ADVANCES IN URBAN CONSTRUCTION AND MANAGEMENT ENGINEERING

Advances in Urban Construction and Management Engineering focuses on the research of urban traffic, city engineering, ecological city and management engineering. The proceedings feature the most cutting-edge research directions and achievements related to Urban Construction. Subjects in the proceedings include:

- Urban development and construction
- Architectural design and urban planning
- Logistics and supply chain management
- Management engineering

The works of this proceedings can promote development of Urban Construction and Management Engineering, resource sharing, flexibility and high efficiency. Thereby, promote scientific information interchange between scholars from the top universities, research centers and high-tech enterprises working all around the world.

PROCEEDINGS OF THE 3RD INTERNATIONAL CONFERENCE ON URBAN CONSTRUCTION AND MANAGEMENT ENGINEERING (ICUCME 2022), GUANGZHOU, CHINA, 22–24 JULY 2022

# Advances in Urban Construction and Management Engineering


*Edited by*

Young-Jin Cha
*University of Manitoba, Canada*



CRC Press
Taylor & Francis Group
Boca Raton London New York Leiden

CRC Press is an imprint of the
Taylor & Francis Group, an **informa** business
A BALKEMA BOOK

First published 2023
by CRC Press/Balkema
4 Park Square, Milton Park / Abingdon, Oxon OX14 4RN / UK
e-mail: enquiries@taylorandfrancis.com
www.routledge.com – www.taylorandfrancis.com

*CRC Press/Balkema is an imprint of the Taylor & Francis Group, an informa business*



*Library of Congress Cataloging-in-Publication Data*
A catalog record has been requested for this book

ISBN: 978-1-032-39018-5 (hbk)
ISBN: 978-1-032-39019-2 (pbk)
ISBN: 978-1-003-34802-3 (ebk)

DOI: 10.1201/9781003348023

Typeset in Times New Roman
by MPS Limited, Chennai, India

*Advances in Urban Construction and Management Engineering – Cha (Ed)*
*© 2023 the Editor(s), ISBN 978-1-032-39018-5*

# Table of contents

*Advances in Urban Construction and Management Engineering – Cha (Ed)*
*© 2023 the Editor(s), ISBN 978-1-032-39018-5*

# Preface

The 2022 3rd International Conference on Urban Construction and Management Engineering (ICUCME 2022), a virtual conference, was successfully held in Guangzhou, China on July 22–24, 2022. The conference mainly focused on the development of urban traffic, city engineering, ecological city as well as management engineering. With the theme "Urban Construction" and "Engineering Management", ICUCME 2022 aspired to keep up with advances in such a constantly changing field.

We were greatly delighted and honored to have Assoc. Prof. Ahmad Safuan A Rashid from Universiti Teknologi Malaysia, Malaysia to serve as our Conference General Chair. There were about 160 individuals worldwide who attended the conference. Divided into three parts, the conference agenda included keynote speeches, oral presentations and online Q&A discussions. Primarily, the keynote speakers were each allocated 30–45 minutes to address their speeches. Then in the next part, oral presentations, the excellent papers we had selected were presented by their authors one by one.

Three distinguished professors were invited to hold their speeches during the conference. Among them, Prof. Biao Wang from Soochow University, China performed a thought-provoking speech on Urban Wind Energy Evaluation with Urban Morphology. To evaluate wind energy potential with study of urban form in a block scale (500 m * 500 m), simple building forms were tested for exploring the impact of building form on wind potential, urban form unit models were then considered to learn the impact of certain urban's form and feature on wind potential, and finally a block model of Beijing was given for a case study on urban wind evaluation. And then Assoc. Prof. Jiwei Wen from Shijiazhuang Tiedao University, China addressed a speech: Research on the Key Techniques of Trenchless Pipe Jacking Construction of Urban Underground Structures. This study analyzed some crucial factors related to friction resistance, and proposed some feasible techniques to reduce that resistance, providing certain references for practical pipe jacking. Their insightful speeches had triggered heated discussion. And every participant praised this conference for disseminating useful and insightful knowledge.

We are glad to share with you that we received lots of submissions from the conference and we selected a number of high-quality papers and compiled them into the proceedings after rigorously reviewing them. These papers feature but are not limited to the following topics: Urban Construction and Application, Smart City Construction, Green Ecological Community, Environment and Equipment Engineering, Urban Traffic Management, etc. All the papers have been checked through rigorous review and processes to meet the requirements of publication.

We would like to acknowledge all of those who supported ICUCME 2022 and made it a great success. In particular, we would like to thank the organizing committee, for its valuable inputs in shaping the conference program and reviewing the submitted papers. The help and contribution of each individual and institution was instrumental in the success of the conference.

Committee of ICUCME 2022

 *ISBN 978-1-032-39018-5*

# Committee members

**Conference General Chair**
Associate Professor Ahmad Safuan A Rashid, *Universiti Teknologi Malaysia, Malaysia*

**Publication Chair**
Professor Young-Jin Cha, *University of Manitoba, Canada*

**Committee Member**
Prof. Bachir ACHOUR, *University of Biskra, Algeria*
Prof. Houpan Zhou, *Hangzhou Dianzi University, China*
Ass. Prof. Hossein Medi, *Imam Khomeini International University, Qazvin-Iran*
Ass. Prof. Parthiban Kathirvel, *SASTRA Deemed University, India*
Assoc. Prof. Norzailawati Mohd Noor, *International Islamic University Malaysia, Kuala Lumpur, Malaysia*
Dr. Nur Mardhiyah Aziz, *Universiti Malaya, Malaysia*
Dr. D. C. Haran Pragalath, *British Applied College, United Arab Emirates*
Dr. Ana Almerich-Chulia, *Universitat Politecnica de Valencia, Valencia, Spain*

*Urban construction and analysis of space system design*

*Advances in Urban Construction and Management Engineering – Cha (Ed)*
 *ISBN 978-1-032-39018-5*

# Analysis of the spatial characteristics of the residential communities of COVID-19 cases based on GIS platform—Taking the central urban area of Wuhan as an example

Zheng Yang, Yiheng Wang, Zhicheng Tang, Yangling Zhao & Luo Guo*
*College of Life and Environmental Sciences, Beijing, China*

ABSTRACT: The COVID-19 pandemic is rampant around the world, dramatically increasing the number of infected people, and greatly affecting the lives of people living in central urban areas. In China, one of the cities most affected by COVID-19 is Wuhan, where the distribution of residential communities with COVID-19 cases is indicative of viral spread. Analyses of spatial characteristics and influencing factors in case communities in the central Wuhan urban area provide indicators for exploring the COVID-19 environmental epidemiology and urban planning. This study used ArcGIS technology and spatial analysis methods such as spatial autocorrelation analysis, kernel density estimation, standard deviation ellipse analysis, and near analysis to explore the spatial distribution data patterns in case communities in Wuhan. Results indicate that 73% of the case communities were located within 150 m of main roads, while 90% of the case communities were located within 220 m of urban roads. The clustered spatial distribution of the case communities indicates a multi-core cluster pattern along the Yangtze River, with northwest to southeast trend, large distribution differences, and prominent spatial imbalance. The transportation network is a significant agent of the spread of the epidemic. At the same time, green space, population density, and gross domestic product (GDP) show no significant effects on the spread of COVID-19. The results of this study can provide some references for basic epidemiological research and urban planning.

## 1 INTRODUCTION

The Corona Virus Disease 2019 (COVID-19) is an acute respiratory infectious disease caused by a novel coronavirus infection. This disease has pandemic traits and significant effects on the health and lives of people worldwide (Brenner 2021; Blumenthal et al. 2020). In epidemiology, the distribution of infectious human and animal diseases, pathogen hosts and vectors, as well as temperature, humidity, precipitation, soil, and facilities, are significant factors in the spread of infectious diseases (Deng 2022). GIS, Remote Sensing (RS), Global Positioning System (GPS), and other advanced technologies and means used in epidemiology can explore the spatial distribution and dynamic changes of infectious diseases (Chaput 2002). Since 1988, the Centers for Disease Control and Prevention (CDC) has used these technologies to monitor malaria, Lyme disease, plagues, filariasis, and epidemic poliomyelitis (Xiao 2005). In 1990, Glass et al. used the GIS analysis and modeling methods to explore the Lyme disease epidemic in Baltimore, USA, compared the development trend of cases with randomly-set references, and conducted a logistic regression analysis to form a risk model and predict the scope of the epidemic. The operation of the system modeled by GIS was verified in the disease scenario in 1991 (Glass 1995).

As COVID-19 ravages the world, GIS technology is also widely used for COVID-19 surveillance and prevention. Nath et al. conducted a risk assessment in different areas of Kolkata according to

*Corresponding Author: guoluo@muc.edu.cn

DOI 10.1201/9781003348023-1

the distribution of confirmed patients (2021). Cao et al. (2021) built the novel coronavirus agent simulation model using GIS technology and an agent model based on individual behaviors and social relations in the city to simulate the viral transmission process. GIS-based visual epidemic mapping and activity tracking of confirmed patients have been widely used in epidemic prevention and control (Peng et al. 2021). However, most COVID-19 pandemic spatial analyses focus on cities, provinces, or regions; few studies focus on residential communities.

This paper analyzes the spatial distribution pattern of residential communities in Wuhan and discusses the factors that can affect the number of residential communities. Spatial analysis of epidemic cases indicates that a greater correlation between influencing factors and case distribution indicates stronger effects of influencing factors. This study of urban epidemiological transmission provides a reference for predicting epidemic trends and urban ecological planning.

## 2 STUDY AREA AND METHOD

### 2.1 *Study area*

The central urban area of Wuhan, the capital city of Hubei Province, is located between 114° 07′ – 114° 38' east longitude and 30°42′ – 30°20′ north latitude (Figure 1). Due to its advantageous geographical position and highly developed transportation system, Wuhan is one of the key logistics hubs in China.

Statistics show that the Wuhan downtown area accounts for only 10% of the entire city area. However, the permanent population accounts for more than 55% of the total population. The number of COVID-19 infections in Wuhan accounted for about 60% of the total number of infections in China. Moreover, the number of infections in the main urban areas accounted for more than 85% of the entire area. From February 4th to 14th, Wuhan recorded 658 communities with confirmed cases, including 436 in main urban areas.

Figure 1. Study area locations.

### 2.2 *Data and methodology*

This study used the 2019 Wuhan SHP file as the data source to select the case communities. A total of 658 communities with confirmed cases were counted in Wuhan from February 4th to 14th, including 436 in the central urban areas. The POI data includes specific address information and latitude-longitude coordinates, which may include deviations, data omissions, and other variables. Thus, the selected data had to be cleaned, invalid data eliminated, and the coordinates and address information had to be corrected. The Wuhan population and GNP data are from the annual statistical yearbook of the Wuhan Statistics Bureau. The Wuhan administrative district map, road traffic map, and other data are from national geographic data sources.

This study used spatial autocorrelation analysis, kernel density estimation, standard deviation ellipse analysis, and other methods of spatial analysis to explore the spatial distribution characteristics of case communities in the Wuhan central urban area. Near analysis was used to analyze the distances between the case communities and the main roads, the relationships between the degree of traffic development, and the number of residential communities. SPSS was used to analyze the effects of green space, road network lengths, population densities, GDP, and other factors on the number of residential communities.

## 3 RESULTS AND ANALYSIS

### 3.1 *Analysis results: Distance between case communities and road networks*

Based on the minimum distance principle, mathematical statistics were used on the distance between the Wuhan case communities and the main roads. The average distance between case community entrances and main roads was 124 m. This distance is expressed by the equation y = 197.921/[1+ (41.006/x) 1.218]/[1+ (x/156.628) 4.537]. Moreover, the $R^2 = 0.9865$ fitting degree shows a highly significant fitting relationship. Most of the case communities are close to major roads. As shown in Figure 2, the overall distance between case cells and roads showed an increasing then decreasing trend, where the number of case communities peaked at about 100 m from the main roads and then decreased continuously. At least 73% of the cases were distributed within 150 m of the nearest road. Moreover, at least 90% of the cases were distributed within 220 m of the nearest road. The number of cases decreased significantly with the increase in distance between roads and case community locations.

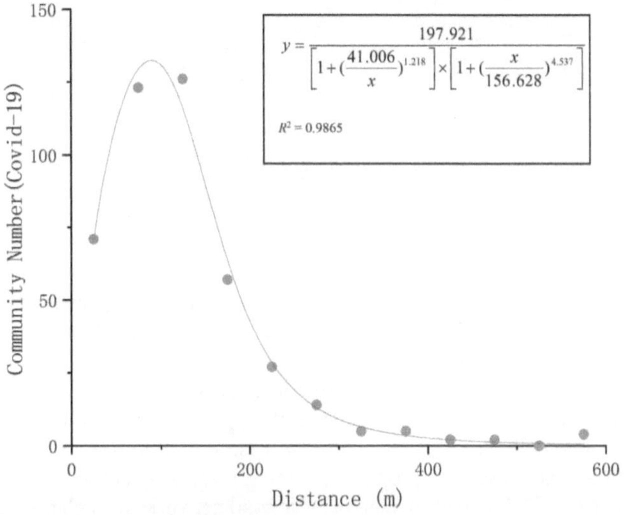


Figure 2. Relationship between community number and traffic.

### 3.2 *Spatial distribution characteristics of case communities*

Figure 3a shows the number of confirmed COVID infection cases in the central Wuhan urban areas. The Moran's I index of the global autocorrelation coefficient of each subdistrict in the central Wuhan urban area was 0.156, with a 99% feasibility (Z-score: 3.001, P-value< 0.003), indicating a positive overall spatial distribution pattern of residential communities of cases, with some degree of agglomeration.

If the elements with no neighbors in the spatial weight are deleted, the Moran's I index is 0.839, indicating significant agglomeration. The average nearest neighbor function was used to analyze the case communities. The 0.704 (less than 1) Nearest Neighbor Index (NNI) indicates a positive spatial correlation in the overall spatial distribution pattern of the case communities, with some degree of agglomeration.

Figure 3b shows the kernel density analysis diagram of case communities in the central Wuhan urban area. The light to dark color indicates the gradual increase in the density of case communities. In addition, the spatial distribution of the case communities along the Yangtze River showed obvious agglomeration characteristics clustered along and scattered around the river. The uneven distribution is discernible.

Directionality is also an important feature of the spatial distribution of the case communities. That is, the aggregation degree of the sample is clearly pointed in a direction. For instance, Figure 3c shows a clear length difference between the long axis and the short axis of the standard deviation ellipse. In addition, the flat rate of the ellipse is large. The spatial distribution direction is indicated by the long axis in the southeast-north direction, while the overall distribution shows a slight north-south trend. The overall space distribution shows that the Wuhan case residential area has a southeast-northwest distribution trend.

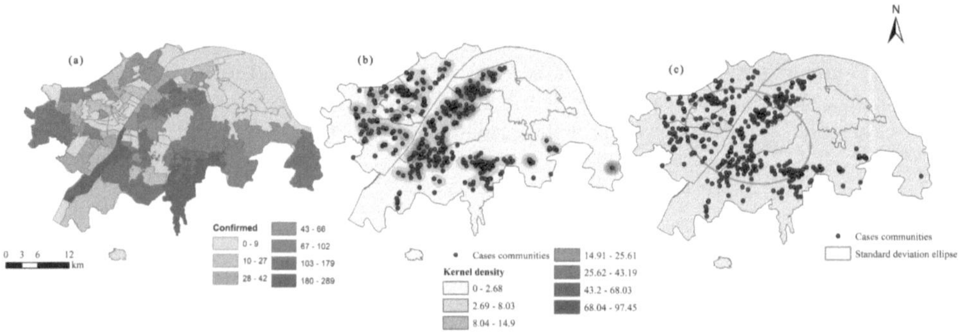


Figure 3. Spatial characteristics of communities with COVID-19 cases.

## 3.3 *Correlation study results*

The overall distribution of infected residential Wuhan communities showed significant clustering toward the city center. This paper explores the influences of social, economic, environmental, and traffic factors on the distribution of infected residential communities.

A parking study explored the effects of the green space service scope on the number of case communities (Tang et al. 2022). Figure 4a shows the park's 15-minute driving distance to the main Wuhan urban area, which intersects with the case communities. Results show 270 case communities within the park's service scope, accounting for 61.9% of the total case communities. Figure 4b shows the 6316.41-km central Wuhan main road network. In each district, the number of case communities covered by green space within a 15-minute drive service area was calculated. If most of the communities were distributed within the park service area, it could prove a correlation between the number of green spaces and cases communities. A Pearson correlation coefficient analysis indicated a correlation coefficient $r$ of 0.5 and a $p$-value of 0.253 (more than 0.05), which means that there is no correlation between the number of case communities and the number of communities within the park service area. As for the relationship between road network length and residential area, the 0.781 correlation coefficient $r$ and the 0.038 $p$-value (less than 0.05) indicated a significant positive correlation.

A Pearson correlation coefficient analysis of the GDP of each district in the Wuhan Yearbook showed a correlation coefficient *r* of 0.077 (close to 09) between the number of case communities in each district and GDP, while the 0.870 *p*-value (more than 0.05) indicated no correlation between the number of case communities in each district and GDP. Moreover, the -0.002-correlation coefficient *r* (close to 0) between population density in each district and the number of case communities, and the 0.997 *p*-value (more than 0.05) indicated no correlation between the number of case communities and population density.

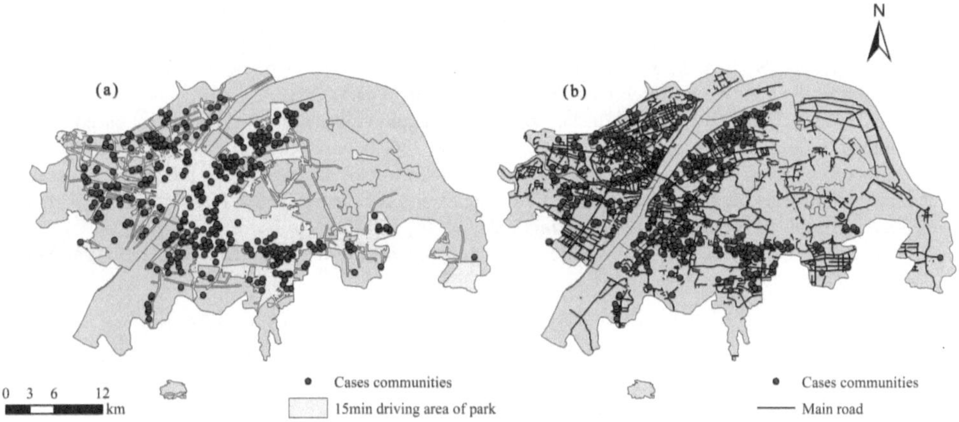


Figure 4. Wuhan Park service areas and road network lengths.

## 4 CONCLUSIONS

In the central Wuhan urban area, at least 73% of the case communities were located 150 m away from main roads. Moreover, at least 90% of the case communities were located within 220 m away from main roads. Most case communities were located approximately 100 m away from main roads. As the distance from the main road increased, the number of case communities first increased and then decreased.

The generally clustered spatial distribution of case communities in Wuhan indicated a multi-core cluster pattern along and scattered around the Yangtze River, with northwest to southeast trend perpendicular to the river. The distribution difference is significant, and the spatial imbalance is discernible.

The factors that may influence the number of case communities in Wuhan include the park service scope, road network lengths, population densities, GDP, etc. No correlation showed between the green space service scope, population density and GDP, and the number of case communities. This may be because the newly confirmed cases used in this paper included data during the containment period in Wuhan, which could not directly and effectively reflect the correlation of the epidemic's natural transmission.

A significant positive correlation showed between case communities and road network length. For instance, Figure 4b shows more case communities in places with dense road networks and fewer communities. To sum up, transportation factors can influence the spread of COVID-19 infection.

## REFERENCES


Blumenthal, D., Fowler, E.J., Abrams, M., & Collins, S.R. (2020). Covid-19 - implications for the health care system. *J. The New England journal of medicine*, 383(15), 1483–1488.

Brenner, M.H. (2021). Unemployment, bankruptcies, and deaths from multiple causes in the COVID-19 recession compared with the 2000–2018 great recession impact. *J. American Journal of Public Health*, 111(11), 1950–1959.
Cao, Z.H., Zhang, J.Q., Yang, M., Jia, P.L., & Deng, S.C. (2021). The city agent model of COVID-19 based on GIS and application: a case study of Guangzhou. *J. Journal of Geo-information Science*, 23(2), 10.
Chaput, E.K., Meek, J.I., & Heimer, R. (2002). Spatial analysis of human granulocytic ehrlichiosis near Lyme, Connecticut. *J. Emerging infectious diseases*, 8(9), 943.
Deng, J.Q., Wu, Q.H., Liu, A.Z. (2006) Analysis About Environmental Factor of SARS Epidemic Based on GIS Logical Information Method. *J. Chinese Journal of Health Statistics*, (04), 301–305.
Glass, G.E., Schwartz, B.S., Morgan III, J.M., et al. (1995). Environmental risk factors for Lyme disease identified with geographic information systems. *J. American journal of public health*, 85(7), 944–948.
Nath, B., Majumder, S., Sen, J., & Rahman, M.M. (2021). A risk analysis of covid infections in Kolkata metropolitan city: a gis based study. *J. GeoHealth*, 5(4), e2020GH000368.
Peng, C., Xie, Y.X., Zhang, H.F., et al. (2021). Application of epidemic prevention and control based on GIS technology. *J. Jiangsu Science & Technology Information*, 38(17), 6.
Tang, Z., Wang, Y., Yang, Z., Jiang, Y., & Guo, L. (2021). Assessing the accessibility of urban green spaces in central Wuhan based on geographic information systems and remote sensing. *J. International Conference on Smart Transportation and City Engineering* 2021, 11(16), 7368.
Xiao, X.X., Xiao, J.G., Xiu, L.C. (2005). Application of geographic information systems in infectious diseases. *J. Modern Preventive Medicine*, 32(10), 3.

*Advances in Urban Construction and Management Engineering – Cha (Ed)*


# Spatial distribution pattern and influencing factors of Tibetan Plateau traditional villages

Luqing Yan*, Haifeng Zhang* & Ru Xu*
*School of Geography Science, Qinghai Normal University, Xining, Qinghai, China*

ABSTRACT: Based on 391 traditional villages on the Tibetan Plateau, the spatial distribution pattern of traditional villages on the Tibetan Plateau and its influencing factors were explored by using mathematical statistics, GIS spatial analysis, and geographic probes, which showed that: 1) Traditional villages on the Tibetan Plateau are cohesively distributed, concentrated and unevenly distributed; 2) Traditional villages have obvious spatial and temporal differentiation characteristics, showing a "single core - double core - single core expansion - double core growth". 3) Among the natural factors, topography has the greatest influence on the distribution of traditional villages; among the socio-human factors, transportation, economic level, and population are the keys to the preservation and continuation of traditional villages; religious beliefs and ethnic culture have a driving role in the protection and inheritance of traditional villages.

## 1 INTRODUCTION

Traditional villages are villages that were formed earlier and have rich tangible and intangible cultural heritage, natural and humanistic historical landscapes, as well as certain historical, cultural, scientific, artistic, social, and economic values that should be protected (Hu 2014). In recent years, with the rapid development of urbanization, industrialization, and modernization of agriculture and rural areas, the number of traditional villages has been decreasing and the phenomenon of cultural loss has continued. In the face of the plight of traditional villages, the Ministry of Housing and Urban-Rural Development and other organizing departments announced the list of traditional villages in China in 2012, and up to now, a total of 6,819 traditional villages have been announced in five batches.

Overseas research on traditional villages started earlier and focused on the cultural landscape of traditional villages (Chen 2010), tourism development (Che 2016), conservation and development (Marschalek 2008), etc. At the end of the 20th century, scholars such as Yin Yongda (1991) and Liu Peilin (1998) conducted the earliest research on the cultural connotation and spatial imagery of traditional villages, and the initial research was mainly based on qualitative analysis and case study analysis. At present, their research is becoming more and more mature, and the main contents are as follows: theoretical connotation and development path of traditional villages (Sun 2017), spatial morphology (Hu 2017), spatial and temporal evolution and formation mechanism (Liang 2018), development, conservation and development of traditional villages (Yang 2018), tourism of traditional villages (Liu 2019), etc. In recent years, geography has exerted its advantages of spatial analysis, combined with time series, and conducted a lot of research on the spatial distribution of traditional villages and their influencing factors (Feng 2017; Li 2020).

At present, studies on the spatial distribution and influencing factors of traditional villages mainly focus on the macroscopic scale, and relatively few studies are conducted on single natural

*Corresponding Authors: 2523008893@qq.com, 675240331@qq.com and 1923762952@qq.com

DOI 10.1201/9781003348023-2

or cultural areas, the study areas are mostly in the eastern and central regions. As the "roof of the world", the Tibetan Plateau's terrain is steep and changeable, and the people of all ethnic groups have created and preserved rich traditional village resources under the harsh natural environment, but few scholars have systematically explored the macroscopic distribution pattern of traditional villages on the Tibetan Plateau. In view of this, this paper takes 391 traditional villages on the Tibetan Plateau as the research object and explores the spatial distribution pattern of traditional villages and their influencing factors, in order to enrich the relevant content of the influence of geographical factors on human beings and provide a reference for the development and protection of traditional villages.

## 2 MATERIALS AND METHODS

### 2.1 *Overview of the study area*

The Tibetan Plateau is located at 26°00′12″N~ 39°46′50″*N*, 73°18′52″N~ 104°46′59″E, with an area of 2572.4 × $10^3$km$^2$, accounting for 26.8% of China's total land area, and an average altitude of about 4385m (Zhang 2002). It is a multi-ethnic area inhabited mainly by Tibetans. The Tibetan Plateau has special natural conditions such as alpine cold and lack of oxygen, strong ultraviolet radiation, etc. The alpine cold and multi-ethnic integration environment of the Tibetan Plateau has nurtured many traditional villages of different ethnic styles (Figure 1), and the unique geographical environment for the study of traditional villages laid the foundation.

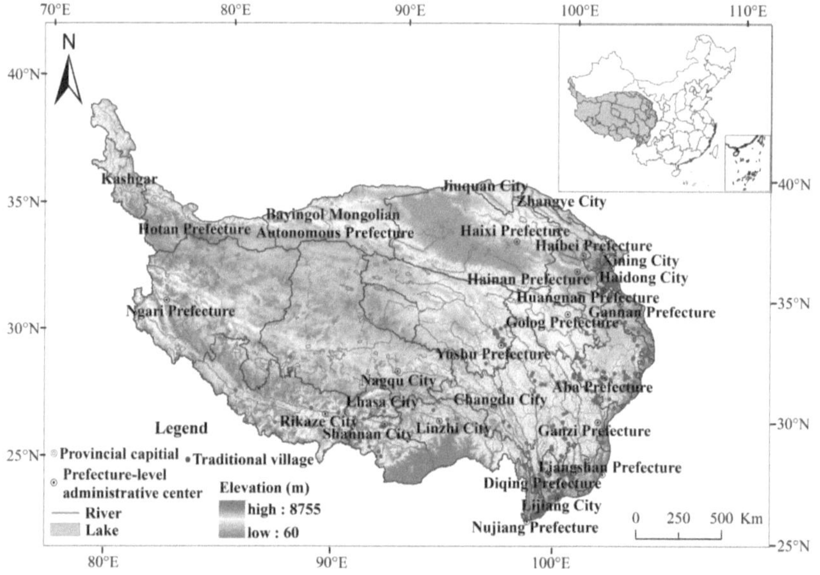


Figure 1. Spatial distribution of traditional villages in the Tibetan Plateau.

### 2.2 *Research methodology*

The methods of nearest point index, imbalance index, and kernel density were used to quantitatively analyze the spatial distribution of traditional villages on the Tibetan Plateau, and the geographic probe (Wang 2017) was used to analyze the factors influencing the spatial distribution of traditional villages, and the methods used and their geographical significance are shown in Table 1.

(1) Nearest neighbor index

Determine the distribution type of the point element. When $R=1$, it means random distribution; when $R<1$, it means coalescent distribution; when $R>1$, it means uniform distribution. Its calculation formula is.

$$R = \frac{\overline{r_i}}{r_E} = 2\sqrt{D} \tag{1}$$

Where: $\bar{r}_i$ is the actual nearest-neighbor distance, $r_E$ is the theoretical nearest-neighbor distance, and $D$ denotes the point density.

(2) Imbalance index

$S$ is between 0 and 1. When $S=0$, traditional villages are evenly distributed in each province and region; if $S=1$, traditional villages are concentrated in one province and region. Its calculation formula is.

$$S = \frac{\sum_{i=1}^{n} Y_i - 50\,(n+1)}{100n - 50\,(n+1)} \tag{2}$$

Where: $n$ is the number of provinces and regions, and $Y_i$ represents the cumulative percentage of the number of traditional villages in each province and region ranked from the largest to the smallest in the study area.

(3) Nuclear density analysis

The kernel density estimates can determine the spatial clustering of traditional villages and identify agglomeration areas. For the data $x_1 x_2, \ldots, x_n$, its calculation formula is.

$$f_h(x) = \frac{1}{nh}\sum_{i=1}^{n} K\left(\frac{x - x_i}{h}\right) \tag{3}$$

Where: $K$ denotes the kernel function, $x_i$ denotes the location coordinates of traditional villages $i$ (i=1, 2, 3, ..., $n$) $n$ denotes the number of traditional villages, and $h$ denotes the search radius of the kernel density function.

(4) Geographical probe

In this paper, we mainly use factor detection in the geodetector to analyze the strength of each factor including the natural and human environment on the spatial distribution of traditional villages on the Tibetan Plateau. $q$ takes the values of [0, 1], and the closer the value of *q is to* 1, the greater the influence of the driver on the spatial distribution of traditional villages, and vice versa, the closer it is to 0, the smaller the influence of the driver on the spatial distribution of traditional villages. The expression of factor detection applied in this paper is.

$$q = 1 - \frac{\sum_{h=1}^{L} N_h \sigma_h^2}{N\sigma^2} \tag{4}$$

where $N$ is the number of grids divided within the study area; $L$ is the stratification of the variable or factor; $N_h$ denotes the number of grids of study cells in the $h$-th stratification $\sigma^2$ is the total variance of spatial heterogeneity within the study area. $\sigma^2$is the variance of spatial heterogeneity of the $h$-th stratum.

## 2.3 *Data sources and processing*

The traditional village data were obtained from the List of Traditional Villages in China (Batch 1–5) promulgated by the Ministry of Housing and Urban-Rural Development of China, from which a total of 391 traditional villages located on the Tibetan Plateau were extracted using ArcGIS software and visualized (Figure 1). geographic data such as DEM, water system and traffic network were obtained from the geospatial data cloud platform. The socio-economic data were obtained from the China Statistical Yearbook, provincial statistical yearbooks, and socio-economic development bulletins.

## 3 SPATIAL CHARACTERISTICS

### 3.1 *Type of spatial distribution*

Based on Equation (1), using the average nearest neighbor distance in the ArcGIS10.2 spatial statistics tool, the average actual nearest neighbor distance is obtained $\bar{r}_i = 0.1668$, theoretical nearest neighbor distance $r_E = 1.0320$, and the nearest neighbor index $R = 0.161603 < 1$, $\bar{r}_i < r_E$ and the Z-score is –31.552657, indicating that the traditional villages on the Tibetan plateau tend to be cohesively distributed.

### 3.2 *Balanced spatial distribution*

Based on equation (2) calculated by Excel, the imbalance index S = 0.524 is close to 1, indicating that traditional villages are not evenly distributed within the Tibetan Plateau. The Lorenz curve (Figure 2) also reveals that the Lorenz curve is curved to a large extent, further indicating that traditional villages on the Tibetan Plateau are unevenly distributed and concentrated in Sichuan, Qinghai and Yunnan provinces.

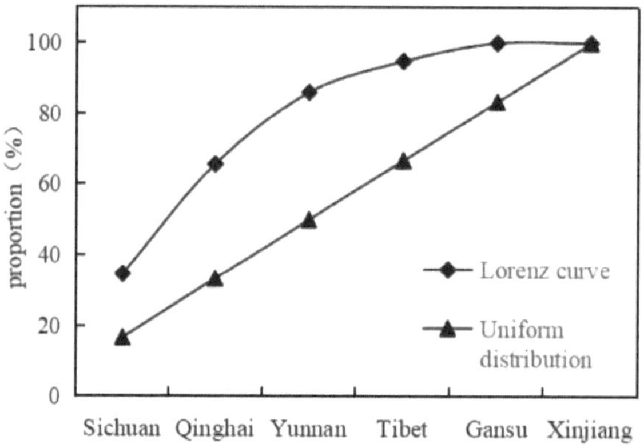


Figure 2. Lorenz curve of the spatial distribution of traditional villages in the Tibetan Plateau.

### 3.3 *Spatial clustering*

Kernel density analysis of traditional villages on the Tibetan Plateau using ArcGIS software Kernel Density tool (Figure 3) revealed that there were obvious regional differences between the 1st–5th batches of traditional villages on the Tibetan Plateau, forming two high-density areas and one secondary high-density area (Figure 3f), and presenting a "single-core-double-core expansion-dual-core growth" spatio-temporal distribution characteristics. By batches, the 1st and 2nd batches of traditional villages initially formed Qinghai Hehuang Valley high-density area and Northwest Yunnan high-density area (Figures 3a, 3b); the 3rd batch further consolidated the core position of the 1st and 2nd batches of high-density area (Figure 3c); the 4th batch of traditional villages of Hehuang Valley high-density area expanded to the periphery, and the dominant position of Northwest Yunnan was weakened (Figure 3d); the 5th batch of Qinghai Hehuang Valley high-density area still occupies In the 5th batch, the Hehuang Valley high-density area still occupies the core position, and the Ganzi-Aba sub-high-density area is formed (Figure 3e).

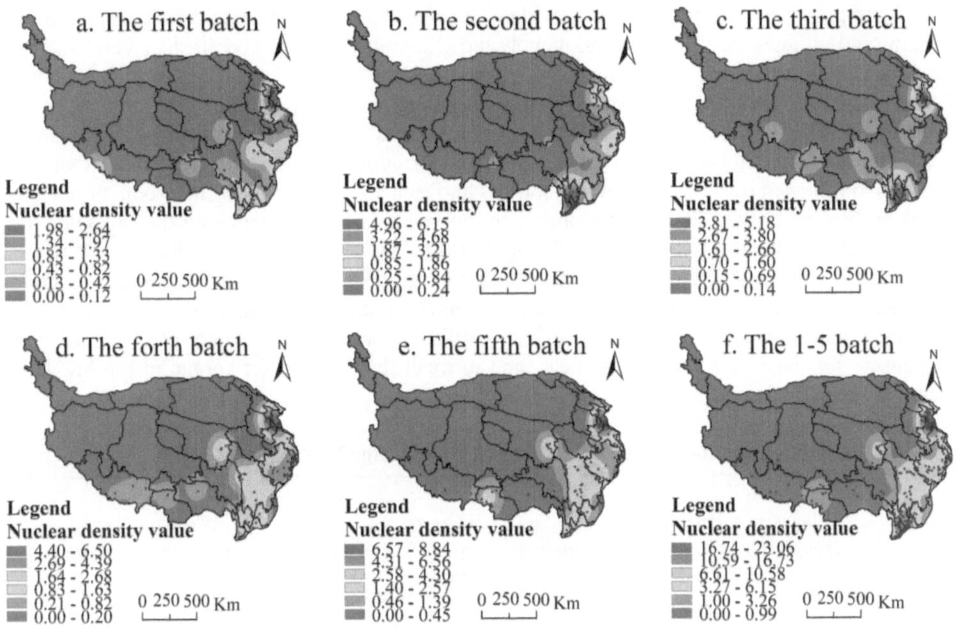


Figure 3. Distribution of kernel density of traditional villages in the Tibetan Plateau.

## 4 INFLUENCING FACTORS

The spatial distribution and evolution of traditional villages are influenced by the combined effect of various factors. Six categories of factors, such as elevation, river network density, slope direction, etc. were selected to analyze their driving mechanisms. The natural interruption point method was used to discretize the data of the above six categories except for social culture into six classes in ArcGIS, and the discrete distribution map of each factor was obtained. Applying the geographic probe, the size of the influence of each factor on the spatial distribution of traditional villages was obtained (Table 1).

Table 1. Geographical influences of spatial distribution factors on the Tibetan Plateau.

| Influencing Factors | Detection factor | Indicators | q-value |
|---|---|---|---|
| Nature | Terrain | Elevation (m) | 0.9342 |
| | River | River network density (km/km$^2$) | 0.7925 |
| | Slope direction | Slope direction | 0.5021 |
| Humanities | Population | Population density (persons/km$^2$) | 0.5378 |
| | Transportation | Road network density (km/100km$^2$) | 0.6624 |
| | Economy | GDP per capita (yuan) | 0.5043 |

Natural environment factors. The natural environment is the limiting factor affecting the distribution of traditional villages on the Tibetan plateau. 1) The influence of altitude is highest at 0.93 (Table 1). Most of the traditional villages on the Tibetan Plateau are located in the mountainous and river valley areas, concentrated in the middle and high-altitude areas with obvious and complex terrain changes from 1000 m to 3500 m (Table 2), which are less affected by external factors and thus easy to preserve. 2) River systems provide water resources for the formation and development of traditional villages. The three rivers, namely the Nujiang, Lancang, and Jinsha rivers, and the Hehuang valley of Qinghai, have become the main clusters of traditional villages on the Tibetan Plateau (Figure 1), and about 96% of the traditional villages are located within 50km of the rivers (Table 3). In addition, the widely distributed pastoral areas in the west and north of the Tibetan Plateau are dominated by animal husbandry, and the distribution of pasture in areas near water systems is also followed by more, near water sources and rich pasture can meet the production needs of plateau herders (Zhang 2019). 3) Slope direction has a certain influence on the site selection and layout of traditional villages on the Tibetan Plateau. The distribution of traditional villages on the Tibetan Plateau is characterized by "distribution by shaded slope", and the number of villages on the shaded slope is 1.3 times higher than that on the sunny slope (Table 4). The villages on the shaded slopes have relatively gentle terrain, sufficient water sources, and lush vegetation to compensate for the lack of water and heat, and to meet the demand for forage in the highland pasture (Zhang 2019).

Table 2. Spatial distribution of villages with different altitudes in the Tibetan Plateau.

| Altitude (m) | Number of villages (pcs) | Percentage (%) | Cumulative share (%) | Altitude (m) | Number of villages (pcs) | Percentage (%) | Cumulative share (%) |
|---|---|---|---|---|---|---|---|
| <1000 | 2 | 0.5 | 0.5 | 3500~4500 | 82 | 21.1 | 98.5 |
| 1000~2500 | 151 | 38.7 | 39.2 | 4500~5000 | 6 | 1.5 | 100.0 |
| 2500~3500 | 149 | 38.2 | 77.4 | >5000 | 0 | 0 | 100.0 |

Table 3. Spatial distribution of villages with different distances from river buffer zones in the Tibetan Plateau.

| Buffer distance (km) | Number of villages (pcs) | Percentage (%) | Cumulative share (%) | Buffer distance (km) | Number of villages (pcs) | Percentage (%) | Cumulative share (%) |
|---|---|---|---|---|---|---|---|
| 0~5km | 119 | 30.5 | 30.5 | 20~30km | 67 | 17.2 | 79.8 |
| 5~10km | 48 | 12.1 | 42.6 | 30~50km | 61 | 15.6 | 95.4 |
| 10~20km | 78 | 20.0 | 62.6 | >50km | 18 | 4.6 | 100.0 |

Table 4. Slope direction of traditional villages in the Tibetan Plateau.

| Slope direction | Number of villages (pcs) | Percentage (%) | Slope direction | Number of villages (pcs) | Percentage (%) |
|---|---|---|---|---|---|
| North | 73 | 18.67 | South | 56 | 14.32 |
| Northeast | 64 | 16.37 | Southwest | 44 | 11.25 |
| East | 38 | 9.72 | West | 39 | 9.98 |
| Southeast | 30 | 7.67 | Northwest | 47 | 12.02 |

Socio-economic factors. Traditional villages are the products of social and historical development to a certain stage. Human social activities promote the formation and development of traditional villages to a certain extent, and their influence is ranked as follows: transportation > population > economy. 1) Transportation provides important conditions for population gathering. determines the degree of communication between traditional villages and the outside world. From the road network density map (Figure 4d), it can be seen that traditional villages on the Tibetan Plateau are mainly clustered in areas where the road network density is at a medium level. In the current era of rapid urbanization, convenient transportation can easily expose traditional villages to foreign

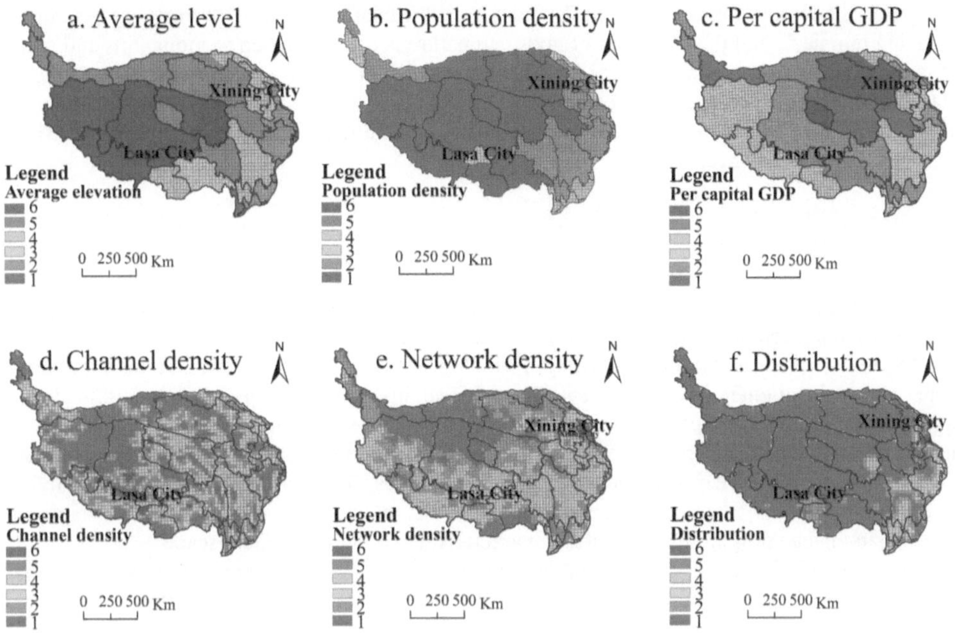


Figure 4. Discrete distribution diagram of factors affecting the spatial distribution of traditional villages.

cultural influences, resulting in the assimilation and destruction of traditional villages; while the medium level of transportation can maintain the accessibility of villages while avoiding the impact of too much communication on villages (Li 2021). 2) The distribution of traditional villages on the Tibetan Plateau is closely related to the distribution of population. The population density map (Figure 4e) shows that there are more traditional villages clustered in the Hehuang Valley in eastern Qinghai. 3) The level of economic development has the weakest influence on the layout of traditional villages, and the level of the economy has a certain influence on the financial investment and protection of traditional villages' development and protection. The urbanization rate and per capita GDP of the five regions with the largest distribution of traditional villages on the Tibetan Plateau (Table 5) indicate that traditional villages are mostly distributed in areas with relatively slow economic development levels, and their development efforts are small, creating conditions for preserving traditional culture and preserving traditional villages.

Table 5. Statistics of social and economic indicators in the Tibetan Plateau.

| City/Region/State | Number of traditional villages | GDP per capita | Urbanization rate/% |
|---|---|---|---|
| Garze | 71 | 37077.71 | 31.01 |
| Haidong City | 59 | 37880.82 | 40.40 |
| Aba | 41 | 50055.50 | 41.49 |
| Lijiang City | 33 | 40893.43 | 47.64 |
| Huangnan Tibet | 30 | 39603.21 | 41.57 |

Social and cultural factors. The Tibetan Plateau is a region inhabited by ethnic minorities, and traditional villages are distinguished from other traditional villages by their special ethnic culture and religious beliefs and form their unique traditional village culture. The Tibetans are the main ethnic minority living on the Tibetan Plateau, and all of them believe in Tibetan Buddhism, and their religious beliefs have a profound influence on the formation and development of traditional villages in the region, for example, the Tibetans have the characteristic of "living next to temples or facing temples", and their worship of gods and nature is often attached to mountains and rocks (Zhang 2019), which is related to the high terrain and harsh natural environment of the Tibetan Plateau. Other ethnic minorities, such as the Hui and Salar, have the custom of "having a mosque in every village", which contributes to the clustering of traditional villages; the Lisu have fish as their god and thus live near rivers.

## 5 DISCUSSION

The study indicates that the spatial distribution of traditional villages on the Tibetan Plateau shows a distribution pattern of "single core - double core - single core expansion - double core growth". The distribution pattern of traditional villages with high altitudes, leaning on mountains and shaded slopes shows the uniqueness of the Tibetan Plateau, and this distribution pattern has certain advantages for the preservation of traditional villages and the development of agriculture and animal husbandry on the plateau. In addition, as a multi-ethnic area, religious beliefs and ethnic culture have a profound influence on the layout of traditional villages on the Tibetan plateau, such as "living around a temple" and "there must be a mosque in a village", which makes traditional villages on the Tibetan plateau present unique plateau characteristics of the village landscape.

The study initially explored the spatial distribution and influence mechanism of 391 traditional villages on the Tibetan Plateau, which is both an understanding of the current situation of traditional villages on the Tibetan Plateau and a reference value for the conservation and rational development and utilization of traditional villages, as well as enriching the case studies of traditional villages in this area. In the future, this study needs to further focus on the internal structure and cultural characteristics of the villages in combination with data and field research.

## 6 CONCLUSION

The spatial pattern of traditional villages and the influencing factors in the Tibetan Plateau were analyzed by using a combination of GIS spatial analysis, mathematical statistics and geographic probes and other methodological tools. The results show that.

(1) The nearest neighbor index $R < 1$ for traditional villages on the Tibetan Plateau is a cohesive distribution; the equilibrium index $S = 0.524$ indicates that the distribution of traditional villages is uneven and spatially divergent.
(2) Traditional villages on the Tibetan Plateau as a whole have the spatial and temporal distribution characteristics of "single core - double core - single core expansion - double core growth". There are two high-density areas in Hehuang and northwestern Yunnan and a secondary high-density area near Longnan City in Gansu with Aba and Ganzi as the core.
(3) From the perspective of influencing factors, topography and rivers are the dominant factors affecting traditional villages on the Tibetan Plateau; transportation is the main social factor affecting the spatial distribution of traditional villages; religious beliefs and ethnic culture promote the transmission of traditional village culture.

## REFERENCES

Bo Y. (2021). Study on spatial pattern and influencing factors of traditional villages in Hunan province. *J. Science of Surveying and Mapping*. 46(04), 150–157.

Buqing L. & Dawei X (2018). Spatio-Temporal distribution and evolution of Hakka traditional villages in Ganzhou. *J. Economic Geography*. 38(08), 196–203.

Chen B. & Nakama Y (2010). A study on village forest landscape in small island topography in Okinawa, Japan. *J. Urban Forestry and Urban Greening*. 9(02), 139–148.

Haipeng Z. & Jie F (2019) Spatio-temporal evolution of settlements and its driving mechanisms in Tibetan Plateau pastoral area: Taking Nagqu county in northern Tibet as an example. *J. Scientia Geographica Sinica*. 39(10), 1642–1653.
Jiangsu L. & Xiaorui W (2020). Spatial distribution characteristics and influencing factors of Chinese traditional villages. *J. Economic Geography*. 40(02), 143–153.
Jinfeng W. & Chendong X (2017). Geodetector: Principle and prospect. *J. Acta Geographica Sinica*. 72(01), 116–134.
Jiuxia S (2017). Traditional village: theoretical connotation and development path. *J. Tourism Tribune*. 32(01), 1–3.
Liguo Y. & Hualong L (2018). The protection and its evaluation system of traditional villages: a case study of traditional villages in Hunan province. *J. Human Geography*. 33(03), 121–128+151.
Marschalek I (2008). The concept of participatory local sustainability projects in seven Chinese villages. *J. Journal of Environmental Management*. 87(2), 226–235.
Peilin L. & Shuangshuang D (1998). Study of landscape-image of Chinese ancient village. *J. Geographical Research*. 1, 32–39.
Tianzhao L. & Peilin L (2019). The protection and tourism development path of ancient villages and old towns under the background of new-type urbanization: A case study of the old town of Xuanzhou in Hunan province. *J. Geographical Research*. 38(01), 133–145.
Xuehui H (2017). Research on spatial morphology of traditional villages in the northeast region. D. Harbin Institute of Technology.
Yafeng F. & Wanyuan Y (2017). Spatial distribution features and controlling factors of traditional villages in Guangdong province. *J. Scientia Geographica Sinica*. 37(02), 236–243.
Yan H. & Shen C (2014). The Concept and Cultural of Traditional Villages. *J. Urban Development studies*. 21(01), 10–13.
Yili Z. & Bingyuan L (2002). A discussion on the boundary and area of the Tibetan Plateau in China. *J. Geographical Research*. 1, 1–8.
Yongda Y (1991). On the model and cultural connotation of Huizhou traditional village Shuikou. *J. Southeast Cultural*. 2, 174–177.
Zheng C (2016). The advantages and disadvantages of tourism development on traditional village style. *J. Tourism Research*. 8(03), 1–3.

*Advances in Urban Construction and Management Engineering – Cha (Ed)*
*© 2023 Copyright the Author(s), ISBN 978-1-032-39018-5*

# Integrated ecological and environmental risk assessment of Chongqing city based on urban expansion

Yifan Pu, Yangling Zhao & Luo Guo*
*College of Life and Environmental Sciences, Beijing, Beijing, China*

ABSTRACT: The expansion of construction land in the process of urbanization will inevitably encroach on arable land, grassland, and other ecological land and damage the structure and function of the ecosystem. Therefore, it is important to identify the ecological risk changes in urban expansion for urban land use planning and risk prevention and control. In this study, a "factor-exposure-impact-response" ecological risk evaluation system was established based on urban expansion in Chongqing, and the spatial and temporal ecological risk evaluation and risk control were conducted for the study area in 1990, 2000, 2010, and 2020. The results show that the scale of urban land in Chongqing is expanding, and the expansion rate and intensity are on the rise; the construction land expansion intensity index for the four jurisdictions during the study period was 1.25 for Northern District, 0.65 for Western District, 0.64 for Southern District and 0.39 for the main urban area; there are significant differences in the size and intensity of construction land expansion in each jurisdiction; temporally, the integrated ecological risk values of Chongqing and its four jurisdictions are on the rise; the risk level of Chongqing evolves from low to medium, and there is a strong correlation between the risk value and the intensity of construction land expansion; spatially, the integrated ecological risk level, the spatial variation of risk values is significantly influenced by the intensity of construction land expansion; the risk level with the largest proportion of area in Chongqing evolves from low to medium; the proportion of each risk level in the four jurisdictions varies significantly and is consistent with the intensity of construction land expansion.

## 1 INTRODUCTION

In recent years, as a result of China's western development policy, the urbanization of various core western cities, represented by Chongqing, has also brought about ecological problems. The expansion of construction land during urbanization will inevitably encroach on arable land, grassland, and other ecological lands, damaging the ecosystem (Adrianna *et al.* 2018). Therefore, it is important to identify the ecological risk changes in urban expansion for urban land use planning and risk prevention and control. In addition, it is necessary to understand the evolution of the environment and its development mechanism and to identify potential ecological safety hazards in a timely manner. And assessing ecological hazards requires an ecological risk assessment. In the field of GIS, research on urban ecological risk assessment mainly includes risk sources and drivers, risk receptors, evaluation endpoints, and systematic evaluation methods. On the one hand, scholars construct mathematical models from the perspective of land use to evaluate ecological risks in the Chongqing landscape in time and space, but lack the analysis of the impact of human factors on ecological risks (Liu *et al.* 2011); on the other hand, the evaluation model of "pressure-state-response" is constructed from the socio-economic perspective to analyze the temporal changes of ecological security in Chongqing, but lacks the analysis of spatial changes (Liu *et al.* 2019).

Based on the ecological environment changes brought about by urban expansion, this paper identifies disaster-causing factors from multiple perspectives, including land use, human economic activities, and natural disasters, take the landscape pattern as the risk receptor, analyzes the change

*Corresponding Author: guoluo@muc.edu.cn

 DOI 10.1201/9781003348023-3

of the integrated ecological risk and each index level in Chongqing, which shows that the ecological risk value of Chongqing City from 1990 to 2020 increased from 0.35 to 0.41 year by year, and the risk level evolved from low to medium. 1990–2000 was a low ecological risk, etc. This study analyses the adverse effects of ecological risks on nature in terms of human-built factors and incorporates human responses to environmental ecological risks into the assessment system (Song *et al.* 2019).

## 2 STUDY AREA AND RESEARCH METHODOLOGY

### 2.1 *Study area overview*

Chongqing (105°11′-110°11′E, 28°10′-32°13′N) is an important core city of the Yangtze River Economic Belt in western China, and it is of great significance to reduce its ecological risk and balancing urban development with environmental protection (Figure 1). From 1990 to 2020, the scale of the built-up area of Chongqing City has been expanding, and the expansion rate and intensity are both on the rise. By 2020, the size of the built-up area further increases to 260 km$^2$, and the expansion rate and intensity are much higher than those of the previous two periods (Xiu *et al.* 2015).

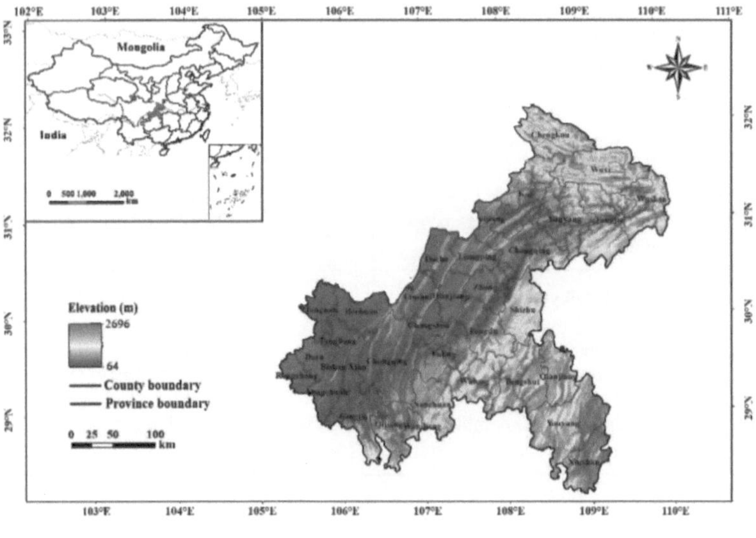


Figure 1. The location of the research area.

### 2.2 *Data source*

This study uses land use data from 1990, 2000, 2010, and 2020 from the Data Centre for Resources and Environmental Sciences of the Chinese Academy of Sciences (http://www.resdc.cn/). The resolution of land use data is 30m×30m. The land use data of RESDC has been used in related research. The spatial elevation data comes from the geospatial data cloud (http://www.gscloud.cn/), with a spatial resolution of 30 m×30 m. Socio-economic data from Chongqing Statistical Yearbook and Sichuan Statistical Yearbook, partly from government work reports.

### 2.3 *Research methodology*

Based on the "factor-exposure-impact-response" framework, this study combined the characteristics of Chongqing to select the indicators of the evaluation system (Yunfeng *et al.* 2020). According to the relationship between the index and ecological risk, the index is divided into positive and negative, and the polarization method is used to standardize the data. Input the standardized data into SPSS 25 to obtain KMO and sig test values, eigenvalues, cumulative contribution rate, and component matrix. Using the first n principal components with eigenvalues greater than 1, the cumulative contribution of principal components reaches 85%, and the specific weights of each indicator are obtained through principal component analysis (Table 1).

Using the weighted summation method, a comprehensive ecological risk evaluation model is constructed with the following equation:

$$\sum_{i}^{n} E_i = E_i \times X_{ij}$$

In the formula, $X_{ij}$ is the specific indicator value, and $E_i$ is the corresponding weight of each indicator.

Table 1. Ecological risk evaluation index system in Chongqing.

| Indicator layer | Factor | The weighting of indicator layers | Total weighting |
|---|---|---|---|
| Factor layer | Population density | 0.1358 | 0.0339 |
| | Percentage of land used for construction | 0.4862 | 0.1216 |
| | Share of agricultural land | 0.0342 | 0.0086 |
| | Road network density | 0.3438 | 0.086 |
| Exposure layer | Aggregation index | 0.4016 | 0.1004 |
| | Sprawl index | 0.0381 | 0.0095 |
| | Separation index | 0.1404 | 0.0351 |
| | Aroma Diversity Index | 0.1325 | 0.0331 |
| | Plaque Density Index | 0.2874 | 0.0719 |
| Impact layer | Vegetation cover | 0.7429 | 0.1857 |
| | Land degradation index | 0.0632 | 0.0158 |
| | Water network density | 0.1939 | 0.0485 |
| Response layer | GDP per capita | 0.3196 | 0.0799 |
| | Green space per capita | 0.5584 | 0.1395 |
| | Woodland area | 0.122 | 0.0305 |

## 3 RESULTS AND ANALYSIS

### 3.1 *Spatio-temporal sequence analysis of urban expansion in Chongqing*

From 1990 to 2020, the scale of the built-up area of Chongqing City has been expanding, and the expansion rate and intensity are both on the rise. By 2020, the size of the built-up area further increases to 260 $km^2$, and the expansion rate and intensity are much higher than those of the previous two periods. The urban expansion intensity of this district has been at a medium level for a long time. Therefore, the Northern construction land expansion intensity index shows "high-low-high", the Western District shows "high-low-low", and the trend is "low-high-low" in Southern District.

The topography of Chongqing is high in the east and low in the west, and the towering terrain of Daqing Mountain in the north prevents the expansion of urban land to the north, so the expansion of construction land to the southwest and southeast is dominant after 2000, which shows a decrease in the intensity of construction land expansion in Northern District and an increase in Southern District (Figure 2). Western District, which has the highest expansion intensity, is located in the alluvial fan plain in the southwest, and the terrain is smooth and more conducive to urban expansion.

In contrast, Nanchuan District in the southeast is a water resource protection zone, which restricts the expansion of local urban land in this direction.

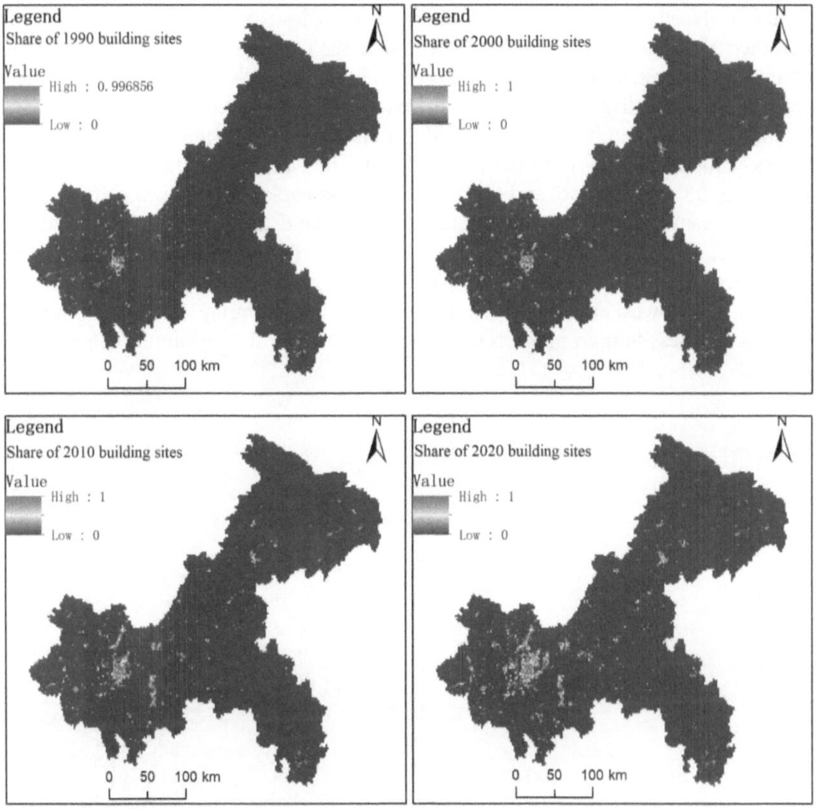


Figure 2. Expansion of construction land in Chongqing and division of circles.

### 3.2 *Temporal change analysis of ecological risk elements in Chongqing*

Analysis of the changes in ecological risk in Chongqing shows that the ecological risk value in Chongqing increased from 0.35 to 0.41 from 1990 to 2020, with the risk level evolving from low to medium (Figure 3). In 2000, the built-up area of Chongqing was 1.6 times larger than that in 1990, and the conversion of a large amount of ecological land to construction land inevitably led to a deepening of the disturbance to the ecological environment caused by human activities, with the risk values of the factor layer, exposure layer and impact layer increasing from 0.058, 0.063 and 0.148 to 0.076, 0.070 and 0.155 respectively. During the transition period from 2000 to 2010, the ecological risk increased from a low level to a medium level, and the risk value further increased to 0.40, which remained at a high level of 0.002. The risk value of the response level decreases slightly and the overall ecological risk is on the rise. In the period 2010–2020, the medium ecological risk level, the risk value increased at a low rate of 0.001 to 0.41.

On the flip side, the structure of urban land tends to stabilize and the risk value of the exposure layer begins to decline, which makes the ecological risk value increase at a slower rate. However, the built-up area expanded by 94 square kilometers during this period, and the rate and intensity of expansion were three times and two times that of the previous period respectively, and the risk to the factor and impact layers still increased, so the ecological environment was most vulnerable at this time.

### 3.3 *Spatio change analysis of ecological risk in Chongqing*

The integrated ecological risk values of all four jurisdictions showed an increasing trend (Figure 3). The ecological risk value of the Northern District changed the least, from 0.292 to 0.327, an increase of 0.036. Northern District has no distribution of low and lower grades in all four time periods, and the risk grade with the largest area share rises from medium to higher, and the whole district is basically covered by high and higher grades by 2020, with the least optimistic ecological environment. Southern District only had a small area of low-grade area in 1990, and the distribution of lower, middle, higher, and higher grades did not change much in the remaining years, and its ecological risk value changed within the middle grade. In Western District, all risk classes were distributed during the study period, and the dominant class evolved from lower to medium, which was related to the conversion of a large amount of arable land and unused land into construction land in the district after 2000. All risk levels are also distributed in the main 5 urban districts, where the low ecological risk level is widely distributed in the area where Daqing Mountain is located, which has the lowest ecological risk due to the high topography, absence of industries, and dense buildings, and less human interference, making the natural conditions such as climate and vegetation good.

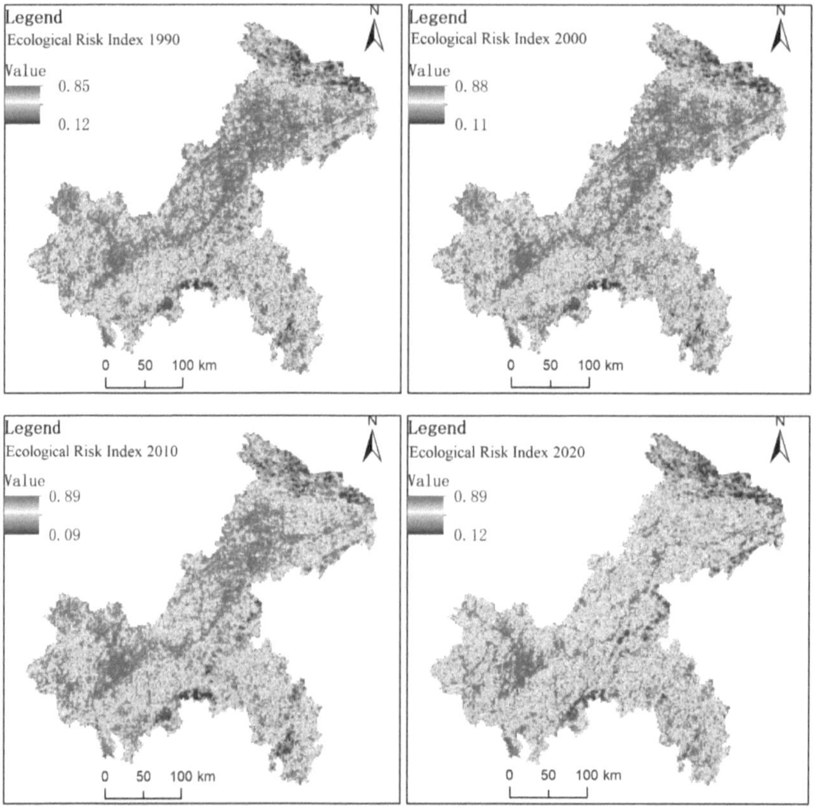


Figure 3. Ecological risk values of indicator layer in 1990–2020.

## 4 CONCLUSION

Ecological risk evaluation is the main direction of ecological security research in China. Based on urban expansion, this paper constructs an index system from multiple perspectives and calculates the integrated ecological risk values of Chongqing and its four jurisdictions in 1990, 2000, 2010, and 2020, and their responses to urban expansion. The results illustrate the spatial and temporal evolution of construction land expansion and ecological risk, reveal a significant correlation between the two, and conclude that urban expansion and the series of human-induced disturbances are one of the reasons for the increasing ecological risk year by year.

Compared with previous studies, on the one hand, the results of ecological risk increasing year by year are consistent with the conclusions drawn from both landscape ecological security; on the other hand, this paper integrates socio-economic and landscape structural elements to analyses not only the ecological risks in Chongqing but also the dynamics and geographical distribution of risks in four geographical regions at the numerical level, in order to reflect in detail the ecological risks in the whole Chongqing region. However, due to the availability of indicators, the 25 indicators alone are not comprehensive enough to evaluate the comprehensive ecological risk, especially since the response part of the indicators is small, and some indicators reflect environmental management (Zhang *et al.* 2020), such as the amount of household garbage harmlessly removed, are not selected because they cannot be spatially compared. At the same time, this paper calculates the comprehensive ecological risk by the quantitative model (Zhang *et al.* 2020), which cannot analyze the unknown factors that are difficult to quantify and the mutual driving relationship behind the data, so only the spatial and temporal evaluation of ecological risk cannot intuitively guide the ecological risk prevention and control.

## REFERENCES


Adrianna Damiana Mastalerz-Kedzie Katarzyna Pośpiech. (2018). Application of Hölder Function to Expansion Intensity of Spatial Phenomena Analysis. Acta Universitatis Lodziensis. *Folia Oeconomica*, 72:16–27.

Liu, M., Chen, L., Gou, Y. & Dong, R. (2011). Assessment of urban ecological risk from spatial interaction models for Lijiang City, *International journal of sustainable development and world ecology*, 18(6): 537–542.

Liu, X. & Chen, H. (2019). *Integrated assessment of ecological risk for multi-hazards in Guangdong province in southeastern China, Geomatics, natural hazards and risk*, 10(1): 2069–2093.

Song, H.M. & Xu, L.Y. (2011). A method of urban ecological risk assessment: combining the multimedia fugacity model and GIS, *Stochastic environmental research and risk assessment*, 25(5):713–719.

Xiu, Y.C., Liu, W.B., Liu, B. & Li, H.X. (2015). An Oilfield Ecological Risk Assessment System Integrating OERAM and GIS, *Chemical engineering transactions*, 46:33–56.

Yunfeng, H. & Ge, G. (2020). Framework and practice of urban landscape ecological risk assessment: a case study of the Tiantan region in Beijing, *Acta Ecologica Sinica*, 21: 78–95.

Zhang, D., Shi, X., Xu, H., Jing, Q., Pan, X., Liu, T., Wang, H. & Hou, H. (2020). A GIS-based spatial multi-index model for flood risk assessment in the Yangtze River Basin, China, *Environmental impact assessment review*, 83:106–197.

Zhang, D., Shi, X., Xu, H., Jing, Q., Pan, X., Liu, T., Wang, H. Hou, H. (2020). A GIS-based spatial multi-index model for flood risk assessment in the Yangtze River Basin, China, *Environmental impact assessment review*, 83:10–13.

*Advances in Urban Construction and Management Engineering – Cha (Ed)*


# Numerical analysis of temperature field in the precast construction stage of high-speed railroad box girder

Xiaowei Tao
*School of Traffic Engineering, Huanghe Jiaotong University, Jiaozuo, China*

Haikuan Liu*
*School of Civil and Hydraulic Engineering, Zhengzhou University, Zhengzhou, China*
*Henan Transportation Research Institute CO., LTD, Zhengzhou, China*

ABSTRACT: To calculate the heat of hydration and internal temperature field of concrete in the precast construction stage of a high-speed railroad box girder and provide guidance for the early thermal insulation and maintenance of box girder concrete, this paper examined the adiabatic temperature rise of concrete, constructed a finite element solid model for box girder and analyzed the variation law of internal temperature and temperature difference in the precast construction stage of the box girder. The analysis results provide a scientific basis for the thermal insulation and maintenance of box girders, which can reduce the early cracking risk of concrete.

## 1 GENERAL INSTRUCTIONS

With the rapid development of high-speed railroad construction in China, a large number of high-speed railroad bridges have been intensively constructed. Common structural forms of the bridge are all applied on high-speed railroads, among which simply-supported box girder, characterized by good bending performance, high torsional stiffness, simple and clear structural forces, and short construction period (Ban 2021; Dong 2010; Liu et al. 2015; Xing 2015), is the most common, accounting for more than 90% of the total bridge mileage (Ban 2021; Xing 2015). Most simply-supported girders are prefabricated in the precast yard, which can strictly regulate prefabrication conditions and guarantee quality. The inherent quality of a box girder is generally determined by the construction control and maintenance state control at the prefabrication stage. Cement generates a large amount of hydration heat in the precast construction stage of box girders. Due to the low thermal conductivity of concrete, heat is not likely to dissipate, and the inner temperature of the structure heats up rapidly but cools down slowly, forming a large temperature gradient that is easy to produce cracks. Cracks will seriously affect the safety, beauty, resistance to seepage, and durability of the structure after emergence. Therefore, to effectively control the internal and external temperature difference and prevent further cracks in construction and maintenance, it is of great importance to explore the change of temperature field in the precast construction stage of the box girder. Taking a 32-meter-long precast box girder of the high-speed railroad as an example, this paper evaluated and studied the internal heat of hydration and the variation law of internal and external temperatures of the box girder in the precast construction stage. The results provide a scientific basis for temperature control and maintenance in the precast construction stage.

## 2 32-METER-LONG PRECAST BOX GIRDER FOR HIGH-SPEED RAILROAD

A 32-meter-long precast box girder of a high-speed railroad was a simply-supported box girder with a single box of equal height, as shown in Figure 1. The top plate, bottom plate, and web plate

*Corresponding Author: lhkuan@foxmail.com


DOI 10.1201/9781003348023-4

of the box girder were partially thickened inward. The top surface of the box girder was 12.2 mm wide. The total length, calculated span, and height of the box girder were 32.6 m, 31.5 m, and 2.60 m, respectively, and the center distance of the cross-bridge directional bearing was 4.4 m. C50 concrete was used as its raw material. A box girder with a dead weight of 822.9 tons consumed 321.93 $m^3$ concrete. Box girder is generally prefabricated with fixed steel formwork and adopts a post-tensioned prestressing system.

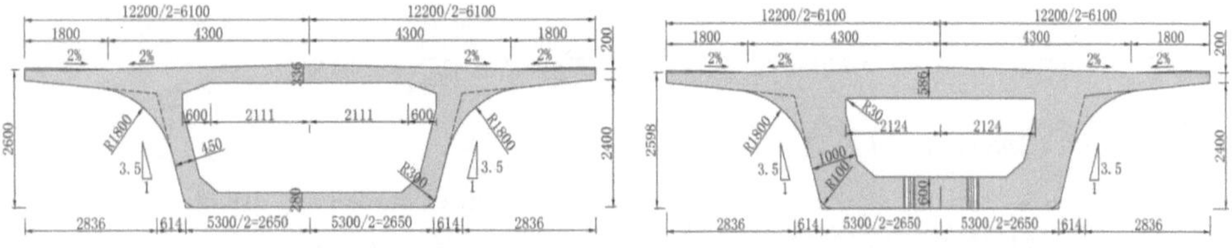


(a) The mid-span cross-section

(b) The support point cross-section

Figure 1. Cross-section of 30m prefabricated box girder (unit: mm).

## 3 MATERIAL PROPERTIES OF C50 CONCRETE FOR BOX GIRDER

### 3.1 *Adiabatic temperature rise test of concrete*

The hydration temperature rise of C50 concrete was measured for 7 d according to the standard concrete adiabatic temperature rise test method (2018, 2020). The adiabatic temperature rise values of concrete at various ages are listed in Table 1. It can be seen from the table that the concrete mostly reached its maximum temperature rise at 45~50 h and that its maximum temperature adiabatic temperature rise of 42.6°C occurred at 168 h. The temperature rise test results of concrete at the early stage provide the accurate heat source for the finite element thermal analysis in the construction process of the box girder, and thus a basis for temperature field analysis.

Table 1. The adiabatic temperature rise of concrete at various ages.

| Concrete age (h) | 1 | 5 | 10 | 15 | 20 | 25 | 30 | 35 | 40 | 45 | 50 | 55 | 60 |
|---|---|---|---|---|---|---|---|---|---|---|---|---|---|
| Temperature rise (°C) | 0.5 | 2.5 | 28.5 | 38.6 | 41.5 | 42.3 | 42.4 | 42.4 | 42.5 | 42.5 | 42.6 | 42.6 | 42.6 |
| Concrete age (h) | 65 | 70 | 75 | 80 | 85 | 90 | 95 | 100 | 105 | 110 | 115 | 120 | 168 |
| Temperature rise (°C) | 42.6 | 42.6 | 42.6 | 42.6 | 42.6 | 42.6 | 42.6 | 42.6 | 42.6 | 42.6 | 42.6 | 42.6 | 42.6 |

### 3.2 *The mechanical and thermal parameters of concrete*

The compressive strength development of C50 concrete for box girders was simulated with reference to the ACI code (2011). The strength development function is expressed as:

$$f_{ck(t)} = \frac{t}{a + bt} f_{ck(28)}$$

where $f_{ck(t)}$—Concrete compressive strength at age t days, in MPa.

$f_{ck(28)}$—Compressive strength of concrete at 28 days of age, in MPa.

$a, b$—Constant cement type.

The test results indicated that the compressive strength of concrete at 3 d, 10 d, and 28 d was 46.1 MPa, 56.3 MPa, and 66.1 MPa, respectively. Based on the measured strength, the least square method was used to regress the cement type constants, $a = 1.655$, $b = 0.958$. Other mechanical and thermodynamic parameters of materials for C50 concrete were calculated, as shown in Table 2.

Table 2. The mechanical and thermodynamic parameters of C50 concrete.

| Physical Characteristics | Values | Physical Characteristics | Values |
|---|---|---|---|
| Modulus of elasticity (MPa) | $3.86 \times 10^4$ | Specific heat capacity $(kJ/kg \cdot °C)$ | 0.97 |
| Poisson's ratio | 0.2 | Thermal conductivity $(kJ/m \cdot h \cdot °C)$ | 10.0 |
| Coefficient of linear expansion | $1 \times 10^{-5}$ | Weight capacity $(kn/m^3)$ | 26.0 |

## 4 THE FINITE ELEMENT MODEL AND CONVECTIVE BOUNDARY

### 4.1 *The finite element model*

According to the symmetry of the box girder, the 1/2 box girder model was established by using Midas/Civil finite element software, as shown in Figure 2, and the temperature boundary conditions were set in the symmetric section. As can be seen from Figure 2, the model was divided into 13,513 units and 6,069 nodes, with a boundary, either a support boundary or a convection boundary, in each face. The support stiffness of the formwork to the box girder during construction was set according to the actual conditions, and the support boundary was set on the symmetry surface to constrain its horizontal displacement.

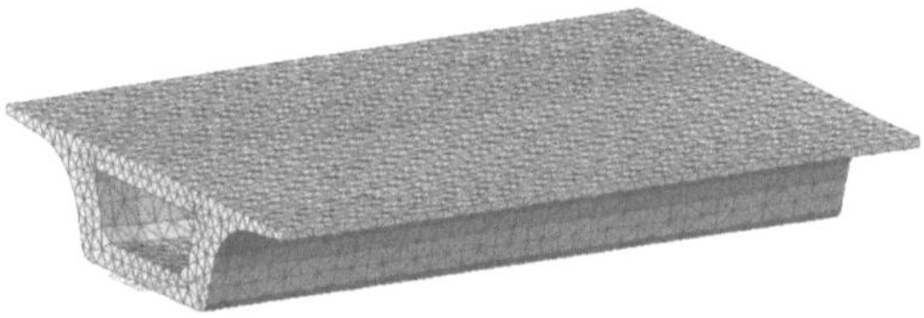

Figure 2. The solid finite element model of the box girder.

### 4.2 *The convective boundary of the finite element model*

In addition to setting the support boundary of the model, the convection boundary of the finite element unit should also be set correctly when the finite element model is used to analyze the hydration heat of the box girder. The combined steel formwork was used for inner and outer formworks in the prefabrication of the box girder, and the formwork and stiffening ribs were converted to steel plates with a thickness of 2 cm. After casting, the top surface of the box girder was covered by a geotextile cloth with a thickness of 2 cm. When the surface of the concrete was covered with formwork or insulation layer, the third type of boundary conditions was adopted for calculation (Liu 2011), but the impact of formwork or insulation layer was considered by the method of equivalent exothermic coefficient $\beta_s$.

$$\beta_s = \frac{1}{(1/\beta) + \sum (h_i/\lambda_i)}$$

where $\beta$—The coefficient of heat release from the concrete surface into the air.

$h_i$—Thickness of the insulation layer.

$\lambda_i$—Thermal conductivity of the insulation layer.

The equivalent exothermic coefficients of different convection boundaries of the box girder were calculated as follows.

(1) Convective heat release coefficient of the top surface: The surface was covered with 2-centimeter-thick geotextile insulation, and the thermal conductivity was 0.673 $kJ/m \cdot h \cdot °C$. When the wind speed at 0 m/s was taken into account, the heat release coefficient to the air was 21.06 $kJ/m^2 \cdot h \cdot °C$, and the calculated equivalent heat release coefficient was 12.95 $kJ/m^2 \cdot h \cdot °C$.

(2) Convective heat release coefficient of the internal mold: The steel template was converted into a 2-centimeter-thick steel plate, and its thermal conductivity was 288 $kJ/m \cdot h \cdot °C$. When the wind speed at 0 m/s was considered, the heat release coefficient to the air was 18.46 $kJ/m^2 \cdot h \cdot °C$, and the calculated equivalent heat release coefficient was 18.45 $kJ/m^2 \cdot h \cdot °C$.
(3) Convective heat release coefficient of the external mold: The steel template was converted to a steel plate with a thickness of 2 cm, and its thermal conductivity was 288 $kJ/m \cdot h \cdot °C$. When the wind speed at 1 m/s was considered, the heat release coefficient to the air was 35.75 $kJ/m^2 \cdot h \cdot °C$, and the calculated equivalent heat release coefficient was 35.66 $kJ/m^2 \cdot h \cdot °C$.

## 5 THE VARIATION LAW OF TEMPERATURE FIELD FOR BOX GIRDER

The variation of the internal temperature of the box girder with age was analyzed, and the temperature field distribution and the variation curve of internal temperature at each age were obtained. The temperature of concrete entering the formwork was considered to be 20°C, and the variation of ambient temperature was simulated as a sinusoidal function $T = 10\sin(\pi t/12) + 25$ according to local meteorological data. In this paper, the temperature variation curves of the feature points on the characteristic sections of the box girder the mid-span, 1/4-span, and supporting point sections were presented, and the difference between internal and external temperatures and the temperature gradient of the structure were also analyzed.

### 5.1 *The temperature variation of the mid-span section*

Eight feature points at different locations were selected for temperature variation analysis for the mid-span section, as shown in Figure 3. Among them, points 1~3 were located at the top edge of the top plate, point 4 at the center of the junction between the web and the top plate, point 5 at the outer edge of the middle and upper part of the web, point 6 at the inner edge of the middle and lower part of the web, point 7 at the outer edge of the junction between the web and the bottom plate, and point 8 at the lower edge of the middle of the bottom plate. The temperature trend at each feature point of the section with age is shown in Figure 4. As can be seen from the figure, the box

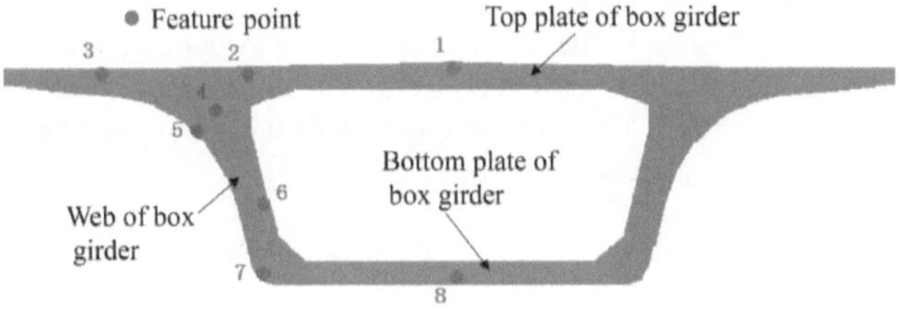


Figure 3. Distribution of feature points of the mid-span section.

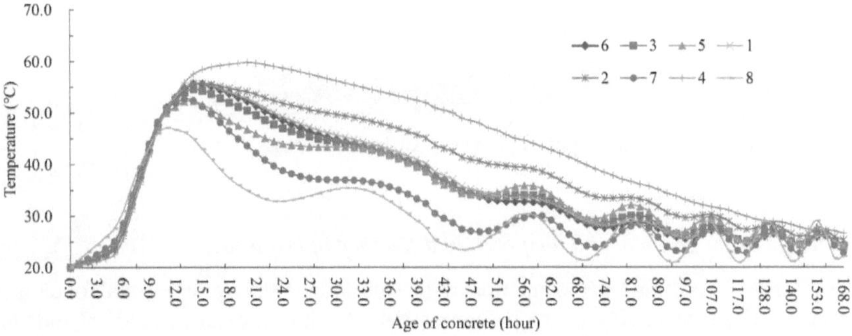


Figure 4. The temperature trend at each feature point of the mid-span section.

girder coagulation was mostly exothermic at the age of 5~15 h. Each feature point of the mid-span cross-section reached its maximum temperature of 47 ~ 59°C at 10 ~ 25 h. The concrete inside the thicker web reached its maximum temperature at 20 ~ 25 h, with a peak temperature of 59°C.

### 5.2 *The temperature variation of the 1/4-span section*

For the 1/4 span section, the distribution of feature points is the same as that of the mid-span section, and the temperature variation trend of each feature point with age is shown in Figure 5. The box girder concrete concentrates the exothermic heat in the process of setting and hardening at the age of 5~15 h. The maximum temperature of each feature point is reached at 10~20 h, and the peak temperature is 48~60°C. The concrete inside the thicker web reaches the maximum temperature at 20~25 h of age, with a peak temperature of 60°C.

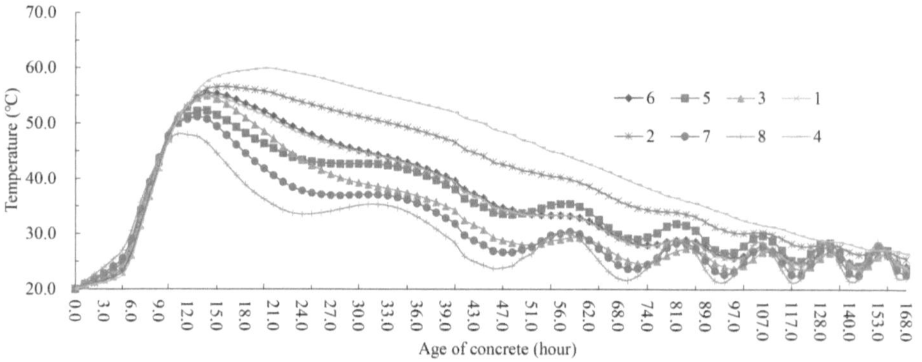


Figure 5. The temperature trend at each feature point of the 1/4 span section.

### 5.3 *The temperature variation of the support point cross-section*

In terms of supporting point cross-section, nine feature points at different locations were selected for analysis, as shown in Figure 6. The temperature trend of each feature point of the supporting point section with age is shown in Figure 7. The concrete of the box girder concrete intensively released heat in the setting and hardening processes at 5~15 h. The temperature reached its maximum at 10~25 h, with a peak value of 50 ~ 61°C. The concrete inside the thicker web reached its maximum temperature at 20~2h, with a peak temperature of 61°C.

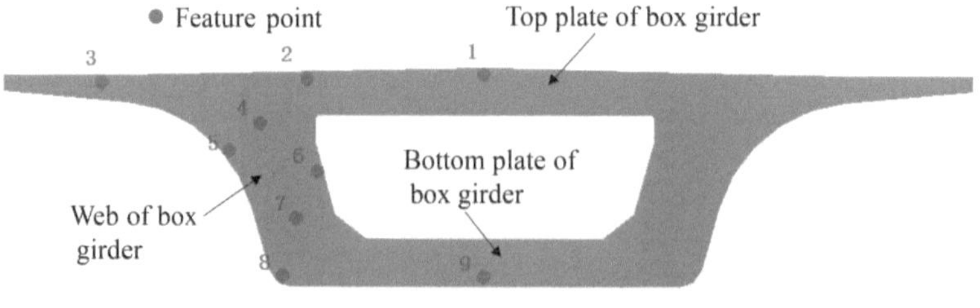


Figure 6. The distribution of feature points of the supporting point cross-section.

### 5.4 *The temperature gradient and temperature difference of box girder*

The feature points 2, 3, 4, and 5 at the junction of the web and top slab were selected for temperature gradient analysis, as respectively shown in Figures 3 and 6. The temperature gradient and temperature difference between inside and outside of the box girder at 20 h, 30 h, 40 h, 60 h, 92 h, and 168 h

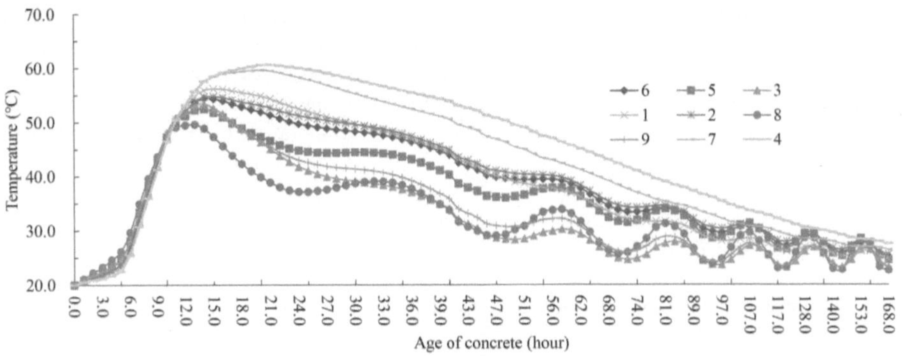


Figure 7. The temperature trend at each feature point of the supporting point cross-section.

for the mid-span section and supporting point cross-section were respectively counted, as displayed in Table 3. The temperature difference and temperature gradient between inside and outside of the box girder reached their maximum around 40 h, indicating that the most critical period for the maintenance and insulation of the box girder was from the completion of concrete placement to 40 h, and that external nourishment and protection measures should be taken in advance to avoid cracks caused by excessive temperature difference.

Table 3. The maximum temperature difference and temperature gradient at the feature points of the box girder.

| The mid-span cross-section | | | | | | | The support point cross-section | | | | | | |
|---|---|---|---|---|---|---|---|---|---|---|---|---|---|
| | Feature points / Temperature (°*C*) | | | | Maximum temperature difference | Maximum temperature gradient | | Feature points / Temperature (°*C*) | | | | Maximum temperature difference | Maximum temperature gradient |
| Age (*h*) | 2 | 3 | 4 | 5 | (°*C*) | (°*C*/m) | Age (*h*) | 2 | 3 | 4 | 5 | (°*C*) | (°*C*/m) |
| 20 | 54.0 | 48.4 | 59.8 | 46.7 | 13.1 | 24.7 | 20 | 53.1 | 46.4 | 60.6 | 47.3 | 14.2 | 24.5 |
| 30 | 49.6 | 39.4 | 56.1 | 43.4 | 16.7 | 24.1 | 30 | 49.6 | 39.1 | 57.9 | 44.4 | 18.8 | 24.9 |
| 40 | 45.0 | 34.2 | 51.9 | 38.6 | 17.7 | 25.1 | 40 | 44.9 | 34.0 | 53.9 | 40.3 | 20.0 | 25.3 |
| 60 | 38.6 | 24.7 | 43.4 | 35.2 | 18.7 | 15.4 | 60 | 39.2 | 30.1 | 45.7 | 37.4 | 15.7 | 15.5 |
| 92 | 30.5 | 24.8 | 34.3 | 27.3 | 9.6 | 13.3 | 92 | 31.1 | 24.5 | 36.7 | 29.3 | 12.1 | 13.6 |
| 168 | 25.2 | 23.6 | 26.5 | 23.9 | 2.9 | 5.0 | 168 | 25.6 | 23.3 | 27.6 | 24.7 | 4.3 | 5.5 |

## 6 CONCLUSION

The adiabatic temperature rise of C50 concrete was measured, and the variation law of the temperature field of the box girder was investigated numerically. The conclusions are drawn as follows:

(1) The concrete of the box girder is released in the setting process at 5~15 h, reaching the maximum temperature of 47 ~ 61°C at 10~15 h. The inner part of the thicker web reaches the highest temperature at 20 – 25 h, peaking at 60 – 61°C. The heat dissipation is slow such as at point 4 and less affected by the external temperature after internal concrete reaches its maximum temperature,
(2) The inner and outer surfaces of the box girder in contact with the steel form have good heat dissipation, and the temperature is significantly affected by temperature fluctuation. During construction, attention should be paid to the impact of sudden temperature changes on the temperature of the concrete surface.

(3) After reaching the peak temperature, the average cooling rate inside the concrete and on the surface is around 5°C/d and 7.5°C/d, respectively. The box girder has a greater cooling rate under natural conditions and thus no additional cooling measures are required. However, in the case of extreme cooling weather or construction in winter, the concrete surface should be well insulated to avoid cracks caused by an excessive temperature difference between inside and outside.


## ACKNOWLEDGEMENT

Funded by the Transportation Science and technology planning project of Henan Province (No.2021J2, No.2015X01-2).


## REFERENCES


2011. *Building Code Requirements for Structural Concrete*. In: INSTITUTE, A. C. (ed.).

2018. *The standard for the construction of mass concrete*. National standard of the People's Republic of China.

2020. Hydraulic concrete test procedure. *Water Resources Standard*.

Ban, X. 2021. Theoretical study on the design of prefabricated 40m simply supported box girder for the high-speed railroad. Ph.D., *China Academy of Railway Science*.

Dong, C. 2010. *Spatial finite element analysis of mechanical properties of pre-stressed simply-supported box girder for a high-speed railroad*. Master, Central South University.

Liu, H. 2011. *Design and temperature control of bridge mass concrete*. Master, Changsha University of Science and Technology.

Liu, Y., Dai, G. & Kang, C. 2015. A review of simply supported girders for high-speed railroads in China. *Journal of Railway Science and Engineering*, 12, 242–249.

Xing, T. 2015. *Research on prefabrication process of simple-supported box girder and its key technology for a high-speed railroad*. Master, Shijiazhuang University of Railways.

*Advances in Urban Construction and Management Engineering – Cha (Ed)*


# Analysis of the effects of different bracing construction schemes for curved girder bridges

Yang Wei*, Chengtao Cui*, Pengfei Xie*, Junbin Zhang* & Jie Liu*
*China Water Conservancy and Hydropower14th Engineering Bureau Co., Ltd, Guiyang, China*

Fangfang Li*
*School of Civil Engineering and Construction, Xi'an University of Technology, Xi'an, China*

ABSTRACT: To study the influence of different bracket construction schemes on the displacement and structural internal force of the curved girder bridge, to ensure the safety of the bridge during construction, and to ensure the reasonable force in the completed state, the main bridge of Erbao Special Bridge with (45 + 70 + 45) m prestressed concrete variable-section continuous girder was used as the engineering background. The finite element software MIDAS/civil was used to model the full cloth full bracket (scheme 1) and the combined bracket (scheme 2), respectively, calculate the bearing capacity of the bracket foundation, analyze the bracket force and deformation during the construction process, and carry out a comparative analysis of schemes according to the calculation results. The results show that option 1 can meet the requirements of the whole construction stage, and there is enough surplus, especially when the beam height is small, using 0.9 m longitudinal spacing, the surplus is large, and the spacing can still be optimized; option 2 can meet the requirements of stress and deformation in the whole stage, but the stress is close to the specification limit, and the load transferred from the steel column to the foundation is too large. The foundation bearing capacity does not meet the requirements and produces certain impact on the underpass structure.

## 1 INTRODUCTION

The types of brackets commonly used in bridge construction are fastener type steel pipe bracket, bowl buckle type steel pipe bracket, portal type steel pipe bracket, and full cloth type full hall bracket (Jang 2008). To ensure the reasonably structural force premise, while saving the project cost and shortening the construction period, it is of certain theoretical significance and practical need to study the force conditions of main girders under different bracket construction.

Numerous scholars in China have carried out related research work. For example, Zhuang Jinping and others (Zhuang 2016) studied the influence law of crossbar step distance, vertical bar cross distance, and initial defects on the stable bearing capacity of tall formwork fastener type steel pipe bracket system by finite element software ABAQUS. Ran Tao and others (Ran 2020) conducted modeling analysis based on numerical simulation means to study the influence law of different scale factors on the stable bearing capacity of the structure. Liang Yan and He Enming took cast-in-place concrete box girder as the research object and compared the bearing performance, structural safety, and economy of the socket type coiled buckle bracket system and bowl buckle bracket system (He 2015; Liang 2016). The comparison of force performance, structural safety, and economy between socketed pan buckle bracket system and bowl buckle bracket system pointed out that the force performance of the uprights of pan buckle bracket is better than that of bowl buckle bracket. The structural safety performance of pan buckle bracket is better than that of bowl buckle bracket, and

*Corresponding Authors: 1191539@qq.com, 254716264@qq.com, 290488303@qq.com, 3428760233@qq.com, 1430588166@qq.com and 2730841253@qq.com

DOI 10.1201/9781003348023-5

it can meet the requirements of safe and civilized construction better. The pan buckle bracket can save the construction period, but shortcomings are to be handled. Shao Hongxing et al., based on the construction and budget cost of steel pipe bracket and full-bore bracket of cast-in-place bridge, analyzed the scheme and main materials used in actual construction, and proposed the billing analysis of different brackets (Shao 2014).

In this paper, a three-dimensional finite element analysis model is established based on the main bridge project of the Erbao Special Bridge by using numerical simulation, and two bracing schemes are proposed: the full-braced bracing scheme and the combined bracing scheme. According to the current highway bridge design specification, the two bracing schemes are analyzed in relation to the specific construction of the (45 + 70 + 45) m prestressed concrete variable-section continuous girder of the Erbao Special Bridge, and whether the force deformation and foundation bearing capacity of the two bracing schemes meet the requirements during the construction process, and the scheme is compared.

## 2 ESTABLISHMENT OF FINITE ELEMENT MODEL

The finite element model of the continuous girder is divided into 278 units and 546 nodes according to the actual construction process, and the radius of curvature of the model plane is 300 m. The finite element simulation model of the whole bridge is shown in Figure 1, and to simplify the calculation, the calculation model is established in scheme 1: full cloth full hall brace with a variable section, longitudinal spacing of brace 0.9 m section, and 6 rows of the brace as the research object can be seen in Figure 2; scheme 2: establishment of portal bracket structure model, the upper beam and bracket self-weight, and other construction loads are loaded on to the model with beam unit load, which can be seen in Figure 3.

Figure 1. Model of the main bridge of Erbao special bridge.

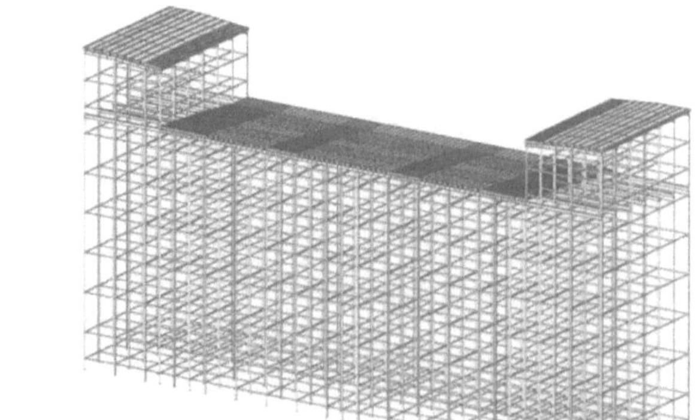

Figure 2. Model of full-span full-span scaffolding.

The full-braced method is a more common construction method in bridge construction, which involves building scaffolds that act as supports at certain intervals in a dense manner to transfer the upper loads to the foundation (Lu 2017; Shao 2014). In addition, the multi-point support during the construction of the full hall support is easy to control the settlement, and the support rebound is small with tension, which is beneficial for the health of the main beam, and the linearity is also easy to control.

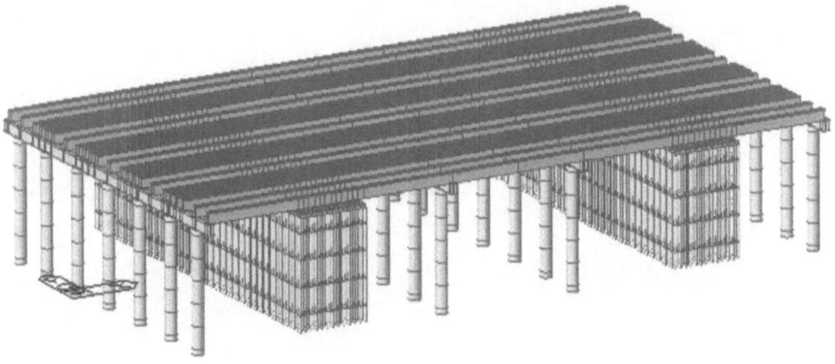

Figure 3. Cross-sectional view of the portal bracket.

Portal brackets are very widely used because of their easy processing, flexible disassembly and assembly, convenient handling, and high versatility (Gui 2018; Zhao 2011).

## 3 BEAM FORCE AND DEFORMATION ANALYSIS

Using the above finite element model, the whole construction process is simulated and the results of stress and deformation diagrams for each construction stage are discussed and analyzed as follows.

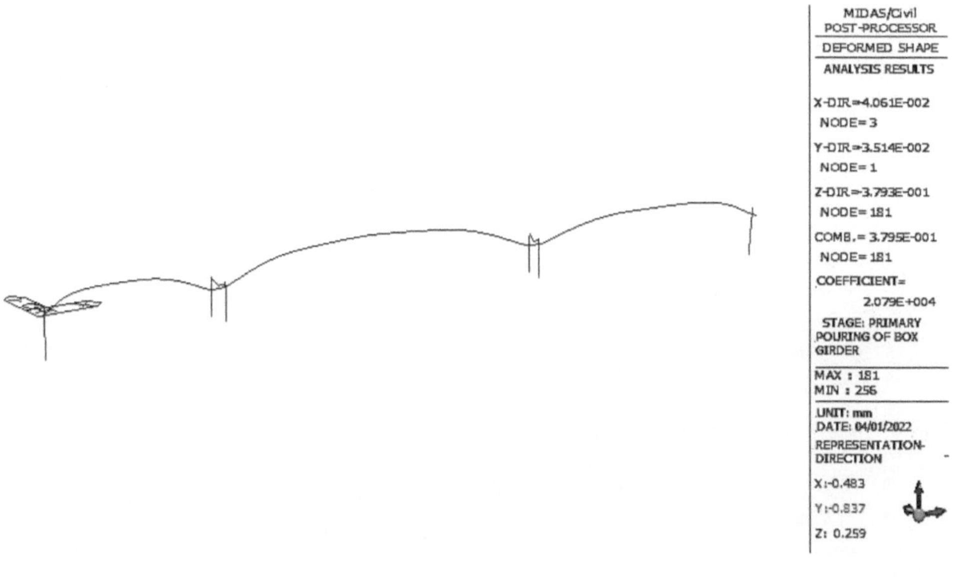


Figure 4. Deformation diagram of the beam one casting.

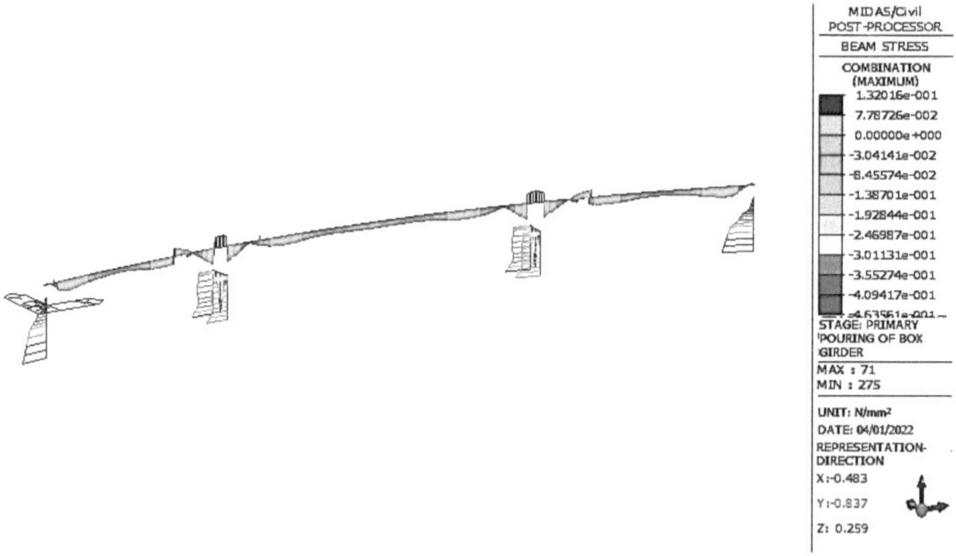


Figure 5. Stress diagram of the beam in one casting.

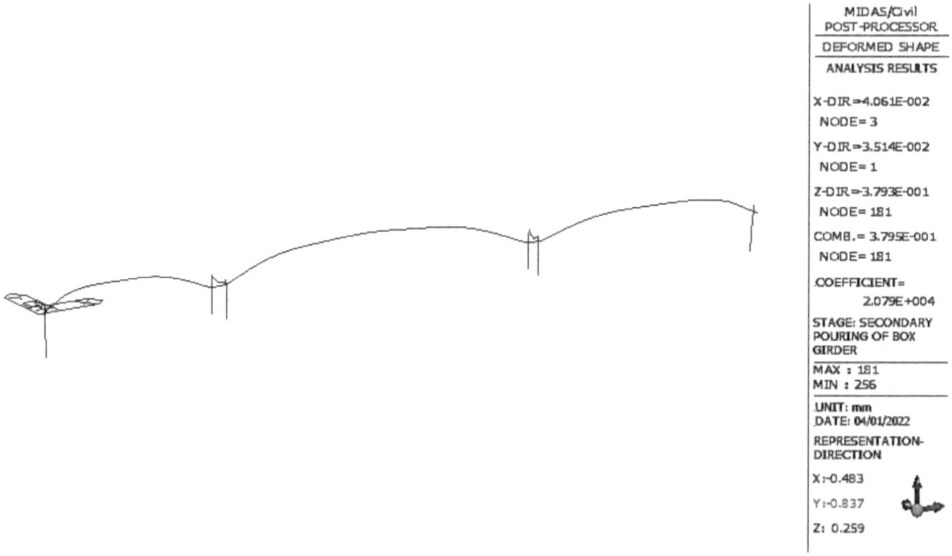


Figure 6. Stress diagram of secondary cast beam.

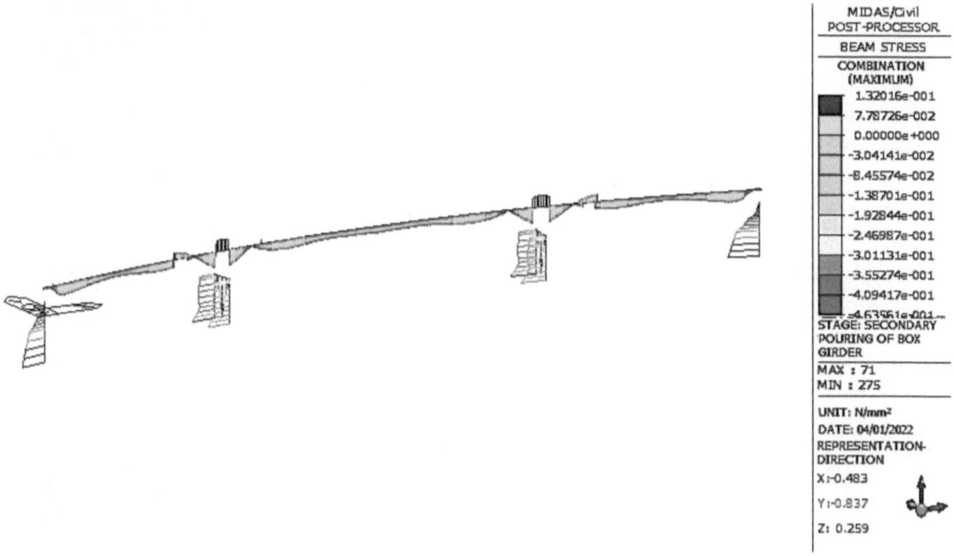


Figure 7. Stress diagram of secondary cast beam.

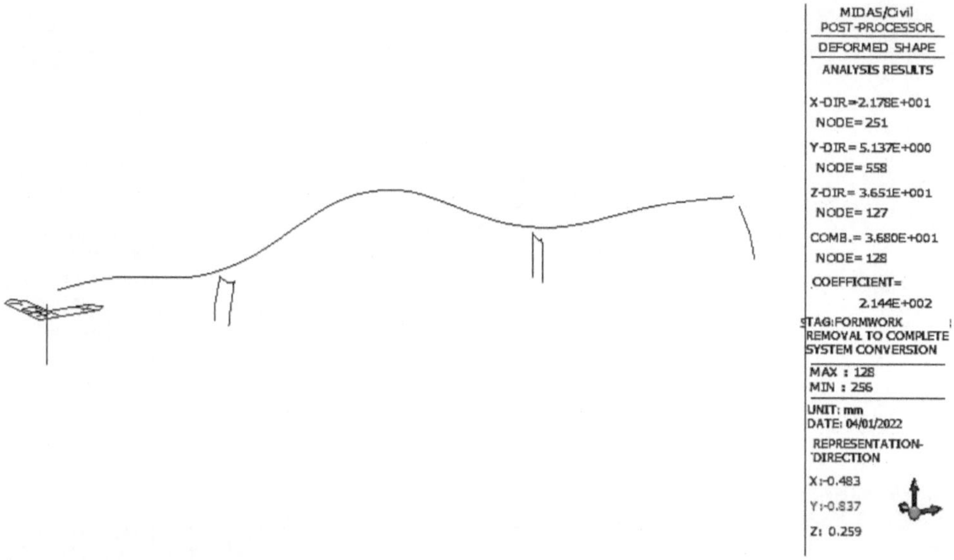


Figure 8. Deformation diagram of the complete system conversion by removing the formwork.

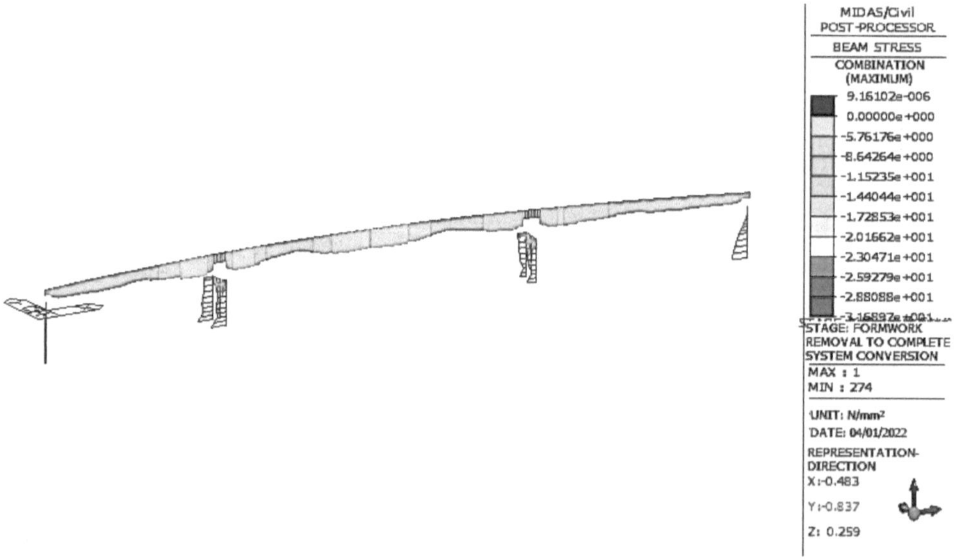


Figure 9. Dismantling formwork to complete the system conversion strain diagram.

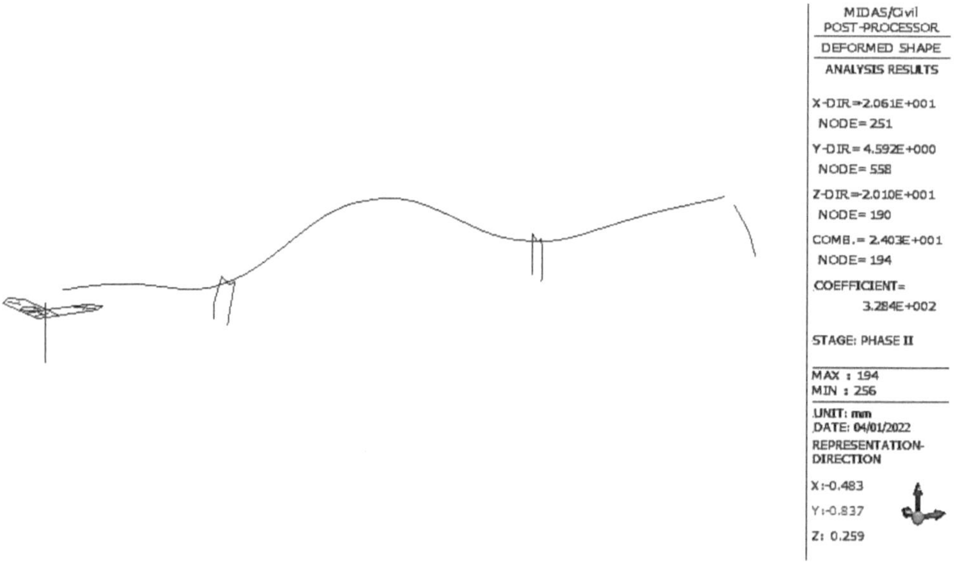


Figure 10. Deformation diagram of the second.

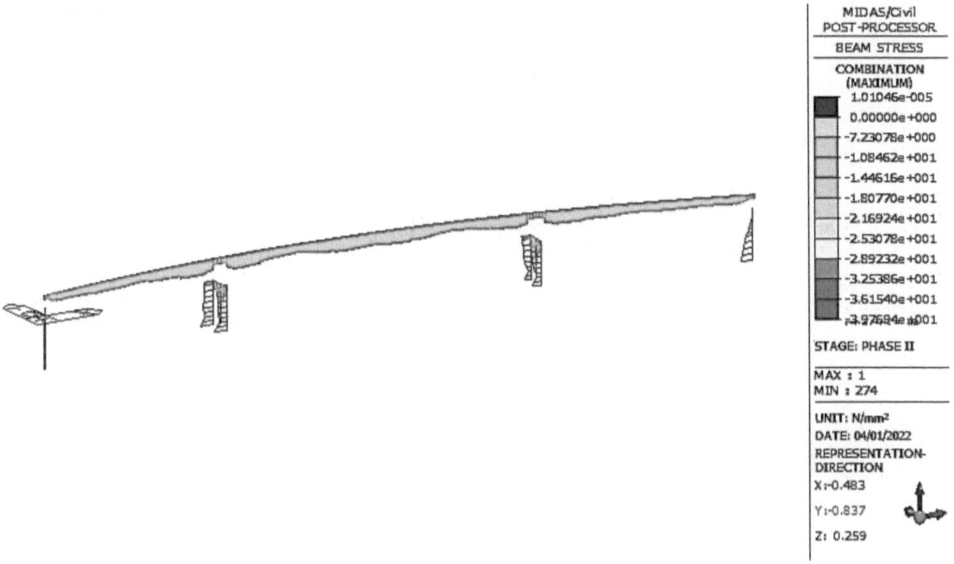


Figure 11. Second phase paving strain diagram phase paving.

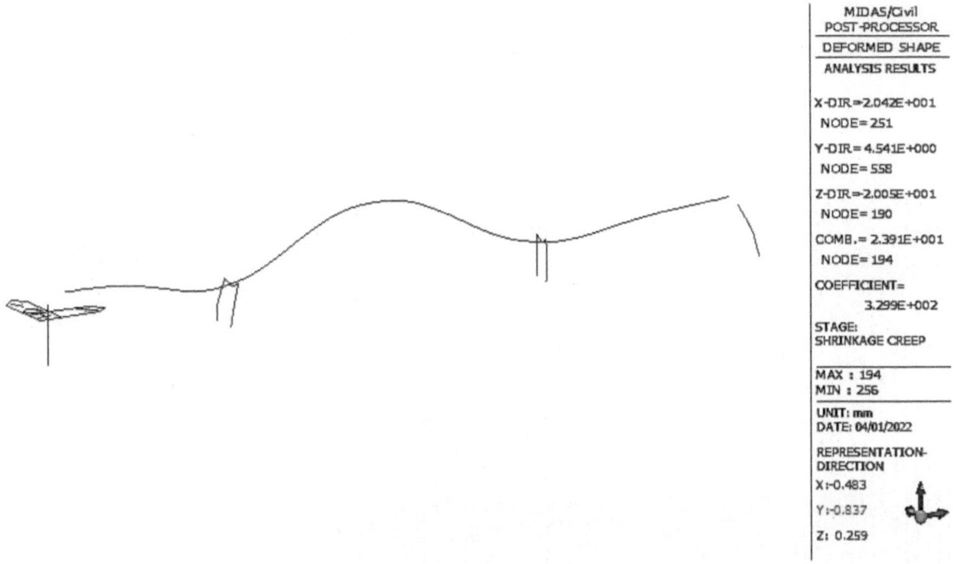


Figure 12. Systolic creep deformation.

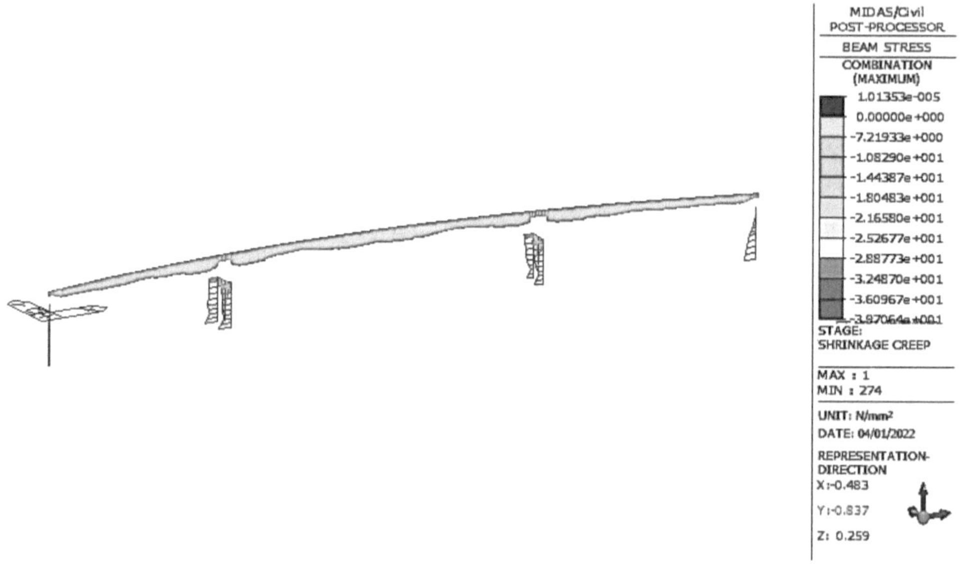


Figure 13. Systolic creep stress diagram.

Figure 4–13 shows the complete construction process, the full bridge stress, and deformation according to the above results and the bracket force analysis.

## 4 BRACKET FORCE DEFORMATION STABILITY ANALYSIS

### 4.1 *Option 1 bracket force deformation analysis*

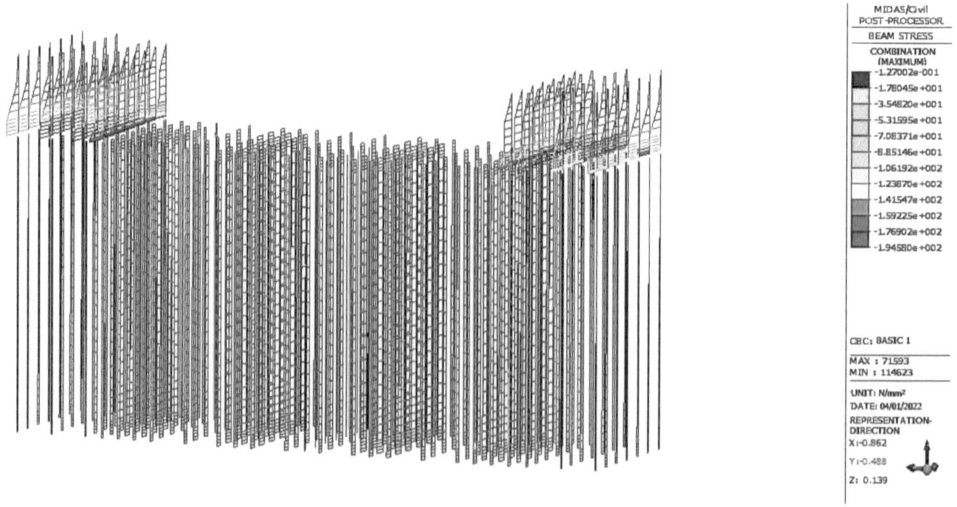


Figure 14. Stress diagram of vertical rod.

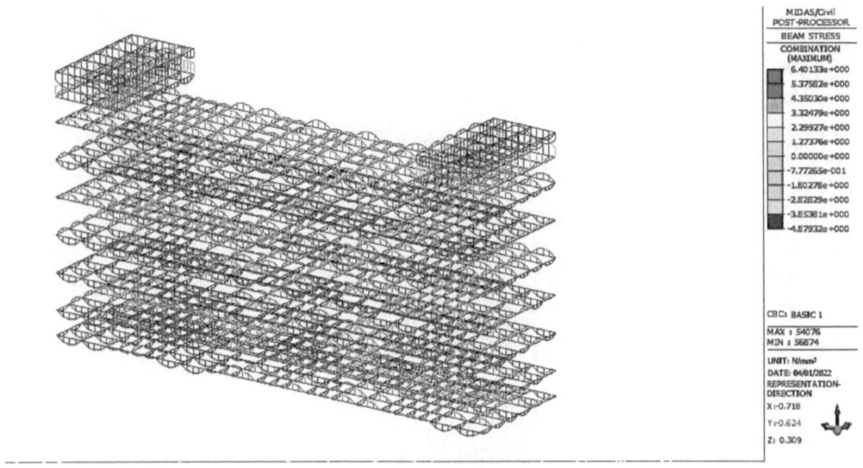


Figure 15. Horizontal rod stress diagram.

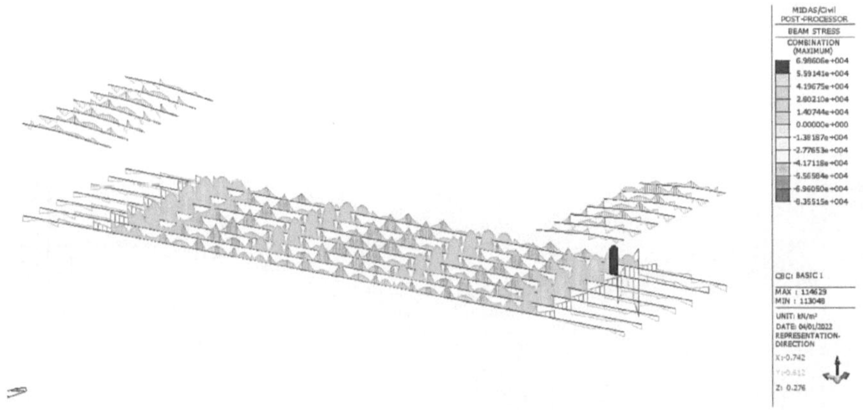


Figure 16. Crossbeam (I-beam) stress diagram.

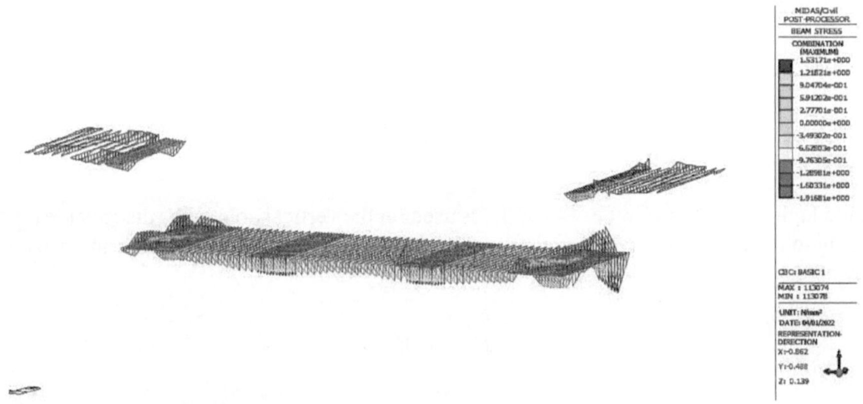


Figure 17. Secondary beam (square wood) stress diagram.

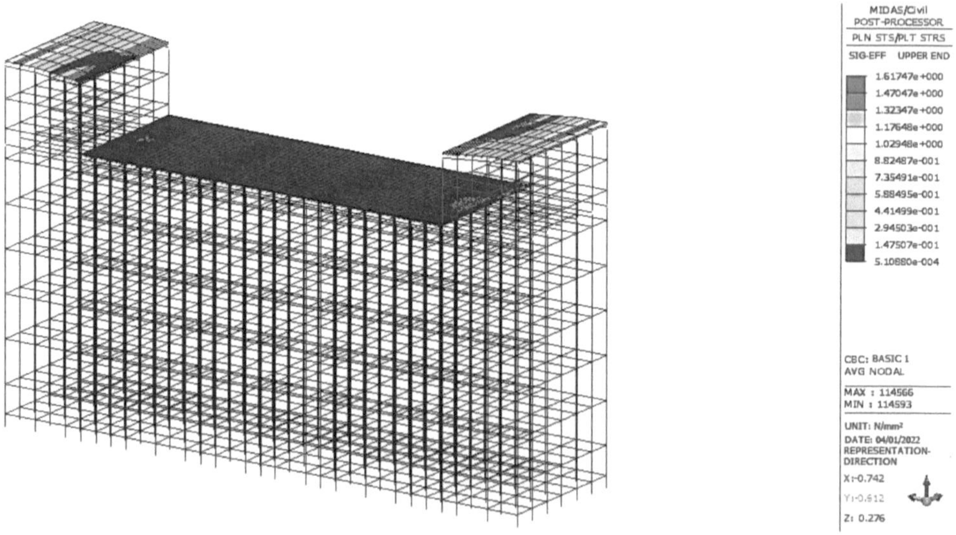


Figure 18. Adhesive plate stress diagram.

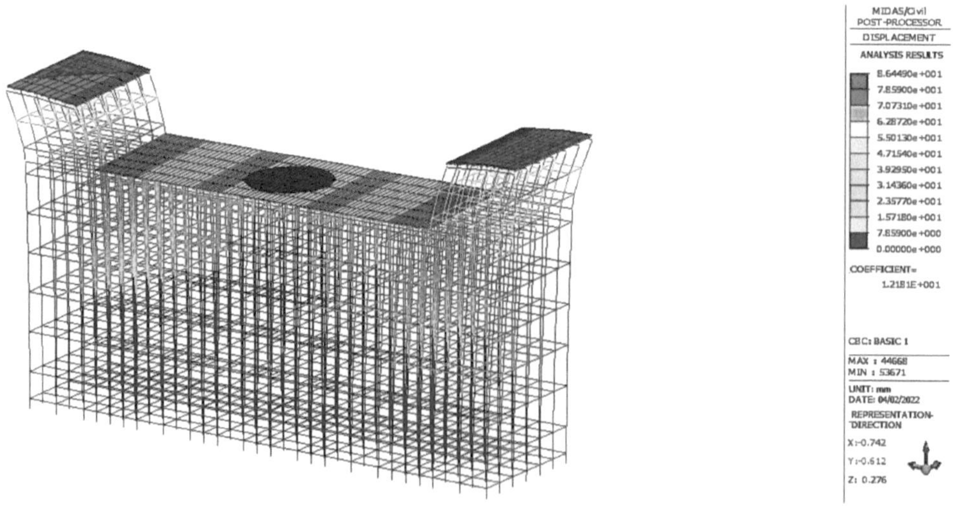


Figure 19. Deformation diagram of full cloth type full hall bracket.

The Q345 high-strength steel pipe Φ48 × 3.2 is used for the vertical pole of the disc buckle bracket. According to the current code of our country, the design value of the stability bearing capacity of the vertical pole is calculated according to the following formula:

$$K_d = \varphi A f \tag{1}$$

From the results of Figures 14–19, it can be seen that the maximum displacement of the bracket is 86.4 mm and the maximum stress is 194.6 MPa, which meets the specification requirements.

## 4.2 *Option 2 bracket force deformation analysis*

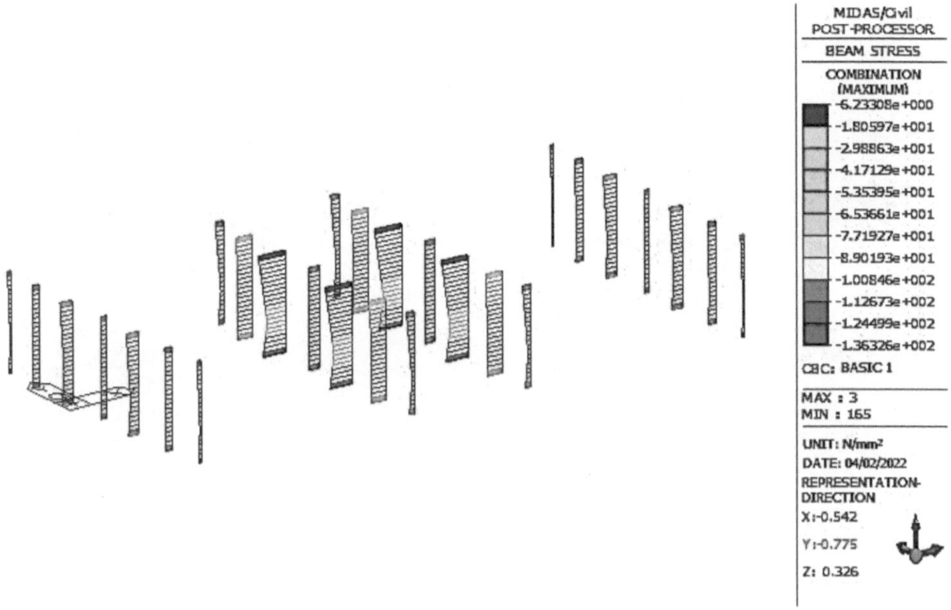


Figure 20. Steel column stress diagram.

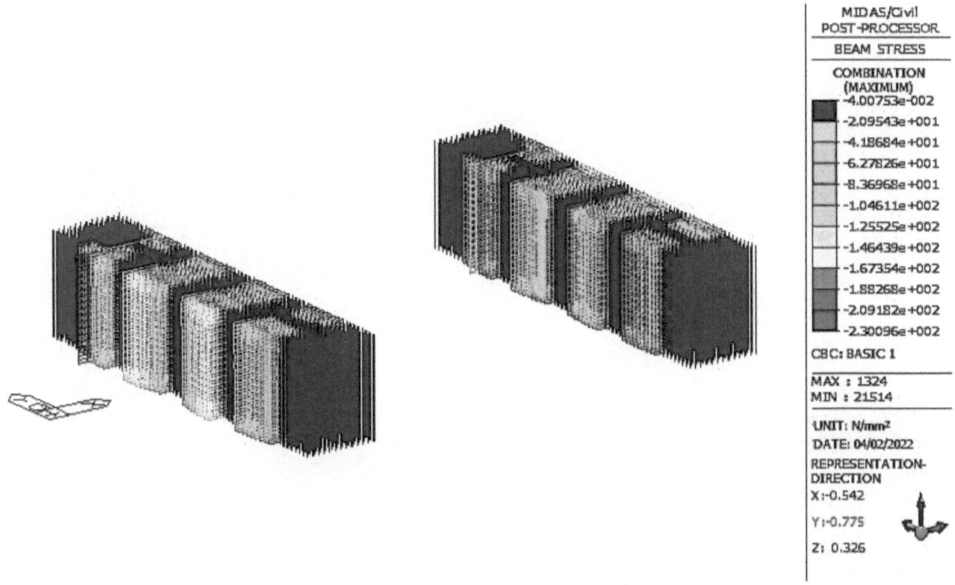


Figure 21. Stress diagram of the column.

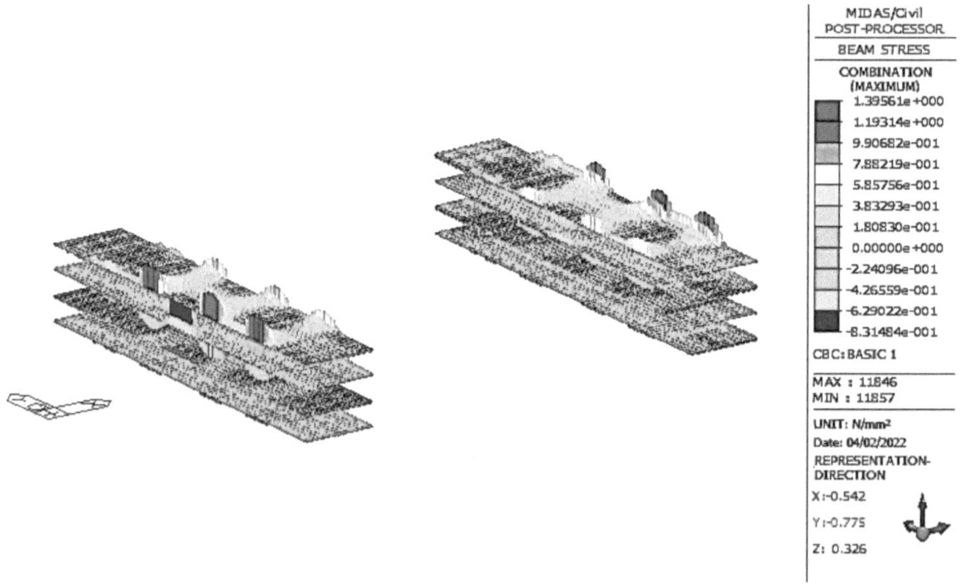


Figure 22. Horizontal bar stress diagram.

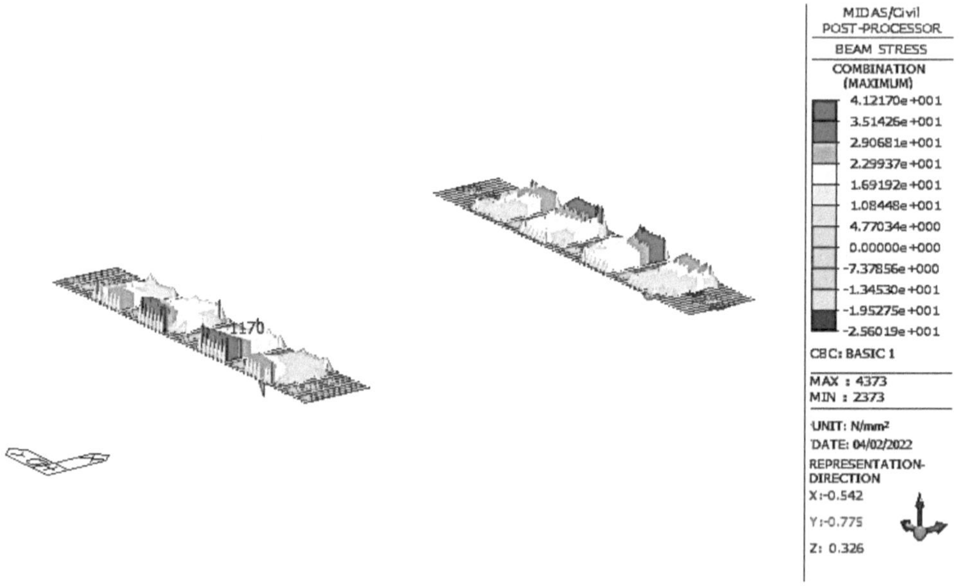


Figure 23. Stress diagram of I-beam No.14.5.

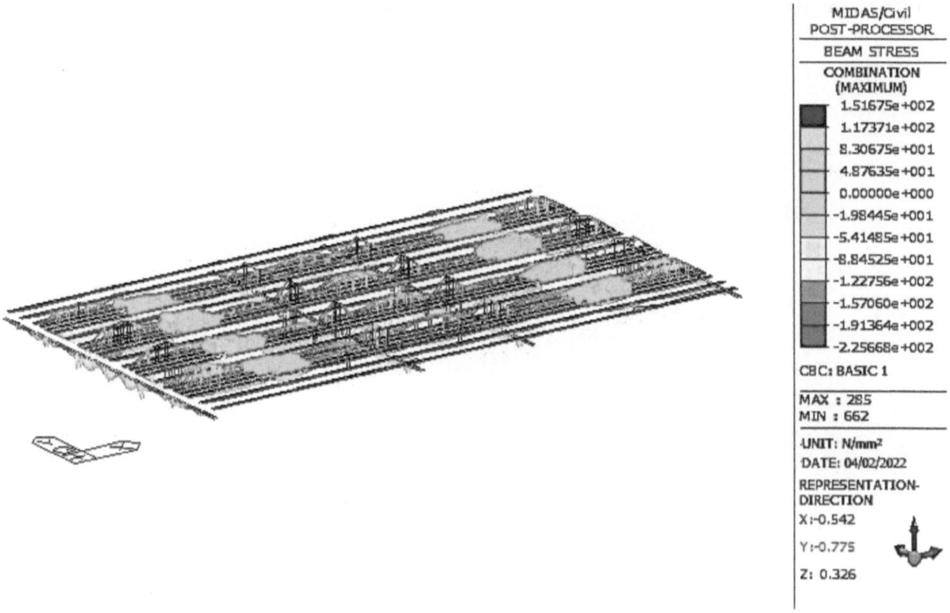


Figure 24. 45 I-beam stress diagram.

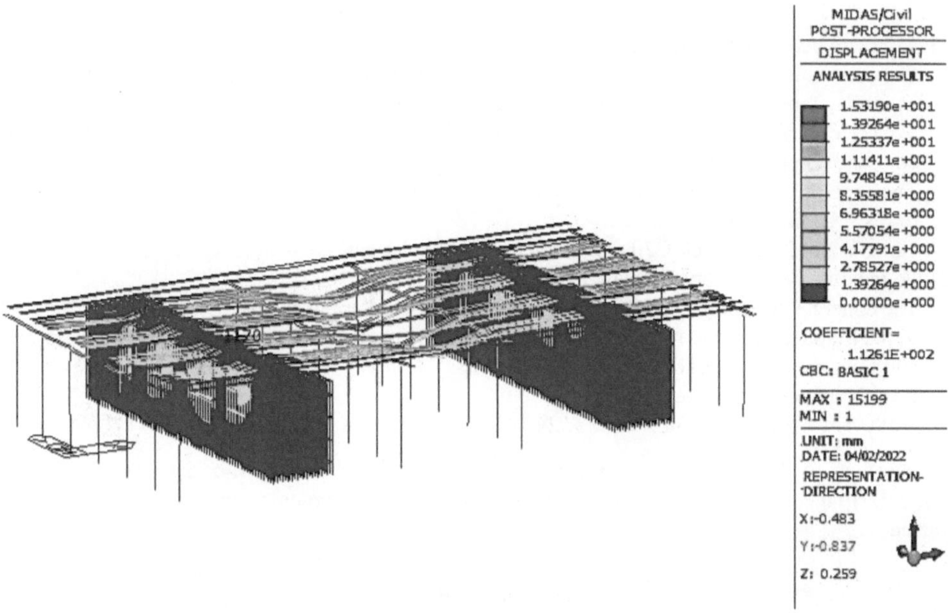


Figure 25. Deformation diagram of portal frame.

From Figures 20–25 stress deformation diagram, it is observed that the maximum displacement of the bracket is 15.3 mm and the maximum stress is 230.1 MPa. The maximum stress will appear at the uprights, which meets the code requirements, but the result is close to the code limit.

## 5 FOUNDATION BEARING CAPACITY CALCULATION

### 5.1 *Program 1 foundation bearing capacity calculation*

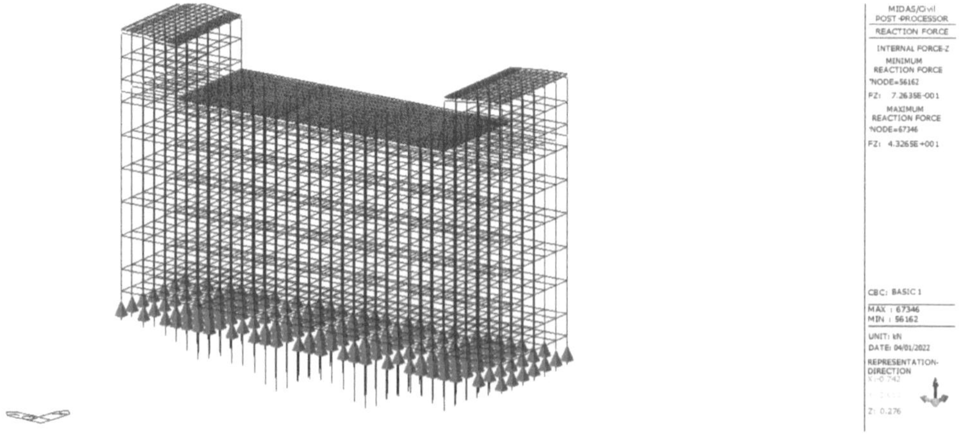


Figure 26. Full cloth type full hall bracket reaction force diagram.

According to the results of the most unfavorable section check: the crossbeam and web bottom uprights are subjected to an axial force of 72.635 kN, the base plate is a 150×150 mm steel plate, and a 25 cm thick C20 plain concrete bedding is laid.

(1) Concrete bedding calculation
Concrete compressive stress = 72.635 × 103/(150 × 150) = 3.33 N/mm$^2$ < 9.6 N/mm$^2$ (C20 concrete compressive strength design value, to meet the requirements)
Concrete shear calculation = 72.635 × 103/(150 × 150)/4 = 0.83 N/mm$^2$ < 1.1 N/mm$^2$ (C20 concrete compressive strength design value, to meet the requirements)
The concrete bedding layer is full, so the bottom of the concrete bedding layer is not considered to resist bending and pulling.
Conclusion: 25 cm thick C20 plain concrete bedding layer meets the requirements.
(2) Foundation bearing capacity calculation
The concrete diffusion angle is taken as 45°, then the area of the base of the upright rod diffused to the top surface of the foundation through the concrete mat is A = 0.65 × 0.65 = 0.4225 m$^2$, and the required foundation bearing capacity is:
P = (N+concrete self-weight)/A
= (90.111 + (0.65 × 0.65 × 0.25 × 24))/0.4225 = 219.28 kPa (KN/m$^2$) ≤ 250 kN/m$^2$

According to the "Highway Bridge Foundation and Foundation Design Specification" (JTJ D63-2007), the foundation bearing capacity meets the requirements, and the support can be erected.

### 5.2 *Option 2 foundation bearing capacity calculation*

According to the calculation to determine the steel pipe column foundation using 1.5 m × 2 m ×0.8 m C30 reinforced concrete independent foundation, the maximum calculated bearing capacity of 250 kPa is obtained, according to the detailed survey report and the actual measurement of the section of the foundation bearing capacity to meet the requirements, no foundation treatment. Considering the reuse of concrete foundation, and not to damage the original municipal road, the foundation is set up with the road above.

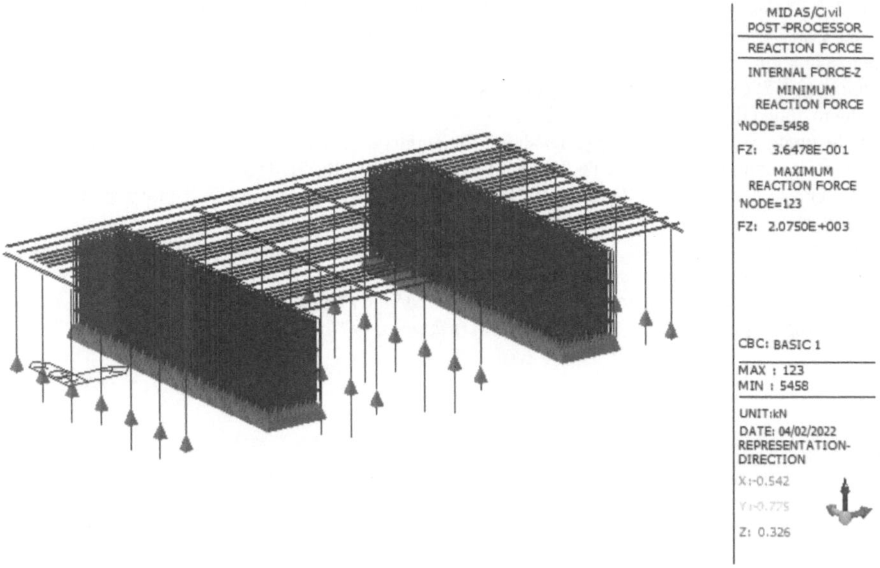


Figure 27. Portal frame reaction force diagram.

A single column to the foundation load F = 820.6 kN

$p = F/A = 820.6/(1.2\times 2) = 341.9$ kPa $\geq 250$ kPa, the foundation bearing capacity does not meet the requirements.

## 6 CONCLUSION

Based on the prestressed concrete variable-section continuous beam, a spatial finite element model was established to analyze the effects under different bracket schemes. The analysis shows that: (1) the maximum displacement of the bracket of scheme 1 is 86.4mm and the maximum stress is 194.6 MPa, which meets the specification requirements; the maximum displacement of the bracket of scheme 2 is 15.3 mm and the maximum stress is 230.1 MPa, and the maximum stress appears at the vertical rod, (2) the foundation bearing capacity of scheme 1 is 219.28 kPa $\leq$ 250 kPa, which meets the requirements. The foundation bearing capacity of scheme 2 is 341.9 kPa $\geq$ 250 kPa, which does not meet the requirements.

From the results of maximum stress and foundation bearing capacity, the bracket construction of the main bridge of Erbao Special Bridge is better by using the full cloth type full hall bracket, and the research results provide the reference significance for the bracket construction scheme of a curved girder bridge.

## REFERENCES

Chen Long.(2013) *Analysis of the factors affecting the force performance of socketed plate buckle type steel pipe formwork bracket* [D]. Changsha: Central South University.

Gui Zhijing, Zhang Jiandong, Feng Xiaonan, Liu Duo (2018). Application of socket type coiled buckle steel pipe support in bridge engineering [J]. *Highway Traffic Science and Technology*, 35(10):76–81.

He Enming, He Xiaohong, Zhu Guanghua, et al.(2015) Comparative analysis of construction cost of socket type coil buckle steel pipe bracket and bowl buckle steel pipe bracket[J]. *Chongqing Construction*,14(11):60–62.

Jiang X.(2008) Experimental and numerical simulation study of a newly inserted plate scaffold[J]. *Journal of Civil Engineering*,41(7): 55–60.

JTJ D63-2007 "*Highway Bridge Foundation and Foundation Design Specification*" [S].
Liang Yan, Gao J, Chen Quanxing. (2016) Preferential analysis of full support system scheme for cast-in-place concrete box girder[J]. *Construction Technology*, 45(24):64–67+87.
Lu Weibing. (2017) Construction technology of cast-in-place concrete continuous box girders by the full-span bracing method [J]. *Traffic World*(18):72–73.
Ran Tao, Li Junde, Wei Zhiguo, Huang Xianzhou, Liu Xingtong.(2020) Research on the structural stability of tall full-bore supports based on scale factors[J]. *Highway*, 65(10):151–155.
Shao Hongxing, Li Shuzhen, Li Weiping. (2014) Costing analysis of common cast-in-place bridge support schemes [J]. *Highway*,59(08):221–223.
Zhao Haisen. (2011) Application of portal frame in cast-in-place box girder [J]. *Science and technology information development and economy*, 21(02):218–220.
Zhuang Jinping, Cai Xuefeng, Zheng Yongqian, et al. (2016) Study on the overall force performance of fastener type steel pipe support system for tall formwork[J]. *Journal of Civil Engineering*, 49(10): 57–87.

*Advances in Urban Construction and Management Engineering – Cha (Ed)*
 *ISBN 978-1-032-39018-5*

# Study on the coupling and coordination of habitat system in Qinghai Province

Shu Chen* & Haifa Jia*
*College of Civil Engineering, Qinghai University, Qinghai, China*

Chengkui Liu
*Qinghai Building and Materials Research Co., Ltd, Xining, Qinghai, China*
*The Key Lab of Plateau Building and Eco-community in Qinghai, Xining, Qinghai, China*

ABSTRACT: With the rapid urbanization and rapid economic development, the habitat system in Qinghai Province shows structural imbalance and disorderly development. This paper adopts the entropy value method and coupled coordination model to measure and evaluate the level of habitat construction in Qinghai Province from 2010 to 2020. It analyzes the degree of coupling and coordination among three systems of socio-economic conditions, living conditions, and natural environmental conditions, and their evolution laws. The research results show that the overall score of habitat construction in Qinghai Province from 2010 to 2020 has increased from 0.19 to 0.83, and the construction level has improved significantly. Among them, the overall score of residential living conditions grew the fastest, with a growth rate of 84%, and the natural environment conditions grew the slowest, with a growth rate of 57%. The coupling coordination degree of the habitat system shows a linear increase from 2010 to 2020, but it is still at the barely coordinated stage until 2020, with a coordination degree of only 0.526. Therefore, Qinghai Province should pay attention to the coordinated development of the habitat environment, continuously improve the natural environment and enhance the quality of life of the masses, and comprehensively improve the overall level of habitat construction.

## 1 INTRODUCTION

Since Dausadias put forward the theory of human settlements in the 1950s, foreign scholars' studies on the coordinated development of human settlement systems have mainly focused on the perspectives of the economy, urbanization, tourism, the natural environment, and so on (Asare et al. 2016; Baccini 2018; Delhi, 2017; Hauser 2015). The degree of coordination of various systems within the human settlement environment is related to the sustainable development of the city. The urban living environment is not only the basic place for people to live but also an important symbol to measure whether the economic development of a region is balanced and whether the material culture is advanced (Bear et al. 2007). In 1993, Professor Wu Liangyong first proposed the establishment of the human settlements discipline in China, which played a key guiding role in the formation and development of the human settlement environment in China. Subsequently, in 1994, the Executive Meeting of the State Council issued *China's Agenda for the 21st Century*, which pointed out that environmental protection, construction security, and population control should be continuously strengthened to realize the sustainable development of environment construction for urban human settlements in China.

By reviewing the existing research (Li et al. 2016), it is found that the current research on the human settlement environment as a complex and diverse system mainly focuses on the external

*Corresponding Authors: 1343540672@qq.com and jiahf2014@qhu.edu.com

DOI 10.1201/9781003348023-6

coordination between the human settlement environment system and the independent system, and rarely analyzes the interaction and coordination of the internal system of the human settlement environment. However, internal coordination is the key to sustainable development.

Qinghai Province, as a major province in the northwest, the research on the coupling and coordination relationship between the human settlement environment system in this specific area is still in a blank state. Therefore, this research constructs a comprehensive evaluation index system for the coupling and coordination of the human settlement environment system in Qinghai Province. Through the comprehensive measurement and evaluation of the level of human settlement environment construction and its coupling and coordinated development in Qinghai Province, it found out the characteristics and existing problems of the construction status of Qinghai Province, to provide guidance and basis for Qinghai Province to formulate policies to improve the human settlement environment.

## 2 OVERVIEW OF THE STUDY AREA

Qinghai Province, the source of the river and the water tower of Asia, is located in the northeastern part of the Qinghai-Tibet Plateau (Jin et al. 2011). It has a special strategic position in terms of geography, homeland security, ecological security, and social stability. The province administers two prefecture-level cities of Xining and Hai Dong, and six autonomous prefectures including Hai Bei, Huang Nan, Hainan, Guo Luo, Yu Shu, and Hai Xi. The terrain is generally high in the west and low in the east, high in the north and south, and low in the middle. The altitude in the west is high and steep, inclined to the east, showing a ladder-like decline. The eastern region is the transition zone from the Qinghai-Tibet Plateau to the Loess Plateau, with a complex topography and diverse landforms. Qinghai Province has always been a relatively backward region in China's economic development level. By the end of 2020, the permanent population of Qinghai Province was 5.92 million, the GDP was 300.592 billion yuan, and the per capita GDP was only 50,819 yuan.

## 3 RESEARCH METHODS AND DATA SOURCES

### 3.1 *Descriptive statistical analysis*

By selecting typical indicators from four dimensions of the natural environment, urban economy, social harmony, and residents' life, this paper analyzes the general situation of the overall development of human settlements in Qinghai Province and finds out the problems existing in the development status of human settlements construction in Qinghai Province.

### 3.2 *Entropy method*

This paper adopts the entropy value method to select 35 indicators from 3 aspects, namely socio-economic conditions, residents' living standards, and natural environmental conditions of Qinghai Province as a whole, to comprehensively evaluate the level of habitat construction from 2010 to 2020.

(1) Standardized treatment of indicators.

$$\text{Negative indicator: } X'_{ij} = \frac{X_{ij}\max - X_{ij}}{X_{ij}\max - X_{ij}\min} + 0.01 \tag{1}$$

$$\text{Positive indicator: } X'_{ij} = \frac{X_{ij} - X_{ij}\min}{X_{ij}\max - X_{ij}\min} + 0.01 \tag{2}$$

(2) Calculate the proportion of indicator j of the evaluation object of evaluation in the year I in the aggregate of the index.

$$y_{ij} = \frac{X'_{ij}}{\sum_{i=1}^{m} X'_{ij}} \tag{3}$$

(3) Calculate the index information entropy value $E_j$ and the information utility value $D_j$.

$$E_j = -K\sum_{i=1}^{m} y_{ij}\ln(y_{ij}) \tag{4}$$

where $K=\frac{1}{\ln(m)}$ m is the year, $k > 0, 0 \leq E_j \leq 1$;

$$D_j = 1 - E_j \tag{5}$$

(4) Calculate the weight of evaluation indicators and the comprehensive development level.

$$W_j = \frac{D_j}{\sum_{i=1}^{m} D_j} \tag{6}$$

$$U_i = \sum_{i=1}^{m} W_j \times X'_{ij} \tag{7}$$

### 3.3 *Coupling coordination degree model*

In this paper, the coupling coordination degree is referenced in the capacity coupling coefficient model in physics. The SPSSAU data scientific analysis platform was selected to evaluate the coupling coordination degree for the comprehensive score. The coupling coordination degree level is divided concerning the coordination degree model of Dongling Song et al (2021), and the criteria are shown in Table 1.

Table 1. Classification standard of coupling coordination degree.

| Coupling coordination D value interval | The rank of harmony degree | Coupling coordination degree |
|---|---|---|
| (0.0~0.1) | 1 | extreme imbalance |
| [0.1~0.2) | 2 | serious imbalance |
| [0.2~0.3) | 3 | moderate disorders |
| [0.3~0.4) | 4 | mild disorders |
| [0.4~0.5) | 5 | endangered disorders |
| [0.5~0.6) | 6 | reluctant coordination |
| [0.6~0.7) | 7 | primary coordination |
| [0.7~0.8) | 8 | intermediate coordination |
| [0.8~0.9) | 9 | well coordination |
| [0.9~1.0) | 10 | quality coordination |

### 3.4 *Data sources*

The research data used in this article are all from *the China Urban Statistical Yearbook*, *Qinghai Provincial Statistical Yearbook*, and *Qinghai Provincial Statistical Bulletin of National Economic and Social Development* from 2010 to 2020. The missing data in individual years are obtained through appropriate calculation according to the fitting curve method.

## 4 RESULTS AND ANALYSIS

### 4.1 *Construction of the evaluation index system of the human settlement environment in Qinghai Province*

Referring to a large number of scholars' research results on human settlement environment science, combined with the characteristics of the economy, social development, and environmental resources in Qinghai Province, a comprehensive evaluation index system of the human settlement environment system in Qinghai Province is constructed based on the principles of scientific nature, comprehensiveness, representativeness and operability, which consists of 35 indicators in the three systems of social-economic conditions, living conditions and natural environment conditions (Table 2).

### 4.2 *Comprehensive measurement and evaluation results of human settlement environment system in Qinghai Province*

According to the actual value of each index from 2010 to 2020, the weight coefficient of each index in the comprehensive evaluation system of human settlements in Qinghai Province is calculated by using the entropy method formulas (1) ~ (6). The results are shown in Table 2. Based on the statistical data and weight coefficients of various indicators in Qinghai Province from 2010 to 2020, the changing trend of the comprehensive development index is calculated by the formula (7) as shown in Figure 1.

As can be seen from Figure 1, the overall score of the level of habitat construction in Qinghai Province from 2010 to 2020 rose from 0.1987 in 2010 to 0.8324 in 2020, and the level of habitat construction improved significantly. Among them, residential and living conditions increased

Table 2. The evaluation index system of the human settlement environment system in Qinghai Province.

| First-grade indexes | Secondary indicators: Name of indicator | Secondary indicators: The nature of indicators | Weights |
|---|---|---|---|
| Social-economic condition | Total retail sales of social consumer goods (100 million yuan) | + | 0.0212 |
| | The proportion of tertiary industry in GDP (%) | + | 0.0300 |
| | Total investment in fixed assets of the whole society (100 million yuan) | + | 0.0198 |
| | per capita GDP (yuan) | + | 0.0157 |
| | Population density (person/km$^2$) | + | 0.0140 |
| | The proportion of housing security to fiscal expenditure (%) | + | 0.0420 |
| | Urban registered unemployment rate (%) | - | 0.0259 |
| | Natural population growth rate (‰) | - | 0.0647 |
| | Total postal services (billion yuan) | + | 0.0465 |
| | Per capita disposable income of urban residents (yuan) | + | 0.0238 |
| | Residents' per capita living consumption expenditure (yuan) | + | 0.0191 |
| | The proportion of expenditure on education (%) | + | 0.0165 |

(*continued*)

Table 2. Continued.

| First-grade indexes | Secondary indicators: Name of indicator | Secondary indicators: The nature of indicators | Weights |
|---|---|---|---|
| Living conditions of residents | Engel coefficient of urban residents (%) | - | 0.0247 |
| | Per capita residential building area ($m^2$) | + | 0.0196 |
| | Urban water using penetration rate (%) | + | 0.0229 |
| | Urban gas penetration rate (%) | + | 0.0190 |
| | Internet broadband access penetration rate (%) | + | 0.0332 |
| | Number of full-time teachers in ordinary colleges and universities (person) | + | 0.0440 |
| | Number of students in ordinary colleges and universities (person) | + | 0.0310 |
| | Number of public library books per 10,000 people (10,000 volumes) | + | 0.0382 |
| | Number of hospital beds per 10,000 people (sheets) | + | 0.0267 |
| | Thousands of health technicians (person) | + | 0.0231 |
| | Number of people insured by medical insurance (10,000 people) | + | 0.0506 |
| | Number of people participating in endowment insurance (10,000 people) | + | 0.0126 |
| Natural environment conditions | The green coverage rate of built-up area (%) | + | 0.0255 |
| | Per capita public green space area (square meters) | + | 0.0159 |
| | Forest cover rate (%) | + | 0.0280 |
| | Harmless treatment rate of domestic waste (%) | + | 0.0119 |
| | The proportion of days with good air quality (%) | + | 0.0114 |
| | Sewage treatment rate (%) | + | 0.0187 |
| | Total sewage discharge (10,000 cubic meters) | - | 0.0541 |
| | Domestic garbage removal volume (10,000 tons) | + | 0.0259 |
| | Drainage pipeline density in a built-up area ($km/km^2$) | + | 0.0397 |
| | Per capita road area (square meters) | + | 0.0597 |
| | Elasticity coefficient of energy consumption (%) | - | 0.0248 |

by 0.2569, socio-economic conditions increased by 0.2388, and natural environmental conditions increased by 0.1381. From the score results, it is found that the overall score of the natural environment system in 2020 is the lowest, only 0.2421, which indicates that the overall construction level of the habitat environment in Qinghai Province is constrained by the natural environmental conditions.

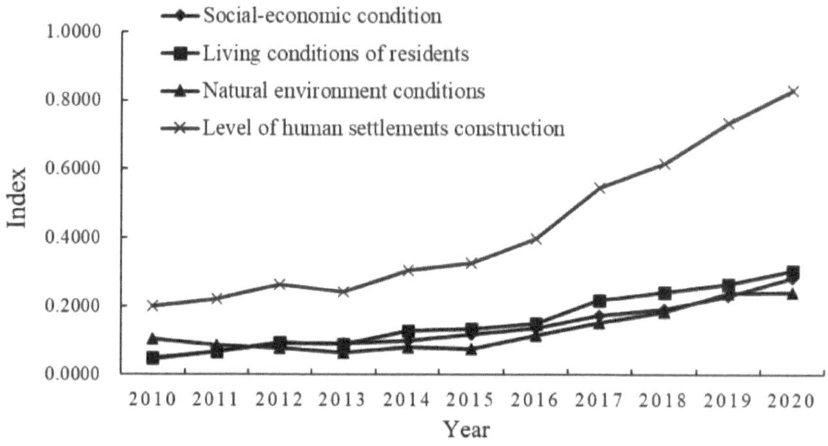


Figure 1. Construction level of human settlements in Qinghai Province and comprehensive scores of each subsystem.

Therefore, Qinghai Province should further strengthen the improvement of the natural environment in the process of future habitat construction and strive to achieve balanced development.

### 4.3 *Analysis of coupling coordination degree*

Based on the comprehensive development index of each subsystem in Qinghai Province, the coupling coordination degree of the three major systems of social-economic conditions, living conditions, and natural environment conditions in Qinghai Province from 2010 to 2020 is calculated by using SPSS software. It can be seen from Table 3 that the coupling coordination degree of each subsystem of the human settlement environment in Qinghai Province is basically on the rise, and further subdivided, which can be divided into three stages: (1) During 2010-2013, the coupling coordination degree of the subsystems of the environment of the human settlement in Qinghai Province increased slowly and remained in the stage of moderate imbalance. (2) From 2014 to 2019, the coupling coordination degree of each subsystem of the human settlement environment in Qinghai Province continued to improve, and the coupling degree transited from mild imbalance to

Table 3. Coupling coordination degrees and types of human settlement environment subsystems in Qinghai Province from 2010 to 2020.

| year | D value of coupling coordination degree | rank of harmony degree | coupling coordination degree |
|---|---|---|---|
| 2010 | 0.248 | 3 | moderate disorders |
| 2011 | 0.271 | 3 | moderate disorders |
| 2012 | 0.296 | 3 | moderate disorders |
| 2013 | 0.282 | 3 | moderate disorders |
| 2014 | 0.316 | 4 | mild disorders |
| 2015 | 0.325 | 4 | mild disorders |
| 2016 | 0.363 | 4 | mild disorders |
| 2017 | 0.424 | 5 | endangered disorders |
| 2018 | 0.452 | 5 | endangered disorders |
| 2019 | 0.495 | 5 | endangered disorders |
| 2020 | 0.526 | 6 | reluctant coordination |

the stage of endangered disorders. (3) From 2019 to 2020, at this stage, the coupling coordination degree of each subsystem of the human settlement environment in Qinghai Province is further improved, reaching the stage of reluctant coordination.

## 5 CONCLUSION

In this study, 35 indicators are selected from 3 aspects: socio-economic conditions, residents' living standards, and natural environmental conditions, and the entropy value method is used to comprehensively evaluate the level of habitat construction in Qinghai Province from 2010 to 2020, and the coupling coordination degree model is used to measure and analyze the coupling coordination of each subsystem of the habitat environment in Qinghai Province, and the following conclusions are obtained: The level of habitat construction in Qinghai Province rose significantly from 2010 to 2020, and the overall score of habitat level increased by 76.12% by 2020. The changes in the coupling coordination degree of each subsystem of the habitat environment in Qinghai Province from 2010 to 2020 show a steady increasing trend, but in 2020 the coupling coordination degree can only reach 0.526, namely, the barely coordinated stage, which indicates the uneven development in the construction process of the habitat environment in Qinghai Province. Therefore, Qinghai Province should continue to promote habitat improvement in the future construction of habitat environment, further enhance the quality of habitat environment, improve the quality of life of the masses, and continuously enhance the sense of access and happiness of the masses. At the same time, the improvement of the natural environment should be strengthened, namely, enhancing the greening coverage of the built-up area, optimizing the energy consumption structure, strengthening the sewage treatment, etc., and comprehensively improving the overall level of the construction of the human living environment.

## ACKNOWLEDGMENTS

This work was financially supported by the Open Fund Project of Key Lab of Plateau Building and Eco-community in Qinghai (Serial number: KLKF-2019-005).

## REFERENCES

Asare K N, Boadu A T, Adimado A A. Evaluation of groundwater and surface water quality and human risk assessment for trace metals in human settlements around the Bosomtwe CraterLake in Ghana [J]. *Earth and Environmental Sciences*, 2016, 5(1):1812.

Baccini P. *The horizontal metropolis between urbanism and urbanization*[M]. Barcelloni Corte: Springer International Press, 2018:111–118.

Bear Eagle, Zeng Guangming, Dong Lisan, etc. Quantitative evaluation of the uncertainty of the coordinated development of urban living environment and economy – Take Changsha City as an example [J]. *Geography*, 2007, 62(2): 397–406.

Delhi P. The effects of tourism in Greek insular settlement and the role of spatial planning [J]. *Knowledge-Economy*, 2017, 8(1): 319–336.

Hauser W. The infrastructure of nature's island: Settlements, networks, and economy of two plantations in Colonial Dominica [J]. *Historical Archaeology*, 2015,19(1):601–622.

Jin Shisheng, Cai Yongli, Liu Jingyi. Research on the Construction of Habitat Environment Based on the Harmonious Development of "San Sheng" –A Case Study of Guide County, Qinghai Province [J]. *Qinghai Social Science*, 2011 (06): 57–60.

Li Bohua, Liu Yan, Liu Peilin, etc. Spatial and Temporal Evolution of Coupling Degree of Human Settlements System in Hunan Province [J]. *Statistics and Decision-making*, 2016 (18): 104–107.

Song Dongling, Ma Yue. Study on the coupling of green water resources utilization rate and high-quality economic development in the Yellow River Basin – Taking Henan Province as an example [J]. *Ecological Economy*, 2021, 37 (05): 14–19 + 42.

*Advances in Urban Construction and Management Engineering – Cha (Ed)*
 *ISBN 978-1-032-39018-5*

# Research on design strategies for renovation of vacant existing office buildings

Changgui Ou* & Yuying Shuai*
*Hunan Non-Ferrous Metals Vocational & Technical College, Zhuzhou, China*

ABSTRACT: Renovation and reusing of the existing office buildings can extend their lifetime, and also make them better adapt to the new office needs. In addition, existing buildings were born with the development of urbanization, and witnessed the historical imprint of the city. Therefore, using comprehensive methods to renovate existing buildings is also meaningful for urban renewal. Through literature research, field research, and practical case analysis, this paper sorts out the spatial characteristics and practical cases type in which existing buildings were regenerated for offices. Then, the focus is placed on the functional organization, building facade, and site design, summarizing reusing design strategies. The function and image of the vacant office buildings are updated to provide design ideas for the realization of new usage of old buildings. The functional renewal and image improvement of existing buildings can provide ideas for reuse.

## 1 INTRODUCTION

The office building is an important part of urban space and one of the largest proportion of public buildings. People spend about 1/3 of their time in the workplace per day. Affected by the epidemic, the boundaries of work and life become blurred. People pay more and more attention to the comforts and feelings of the workspace. Meanwhile, the old buildings do not match the increasing demands.

Furthermore, affected by the epidemic and being uncompetitive with new projects, some existing office buildings keep a high vacancy rate for a long time, resulting in a waste of construction resources. Taking Changsha, the second largest city in central China, as an example, the vacancy rate of class 5A office buildings has been higher than 30% in the past two years. Under similar conditions, the office space absorption capacity of first-tier cities is slightly better than that of second-tier cities. However, the rigidity of the original office area in the city could not be ignored.

From the perspective of environmental protection and resource conservation, these vacant office buildings still have great potential value. However, they are facing the crisis of shortened serving life, because of the mismatch of the current burgeoning demands, lack of space adaptability, the high operation energy consumption, and the unimproved internal environment. Therefore, renovation by subtle intervention can efficiently use functional space, enhance the work efficiency and communication of employees, meet people's multi-dimensional health needs, and rejuvenate the vitality of the buildings.

## 2 SPATIAL CHARACTERISTICS OF OFFICE BUILDINGS

At present, office buildings are distributed in clusters, and the types include (1) high-density, high-volume-ratio office clusters, such as headquarters buildings, CBD; (2) low-density, low-volume

*Corresponding Authors: 272201546@qq.com and 286306262@qq.com

 DOI 10.1201/9781003348023-7

ratio office clusters, such as office parks. The first type is mostly high-rise or super-high-rise buildings, which have a large internal space and free facade, and commonly use frame or frame-core tube structures. The second type is mostly multistory buildings, which contain various forms of interior space composition, and there is a certain connection between independent buildings. The multistory office cluster maintains an open and inclusive attitude towards the external environment and has great potential for renovation.

The internal spatial structure of office buildings can be divided into single modules, open spaces, unit-typed, and mixed-typed ones. The types of offices are shown in Figure 1. Office in a single module connects the individual rooms by one-sided or two-sided corridors. From the perspective of use and equipment operation, it is uneconomical and space-constrained, ignoring the individual needs of employees. Open space office is similar to the mixed-typed, with a larger interior space. They are divided into flexible items such as movable furniture, plants, and decorative partitions, to facilitate communication and collaboration between work teams. These office buildings are more efficient than the office in a single module. The unit-typed office combines living and working scenes, with a similar spatial structure to the residence. The atmosphere of a unit-typed office is more active and suitable for small-scale work teams (Cheung 2022).

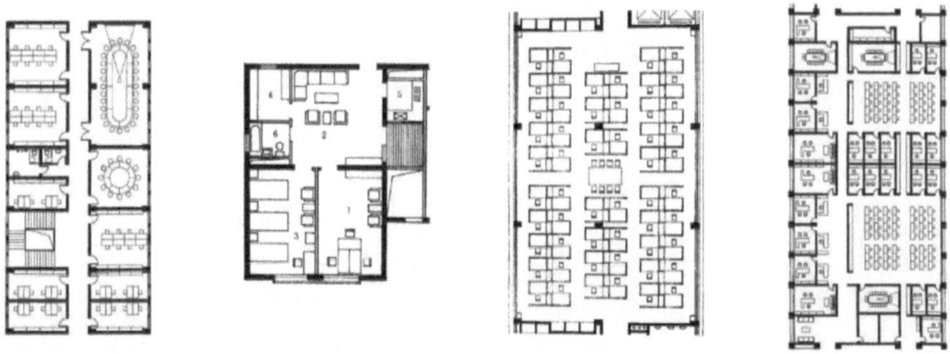

Single module space | Unit-typed space | Open space | Mixed-typed space

Figure 1. Types of office building interior spaces.

For being conveniently pre-emptive, the existing office buildings have geographical advantages in the city, but now part of them has become an inefficient stock space. They are constrained by the lack of flexible and variable functional spaces, the aging of the building envelope, the outdated ventilation and lighting equipment, and the backwardness of intelligent management. In addition to achieving higher economic benefits, working also provides staff with a pleasant shared culture and public life experience. Office mode and needs growing diversity are inevitable. Especially after the 2019 epidemic, a growing number of companies are transitioning to "hybrid mode" in their daily work, which has affected working function. It's not enough to just superimpose new functions. Health, environmental protection, sustainability, and humanization have become essential characteristics of office buildings.

## 3 CLASSIFICATION: RENOVATION OF EXISTING BUILDINGS FOR OFFICE USE

There are 3 types of reusing existing buildings as workplaces: reconstruction of existing office buildings, industrial heritages, and others (commercial, residential and cultural buildings). Retaining structural components of existing buildings or optimizing and upgrading the stock space with light intervention could extend the vitality of vacant space. Due to the different functional combinations, structures, site characteristics, and urban positioning of existing buildings, the reconstruction methods of buildings are different. Examples are depicted in Table 1.

Table 1. Existing buildings reconstruction for working.

| Classification | Projects | Picture | Change of function | Reconstruction scheme |
|---|---|---|---|---|
| Office Building Renovation | The Droogbak transformation (Amsterdam Netherlands) |  | Historic office → contemporary office | Spatial, functional, and technical modifications |
| Industrial heritages to office | Pearl River Brewery Phase II (Guangzhou, China) |  | Industrial heritages → Commercial-cultural project | 1. Turn the old factory into an office 2. Landscape Revitalization |
| Others building to office | Ningbo Academician Center (China) |  | Historical school → Academicians for scientific research | 1. Functional business & cultural venue & rural tourism update 2. Natural ecological renewal |

## 4 DESIGN STRATEGIES FOR REGENERATION OF EXISTING BUILDINGS

The renovation of existing buildings should not be limited to meeting the basic working demands, but follow the trends of variable space. The renovation should focus on optimizing the spatial structure, creating a pleasurable interior and exterior environment, reducing building energy consumption, and enhancing the external image of architecture. For the existing buildings located in the old town, they carry the life imprint of neighbors, and rejuvenating historical buildings is meaningful for urban renewal and regional context continuation. In the process of renovation, attention must be given to stitching the interface between existing buildings and the city (Li 2020), and the site design should be fully considered.

### 4.1 *Spatial reorganization*

For existing buildings that do not encourage communication with a single module and small bay, there are methods to reuse the buildings that fill multi-functional space, break down the barriers of "cubicles", blur the boundary of space, expand space horizontally, and release the "third space" between the independent office and public space (Jan 2002). The horizontal expansion makes the room more convenient for multiple uses, however, it is not practical to simply superimpose a workspace and activities venues. These piled-up "third spaces" cannot really promote communication and collaboration, but are more likely to be idle or even become negative spaces.

For existing buildings with closed sight lines and a single function, the usable area can be increased by expanding vertically and adding upward or underground spaces. Coffee bars, fitness rooms, and bookstores, these "leisure warehouses" scattered in the space can provide brain power stations for knowledge workers. By inserting the atrium, increasing blowing space, and partially demolishing or adding, existing buildings establish a new connection in multiple dimensions of traffic organization, sight, sound, and light. These vertical expansions create visual focus, enrich spatial levels, and activate spatial rhythm.

### 4.2 *Update building skin*

Transforming building skin is the regeneration of envelope structure, at the same time affecting the lighting, ventilation, the permeability of urban interface, and architectural image. A comfortable indoor environment requires soft lighting, natural ventilation, and anti-noise. For example, in Changsha, which is hot in summer and cold in winter, skin renovation should focus on shading, heat insulation, and natural ventilation throughout the entire building life cycle. According to local conditions, vertical greening, ventilation green corridors, and roof gardens are integrated into the skin to promote the penetration of natural elements such as wind and sunlight into the interior such as winds and sunshine, and show ecological value (See for example Figure 2). Using ecological technology reduces the canyon effect in high-density environments, increases activity spaces in the city, and is beneficial to reduce overall energy consumption. The original structure remains as the skeleton, and connection with the new skin must fully consider its bearing capacity and actual loss. The design of the new skin should respect the actual, and select appropriate materials, constructions, and structures to ensure sustainable operation of the entire building life cycle (Zhang 2020).

Figure 2. Communal terraces at the Biotope let fresh air and sunlight into interior spaces. (Lille, France)

### 4.3 *Expansion surrounding site*

The regeneration of office buildings combines new with old at the urban interface. Transformation basically maintains the original building shape and the existing skyline of the city. The renewal of surroundings such as greening, pavement, roads, and parking systems directly affects the indoor environment. The first-floor space is open to the surroundings, which integrates into the 15–20 minutes comfortable life circle of the city (Gensler 2021), and becomes the vitality center of the community (Figure 3). The site design should minimize hard paving, and increase vibrant greening, water bodies, rain garden, and ecological parking lots. It can improve the quality of the surrounding environment, reduce the urban heat island effect, and optimize the field microclimate. From the perspective of the urban interface, the existing office buildings bear the life marks of people. The renewal design should cater to the original environment and continue the characteristics and cultural context, so as to inject vitality into the sleeping blocks. (Figure 4)

Figure 3. Surrounded by a multitude of shared open spaces for greater social opportunities. (Hangzhou, China).

Figure 4. The montparnasse tower facade reflects the Eiffel tower. (Paris, France).

## 5 CONCLUSION

Reusing the existing buildings is a process of constantly improving the office environment, which has practical significance in saving resources and activating stock buildings. Due to differences between the actual structure, functional positioning, and site conditions of the buildings, the design should fully respect the existing conditions, and consider the variability and adaptability of the building's life cycle. These three design strategies for the reconstruction of existing buildings proposed in this paper aim to balance the interaction between buildings, environment, and people, and create a physically and mentally pleasing working environment, and a healthy and sustainable workspace. Furthermore, it has enlightenment for urban renewal activities. Instead of collaging elements to continue the urban context, it is more sensible to extend the vitality of the existing buildings to show the deep-seated appeal of buildings.

## REFERENCES

Cheung, F. (2022). Designing holistic well-being in the age of hybrid working. *J. Architectural Worlds*. 37(1), 35–41.

Gensler. (2021). *Gensler design forecast 2021*. R. https://www.gooood.cn/gensler-design-forecast-2021.htm

Jan. G. (2002). *Life between Buildings*. M. China Architecture & Building Press. 203, 133–197.

Li, J.L. (2020). *Research on skin design strategy in the transformation of existing office buildings based on subtle intervention*. D. Hunan University.

Zhang, J. (2020). *Research on low carbon design strategy of Japanese office buildings*. D. Xi'an University of Architecture and Technology.

*Advances in Urban Construction and Management Engineering – Cha (Ed)*


# Spatial distribution of cultural landscape and its impact on land use pattern in Qinghai-Tibet Plateau: Tibet

Huaigan Wang, Jiayu Kuang, Lanjin Bai & Luo Guo*
*College of Life and Environmental Sciences, Beijing, China*

ABSTRACT: China's Qinghai-Tibet Plateau is a crucial biological barrier with significant ecological protection implications. It is a characteristic region of Tibetan Buddhist culture as well. The pattern of living, the produced approach, and the ecological surroundings are all impacted by culture. This study utilized remote sensing, geographic information technology, spatial statistical methods, and data from five years between 2015 and 2020 to examine the spatial distribution, land utilization pattern change, and landscape pattern change of Tibet's cultural landscape. The findings revealed that: (1) The northeast slope was the greatest, the altitude was 3,900–4,099 m, and the primary slope of temples was 2–6°. (2) Between 2015 and 2020, the amount of unused land in Tibet and around its temples rose, while the amount of grassland dramatically reduced. Most of the diminished grassland was transformed into unused land and forest. (3) The population was concentrated and there was a lot of cultivated land near temples. Grasslands, cultivated lands, and forests were more abundant around temples than in residential areas. In temples, the grassland vegetation is superior to that in residential areas. (4) The human influence in the region surrounding temples grew yearly, and the cultural landscape of the temples increased the diversity and richness of land use, which also increased the proportional differences between different types of landscapes and the degree of fragmentation. This study contributes to the sustainable development of the Qinghai-Tibet Plateau by providing a theoretical foundation and support for illuminating the interaction between humans and land as well as the sensible distribution of land use.

## 1 INTRODUCTION

The Qinghai-Tibet Plateau is a region susceptible to global change. Since the 1970s, due to inappropriate development and pursuit of high consumption and other factors, ecological deterioration, resource depletion, population growth, food availability, and other issues are becoming more severe in this region (Li 2015). In this context, landscape ecological research has been prioritized. Land use/land cover change (LUCC) is one of the primary components and significant causes of global environmental change; it is a process involving the close cross-combination of natural and human variables. LUCC shows the significant changes in the surface landscape structure, and its process reflects the restrictions and modifications of the natural environment, the modifications of the social and economic environment, and technology advancement. The Intergovernmental Panel on Climate Change Landscape pattern index is an ecologically significant indicator that may reflect the spatial variability of landscape and its fragmentation of it, which may then influence several ecological processes.

The cultural landscape transports regional culture. The distinctive natural and humanistic geography of the Qinghai-Tibet Plateau affected the introduction, creation, spread, and dispersion of Buddhism, Buddhist traditions and teachings, temple architecture, etc. In the meantime, culture has affected people's lifestyle and production, so it has changed their living environment. In contrast

*Corresponding Author: guoluo@ muc.edu.cn

 DOI 10.1201/9781003348023-8

to other elements of Tibetan studies, the study of the natural, human, and geographical environment of the Qinghai-Tibet Plateau has not been addressed seriously for a long time (Wang 2010). By establishing the geographic information system of the cultural landscape of Qinghai-Tibet Plateau, the study analyzed the spatial distribution characteristics of temples and residential areas in Qinghai-Tibet Plateau, contributing to the sustainable development of the region.

## 2 STUDY AREA AND METHOD

### 2.1 *Study area*

In the southwest part of the Qinghai-Tibet Plateau, Tibet Autonomous Region is located, with an average height of nearly 4,000 m, in latitudes ranging from 26°50′ to 36°53′ north and longitudes of 78°25′ to 99°06′ east. It is regarded as the "Roof of the World" and governs Lhasa, Shigatse, Qamdo, Linzhi, Shannan, Naqu, and Ali, with a total area of over 1.22 million square kilometers, including 650,000 hectares of grassland; cultivated land is concentrated in valleys and river basins in southern Tibet, with a total area of 360,000 hectares (Deji et al. 2011). Tibet's land resources are characterized by a vast quantity of unused land, which accounts for 30.71 % of the total land area. The Northwest climate is chilly, whereas the southeast climate is warm and humid, demonstrating a zonal transition from southeast to northwest. Around 46,000 monks are residing in over 1,700 Tibetan Buddhist temples, which are widely dispersed and numerous.

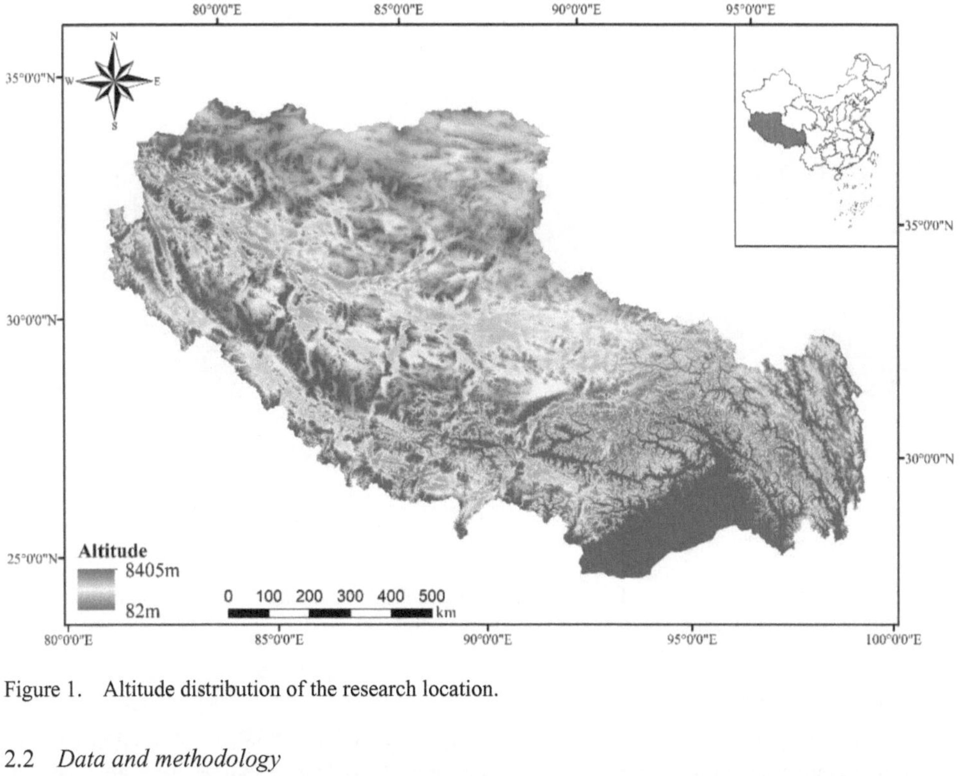


Figure 1. Altitude distribution of the research location.

### 2.2 *Data and methodology*

This study employed the land use data for two periods from 2015 to 2020 from the Data Center for Resources and Environmental Sciences of the Chinese Academy of Sciences, with a 1 km resolution, because land use in the study region changed little before 2015. To represent the dynamic changes in local land resources, land use may be categorized into eight categories: cultivated land, forest, high-density grassland, medium-density grassland, low-density grassland, water area, construction land,

and unused land. The spatial location of temples was determined by field survey, GPS positioning, and State Administration of Religious Affairs statistics.

The slope is subdivided into six gradients: 0–2°, 2–6°, 6–10°, 10–14°, 14–18°, and 18–49°, while the aspect is subdivided into north, northeast, east, southeast, south, southwest, west, and northwest. The altitude is separated into eight degrees: 2000–3499 m, 3500–3699 m, 3700–3899 m, 3900–4099 m, 4100–4299 m, 4300–4499 m, 4500–4690 m, and altitudes more than 4700 m. The temples were statistically assessed based on their gradients. Using a land transfer map, the distribution of land use categories in this region from 2015 to 2020 was examined. The aggregation index (AI), contagion index (CONTAG), patch density (PD), landscape shape index (LSI), and Shannon biodiversity index (SHDI) were then computed for the 1–3 km temples buffer zone.

## 3 RESULTS AND ANALYSIS

### 3.1 *Distribution characteristics of temples*

According to Figure 2, there were the most temples with a slope of 2–6°, accounting for 38.20%, followed by 0–2°. The average gradient of the distribution of temples is 4.52°, indicating that many Buddhist temples were constructed on moderate slopes. The slopes of temples were more northeast and upper northwest, and less west and northwest. 67.58% of the land in Tibet is above 4,700 m above sea level, but only 7.08% of the temples were distributed in this area, and the majority were located at 3500–4300 m, a relatively low altitude. In Tibet, the renowned Jokhang Temple, Tashilhunpo Temple, and Sangye Temple are all situated at comparatively low elevations. Temples at high heights, such as Rongbu Temple atop the mountain, are more than 5,000 m above sea level.

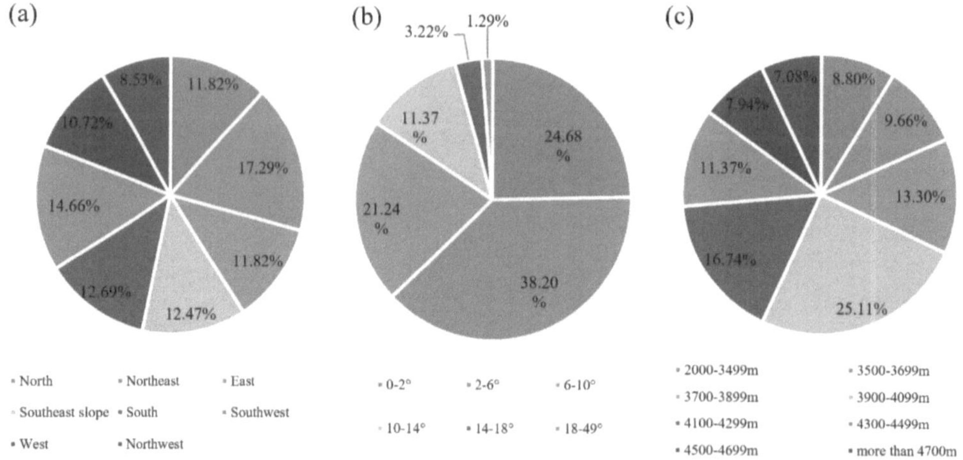


Figure 2. Research location temples slope, slope direction, and altitude distribution. (a. Distribution of slope direction. b. Distribution of slope gradient. c. Distribution of altitude).

### 3.2 *Analysis of land use change*

Grassland was the prominent landscape type in the Tibet Autonomous Region, followed by unused land, which comprises more than 80% of the entire area. Grassland resources were abundant in the Tibet Autonomous Region. There was 70.01% grassland in 2015, but that number dropped to 46.02% by 2020 when it would account for only 46.02% of Tibet's total geographical area. Tibet Autonomous Region's grassland fell by 23.99% between 2015 and 2020. There has been an 18.44% rise in the amount of unused land since 2015. It was mostly unused land and forest that were transformed, while a limited quantity of forest was used for construction land and water areas.

Most of the increased forest was converted from high-density grassland, accounting for 8.3% of the forest area in 2020. From 2015 to 2020, the water area grew by 3.63 %, mainly from grassland and unused land conversion. It shows that after the implementation of the project of returning grazing land to grassland, the area of forestland has increased, and the increase of construction land indicates that the economic development of the Tibet Autonomous Region has accelerated (Dai et al. 2013; Liu et al. 2011). The Qinghai-Tibet Plateau is an ecologically fragile region. The degradation of the grassland vegetation and soil conditions in the Tibet Autonomous Region has decreased grassland density and serious grazing problems. Overgrazing and other human factors have also had an impact on plant community structure and composition (Zhou et al. 1987).

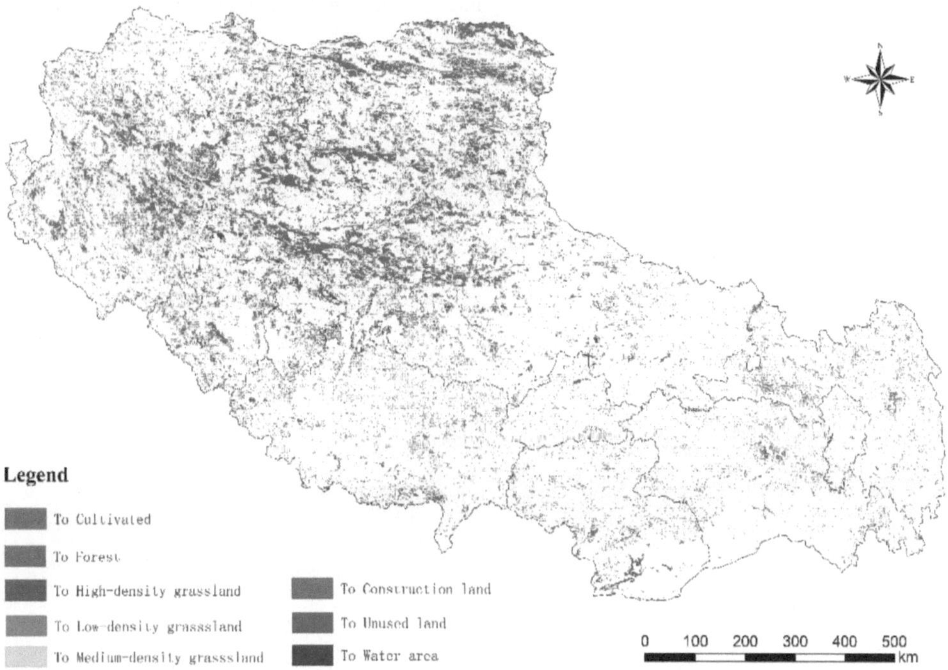


Figure 3. Land use transfer in Tibet Autonomous Region from 2015 to 2020.

### 3.3 *Land use change in the 1–6 km buffer zone around temples and residential areas from 2015 to 2020*

Based on a comparison of land use categories in Figure 4, it can be shown that as buffer space increases the proportions of cultivated land and construction land steadily decrease. Between residential areas and temples, temples have twice as much cultivated land and more forest. In addition, there were more high-density and medium-density grassland in temples, while there was more low-density grassland in residential areas. According to this study, many temples were in locations where cultivated areas and densely populated. The grassland vegetation around the temple is well-protected and in high health.

### 3.4 *Analysis of temples buffer zone landscape pattern index from 2015 to 2020*

According to Figure 5, the degree of landscape aggregation and sprawl increased from 2015 to 2020. With the increase of SHDI, the degree of landscape diversity around temple sites increased, the land use became richer, and the proportion of landscape types varied greatly. The increasing PD suggests a more fragmented landscape, a reflection of the extent to which the landscape has been disturbed by humans. Since humans have been more involved in the natural world, this demonstrates that human influence has grown. Landscape index LSI and maximum patch number LPI are dropped, showing that grassland, the primary land use type, is losing its advantages at the patch level.

Increased buffer area reduced SHDI, PD, CONTAG, and LSI, all of which suggested a reduction in landscape variety, fragmentation, and form complexity. LPI and AI increased, indicating that the farther away from the temple site, the less human intervention, and the dominant land type – grassland increased.

In general, the degree of manufactured influence around the temple site increases year by year. The cultural landscape of the temple increases the diversity of the surrounding landscape, enriches the land use, increases the proportion difference of various landscape types, and increases the degree of fragmentation. The surrounding grassland environment has been severely damaged, and this devastation was worsened between 2015 and 2020.

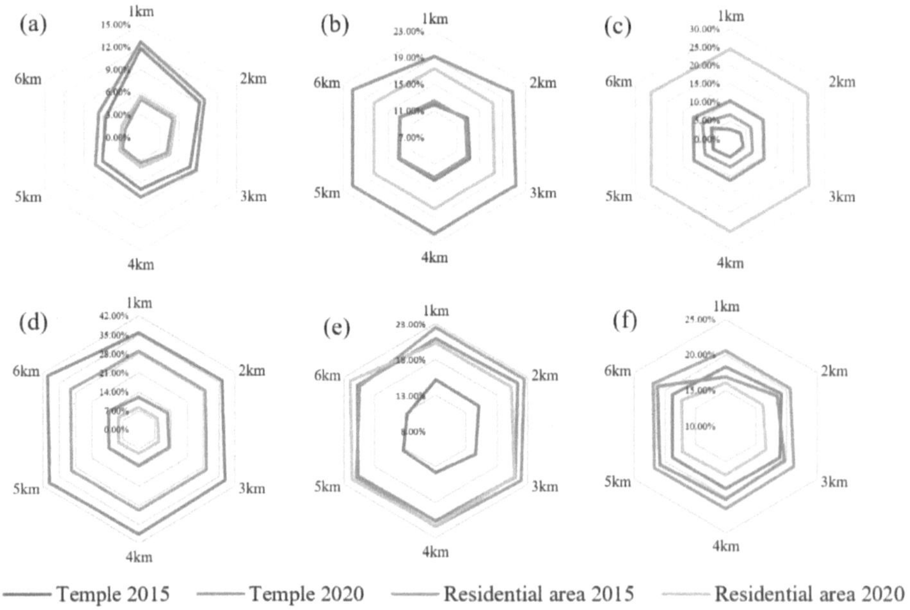


Figure 4. Comparative analysis of land usage in the 1 to 6 km buffer zone between temples and residential areas (a. Cultivated land. b. Forest. c. Unused land. d. High-density grassland. e. Medium-density grassland. f. Low-density grassland).

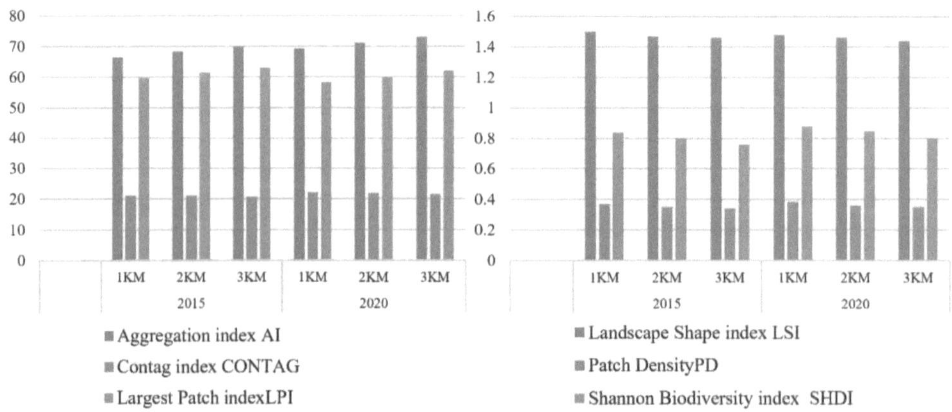


Figure 5. Landscape pattern index of temples 1-3 km buffer zone from 2015 to 2020.

## 4 CONCLUSIONS

Using remote sensing, GIS, and spatial statistical methodologies as well as data from 2015 to 2020, this research examined how Tibet's land use, temple landscape distribution, and surrounding landscape patterns have changed. The results are as follows.

(1) There is a dense concentration of temple slopes between 3,900 and 4,099 m, with an average slope of 4.52°. The northeast and southwest slopes were the most prominent.
(2) From 2015 to 2020, land use changed in the same way in both temples and Tibet. The unused land increased significantly, mainly from grassland conversion, and grassland degradation. There was more forest, mainly from high-density grassland conversion.
(3) The proportion of cultivated land, forest, and construction land around the temples is higher than the average. The closer to the temple, the more obvious this trend become. There are more high-density and medium-density grasslands near the temple, whereas low-density grasslands are near the residential areas. The grassland near the temple is in better condition than that around the residential area.
(4) The human influence in the region surrounding temples grew yearly. In addition to enriching the land use and increasing the degree of fragmentation, the temple cultural landscape promoted the variety of the surrounding terrain. The surrounding grassland had been severely damaged, and this influence was worsened between 2015 and 2020.

## ACKNOWLEDGMENTS

This research was financially supported by the Second Tibetan Plateau Scientific Expedition and Research (STEP) program (Grant No. 2019QZKK0308).

## REFERENCES

De Ji., Qiaosheng Wu. (2011). Mineral Resources and sustainable development in Tibet. *China Land and Resources Economy*., 09(24):15–17+54.

Honglai Liu., Weihua Lu., Chao Chen. (2011). Advances in grassland degradation succession and diagnosis. *Acta Agrestia Sinica.*, 05(19):865–871.

Kaidui Wang. (2010). On the Influence of Geographical Environment on Tibetan Buddhis. *Journal of Guangzhou Institute of Socialism.*, 08(1):58–62.

Rui Dai., Zhihong Liu., MengJun Lou., Jin Liang., Mingyang Yu. (2013). Spatial-temporal characteristics of grassland degradation in Nagqu region, Northern Tibet. *Acta Agrestia Sinica.*, 01(21):37–41+99.

Shichen LI., Yili Zhang., Fanneng He. (2015). Reconstruction of cultivated land spatial pattern and its spatial-temporal changes in Qinghai and Tibet during the past 100 years. *Progress in Geography.*, 34(2):197–206.

Xingmin Zhou., Qiji Wang., Yanqin Zhang., Xinquan Zhao., Yapin lin. (1987). quantitative analysis of vegetation succession in alpine meadow under different grazing intensities. *Journal of Plant Ecology and Geobotany.*, 04:276–285.

*Advances in Urban Construction and Management Engineering – Cha (Ed)*
*© 2023 Copyright the Author(s), ISBN 978-1-032-39018-5*

# Research on lighting layout design of urban tunnel based on DIALux evo

Feng Du*, Zhiqiang Dai* & Feiyang Zhang*
*Fujian University of Technology, China*

ABSTRACT: There are huge differences between urban tunnels and highway tunnels in tunnel height, the speed limit of vehicles, traffic volume, etc. Therefore, the standards and experiences of highway tunnels cannot be copied in tunnel lighting design. In this paper, based on the characteristics of urban tunnels and the layout of tunnel lighting lamps, the software DIALux evo is used to establish tunnel models, and the control variable method is used to simulate different layout methods. Taking the threshold zone as an example, four lighting evaluation indexes are compared and analyzed. On the premise of ensuring safety, the advantages and disadvantages of different lighting layout methods are analyzed from two aspects of energy efficiency (Du 2019) and comfort. Finally, it is concluded that the energy efficiency of lighting with a symmetrical arrangement on both sides of general urban tunnels is slightly lower, but the comfort is improved, which has a high application value. The research results can provide a basis for optimizing the lighting design scheme of urban tunnels.

## 1 INTRODUCTION

### 1.1 *Lighting evaluation index*

Traditional lighting evaluation indexes include average illuminance of pavement, total uniformity of pavement illuminance, and longitudinal uniformity of pavement centerline illuminance. Based on the characteristics of urban tunnels, this paper supplements the research on the disability glare index. Among them, the average illuminance of pavement is an index to evaluate energy saving, while the other three indexes represent the level of lighting comfort. The average illuminance of pavement is an important factor affecting whether obstacles can be seen, so it is required that the average illuminance of pavement should be quite high. The illuminance of each lighting zone should look up the table according to the design speed and traffic volume. For details, please refer to the *Detailed Rules for Lighting Design of Highway Tunnel* (Standard Code: JTG/T D70/2-01-2014). The total illuminance uniformity is the ratio of the minimum illuminance to the average illuminance on the pavement, and the longitudinal illuminance uniformity is the ratio of the minimum illuminance to the maximum illuminance on the lane centerline. The requirements of highway tunnel lighting design for total pavement illuminance uniformity and pavement illuminance uniformity are shown in Table 1. In tunnel lighting, CIE adopts relative threshold increment (TI) to explain the decline of human visual function caused by glare. CIE's *Guide for the Lighting of Road Tunnels and Underpasses* (CIE 2004) also puts forward specific requirements for TI. The maximum TI of the threshold should not be greater than 15%, and its calculation is shown in Formula (1) and Formula (2).

$$TI = 65\frac{L_v}{L_r^{0.8}} \quad L_r \leq 5\ \text{cd/m}^2 \tag{1}$$

*Corresponding Authors: 59154586@qq.com, 1768949893@qq.com and 1185796323@qq.com


DOI 10.1201/9781003348023-9

$$TI = 65\frac{L_v}{L_r^{1.05}} \quad L_r \geq 5\ \mathrm{cd/m^2} \tag{2}$$

where $L_r$ is the average illuminance of the pavement; $L_v$ is Illumination of a light curtain formed by all lamps in the driver's field of vision.

Table 1. Requirements of total pavement illuminance uniformity and pavement illuminance uniformity.

| Design traffic flow N/[veh/(h·ln)] | | Total illuminance uniformity of pavement | Total illuminance uniformity of pavement |
|---|---|---|---|
| Single-lane traffic | Two-lane traffic | | |
| ≥1200 | ≥650 | 0.4 | 0.6 |
| ≤350 | ≤180 | 0.3 | 0.5 |

### 1.2 *Characteristics of urban tunnels*

Compared with highway tunnels, urban tunnels have the characteristics of a small tunnel radius, high illumination of tunnel external environment at night, large traffic flow, and small speed limit. Based on these characteristics, this paper makes some improvements to the lighting design of urban tunnels: because of the small radius of the tunnel, the tunnel height is set to 6m; due to the large traffic flow and low vehicle speed, the traffic flow is set at 1200/[veh/(h·ln)] and the speed limit is set at 60 km/h; due to many small vehicles, the equivalent light curtain is set at a height of 1.26 m. Finally, the weather is set to cloudy to simulate the high-brightness lighting environment outside the tunnel at night.

### 1.3 *Layout form of lighting fixtures*

Reasonable arrangement of tunnel lighting fixtures is an effective way to meet the requirements of lighting quality and improve lighting efficiency. At present, the Chinese tunnel lighting design standard stipulates four kinds of lamp arrangement forms, namely, the midline, the midline is sideways, both sides are staggered, and both sides are symmetrically arranged (Li 2019). However, there are no specific requirements for relevant parameters, and different layout forms have their advantages and disadvantages. An appropriate layout method should be adopted in combination with the actual situation. The arrangement of lamps and lanterns is shown in Figure 1.

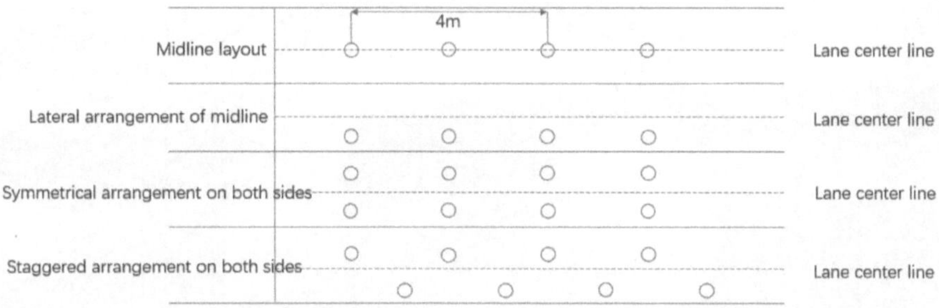


Figure 1. Schematic diagram of lamp layout.

## 2 SIMULATION DESIGN OF TUNNEL LIGHTING BASED ON DIALUX EVO

### 2.1 *Tunnel model building*

After determining the parameters of the tunnel, it is necessary to complete the establishment of the tunnel model. The establishment of a tunnel model mainly includes three parts: model establishment, lighting arrangement, and calculation area establishment.

In the modeling part, the tunnel section drawn by CAD is imported into the outdoor scene of DIALux evo, and the tunnel model is initially created through the Boolean operation. Then, the entrance area of the tunnel needs to be added with standard windows to ensure that sunshine calculation can be introduced. As the final lighting effect of the tunnel is determined by the direct light of the lighting lamps and the reflected light of the inner wall of the tunnel, it is necessary to endow the completed tunnel pavement, side walls, and vault surfaces with the same or similar materials as the reality. The side wall of the tunnel is an ordinary cement wall, the reflection coefficient is adjusted to 27%, the pavement is asphalt concrete, the reflection coefficient is 18%, and the tunnel vault is a dark mixed roof soil slab, reflecting 10% (Shi 2020).

In the part of lighting arrangement, Samsung Aurora Star series LED tunnel lamps are used for lighting arrangement at the threshold zone, and the maintenance coefficient of the lamps is 0.7. The distance between the four lighting modes is 4 m. The lights in the center line layout mode are installed just above the center line of the driveway, with a single light power of 240 W and an installation height of 6 m. The lamps and lanterns arranged in the way of center line deviation are installed 1m away from the center line of the carriageway, and the installation height is 5.5 m. The lamps arranged symmetrically on both sides and staggered on both sides are installed on both sides of the middle line of the carriageway, with an installation height of 5 m and a single lamp power of 120 W.

Finally, the calculation area is set in the tunnel model to calculate the tunnel lighting evaluation index. Set a calculation area of 8 × 20 m on the road surface of the threshold zone to calculate the average illuminance and other indicators. In the calculation area of the maximum TI, considering that the vehicles driving in urban tunnels are mostly small, the sight height of small vehicles is 1.26 m, and the driver's gaze angle is about 1°, so the equivalent light screen is set at the horizontal height of 1.26 m, forming an angle of 89 with the ground. In DIALux evo, the light received by the equivalent light screen from all directions of the tunnel is different from the actual feeling of the driver. Therefore, it is necessary to build an opaque and non-reflective baffle with an area parallel to the equivalent light screen and the same size to control the interference on the back of the light analysis surface (Zheng 2021). The tunnel model is shown in Figure 2.

Figure 2. Tunnel model.

## 2.2 *Simulation of different lighting modes*

The purpose of this paper is to analyze the proper lighting arrangement of urban tunnels. By using the research method of control variables, four lighting arrangements are simulated and calculated under the conditions of controlling the lamp type, lighting spacing, lamp power, and being in the same tunnel. Firstly, three kinds of layout modes, symmetrical on both sides, staggered on both sides, and offset on the center line, are compared and analyzed. When the deflection angles are 0, 10, 20, 30, 40, 50, 60, 70, and 80, the reasonable angles of these three layout modes are obtained by simulation calculation. Then, the software adjusts the four lighting modes to a reasonable angle and gets the data of the evaluation index of the threshold zone currently. Finally, the lamp power is reduced to 40%, and the weather ambient exterior of the tunnel is set to be cloudy to simulate the high illuminance ambient exterior of the urban tunnel at night and get the data of the corresponding evaluation index affected by the external light source.

# 3 RESULT ANALYSIS

## 3.1 *Analysis of deflection angle*

As shown in Figure 3, the evaluation index of the threshold zone lighting changes with the deflection angle.

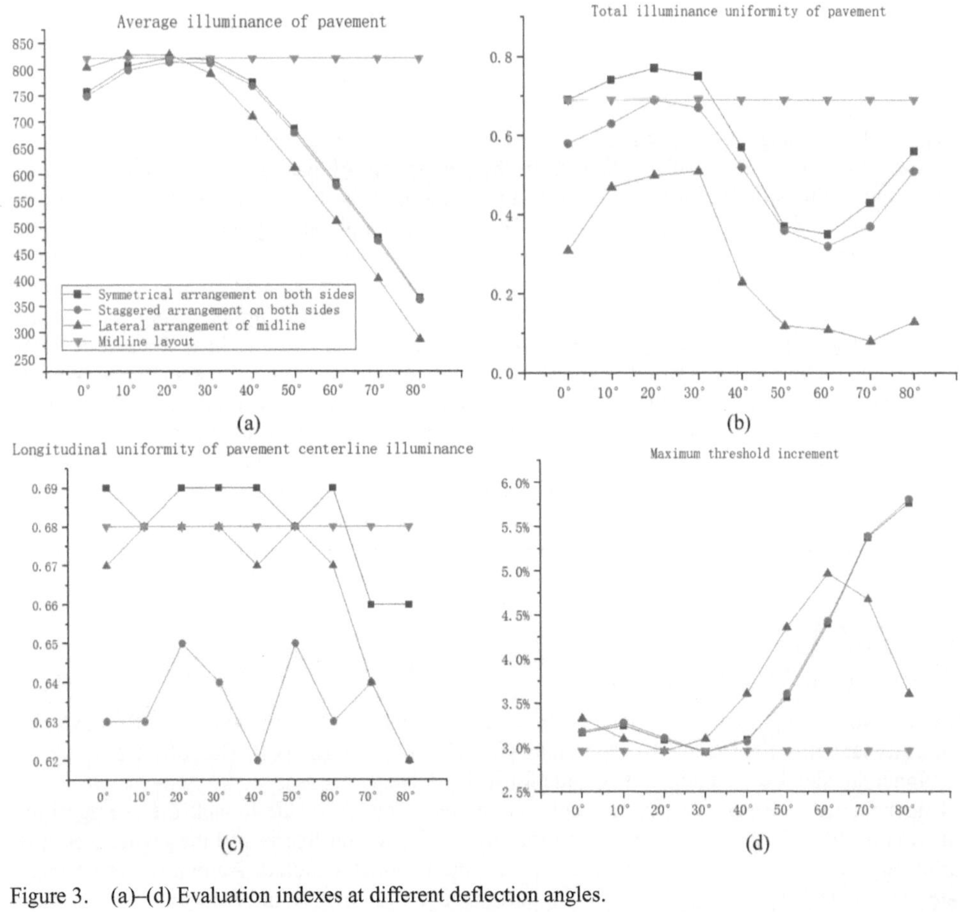


Figure 3. (a)–(d) Evaluation indexes at different deflection angles.

The curve in Figure 3(a) shows the law of first increasing and then decreasing with the increase of the installation angle of lamps, and the maximum value is between 20 degrees and 30 degrees. Among them, the curves of bilateral symmetry and bilateral staggered mode coincide, which meet the requirement of greater than 550 lx between 0 and 70 degrees, and the curve value is greater than that of the midline offset mode after being greater than 30 degrees. The curve value of the midline deviation mode is larger between 0 and 30 degrees, and it meets the specification between 0 and 70 degrees.

The curve in Figure 3(b) shows the law of first increasing and then decreasing, reaching the maximum value between 20 degrees and 30 degrees. The symmetrical and staggered curves on both sides are the same, and the symmetrical pattern on both sides has the largest value, which can meet the specification of greater than 0.4 from 0 to 80 degrees. The curve of the midline deviation mode is lower, only between 10 and 30 degrees, which is greater than 0.4.

Figure 3(c) curve shows a steady trend, with a slight fluctuation, and the change rate reaches the highest at 60 to 80. Among them, the curve with bilateral symmetry is the highest, followed by the deviation of the midline, and finally, the bilateral staggered arrangement. All three arrangements can meet the specification of greater than 0.6 in the range of 0–80 degrees and all three arrangements are stable in the range of 0–60 degrees. The symmetrical arrangement on both sides drops sharply at 60 degrees, the staggered arrangement on both sides and the deviation of the midline drops sharply at 70 degrees, and the maximum value of the three lamp arrangement methods is between 20 and 30 degrees.

In the curve of Figure 3(d), all three modes meet the requirement of less than 15% in the range of 0 degrees to 80 degrees. The maximum TI value of bilateral symmetry and bilateral staggering is the same, and rises with the increase of deflection angle, reaching a stable level between 70 and 80 degrees. The midline deviation increases from 0 to 60 degrees and decreases at a faster rate from 60 to 80 degrees. Within the reasonable deflection angle range, the bilateral arrangement curve is better than the midline lateral arrangement.

The deflection angle does not need to be considered in the central line layout. According to the comprehensive analysis of the data, from the perspective of average illumination, the value of the central line side layout is the highest, indicating that the central line side layout is more energy-saving and has higher light efficiency. When the lamps are arranged based on a reasonable deflection angle, the illuminance difference among the four modes is very small. As for the total uniformity of pavement illuminance, the value of symmetrical arrangement on both sides is much larger than that of other arrangements, which can make the tunnel illuminance uniform. In the aspect of longitudinal uniformity of pavement illuminance in the middle line, the symmetrical arrangement on both sides is still the largest, indicating that the difference between the maximum illuminance and the minimum illuminance in the middle line is the smallest, indicating the highest comfort. The maximum TI measures the degree of disability glare, so the smaller the value (Zhang 2021), the better. The maximum TI of midline deviation and midline layout in the table is the smallest, but this data is not within the reasonable deflection angle range. It can be seen from 0 degrees to 60 degrees that the symmetrical arrangement of lights on both sides is obviously the smallest, so the staggered arrangement on both sides can effectively reduce the driver's disability glare.

### 3.2 *Introducing daylight calculation*

DIALux evo is used to set the weather to cloudy day to simulate the light source ambient exterior of the city tunnel at night, and 60% of the intensified lighting is turned off. The software simulates the tunnel threshold zone again and gets the following table.

Under the influence of the external light source environment of the tunnel, the average illumination of the light distribution mode on the side of the central line is still the highest, but the uniformity of the total illumination is lower than the national standard. Among the other three ways, the bilateral symmetry way is the highest average illuminance and the smallest maximum

Table 2. Calculation results of different lamp arrangement forms

| Arrangement mode of lamps | Average illumination (lx) | Total illuminance uniformity of pavement | Longitudinal uniformity of pavement centerline illuminance | Maximum TI (%) |
|---|---|---|---|---|
| Symmetrical arrangement on both sides | 475 | 0.55 | 0.14 | 22.28 |
| Staggered arrangement on both sides | 472 | 0.59 | 0.16 | 22.29 |
| Midline layout | 474 | 0.49 | 0.15 | 22.31 |
| Lateral arrangement of midline | 477 | 0.36 | 0.15 | 22.17 |

TI, and the bilateral staggered arrangement way is the highest total illuminance uniformity and longitudinal illuminance uniformity of pavement centerline.

## 4 CONCLUSIONS

(1) From the point of view of lighting energy efficiency, the lighting efficiency of the center line with one side is the highest, followed by the symmetrical arrangement on both sides, and the staggered arrangement on both sides is the lowest. The reasonable deflection angle of both sides of the layout is larger than that of the midline layout and the midline lateral layout and reaches the maximum value between 20 degrees and 30 degrees.

(2) In terms of lighting comfort, the total uniformity of pavement illuminance and the longitudinal uniformity of pavement centerline illuminance are the largest values of bilateral symmetrical lighting, and the maximum TI is slightly larger than that of midline layout and midline offset layout, but both are much lower than the 15% specification. Therefore, the tunnel lighting comfort of symmetrical arrangement on both sides is the highest.

(3) The influence of external lighting environment of the urban tunnel at night has little influence on the energy-efficiency index but has a great influence on the comfort index. Among them, the highest lighting efficiency is the arrangement with the midline on one side, and the most comfortable arrangement is the arrangement with bilateral symmetry.

(4) Comprehensive comparison shows that the evaluation indexes of the four lighting arrangements can meet the requirements of lighting design at most deflection angles, and the lighting arrangements on both sides have larger reasonable deflection angles. Under the condition that the difference in average illuminance on the pavement is small and meets the design requirements, the comfort index of symmetrical lighting on both sides is more prominent. Therefore, the symmetrical lamp arrangement on both sides has more application value in urban tunnel lighting design.


## ACKNOWLEDGEMENTS

This research was supported by the Natural Science Foundation of Fujian Science and Technology Department (Grant No. 2019J01787) and Fujian Tonglian intelligent illumination intellectual property research and development (Grant No. GY-H-21203).

## REFERENCES


CIE: 88-2004: Guide for the lighting of road tunnels and underpasses[S], CIE, 2004.

Feng Du: *Research on visual adaptation curve of highway tunnel lighting under light color change* [D], Chongqing, 2019.

Industry Recommended Standards of the People's Republic of China. JTG/T D70/2-01-2014. Detailed Rules for Lighting Design of Highway Tunnels [S]. *Ministry of Transport of the People's Republic of China*. 2014.

Kairong Shi. Study on the influence of tunnel lighting fixture layout on lighting effect [J] *Transportation energy conservation and environmental protection*, 2020,16 (04): 134–137.

Shanghui Li, Chengjiang Zhang. Study on typical lighting parameters of highway tunnel based on DIALux [J] *Journal of Fujian Institute of engineering*, 2019,17 (06): 580–585.

Xuan Zheng, Zhi Li, Xue Li. Glare simulation analysis of highway tunnel lighting fixtures [J] *Journal of lighting engineering*, 2021,32 (03): 172–178.

Yijun Zhang, Zheng Zhang, Li Li, Hu Zhou. Study on glare index of highway tunnel under symmetrical light distribution illumination [J] *Transportation energy conservation and environmental protection*, 2021, 17 (03): 115–118.

*Advances in Urban Construction and Management Engineering – Cha (Ed)*


# Urban spatial elements for walkable neighborhood—A case study of Jindi District in Hunnan, Shenyang

Danyang Li*, Rui Chen*, Mingyuan Ma* & Ou Pan*
*School of Architecture and Planning, Shenyang Jianzhu University, China*

ABSTRACT: With Jindi District in Hunnan Shenyang as a study case, this paper obtains data on walking behavior, social attribute of residents, and urban spatial elements through questionnaire, and site survey, and makes an analysis of physical space morphology from convenience, connectivity, comfort, and security with methods of correlation analysis and logistic regression model. It is found that the density of community service facilities, types, the density of educational facilities, types, green space density, the bus transfer station, and other factors have different degrees of influence on space convenience, connectivity, comfort, and safety. Among these factors, functional layout, path optimization, space comfort, and safety are put forward.

## 1 INTRODUCTION

The urban residential street is an important place to carry residents' daily life, and walking is a common green way to travel in residents' daily life. In recent years, in the process of urban renewal, the street space has become functional rapidly, the pedestrian space has become a temporary parking lot, the lack of pedestrian facilities, the privatization of public space and other factors have led to the continuous deterioration of the pedestrian environment in residential areas (Chen 2017; Lu 2012; Tan 2010). Therefore, the street space of urban residential areas in the era of motor traffic will become an important issue for urban renewal, considering the daily walking space scale and environmental requirements of residents (Ewing & Handy 2012).

The residential space suitable for walking is affected by the climate, urban spatial pattern, cultural customs, etc. in different regions. Due to the great differences in urban development models adopted by different countries and regions, there are great differences in the cognition of walking suitable in different countries and regions. The academic community also disagrees with this recognition, and the evaluation standard of urban space suitable for walking is not complete (Millington & Thompson 2009). Therefore, it is necessary to carry out localization research in China and explore the suitable walking space for urban residential areas in China.

Considering that domestic research pays little attention to the influence of residential street space on residents' walking, it is necessary to explore the walkable space environment suitable for the north. Taking Jindi Community in Hunnan, Shenyang as the research object, this paper focuses on the spatial form of residential streets, the frequency and behavior level of residents' walking trips, and explores the spatial characteristics to meet people's walking needs.

*Corresponding Authors: lee_dy@126.com, 1813331098@qq.com, 56356090@qq.com and 1873802194@qq.com

DOI 10.1201/9781003348023-10

## 2 ON-SITE INVESTIGATION, DATA COLLECTION, AND RESEARCH

### 2.1 *Research objects*

Jindi community is mature and all kinds of functional facilities are perfect. Five residential areas in Jindi Community, Hunnan, and Shenyang are selected as the research objects. For the convenience of research, the buffer radius of 500 meters of residential areas with a similar physical environment and community psychological identity is selected as the research scope.

### 2.2 *Research steps*

Selecting appropriate influencing factors, the relationship model between walking spatial features and walking behavior in residential areas is established.

This paper analyzes the influence of walking space elements on walking behavior from four aspects: convenience, connectivity, comfort, and safety.

This paper discusses the significance of the model results to the optimization of pedestrian space form in residential areas and puts forward optimization strategies.

### 2.3 *Data sources*

The data of this paper mainly includes two parts: one part is a questionnaire survey, 20-30 questionnaires are distributed in each residential area, 125 questionnaires are distributed in total, and 95 valid questionnaires are screened; The other part is the field survey data, which provides information according to the field survey and map and obtains relatively accurate data by combining CAD calculation.

### 2.4 *Research focus*

#### 2.4.1 *Walking behaviors of residents*

In this paper, walking behavior refers to the walking behavior that occurs in the walking space of urban residential areas, that is, the behavior of residents leaving their homes for outdoor activities by walking is walking behavior (Tan 2010). The survey found that the walking travel of residents in the residential area mainly includes commuting walking, shopping walking, and leisure walking, and the walking frequency is taken as the main indicator to measure the walking behavior (Chen 2012). The study approximately believes that high walking frequency represents a large amount of walking volume of residents, and to a certain extent, it indicates the appropriate walking level of walking space (Chen 2017; Ewing & Handy 2012).

#### 2.4.2 *Social attributes of residents*

The follow-up study of the impact of walking space form on walking behavior focuses on factors such as age, gender, occupation, education, and ownership of private cars while excluding irrelevant factors.

#### 2.4.3 *Form of walking space*

The pedestrian space form studied in this paper refers to the combination of a series of elements related to land use, transportation system, and urban design, including convenience, connectivity, comfort, and security. Among them, convenience, and connectivity study the impact of macro land use and distribution on walking behavior, while comfort and security mainly evaluate the impact of pedestrian space form on walking behavior from a micro perspective (Chen 2012; Chen 2017; Handy S L 2002; Koohsari M J 2016).

## 2.5 *Research methods*

Through SPSS software, this paper analyzes the correlation between commuting walking frequency and influencing factors of walking spatial form, calculates the Pearson correlation coefficient, and preliminarily screens influencing factors of walking spatial form variables related to walking behavior.

Logistic regression analysis is carried out to generate the basic model of walking frequency using the reclassified walking frequency as the dependent variable and the residents' social attributes as the independent variables. At the same time, to investigate the influence of different factors in residents' social attributes on their walking behavior, the regression reserved variables were removed from the basic model one by one, and the pseudo $R^2$ changes were observed to obtain the basic model of residents' social attributes.

From Pearson correlation analysis, the variables of walking space morphology with walking frequency less than 0.1 are selected and incorporated into the basic model of residents' social attributes in the previous step. On the one hand, the influence of this variable on walking frequency is measured by the added value of pseudo-$R^2$, and then the key indicators of convenience, connectivity, comfort, and safety are extracted to analyze the elements of walking space morphology in residential areas suitable for walking. On the other hand, based on the model analysis results, the factors are sorted according to the added value of pseudo $R^2$, and the key indicators are selected for analysis (Tables 1 to 6).

Table 1. Fitting results of commuter walking model.

| | | | B (Sig. < 0.1) | | |
|---|---|---|---|---|---|
| Personal social attributes | Pseudo $R^2$ added value | LRT | Model 1 (frequent/less travel) | Model 2 (frequent/less travel) | Model 3 (frequent/less travel) |
| Age | 0.012 | 0.001 | 0.430** | 0.230** | – |
| Professional | 0.205 | 0.000 | 4.890** | 3.850** | – |
| Private Car | 0.445 | 0.000 | −5.565** | −4.449** | −1.156** |

Notes: ILR is an index reflecting authenticity, which belongs to a composite index reflecting sensitivity and specificity at the same time. It is a general rule for seeking test methods. It can be set that the random sample composed of $n$ observation values $(x_1, x_2, \ldots, x_n)$ comes from the density function f $(x; \theta)$, where $\theta$ is an unknown parameter. "**" indicates that the value of sig. is less than 0.05; "*" indicates that the value of sig. is between 0.05 and 0.1. The basic fitting model is: ln(pi/p0) = βi + βig1Xg2 + βig2Xg3 + βig3Xg4, where Xg2 refers to age, Xg3 refers to occupation, and Xg4 refers to private car; i = 1, 2, and 3. Trip 1 is a general trip, trip 2 is more frequent, and trip 3 is frequent; $\beta$ign is the corresponding b-value coefficient in parameter estimation.

Table 2. Key spatial variables affecting commuter walking.

| Spatial variables | The density of the plots' entrance and exit | Street network density | Street side length | Educational facilities distance | Intersection density | Bus transfer density | Service facility density | Service facilities types | Commercial density along the street interface |
|---|---|---|---|---|---|---|---|---|---|
| Pseudo $R^2$ added value | 0.015 | 0.014 | 0.013 | 0.013 | 0.011 | 0.010 | 0.009 | 0.007 | 0.005 |

Table 3. Fitting results of shopping walking model.

| Personal social attributes | Pseudo $R^2$ added value | LRT | B (Sig. < 0.1) Model 1 (frequent/less travel) | Model 2 (frequent/less travel) | Model 3 (frequent/less travel) |
|---|---|---|---|---|---|
| Age | 0.017 | 0.016 | 0.455** | 0.358** | – |
| Professional | 0.030 | 0.070 | −0.558** | −0.400** | −0.334** |
| Private car | 0.075 | 0.000 | −2.324** | −2.183** | −0.998** |

Notes: "**" indicates that the value of sig. is less than 0.05; "*" indicates that the value of sig. is between 0.05 and 0.1. The basic fitting model is: ln(pi/p0) = βi + βig1Xg2 + βig2Xg3 + βig3Xg4 + βig4Xg5 + βig5Xg6 + βIg6xg7, where xg2 is gender, xg3 is age, xg4 is occupation, xg5 is education, xg6 is family population, and xg7 is the family private car; i = 1, 2, and 3. Trip 1 means normal travel, travel 2 means more travel, and travel 3 means frequent travel; $\beta$ign is the corresponding b-value coefficient in parameter estimation.

Table 4. Key spatial variables affecting commuter walking.

| Spatial variables | Cultural and recreational facilities' density | Street network density | Intersection density | Street network density | Street side length | The density of the plots' entrance and exit | Commercial density along the street interface | Service facilities types | Walkable area |
|---|---|---|---|---|---|---|---|---|---|
| Pseudo $R^2$ added value | 0.060 | 0.058 | 0.058 | 0.054 | 0.048 | 0.045 | 0.043 | 0.039 | 0.038 |

Table 5. Fitting results of leisure walking model.

| Personal social attributes | Pseudo $R^2$ added value | LRT | B (Sig. < 0.1) Model 1 (frequent/less travel) | Model 2 (frequent/less travel) | Model 3 (frequent/less travel) |
|---|---|---|---|---|---|
| Age | 0.017 | 0.001 | 0.251** | 0.429** | 0.356** |
| Professional | 0.105 | 0.012 | −3.677** | −1.688** | −1.711** |
| Private car | 0.022 | 0.015 | – | −1.159** | – |

Notes: "**" indicates that the value of sig. is less than 0.05; "*" indicates that the value of sig. is between 0.05 and 0.1. The basic fitting model is expressed as ln(pi/p0) = $\beta$i + $\beta$ig1Xg2 + $\beta$ig2Xg3 + $\beta$ig3Xg4 + $\beta$ig4Xg5, where xg2 is age, xg3 is occupation, xg4 is family population, and xg5 is the family private car; i = 1, 2, and 3; trip 1 is general travel, trip 2 is more travel, trip 3 refers to frequent travel; $\beta$ign is the corresponding b-value coefficient in parameter estimation.

## 3 DATA ANALYSIS

### 3.1 *Analysis of key factors affecting commuting walking*

The density and types of service facilities in the convenience evaluation factor can meet the living needs of residents in the process of commuting, such as shopping, living payment, etc. The density and distance of educational facilities reflect the convenience of parents to pick up and send their children and play a decisive role in the way parents pick up and send their children. The short

Table 6. Key spatial variables affecting recreational walking.

| Spatial variables | The green open space density | cultural and recreational facilities' density | Intersection density | Street network density | Street network density | The density of the plots' entrance and exit | Length and density of sidewalks above 8M | Commercial density along the street interface | The density of the plots' entrance and exit |
|---|---|---|---|---|---|---|---|---|---|
| Pseudo $R^2$ added value | 0.051 | 0.043 | 0.030 | 0.028 | 0.025 | 0.020 | 0.016 | 0.010 | 0.051 |

distance between bus stops is conducive to walking conversion and promotes residents' walking commuting.

The connectivity evaluation factor is the high density of the middle street road network and the high frequency of pedestrian travel. Therefore, strengthening street connectivity is conducive to convenient travel, especially convenient access to educational facilities and bus transfer stations, which can facilitate residents' walking.

Among the comfort evaluation factors, rich Street businesses are conducive to residents' shopping during commuting. The high density of entrances and exits of the plot can provide residents with a variety of travel routes, shorten the distance of residents to their destination, and promote walking.

### 3.2 *Analysis of key factors affecting shopping walking*

The diversified social service functions of residential areas can meet the multiple purposes of residents' one-time travel. The density and distance of cultural and recreational facilities have a significant impact on shopping walking. When commercial facilities are combined with cultural and leisure facilities, residents' travel frequency is greatly increased.

The road network density in the connectivity evaluation factor has a positive impact on walking travel, that is, the larger the street road network density, the smaller the block size, while the shopping walking path is more convenient.

The high density of entrances and exits of the plot provides residents with a variety of path choices for travel, thereby promoting pedestrian travel, especially at the entrances and exits of the community with similar distances, shortening the distance from residents to shopping places. The wide sidewalk can accommodate more service facilities and improve the shopping experience.

### 3.3 *Analysis of key factors affecting leisure walking*

Diversified service facilities reflect the diversity of spatial functions and social activities. The density of cultural and recreational facilities is an important factor that leads to leisure sexual behavior. Green space provides a basic space guarantee for leisure behavior.

There is no clear destination for daily leisure walking, and the path selection is relatively random. Therefore, the high permeability street network provides residents with more path choices, thereby enriching the walking experience.

The survey found that the small square space on the corner will become a fixed activity place in the community over time, attracting some residents to come here every day. The continuous and abundant commercial distribution along the street attracts leisure walking activities. The high density of entrances and exits of the plot can provide residents with more path choices.

In the safety evaluation factors, traffic flow and vehicle speed cause the contradiction between people and vehicles in the residential area, which affects the choice of residents' travel paths and leads to a lower frequency of walking trips.

### 3.4 *Analysis of the morphological elements of walking space in residential areas*

This study classifies the influence degree of the three types of walking by the key indicators. It is found that the density of facilities, the density of the street network, and the richness of the commercial interface along the street are the key indicators that affect the three types of walking behavior; the type of facilities and the density of cultural and recreational facilities have an impact on the two types of walking; the density of cultural facilities, the density of educational facilities, the density of green space, the density of bus transfer stations, the distance between cultural and recreational facilities, the distance between educational facilities, the walkable area and the density of the length of sidewalks above 8 m only affect the first kind of walking.

## 4 WALKING SPACE OPTIMIZATION STRATEGY

### 4.1 *Improve the convenience of walking space*

We should optimize the layout of functional spaces and implement mixed layouts of functions in combination with public transportation. Public transportation hubs and various facilities can guide shopping behavior during commutes and improve the frequency of residents' commuting, shopping, and walking trips.

The centralized layout of diversified and high-density community service facilities can better meet the living needs of residents and meet shopping needs while commuting and leisure.

### 4.2 *Optimize the connectivity of walking paths*

Small scale blocks can not only increase the convenience of walking but also create rich Street View space. The planning and design take the density of the street network as the main index and measure the road connectivity of the residential area in combination with the number of road intersections and the number of entrances and exits of the plot.

### 4.3 *Create comfortable walkways*

Governments should increase the width of some pedestrian walkways and provide continuous and diverse areas for pedestrian activity; reduce physical walls along the street and increase the density and continuity of commercial stores along the street, and enhance the mobility of green spaces by increasing the number of open spaces.

### 4.4 *Ensure the safety of the walking space*

It is found that the traffic flow and traffic speed in the residential area have a certain influence on leisure walking. In the planning, people and vehicles should be separated as much as possible, and the speed should be limited as much as possible to provide a good safety guarantee for residents' leisure walking.

## 5 CONCLUSION

This study focuses on the influence of individual social attributes of residents and the spatial form of walking in residential areas on walking behavior but does not consider the influence of perceived environmental characteristics of pedestrians on walking behavior, and the results will be somewhat biased. For example, the distance perception of commercial facilities within the same distance may be different because individuals perceive the walking environment differently, that is, the real accessibility is inconsistent with the perceived accessibility, and the variables selected in the study are all objective spatial variables, so there will be some errors. In addition, there is another possibility of the relationship between spatial form and walking behavior, that is, residents' self-choice behavior rather than physical space affects walking behavior.

## ACKNOWLEDGEMENTS

This research was supported by the Scientific Research Project (Youth Project) of the Liaoning Provincial Department of Education (Project title: Research on the Form of Pedestrian Space in Urban Residential Areas Based on Behavioral Intervention Theory—A Case Study of Typical Residential Areas in Shenyang) (Grant No. lnqn202002); and 2021 Undergraduate Innovation and Entrepreneurship Training Program of Shenyang Architecture University (Project title: Research on the Optimization Strategy of Pedestrian Space in Urban Old Residential Areas Under the Background of Healthy City).

## REFERENCES

Chen Qianhu, Fang Liyan, Deng Yiling. (2017). *A Study on Neighborhood Compound Environment and Walking Behavior from the Perspective of Heterogeneity: Taking Hangzhou As an Example*, City Planning Review. 9, 48–57.

Chen Yong, He Ning. (2012). Analysis of Walkable Environment and Influential Factors in Rail Transit Station Areas:Case Study of 12 Neighborhoods in Shanghai, *J, Urban Planning Forum*. 96–104.

Chen Yong, Wang Quanyan, Xi Wenxin. Planners. (2017). *Analysis on the Influence of Block Space form on Residents' Pedestrian Traffic*, 2, 74–80.

Handy S L, Boarnet M G, Ewing R, etal. (2002). How the Built Environment Affects Physical Activity: Views from Urban Planning, *J, American Journal of Preventive Medicine*. 2, 64–73.

Koohsari M J, SugiyamaT, Mavoa S, et al. (2016). Street Network Measures and Adults' Walking for Transport: Application of Space Syntax, *J, Health & Place*, 38, 89–95.

Lu Yintao, Wang De. (2012). Walkability Measuring in America and Its Enlightenment, *J, Urban Planning International*, 1, 10–15.

Millington C, Thompson C W, Rowe D, et al. (2009). Development of the Scottish Walkability Assessment Tool (SWAT), *J, Health & Place*, 15: 474–481.

Reid Ewing, Susan Handy (2012), Translated by Jiang Wenjing, Proofread by Zhou Jiangping, Measuring the Unmeasurable: Urban Design Qualities Related to Walkability, *J, Urban Planning International*, 5:43–53.

Tan Shaohua, Guo Jianfeng, Jiang Yi. (2010).Impacet of Human Settlements on Public Health: New Frontier in Urban Planning Research, *J, Urban Planning Forum*, 4: 66–70.

Xu Leiqing, Song Haina, Huang Shuqing. (2017). Reflection on Community Micro-regeneration undere the Background of Innovative Social Governance: Two Practice Cases by 408UDR Lab, *J, Urban and Rural Planning*, 4: 43–51.

*Advances in Urban Construction and Management Engineering – Cha (Ed)*


# Research on the application of UAV 3D aerial surveying technology in landscape design—Taking the landscape design of the southern ecological park of Shenyang Jianzhu University as an example

Wei Dai* & Haoyu Lan*
*School of Design and Art, Shenyang Jianzhu University, Shenyang, Liaoning, China*

ABSTRACT: The application of 3D aerial surveying technology has gradually sprung up in recent years with the development of unmanned aerial vehicle (UAV) photography and mapping technology. 3D aerial surveying technology has been used in urban planning and other large-scale spatial design, but only possessed a single account. In landscape design, this technology has a greater application prospect and will bring a revolution to landscape design work in a new way. The pursuit of efficiency in modern society can make the application of this technology attract even more attention. Faster modeling speed and more accurate model data will greatly improve efficiency, so the technology can promote a new round of innovation in design methods while conforming to the trend of technological development and policy requirements. This paper reviews research on 3D aerial surveying technology applications in various fields in China in recent years. On this basis, we propose the technical route and specific method of applying the UAV 3D aerial surveying technology to landscape design and take the landscape design of the southern ecological park of Shenyang Jianzhu University as an example to elaborate.

## 1 INTRODUCTION

With the rapid development of modern technology, the acquisition of geographic information data has become increasingly diversified, and the requirements for the timeliness and accuracy of data have gradually increased. The traditional mapping method not only consumes much time but also requires secondary modeling, even though it is accurate. It cannot accurately restore the actual situation of the site. In recent years, with the mature development of UAV technology, 3D aerial surveying and modeling by UAV have gradually emerged. The UAV is a vehicle with power wheel drive, not carrying people on board. It is controlled by remote control equipment and can be used repeatedly (Li 2019).

UAV 3D aerial surveying technology consists of UAV photogrammetry, UAV control system, airborne camera, data processing program, and many other technologies. From both a hardware and software perspective, it can be divided into hardware systems and software systems. The hardware system of the current mainstream UAV is mainly composed of four systems: power system, remote control system, ground station (monitoring system), and data transmission system. The software system includes ground station software, aerial film processing software, and data processing software. At present, there are open source and non-open-source software for ground station software. Most of the ground station software has a route planning function, which is operated by the ground station controller to plan the corresponding route according to the mission requirements to obtain data. The aerial image processing software processes the raw data obtained by the UAV to generate a usable data package. The data processing software generates relevant analysis maps from the collated data packages or publishes and shares the data online.

*Corresponding Authors: 14106361@qq.com and 15500009686@163.com


DOI 10.1201/9781003348023-11

Compared with other traditional surveying and mapping methods, UAV 3D aerial surveying has the characteristics of a short measurement period, high working efficiency, and saving workforce and material resources. Therefore, this technology is now widely used in various industries. This paper provides a brief review of the current research status of domestic UAV 3D aerial surveying systems, focusing on its key applications in landscape design and further prospecting its future.

## 2 CURRENT STATUS OF DOMESTIC RESEARCH IN RELATED FIELDS

In recent years, China's UAV-related industries have developed rapidly, and civilian UAV companies led by DJI have also developed more industry-specific UAV products. After years of rapid development, UAVs with three-dimensional aerial surveying and remote sensing functions have also entered the civilian sector. These products have the characteristics of low cost, high resolution and timeliness. At present, UAV technology has been widely used in surveying mapping, urban planning, meteorological observation, and other fields. Lu Huanhuan et al. (2021) used UAV aerial surveying technology to establish a high-precision three-dimensional model and conducted in-depth research in this direction. Dong Jiaojiao et al. (2017) conducted an in-depth discussion on the application of UAV photogrammetry in urban planning. Hong Haiyang et al. (2015) conducted an in-depth discussion on the construction of a landscape information model based on visualization technology. And Wei Lai used UAVs to conduct 1:500 measurement tests and analysis in strip maps. Thus, UAV aerial surveying has relevant practical projects for practical exploration in urban construction, planning, agriculture, and territorial development. With the development of technology, the rapid and extensive application of UAV technology will be recognized by the public.

In the process of landscape design, the traditional way of acquiring topography and landforms has certain lag and limitations and is currently mainly using field measurement and inspection, which has the disadvantage of high cost and long period. The current application of satellite remote sensing image measurement and other means has the characteristics of low photo resolution and high cost (Figure 1), which is limited in the application of the actual project. With the rise of UAV

(a) Satellite image of Shenyang Jianzhu University in 2012

(b) Satellite image of Shenyang Jianzhu University in 2021

Figure 1. Google Earth intercepted part of the image data of Shenyang Jianzhu University (2012 and 2021 change comparison).

aerial surveying technology, the use of UAVs to conduct 3D aerial surveying of the site, not only quickly understands the site overview, but also generates the 3D model of the site through the data processing for a detailed analysis. At present, the domestic application of UAV 3D aerial surveying model reconstruction technology to landscape design is in the exploration stage.

## 3 TILT PHOTOGRAPHY AERIAL TRIANGULATION AND STEREO ACCURATE 3D MODELING TECHNOLOGY RESEARCH

### 3.1 *3D modeling technology route comparison*

There are two major technical routes for 3D modeling: traditional manual modeling (Figure 2) and fully automated modeling by tilt photography.

Traditional manual modeling is time-consuming, and the modeling speed is about 0.02 square kilometers/day/person. In contrast, tilt photography modeling only requires flight aerial surveying, and the subsequent steps are generated fully automatically by a computer. The modeling speed is about 8 square kilometers/day/person, which is 400 times more efficient than manual modeling.

### 3.2 *Overview of tilt photography techniques*

By carrying one or more sensors (i.e., cameras, currently more industry-use UAVs on the market are mostly five-lens camera modules) on one vehicle, the technology acquires images from different angles, such as vertical and oblique, at predetermined mission heights and locations to obtain more complete and accurate information about ground objects. In tilt photography, the positive film is the mission photo taken from the vertical ground angle, and the oblique film is the image acquired by the lens forming a certain angle with the ground. A positive film is a group, and an oblique film is a group of four (Song 2016).

### 3.3 *3D reconstruction technology realization path*

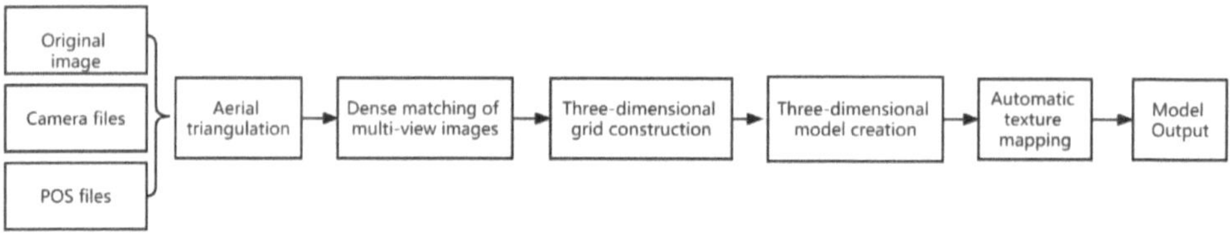


Figure 2. Tilt photography 3D aerial surveying and modeling process.

(1) Raw data acquisition. Tilt photography modeling generates pos files from the images taken by the UAV in the designated mission area (raw images), specific parameters such as camera focal length and sensitivity (camera files), and the position (latitude and longitude coordinates) and height (relative height) of the camera when it was taken.
(2) Aerial triangulation and matching. The above three data are subjected to aerial triangulation (the pose of the camera in the air is solved so that the position of each pixel of the connection point of the photo is calculated). The input image resolution results are then matched to form a 3D dense point cloud.
(3) 3D mesh construction and texturing. After the first two steps are completed, the computer generates a white model from the calculated point cloud data, and then automatically performs texture mapping.
(4) Generate a 3D model. 3D reconstruction work is completed, and the image stitching is finished. At this stage, we can view the high-precision 3D model through the viewing software, you can also generate other desired model data through the model software (such as orthophoto and elevation map).

## 4 APPLICATIONS OF 3D AERIAL SURVEYING TECHNOLOGY FOR LANDSCAPE DESIGN OF THE SOUTHERN ECOLOGICAL PARK OF SHENYANG JIANZHU UNIVERSITY

### 4.1 *Area conditions in data acquisition task*

The project is located on the southwest side of Shenyang Jianzhu University campus in Hunnan District, Shenyang City, Liaoning Province (Figure 3). The site has undergone several rounds of expansion, new construction, and reconstruction projects in recent years. The original drawings have been unable to feedback its true topography. The area is also a typical representative of the design area plots in the process of landscape design. The selected acquisition point is Shenyang Jianzhu University Ecological Park and part of the Science and Technology Park, including C5, B4, Huizhou-style houses, TenthPrince's Mansion, University Hospital, Eighth Prince's Academy, Sino-German Energy Conservation Demonstration Center, and other buildings. The positioning control point is E123.505 621°, N41.736829°, and the aerial surveying area is about 42,000 square meters. As the UAV takeoff and landing area needs open space, the second sports field of the school at 0.1 km to the northeast of the mission area was chosen as the takeoff and landing point for this aerial surveying. On the day of the aerial surveying, the environment was good, with wind force 2 and suitable light conditions. The flight crew held a UAV pilot license issued by the Civil Aviation Administration, which is a legal flight activity.

Figure 3. Aerial view of the southwest corner of Shenyang Jianzhu University by UAV.

### 4.2 *Flight route mission planning and acquisition*

The flight was conducted with a DJI brand flying vehicle, model Elf 4V2.0. The ground station software is DJI Terra, which can be used for route planning and data processing. This aerial surveying uses software for flight setup, with the flight altitude setting to 80 m, the positive flight speed to 3 m/s (maximum flight speed 7.9 m/s), and the oblique flight speed to 3.4 m/s (maximum flight speed 7.9 m/s). The mission area was a triangular area of about 350m north to south and 350m east to west.

Since various factors in the flight process can affect the accuracy of its route flight, to ensure that the modeling of the project runs smoothly, the parameters of the UAV were set to 75% overlap in the side direction, 75% overlap in the side direction (tilt), 85% overlap in the heading, 85% overlap in the heading (tilt), 256° for the main route angle, and 20 meters for the routing margin. The higher overlap rate can ensure a smooth 3D modeling task.

The DJI Terra ground station software was used to set the flight parameters, and a laptop was used to connect to the flight remote control in the field. The ground station software can observe the parameters of the aircraft in real-time. The flight had 6 landings and takeoffs, 11,148 meters of route length, and a total flight time of about 1.8 hours. The total number of navigation points was 92, and the total number of photos was 1038. The camera photo ratio was 3:2, and the pitch angle of the tripod head (tilt) was $-60°$ (Figure 4).

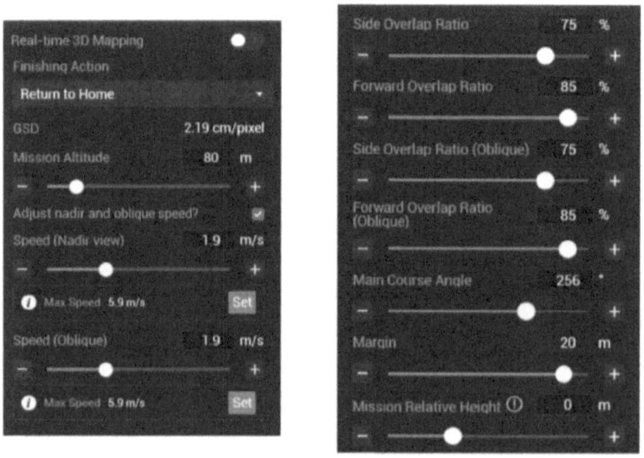


Figure 4. Parameter settings for the aerial surveying task.

## 4.3 *Multi-view image 3D reconstruction*

The 1,038 photos collected by the UAV according to the predetermined route mission were input into the Smart3D software system for air-triple calculation and processing of the multi-view images.

(1) Raw data acquisition. At the end of the flight, a total of 1,038 photos were taken (Figure 5). After filtering, in the Smart 3D software, a new project file was created and the organized photos were imported into the block. Through the system setting, the raw photo data contains basic information such as geographic location information, orientation angle, height, and camera angle. at the time of the shooting. After the image is recorded into Smart 3D software, the system

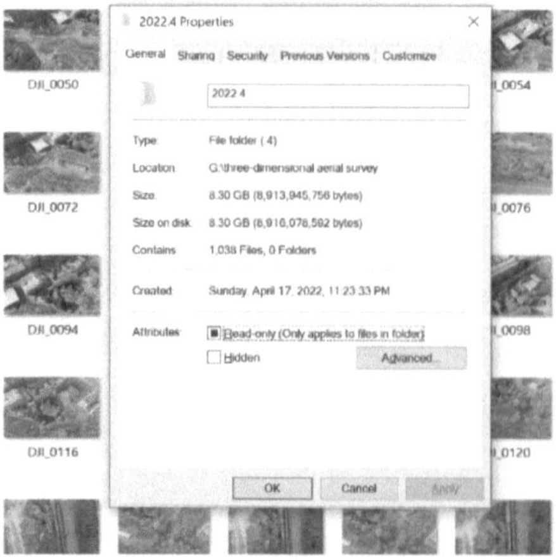


Figure 5. Partial display of the aerial surveying file.

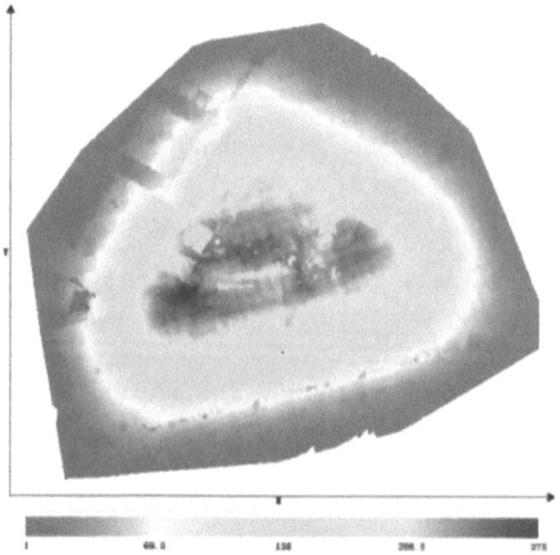

Figure 6. Top view of the number of images of the scene (The color indicates the number of images in the area).

will automatically read the photo information and generate the scene coverage information (Figure 6), and camera positioning error estimation (Figure 7).

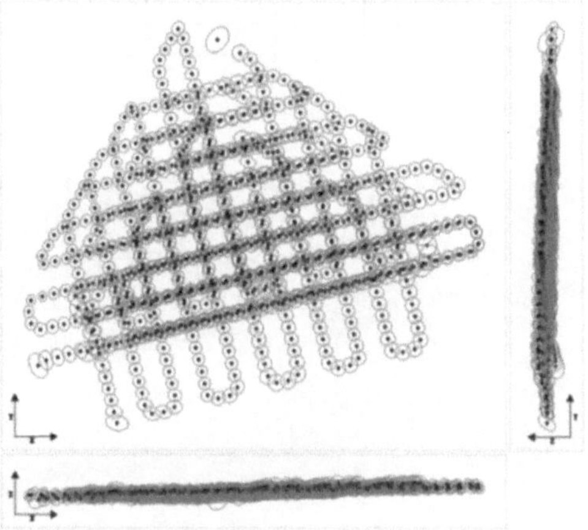

Figure 7. Image position uncertainty.

(2) Aerial triangulation and matching. The system performs camera calibration on the photos (Figure 8), calculates the data for aerial triangulation, and presents the point cloud image after calculation based on coefficients and correlation matrix in the same reference coordinate system.

| Calibration result | | | | | | | | | |
|---|---|---|---|---|---|---|---|---|---|
| | Focal length (mm) | The equivalent focal length of 35 mm (mm) | Main point X (pixels) | Main point Y (pixels) | K1 | K2 | K3 | P1 | P2 |
| Original Value | 8.8 | 24 | | | | | | | |
| Optimization value | 8.83147 | 24.0858 | 2744.77 | 1844.53 | 8.88032e-05 | -0.0056679 | 0.00667996 | -0.00122492 | 0.00128393 |
| Difference between original and optimized values | 0.03147 | 0.0858 | | | | | | | |

Figure 8. Calibration results of the camera.

(3) 3D mesh construction and texture matching. The system constructs mesh based on the generated aerial triangulation images to generate image connection points and optimize the errors (Figure 9) and then assigns corresponding texture information to the 3D white model for mapping (Figure 10). The accuracy can be selected as high, medium, or low to match different modeling requirements, and the lower the accuracy, the faster the modeling speed.

(4) The 3D model was generated by submitting the production project in Smart3D software (Figure 11), where the rough elevation coloring data can also be viewed (Figure 12).

(5) Combined with GIS software such as Global mapper, the model can be visualized and interacted with. The area and distance measurement of the area can be carried out, and the analysis of sunlight can also be carried out, etc. It provides basic data for future elevation mapping, slope measurement, and line of sight analysis, and provides important help for the preliminary research work on landscape design, as shown in Figures 13 and 14.

| Generated connection points | | | | | |
|---|---|---|---|---|---|
| Number of points | The median number of images per point | The median number of points per image | Median reprojection error (pixels) | Reprojection error RMS (pixels) | RMS of the distance to the light (m) |
| 26,8304 | 4 | 1,536 | 0.39 | 0.51 | 0.0948 |

Figure 9. Number and quality of image connection points.

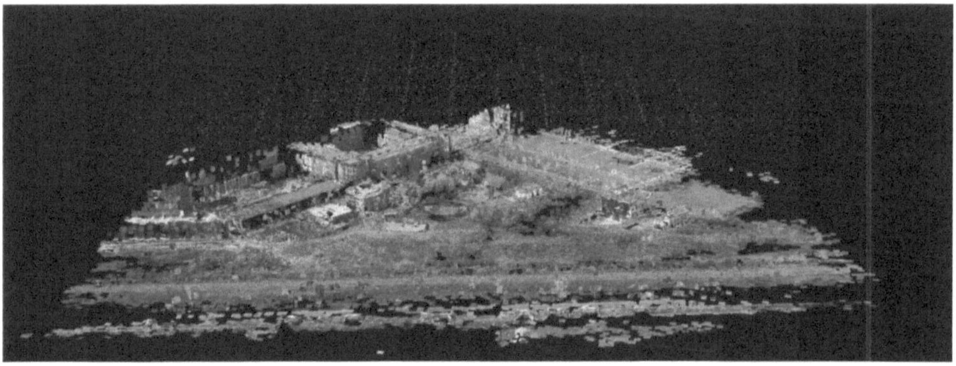

Figure 10. Texture generation.

Figure 11. 3D model generation.

Figure 12. Elevation coloring data.

Figure 13. Contour plotting.

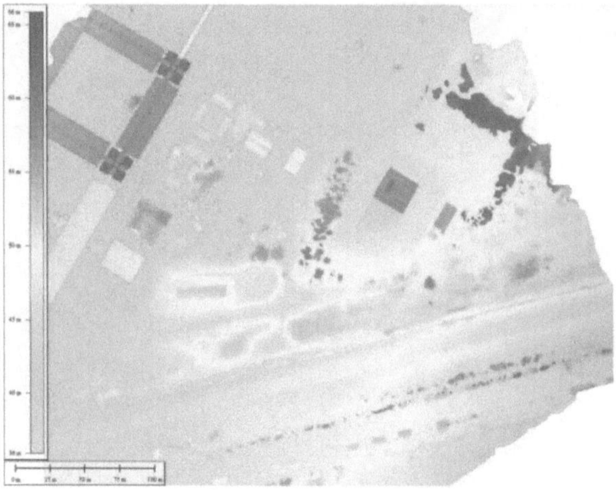

Figure 14. Reconstructed digital elevation model.

### 4.4 *Advantages and innovation points of 3D aerial surveying images in landscape design*

#### 4.4.1 *Advantages of 3D aerial surveying in landscape design*

Using drones for 3D reconstruction of the site can make the base data we collect at the beginning of the design more accurate, providing convenient and efficient underlying data for the landscape design work, which can accurately describe the current state of the base.

The original conventional surveying and mapping method is time-consuming and the workload of personnel is high, resulting in high costs. The use of UAV 3D aerial surveying and modeling technology can greatly shorten the survey time, in good weather conditions, an aerial surveying point a working day to complete the task, and the original high-risk point mapping task using aerial surveying becomes more efficient, which is important for the rapid development of landscape design work.

UAV flight height can be controlled between 20 and 500 meters, according to the task needs and the current situation of the site to determine the safe flight height, the lower the operating height the higher the accuracy, theoretically can reach sub-meter accuracy, orthophoto high definition to obtain information more intuitive. The use of three-dimensional aerial surveying means of drones effectively solves the problem of data surveys in areas not convenient for personnel to enter.

The original landscape design of the pre-analysis work is mainly done by on-site research and filming, measurement, and requesting relevant information from the relevant departments, the time cost, and labor cost are high, and cannot accurately grasp the entire base of the current topography. Therefore, compared with the limitations of traditional channels, 3D aerial surveying has irreplaceable advantages. Figure 15 shows the 3D topographic information of the southern ecological park of Shenyang Jianzhu University.

#### 4.4.2 *Innovation points of 3D aerial surveying application in landscape design projects*

In the landscape design project of the southern ecological park of Shenyang Jianzhu University, the use of three-dimensional aerial surveying image reconstruction by drones made the research work more convenient. High-precision aerial surveying images allow designers to accurately grasp the location information of various structures within the base, providing an important basis for site analysis. Compared with the traditional manual research method, the UAV 3D aerial surveying can provide a broader perspective and break the limitations of the previous manual research, for example, Figure 16 shows the difference in visual breadth between manual research and 3D aerial surveying.

Figure 15. Acquisition of 3D terrain information of the southern ecological park of Shenyang Jianzhu University.

Figure 16. Manual research view (left) UAV 3D aerial surveying view (right).

The rapid development of UAV photography in recent years has changed the previous manual research methods. UAV photography brings a higher visual angle and can observe the current situation of the base from a bird's eye view. The limitations of UAV photography are also obvious, the UAV only takes pictures, and the pictures cannot show the whole picture of the current situation in the base, while the 3D aerial surveying images through the UAV can show the whole picture of the current situation in the base from all angles, and can give more reference indicators for the design work. For example, Figure 17 shows the gap between the single flight aerial photograph of the environment around the seedling cultivation shed in the base and the 3D modeling in dynamic analysis. A single aerial photograph can only express what structures are in the base, but cannot express the height, volume, and another three-dimensional visual expression, and requires multiple aerial photographs for comprehensive analysis to draw relevant conclusions. In 3D aerial modeling, we can observe the structures from multiple angles, and analyze the building height, volume, surrounding environment, etc.

3D aerial surveying and modeling can not only be used for small-scale research but also for macro-scale research. For example, by combining the full-color 3D model of Figure 12 and the 3D terrain model of Figure 15 for the analysis of the macro-scale 3D model of the southern ecological park of Shenyang Jianzhu University, the following geomorphological characteristics of the area were obtained.

Figure 17. Single aerial photo (left panel) and 3D modeling dynamic analysis (right panel).

The southern ecological park of Shenyang Jianzhu University is mainly composed of ancient buildings and sloping land, in which the terrain mainly shows a general trend of low in the northwest and high in the southeast, and presents a pattern of more in the north and less in the south in terms of building distribution. In terms of topography and geomorphology, there is a water body with an area of about 1,800 square meters in the middle of the site (Figure 18), and the height difference within the site is about 10 meters, with the highest point located on the soil slope in the southeast and the low point located near the basic experiment center in the southwest. On the road, there is a campus main road on the north side, and there is no obvious road interspersed on the south side near the railroad, which is analyzed to be a barren land (Figure 19), and the landscape layout is more deprived on the south side.

Figure 18. Area measurement of the 3D model.

Figure 19. High-definition plane image of Southern Ecological Park, Shenyang Jianzhu University.

In the previous landscape design rendering, it is necessary to model the surrounding environment of the project to facilitate the study of the fit between the design plan and the project's peripheral environment or the impact of the project on the periphery, and to facilitate the party A to accurately understand the relationship between the project and the surrounding environment when the result enters the rendering stage. At present, there are two main methods of solving the surrounding environments of a project. The first method is environmental modeling, which involves gathering data on the built environment surrounding the project, then simulating and modeling it on a computer. The second is post-processing on render in the final stage of rendering, adding environmental pictures by Photoshop.

Modeling with help of UAV 3D aerial surveying can introduce the base environment directly into the design in an innovative way. The 3D model can be directly used in the design software, which can simulate planting trees, placing buildings, placing people, and so on in the model. It also interacts with other design software to modify the original base of the project. Using this method, the design environment can be introduced into the design process in real-time, significantly improving the efficiency of landscape design. After the design is completed, the rendering effect can be done directly without modeling the surrounding environment, which can significantly save the drawing time. Figure 20 shows a comparison of the landscape design of the seedling cultivation shed and its surrounding environment in the southern ecological park of Shenyang Jianzhu University. The left picture is the 3D model before the landscape design, and the right picture is the 3D model after the landscape design.

(a) Base of the original 3 D model

(b) Landscape design scheme model

Figure 20. Comparison between the current base 3D model and the design scheme 3D model.

## 5 CONCLUSIONS

In landscape design, a lot of preliminary analysis of the base is needed, but the original analysis method cannot fully grasp the comprehensive information of the base, which leads to the restriction in terms of data acquisition. It cannot accurately obtain accurate design information, which brings obstacles to the development of landscape design work. The application of 3D aerial surveying not only provides a more convenient way for the preliminary research work of landscape design but also directly imports the application of design software, which is an innovative design method for landscape design work.

In future landscape design, this technology can be used to make data collection work more efficient. Its acquisition accuracy will also reach the centimeter level to meet the practical needs of landscape design. In future research, the integration of UAV 3D navigation and landscape design will also become a research hotspot. With the popularization of UAVs and the reduction of the cost of software and hardware, it will be an inevitable trend for UAV 3D aerial surveys to replace the traditional measurement method.

## REFERENCES


Dong Jiaojiao, Wang Lin. Unmanned aircraft photogrammetry and its application in urban planning [J]. *Chinese and foreign architecture*, 2017(08):103–107.

Hong, Ocean. *Research on the construction of landscape information model based on visualization technology*[D]. Northeastern Agricultural University, 2015.

Li Yanxue, Li Fu, Zhu Shu, Xu Dawei. Application of multi-view image 3D reconstruction technology in rural landscape design [J]. *Journal of Northeast Forestry University*, 2019, 47(12): 84–89.

Lu Huanhuan, Dong Chuansheng, Li Zongbao, Wu Shaohua, Zhang Zhifang,Liu Peisen. *Analysis and research on the establishment of high-precision three-dimensional model by aerial surveying technology of unmanned aircraft* [C]//. Proceedings of the Twenty-second East China Six Provinces and One City Survey and Mapping Society Academic Exchange Conference (I). [publisher unknown], 2021:117–119.

Song Wenping. *Research on UAV aerial surveying system integration and image post-processing related issues*[D]. Chang'an University, 2016.

*Advances in Urban Construction and Management Engineering – Cha (Ed)*


# Research on the satisfaction evaluation of the public space of the old unit community based on POE—Taking Xiaodongmen railway community in Wuhan as an example

Fan Zhang*, Gewei Zheng* & Yanhong Zheng
*Department of Art and Design, Hubei University of Technology, China*

ABSTRACT: Under the new background of urban stock planning, the transformation of old communities has received more policy support. Because of the problems of lack of activity space and inactivation of public space in most communities, to effectively awaken the vitality of public space and effectively evaluate the user satisfaction of public space in old communities, this paper focuses on the construction of the satisfaction evaluation model of the public space of the old unit community in Wuhan. The model is based on the POE evaluation method and uses the knowledge of statistics and other disciplines to summarize the residents' evaluation opinions on the public space. In this paper, the commonly used analytic hierarchy process (AHP) and questionnaire survey methods are used in the POE evaluation method and combined with the basic situation of the transformation of old communities in China. The evaluation model of public space use in old communities is constructed, and many survey data on public space and residents' public life in old communities are summarized and tabulated. The survey of community satisfaction can effectively solve the problem of inactivation of public space, and to a certain extent, provide a reference for the transformation of public space in old communities of the same type in other cities.

## 1 INTRODUCTION

The basic unit of the urban organism is the community. If we intend to enhance the vitality of social development, we need to continue to deepen and pay attention to the scientific, refined, and intelligent development of the community. According to the policy document issued by the state, during the "14th Five-Year Plan" period, the country is estimated to have completed all the transformation projects of old urban areas in 219,000 cities and towns completed by the end of 2000. As of 2019, there are 59.000 new residents moved in, benefiting 10.88 million families. Given the renovation effect of old residential areas, supporting facilities such as water, electricity, roads and gas have been improved, and public service facilities such as old-age care, nursery, and convenience stores are increased according to local conditions, eliminating many potential safety hazards, and improving the living conditions and environment of residents in old residential areas. However, some problems have also been witnessed. There are many issues, such as the lack of a unified management department, a large funding gap, problems with the community autonomy function, insufficient participation from residents, etc. To further promote the transformation of old residential areas in cities and towns, planning guidance and coordination should be strengthened. The diversification of residents' material lifestyle and the improvement of living standards also brings a tense, crowded, noisy, and polluted living environment. The problems existing in the living environment, public service, community construction, and cultural characteristics of old residential areas affect the people's sense of access, happiness, and security. Residents enjoyed the spiritual life, were eager to improve the environmental quality of urban living space, and gradually

*Corresponding Authors: 1351783156@qq.com and 40498660@qq.com

DOI 10.1201/9781003348023-12

participated in urban public life., From the perspective of POE, this paper analyzes the public space of the old community, studies the residents' feelings and practical experience behavior of the public space of the old community, and analyzes the use demand of the public space quantitatively, trying to provide a reference for the transformation of the unit-type old community public space to a certain extent.

## 2 PURPOSE AND SIGNIFICANCE OF THE SURVEY

For the research on the public space of the old community of Wuhan, we should grasp the real experience and actual needs of the residents for the public space, and understand their satisfaction with the community using this study. The overall satisfaction of Wuhan residents with their communities is a comprehensive index. It reflects the cumulative effects of subtle feelings observed at multiple levels over time (Geng et al. 2007). POE takes the value orientation of user groups as the starting point of evaluation, takes the evaluation results as a basic standard to judge the rationality of design, establishes the analysis based on investigation and research, pays attention to the reliability of data collection and the scientificity of analysis methods, and introduces more objective factors for the renewal of public space in old communities (Zhao 2007). Through the research and data analysis of the actual needs of the community in daily life and the scientific and objective weight analysis of analytic hierarchy process (AHP), which is commonly used in POE post-evaluation methods, we can further explain the satisfaction of the community. This can also aid in the identification of the key factors affecting community satisfaction to evaluate the rationality of the community, strengthen the vitality of the community, and improve the satisfaction of community residents. The AHP, which is commonly used under the POE research method, can be used to evaluate the suitability characteristics of public space in old residential areas in Wuhan, and the combination of subjective and objective, quantitative and qualitative analysis is selected for analysis. On this basis, this paper puts forward the principle of selecting the index system, classifies the index system, establishes the corresponding index system for the suitability of public space in residential areas, classifies and calculates the indexes of the survey results and the actual questionnaire, and obtains the corresponding index weight. This step can provide data and technical support for the later analysis of the unit-type old community (Wang 2016).

This paper starts with the public space of a typical old community of unit system, such as Xiaodongmen Railway Community in Wuchang District of Wuhan City, puts forward reasonable suggestions to relevant departments from the perspective of post-occupancy evaluation, explores the satisfaction evaluation index of environmental elements of the old community of unit system, enhances the vitality of the community, meets the needs of residents, and provides a reference value for the transformation of the old community.

## 3 RESEARCH PROCESS

### 3.1 *Object of study*

(1) Definition of the old community of unit type
Urban old communities can be divided into three types according to the space type, namely, old communities of street type, old communities of commodity housing type, and old communities of unit courtyard type. The objects of this survey mainly focus on the old community of unit type. Unit-type old residential area refers to the residential area for workers, which is built with the establishment of factories in the process of realizing the social and economic organization model and has the characteristics of a "unit-run society" (Mei 2020).

(2) Characteristics of the unit-type old community
The old community of unit type covers a large area, and many commercial complexes have been implanted into the old residential area of the unit system during the blank period when the management of state-owned enterprises has withdrawn from the socialized management

(Mei 2020). The problems faced by the old residential areas of the unit system are as follows: the aging of infrastructure, the disrepair of architectural style, the old facade of buildings, the weakening of public places of activity, and the aging trend of the population. Compared with the old communities of street type and commercial housing type, the old communities of the unit system mainly take the unit as the core, relying on the collective housing distribution system of collective units.

### 3.2 *Analysis of investigation status*

The investigation site is located in Xiaodongmen, Wuhan City, north of the intersection of Zhongshan Road and Minzhu Road, along the west side of Zhongshan Road in Wuchang District. The Xiaodongmen Railway Community under the New Minzhu Road Community in Wuchang District is mainly selected for investigation. New Democracy Road Community is a combination of the original installation community and East Democracy Road Community, including the railway community, with a total area of about 1.6 to 1.8 sq. km. A network questionnaire survey was conducted on the use of micro-renewal of public space in the old community of unit system, and the respondents were the permanent residents of the old community of unit system in Xiaodongmen Railway District of Wuchang District, and 112 valid questionnaires were collected. According to the data analysis of the valid sample based on the user's situation, the ratio of male to female is close to 3:7 (27.68%:72.32%). In terms of age, the residents aged 41 to 65 account for 44.64%. Most of the respondents are residents of old communities who have lived for more than two years, and they have a clear understanding of the overall situation of old communities. According to the functionality, the public space of the old community of the unit system is divided into four categories for investigation.

(1) Use of leisure and entertainment space
The results of the questionnaire show that there are some problems in the Xiaodongmen Railway District, such as a lack of cultural characteristics and insufficient public venues. In addition to gate ball venues for teenagers, this venue is primarily used on weekdays for residents to dry off. The old community's small sports venues are generally acceptable to 35.71% of the residents. However, 8% of the residents are not satisfied with the old community's sports venues, and 22.32% of the residents have no sports facilities in the old community.

(2) Use of living public space
In recent years, old communities have been slightly renovated, and the drying space is a powerful project vigorously renovated by the government. Most of the drying points are located near the landscape flower beds at the bottom of residential buildings, and some larger old communities will be divided into specific areas to create drying space, which is also a strong support point for residents' satisfaction with community renovation. In the Xiaodongmen Railway Community, 39% of residents are not satisfied with the public drying space, 37.5% believe that it is general, and only 8.04% are satisfied with the public drying space. Residents of the community are most satisfied with the proximity of the shopping mall and the convenience of shopping, while only 38.13% of them are most satisfied with the nearby work unit. According to the feedback from the residents, the biggest problem in the community is the serious shortage of parking spaces, which is very inconvenient for many residents with private cars, and the lack of public service facilities cannot cover the daily life of residents.

(3) Use of an ornamental landscape
The community entrance landscape is the facade project of the community, and it is also the only way for every resident. According to the questionnaire survey data, 77.68% of the residents think that a comfortable space scale is the most important.

(4) Use of cultural education space
A total of 78.45% of the residents believe that the community should be imbued with culture, and 57.8% of the residents in other old communities that have been transformed reported feeling a strong cultural atmosphere, 20% were unable to detect it, and 22% had no obvious sense of it.

Table 1. Analysis of the current situation of investigation and research.

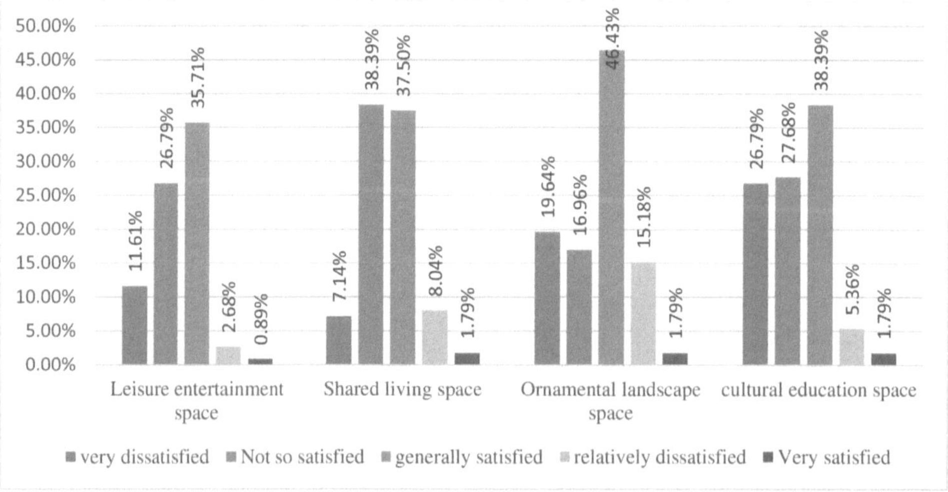


(Drawn by the authors)

### 3.3 *Summary of survey data*

According to the questionnaire data, the results show that 55.63% of the people are satisfied with the Xiaodongmen Railway Community where they live, and only 1.88% of the people have a poor overall impression of the community; 45.5% of the residents are generally satisfied with the public square of Xiaodongmen Railway District; 21.43% of the residents are not satisfied with the public square space of the old community. In the community public square, 48.2% of the residents passed by here to rest, 37.5% of the residents basked in the sun and chatted, 34.8% of them did an exercise with equipment, and most of them did activities between 19:00 and 21:00. In general, Xiaodongmen Railway District residents are satisfied with the sanitation and microclimate conditions, as well as with the quality of night lighting in public spaces, community pavements, and rest areas. Xiaodongmen Railway District do not have too many innovative points and memory points, the visual effect is general no cultural propaganda can only meet the ordinary living needs of residents. The accessibility of the community is good and the purchase of goods is convenient, but the scale of public space is cohesive and slightly small, and 26.8% of the old community space is very small, which makes people feel depressed, and the number of activity space is small, which cannot meet the needs of all residents. There will be a problem of overlapping public space sites in the same period.

## 4 SATISFACTION EVALUATION OF PUBLIC SPACE IN OLD COMMUNITY BASED ON POE

### 4.1 *Definition and research status of POE*

The POE method is a systematic evaluation study on the facilities or outdoor space built and used for a period of practice by using part of the system data and research methods from the user's point of view. It refers to the scientific data and information collection of the planning and design projects built after a certain period in a systematic and normative way. To collect and analyze effective evaluation, an evaluation group composed of psychologists, sociologists, designers, and technical specialists in some fields (such as evaluation libraries, and experts in library management) focuses on the investigation of behavioral psychology, using scientific instruments, psychological questionnaires, on-site observation, interviews, and other methods. Through the collection of reliable information, the project designer will be able to determine how well the project design meets the needs of the user groups, and verify the dynamic effects (physical and psychological) of the project users (individuals, groups, and institutions). The project design will be compared to the original design objectives, the design problems will be identified, and a scientific reference for a similar project design will be provided (Luo & Lu 2004).

POE originated in Europe and the United States in the 1960s and has gradually developed from case studies to all kinds of public buildings and urban open spaces. Its research methods have entered a practical stage and attracted wide attention from academia. Its research scope has expanded from users' psychological feelings and behaviors to site profiles, surrounding environment, owners' needs, design procedures, and social and historical backgrounds. These research programs are also gradually expanding to developing countries. The POE evaluation model can form a systematic comprehensive attribute evaluation system, which can measure the physical environment on the one hand, and evaluate the psychological measurement on the other hand.

The research method of POE is maturing regularly. The POE method is introduced into the design. Through the investigation and analysis of the design, it can provide some objective influence on the design project. The POE research method provides scientific and effective design decisions.

### 4.2 *Evaluation index system*

(1) Statistics of questionnaire data

In this paper, according to the purpose and content of the study, a questionnaire was made and distributed to the permanent residents of Xiaodongmen Railway Community in Wuchang District of Wuhan City, and the spatial data were statistically analyzed by EXCEL software. Five influencing factors, including old community public squares, old community sports venues, old community public drying space, the number of activity spaces in old communities, and the scale of community public space, were extracted for SPSS analysis of satisfaction (as shown in Table 1).

Table 2. SPSS analysis report on functional satisfaction of unit-type old community space.

| Project | Average value | Standard deviation | Satisfaction with public square space | Satisfaction with small sports venues | Satisfaction with public drying space | Satisfaction with the number of activity spaces | Satisfaction of public space scale |
|---|---|---|---|---|---|---|---|
| Satisfaction with public square space | 3.13 | 1.44 | 1 | | | | |
| Satisfaction with small sports venues | 3.21 | 1.67 | 0.69 | 1 | | | |
| Satisfaction with public drying space | 2.80 | 1.20 | 0.44 | 0.43 | 1 | | |
| Satisfaction with the number of activity spaces | 2.47 | 0.96 | 0.14 | 0.07 | 0.20 | 1 | |
| Satisfaction of public space scale | 2.13 | 1.13 | 0.06 | −0.01 | 0.05 | 0.37 | 1 |

(Drawn by the authors)

(2) Establishing a hierarchy evaluation structure model using ATP
In this paper, the AHP is used to construct the evaluation index of the public space of Xiaodongmen Railway District in Wuhan. Given the complexity and diversity of evaluation indicators, this paper classifies them and evaluates them according to certain scientific, comprehensive, and horizontal comparability. Specifically, it is divided into three levels. The first level is to evaluate the satisfaction and demand for the use of public space in the old community of unit type. The second level involves the evaluation of the functional satisfaction of the public space in the old community of the unit system, which is divided into leisure and entertainment space, living public space, ornamental landscape space, and cultural education space (X1, X2, X3, and X4). The third level is the indicator layer. Indicators are the middle layer of concretization and operationalization, which are marked as A1 to A8 in the model, and the specific indicators are shown in Table 3.

Table 3. Satisfaction evaluation of space use in old communities.

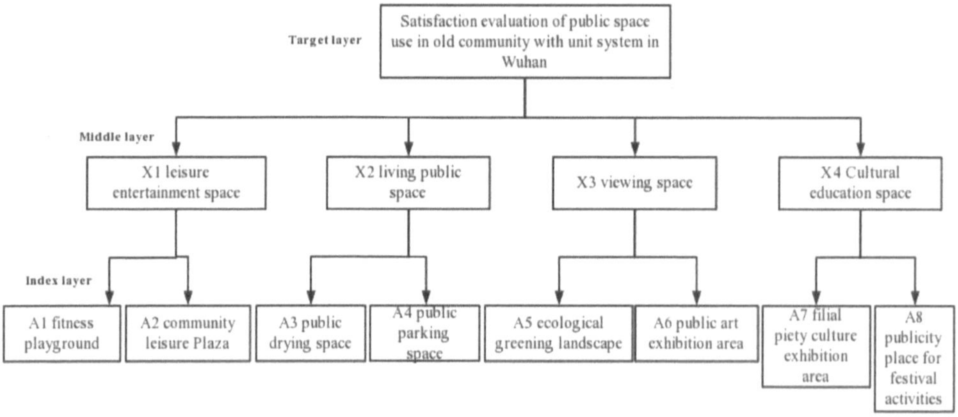


(Drawn by the authors)

The selected three-level evaluation index of community satisfaction includes index layers: A1 (fitness equipment and facilities), A2 (community leisure square and activity center), A3 (public drying space), A4 (public parking space), A5 (ecological greening landscape), A6 (public art sketches), A7 (historical integrity and filial piety culture), and A8 (festival activities publicity).

According to the comprehensive data of the questionnaire, it can be concluded that the importance of residents to public spaces is ranked as follows: A2 > A3 > A4 > A1 > A5 > A6 > A7 > A8.

### 4.3 *Calculation of comprehensive scores*

After the hierarchical model is established, the pairwise comparison is used to construct the judgment matrix, to clarify the weight relationship between the scheme layers, and 1 to 9 is used to assign the weight. The meanings of the scale levels 1 to 9 are shown in Table 3. For the influencing factors A1, A2, and A8 obtain the important proportion of each factor to the target through pairwise comparison and form a judgment matrix A (see Table 5). The next step is to solve the problem of calculating the ranking weights of the eight influencing factors. The judgment matrix A is normalized by column in EXCEL and the consistency test is carried out. If the test value $CR \leq 0.1$, it means that the judgment matrix is correct. If the one-time test value $CR > 0.1$, it means that the judgment matrix is wrong and needs to be corrected appropriately. The one-time test formula is $CR = CI/RI$. According to the RI order of the AHP, the RI value is 1.41. The largest eigenvalue

Table 4. Implications of the 1 to 9 scales.

| | |
|---|---|
| 1 | Two elements are of equal importance compared to each other. |
| 3 | One of the two elements is slightly more important than the other. |
| 5 | One of the two elements is significantly more important than the other. |
| 7 | One of the two elements is more important than the other. |
| 9 | One of the two elements is more important than the other. |
| 2, 4, 6, and 8 | The intermediate value of the above adjacent judgment. |
| Corresponding to the reciprocal of the above values | The above description in which one factor is less important than another. |

(Drawn by the authors)

of the matrix I $\lambda_{max} = \sum_{i=1}^{n} \frac{[AW]_i}{nw_i} = 8.48$, CI = (λ−n)/(n−1) = 0.05, CI = 0.05 < 0.1, indicating that the judgment matrix is correct.

## 4.4 *Analysis of POE evaluation results*

According to the weight values (as shown in Table 5) obtained from the hierarchical analysis based on the questionnaire data and the normalization of the satisfaction scale scores in Table 4, among the eight influencing factors, the residents of Xiaodongmen Railway Community consider that the community leisure square and public drying space are more important, while the green landscape type is less concerned. Residents have a greater demand for leisure and entertainment space and living public space in the old community of unit system, and less demand for ornamental landscape space and cultural education space.

Table 5. Weight relation of judgment matrix.

| A | 1. Fitness equipment facilities | 2. Community leisure square | 3. Public drying space | 4. Public parking space | 5. Ecological greening landscape | 6. Public art sketches | 7. Filial piety culture publicity site | 8. Publicity of festival activities |
|---|---|---|---|---|---|---|---|---|
| 1. Fitness equipment facilities | 1 | 1/5 | 1/4 | 1/3 | 2 | 3 | 4 | 5 |
| 2. Community leisure square | 5 | 1 | 2 | 3 | 4 | 5 | 6 | 7 |
| 3. Public drying space | 4 | 1/2 | 1 | 2 | 3 | 4 | 5 | 6 |
| 4. Public parking space | 3 | 1/3 | 1/2 | 1 | 3 | 4 | 5 | 6 |
| 5. Ecological greening landscape | 1/2 | 1/4 | 1/3 | 1/3 | 1 | 2 | 3 | 4 |
| 6. Public art sketches | 1/3 | 1/5 | 1/5 | 1/4 | 1/2 | 1 | 2 | 3 |
| 7. Filial piety culture publicity site | 1/4 | 1/6 | 1/5 | 1/5 | 1/3 | 1/2 | 1 | 3 |
| 8. Publicity of festival activities | 1/5 | 1/7 | 1/6 | 1/6 | 1/4 | 1/3 | 1/3 | 1 |
| Sum | 14.28 | 2.79 | 4.65 | 7.28 | 14.08 | 19.83 | 26.33 | 35.0 |

(Drawn by the authors)

Table 6. Normalized weight value by column

| Z | 1. Fitness and sports venues | 2. Community leisure square | 3. Public drying space | 4. Public parking space | 5. Ecological greening landscape | 6. Public art exhibition area | 7. Honest and filial piety culture exhibition area | 8. Publicity venues for festival activities | w | AW |
|---|---|---|---|---|---|---|---|---|---|---|
| 1. Fitness equipment facilities | 0.07 | 0.07 | 0.05 | 0.05 | 0.14 | 0.15 | 0.15 | 0.14 | 0.10 | 0.88 |
| 2. Community leisure square | 0.35 | 0.36 | 0.43 | 0.41 | 0.28 | 0.25 | 0.23 | 0.20 | 0.31 | 2.77 |
| 3. Public drying space | 0.28 | 0.18 | 0.22 | 0.27 | 0.21 | 0.20 | 0.19 | 0.17 | 0.22 | 1.93 |
| 4. Public parking space | 0.21 | 0.12 | 0.11 | 0.14 | 0.21 | 0.20 | 0.19 | 0.17 | 0.17 | 1.49 |
| 5. Ecological greening landscape | 0.04 | 0.09 | 0.07 | 0.05 | 0.07 | 0.10 | 0.11 | 0.11 | 0.08 | 0.66 |
| 6. Public art sketches | 0.02 | 0.07 | 0.04 | 0.03 | 0.04 | 0.05 | 0.08 | 0.09 | 0.05 | 0.43 |
| 7. Filial piety culture publicity site | 0.02 | 0.06 | 0.04 | 0.03 | 0.02 | 0.03 | 0.04 | 0.09 | 0.04 | 0.32 |
| 8. Publicity of festival activities | 0.01 | 0.05 | 0.04 | 0.02 | 0.02 | 0.02 | 0.01 | 0.03 | 0.02 | 0.21 |

(Drawn by the authors)

## 5 CONCLUSIONS

From the perspective of POE, from the perspective of community residents' satisfaction, based on the questionnaire survey data, this paper compares the weights of four types of functional public space in the old unit community of Wuhan. It introduces the AHP, wherein residents fill in questionnaires, construct the structural index system of public space in the old community, and quantifies the actual needs of residents into the corresponding weights of the index system. Then the specific score of the scheme layer is weighted, and the final maximum score is the best. On this basis, a three-level community evaluation index system is established, in which the weight of the second-level index to the first-level index is based on the contribution of each main factor, and the weight of the third-level index is based on the second-level index.

This paper uses the questionnaire data of the old community residents in Wuhan to make an empirical analysis and obtains the final evaluation model. By evaluating the satisfaction of the public space in the old community, the satisfaction level evaluation is obtained, and then the quantitative evaluation is carried out. POE is a concept and method, which requires the design of the project to fully consider the needs of users. This evaluation, however, is not comprehensive, since, in the actual community transformation, we must consider all parties' interests, ensure user feedback, provide ideas, and reference value for our research on the old public space of the community, and ensure that the evaluation results are more scientific and objective.

## REFERENCES


Geng Jinhua, Gao Qisheng, Zhang Siying. Community Satisfaction Evaluation System Based on AHP and Factor Analysis_Geng Jinhua [J]. *Journal of System Management*, 2007(6): 673–677.

Luo Lingling, Lu Wei. The International Trend of POE Research and the Realistic Thinking of Its Introduction into China _ Luo Lingling [J]. *Journal of Architecture*, 2004 (8): 82–83.

Mei Lei, Huang Zeliu, Wang Qi, Bai Yu. Study on the reconstruction of unit system old residential area – taking the reconstruction of Tongda community in Qingshan District, Wuhan, Wuhan as an example [J]. *Urban-rural development*, 2020(20): 44–47.

Wang Hong. *A study on the aging adaptability of the external public space of the aged community in Chengdu based on AHP* [ D ] . Southwest Jiaotong University, 2016.

Zhao Donghan. Post-use evaluation of the development characteristics of POE abroad and its applicability in China_Zhao Donghan [J]. *Journal of Peking University* (Natural Science Edition), 2007(6): 797–802.

*Advances in Urban Construction and Management Engineering – Cha (Ed)*


# Analysis of spatio-temporal variation characteristics of environmental carrying capacity in Shigatse area of Qinghai-Tibet Plateau

Ya Tu, Yutao Li, Huiting Zou & Luo Guo*
*College of Life and Environmental Sciences, Beijing, China*

ABSTRACT: Shigatse is the second largest city in the Tibetan region with rapid economic development and the major task of building a firm national ecological security barrier. This research used remote sensing data, field research data, and statistical data from 2010 and 2015 to analyze the changes in land use patterns in the Shigatse region, and analyzed the changes in ecological carrying capacity with principal component analysis. The research results are as follows. 1) The most significant type of land use transfer was "grassland → unused land", with the transferred area accounting for 40.70% of the total transferred area. 2) From 2010 to 2015, the overall trend of the environmental carrying capacity was weakening, mainly the strong and stronger carrying capacity areas were decreasing, and the medium or weaker carrying capacity areas were increasing. 3) The natural factors affecting the environmental carrying capacity were mainly climate and terrain, while the human factors were mainly unreasonable land resource development and unbalanced agricultural and animal husbandry industry development. On this basis, this research provided suggestions and countermeasures for ecological environmental protection in the region and a reference basis for the sustainable development of the Qinghai-Tibet Plateau region.

## 1 INTRODUCTION

Environmental carrying capacity refers to the limit of the ability of the regional environment to support human social and economic activities. The status of environmental carrying capacity has a direct impact on the development of regional ecology, economy, society, and other aspects, and has an important influence on regional economic development and sustainable development. At present, the hot issues of environmental carrying capacity research are mainly on the comprehensive evaluation index system, evaluation methods, and evaluation models of different scales and different fields. The current study of the environmental carrying capacity of the Shigatse region lacks systematic environmental quality evaluation and analysis of the elements affecting environmental quality.

As the second largest city in Tibet, Shigatse is experiencing rapid economic development and is also charged with the major task of building a firm national ecological security barrier, so it is of great ecological significance to conduct an environmental carrying capacity assessment of the city. Exploring the main factors affecting and restricting the ecological and environmental protection of the region and the protection countermeasures can provide a scientific basis and strong guarantee for the ecological environment and socio-economic development of the Tibetan Plateau region.

*Corresponding Author: guoluo@muc.edu.cn

 DOI 10.1201/9781003348023-13

## 2 STUDY AREA AND METHOD

### 2.1 *Study area*

Shigatse is located in the southwestern part of the Tibetan Plateau (27.13°31.49°N, 82.10°90.20°E) (see Figure 1). There are currently one municipal district and 17 counties under the jurisdiction of the city. The altitude of the region ranges from 1,463 to 8,619 m above sea level, and is typical of the mountainous climate of the plateau, with complex and varied terrain, and is the birthplace of the Yarlung Tsangpo River, the largest river in Tibet. The region has a predominantly rural population, accounting for 77.75% of the population, with many ethnic minorities. Since the peaceful liberation of Tibet, Shigatse has experienced rapid economic development.

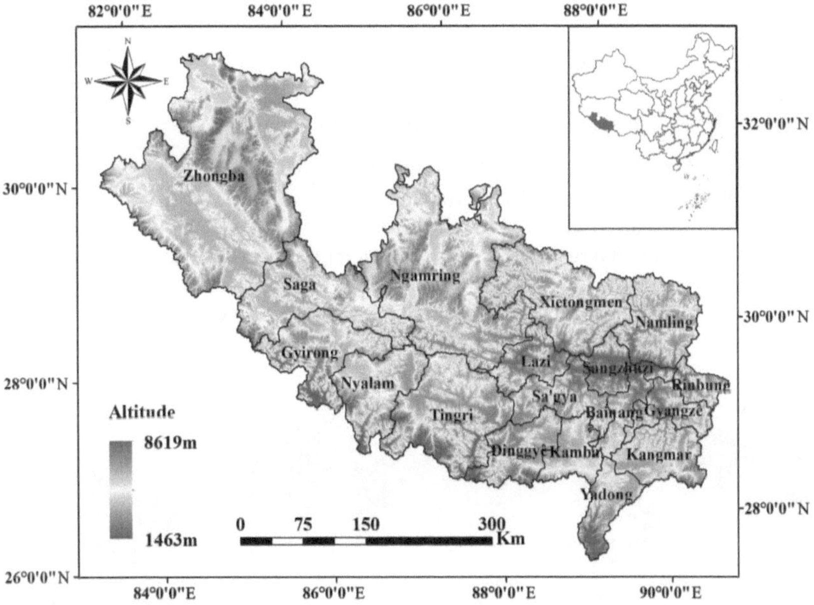


Figure 1. Study area.

Table 1. Principal component analysis results of environmental carrying capacity evaluation indexes.

| | 2010 | | | 2020 | | |
|---|---|---|---|---|---|---|
| Principal component | Eigenvalue | Cumulative contribution rate | Final weight | Eigenvalue | Cumulative contribution rate | Final weight |
| F1 | 5.48 | 42.16 | 0.48 | 5.24 | 32.07 | 0.47 |
| F2 | 3.47 | 68.87 | 0.31 | 3.20 | 59.18 | 0.29 |
| F3 | 1.29 | 78.77 | 0.11 | 1.43 | 72.29 | 0.13 |
| F4 | 1.10 | 87.23 | 0.10 | 1.17 | 84.83 | 0.11 |

### 2.2 *Data and methodology*

We used Landsat TM/ETM remote sensing imagery from 2010 and 2015, and classified the local land types into six primary types including farmland, forestland, grassland, water, building land, and unused land. Based on the evaluation procedure of "ecological problems – environmental factors – evaluation indicators", combining the natural characteristics, 13 indicators were selected from topography, surface, meteorology, and human interference to build the evaluation of the ecological and environmental carrying capacity of Shigatse. The 13 indicators were the proportion of building land, grassland area, annual radiation intensity, land utilization, terrain features, annual precipitation, vegetation cover, average annual temperature, plateau tundra rate, population density, and desertification of the land, animal husbandry output value and GDP per capita. The results were classified as weak, weaker, medium, stronger, and strong. Principal component analysis in ArcGIS was used to extract the principal component factors affecting the changes in ecological carrying capacity. According to the principle that the cumulative contribution of variance was greater than or equal to 85%, the number of principal components P was determined, and the factor loadings of the principal components and the scores of each factor were derived (Liu 2018) (Table 1).

## 3 RESULTS AND ANALYSIS

### 3.1 *Analysis of land use change in Shigatse from 2010 to 2015*

From 2010 to 2015, there were 10 types of land use transfer, and the total area of land use transfer was 86 $km^2$, of which the most significant transfer type was "grassland → unused land", which accounted for 40.70% of the total transfer area. This type of land use change was concentrated in Angren County, indicating that there was a serious problem of grassland degradation in this region in recent years. The transfer area of "grassland → water" and "water → unused land" accounted for 24.42% and 18.61% of the total transfer area respectively, which was relatively scattered in spatial distribution.

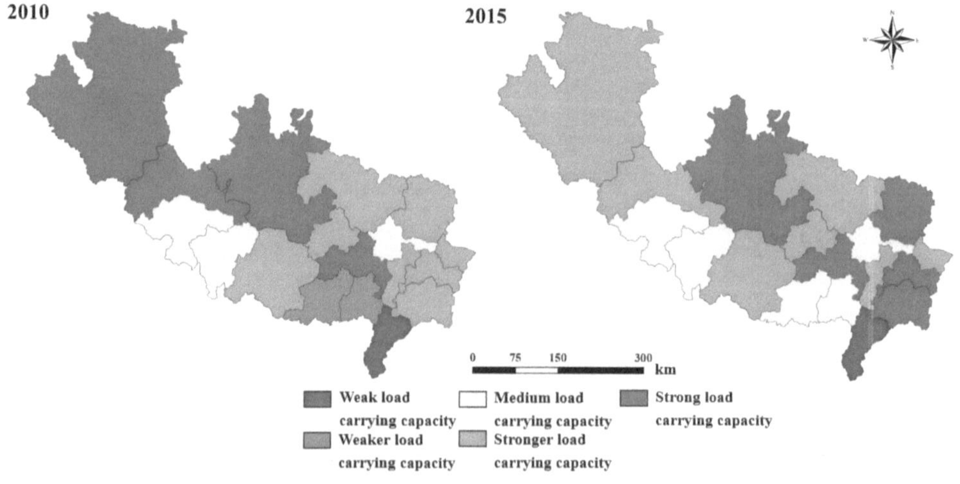


Figure 2. The environmental carrying capacity of Shigatse from 2010 to 2015.

### 3.2 *Spatial and temporal characteristics of the environmental carrying capacity of Shigatse*

From 2010 to 2015, the environmental carrying capacity developed towards a weakening trend, in which the strong and stronger carrying capacity areas decreased significantly, and the medium or weaker carrying capacity areas increased (Figure 2). In 2010, the environmental carrying capacity showed a spatial pattern of gradual weakening along the northwest-southeast, with the strong carrying capacity area in the northwest, the stronger carrying capacity area in the east, the medium carrying capacity area in the southwest, and the weaker and weak carrying capacity area in the south. In 2015, compared with 2010, the overall environmental carrying capacity decreased and the distribution pattern showed a weakening along the northeast-southwest direction. The total area of the strong carrying capacity area decreased to 34.09%; the stronger carrying capacity area increased to 48.07%; the medium carrying capacity area increased to 15.65%; there was no weaker carrying capacity area; the weak carrying capacity area did not change, only in Yadong County.

Table 2. The principal component scoring coefficient matrix in 2010 and 2015.

| | 2010 Principal Component Score Matrix | | | | 2015 Principal Component Score Matrix | | | |
|---|---|---|---|---|---|---|---|---|
| Index | 1 | 2 | 3 | 4 | 1 | 2 | 3 | 4 |
| The proportion of building land | 0.237 | 0.04 | 0.15 | −0.167 | −0.17 | −0.186 | 0.155 | −0.042 |
| Grassland area | 0.113 | −0.049 | −0.213 | 0.04 | −0.157 | −0.037 | −0.065 | −0.082 |
| Annual radiation intensity | −0.062 | −0.01 | 0.137 | 0.324 | 0.172 | −0.063 | 0.194 | 0.19 |
| Land utilization | −0.197 | 0.019 | 0.058 | −0.082 | 0.21 | 0.023 | −0.069 | −0.008 |
| Terrain features | −0.003 | 0.295 | −0.101 | −0.086 | −0.017 | 0.297 | −0.072 | −0.032 |
| Annual precipitation | −0.053 | 0.167 | 0.119 | −0.076 | 0.072 | −0.065 | 0.325 | −0.149 |
| Vegetation coverage | 0.011 | 0.295 | −0.133 | −0.044 | −0.015 | 0.302 | −0.100 | −0.026 |
| Average annual temperature | 0.065 | 0.305 | −0.11 | −0.083 | −0.1 | 0.307 | −0.013 | −0.11 |
| Plateau tundra rate | 0.189 | 0.11 | −0.15 | 0.204 | −0.211 | 0.128 | 0.028 | 0.049 |
| Population density | −0.234 | 0.042 | −0.357 | −0.062 | 0.276 | −0.005 | −0.359 | 0.057 |
| Desertification of the land | 0.162 | −0.032 | 0.051 | −0.167 | −0.04 | 0.013 | −0.163 | 0.478 |
| The animal husbandry output value | 0.09 | −0.114 | −0.027 | 0.765 | 0.085 | −0.097 | 0.082 | 0.589 |
| GDP per capita | 0.089 | −0.123 | 0.686 | −0.009 | −0.12 | −0.071 | 0.636 | 0.001 |

### 3.3 *Analysis of driving factors*

The results of the analysis of the 13 evaluation indicators through the principal component analysis tool in ArcGIS are shown in Table 2, which shows that the eigenvalues of the four principal components before 2010 are 5.48, 3.47, 1.29, and 1.10, respectively, and the eigenvalues of the four principal components before 2020 are 5.24, 3.20, 1.43, and 1.17, respectively, all of which are greater than one, and the cumulative contribution rate is more than 80%, which indicates that the first four principal components could represent the original indicators for the evaluation and analysis of the environmental carrying capacity of Shigatse.

Table 3 presents the principal component loading matrix and the principal component score coefficient matrix. It shows that the first principal component in 2010 has a large positive correlation with the degree of desertification of the land, grassland area, average annual temperature, vegetation cover, the proportion of building land, plateau tundra rate, GDP per capita, and animal husbandry output value. There is a significant negative correlation between annual radiation intensity, land utilization, annual precipitation, terrain features, and population density. These indicators reflect the state of resource consumption for the natural, social, and economic development of Shigatse, reflecting that the rapid development of the quality of life of the inhabitants has come at the cost of environmental pollution and resource consumption. Population density has the greatest influence on the first principal component, with a loading coefficient of −0.234. It is the most important indicator affecting the environmental situation; the second principal component has a more obvious positive correlation with the average annual temperature, terrain features, and vegetation cover, and a more obvious negative correlation with the Animal husbandry output value and GDP per capita, etc. The terrain features and the animal husbandry output value in the first principal component in 2020 have no change in influence compared with those in 2010, but the remaining 11 indicators have all shifted in positive and negative correlations. Among them, the plateau tundra rate has the greatest influence on the first principal component, with a loading coefficient of −0.211, making it the most important indicator affecting the environmental condition of Shigatse; the second principal component has a more obvious positive correlation with the average annual temperature, terrain features, and vegetation cover. The second principal component has a more obvious negative correlation with the proportion of building land, animal husbandry output value, and GDP per capita.

## 4 DISCUSSION

### 4.1 *Influences of natural factors on changes in environmental carrying capacity*

The natural environmental factors that have the greatest impact on the environmental carrying capacity of Shigatse include topographical factors, climatic factors and natural disaster factors. From the topographic factors, the topography of Shigatse is complex and diverse, and the high-altitude areas are mainly located in the north and south. From 2010 to 2015, the weak environmental carrying capacity area was mainly located in the south, because the climate was cold and dry, with many hailstorms, snowstorms, high winds, and serious freeze-thaw erosion. In addition, the degradation of grasslands was becoming more serious and has weakened the climate regulation function, while making its ecosystem regulation function, soil and water regulation function, and biological shelter function decline significantly. In the northern region, although the altitude was also higher, the territory was densely covered with large and small lakes, lake basin plains, and flat dams and beaches, with better land protection, sufficient sunshine, and high annual radiation intensity, the overall ecological carrying capacity was stronger and the ecological environment was in good condition. Therefore, the environmental carrying capacity of different regions is affected by various factors to different degrees, but in general, it is mainly influenced by factors such as topography, precipitation, and radiation intensity.

In terms of climatic factors, the highland subalpine arid and semi-arid monsoon climate of Shigatse is the main cause of recurrent droughts, high winds, and floods in the region, and natural disasters are frequent in the region, resulting in a more fragile regional ecological environment. The soil in the southeast is porous, which exacerbates the desertification of the land in the southeast, making it more vulnerable to climate impacts, floods, mudslides, and other problems, resulting in the environmental carrying capacity of the southeast at a weaker level for a long time; the northwestern region is more protected and has more plains, and is less affected by natural disasters brought about by climate. The northwestern region, with better land protection and more plains, is less affected by natural disasters and has a stronger overall ecological carrying capacity. However, it is also affected by climate warming, and erosion phenomena such as glacier retreat and freeze-thaw landslides are common, which also brings great challenges to the ecological protection of Shigatse.

### 4.2 *Anthropogenic influences on changes in environmental carrying capacity*

The influences of anthropogenic factors on the ecological carrying capacity of the Shigatse are mainly in the form of unreasonable land resource development and unbalanced agricultural and animal husbandry industrial development. Areas with weaker carrying capacity, such as Jilong County, Shigatse City, and Yadong County, are mostly located in river valley areas with good natural conditions, leading to the rapid expansion of construction land, while construction wastes a lot of resources and generates a lot of waste during the construction process, and also encroaches on some arable land, which makes the contradiction between the expansion of construction land and the protection of arable land more and more prominent. In terms of agricultural and livestock industry development, the environmental carrying capacity of all counties decreased between 2010 and 2015 as their agricultural and livestock industry output increased. Large-scale agricultural and livestock development has brought huge challenges to the ecological environment. The rapid development of agropastoralism has led to the predatory exploitation of grassland resources and long-term over-eating by livestock, while excessive trampling by livestock has also made grassland increasingly compact, resulting in degradation of grassland, prominent contradictions between grass and livestock, and long-term overload of grassland; the decrease in the permeability of the soil has exacerbated grassland degradation, resulting in a vicious cycle (Wang 2021).

### 4.3 *Suggestions for sustainable development*

Environmental protection is a prerequisite for sustainable social development, as well as for economic development and the continuation of human civilization (Cao et al. 2018). According to the work report of the Shigatse government, the government was aware of the environmental pressure and the urgency of environmental protection in the region and has gradually carried out related work. In the future, the government should pay more attention to the problem of grassland degradation, reduce the pressure on grassland by setting reasonable livestock carrying capacity, strictly prohibiting overgrazing and implementing fencing; strengthen the management of forest areas, expand afforestation areas and implement forest closure policies; enhance the concept of ecological environmental protection in forest areas for all people, and encourage all sectors of society to strengthen their support and participation in forest area protection (Pimm 2021). Sustainable development should also be achieved by strengthening the conservation of biodiversity, introducing specific legislation for nature reserves and parks, and enhancing public education (Cheng et al. 2019). The government should promote the green development of industries, introduce advanced technological tools, implement projects for the sustainable use of resources, and form a resource-saving and environmentally friendly industrial structure.

## 5 CONCLUSION

The research selected 13 indicators to construct an evaluation index system for the ecological and environmental carrying capacity of Shigatse and used the principal component analysis method to analyze the environmental carrying capacity. The results of the study showed that, among the different land use types in Shigatse, the area of water, forest, building land, and unused land increased, while the area of farmland and grassland decreased. In terms of environmental carrying capacity, Dingjie County, Gangba County, Jilong County, Nyalam County, Shigatse City, and Yadong County have a weaker carrying capacity. The environmental carrying capacity was consistent with the ecological and socio-economic characteristics of each region. The evaluation method provided a scientific basis for the development of regional resources and ecological environmental protection and offered suggestions for the construction of sustainable development in the Shigatse region.

## ACKNOWLEDGMENT

This work was financially supported by the Second Tibetan Plateau Scientific Expedition and Research (STEP) program (Grant No. 2019QZKK0308).

## REFERENCES

Cao, J., Li, M., & Qi, F. (2018). *Comparison of social-ecological resilience between two grassland management patterns driven by grassland land contract policy in the Maqu*, Qinghai-Tibetan Plateau. Land Use Policy, 74.

Cheng, F., Su, F., Chen, M., Wang, Q., Jiang, H., & Wang, X. (2019). An evolving assessment model for environmental carrying capacity: a case study of coral reef islands. *Journal of environmental management*, 233, 543.

Liu, Y., Zhang, J., Wang, S., Yan, W., & Zhao, A. (2018). Assessment of Environmental Carrying Capacity Using Principal Component Analysis. *Journal of Geoscience and Environment Protection*, 6(3): 54–65.
Pimm, S.L. (2021). What is biodiversity conservation?: This article belongs to Ambio's 50th Anniversary Collection. Theme: Biodiversity Conservation. *Ambio*, 50(5).
Wang, M., & Feng, C. (2021). The win-win ability of environmental protection and economic development during China's transition. *Technological Forecasting & Social Change*, 166.

*Advances in Urban Construction and Management Engineering – Cha (Ed)*
 *ISBN 978-1-032-39018-5*

# Steel grid roof bearing capacity checking

Xiaoyun Jiang*
*Lanzhou Jiaotong University, Lanzhou, China*

ABSTRACT: Load type, temperature effect, and construction technology are one of the main reasons for the deterioration of steel structures. Based on the influence of load type, temperature effect, and construction technology on steel structure, the bearing capacity of steel grid roof is explored, and the influence degree of load type, temperature effect, and construction technology on the bearing capacity of steel grid roof is mainly analyzed. The results show that the bearing capacity of the structure meets the requirements of the code and that chloride ion erosion accelerates the formation of the peritoneal cracks.

## 1 INTRODUCTION

With the rapid growth of China's economic aggregate, all kinds of new structures are springing up like bamboo shoots after a spring rain, among which the most common, beautiful shape, and excellent structural performance of steel structure is the most widely used. As one of the basic forms of steel structure, the steel grid is also widely used in various building structures (Chu 2009; Chen 2020).

In this paper, steel grid roof as the research object, based on a stadium project, steel grid roof carrying capacity research. The total construction area of the project is nearly 40,000 square meters. The span from the E-axis to the N-axis of the steel grid roof is 67.2 meters. The structure of the roof is a double-layer square pyramid (a three-layer grid frame is used locally), and the support type is lower chord column point support. The safety level of the grid structure is level ONE, the designed service life is 50 years, and the seismic fortification intensity is 8 degrees.

## 2 MODEL AND EXPERIMENT

(1) Purpose of the test
By establishing the finite element model of the steel grid roof, the bearing capacity of the steel grid roof is analyzed (Deng 2006; QIN 2020).
(2) Test content
Using CAD software series of Tongji University -3D3S steel structure-Space structure design software, the current load of the steel grid structure (upper chord constant load: 1.0 $kN/m^2$, bridle load: 18 kN, temperature $\pm 30°$) was modeled and analyzed under the action of load temperature.
(3) Steel grid roof model
   1) Main parameters of the model
   The main specifications of the rod are $75.5\times4$, $88.5\times4$, and $114\times4$, using Q355B straight seam welded steel pipe. The main specifications of the bolt ball are 120, 140, and 350. The bolted ball is made of GB/T699-2015 medium 45 steel hot forging; The main specifications

*Corresponding Author: 25641176@qq.com

of welding hollow ball are WSR450×18, WSR650×25, WSR700×28, WSR750×30, made of Q355B steel; Q355B steel is used for bearing, sleeve and cone head of sealing plate corresponding to steel pipe. A metal roof has been adopted for the grid. The specific approach is (1) 0.9 mm vertical lock edge aluminum magnesium manganese alloy roof panel 65 to 400; (2) H81 high strength aluminum alloy bearings with heat insulation pads; (3) waterproof layer with 1.5mm TPO waterproof roll; (4) thermal insulation cotton: 50+50 mm thermal insulation rock wool, bulk density 120 kg/m$^3$; (5) sound-absorbing layer: 100 mm glass fiber sound-absorbing cotton, with a bulk density of 24 kg/m$^3$; (6) steam insulation layer: 0.3 mm steam insulation film; (7) bottom layer: 0.7 mm aluminum plated perforated steel PLATE HV-840, perforation rate is 23%, and the back is sprayed with plastic; (8) secondary purlin: 150 × 100 × 4.0 mm rectangular steel pipe, Q235B, epoxy Zinc-rich primer 2 lines 70 μm + Epoxy iron intermediate paint 2 lines 105 μm+ fire retardant paint @ ≤ 1550; (9) main purlin: 200 × 100 × 4.0 mm rectangular steel tube, two lines of Epoxy Zinc-rich primer, 70 μm + two lines of epoxy iron intermediate paint, and 105 μm+ Fire retardant paint (Shi 2007).

2) Model building

Taking the steel grid roof of the project as an example, a 1:1 model is established. The structure layout is shown in Figure 1, and the finite element model is shown in Figure 2.

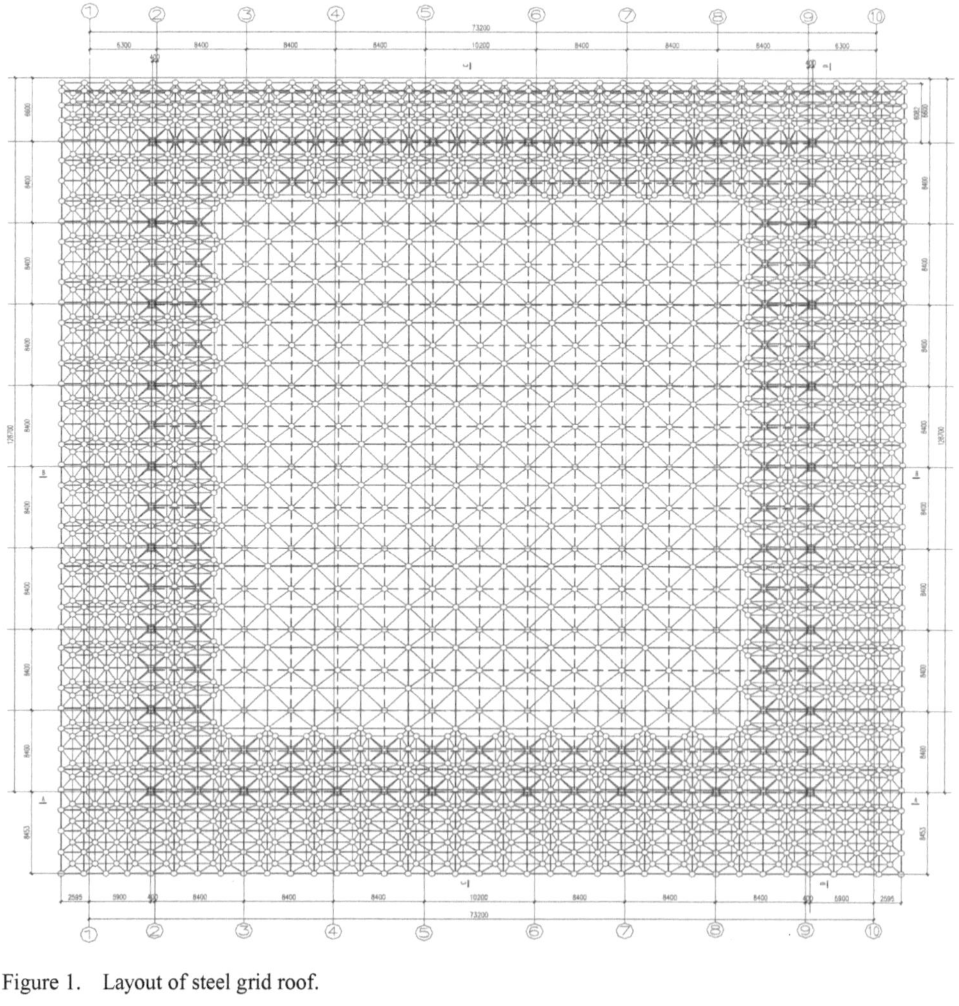

Figure 1. Layout of steel grid roof.

Among them, (1) the structural information is the total number of nodes: 2201; the total number of supports: 30; the total number of units: 8582; the type of material: 1; the type of section: 14; the load condition: 5. (2) Calculation parameters (dynamic characteristic calculation) mode number: 30; mode type: characteristic vector. (3) Linear calculation, beam element attribute: general beam element (Euler beam), beam torsional moment of inertia: a polar moment of inertia, considering P-$\delta$/second order effect: none. (4) Structural importance coefficient: 1.100; support critical angle: 15.000 (Xu 2013; Xie 2022).

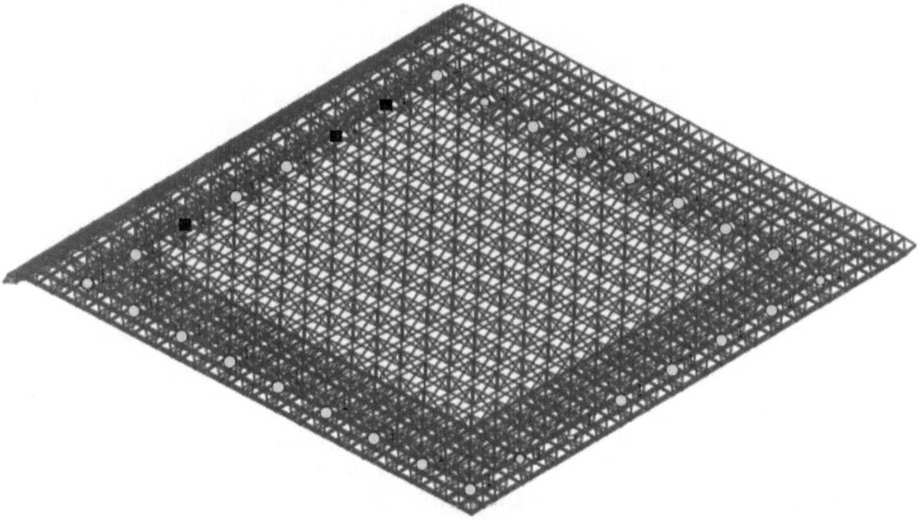

Figure 2. Schematic diagram of grid calculation (circle represents support and the number represents the node number).

(4) Structure bearing capacity check calculation

According to the *Spatial Grid Structure Technical Specification* (Standard ID: JGJ7-2010) and *Steel Structure Design Standard* (Standard ID: GB50017-2017) and other specifications, the bearing capacity checking calculation analysis. This time, the CAD software series of Tongji University – 3D3S steel structure – space structure design software is used to check the bearing capacity of steel grid structural components. Due to construction reasons, the method of high-altitude bulk is adopted in the construction process, and pre-arching is not considered in the construction process, resulting in geometric deviations in horizontal and vertical positions of the steel grid. The spherical coordinates of the nodes in the steel grid model are established by using the spherical coordinates measured in the *Steel grid Measurement Report of the Project Gymnasium* issued by a company.

1) Structure calculation principles
   a) Structural analysis and checking methods of structures or components shall comply with the provisions of the current national design code.
   b) The calculation model used for structural analysis and checking of structures or components shall conform to the actual stress and structural conditions of the structure.
   c) The standard value of the function on the structure shall be based on the actual field investigation and the General Code for Engineering Structures (Standard ID: GB55001-2021).
2) The standard value of material strength shall be set according to the actual condition of the component and the obtained test data according to the following principles:
   a) When the type and performance of materials meet the requirements of the original design, the value can be set according to the original design standard value;

b) When the type and performance of the material are inconsistent with the original design or the material performance has been significantly degraded, the value shall be taken according to the measured data and the provisions of the current relevant testing technical standards.

3) The geometrical parameters of the structure or component should be measured and determined according to the actual deformation, construction deviation, cracks, defects, damage, corrosion, and other influences of the structure.

(5) Structure bearing capacity checking (not calculated earthquake action)
During checking calculation, the component layout in the calculation model is established by an on-site investigation. The load combination is following the *General Code for Engineering Structures* (Standard ID: GB 55001-2021) and *Unified Standard for Reliability Design of Building Structures* (Standard ID: GB 50068-2018). The dead load of the steel grid roof of the basketball stadium in this project is determined according to the actual on-site investigation and design drawings. Roof dead load is calculated according to the metal roof deepening design drawing of Zhejiang Zhongnan Construction Group Steel Structure Co., Ltd., and the general description of metal roof deepening design.

## 3 RESULT ANALYSIS

### 3.1 *Checking of structural bearing capacity (excluding seismic action)*

(1) Load investigation
During checking calculation, the component layout in the calculation model is established by an on-site investigation. The load combination shall be following the *General Code for Engineering Structures* (Standard ID: GB 55001-2021) and *Unified Standard for Reliability Design of Building Structures* (Standard ID: GB 50068-2018). The dead load of the steel grid roof of the basketball stadium in this project shall be determined according to the actual on-site investigation and design drawings. The structural layer of the metal roof is calculated and the live load of the roof is calculated according to the value of the roof.

(2) Checking the bearing capacity of structural members
Through CAD software series -3D3S steel structure-Spatial structure design software ofTongji University, the bearing capacity of steel grid structural members of the basketball stadium of this project is checked and calculated, as shown in Figure 3. The checking results show that the stress ratio of each member is less than one, and the stress ratio and slenderness ratio meet the requirements of the specification.

The checking results in Figure 3 show that the stress ratio of each member is less than one, the stress ratio and slenderness ratio meet the requirements of the code, and the bearing capacity of the structure meets the requirements.

### 3.2 *Deformation analysis*

Through three on-site measurements, the maximum deformation position of the grid frame is the 5–6/J axis. The temporarily fixed plate at the support was removed on May 10, and the measurement was carried out again on May 12. The final deformation data shall prevail on May 12. The existing deformation value of the 5–6/J axis is 273 mm (the height difference between 2/J axis support and 9/J axis support is 28 mm, and the 5–6/J axis is the span, and the 1/2 height difference of 14 mm has been deducted during the calculation). CAD software series of Tongji University -3D3S steel structure – Space structure design software is used to analyze the current load of the steel grid structure (upper chord constant load: 1.0 kN/m2, bridle load: 18 kN, temperature ±30°C). The maximum deflection of the grid under the current load is the 5–6/J axis, and the maximum deflection is 76.0 mm, as shown in Figure 4.

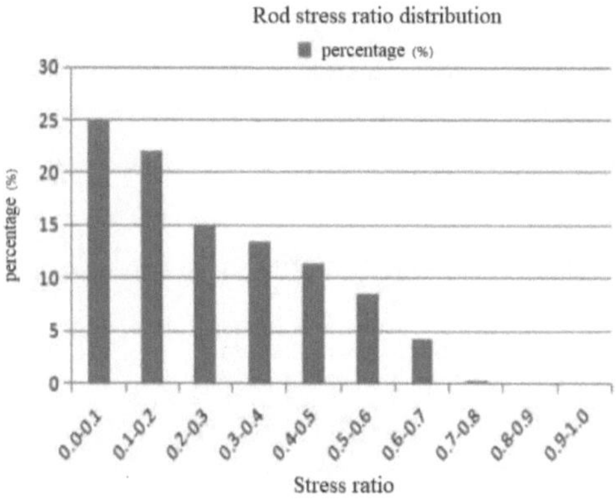


Figure 3. Rod stress ratio distribution (The test results show that the maximum stress ratio is 0.77, and the structure can meet the calculation requirements of bearing capacity).

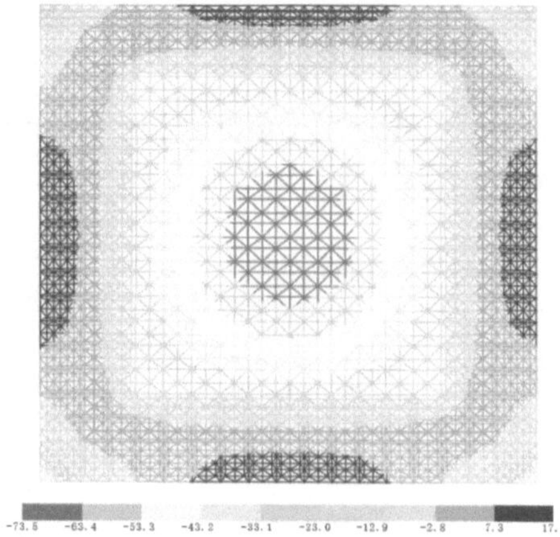

Figure 4. Maximum positive displacement: Uz (unit: mm).

## 4 CONCLUSIONS

Based on the model analysis, this paper studies the influence of load type, temperature effect, and construction technology on steel structures. The main conclusions are as follows:

(1) Through CAD software series -3D3S steel structure-spatial structure design software of Tongji University, the bearing capacity of steel grid structural members of the basketball stadium of this project is checked and calculated, as shown in Fig. 3. The checking results show that the stress ratio of each member is less than 1, and the stress ratio and slenderness ratio meet the requirements of the specification.

(2) From the conclusion of the deformation inspection, the deformation of the grid meets the requirement of calculating allowable deflection value under the current load during testing.

## REFERENCES

Chen shengmin (2020) Optimization and construction quality control of large-span bolted ball steel grid [J]. *Construction mechanization*, 41(10):53–55.

Chu Mingjin, Feng Peng, Hou Jianqun, Ye Liping, Liu Yansheng (2009). Experimental study on seismic performance of multi-storey residential wall with steel mesh frame and concrete composite structure [J]. *China civil engineering journal*, 42(07):36–45.

Deng Chao, Chen Zhongfan (2006). Experimental Research on Concrete Composite Wall with Steel Mesh Frame [J]. *Jiangsu Architecture*, (05):17–19.

Qin Lijie (2020). Analysis on Construction Quality Control of Bolt-ball Steel Grid Structure [J]. *Installation*, (03):62–64.

Shi Shengdong, Chen Zhongfan(2007). Research on self-compacting fly ash concrete for Composite Wall with Steel Mesh Frame [J]. *New Building Materials*, (07):61–64.

Xie Di, Xu Meng (2022). Application of BIM technology in construction of Irregular grid ball Joints [J]. *Intelligent Building and Smart City*, (04):113–115.

Xu Ming, Yao Yongqin, Zhu Xiaoli, Chen Zhongfan(2013). Experimental Study on flexural Performance of concrete floor slab with steel mesh Frame [J]. *Concrete*, (01):117–119+126.

*Advances in Urban Construction and Management Engineering – Cha (Ed)*
*© 2023 Copyright the Author(s), ISBN 978-1-032-39018-5*

# The influence of mud index on the wear of cutter head and cutting tool

Qing Yang*, Xiangchuan Yao* & Wen Liu*
*CCCC Second Navigation Engineering Bureau Co., Ltd., Wuhan, Hubei Province China*

Yang Chen*
*School of Civil and Hydraulic Engineering, Huazhong University of Science and Technology, Wuhan, Hubei Province, China*

ABSTRACT: The good performance of the cutter is a prerequisite to ensuring the smooth tunneling of the shield. The research on reducing the wear of the cutter head and cutting tool by optimizing the setting of the tunneling parameters has matured, while the effect of mud quality on wear reduction of the cutter head and cutting tool is only explained qualitatively. The type of mud index that can reduce wear on the cutter head and cutting tool is a problem that needs to be addressed in slurry shield engineering. Based on the river-crossing tunnel project of the Weisan Road in Nanjing, the corresponding research on the influence of mud index on the wear of cutter head and cutting tool is carried out in this paper. The results show that increasing the weight and viscosity of mud helps slow down the wear of the cutter head and cutting tool. However, the mud with high weight and viscosity will increase the load of the slurry treatment system as well as the construction cost and risk. In the actual construction process, the control of the mud index is first to adapt to geological conditions and ensure the stability of the excavation face, under which the mud index can be appropriately improved to slow down the wear of the cutter head and cutting tool. The research proposed in this paper, with certain guiding significance, provides reference experience for similar projects in the future.

## 1 INTRODUCTION

With the rapid development of the economy, the continuous improvement in the rate of urbanization, and the significant increase in urban population density, the development of urban underground space has become an important means to alleviate traffic congestion and promote the development of sustainable cities. As the main construction method for the development and construction of underground space, the shield method has been widely used in urban road traffic or rail transit construction due to its advantages of high construction efficiency, small surface disturbance, and safe working surface. In the construction of urban underground space, the construction conditions are complex and the risks are high. In the process of shield tunneling, the wear of the cutting tool is inevitable. After the tool is worn, it must be replaced in time. If the cutting tool is seriously worn and not replaced in time, it may cause the wear of the cutter head, further leading to unacceptable accidents such as the failure of the main bearing seal, which will seriously affect the progress of the project and even fail the project. The wear of the cutter head and cutting tool not only brings economic losses but safety risks. Therefore, reducing the wear of cutting tools and cutter head is one of the research projects focuses and difficulties in the engineering field.

Many studies on the wear of cutting tools have been carried out in China and abroad, and achieved some results. Relevant research mainly focuses on tool cutting and wear mechanism, cutting tool arrangement, and indoor similar model experiments, the main purpose of which is to design the

*Corresponding Authors: hust.qingyang@gmail.com, 605866412@qq.com, 379047775@qq.com and 605866412@qq.com


DOI 10.1201/9781003348023-15

cutter head for shield machines and predict the wear of cutting tools (Cao 2017; Pu 2011; Song 2009; Zhang 2008). Reasonable mud performance helps reduce the wear of the cutter head and cutting tool, which has been widely recognized by colleagues in the industry, but few focus on improving the mud performance to alleviate the damage to the cutter head and cutting tool. The mud film formed on the excavation surface has the characteristics of lubrication, which is beneficial to reducing the friction between the tool and the stratum, further abating the wear of the cutting tool, and prolonging its service life. Appropriate mud can make the cutting tool surface adsorb bentonite particles, reduce the direct contact between the cutting tool and the slag particles, and thus alleviate the wear of the cutting tool. In the current research, the influence of mud on the anti-wear effect of the cutter head and cutting tool is only a qualitative conclusion based on engineering experience, and quantitative research results are few. The special research on the influence of mud index on the wear of cutter head and the cutting tool has not been reported publicly, and what kind of mud index can reduce the wear of cutter head and cutting tool is one of the problems worthies of attention in mud shield engineering.

In this paper, considering the Nanjing Weisan Road crossing-river tunnel project and the actual engineering stratum environment on site, an effective test system is designed in terms of the construction technical problems related to slurry shield construction as well as the cutting principle of cutter head props and soil. In addition, the influence of the mud index on the wear of the cutter head and cutting tool is studied with the help of the shield machine model test.

## 2 RESEARCH ON CUTTING WEAR MECHANISM AND LAYOUT OF CUTTING TOOLS

### 2.1 *The cutting mechanism of the scraper*

The basic cutting process of the scraper is as follows. The scraper generates stress and deformation for the excavated soil through the cutting effect of the blade and the pushing effect of the rake face. The cutting effect of the blade makes the stress of the soil in the cutting layer exceed the strength of the soil, separating the soil in the cutting layer along the blade direction. The pushing effect of the rake face makes the separated soil deform and separate from the parent body to form soil debris that enters the aperture with the front of the scraper, so the cutting tool has both the cutting function and the loading function (Min 2018; Yang 2015).

The deformation of soil in the cutting area can be divided into three deformation regions as shown in Figure 1.

The first deformation area (I) refers to the deformation of the cutting layer near the rake face under the extrusion effect of the cutting tool. For plastic materials, it mainly refers to the slip deformation along the shear surface.

The second deformation area (II) refers to the extrusion friction deformation between the cutting chips and the rake face during the discharge.

The third deformation area (III) refers to the deformation zone generated inside the palm surface near the forefront, which is mainly formed by the extrusion and friction that the blunt circular part of the scraper and the rear knife apply to the palm surface.

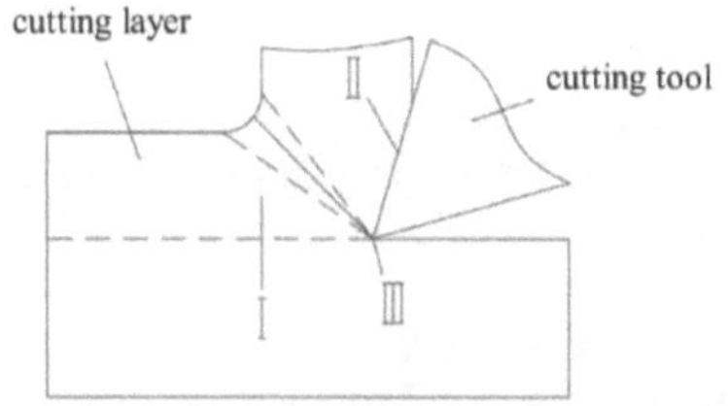


Figure 1. Deformation of cutting soil.

The failure form of muck in scraper cutting is related to the formation properties, cutting tool parameters (front and rear angles), and cutting thickness. Some scholars have corresponded and summarized the common muck flow failure form to the scraper cutting mechanism, including flow-type cutting, shear-type cutting, fracture-type cutting, and exfoliation-type cutting, as shown in Figure 2.

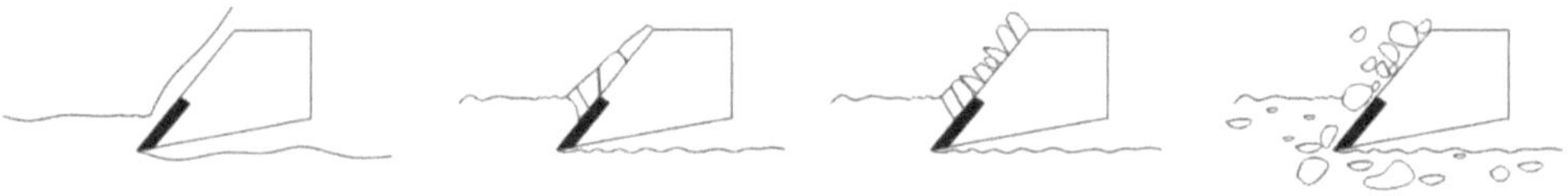

(a) Flow-type cutting (b) Shear-type cutting (c) Fracture-type cutting (d) Exfoliation-type cutting

Figure 2. Four different cutting mechanisms.

Flow-type cutting mainly occurs in the silt stratum, water-rich clay stratum, silty soil stratum, and silty sand stratum. The formation strength is low. With the movement of the scraper, the soil produces continuous shear deformation from the blade, and the residual soil flows continuously along the rake face of the tool. Under this condition, the cutting resistance of the cutting tool is very small and stable. The cutting tool mainly occurs in normal wear, is not easy to chip, and has a long life.

Shear-type cutting mainly occurs in clay and silty soil layers with relatively high strength. The decrease in the rake angle of the cutting tool will increase this trend. Due to the relatively high formation strength, the soil will first produce compression deformation during cutting, and then shear deformation along a plane from the blade, finally destroying and falling off.

Fracture-type cutting is a further development of shear-type failure, which occurs in clay and silty soil formations with low water content and higher strength, and gravel formations with better filling and high cementation strength. When cutting, the soil first generates compression deformation and maintains certain stability. With the cutting, the soil cracks and destroys by the blade, and the formed slag soil is in the form of a small lump. Under this condition, the cutting resistance acting on the scraper is pulsating, causing the cutter head and even the shield to vibrate. The frictional resistance between the shield and the formation is usually small, which is easy to cause the shield to rotate.

Exfoliation-type cutting occurs in ordinary gravel and sandy pebble formations. With small soil cohesion and large particle size, the cutting tool is less in the true sense of cutting, but the particles are stripped from the original formation, so the blade alloy is easy to crack.

The flow failure model of soil is related to the mechanical state when a failure occurs. The flow type, shear type, and fracture type belong to the plastic failure caused by shear deformation, and the exfoliation type belongs to the brittle failure caused by tensile deformation. The flow pattern of the muck will change with the soil quality, cutting angle, cutting speed, or cutting thickness. When the moisture content of the soil is low, if the cutting speed increases, the soil will change from fracture failure to flow failure. When the cutting angle of the scraper increases (the rake angle decreases) or the cutting thickness increases, the soil will change from flowing failure to fracture failure.

### 2.2 *Wear mechanism of scraper*

In terms of the cutting tools, the process from wear to damage can generally be divided into three stages (Wu 2020).

The first stage is the initial wear stage. After the tool is made, the rake face and the flank face are not completely smooth, with many very sharp and small bumps on the surface. When these bumps are subjected to cutting friction and chip impact, they will be flattened quickly, which is the initial wear stage.

The second stage is the normal wear stage. With the increase of cutting time, a wear band will gradually form on the flank of the cutting tool. Moreover, the wear amount will increase with time increase, and the wear on the surface of the cutting tool will intensify. This stage lasts for a long time, which is the normal wear stage.

The third stage is the rapid wear stage. When the cutting tool enters the later stage of normal wear, the wear on the surface of the cutting tool increases gradually, which will lead to sharp wear of the cutting tool, and the cutting tool will collapse. At this time, the cutting tool will fail and must be replaced.

The wear of the scraper is related to the formation properties, muck flow pattern, cutting tool parameters, and tunneling parameters. It is of great significance to analyze the wear process of the cutting tool and put forward the wear mechanism, thus summarizing the reasonable scraper parameters and tunneling parameters under different strata to reduce the scraper wear.

When the scraper is worn normally, the wear line is a gentle arc. With the development of wear, the blade becomes blunter and blunter, and the blade body is gradually damaged. Normal wear usually develops slowly, and the loss to the tool is small. In silt, clay, and silty soil layers, the wear of the cutting tool is mainly normal, and the soil is characterized by flow and shear failure. In this kind of stratum, a large rake angle and small rear angle can be selected, and the speed will be relatively high so that the thrust resistance and thrust torque can be reduced to improve efficiency.

Scrapers in silty and fine sand strata also show normal wear. However, due to the high quartz content, the secondary wear of the relief surface of the scraper is serious, and the rear corner should be appropriately increased. The secondary wear of the relief surface is more prominent in the gravel and sandy cobble formations, which leads to insufficient blade embedding depth and alloy shedding in severe cases.

Due to the large particle size and high strength of gravel and sandy pebble strata, the soil is mainly fracture-type failure and exfoliation-type failure. The performance of the cutting tool is more abnormal wear, including the secondary wear of the uncut stratum and the soil silo on the cutting tool, the alloy cracking caused by the tool exfoliating the pebbles or gravel, and the impact damage to the alloy when the pebbles fall.

When tunneling in such strata, exfoliation-type cutting should be formed as far as possible, and the frontal impact of the blade and the pebble, which is caused by the tool cutting the formation directly, also should be avoided as much as possible. When designing the cutting tool, the rake angle should be reduced, and the size of the alloy section should be increased. Moreover, the head should not be too sharp so that the impact resistance can be enhanced. At the same time, giving full play to the role of the advanced knife can effectively reduce the force of the cutting tool, reduce the probability of abnormal wear of the tool and prolong its life.

### 2.3 *Research on the arrangement of the cutting tool*

At present, the arrangement of scrapers includes Archimedes' spiral arrangement and concentric symmetrical arrangement. Usually, the Archimedes spiral arrangement method is used in the design of the shield machine cutter head. In this method, the overturning moment of the cutter head is 0, and the stress condition is the best (Li 2021; Ji 2009; Zhang 2021).

Archimedes spiral is known as the constant velocity spiral. When a moving point moves in a straight line at a uniform speed along the polar diameter, and the polar diameter itself rotates at a uniform angular velocity, the trajectory of the moving point is called Archimedes spiral.

The polar coordinates of the Archimedes spiral are described as:

$$\rho = \rho_0 + \alpha \cdot (\theta - \theta_0) \tag{1}$$

where $\rho$ is the polar diameter, $\rho_0$ refers to the initial value of the polar diameter, $\alpha$ is the constant coefficient, $\theta$ is the polar angle, and $\theta_0$ is the initial value of the polar angle.

The constant coefficient $\alpha$ is:

$$\alpha = \frac{N \cdot b}{2\pi}$$

where $b$ is the width of the cutting tool, and $N$ is the number of spokes of the cutter head.

Then the expression of the Archimedes spiral arrangement of the cutting tool on the cutter head is:

$$\rho = \rho_0 + \alpha \cdot (\theta - \theta_0) = r_1 + \frac{N \cdot b}{2\pi}(\theta - \theta_1) \tag{2}$$

where $r_1$ and $\theta_1$ are the installation radius and polar angle of the first scraper.

## 3 DESIGN OF MODEL TEST SYSTEM OF SHIELD MACHINE

Although real propulsion parameters and cutting tool parameters can be obtained in on-site excavation, some on-site propulsion parameters are affected by several irrelevant factors. As an example, the thrust torque is affected by friction forces in the stratum, traction forces in the rear support, friction torque in the cutter head, etc. This does not fully reflect the cutting force and cutting torque required to cut the material. The parameters of the cutter head and the cutting tool cannot be changed under the same engineering and geological conditions, which leads to certain limitations to study the cutting mechanism and law only relying on field data. Although the model experiment is not completely equivalent to the real shield tunneling, its simplicity and repeatability make up for the lack of site tunneling. The combination of field excavation and model experiments helps systematically study the geological adaptability of the cutter head and the cutting tool.

### 3.1 *Design principle*

In this test, to study the influence of a single variable of mud index on the wear of the cutter head and cutting tool, a simplified cutter head model of the shield machine is designed, as shown in the figure below. The cutter head is driven by a frequency conversion motor plus reducer, and the speed is 20 to 50 rpm.

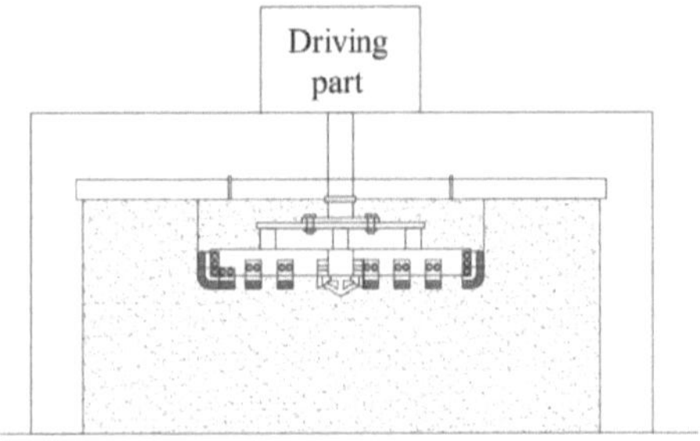


Figure 3. Schematic diagram of the system.

Figure 4. Model entity.

### 3.2 *Integration of the system*

#### 3.2.1 *The arrangement of the cutter head and cutting tool*

There are four spokes on the model cutter head, and the diameter of the cutter head is 1,000 mm. The flange connection is used between the cutter head and the drive part.

A total of 22 front scrapers, 14 edge scrapers, and 1 fishtail knife are set on the blade spoke. The height difference between the fishtail knife and the scraper is fixed to 30 mm, and the blade height of the scraper is 50 mm. The positioning steel plate of the cutting tool, with a thickness of 15 mm, is welded on the spoke, and the scraper and the spokes are connected by bolts.

The structure of the cutter head and the arrangement of the cutting tool are shown in Figures 5 and 6.

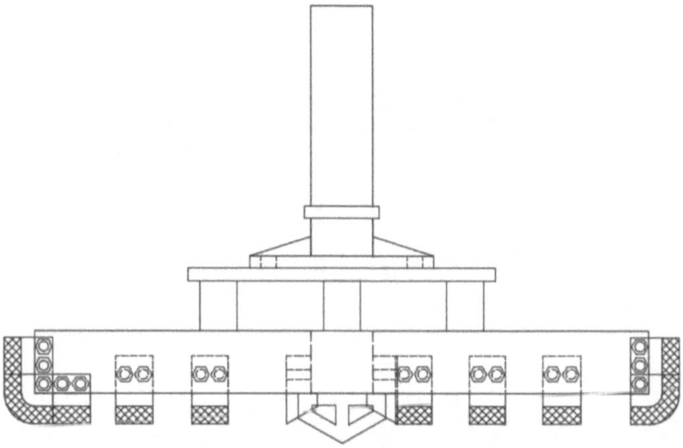

Figure 5. Structure of the cutter head.

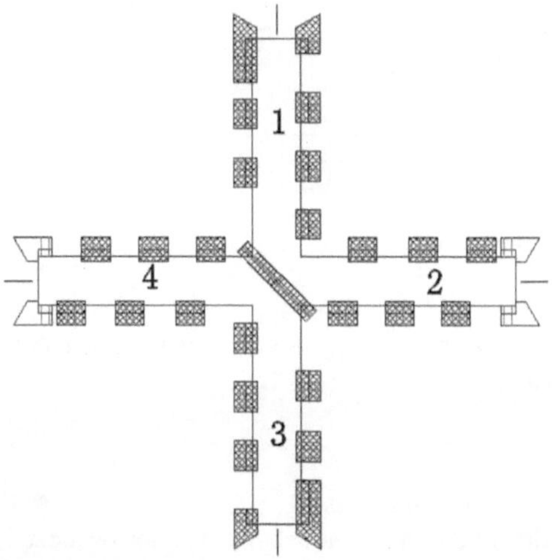


Figure 6. The layout of the spoke scraper.

3.2.2 *The design of the model scraper*

The structure of the front scraper is shown below.

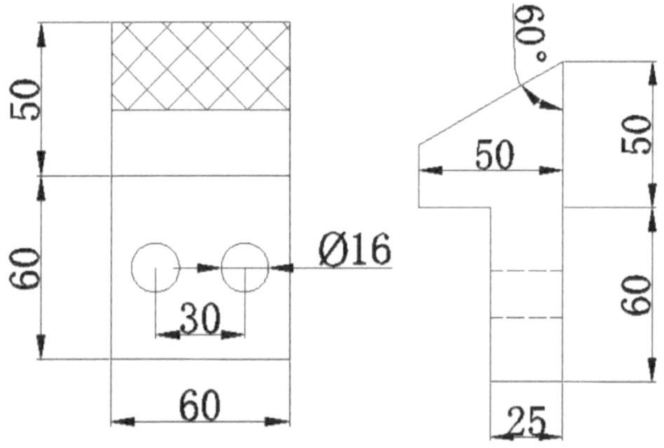


Figure 7. Structure diagram of the front scraper.

The structure of the edge scraper group is shown in Figure 8. The edge scrapers of different spokes can be composed of E-1, E-2, and E-3.

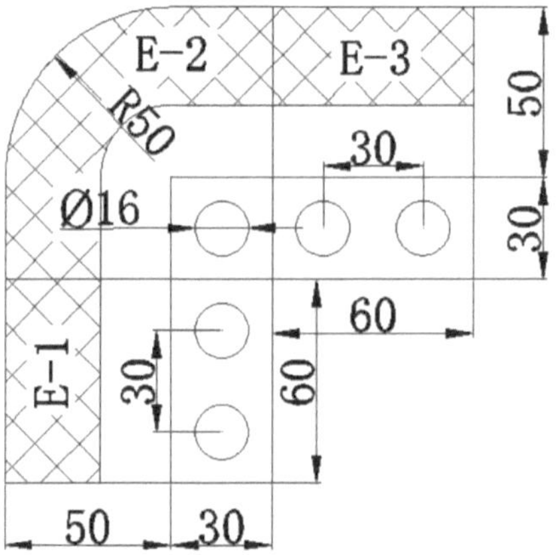


Figure 8. Structure diagram of edge scraper group.

## 4 EXPERIMENTAL RESEARCH ON THE INFLUENCE OF MUD INDEX ON THE WEAR OF THE CUTTING TOOL

### 4.1 *Experimental design*

4.1.1 *The configuration of the test soil sample*

Affected by the size limitation of the shield model machine, the oversized particles must be treated when the test formation is configured. At present, there are generally three methods for handling over-sized particles at home and abroad: (1) exclusion method; (2) equivalent substitution method; (3) similar gradation method. The exclusion method is to eliminate the oversized particles and take the remaining part to calculate the content of each particle size group, which will increase the content of fine particles and change the properties of coarse-grained soil. Therefore, this method is generally not used except for coarse-grained soil with very small particle sizes. For the equivalent substitution method, the coarse particles below the maximum particle size allowed by the minimum size of the model box are replaced by proportional equivalents. Although the replaced coarse-grained soil grades maintain their original fine particle contents, its uniformity coefficient changed. Relevant tests have confirmed that the samples prepared by the equivalent substitution method are more in line with the actual situation than the exclusion method. The similarity grading method is to reduce the particle size of the original coarse-grained soil according to the determined maximum allowable particle size in equal proportion to the geometric similarity principle, that is, the particle curve is translated according to a certain geometric simulation scale. Although this method keeps the non-uniformity and curvature coefficient unchanged, the fine-grained content increases, which changes the engineering properties of the original coarse-grained soil, especially the permeability of the soil. In the test, according to the main objectives of the test, the key mechanical indexes should be selected as the main control conditions to achieve the purpose of correct key mechanical performance.

Based on the above analysis, the soil samples are prepared by an equivalent substitution method.

In this test, natural sand with single particle size and slurry separation slag sample at the construction site are used to configure the test formation to simulate the wear of the cutting tool in the 5-2 gravel formation. The test formation is shown in Figure 9, and its particle gradation curve is shown in Figure 10.

Figure 9. Test formation.

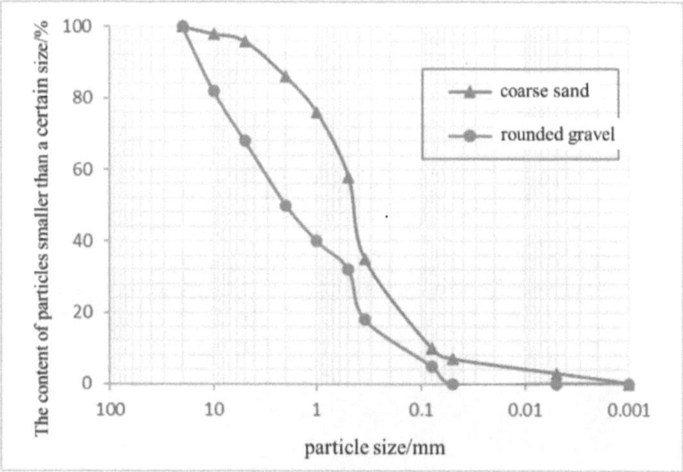


Figure 10. Particle gradation curve of test formation.

#### 4.1.2 *Design of experimental parameters*

This test is to study the influence of mud index on the wear of the cutting tool. The control parameter of the test is the mud index. A total of four groups of tests are designed.

Table 1. Design of experimental parameters.

| Stratum | Mud indices | | | |
|---|---|---|---|---|
| Five-2 gravelly sand | Freshwater | Weight: 1.10 g/cm$^3$<br>Viscosity: 29 s (Marsh) | Weight: 1.15 g/cm$^3$<br>Viscosity: 35 s (Marsh) | Weight: 1.20 g/cm$^3$<br>Viscosity: 40 s (Marsh) |

#### 4.1.3 *Test steps*

(1) Prepare test. Clean the soil box, fill the test soil layer, and add water or mud to the soil layer to saturation.
(2) Start the test. Each group of tests runs for 72 hours, and the change in the wear of the cutting tool is recorded every 24 hours.

Figure 11. Test preparation.

Figure 12. The measurement for the wear of the cutting tool every 24 hours.

Since the test soil box is an open structure, a large amount of heat will be generated during the test operation, which will lead to water evaporation. Therefore, during the test process, it is necessary to add fresh water or mud to the test soil layer in the soil box.

### 4.2 *Test results*

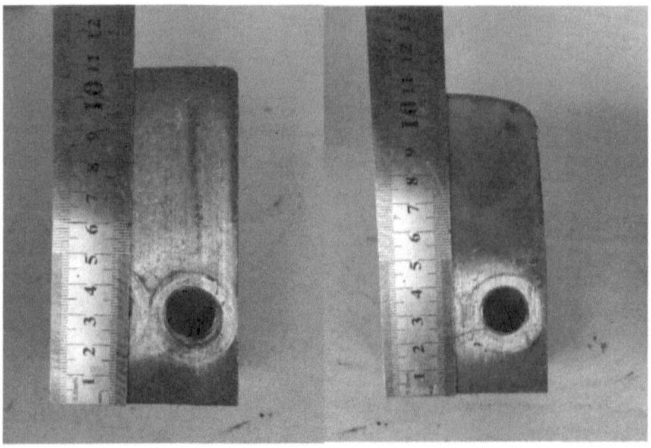

Figure 13. The wear measurement of the cutting tool.

Table 2. Statistics of the wear test on the cutting tool.

| Control index | The maximum wear (mm) | | | |
|---|---|---|---|---|
| | 24 hrs | 36 hrs | 48 hrs | 72 hrs |
| Freshwater | 2 | 4 | 7 | 11 |
| Weight:1.10 g/cm$^3$ Viscosity: 28 s | 2 | 3 | 5 | 8 |
| (Marsh) Weight:1.15 g/cm$^3$ Viscosity: 35 s | 1 | 3 | 5 | 7 |
| (Marsh) Weight: 1.20 g/cm$^3$ Viscosity: 40 s (Marsh) | 1 | 2 | 3 | 5 |

It can be seen from the test statistics in Table 2 that with the increase of time, the maximum wear value of the cutting tool gradually increases, and the maximum wear value of the cutting tool in the mud in the same period is smaller than that in the freshwater. Moreover, with the increase in mud weight and viscosity, the maximum wear value of the cutting tool in the same period also shows a downward trend.

## 5 CONCLUSIONS

Based on the Nanjing Weisan Road crossing-river tunnel project, the influence of mud index on the wear of cutter head and cutting tool during slurry balanced shield construction is studied in this paper. Moreover, by investigating the cutting wear mechanism of the cutting tool in detail and considering the actual engineering stratum environment, an effective test system is designed. Several groups of field tests are carried out based on the test system, obtaining sufficient sample data. Through comparatively analyzing the experimental data, the following conclusions are drawn:

(1) Compared with fresh water, the mud with a lubricating effect can make the formation particles and the surface of the cutting tool adsorb bentonite (clay) particles, reducing the direct contact between the cutting tool and the slag particles, and significantly relieving the wear of the cutter head and cutting tool;
(2) Increasing the weight and viscosity of the mud helps ease the wear of the cutter head and cutting tool;
(3) The mud with high weight and viscosity can effectively reduce the wear of cutting tool, which will simultaneously increase the load of the slurry treatment system and construction costs and risks;
(4) In the actual construction process, the control of the mud index is first to adapt to geological conditions and ensure the stability of the tunneling face, under which the mud index can be appropriately improved to abate the wear of the cutter head and cutting tool.

## REFERENCES

Cao Li, Qiu Gong. (2017). Research on shield tool wear prediction model based on propulsion parameters[J]. *Modern tunneling technology*, 54(03):155–161.
Ji Guang-bin. (2009). Design study of shield cutterhead[D]. Tianjin University.
Li Feng, Li Shuang-xi. (2021). Wear and wear reduction measures of shield cutting tools[J]. *Mechanical Design*, 38(S1):126–130.
Min Fan-lu, Bai Yu-xin. (2018). Research progress of shield cutting tools and case analysis of abnormal wear[J]. *China Journal of Highway and Transport*, 31(10):47–58.

Pu Yi, Liu Jian-qin. (2011). Research on cutterhead tool layout method of EPB shield machine[J]. *Chinese Journal of Mechanical Engineering*, 47(15):161–168.

Song Yun. (2009). *Research on cutterhead selection and design theory of shield machine*[D]. Southwest JiaoTong University.

Wu Jun. (2020). *Study on the interaction and wear between shield cutting tools and geotechnical mechanics*[D]. Beijing JiaoTong University.

Yang Hui. (2015). *Cutting mechanism analysis and multi-objective optimization design of shield machine tools*[D]. Hunan University.

Zhang Ming-fu, Yuan Da-jun. (2008). Dynamic wear analysis of shield tools in sand pebble formation[J]. *Chinese Journal of Rock Mechanics and Engineering*, (02):397–402.

Zhang Zhao-huang, Syed Naveed Haider Naqvi. (2021). Design and analysis of shield cutterhead structure[J]. *Journal of China Institute of Water Resources and Hydropower Research*, 19(03): 342–349.

*Advances in Urban Construction and Management Engineering – Cha (Ed)*


# Technical and economic research on environmental engineering projects

Pengxiang Liao*
*Guangdong University of Science and Technology, Dongguan, China*

ABSTRACT: As a prominent part of new energy, environmental engineering has an important impact on the global economy. Since the double carbon target was written into the government report of the two sessions in 2021, environmental engineering has been raised to a new level. Environmental engineering is a branch of environmental science. The discipline field is still developing, and its importance is also increasing. Based on relevant theory, this paper analyzed the evaluation model of environmental engineering projects by using relevant indicators.

## 1 INTRODUCTION

With the increasingly serious global environmental problems, new energy is an important theme at present. The proposal of sustainable development has brought environmental governance into a new era. As a component of new energy, environmental engineering has become a problem that needs attention. Environmental engineering has a long history, but the formation of a systematic discipline branch is one of the key high-tech developments in the 21st century. According to the relevant experience in the past, among the influencing factors of environmental engineering, the prior control, including initial design, directly affects the economic effect of environmental engineering. For technical analysis of environmental engineering, it is necessary to conduct a comprehensive analysis from the qualitative and quantitative perspectives, compare and demonstrate different technical policies, technical plans, and technical schemes, and finally evaluate different technical schemes to meet the cost-benefit principle (Bohumil 2017), thereby achieving the best combination of technology and economy. Environmental engineering should not only consider the cost but also ensure the quality. At the same time, it also needs to be improved in technology. The technology of environmental engineering design should be closely linked with economic benefits, so corresponding financial indicators are required. Through financial indicators, the technical and economic analysis of environmental engineering is quantified, and the optimal scheme is selected.

## 2 DEVELOPMENT OF ENVIRONMENTAL ENGINEERING AND THE CURRENT ENVIRONMENTAL SITUATION IN CHINA (TAKING SURFACE WATER QUALITY AS AN EXAMPLE)

China's environmental engineering market started late and is not very mature at present, but it is also in continuous development and improvement. The more important time point is the promulgation of the 12th Five Year Plan for national environmental protection, which puts forward higher requirements for environmental quality and protection, and brings certain standards for the quality and benefits of environmental engineering. In this case, the environmental protection work has ushered

*Corresponding Author: 303777914@qq.com

DOI 10.1201/9781003348023-16

in an opportunity. Many environmental protection companies have begun to set up and carry out their work in an orderly manner. Companies that do not pay attention to quality and economic effects will also be eliminated by the market. At the beginning of the concept of environmental protection, environmental protection only existed in urban construction as a subsidiary concept. With the continuous improvement of the concept of environmental protection, environmental engineering has become a theme of project engineering. Combined with supporting projects such as electric power, water conservancy, mechanical automation, and network informatization, environmental engineering is becoming increasingly important in urban construction. At the same time, with the reform and development of China's professional competence authentication system, environmental engineering has also participated in the joint release of the registration system of Interim Provisions for the environmental protection engineer system (Zhao 2013). Only by taking the examination and obtaining the professional qualification certificate, can the personnel who have obtained the certificate can engage in environmental engineering work in relevant units, which means that the demand and development of environmental engineering projects have reached a new height.

The development of environmental engineering has radically changed the environmental situation in China. Surface water and ambient air are used as examples in this paper to examine the effects of environmental engineering.

Table 1. The proportion of surface water quality categories from January to December 2021.

| Water quality category | Proportion | Main pollution indicators |
|---|---|---|
| Class I | 6.9% | Chemical oxygen demand, permanganate index, and total phosphorus |
| Class II | 49.2% | |
| Class III | 28.8% | |
| Class IV | 2.2% | |
| Class V | 11.8% | |
| Inferior to Class V | 1.2% | |

From January to December, among 3641 national surface water assessment sections, the proportion of sections with excellent water quality (Class I–III) was 84.9%, an increase of 1.5 percentage points compared with 2020; the proportion of sections inferior to class V is 1.2%, which meets the water quality target of 2021. The main pollution indicators are chemical oxygen demand, permanganate index, and total phosphorus.

## 3 RELEVANT FACTORS TO BE CONSIDERED IN TECHNICAL AND ECONOMIC EVALUATION OF ENVIRONMENTAL ENGINEERING

In the environmental engineering scheme, the technical and economic analysis and evaluation should be considered from the process point of view, according to the project process, and should be analyzed from the perspective of cash flow. To grasp the direction of cash flow, we can analyze it from the perspectives of capital preparation, capital inflow, capital operation, and capital outflow, and guide the technical and economic evaluation of environmental engineering by evaluating the use of cash.

### 3.1 *Preparation of environmental engineering funds*

Environmental engineering is a long-term project, also a capital construction project, and a livelihood project. Compared with general construction projects, it has the characteristics of large investment, long cycle, difficult design, and high maintenance cost. The long battle front requires long-term capital investment, which is also the limiting factor for environmental engineering to do

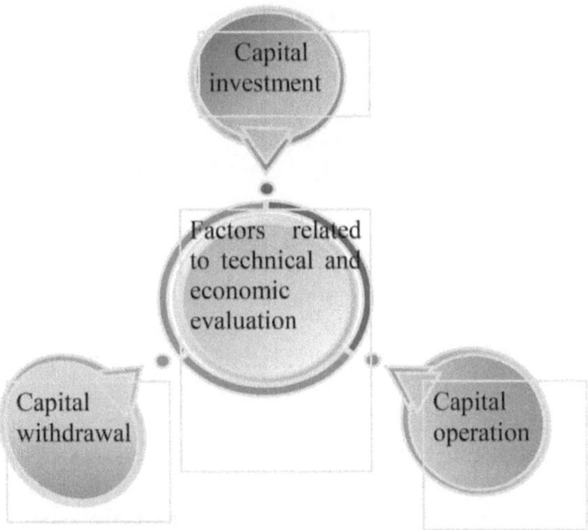


Figure 1. Relevant factors of technical and economic evaluation in environmental engineering.

well, and determines that environmental engineering requires high capital investment. The fund is the key point, but the preparatory work is the difficulty. At present, the fund-raising of environmental engineering is mainly reflected in government funds. Due to the relatively large difficulty of private financing, the current environmental engineering relies more on government funds, and the loans of financial institutions such as the world bank also need to be led by the government. The preparation of funds should be diversified, with a single way and simple channels, which will have a great impact on the efficiency and effect of fund-raising, and even cause the rupture of the capital chain. The orderly preparation of funds will improve the effect of technical and economic evaluation of environmental engineering (Ren 2017).

### 3.2 *Investment in environmental engineering*

The investment of environmental engineering funds determines the efficiency of technical and economic evaluation of environmental engineering. The quality of environmental engineering depends on the investment of funds. The investment of funds needs policy support and a reasonable design of environmental engineering treatment scheme. The social effect is an index of technical and economic analysis of environmental engineering, which covers the quality and social response of environmental engineering. Taking air pollution as an example, to reduce and improve air pollutants, it is first necessary to design a reasonable environmental engineering implementation plan, set up expected goals based on funds, mainly for social and economic settings, and need to add the satisfaction of the masses to monitor its efficiency. However, the most fundamental thing is a reasonable plan and capital investment, this will directly affect the technical and economic evaluation results of environmental engineering.

### 3.3 *Fund operation of environmental engineering*

With the investment of funds and the reasonable design of the scheme, we need to strictly implement it and revise the scheme during the implementation. In the process of environmental engineering construction, much mechanical equipment and workforce are required, which is also the main direction of fund application. The technical and economic evaluation of mechanical equipment is mainly focused on the use efficiency and use effect, and the inverse index is mainly reflected in some maintenance costs. The first impact of mechanical equipment on environmental engineering is the use frequency. Within the normal working capacity of mechanical equipment, the use frequency should be maximized, because for mechanical equipment, if it is not used, the equipment is still depreciated. Without time running in, the operating capacity of the equipment will decline. The second aspect is the use effect of the equipment. The main objective of environmental engineering is to ensure the quality and quantity of construction. Many pieces of equipment may be disposed of after the work has been completed. Under the same effect, select cost-effective equipment, increase revenue, reduce expenditure, and do a good job in capital operation.

### 3.4 *The outflow of environmental engineering funds*

After the investment and operation of environmental engineering, it is also necessary to consider its capital outflow. Capital is not stagnant, but in constant flow, so it also needs to flow out. For example, recruitment for construction work, purchase of large equipment, and daily expenses are all outflow methods. Outflow needs to reflect the application effect of funds. With outflow, the capital flow of the entire environmental project is complete. If there are individual problems in the middle, there may be fractures. Technical and economic analysis is a systematic analysis of all the cash flows of the environmental engineering project design, and an analysis of the economic benefits of each project in environmental engineering. After all, the types of projects covered in environmental engineering are complex, and their flows are different.

## 4 ANALYSIS OF RELEVANT FINANCIAL INDICATORS FOR TECHNICAL AND ECONOMIC ANALYSIS OF ENVIRONMENTAL ENGINEERING

At present, there are qualitative and quantitative methods for the technical and economic evaluation of environmental engineering. Quantitative analysis is mainly conducted from the perspective of indicators. We will conduct theoretical analysis from the perspective of project structure and the overall project.

### 4.1 *Project-based economic analysis*

According to the cost-benefit principle, the project-based economic analysis mainly refers to the economic analysis based on various indicators, including both static indicators and project dynamic indicators. In environmental engineering, the economic benefit index without considering the time value of funds is called the static evaluation index, which mainly includes a static investment payback period and investment return rate. The advantages are clear and intuitive economic significance, and simple calculation; to some extent, it reflects the advantages and disadvantages of the investment effect; it can be applied to various investment scales, but the time value of funds is not considered, so it is impossible to correctly identify the advantages and disadvantages of the project. Commonly used dynamic evaluation indicators are net present value, internal rate of return, dynamic investment payback period, etc. The two accounting methods have their advantages and disadvantages, which should be combined to consider the time value of money as much as possible. The main models are listed as follows:

(1) Payback period model (Bogatkina 2020). The payback period refers to the time required to recover the project investment with the net income of the project, which can be divided into the investment recovery period including the construction period, and the investment recovery period excluding the construction period according to the actual situation. The investment recovery period calculated by this model is the investment recovery period including the construction period.

The calculation for the payback period is:

$$\sum_{t=0}^{n}(CI - C0)_t = 0 \tag{1}$$

where $CI$ is cash inflow; $C0$ is cash outflow; $(CI - C0)_t$ is net cash flow in year $t$.

(2) Internal rate of return. The internal rate of return (IRR) of the project refers to the discount rate that can make the sum of the present value of the net cash flow of each year in the whole calculation period of the project zero. It is an indicator to examine the profitability of the project.

(3) Net present value (NPV). NPV is the sum of the present values of net cash flows in each year of the project calculation period calculated according to the set discount rate (*I*).

The calculation for NPV is:

$$NPV = \sum_{t=0}^{n} (CI - C0)_t * (1 + i)^{-t} \tag{2}$$

The financial net present value is an absolute indicator to evaluate the profitability of the project, which reflects the present value of the excess profits that the project can obtain in addition to meeting the profits according to the set discount rate. If it is greater than or equal to zero, it means that the project meets the evaluation requirements (Gao 2021).

### 4.2 *An overall analysis of capital income*

To conduct a comprehensive analysis of environmental engineering projects from a technical and economic perspective, many factors must be considered, including investment, risk tolerance, income status, etc. A suitable model can be suggested, dependent and independent variables can be defined, and statistical data analysis can be employed to determine the relationship between them. Consequently, relevant assumptions can be made: (1) Investment funds are positively correlated with environmental engineering economic evaluation. (2) Risk tolerance is primarily determined by the current state of fundraising. (3) Income is positively correlated with the economic evaluation of environmental engineering.

## 5 CONCLUSION

As a discipline branch of new energy, environmental engineering has been accelerating its development due to the continuous innovation of technology, and its relationship with technical and economic evaluation has been developing. Environmental problems are not only governance problems but also social problems, which can better reflect economic effects. Therefore, it is necessary to conduct technical and economic analysis on environmental engineering. Through static and dynamic financial indicators, considering risk and other factors, building relevant models can directly reflect the technical and economic situation of environmental engineering. In the future, it will be analyzed quantitatively through empirical analysis of industries or enterprises.

## REFERENCES


Bogatkina Yu. G., Eremin N.A. The Methodology for Economic Evaluation of Oil and Gas Investment Projects in Kazakhstan[J]. *Neftyanoe khozyaystvo-Oil Industry*, 2020, 1.

Bohumil Hlavenka, Roman Kubík. Expert System for Economic Evaluation of Technological Projects [J]. *Management: Journal of Contemporary Management Issues*, 2017.

Gao Xueping. Establish Economic Evaluation Model of Investment Project by Using Excel [J]. *China Management Informatization*, 2021, 24 (01): 96–99.

Ren Shengli. Technical and Economic Analysis of Environmental Engineering Projects [J] *Research on Urban Construction Theory* (electronic version), 2017 (11): 232.

Zhao Anna. Analyze the current situation and new development of environmental engineering design in China [J] *Intelligence*, 2013 (14): 285–286.

*Advances in Urban Construction and Management Engineering – Cha (Ed)*


# The application of traditional Chinese gardening techniques in the planning of "Park City" in Luxian, Sichuan Province

Xuehong Zhou*
*Sichuan Institute of Territorial Spatial Planning, China*

ABSTRACT: The concept of "park city" is a new concept of urban planning and construction proposed in the context of ecological civilization. After nearly four years of practical exploration of park city construction in Chengdu, Sichuan Province, the city has vigorously optimized the layout of green space systems, built greenways around the city, continuously improved the service quality and equalization of public service product supply, created various park city scenes to meet people's aspirations and needs for a better life, and promoted urban development transformation. A variety of traditional Chinese gardening techniques are discussed in this paper, including division of scenery, inscriptions, interspersed scenery, and punctuated scenery. The purpose of this paper is to explore how these traditional techniques can be incorporated into the planning and construction of park cities in Luxian, Sichuan Province, to address key concerns. We do this by drawing on the Chengdu Park city planning and construction model. For small towns in less developed areas of western China, this paper offers a useful reference on how to apply the concept of park cities.

## 1 INTRODUCTION

The park city concept originated in February 2018. In the past four years, Chengdu City in Sichuan Province has actively explored the connotation of the concept and the implementation path of the park city from both theoretical and practical levels and achieved remarkable results. For many small towns located in the west with excellent resource and environmental backgrounds, the current situation is mainly faced with the problems of low level of development of new urbanization, the outflow of production factors, weak industrial economic base, and low urban competitiveness. With the park city concept as a guide, high-quality development as the objective, small town economic development as the support, and small-scale progressive promotion as the guide, we can learn from Chengdu's planning and construction paradigm. To define and highlight the characteristics of western small towns, park city planning needs to combine local practice and in-depth research.

## 2 PLANNING RESEARCH ON "PARK CITIES" IN CHENGDU

In 2019, the Chengdu Tianfu Park City Research Institute completed the Chengdu City Park City Plan (2018–2035), a new research topic to achieve high-quality urban development in the context of ecological civilization. The plan conducts a series of studies on the connotation, characteristics, index system, and construction paths of park cities. Combined with Chengdu's practice, the plan takes the lead in proposing the interpretation of the theoretical system of park cities in Chengdu under the guidance of Xi Jinping's thought of socialism with Chinese characteristics in the new era and completes the park city planning paradigm and construction samples by combining the

*Corresponding Author: 313631774@qq.com


DOI 10.1201/9781003348023-17

judgment of the basic conditions for building park cities in Chengdu with the requirements of the new round of urban master planning. The plan clearly defines the development goal of Chengdu as a beautiful and livable park city by the middle of this century and forms twenty planning strategies for building a park city around the four dimensions of "people, city, environment and industry" (Wu 2019). The six park scenes are built according to the park-like requirements of "accessible, participatory, scenic and landscaped". The Chengdu Park city planning paradigm focuses on building a park city theoretical system and exploring a park city with the characteristics of the western Sichuan plain, which provides a certain method and experience for the practical path of planning and construction of beautiful and livable park cities in megacities in the plain area of China.

## 3 THE APPLICATION OF TRADITIONAL GARDENING TECHNIQUES IN THE PLANNING OF A "PARK CITY" IN LUXIAN, SICHUAN PROVINCE

### 3.1 *Overview of Luxian county*

Luxian, also known as Dragon City, is located in Luzhou City, Sichuan Province. It is located at the confluence of the Tuo River and the Yangtze River, the core hinterland of the twin-city economic circle in the Chengdu-Chongqing region, and the node of the southwest sea route. Luxian has a long history and was established in 135 B.C. It has a diverse topography and rich natural resources, is known as the "hometown of Chinese dragon culture" and is a national cultural heritage county. It covers an area of 1,525.3 square kilometers, with 19 towns and one street under its district (see Table 1). According to the population statistics of the Seventh Poo, the resident population of Luxian in 2020 is 764,400, with an urbanization rate of 41.22%. The county is conveniently located with an urban population of 120,000. Located 230 km from Chengdu and 130 km from Chongqing, it is an important node at the intersection of the 2-hour economic circle of Chengdu and the 1-hour economic circle of Chongqing.

Luxian County has a high-quality natural ecological environment. "The mountains and waters in the south of Shu are the most beautiful in Luchuan, with the Jade Toad in the drunkenness", which is a beautiful picture of Luxian County in the North Song Dynasty, written by Huang Tingjian. In the south-east of the county, there is the Jade Toad Mountain with its lush forests, the Seti River and the Jiuqu River running through the center of the county, the Jiuqu River Central Park, the Xiaolu River Botanical Garden, the Seti River Riverfront Park, and the street green areas and small pleasure gardens, and the Qinglong River and Xiaolu River are rich in a rural wilderness. With its pleasant climate and rich biodiversity, the county presents a natural and harmonious beauty that combines farmland and gardens with urban paths.

Since 2020, Taking advantage of the planning paradigm and construction experience of Chengdu city park city, Luxian has created a national garden city, combining it with the pilot project of the provincial sponge city and livable county. Following the park city construction concept, the "Millennium Song rhyme Dragon City, happy park city" project focuses on preparing park city planning and pilot construction.

### 3.2 *Key issues that need to be clarified in the planning of a "park city" in Luxian county*

#### 3.2.1 *A comprehensive understanding of the connotations and extensions of a county-scale park city*

A "park city" is an urban-rural green space system and park system, a park-like urban-rural ecological pattern and landscape as a basic and pre-configuring element of urban-rural development and construction, and an optimal and harmonious relationship between "citizen-park-city" and "park". The new concept of urban and rural habitat construction and the ideal city construction model, which takes the optimization and harmony of the relationship between the "citizen-park-city" as an important element in creating a better life, and provides more high-quality ecological products to

Table 1. Luxian County's land use structure in 2020.

| Type of land use | | Area (Square kilometers) | Proportion (%) |
|---|---|---|---|
| Cropland | | 759.7 | 49.81 |
| Garden | | 41.7 | 2.74 |
| Woodland | | 304.3 | 19.95 |
| Grassland | | 1.3 | 0.09 |
| Wetlands | | 0.5 | 0.03 |
| Urban and rural building sites | City | 24.1 | 1.58 |
| | Village | 112.5 | 7.38 |
| Regional infrastructure sites | | 16.4 | 1.07 |
| Other building sites | | 86.6 | 5.68 |
| Terrestrial waters | | 54 | 3.54 |
| Other lands | | 124.2 | 8.14 |
| Total | | 1525.3 | 100 |

meet the growing needs of the people for a beautiful ecological environment. The resource carrier of park city construction is the urban ecological environment and urban green space system. For cities and towns of different scales and different topographical conditions at different stages of development, the connotation and extension of park city planning and construction should be different and focused. As a form of ecological civilization on an urban scale, Chengdu has the characteristics of a mega-city park city in the plains, while the county should have a different positioning from that of a mega-city. The construction of county park cities is more conducive to the creation of an overall distinctive environment and should be based on an idyllic city and countryside, with the earth's landscape as the substrate, and the urban green space system, urban and rural ecological pattern, and urban cultural landscape should be systematically optimized to improve spatial quality and meet people's needs for a better life.

3.2.2 *Accurately grasping the content and depth of county-scale park city planning*

As a new type of planning, Chengdu has summarized the park city planning examples suitable for Chengdu based on the local development reality, but for the county scale, the technical path of park city planning is bound to be different due to the different environmental background and different economic base. It is necessary to analyze the background, connotation, and significance of park cities, and to summarize the content and depth of park city planning on smaller scales, based on the technical paths of park city planning in Chengdu. Based on the analysis of the basic characteristics and current problems of the construction of park cities in Luxian County, it is necessary to focus on the county, take the point as a whole, and propose the target positioning of the construction of park cities in Luxian County at different times, study the general layout and construction strategy of the construction of park cities in Luxian County and form an action plan for the construction of park cities to strengthen the suitability of the content of park city planning at the county level and the practical operation of the action plan.

3.2.3 *Precisely determining the focus of county-scale park city planning and construction*

The economic base of counties in the western region is generally weak, while park cities require the implementation of many projects with large funding requirements, making it difficult to promote full implementation in parallel. According to the demand for park city construction in Luxian County, the five-year action project pool, which was initially completed, includes 44 projects in five categories: building a park for the whole area, improving public services, optimizing urban form, building a management platform, and constructing a cultural scene, with a total investment of about 8 billion yuan. Funding and land elements guarantee will become the main constraint for the smooth implementation of the county-level urban park city plan. It is necessary to combine the actual economic situation of the county, carefully study from the perspective of operability, thoroughly sort out the stage objectives of the park city planning and construction, highlight the key points, and face the implementation, first easy and then difficult, accurately determine the implementation time sequence of the project according to the actual financing, achieve gradual and progressive, rolling development.

### 3.3 *The use of traditional gardening techniques in the planning of a "park city" in Luxian*

In classical gardens in China, framing, borrowing, interlacing, and barricading are common gardening techniques. Through the analysis of different scenes, a different perception of space is created. In Luxian park city planning and construction practice, based on the Luxian scale and the actual development of Luxian, the traditional gardening concept is introduced in the planning, taking the Luxian green space system and natural ecological pattern as the resource base, focusing on the optimization of urban functions, using traditional Chinese gardening techniques such as dividing scenes, inscriptions, interspersed scenes and pointing scenes, incorporating panoramic structures, scene-like compositions, converging veins and contrasting symbols into the county-scale The park city plan is prepared to highlight the characteristics and focus of Luxian's park city and is geared towards implementation, strengthening the operability of the planning scheme, guiding Luxian in the direction of a park city in an orderly manner, and building a park city with the qualities of a beautiful landscape dragon city.

#### 3.3.1 *Sensitization: a split-scene – panoramic structure*

The landscape is divided into many spaces to obtain a garden within a garden, a scene within a scene, and to make the garden scene virtual, half virtual, half-real or real in real, real in virtual, with rich scenery and variable spatial processing techniques (Zhang 2019). In 2020, the city will have a greening coverage of 38.12%, a green space rate of 34.05%, and a per capita park area of 12.65 square meters, making it a good ecological base for park city planning.

Luxian's park city planning breaks the traditional framework of green space system planning, taking the scenic design of "park + landscape" as a guide, using the "landscape" gardening technique, using landscape as a natural boundary, breaking the traditional pattern of green space system planning in which, the green space system and urban functions are laid out independently. The traditional pattern of the independent layout of the green space system and urban functional groups is broken down by integrating the functional groups of the county into the natural landscape, forming a panoramic structure, forming a space with a city in the garden, a city in the landscape, an interplay between reality and reality, and a rich and varied landscape depth. Combining the characteristics of the landscape pattern, road traffic, and the layout of urban functions in Luxian city, the complex of landscape space with urban park space and urban function space is strengthened, and the Luxian County, with Yuzhan Mountain, Seti River, and Jiuqu River as the core, the wedge-shaped green ecological corridor and greenway as the skeleton vein, and the neighborhoods of Longbouqiao Park, Chengbei Park, Jiuqu River Central Park and Wetland Park as scenic areas, is built. The spatial structure of the park city is "one mountain, two rivers, six veins, and eight scenic spots". According to the leading functions of each group linked by the landscape parks, the county is divided into eight functional scenic spots, mainly including the scenic spot of the King of Hundred Herbs and Medicine, the scenic spot of the Liquor Scented Deer Creek, the scenic spot of the Dragon Lake Spring Light, the scenic spot of the Jinping Flower Sea, the scenic spot of the Jiuzhu Goblet, the scenic spot of the White Pagoda Learning, the scenic spot of the Longbao Green Rhythm and the scenic spot of the Jade Toad Moonlight.

#### 3.3.2 *Landscaping: Thematic scene – scene composition*

In park city planning, according to the characteristics and environment of the urban landscape, the theme of landscape design is "park + culture", combining the scene requirements of different scenic areas, using artistic techniques such as pagodas, large scenic stone carvings, and native plant shapes to point out the cultural characteristics of the scenic area and render a unique mood. The landscape of the different scenic areas of the city is sublimated by the artistic techniques of pagodas, large scenic stone carvings, and native plant forms. Luxian has a long history of dragon culture, and is known as the "hometown of Chinese dragon culture". There are more than 140 dragon bridges in the county, 46 of which are national key cultural relics, the most representative of which is the Dragon Brain Bridge, built during the Ming Hongwu period and comparable to the Zhaozhou Bridge and Lugou Bridge. These dragon bridges are dominated by stone slab girder and arch bridges, all carved with dragons and spirit animals with exquisite quality and subtlety, representing the peak of human medieval stone dragon carving culture. In Luxian Park City planning, based on a deep interpretation of Luxian dragon culture and shaping diverse dragon cultural IP, the theme of the site is used to root the spiritual core of the park city, linking history and the future, and carrying the soul of the physical form of the park city.

In Luxian's park city planning, based on the charm of the Song Dynasty Dragon City theme, the dragon culture connotation should be fully integrated into the city scenic planning, and the dragon culture connotation of "dragon shape, dragon decoration, dragon dance, dragon poetry" should be explored and visualized in the planning of different scenic spots to create a special scenic spot with dragon culture connotation, and to highlight the distinctive city culture with a theme park.

#### 3.3.3 *Accessible: the pulse of the mezzanine – convergence style*

In the garden landscape treatment techniques, a clip scene is a kind of landscape composition with control, it can not only show a specific mood, convey the solemn, far-reaching, forward, exploratory, and other infectious, but also can strengthen the design conception mood, highlight the status of the end scene, and can induce, organization, the convergence of sight, so that the scene view space directed extension, until the culmination of the end scene (Xin 2020). Luxian County is surrounded by mountains and water, the terrain is flat, the road network is low density and residents travel mainly on foot and by non-motorized vehicles. Some of the current roads have already formed some pinch scenery effects, but the visibility and guidance between the county and the mountains, scenery, and parks are poor. In the Luxian park city planning, the concept of "park + traffic" was introduced to meet the function of urban road traffic, based on the traffic organization of the scenic city space, systematic research from the level of vehicle lines, pedestrian lines, urban touring lines, and static traffic to carefully sort out the visual relationship between the current city roads and the mountain, water, and scenic gardens. In addition, all the roads in the county that are visible to Yuqan Mountain are taken as the leading viewing space for organizing space and guiding tours, and Yuqan Mountain is taken as the end view. The building height and massing of the buildings are controlled and guided reasonably so that the planar space is open and well-defined, the three-dimensional space is pleasant and friendly, and the visual relationship is transparent and focused. The street interlayer is used to guide the viewer to Jade Toad Hill.

#### 3.3.4 *Visualization: pointed scenery – a sign of the pairing style*

In classical Chinese gardens, point-scenes refer to the use of embellishment to decorate spots or scenery, highlighting the features and mood of the scene and adding charm and color to the landscape (Xiao 2011). The most important physical representation in the city is the building, which is also the most important element of the punctuated landscape in park city planning. In recent years, as the urban area of Luxian County has continued to expand, buildings of different construction eras, building quality, and construction styles have been mixed, and the waterfront shoreline of the Seti River and Jiuqu River has been encroached upon by high-rise buildings lacking in character, and the relationship between people, the city and the natural landscape, which is interdependent, has been artificially fragmented, with the city entering the landscape and the city's sense of historical place and spatial recognizability diminishing.

In the planning of Luxian Park City, the concept of "park + architecture" visualization is used as a guide to highlight the iconic landscape features of the key buildings or structures in the eight scenic areas and to strengthen the role of architecture in the landscape control of the county park city so that the architectural space system and the park. The city presents a naturally harmonious and unified relationship between the architectural space system and the park city. Different scenic areas undertake different urban functions, and the elements of the dotted landscape are different. Both the Bacao Medicine King Scenic Area and the Liquor Deer Creek Scenic Area are clusters of urban industrial functions, and the main architecture of the scenic spots should be large-volume traditional-style low-rise industrial factory buildings. The Long Lake Spring Landscape and the Jinping Flower Landscape are the main residential groups in the city. The main landscape buildings can be planned and laid out as the Long Lake Pavilion on the Long Lake Island and the Range Pavilion on the top of Jinping Mountain as the scenic elements of the landscape, which can become the core landscape of the landscape and become the opposite architecture to break the rich skyline, with deep scenery and interesting views. The main function of the Quliu Goblet scenic area is to integrate the old city, which should be designed in key areas of high-rise buildings as a symbol of the scenic area. The old city walls, the battlements, and the battlements are all in sight, and the beauty of the architecture and the landscape are in harmony.

## 4 CONCLUSIONS

The German philosopher Heidegger once said, "The essence of life is a poem, and man is supposed to dwell poetically on the earth." The park city is a new concept of urban and rural habitat construction, a new construction mode to achieve high-quality urban development, and a new model of urban park and green space system, ecological pattern, historical culture, and traditional style as the basic, elements of urban development and construction. For small towns in less developed areas in the west, traditional planning concepts, a relatively weak economic base, and limited resources are important factors limiting the character of the town. Luxian's park city planning uses traditional gardening techniques for creative exploration and a small-scale urban practice, which seeks to explore the excellent ecological foundation of the small town and reshape it from a characteristic research perspective. By focusing on landscape, culture, transportation, and architecture, the project aims to fully explore the excellent ecological environment of the small town and reshape it through planning. The study has proposed related strategies in four aspects – landscape, culture, transportation, and architecture – to achieve integration of city, site, park, and people.

## REFERENCES


Blackbird in the night sky. *Park city series talk (I): background, connotation and conceptual characteristics of park city* [W].

Chengdu City Park City Construction Leading Group. *Park city: a theoretical exploration of a new model of urban construction*. Sichuan People's Publishing House: 2019.4 http://www.360doc.com/content/18/1204/18/47586630_799295895.shtml.2018-12-4

Wu Yan. Wang Zhongjie. The connotation of park city concept and suggestions for planning and construction in Tianfu New Area[J]. *Pioneer* 2019(4):27–29.

Xiao Yongqin. The application of Chinese traditional literature in gardens [J]. *Anhui Agricultural Science*. 2011: 623–625.

Xinseibo. Between contemplation and experience – a trial analysis of Dong Yu Gan's gardening practice [J/OL]. *New Architecture* 2020 (3).

Zhang Yanlin. Research on the application of the design theory of "Yuanye" in modern landscape [J]. *Journal of Sanmenxia Vocational and Technical College* 2019(12):141–146.

*Advances in Urban Construction and Management Engineering – Cha (Ed)*


# Discussion on the design points of railway across the urban spillway

Zheng Feng & Wenbo Wang*
*China Railway First Survey and Design Institute Group Ltd., Lanzhou, China*

ABSTRACT: In recent years, the water authorities on the railway across the urban area of the river floodway related requirements tend to be stricter, such as the water department simply requires that the railway structure in the river channel water resistance ratio should not be greater than 5%, special conditions shall not be greater than 8%. For most of the floodways, a span is required, which shall not be within the river management scope of the construction of piers and other structures. The actual extent of most rivers is smaller, while the delineated river management area (urban blue line) is much larger than the actual river extent. This caused the railway bridge span to be too large and substantially increased the construction investment, resulting in unnecessary waste. This paper takes the new Tianshui-Longnan railway as an example, combines the problems encountered in the bridge hole span arrangement and flood control evaluation, discusses the design points of the railway and river crossing, and systematically analyzes how to balance the relationship between the railway bridge span arrangement and flood control safety, providing reference significance for the railway alignment selection as well as bridge span design.

## 1 INTRODUCTION

"Floodproofing Law of The People's Republic of China" was revised in 2017 (National People's Congress 2016). In 2019, the State Council promulgated and implemented the "Regulations of the People's Republic of China on the Management of River Courses (State Council of the PRC2019)." In 2021, the Ministry of Water Resources issued "Guidelines on the formulation of flood control assessment report within the management scope of river course (MWR 2021)", the water authorities tend to be stricter on the requirements related to the railway crossing the river floodway in the urban area, and many railway projects have not received approval of flood control evaluation, such as Zhongwei to Lanzhou high-speed railway, Lanzhou to Hezuo railway, and Tianshui to Longnan railway. However, at the same time, the water authority has problems with the approval stage and unclear scope of approval in the process of railway engineering approval. There is a greater impact on the scheme and cost of railway projects. Taking Tianshui to Longnan railway as an example, this paper discusses the design points of the railway and river crossings in combination with the problems encountered in bridge hole span arrangement and flood protection evaluation and systematically analyzes how to balance the relationship between railway bridge span arrangement and flood protection safety to provide reference significance for railway alignment selection as well as bridge span design.

## 2 WATER AUTHORITIES ON THE APPROVAL REQUIREMENTS FOR RAILWAY CROSSINGS OF RIVERS

In Gansu Province, for example, the Provincial Water Resources Department issued at the end of 2017, "Gansu Province, the notice of the administrative licensing and information announcement

*Corresponding Author: 412049210@qq.com

 DOI 10.1201/9781003348023-18

system for river-related construction projects," the document requires that "the construction unit or individual of the river-related construction project must, following the river management authority, submit the construction program of the river-related project to the competent river authority with jurisdiction to review and agree. Without the review and consent of the competent river authorities, the construction unit shall not start construction." Therefore, the following procedures are currently followed for approval related to traffic engineering across rivers.

The construction stages are as follows: (1) prepare a feasibility study of the project, (2) obtain approval from National Development and Reform Commission, (3) prepare preliminary design, (4) obtain approval from relevant ministries, (5) design and approve construction drawings, and (6) construct the project. The approval of the river crossing is concentrated in the preliminary design or construction drawing design stage. Generally, a third-party agency prepares a flood control evaluation report for river-related projects that are submitted to the appropriate authority of the river authorities. After the expert review meeting has been held, the report is completed. The competent authorities approve river-related projects based on the opinions of the expert review meeting.

## 3 COMMON PROBLEMS IN THE ADMINISTRATIVE APPROVAL PROCESS

### 3.1 *The impact of different design stages to declare program approval*

There is no document in which the design phase should be declared to the competent river authorities to make specific provisions, highway projects in the feasibility study stage that the program declaration, there are also some projects in the preliminary design stage declaration; but the railway project in the preliminary design or even construction drawings stage are required to declare the program.

Declaration in different design stages will have different degrees of impact on the project investment control. China stipulates that the construction design budget in principle shall not exceed 1.1 times the preliminary design estimate, and shall not exceed 1.2 times the feasibility study estimate. In the actual project, most of them are controlled according to the construction drawing budget not exceeding the estimation of the feasibility study. Therefore, declaring the program in the preliminary design stage or even construction design stage to the competent water authorities may cause significant changes in the program, a substantial increase in investment, or even lead to project investment exceeding the control requirements, or to re-declare the feasibility study.

As an example, the preliminary design and construction drawings for the Zhongwei-Lanzhou high-speed railway have been completed. In the review of the program organized by the government departments of Gansu Province and Ningxia Autonomous Region, as well as the Water Resources Departments of the two provinces participating in the meeting, the railway corporation did not raise objections. Nevertheless, before the start of the project, the construction unit informed the competent water authorities that flood control evaluation and review were needed, the review was passed before construction began. In the follow-up review process, the requirement to increase the bridge span resulted in multiple changes to the design, increasing investment of tens of millions of yuan.

### 3.2 *There is no clear regulation on what size of river channel needs to be approved*

The Provincial Water Resources Department printed the following requirements: "River-related construction projects, whether they are built by units or individuals, must comply with the river management authority." The river-related engineering construction program must be approved by the competent river authorities with authority. However, there is no clear requirement on what size of river needs to be evaluated and approved for flood control, which leads to different implementation of the policy by the administrative authorities at the county level along a project, and even a few meters wide creek also needs to be prepared for flood control evaluation report requirements, resulting in a greater waste of resources.

Based on the above two problems, the Tianshui-Longnan railway is an example discussed in the next paragraphs in relation to specific bridgework points across the river program design, according to the review conclusions to give relevant recommendations.

## 4 DESIGN POINTS OF CROSSOVER BETWEEN TIANSHUI-LONGNAN RAILWAY AND RIVER

### 4.1 *Overview of Tianshui to Longnan Railway*

The Tianshui to Longnan railway is located in the southeast of Gansu Province, in the cities of Tianshui and Longnan. The new mainline has a length of 208.207 km, which includes 51.261 km in Tianshui City and Longnan City, 156.991 km (FSDI2021), 24.515 km/40 bridges, and 162.618 km/25 tunnels, a bridge-to-tunnel ratio of 89.88%. The construction of Tianshui station contact line 2.4 km, tunnel 0.506 km/1 bridge, 1.682 km/2 bridges; and Longnan West station relief line 3.177 km, 2.143 km/1 bridge. Reconstruction of existing Tianshui to Pingliang railway main line 1.413 km, tunnel 0.407 km, and reconstruction of Tianshui to Pingliang railway under construction to relieve the line 0.156 km, all for the construction of the bridge. Anhua Tunnel (L-19.980 km) is the longest tunnel on the line, and West Hanshui Special Bridge is the highest bridge (H-98 m) (FSDI 2021).

### 4.2 *Railway crossing major river profile*

As discussed in Section 2.2, the Provincial Water Resources Department has not made clear requirements on what size of river needs to be evaluated and approved for flood control, resulting in different requirements for each administrative region of the Tianlong Railway, the main river crossed by the railway is as follows.

Table 1. Spillway statistics of Tianshui to Longnan railway.

| Spillway | Width of the existing spillway (m) | Fortification standard for flood protection embankment (year) | Design span |
|---|---|---|---|
| Yingchuan | 98 | 20 | 32+24 m T-beam +48 m box girder |
| Dongshui | 30 | 10 | 2-32 m T-beam |
| Baijia | 80 | 20 | 4-32.6 m T-beam |
| Liziyuan | 180 (along the river) | / | 2-32+24+2-32 m T-beam+48 m box girder |
| Mayan | 130 | / | 5-32 m T-beam |
| Nianzi | 28 | 10 | 32 m T-beam |
| Dong | 80 | 20 | 24+3-32 m T-beam |
| Weier | 5 | 20 | 2-32 m T-beam |
| Lu | 11 | 20 | 32 m T-beam |
| Chenzhuang | 18 | 20 | 2-32 m T-beam |
| Nan | 250 | 20 | 5 × 60 m box girder |
| Liangshui | 25 | 20 | 32 m T-beam |
| Xihanshui | 130 | 10 | 59+105+2 × 136+105+59 m continuous rigid-frame beam |
| Pinluo | 30 | 10 | 60+100+60 m continuous box girder |
| Beiyu | 18 | 10 | 3-32+24 m T-beam |
| Youfang | 6 | 10 | 32 m T-beam |
| Long | 13 | 10 | 32 m T-beam |
| Gouba | 30 | 20 | 48+40+64 continuous box girder |

It should be noted that from the above table, Tianlong railway crosses Nianzi and Pinluo many times, Liziyuan, Nianzi for the bridges in the direction of the river. As the railway bridge fortification standard is much higher than the flood control project fortification standard, and the bridges are higher. The flood control evaluation report prepared by a third party concluded that the railway bridge hole span arrangement meets the flood control requirements. However, the author participated in the expert review meeting organized by the Department of Construction for flood protection evaluation, only two of the above bridge span layouts successfully passed the review, the rest are recommended to adjust the bridge span. The following will be a typical bridge crossing for a detailed description.

## 4.3 *Xihan bridge*

Xihan bridge is located in Cheng County, Longnan City, spanning the downstream section of the West Hanshui River, 39.1 km from the upstream bridge hydrological station and 36.9 km from the downstream Tanjiaba hydrological station. The bridge is equipped with a 100-year encounter standard and is calibrated for a 300-year encounter. The total length of the bridge is 647.45 m, and the hole span arrangement is (59+105+2 × 136+105+59 m) rigid continuous beam + 1-32 m simple-supported box girder, the angle of intersection between the bridge and the river is 60 degrees, the highest pier height is 93 m.

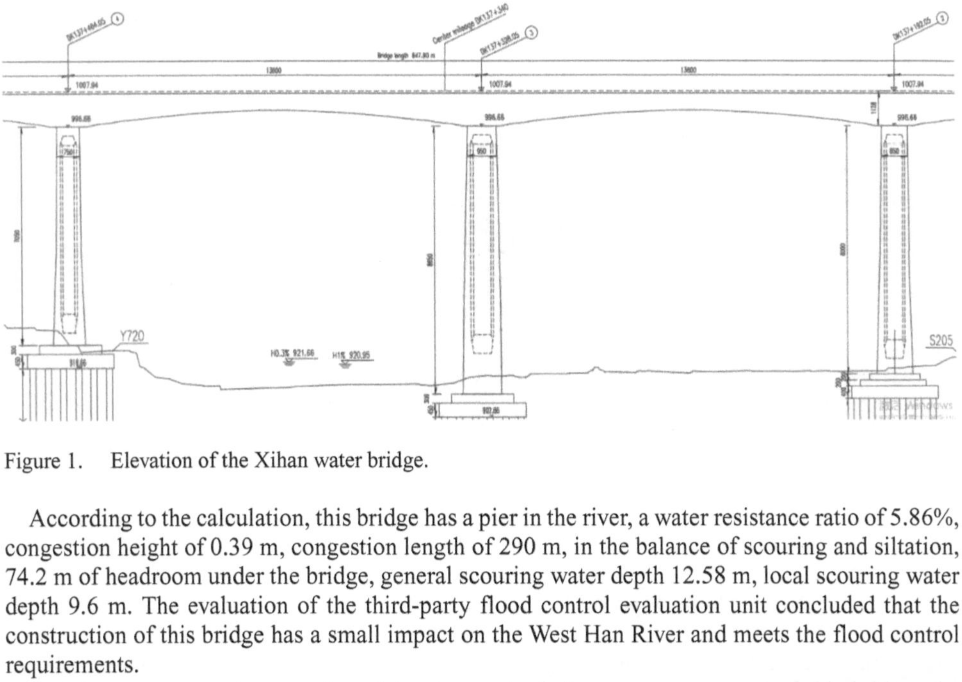


Figure 1. Elevation of the Xihan water bridge.

According to the calculation, this bridge has a pier in the river, a water resistance ratio of 5.86%, congestion height of 0.39 m, congestion length of 290 m, in the balance of scouring and siltation, 74.2 m of headroom under the bridge, general scouring water depth 12.58 m, local scouring water depth 9.6 m. The evaluation of the third-party flood control evaluation unit concluded that the construction of this bridge has a small impact on the West Han River and meets the flood control requirements.

But the program review meeting did not pass the hole span arrangement of this bridge, the participating experts requested that no piers should be set in the river. The reason is that the river pier will scour the river bank, and there are safety hazards. If the meeting requirements are not set in the river pier, then the bridge structure needs to be changed to a cable-stayed bridge or suspension bridge, the estimated investment in this bridge needs to increase by tens of millions of yuan, while the bridge is high, wind load and bridge stiffness lead to challenges, and thus the design is difficult.

The reason experts proposed to increase the bridge span is more far-fetched, the Reynolds number of the natural river is large, the vortex at the abutment is quickly transformed into thermal energy dissipation, and the impact on the river bank more than a hundred meters away is very small, adjust the bridge structure to increase the cost is too much.

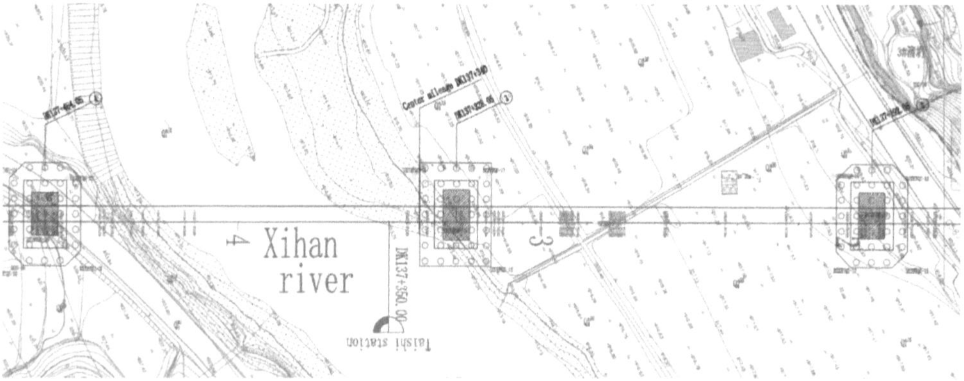


Figure 2. Plan of the Xihan water bridge.

### 4.4 *The Longgou bridge*

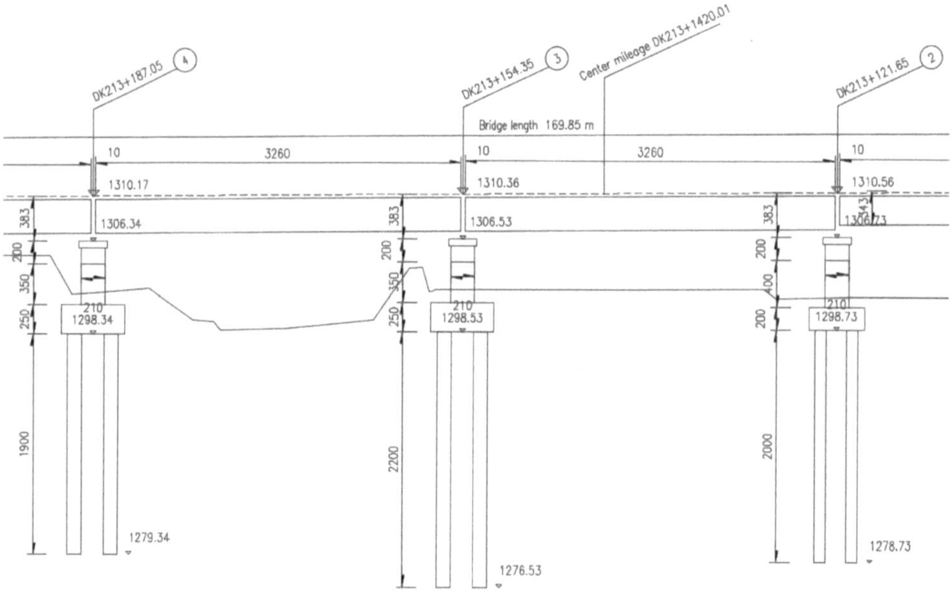


Figure 3. Elevation of the Longgou bridge.

According to the calculation of this bridge in the river without piers, the water resistance ratio is 0, there is no congestion under the bridge clearance value of 2.1 m, and a general scour water depth of 5.91 m is reported. Third-party flood assessment unit evaluation concluded that the construction of this bridge on the Long Ditch impact is small, to meet the flood control requirements.

However, the program review meeting also did not pass the hole span arrangement of this bridge, the participating experts requested that the railway bridge span should not be controlled according to the current river width, but should be controlled according to the river management scope. However, the river management scope of this ditch is 30 m wide, which is much larger than the current river width of 13 m. If the bridge span is controlled according to the river management scope, then it is necessary to increase the span of a simple beam or adjust the bridge structure to a continuous beam, which will increase the beam height and in turn compress the flood net height.

Figure 4. Plan of Longgou bridge.

The reason for increasing the bridge span is also far-fetched, the scientific nature of the river management scope is questionable, the current river has been flooding for decades, and the flood control evaluation report also points out that the bridge span meets the flood control requirements. Adjustment of the bridge span needs to synchronize the elevation of the line, the cost increases more.

### 4.5 *Yingchuan bridge*

DK22+517 with Yingchuan bridge hole span: 6–32 m simple-supported T-girder + 1–24 m simple-supported T-girder + a (48+80+48) m continuous beam + 2-32 m simple-supported T-girder + a (40+64+40) m continuous beam + 13–32 m simple-supported T-girder, the total length of the bridge 1049.94 m, continuous beam side span of 48 m, and 24 m simple-supported beam over the Yingchuan River.

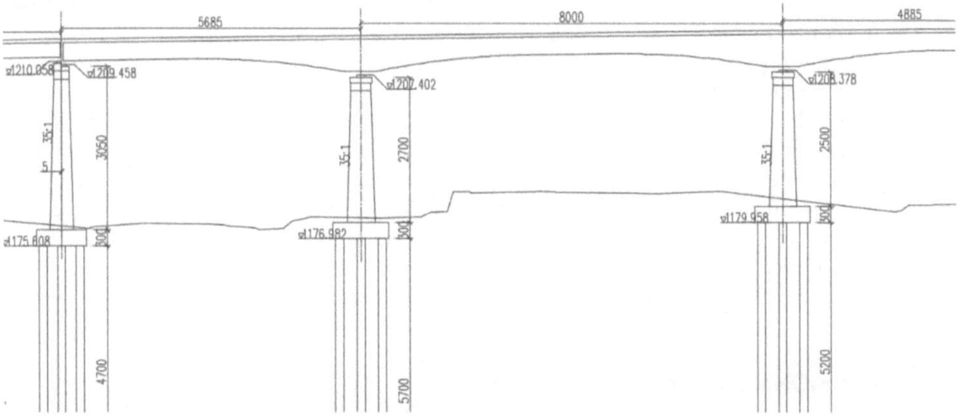


Figure 5. Elevation of Yingchuan bridge.

According to the calculation, the bridge has 1 pier in the main river channel, three piers in the river management area, 13.32% water resistance ratio, 0.61 m congestion height, 136 m congestion length, and 24.83 m clearance under the bridge, and 3.93 m general scouring water depth. The evaluation of the third-party flood control evaluation unit concluded that although the water resistance ratio of this bridge is large, the bridge hole span arrangement is calculated to meet the

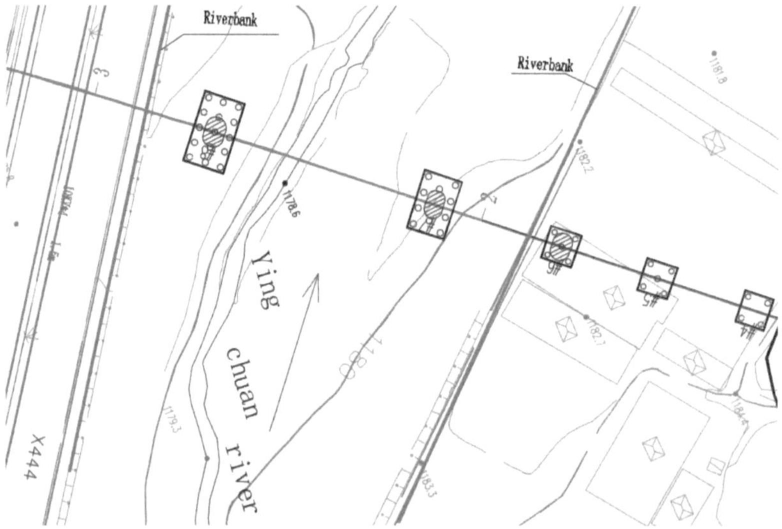


Figure 6. Elevation of Yingchuan bridge.

flooding requirements. Only the construction damage to the shore protection project according to the original standard repair.

However, the program review will not pass this bridge hole span layout, the experts insisted that the water resistance ratio should not be greater than 5%, and can be relaxed to 8%. However, if the bridge span across the Yingchuan River is increased according to the above requirements, it will also affect the span of the adjacent bridge across the X444 road, making the structural design more difficult and significantly increasing the investment.

From the above analysis of typical bridges crossing rivers and the evaluation results, we can get that at present, the water authority only does administrative approval, and the flood protection evaluation is prepared by a third-party company with hydrological qualification, and when the flood protection evaluation report is completed, the water authority will hold a flood protection evaluation expert review meeting, and the participating experts will make a decision on whether the bridge proposal is feasible. However, according to the author after participating in different projects of the flood control evaluation expert review meeting, the participating experts do not pay attention to the evaluation conclusions made by the third-party evaluation agency, but only the water resistance rate as the main criteria for determining whether the program is established, which resulted in the flood control evaluation that concluded that the bridge span layout is feasible, but cannot pass the expert review phenomenon, and ultimately a large number of bridge spans need to be adjusted, and even individual lots. This resulted in a large number of bridge spans needing to be adjusted and even individual lots being re-routed, significantly increasing the construction cost.

## 5 CONCLUSIONS AND RECOMMENDATIONS

This paper takes the Tianshui-Longnan railway as an example, analyzes the typical bridge crossing floodway, analyzes how to balance the relationship between railway bridge span arrangement and flood safety, provides reference significance for railway selection as well as bridge span design, and draws the following conclusions:

(1) There is no document on which design stage should be made to the competent water authorities to make specific provisions for program declaration, and because of the relative independence of China's railway projects, designers pay insufficient attention to the issue, railway projects

mostly in the preliminary design or even construction drawings stage for program declaration if the program does not pass the evaluation, resulting in major changes in the program, a substantial increase in investment, or even increase in project investment exceeding the control requirements, and re-declaring the feasibility study. Therefore, it is recommended that railway projects in the feasibility study stage communicate timely with the water authorities, which makes it easier for key bridge programs to obtain approval.

(2) In Gansu Province, for example, the Department of Water Resources did not make clear requirements on what size of river channel needs to be evaluated and approved for flood control, which led to different implementation of policies by the water administrative departments at the county level along a project, and even the requirement to prepare flood control evaluation reports for streams several meters wide, resulting in a large waste of resources. On this point, it can be proposed to the Department of Water Resources. The Department needs to prepare a flood assessment report on the size of the floodway (catchment area) to make quantitative requirements.

(3) There is no unified regulation on the delineation of river management scope, and the delineation of river management scope by the water authorities at the county level is arbitrary, resulting in most of the river management scope being much larger than the current river. And the bridge program for the review of the river management scope to control the bridge span results in a lot of unnecessary waste of resources. It is suggested that the higher water authorities should make scientific and reasonable unified regulations on the delineation of river management scope, balance the relationship between railway bridge span arrangement and flood safety, and make the healthy development of China's engineering construction.

(4) The experts attending the flood control evaluation meeting did not pay attention to the evaluation conclusion made by the third-party evaluation agency, but only took the water blockage rate as the main judgment criterion for the establishment of the program, resulting in the phenomenon that the flood control evaluation concluded that the bridge span arrangement was feasible, but could not pass the expert evaluation. It is suggested that the water authorities should openly solicit a pool of experts, absorb more comprehensive experts into the pool, and optimize the evaluation criteria.

## REFERENCES

China Railway First Survey and Design Institute Group Ltd. *General description of construction drawing design for Tianshui to Longnan Railway* [R]. Xian: China Railway First Survey and Design Institute Group Ltd, 2021.

China Railway First Survey and Design Institute Group Ltd. *Special section of bridge and culvert for construction drawing design of Tianshui Longnan Railway* [R]. Xian: China Railway First Survey and Design Institute Group Ltd, 2021.

Ministry of Water Resources of the People's Republic of China. *Guidelines on formulation of flood control assessment report within the management scope of river course*. SL/T808-2021 [S]. Beijing: Ministry of Water Resources of the People's Republic of China, 2021.

National People's Congress. *Flood proofing law of The People's Republic of China* [S]. Beijing: National People's Congress, 2016.

State Council of the PRC. *Regulations of the People's Republic of China on the Management of River Courses*. [S]. Beijing: State Council of the PRC, 2018.

*Advances in Urban Construction and Management Engineering – Cha (Ed)*
 *ISBN 978-1-032-39018-5*

# Research on university architectural design from the perspective of epidemic prevention

Yuwei Huang & Guanghu Jin*
*Architecture, Yanbian University, Yanji, China*

ABSTRACT: Since December 2019, cases of pneumonia caused by the new coronavirus infection have been detected in many places in China. Prevent the spread of the new coronavirus in the inner space of the university campus and avoid cross-infection between teachers and students. In view of the existing problems in the epidemic prevention of universities and summarizing the shortcomings in the control of university buildings during the epidemic period, this article points out the applicable methods in the five aspects of space facility layout, ventilation system, 0drainage system, traffic flow arrangement, and intelligent epidemic prevention system. The three levels of "cutting off the transmission route of the virus," "blocking the source of infection," and "emphasizing on susceptible people" are aimed at improving the environmental safety of colleges and universities in terms of epidemic prevention.

## 1 INTRODUCTION

The outbreak of new coronary pneumonia in 2019 is the second large-scale respiratory infectious disease in China after SARS in 2003. (Zhang et al. 2020) In the past seventeen years, most of the architectural field's response to this large-scale infectious disease problem has stayed in medical buildings. There has not been a complete exploration and thinking in some public spaces, such as colleges and universities. However, the personnel in the university are relatively mixed, and large-scale university students will gather in a high density in a limited area and have close contact with each other. As an important part of the public space, university buildings should also bear the responsibility of epidemic prevention and control risks.

However, university buildings are different from the buildings of primary and secondary schools. They have a long history and may have deficiencies in epidemic prevention and control at the beginning of their establishment. Specifically embodied in:

(1) Spaces such as the stairs and elevators of the teaching building are highly intensively flowed when going to and from classes, and traffic flow is generally congested. The lack of clear temporary storage areas for takeaways in student apartments, teaching buildings, and office areas can easily lead to potential infection risks caused by the concentration of teachers and students during the peak meal period, resulting in high population density.
(2) The frequency of classroom use is too high, and it is difficult to ventilate in time
(3) Staff such as takeaway couriers need to be separated from the traffic flow between teachers and students to a certain extent.
(4) High-density high-rise teaching buildings and apartments and high-density stacking in the vertical direction are likely to cause water pollution. (Wang 2019)

These problems involve the various systems and scales of university buildings, and a complete and complete system of epidemic prevention and control for university buildings needs to be established.

*Corresponding Author: jinguanghu@ybu.edu.cn

 DOI 10.1201/9781003348023-19

## 2 TEXT

For an infectious disease such as new coronary pneumonia, if you want to prevent it, you must first understand the three mechanisms of the infectious disease: the source of infection, the susceptible population, and the route of transmission. Prevent from the three general directions of the mechanism:

Table 1. Three major directions in the mechanism of infectious disease prevention.

| | Epidemic prevention system | | |
|---|---|---|---|
| Mechanism | Source of infection | Transmission route | Susceptible population |
| Epidemic prevention methods and measures | Closed management of building zoning | Blocking transmission routes intelligent monitoring of public space | Zoning management of susceptible population |

Based on these three directions, we now propose five suggestions and analysis explanations for the prevention of epidemics in colleges and universities:

### 2.1 *Layout of epidemic prevention space facilities in university buildings*

The new crown virus is currently spread mainly through droplets and contact. People may produce droplets when talking, sneezing, and coughing. Public transportation spaces such as stairs and elevators in colleges and universities are in a closed environment, and there are great hidden dangers.

(1) Teaching buildings generally have two or more stairs, so first of all, the upward and downward flow of the stairs should be separated. Only one staircase is allowed to go up or down to reduce the possibility of infection caused by human interaction. At the same time, the entrances and exits are controlled by management personnel, and it is forbidden to use one entrance and exit to disperse the flow of people as much as possible.

(2) The elevator has strong airtightness and lacks natural ventilation. The only opportunity for mechanical ventilation is through the gap between the upper and lower elevators and between the elevator door switches. It is difficult to ensure the timely evacuation of viruses and bacteria to the outside world. Therefore, installing an infrared temperature measurement system in the elevator room is necessary to ensure timely prevention when infection occurs in the internal space.

(3) The frequency and population base of college students ordering takeout are very large.

According to sampling statistics, the peak period of students taking and delivering food is basically the same as the peak period of students entering and leaving the apartment. Delivery staff usually deliver food delivery to students in the hall of the student apartment. The student apartment needs to set up a takeout storage area near the first floor. It should not be set in the hall on the first floor. The hall on the first floor is the main entrance for students to enter the apartment. There is a dense flow of people and should be avoided. Takeaways have close contact with students entering and exiting. They can be set up near the entrance of the first level, and a simple takeaway reception area can be set up to solve the unnecessary risk of infection caused by taking and delivering food during the epidemic.

### 2.2 *College building ventilation system settings*

From the perspective of the spread of the new coronavirus, a good ventilation system can effectively prevent the spread of the virus. Droplets and sneezes are mainly spread through aerosols. Maintaining air circulation can effectively inhibit the spread of new coronaviruses through aerosols.

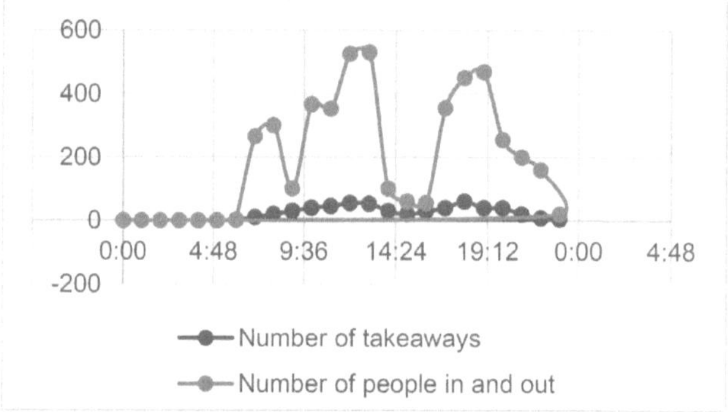


Figure 1. Statistics on the number of takeaways and the number of people in and out of a college.

(1) Due to the large area of public classrooms, it is more difficult to effectively achieve air circulation, especially considering that indoor air conditioning or heating may be turned on in winter Public buildings mainly use central air conditioning, and central air conditioning needs to be reformed: The HEPA filter is installed at the air inlet of the fan coil, and chlorine dioxide or ultraviolet lamp disinfectant is added at the air outlet to filter and kill viruses. The other is to increase fresh air supplement and exhaust measures. In addition, the vents should be set at a low place to ensure that fresh air from the outside can enter, and the exhaust vents should be set at a high place to circulate the air exhaled by people to avoid sinking and causing second-degree infection. (Long & Xue 2020) as the picture shows:

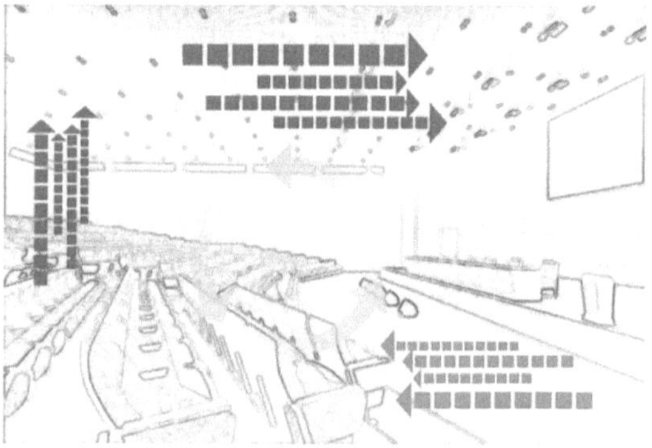

Figure 2. Analysis of gas flow in large conference room.

(2) The student canteen should use a combination of local ventilation and comprehensive ventilation. Partial ventilation ensures that the cooking fume in the kitchen is processed outside the canteen. Comprehensive ventilation ensures that when teachers and students have meals, the indoor air is changed regularly. (Ma 2020) When dining, a transparent barrier can be set up around the dining table to avoid the spread of droplets between people and ensure that the air circulates upwards.

## 2.3 *Discharge of sewage system in university buildings*

University buildings, whether they are teaching buildings or student apartments, are vertically distributed with high density. Therefore, sewage treatment and water quality safety are important points that cannot be ignored during epidemic prevention.

(1) The cleaning staff should check the smoothness of the drainage pipes in time and regularly check the drainage pipe system of the whole school.
(2) In the drainage process, it is necessary to pay attention to the timely outflow of sewage and the problem of air flowing in and out of the water flow. Therefore, the water seal should be strengthened, and the water seal of disinfection water should be adopted.
(3) The use of secondary water after sewage treatment should be reduced, and the intensity of sewage treatment should be strengthened.

### 2.4 *Traffic flow arrangement in colleges and universities*

We are now investigating the traffic flow satisfaction of a school during the epidemic. Using the 4-point Likert scale, four points are awarded for answering "very satisfied," three points for answering "satisfied," two points for answering "unclear," and one point for answering "unsatisfied." Then take the average to get the following statistical chart:

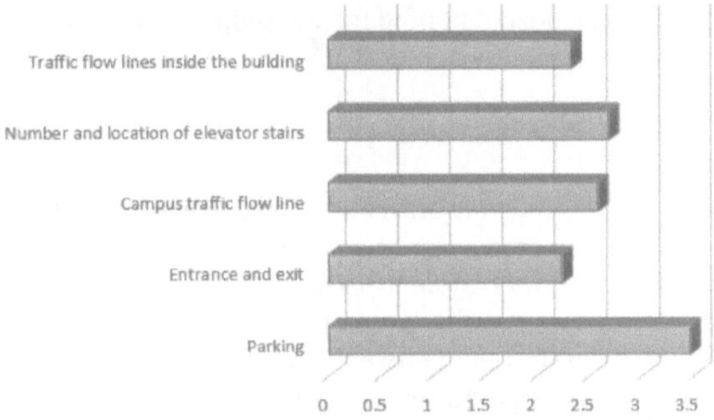


Figure 3. Survey on satisfaction of traffic flow line in a university.

According to statistics, during the epidemic, teachers and students had the lowest satisfaction with the entrance and exit design and the internal traffic flow of the building and slightly lower satisfaction with the campus traffic flow. In this regard, the following suggestions are made for the traffic flow of the three main groups of students, teachers, and staff:

(1) The staff traffic flow line mainly refers to the activity route of the express delivery staff, cleaning staff, canteen staff, etc. The delivery staff, cleaning staff, and canteen staff needs to transport a large number of items, so the flow of people on the line cannot pass. It can't intersect with the teacher-student route while avoiding takeout, food contamination, and the spread of garbage germs. Therefore, the transportation route of the staff should avoid the main entrances and exits of the campus, and enter and exit the back door of the school separately. The flow of teachers and students is small, which not only improves the work efficiency and speed, but also avoids unnecessary infection.
(2) Teachers usually drive in and out of the campus and need to be separated from the dense crowds. Teachers have a large audience. Once they are infected and carry germs along the route, they will likely cause a large-scale infection. Teachers should choose the side entrance of the campus or the teaching building. Compared with the back entrance, the side entrance has a certain degree of traffic convenience and avoids the flow of people at the main entrance.
(3) Students often enter and exit teaching buildings and student apartments, and they account for the largest proportion of the people on campus. They need to occupy the main entrances and exits to ensure the convenience of students entering and exiting the target building, making it easy for students to identify and find, and it is helpful to evacuate dense crowds.

The separation of the three routes will not only greatly improve traffic efficiency during the epidemic, but also avoid the scope of pathogen infection.

### 2.5 *Intelligent epidemic prevention system in university buildings*

The advent of the technological era and the flexible use of the Internet IT industry in all walks of life indicate that humans no longer need to use a lot of manpower and material resources to complete high-precision epidemic detection.

(1) Set up an intelligent detection and video surveillance area at the entrance area of the campus building, and perform infrared detection on the flow of people in and out in a timely manner. And it is bound to a personal mobile phone, through the mobile phone positioning system, to count the flow of people in and out of place and time.
(2) Intelligent early warning is given to teachers and students with abnormal body temperature (Zhao 2016). It can immediately arrange an isolated campus space in the system for the isolation, report to the superior to ensure "person-to-person," calculate the time of quarantined people, and eliminate the early warning of suspected diagnosis in time. (Fu 2019; Wang & Tao 2019)
(3) The intelligent epidemic prevention system should also include statistics on materials and medical supplies and deliver materials to quarantined people regularly and quantitatively every day. The system needs to be updated in time for warehouse storage to facilitate statistics.

## 3 CONCLUSION

It is conceivable that in the future, epidemic prevention and university buildings will be combined to develop together. Not only the new crown virus pneumonia but also any infectious disease can use this set of building epidemic prevention systems to ensure that when an infectious disease breaks out, it will be protected in time. University buildings are equivalent to a community. China has some explorations and discoveries in community epidemic prevention, but there is little research on the combination of university buildings and epidemic prevention. On the one hand, university life is closer to group life than community residents, and infection isolation needs further strengthening. On the other hand, the time period for people from colleges and universities to flow into and out of the same place is similar, and it is easier for frequent crowds to gather. Therefore, the combination of university buildings and epidemic prevention requires continuous exploration and joint efforts in various ways to overcome the epidemic.

## REFERENCES


Fu, G. (2019) *Safety Management*. Science Press, Beijing.

Long, H., Xue, K. (2020) *Discussion on the Architectural Design and Epidemic Prevention Plan of Large-Space Public Buildings under the Background of Healthy City—Taking Large Gymnasium Buildings as an Example*. Shanghai Urban Planning Review,2: 31-37.

Ma, J. (2020) *Guidelines for the Prevention and Control of Novel Coronavirus Pneumonia in Primary and Secondary Schools*. People's Publishing House, Beijing.

Wang, S. (2019) *Construction engineering quality and safety management*. Huazhong University of Science and Technology Press, Wuhan.

Wang, Y., Tao, B. (2019) *Theory and Application of Smart Construction Site*. China Architecture & Building Press, Beijing.

Zhang, Z., Gao, W., Lu, Y., Cheng, C., Li, Y., Mei, S. (2020) Analysis of the Infection Status of Close Contacts of Patients with New Coronary Pneumonia. *Modern Preventive Medicine*,47: 4516–4518,4522.

Zhao, X. (2016) *Engineering Technical Guide for Building Equipment Monitoring System*. China Architecture & Building Press, Beijing.

*Advances in Urban Construction and Management Engineering – Cha (Ed)*


# Study on the local prototype of landscape

Hongyuan Xiao*
*Guilin University of Technology, Guilin, China*

ABSTRACT: The mountain-water landscape as the primary type of Chinese classical garden, promoted the emergence and prosperity of Chinese landscape culture. Unfortunately, the development of the mountain-water landscape is gradually declining at present. The locality is the basis of protecting the essential characteristics of the landscape and sustainable development, and the traditional landscape is the specific embodiment and carrier of the locality. Combining the local Chinese landscape and from the perspective of prototype theory, this paper analyzes the local prototype, representation, and generation mode of landscape to explore its local construction methods. Finally, by analyzing the local development process of landscape, three methods of landscape local prototype construction are proposed: local prototype presentation method, local prototype reproduction method, and local prototype deepening method.

## 1 INTRODUCTION

The mountain-water landscape is the landscape that Chinese people have explored for thousands of years. It has accumulated and settled into a local landscape different from the western landscape and is a kind of landscape image based on the natural landscape, landscape reconstruction, and artificial reproduction according to the different degrees of understanding of landscape architects. Thus, it is inevitably integrated into the ideology and emotional experience of landscape architects. It is a pity that "landscape" seems to be less local at present, which is difficult to be recognized by the public and is gradually declining in the development of modern landscape or even at the risk of disappearing (Jin 2018).

Currently, the mountain-water landscape mostly focuses on the physical landscape image, ignoring the research on thinking and psychological consciousness and the local inheritance and development of the mountain-water landscape under ideology. Although the two are established and developed on different theoretical bases, they are essentially the same. They reflect people's long-term landscape consciousness, mainly manifested in vision, sensation, and perception (Chen 2020; He 2020). The collective subconscious is regarded as the "model" in landscape construction, which can not only serve as the original consciousness of landscape, but also produce different subsequent landscapes and become the thinking paradigm of local inheritance and development. Inspired by Jung's archetype theory, this paper studies the locality of the mountain-water landscape by quoting part of the archetype theory and analyzes the local archetype of the landscape from the form, representation, and generation mode of landscape locality. Then, according to the landscape local prototype paradigm established in the framework, the construction method of the local landscape prototype is elaborated so that the local promotion of landscape becomes a trace to follow.

*Corresponding Author: 619586839@qq.com

DOI 10.1201/9781003348023-20

## 2 THE LOCALISM OF MOUNTAIN-WATER LANDSCAPE

The mountain-water landscape is part of the Chinese classical garden. After long development and evolution, it seems to have become an indispensable part of the construction of Chinese gardens in past dynasties, as a soul bond throughout the ancient times, but the local landscape in its long history has also changed.

The mountain-water landscape has gone through a long development and evolution. Three concepts directly affect the construction of the Chinese landscape. In addition to "the unity of nature and man" the realm of immortals and "landscape and virtue" also directly affects the local landscape garden party. Confucius endowed landscape with benevolence and wisdom, which is typical of Chinese landscape culture, and laid the foundation for our directional thinking of mountains and water. In the Analects of Confucius, it is said, "when one is cold, one knows pines and cypresses and then tunes them," which expresses the aesthetic concept of "virtue by landscape" in Confucianism. "Virtue by landscape" manifests the beauty of life in the original consciousness of human beings and lays a local foundation different from other gardens in other countries (Sun 2022; Song 2022).

### 2.1 *The localism of the northern mountain-water landscape*

In ancient times, there were local differences in landscape. In the northwest plateau, people revere the natural mountains and rivers resulting from ignorance. They believed that the tall mountains were the body of the god, and the water body was the representative of the god, all of which were based on mythological stories. People construct buildings according to the appearance of tall mountains and regard artificial lakes as the habitat of gods, which is called "Lingzhao" in the myth of Kunlun in the northwest plateau. With the spread of the myth, local people on the east coast created Penglai mythology, as shown in Figure 1, which is different from the Kunlun myth. The difference lies in the importance of water, which also reflects that people in the northwest plateau and the east coast have different ideologies and sense of place, giving rise to different myths (Han 2015).

Penglai mythology flourished when the first emperor of Qin unified China, and the pursuit of immortality deepened. At this time, Penglai mythology became the mainstream of Chinese garden culture. The First Emperor of Qin directly moved Penglai fairyland into his palace. He built a rockery in the palace, symbolizing Penglai mountain, which is the restoration of the prototype of Penglai fairyland on the east coast (Meng 2017; Xia 2017). Later, Martial Emperor followed Emperor Qin's will and built three islands in the Taiye Pond in the Jianzhang Palace, representing Penglai, Yingzhou. Since then, the Penglai mythology "one pool and three mountains" has become an enduring gardening model of the Chinese classical garden economy.

### 2.2 *The localism of southern the mountain-water landscape*

At the beginning of the mature stage of Chinese classical gardens, the gardens of the Song Dynasty played a connecting role. In the two Song Gardens, northern Song Gardens are mostly concentrated in the north. As the political center of the Southern Song Dynasty moved southward, gardens at that time were mostly concentrated in Jiangsu and Zhejiang regions. The southern and northern Song Dynasty presents different local characteristics.

With the prosperity of commerce in the Northern Song Dynasty, significant changes occurred in the social structure of cities and the life of the citizens. Many popular activities in the royal family also began to flourish among the general public. In the Southern Song Dynasty, the phenomenon of urbanization of these literati became more apparent.

Combined with the influence of southern Song Dynasty culture, the ceremonial sense of northern imperial gardens gradually declined, and the aesthetic structure gradually refined, reflected in the spatial layout, internal architecture, stacking of mountains, water management, and other aspects (Zhang 2016).

All of these are based on changes in local feelings and ideologies in the Northern and Southern Song Dynasties. Under the political background of the Song Dynasty, the perception and yearning for the natural landscape were transformed into the desire for artistic aesthetics, which had universal behavioral experience and emotional appeal. The literati of the Song Dynasty reflected their ideology through landscape painting, which could reflect not only the painter's pursuit and desire for the natural landscape, but also the ideal blueprint of gardens in their heart. Landscape painting peaked in the Northern Song Dynasty, and the paintings tended to reflect the panoramic style. The broad vision and flat layout all reflected Pingyuan's aesthetic ideology. Although the aesthetic taste of landscape formed in the late Northern Song Dynasty, it did not reach the artistic conception of "heaven and earth in pot." In the Southern Song Dynasty, the landscape portrayal of the "big scene" in the Northern Song Dynasty was changed from magnificent and sad to exquisite and graceful, making the landscape painting a concentrated universe of heaven and earth. In addition, it changed from the panoramic layout of "viewing the small from the big" to "viewing the big from the small," which increasingly showed a clever and subtle implication. This implication also served as a link through the inheritance and development of gardens in the Southern Song Dynasty, showing the localism of gardens in the south of the Yangtze River, which was different from the northern gardens represented by the Northern Song Dynasty (Jiang 2013; Lu 2013; Shen 2013).

### 2.3 *The localism of Lingnan mountain-water landscape*

For more than 2000 years, Lingnan culture originated from the Central Plains and is one of the veins of Yan-Huang culture. However, due to its unique geographical location and social and historical background, it has adopted and integrated the cultures of Wu Chuyue, Han Miao Yao, and other overseas cultures, showing its unique local characteristics.

Lingnan is the earliest place where Chinese culture met with the world culture, and its garden culture is the first place to accept the influence of foreign culture. Chinese history and western culture often collided during the Ming and Qing Dynasties. Two different ideologies can be transformed into cultural collisions, reflected in painting, gardens, architecture, and other aspects. It includes the collision of western geometric thinking and Chinese landscape painting artistic conception, forming a new style of painting of "Western learning spreading to the East" (Li 2016). The important influence of the new painting style lies in the folk, which is a form of ascending gradually. The literati class has a traditional strict standard for painting, while the folk is more willing to accept the exquisite quality of the Western painting. All these make Lingnan gardens show tolerance to Western culture in painting, architectural form, and material technology, which is different from the royal majesty in the north and reflects the market ecology. This can also be reflected in the ideology of different classes, the characteristics of the southern port cities with heavy trade functions, the "ring grave" and coral stone Islamic tombs in the Tang Dynasty, the hundreds of western-style garden architecture in the Ming and Qing Dynasties, and the first introduction of "modern parks" in the Republic of China (Chen 2019).

All reveal the different local characteristics of Lingnan landscape garden, Jiangnan landscape garden, and Northern landscape garden.

## 3 THE PROTOTYPE OF THE MOUNTAIN-WATER LANDSCAPE

Archetype is a term in analytical psychology. Its original meaning refers to the characters, matrix, institutions, and styles repeated in works. Carl Jung, a Swiss psychologist, defined an archetype as a deep-seated transcendental idea and psychological schema accumulated over a long period (Zhao 2014). He thinks that there is a mental structure under the conscious mind, and this mental structure is universal. This view is applied to Aldo. In Rossi's architectural typology, he believes that the city is the collective memory of living in the city, and these memories, in turn, influence the shaping of the city. Collective memory is extended from the "collective unconsciousness" of Jung's archetypal theory. Everyone has collective memory beyond personal consciousness, which has essential similarities (Liu 2012).

### 3.1 *Prototype activation localism*

"Prototype" has always existed in Chinese gardens. The prototype of "one pool and three mountains" originated from the Penglai myth and can be seen in some classic landscape gardens, such as summer resorts until the Ming and Qing Dynasties. It can also be seen in some contemporary landscape compositions. Whether abstract or intuitive, archetypes are always present in the subconscious, a consciousness known in Jung's theory as the "collective unconscious." When a local scene appears, the archetype is activated, awakening the collective unconscious. As an innate ideology, a prototype can be represented through images (Jin 2012; Xiang 2012). However, due to the differences in each person's experience, the image of the sense of place triggered will also be different. It has both elements of prototype and acquired experience. By combining their own aesthetic and experiential experience with landscape ideal, planners form the unity of scene, that is, the establishment of the ideal dwelling place. They place themselves between mountains and rivers. However, due to the difference in personal consciousness and the change of social consciousness of the times, they have different materializations of local images generated. In the interaction with the landscape, experiencers induce the images of a sense of places, such as a sense of local identity, sense of local dependence, and sense of deep-rooted place, from emotion and psychology, and then the feelings become local (Huan 2015; Wang 2015).

### 3.2 *Local prototype analysis of the mountain-water landscape*

The key to the archetype of locality is the influence of people's original consciousness (collective memory) on the construction of emotions in different places People gradually form emotional feelings in the interaction with the ground. In this process, people's original consciousness has influenced the subjective perception and emotional feelings of the place. The sense of place is constantly strengthened and modified, which generates a personal consciousness and constitutes the local prototype.

The meaning of local prototype of the mountain-water landscape has been developing in the direction of natural scenery for thousands of years. The unique flavor of "natural landscape" is not only because of its excellent natural resources, but also because of the admiration for the fairyland of Penglai Three Mountains, the exploration of heaven and the blessed land, and the expectation for the holy land of the humanistic holy land. During this time, the meaning of its prototype had already stretched by time. Under the "landscape" skeleton history precipitated in the heart, rooted in the foot by forming the "collective unconscious," which gradually formed the regional culture and life of philosophizing, spirit, and the humanistic accomplishment of emotions. Fairy mountains, caves, and the land of idyllic beauty are what to pursue, which bears the thinking of accommodating heaven and earth and broad-mindedness of the Chinese nation for thousands of years (Jin 2014; Li 2014).

### 3.3 *Local prototype representation of the mountain-water landscape*

An archetypal image is the representation of the archetypal image in consciousness, and it is a typical form of continuous psychological experience. As the experience deposit of the innate ancestor, the prototype can only be perceived by experiencing some scene stimulation. It belongs to the content of consciousness, so it cannot be materialized and visualized. However, western psychology puts forward that image can express not only inner experience with the image of external things, but also has symbolic significance and is a symbolic language. Therefore, the way of archetypal representation is the original image, which has both archetypal elements and acquired experience elements. The consciousness of it makes individuals have feelings connected with the experience of ancestors (Silvano 1976).

The sense of place reflects the locality of the mountain-water landscape. It is a psychological product of place generated by individuals and groups relying on experience, memory and instinct. It is a conceptual thing generated in the brain of participants, a part of the landscape creation conception of designers, and exists in the experience process of participants. When the landscape stimulates the generation of the sense of place, it is remembered and passed down and used as the creation and expression of people to guide the place. Its essence is a mental image, which can be shaped by structure, form and other elements of material space. For example, Japan has a local attachment to mountains, rivers, lakes and seas. It uses fine sand and gravel in the landscape to represent "lakes and seas" and well-formed stones to represent "mountains and rivers" This local prototype of the landscape is represented as a Japanese rock garden model (Zhang 2014; Zou 2014).

### 3.4 *Local generation of the mountain-water landscape*

The process of local generation can generalize the sense of place generated by the interaction between adults and the landscape. The sense of place is constantly strengthened and accumulated to form a locality, which is manifested in two ways. On the one hand, a human's deep original consciousness influences the interaction between human and landscape, producing a sense of place, which condenses into landscape practice experience and psychological representation. On the other hand, in the interaction process between humans and the landscape, the human's subconscious will influence behavior and cognition, resulting in different experiences with the landscape, which will influence the local landscape and precipitate into conventional formal rules, as shown in Figure 1.

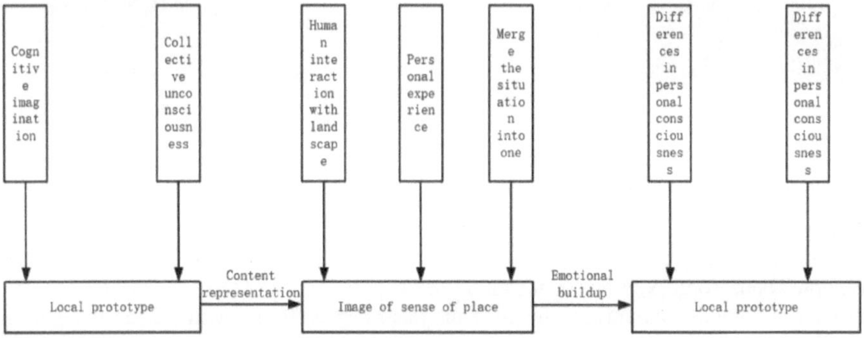


Figure 1. Landscape local prototype activation process.

It traces the influence of deep consciousness on the locality. For example, the idea originates from myth and religious belief, forming people's pursuit of "the ideal realm of immortality" and creating the layout mode of "Three mountains in a pool." The landscape construction pattern of one pond and three mountains is a metaphor for feudal rulers' ideal realm of "immortality" in traditional Chinese gardens, reflecting people's strong pursuit and ideal. Such local factors are influenced by deep consciousness (Zhao 2014).

## 4 LOCAL PROTOTYPE CONSTRUCTION METHOD OF MOUNTAIN-WATER LANDSCAPE

### 4.1 *Local prototype presentation method*

The local representation method is the most direct method, which replicates and processes the image prototype induced by locality. Qinghui Garden, as one of the four major gardens in Lingnan, is different from Jiangnan Garden and the northern garden in its locality, which is influenced by the ideology of the western garden and its terrain. Lingnan region is mainly hilly and mountainous, which is the formation of coastal culture. Under the influence of the political background at that time, the garden has the regular rectangular water courtyard of a European classical garden, the geometric system with the characteristics of a western garden, and the combination of a boat courtyard and garden pavilion with the characteristics of southern water town, as shown in Figure 2. It includes original geometric roads, geometric flower base, extracted Roman style letter and ionic column, as shown in Figure 3, and then inlaid with traditional Chinese painting, integrating the local Chinese and western (Meng 2010; Wang 2010). First, from the prototype of western gardens, geometry is still the classical local representation of western gardens, awakening people's memory of traditional western gardens. Second, starting from the surrounding environment and the prototype landscape construction mode of the garden, relevant geometric elements are extracted and combined with Chinese elements to produce the unique local image representation of Lingnan garden, which is different from the northern garden and Jiangnan Garden. Finally, the distinctive landscape image is produced.

Figure 2. Real scene of Qinghui Garden.

Figure 3. Real scene of Qinghui Garden.

### 4.2 *Local prototype reproduction method*

The local representation method is not the same as the representation method. It is a selective transformation of local archetypes based on memory to create local images. The key to this method lies in people's behavioral experiences and emotional experiences, and selecting the local archetypes left in memory is more conducive to forming the local representation with selection significance based on the original image of the place. In the Song Dynasty, there was the "banquet and shooting" activity in the garden. As the top style of life in the Northern Song Dynasty, it developed into a pure art form, which contains the influence of ancient people's behavioral experiences and emotional experiences. After Taizu, "Sheyan" gradually evolved into a language symbol of "seeking oneself," which is also related to the importance and popularity of "Sheyan" This reflects the changes in garden life during the Song Dynasty and the overall development of the line of "poetic and pictorial." (He 2017). After the Southern Song Dynasty, the garden formed the transformation of "Lin 'an" as the center of the Jiangnan Garden. The development of Chinese landscape gardens turns to more of the characteristics of Jiangnan Gardens, which are "delicate, elegant and elegant, profound and profound, and full of illusion," and gradually away from the plain, magnificent, powerful realm to show the "Jiangnan Garden" based on the local representation.

The key to this method is accumulating people's emotional experiences and ideas, and the second "selection" of the landscape prototype remains in memory. This choice is based on reconstruction, and the deep archetype reappears in consciousness, giving a clearer image. "Archery hall" in the Northern Song Dynasty was an indispensable part of the royal garden. Later, due to the political collapse of the Southern Song Dynasty, the royal family was wandering, the new dynasty established multiple burdens and economic pressure, and the royal family did not want to see the homeland scene again. However, with the memory of garden makers, they still retained the northern Song Dynasty garden construction. They made a "secondary selection" of its prototype to maintain this symbol, making it a unique local landscape garden of the Song Dynasty (Yu 2021).

### 4.3 *Local prototype deepening method*

When the scene is unified, the realistic environment stimulates the generation of images with a sense of place. This method of obtaining local images through primary, secondary, and creative development is called local deepening. This method is from abstract to concreteness and then from concreteness to abstraction.

The karst landform features in the eight scenic spots of Guilin provide the material basis for the rapid accumulation of Guilin culture. The beautiful peaks have become the first choice for constructing the Guilin landscape, as shown in Figure 4.

Figure 4. Guilin landscape.

Although their morphological characteristics are different, their same nature and constituent elements trigger the image of place sense. Their repeated appearance makes the experiencers deeply remember them and constitute the iconic image of the Eight Scenic spots of Guilin. Combining mountain scenery, water landscape, weather, and cultural landscapes with appropriate scale creates Guilin's unique eight scenic spots (Xiao 2018).

Guilin mountain-water landscape shows the artistic conception of the beauty of the Chinese landscape with its unique natural environment advantage, which is mostly divided into the external and internal landscape. The latter landscape is more concentrated in Guilin's unique karst landform mountain and cave landscape. The outward-oriented landscape is dominated by natural mountains, water, lakes, and other large landscape. Landscape elements are more scattered. Mountains are different combined landscapes, such as point groups and surface (rivers), combined with meteorological modification, including deep and flat. From the city, you can see all the mountains and rivers in all directions (Liu 2021; Wu 2021, 2021; Zheng 2021). However, Guilin, as the land of ancient times, becomes a place for poets to express their emotions. The mountains, water, urban layouts, and surrounding buildings of Guilin become the important carrier. "The water likes a QingLuo belt, and the mountain likes a jade hairpin" depicted that the Li River flows through the city like a woman's Qingluo belt, and the mountain likes a jade hairpin of beauty. It also highlights the difference between the real mountain and water landscape in Guilin and the royal garden, and the Jiangnan Garden.

Guilin mountain-water landscape is based on the ideology of the landscape architect. Guilin landscape has carried out different practices; that is, the "coding" and "decoding" way, which is different from other gardens, is called the prototype transformation method. It is mainly the transformation of images. People create different images through different landscape practice experiences. The landscape is built based on the same landscape prototype, but with the acquired differences, the same prototype will produce different images to represent.

## 5 CONCLUSION

The locality is the core concept of human geography. It has been introduced into landscape research in recent years and is an important attribute to maintaining landscape inheritance. In the local research of traditional landscape, the core is to study solid landscape images, ignoring the local inheritance and development of landscape under ideology. With the prototype theory as a new perspective, the whole landscape of China is regarded as the research object to interpret the local landscape. It is a new attempt to study landscape:

(1) Sort out the Chinese landscape and interpret the local development process of the landscape.
(2) According to the archetypal theory of psychology, with the clue of "collective unconsciousness" (collective memory) and "cognitive imagination," we scientifically analyze the local archetypal representation and generation mode of the landscape.
(3) According to the paradigm of the local archetype of landscape, this paper puts forward the construction method of psychological level. By analyzing the local development process of landscape, it puts forward the methods of local archetype presentation, local archetype reproduction, and local archetype deepening.

## REFERENCES


Chen, W. X. (2019). *Spatial landscape logic in Lingnan Gardens from the perspective of landscape painting* (ph. D. dissertation, South China University of Technology).

He Yuhan & Chen Zhanchuan. (2020). On the embodiment of landscape culture in The construction of Chinese landscape architecture. *Guangdong Landscape Architecture* (04), 80–84.

Han, Y. (2015). *The influence of Penglai Xianhua Landscape Mode on Chinese classical Gardens* (Master's thesis, Hebei University of Technology).

Huanzhendan & Wang Yanping. (2015). Similarities and differences between the sense of place and locality and their mutual transformation. *Tourism Research* (02), 64–68.

He, Xiaojing. (2017). *Image and Presentation: a study on the origin and Flow of Jiangnan Gardens in the Southern Song Dynasty* (Doctoral dissertation, China Academy of Art).

Jin, B. (2018). The formation of Chinese traditional garden culture and values. Landscape Architecture (01), 79–82.

Jiang Junhao, Shen Shanshan & Lu Shan.(2013). The Characteristics of garden art in the Southern Song Dynasty from the changes in gardens in the two Song Dynasties. *Chinese Gardens* (04), 104–108.

Jin Yunfeng & Xiang Shuping. (2012). Prototype Activation history: Historical space design in landscape architecture. *Chinese Gardens* (02), 53–57.

Li, Y. X (2016). *Study on the interaction between Jiangnan landscape and Lingnan Landscape* (Master's thesis, Zhejiang A&F University).

Liu, Q. Y. (2012). *Research and application of architectural prototype theory* (Master's thesis, Chongqing University).

Li Tao & Jin Yunfeng. (2014). Landscape Design Strategy: urban space construction based on the prototype of "landscape." Chinese Society of Landscape Architecture.(EDS.) 2014 Annual Conference proceedings of Chinese Society of Landscape Architecture (Volume 1) (pp. 46–50). *China Architecture and Architecture Press*.

Liu Zonglin, Zheng Wenjun, Wu Manni & Wu Liulan.(2021). Landscape language analysis of eight Scenic spots in Guilin. *Journal of Guilin University of Technology* (01), 230–237.

Meng Pei & Xia Han. (2017). The garden pattern of "One Pool and three Mountains" in Chinese classical Gardens. *Research in Fine Arts Education* (22), 62.

Sun Songlin & Song Shuang. (2022). Contemporary public garden construction practice based on Chinese landscape culture. *Landscape Architecture* (05), 59–66.

Silvano A.(1976). Creativity: the Magic Synthesis. New York Basic Books inc.

Wang Chenghui & Meng Haoliang.(2010). Qinghui Garden in Shunde, China. *Guangdong Landscape Architecture* (05), 23–26.

Xiao Zhiyuan. (2018). *Aesthetic connotation and ink expression of Guilin's "Eight Scenes"* (Master's Thesis, Guangxi Normal University).

Yu, Qing-miao. (2021). *A comparative study of Garden aesthetics in the Song Dynasty* (Master's thesis, Heilongjiang University).

Zhang P. (2016). *Research on the Discourse of gardens in the Song Dynasty* (Doctoral dissertation, Beijing Forestry University).

Zhao Jianwei. (2014). *Interpretation of artistic symbolism from the perspective of Jung's archetypal Theory* (Master's thesis, Shanghai Normal University).

Zhang Lei & Zou Guangtian. (2014). Landscape image creation based on prototype theory. *Chinese Landscape Architecture* (05), 40–43.

# Evaluation of the synergistic development of urbanization and ecological environment in Chengdu-Chongqing twin cities economic circle

Junxiang Sun*, Bangguo Xu*, Yongqi Zhu*, Xinru He* & Zhenxiang Liu*
*College of Architecture and Urban and Rural Planning, Sichuan Agricultural University, China*

ABSTRACT: As of 2019, although the urbanization rate of the Chengdu-Chongqing dual-city economic circle was as high as 53.8 %, it was far from the quality of urbanization development in the Yangtze River Delta, Beijing-Tianjin-Hebei region, and other places, and the investment in ecological environment governance also lagged behind the national average. Accordingly, based on the perspective of coordinated regional development, the paper takes 16 cities in the Chengdu-Chongqing Economic Circle as research objects. With the data from the statistical yearbooks of each city in 2011, 2016, and 2020, we used Excel software to study the development level of urbanization, ecological environment, and the coupling coordination degree. Conclusion: The urbanization level of the Chengdu-Chongqing economic circle shows a fluctuating trend of decreasing and then increasing during the period of study, with severe differentiation between the two grades; the ecological environment development has consistency and growth on the whole; the coupling degree of each city presents a remarkably positive trend of increasing.

## 1 INTRODUCTION

On 3 January 2020, the Central Finance and Economics Commission proposed a strategic plan for the Chengdu-Chongqing twin-city economic circle, emphasizing that high-quality development and ecology should have two hands (Xi 2020). On 20 October 2021, the State Council of the Central Committee of the Communist Party of China (CPC) issued *the Outline of the Construction Plan for the Chengdu-Chongqing Twin-City Economic Circle* (the State Council 2021).

Nowadays, three main aspects of domestic and foreign research are as follows. The first research on the measures of the level of urbanization and eco-environment synergy from the perspective of the green development index (Lin 2019), and urban-rural integration development (Ji 2020), focuses on coastal urban clusters. The second research on the influencing factors of urbanization and ecological environment synergy focuses on the polycentric pattern (Huang 2020) and the degree of urbanization (Zhou 2020). The third research is on the coupled and coordinated development of urbanization and the ecological environment. Coupled dynamic simulation model (Cui 2019), "human-land" coupling model (Ren 2019), far- and near-range coupling model (Ren 2020), and so on, while the interaction analysis of the study object is lacking.

In conclusion, urbanization and ecological environment coordinated development research is primarily concentrated in the eastern coastal region, while few papers focus on the Chengdu-Chongqing region. Facing the double pressure of economic development and ecological protection

*Corresponding Authors: 2290201196@qq.com, 549431358@qq.com, 1986685806@qq.com, 2057435245@qq.com and 1653248860@qq.com

DOI 10.1201/9781003348023-21

in the Chengdu-Chongqing economic circle, improving the coordinated development of urbanization and the ecological environment has become an urgent choice to build the twin-city economic area in the Chengdu-Chongqing region.

## 2 RESEARCH AREA AND EVALUATION SYSTEM

### 2.1 *Definition of the research subject*

According to *the Outline of the Construction Plan of Chengdu-Chongqing Regional Twin-City Economic Circle*, the specific scope of the Chengdu-Chongqing Twin-City Economic Circle consists of 27 districts (counties), including Yuzhong District and Jiangbei District in Chongqing, and 15 cities, including Chengdu City and Mianyang City in Sichuan Province. The paper takes Chongqing as the research unit, of which there are 16 units.

### 2.2 *Data sources*

The whole data presented here are from *the Sichuan Statistical Yearbook* of 2020, 2016, and 2011, and *the Chongqing Statistical Yearbook* of 2020, 2016, and 2011.

### 2.3 *Indicator evaluation system*

Urbanization indicators consider four dimensions: population urbanization, economic urbanization, social urbanization, and spatial urbanization. Eco-environmental system indicators consider four dimensions: eco-scale, eco-quality, eco-pressure, and eco-governance.

Regarding the comprehensive representativeness and accessibility of the indexes, the paper establishes the evaluation system of urbanization and eco-environmental indicators of the Chengdu-Chongqing twin-city economic area, as shown in Table 1, by referring to the studies of Fang (2017) and Gao (2016).

## 3 ANALYSIS OF THE DEVELOPMENT LEVEL AND INFLUENCING FACTORS OF URBANIZATION AND ECOLOGICAL ENVIRONMENT

### 3.1 *Research method: Entropy method*

The paper applies the entropy value method and refers to Chen (2010) and Lu (2009) to quantitatively analyze the combined urbanization and ecological environment scorings, and the main steps are as follows.

(a) Normalization of the original data matrix, namely, construction of matrix and data standardization of original index data matrix $A = (X_{ij})_{m\times n}$;

Positive indicators:

$$y_{ij} = \frac{x_{ij-\min x_{ij}}}{\max x_{ij-} \min x_{ij}}; \text{ Negative indicators: } y_{ij} = \frac{\max x_{ij-}x_{ij}}{\max x_{ij-} \min x_{ij}}; \quad (1)$$

(b) Calculate the proportion and entropy of each index

The proportion of index j of the i city:

$$p_{ij} = \frac{y_{ij}}{\sum_{i=1}^{m} y_{ij}} \quad (2)$$

Index j entropy:

$$e_j = -\text{k}\sum_{i=1}^{m} p_{ij} \ln p_{ij}, \text{ where K} = \frac{1}{\ln m}, k > 0 0 < e_j < 1 \quad (3)$$

Table 1. Evaluation system of new urbanization and ecosystem development level indicators.

| Target level | Guideline level | Indicator layer | Indicator attributes |
|---|---|---|---|
| **New Urbanization** | Population Urbanization | $X_1$ Population density (persons/km$^2$) | Positive |
| | | $X_2$ Urbanization rate of the resident population by city (state) (%) | Positive |
| | Economic Urbanization | $X_3$ Per capita gross regional product of each city (state) (yuan) | Positive |
| | | $X_4$ Percentage of secondary and tertiary industries in GDP (%) | Positive |
| | Social Urbanization | $X_5$ General public budget expenditure by city (state) (RMB) | Positive |
| | | $X_6$ Total number of insured persons in each city (state) (million) | Positive |
| | Spatial Urbanization | $X_7$ Area of built-up area (square kilometers) | Positive |
| | | $X_8$ Urban road area per capita (square meters) | Positive |
| **Ecological Environment** | Eco-scale | $Y_1$ Built-up area covered by greenery (hectares) | Positive |
| | | $Y_2$ per capita water resources (tons) | Positive |
| | Eco-quality | $Y_3$ urban greening coverage (%) | Positive |
| | | $Y_4$ Annual average relative humidity | Positive |
| | Eco-pressure | $Y_5$ Per capita wastewater emission (ton) | Negative |
| | | $Y_6$ Sulfur dioxide emissions per capita (tons) | Negative |
| | Eco-governance | $Y_7$ Fixed waste disposal (% or tons) | Positive |
| | | $Y_8$ Wastewater treatment volume (tons) | Positive |

Redundancy of entropy:

$$h_j = 1 - e_j;\ \text{Weighting factor: } w_j = \frac{h_j}{\sum_{j=1}^{n} h_j} \tag{4}$$

(c) Calculate the overall score:

$$S_i = \sum_{j=1}^{n} w_j * y_{ij} \tag{5}$$

The processing results of urbanization and ecological environment development level calculation are available in Figure 1 and Figure 2 below.

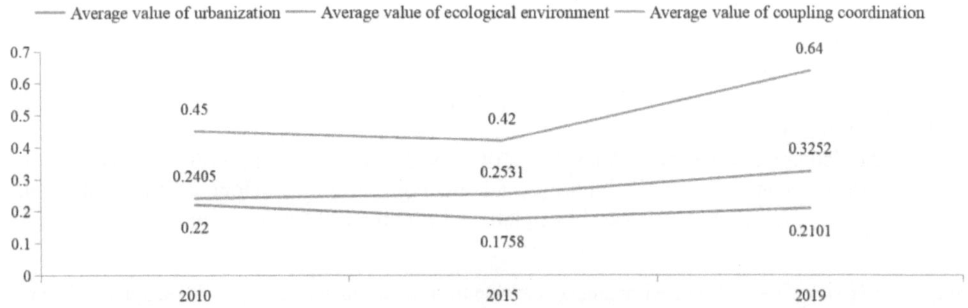


Figure 1. Average values of each score of Chengdu-Chongqing twin-city economic area in 2010, 2015 and 2019. Data source: entropy method and coupled coordination model calculations, as compiled.

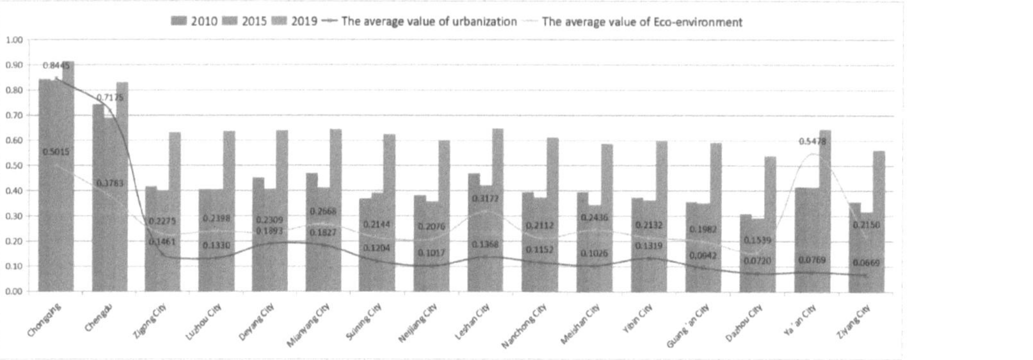


Figure 2. Comprehensive analysis of the coupling and coordination degree of urbanization and eco-environment in the Chengdu-Chongqing economic circle. Data source: calculated by the coupling coordination degree model and collated.

### 3.2 *Level of urbanization development*

The average value of urbanization level is available in Figures 1 and 2. Figure 1 reveals that the overall meaning of urbanization in the twin-city economic circle of the Chengdu-Chongqing region in 2010, 2015, and 2019 fluctuates by decreasing and then increasing. The mean value decreases from 0.2200 to 0.1758 from 2010 to 2015. As of 2019, it is up to 0.2101.

Table 2 classifies the mean urbanization scores into four echelons according to high and low. From Figure 2, Chongqing and Chengdu are in the first echelon with the mean value of 0.8445 and 0.7175, respectively, and are firmly in the first and second positions. However, there is a big difference between Chongqing and Chengdu and Deyang, Mianyang, Zigong, Leshan, Luzhou, and Yibin. Suining, Nanchong, Meishan, and Neijiang are in the third tier, with a lower urbanization level. Guang'an, Dazhou, Ya'an, and Ziyang are in the fourth echelon with lower ratings and the lowest overall urbanization level.

Table 2. Classification of urbanization development level in the Chengdu-Chongqing twin-city economic area.

| >0.5000 | >0.1300 | >0.1000 | <0.1000 |
|---|---|---|---|
| **The first echelon** | The second echelon | The third echelon | The fourth echelon |

### 3.3 *Level of ecological development*

The mean value of the ecological level calculation is in Figure 1, which indicates an overall trend of good growth. It increases from 0.2405 to 0.2531 from 2010 to 2015. However, the ecological level is low and growing slowly. As of 2019, it raises substantially to 0.3252.

Here, the ecological environment level is still ranked into 4-levels according to the size of the mean value as in Table 3. In Figure 2, Chongqing and Ya'an belong to the first level, while Chongqing decreases significantly annually. Chengdu and Leshan are in the second rank,

Table 3. Classification of eco-environmental development levels in the Chengdu-Chongqing twin-city economic area.

| >0.5000 | >0.3000 | >0.2000 | >0.1000 |
|---|---|---|---|
| **Level 1** | Level 2 | Level 3 | Level 4 |

while Chengdu's urbanization and eco-environmental standards remain in good condition. Level 3 includes Mianyang, Deyang, and other ten cities. Then, Guang'an and Dazhou belong to the fourth rank, with a slow growth rate.

## 4 EVALUATION OF COUPLING COORDINATION DEGREE BETWEEN URBANIZATION AND ECOLOGICAL ENVIRONMENT

### 4.1 *Research method*

To construct the coupling coordination model concerning the method of Liao (1999) and Fang (2016):

$$C = \sqrt{\frac{U_1 * U_2}{\left(\frac{U_1+U_2}{2}\right)^2}} \quad T = \alpha U_1 + \beta U_2; \tag{6}$$

$$D = \sqrt{C * T}; \tag{7}$$

U1 and U2 are the combined scores of urbanization and the ecological environment separately. C is the coupling degree value, and $C \in [0, 1]$. When $C \rightarrow 1$, the coupling degree is maximum, which means the two systems promote each other at a high coupling level, otherwise becoming lower. T is the comprehensive evaluation index of urbanization and the ecological environment. $\alpha$ and $\beta$ are coefficients to be determined, taking $\alpha = \beta = 0.5$. D is the coupling coordination degree value, and $D \in [0, 1]$, which is categorized by declining, excessive, and coordinated development according to the degree of coupling coordination in Table 4.

Table 4. Classification of coupling coordination degree between urbanization and ecological environment.

| 0~0.19 | 0.20~0.39 | 0.40~0.49 | 0.50~0.59 | 0.60~0.79 | 0.80~1.00 |
|---|---|---|---|---|---|
| Severe disorders | Moderate disorders | Minor disorders | Reluctant coordination | good coordination | Quality coordination |
| Involution form | | Transition form | | Coordinated development type | |

### 4.2 *Coupling coordination degree between urbanization and ecological environment*

The results of the coupling coordination degree calculation are in Figure 2. The trend is fluctuating upward overall. From 2010 to 2015, the coupling coordination degree declined from 0.45 to 0.42, and more cities belong to the slightly dysfunctional type. As of 2019, Chongqing and Chengdu are significantly higher than other cities, rising to 0.64. Chongqing belongs to the rising quality coordinated category, while Chengdu rises from good coordinated to the quality coordinated county.

Except for Ya'an and Suining, coupling coordination declined in 2010 and 2015. Six cities, including Zigong, Luzhou, and Deyang, are slightly dysfunctional. The rest eight other cities, such as Suining and Neijiang, are moderately dysfunctional and of declining type. In 2019, the coupling coordination degree improved significantly, excepting Meishan, Guang'an, Dazhou, and Ziyang, the coupling coordination degree of the remaining ten cities increased substantially, all of which are greater than 0.60 into well-coordinated type.

## 5 CONCLUSION

Firstly, the urbanization development level of the Chengdu-Chongqing economic circle tends to fluctuate by decreasing and then increasing. With heavy differentiation between the two categories, the growth has been unstable and slow in the last decade. From Figure 4.1, the average urbanization value of Chongqing and Chengdu is much higher than 0.7, while that of the other fourteen cities is not more than 0.2, and even the average value of four cities is lower than 0.1, so the polarization is severe.

Second, the ecological environment development level of the Chengdu-Chongqing economic circle is generally consistent and steadily growing. The ecological environment has improved significantly as of 2019. Except for individual cities such as Chongqing, all show a promising trend of growth year by year. Only Chongqing and Ya'an are slightly greater than 0.5 in this research. The other cities are stable around the 0.1 to 0.3 range, and the ecological environment is in the middle to upper but not superior, possessing great room for development.

Finally, as of 2019, the coupling coordination degree of the Chengdu-Chongqing economic circle presents a promising trend of a significant increase. With high coupling coordination, Chongqing and Chengdu are positioned in the first and second place to promote the development of the entire economic circle. In 2010 and 2015, each of the cities was moderately or slightly dysfunctional except for the coordination of Chongqing and Chengdu. The coupling coordination of all cities improved significantly in 2019. The previously dysfunctional cities all turn into coordinated, most of which turn into good coordinating ones. In addition, the Chengdu-Chongqing economic circle has considerable gaps compared with developed city clusters and has high future development potential.

## REFERENCES

Chen Mingxing, Lu Dadao, Cha Liangsong. A comprehensive evaluation of the impact of urbanization on resources and environment in China (in English) [J]. Journal of Geographical Sciences, 2010, 20 (01):17–30.

Chongqing Statistical Yearbook 2011 Editorial Committee Chongqing Statistical Yearbook 2011 Editorial Committee. Zheng Zibin, Li Taoming, eds, Chongqing Statistical Yearbook, China Statistics Press, 2011, 7, Yearbook.

*Chongqing Statistical Yearbook 2016* Editorial Committee *Chongqing Statistical Yearbook 2016* Editorial Committee. Zhang Fumin, editor-in-chief of Tong Zesheng, Chongqing Statistical Yearbook, China Statistical Press, 2016, 6, Yearbook.

*Chongqing Statistical Yearbook-2020* Editorial Committee *Chongqing Statistical Yearbook-2020* Editorial Committee.Yang Hongyi, Li Taoming, editors-in-chief, Chongqing Statistical Yearbook, China Statistics Press, 2020, 5, Yearbook. DOI:10.38772/y.cnki.ycqtj.2020.000001.

Cui Xuegang, Fang Chuanglin, Liu Haimeng, Liu Xiaofei, Li Yonghong. Research progress on coupling dynamic simulation theory and method of urbanization and ecological environment [J]. Geography, 2019, 74 (06): 1079–1096.

Fang Chuanglin, Liu Haimeng, Li Guangdong. International progress and overall evaluation of research on the interactive coupling effect of urbanization and ecological environment (in English)[J]. Journal of Geographical Sciences, 2016, 26(08):1081–1116.

Fang Chuanglin, Zhou Chenghu, Gu Chaolin, Chen Liting, Li Shuangcheng. Theoretical analysis of the coupling mechanism and coercive effect of urbanization and ecological environment in mega-city cluster areas (in English) [J].Journal of Geographical Sciences, 2017, 27 (12):1431–1449.

Gao Xincai, Yang Fang. Measuring the coordination degree of urbanization and ecological environment coupling in Northwest China[J]. Urban Issues, 2016(12):26–33. DOI:10.13239/j.bjsshkxy.cswt.161204.

Huang Dongmei, Liu Xiaoyu, Zheng Qingchang, Liu Jun. Study on the effect of multi-center pattern on the coupling and coordinated development of urbanization and the ecological environment - - A case study of two metropolitan areas in Fujian Province[J]. Ecology Journal, 2020, 40 (21): 7886–7896.

Liao Chongbin. Quantitative Evaluation and Classification System of Coordinated Development of Environment and Economy - - A Case Study of Urban Agglomeration in Pearl River Delta[J]. Tropical Geography, 1999 (02): 76–82.

Lin Weibin, Su Jian, Zhang Qihui. Measurement of Green Development Level – A Construction of Green Development Index[J]. Learning and Exploration, 2019 (11): 106-113 + 2.

Lu Tianchao, Kang Kai. Application of entropy method and analytic hierarchy process in weight determination[J]. Computer programming skills and maintenance, 2009 (22): 19–20 + 53.

Ren Yawen, Cao Weidong, Zhang Yu,Su Hefang, and Wang Xuewei. Spatial-temporal coupling characteristics of urbanization and ecological environment in the three major urban agglomerations of the Yangtze River Economic Belt [J]. Resource and environment in the Yangtze River Basin, 2019, 28 (11): 2586–2600.
*Sichuan Statistical Yearbook-2012* Editorial Department *Sichuan Statistical Yearbook-2012,* Editorial Department.Xiong Jianzhong, editor-in-chief, Sichuan Statistical Yearbook, China Statistics Press, 2011, 8, Yearbook.
*Sichuan Statistical Yearbook-2016* Editorial Committee *Sichuan Statistical Yearbook-2016* Editorial Committee. Xiong Jianzhong, Sichuan Statistical Yearbook, China Statistical Press, 2016, 4-5, Yearbook. DOI: 10.41462 / y.cnki.ysctn.2017.000001.
*Sichuan Statistical Yearbook-2020* Editorial Committee and Editorial and Publishing Staff *Sichuan Statistical Yearbook-2020* Editorial Committee and Editorial and Publishing Staff. Xiong Jianzhong, Zeng Junlin General Editor, Sichuan Statistical Yearbook, China Statistics Press, 2020, 4-5, Yearbook. DOI:10.41462/y.cnki.ysctn.2020.000001.
The State Council of the Central Committee of the Communist Party of China issued *Outline of the Plan for the Construction of Dual-City Economic Circles in Chengdu and Chongqing* _ Chinese Government Network ( www.gov.cn ) 2021-10-2107: 19 Source: Xinhua
Xi Jinping presided over the sixth meeting of the Central Economic Commission _ China Government Network (www.gov.cn) 2020-01-0320: 26 Source: Xinhua
Zhou Zhengzhu, Wang Junlong. Study on the coupling coordination relationship between urbanization and ecological environment in the Yangtze River Economic Belt [J]. *Urban problems*, 2020 (04): 21 – 32. DOI: 10.13239 / j. bjsshkxy. cswt. 200403.

*Advances in Urban Construction and Management Engineering – Cha (Ed)*
*© 2023 Copyright the Author(s), ISBN 978-1-032-39018-5*

# Spatial distribution of villages and change of landscape in Guizhou Province, China

Hong Ye, Yuanxin Chen, Yangling Zhao & Luo Guo*
*College of Life and Environmental Sciences, Beijing, China*

ABSTRACT: Guizhou is a multi-ethnic province with the largest number of ethnic minority villages in China. Ethnic minority villages in Guizhou have the characteristics of traditional settlements, such as ancient history and cultural accumulation, as well as their own local culture and ethnic flavour, which have a significant impact on the adjacent land use structure and landscape spatial pattern. Based on land use data from 1990 to 2020, this paper explores the spatial distribution of ethnic villages, the dynamics of land use, and landscape spatial patterns in Guizhou Province. The results show that the distribution of ethnic villages is more concentrated in the southeastern region, with 41% of the villages located in the range of 650–900 m, 35% in the range of 0–4 degrees of slope, and 46% facing south. There are obvious similarities and differences in land use patterns among the ethnic villages in the buffer zone. For example, ethnic villages have a greater impact on the conversion of arable land, forestland, and grassland in the 2 km buffer zone and even higher impact on building land and water bodies in the 1 km buffer zone. The changing trends of the landscape pattern index of each ethnic village are different. The degree of fragmentation in the 0–1 km buffer zone of ethnic villages is low, while the degree of fragmentation of patches within 1–2 km is high, and the degree of fragmentation of patches within the 2–3 km buffer zone is the highest. Studying the spatial distribution and landscape pattern changes of ethnic villages in Guizhou Province provides a theoretical basis for local ecological planning and policy management and is of great significance in promoting the coordinated and sustainable development of the natural, economic, and social aspects of ethnic villages.

## 1 INTRODUCTION

Guizhou Province is a multi-ethnic and multi-cultural province, and its ethnic minority areas occupy 55% of the province's area. Minority villages are the main places for the production and life of ethnic minorities. They have become the cradle of national cultural heritage because of their long history, relatively complete original ecological and cultural system, rich material and intangible heritage, and distinctive local characteristics (Yang 2011). In recent years, the proportion of the tourism economy of ethnic minority villages in the economic development of Guizhou Province has gradually increased, indicating that ethnic minority villages are critical to national development and Guizhou Province (Li et al. 2021, He et al. 2018). Therefore, exploring the sustainable development of minority villages in Guizhou Province is urgent.

Land use/land cover change (LUCC) is an important content of global climate change and global environmental change research (Mooney et al. 2013; Sterling et al. 2013). As the most direct signal that characterizes the impact of human activities on the earth's ecosystem, it is the link between human social and economic activities and the natural ecology (Mooney et al. 2013). In the past few decades, with the social and economic development, the ecosystem functions, and

*Corresponding Author: guoluo@muc.edu.cn

 DOI 10.1201/9781003348023-22

structures of ethnic minority villages in Guizhou Province have undergone considerable changes (Xiao et al. 2020). Past studies have mainly focused on the spatial distribution and influencing factors of specific regions or characteristic ethnic villages (Ma & Tong 2022; Xiao et al. 2020; Yihong et al. 2022; Zheng et al. 2022). However, with provinces as the research unit, few literature studies analyze the overall spatial distribution characteristics of ethnic minority villages from the perspective of land use and landscape pattern and clarify the influencing factors of changes. Hence, this study takes the ethnic villages in Guizhou as the research area and focuses on analyzing the geographic spatial attributes of villages in Guizhou Province and the changes in surrounding land use and landscape patterns from 1990 to 2020 to provide suggestions for sustainable development of ethnic areas.

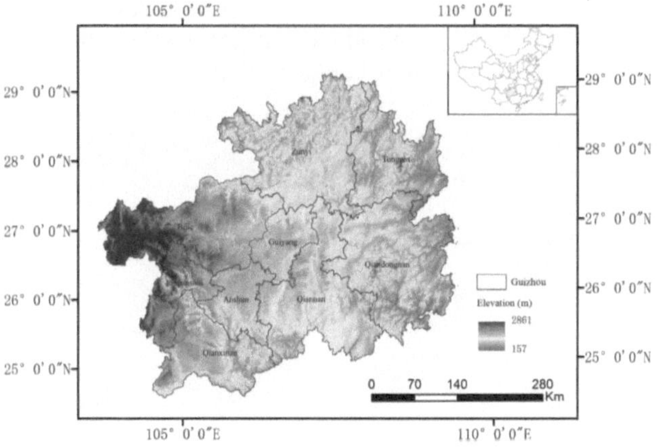


Figure 1. Research location map and altitude distribution.

## 2 STUDY AREA AND METHOD

### 2.1 *Study area*

Guizhou Province (24°37′-29°13′N, 103°36′-109°35′E) is in the southwest of China and belongs to the highland mountainous region (Figure 1). The terrain of Guizhou is high in the west and low in the east. According to the seventh national census of Guizhou Province, the population of ethnic minorities accounts for 36.44%. Guizhou contains 56 ethnic minorities, 18 of which are hereditary. There are three ethnic autonomous prefectures and 11 ethnic autonomous counties, and the autonomous areas of ethnic minorities cover an area of 97,800 sq. km, accounting for 55.5% of the province's land area.

### 2.2 *Data and methodology*

The land use data used for the study area is based on land use/land cover for 1990, 1995, 2000, 2005, 2010, 2015, and 2020 from the Chinese Academy of Sciences Resource Environmental Data Center (RESDC). Digital Elevation Model (DEM) data were sourced from the National Geographic Data Cloud. Ethnic village points are from ethnic villages with ethnic characteristics released by the Guizhou Provincial Ethnic and Religious Commission and the State Ethnic Affairs Commission in October 2020.

In ArcGIS, 0–1 km, 1–2 km, and 2–3 km buffer zones were established based on ethnic village points, and land-use changes within the buffer zones were analyzed from 1990 to 2020. Based on FRSTATS 4.2, this paper selects the patch density (PD), largest patch index (LPI), landscape shape index (LSI), contagion index (CONTAG), the proportion of like adjacency (PLADJ), interspersion juxtaposition index (IJI), landscape division index (DIVISION), Shannon's diversity Index (SHDI), Shannon's evenness index (SHEI), and aggregation index (AI) to analyze the landscape pattern characteristics of ethnic villages.

## 3 RESULTS AND ANALYSIS

### 3.1 *Distribution characteristics of villages in Guizhou Province*

Ethnic villages in Guizhou Province are in the altitude range of 174–1656 m, with an average altitude of 783 m, and only seven ethnic villages are located at altitudes greater than 1400 m, indicating that very few ethnic villages are located at high altitudes. In general, the altitude of ethnic villages shows a normal distribution pattern, with about 41% of the villages located in the altitude range of 650–900 m, and the number of ethnic villages decreases as the altitude increases and decreases.

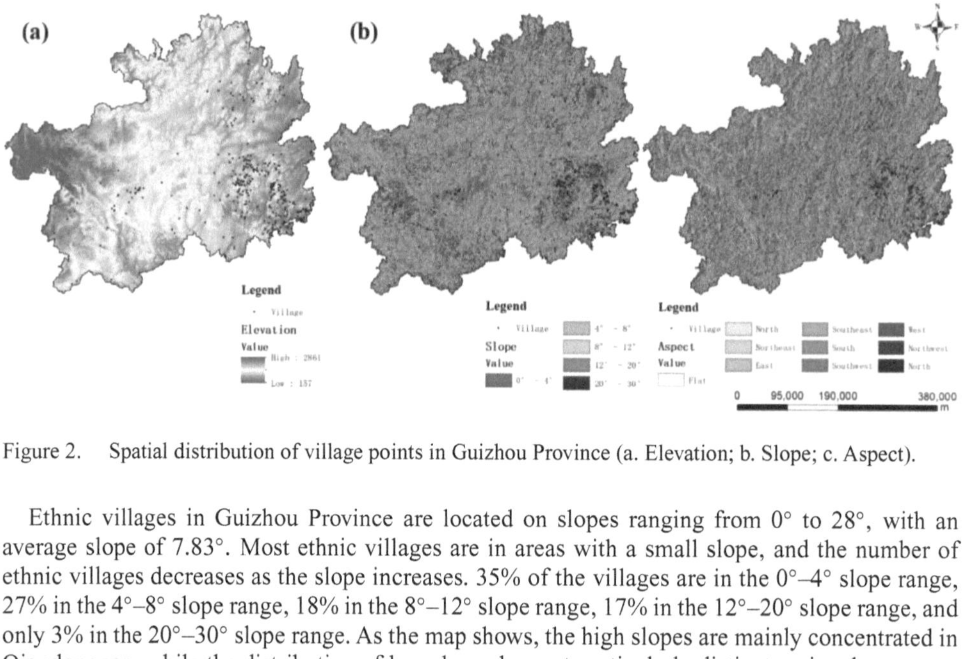


Figure 2. Spatial distribution of village points in Guizhou Province (a. Elevation; b. Slope; c. Aspect).

Ethnic villages in Guizhou Province are located on slopes ranging from 0° to 28°, with an average slope of 7.83°. Most ethnic villages are in areas with a small slope, and the number of ethnic villages decreases as the slope increases. 35% of the villages are in the 0°–4° slope range, 27% in the 4°–8° slope range, 18% in the 8°–12° slope range, 17% in the 12°–20° slope range, and only 3% in the 20°–30° slope range. As the map shows, the high slopes are mainly concentrated in Qiandongnan, while the distribution of low slopes has not particularly distinct regional or county characteristics.

The southeastern and southern slopes are the main aspects of the villages, accounting for 17% and 15%, respectively. On the other hand, the northern slope accounts for only 7%. In particular, the distribution of aspects in Qiandongnan is closely related to the climate and four-season wind of Qiandongnan. In general, the distribution of ethnic villages in the aspects is relatively scattered and uniform, with no obvious pattern to follow.

### 3.2 *Analysis of land use change in villages*

The results of land use transfer in Guizhou Province from 1990 to 2020 are shown in Figure 3. The ethnic villages in Guizhou Province are mainly characterized by the mutual transformation of cultivated land, forestland, and grassland. Over the past 30 years, the growth rates of cultivated land in the 1 km, 2 km, and 3 km buffer zone around the ethnic villages were 1.33%, 7.06%, and 6.40%, respectively; the growth rates of forestland were –0.76%, –3.81%, and –1.87%, respectively; the growth rates of grassland were –0.17%, 9.38%, and –5.68%, respectively. Among them, the growth rate of cultivated land and grassland showed a trend of increasing first and then decreasing with the increase of buffer zone, while the forestland was the opposite, showing a trend of first decreasing and then increasing. However, they all reach extreme values around 2 km. The changes in cultivated land and grassland are more obvious. In terms of geographical distribution pattern, the change of cultivated land in ethnic villages in southeastern and southern Guizhou is mainly manifested in the mutual transformation of cultivated land and forestland, accompanied by a small amount of mutual transformation between cultivated land and grassland. Although there is a large amount of forest-grass conversion in southwestern Guizhou, there are few ethnic groups in this area; that is, ethnic villages have little impact on forest-grass conversion in land use.

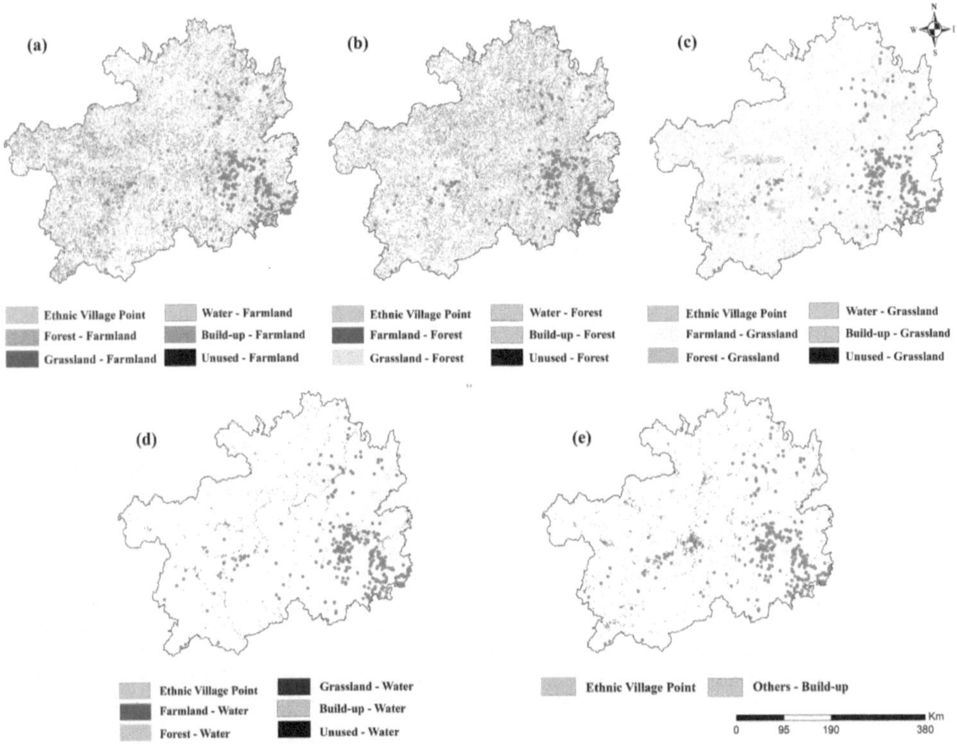


Figure 3. Land use transfer map of village sites in Guizhou Province (a. Cultivated land conversion; b. Forestland conversion; c. Grassland conversion; d. Water body conversion; e. Build-up land conversion).

The ethnic villages show the impact on the conversion of water bodies and building land, as shown in Figures (d) and (e). In the past 30 years, the increasing rates of the water body area of ethnic villages within the buffer zone of 1 km and 2 km have been 1915% and 926%, respectively. At the same time, the growth of build-up land in ethnic villages within the buffer zone of 1 km and 2 km is 1087% and 573%, respectively, reflecting the rapid development of the economy and urban construction in Guizhou Province. From the perspective of geographical distribution, there are many water body transformations at the junction of southeastern and southern Guizhou and around ethnic villages in northeastern Guizhou, mainly manifested in the transformation of forestland and cultivated land into water bodies. There is also a small amount of other land-use types converted into water bodies in ethnic villages in central Guizhou. In addition, the transfer-in and transfer-out of construction land significantly occurred in central Guizhou, but the transfer-out area was much smaller than the transfer-in area. Overall, there is no transfer of build-up land area in almost all ethnic villages.

### 3.3 *Analysis of landscape changes in villages*

The analysis of the 0–1 km buffer zone of ethnic villages in Guizhou Province shows that PD increases slightly from 1990 to 2020, and LPI remains unchanged, indicating that patchiness, fragmentation remains, and the area of large patches of the dominant land use type unchanged and remains unchanged. LSI first increases and then shows a decreasing trend. CONTAG is less volatile and decreasing, indicating that the landscape within its boundaries is fragmented. PLADJ is stable and fluctuating, indicating that the degree of aggregation of patches is stable. IJI is increasing and fluctuating from 2015 to 2020. DIVISION remains unchanged, and AI increases, indicating a high degree of aggregation among patches within 1 km and no fragmentation trend. SHDI remained unchanged from 1990 to 2015 but increased from 2015 to 2020. The increasing trend of the SHEI indicates a more decentralized land use pattern within the village.

Table 1. Landscape pattern index of different buffer zones.

| Buffer | Time | PD | LPI | LSI | CONTAG | PLADJ | IJI | DIVISION | SHDI | SHEI | AI |
|---|---|---|---|---|---|---|---|---|---|---|---|
| 0–1 km | 1990 | 0.0022 | 99.6046 | 2.3433 | 98.5193 | 99.2373 | 53.1051 | 0.0079 | 0.0298 | 0.0166 | 99.4474 |
| | 1995 | 0.0022 | 99.6046 | 2.3423 | 98.5191 | 99.2377 | 53.0104 | 0.0079 | 0.0298 | 0.0166 | 99.4480 |
| | 2000 | 0.0022 | 99.6046 | 2.3444 | 98.5185 | 99.2373 | 53.1400 | 0.0079 | 0.0298 | 0.0166 | 99.4480 |
| | 2005 | 0.0022 | 99.6046 | 2.3444 | 98.5184 | 99.2368 | 53.2800 | 0.0079 | 0.0298 | 0.0167 | 99.4470 |
| | 2010 | 0.0022 | 99.6046 | 2.3449 | 98.5178 | 99.2366 | 53.4487 | 0.0079 | 0.0298 | 0.0167 | 99.4468 |
| | 2015 | 0.0023 | 99.6046 | 2.3459 | 98.5120 | 99.2362 | 54.7273 | 0.0079 | 0.0300 | 0.0167 | 99.4465 |
| | 2020 | 0.0023 | 99.6046 | 2.3309 | 98.5028 | 99.2424 | 56.4926 | 0.0079 | 0.0301 | 0.0168 | 99.4431 |
| 1–2 km | 1990 | 0.0042 | 98.8622 | 4.6179 | 96.7281 | 98.3031 | 45.0431 | 0.0226 | 0.0709 | 0.0364 | 98.5146 |
| | 1995 | 0.0042 | 98.8622 | 4.6174 | 96.7286 | 98.3033 | 45.0142 | 0.0226 | 0.0709 | 0.0364 | 98.5147 |
| | 2000 | 0.0043 | 98.8622 | 4.6215 | 96.7228 | 98.3016 | 45.4433 | 0.0226 | 0.0710 | 0.0365 | 98.5131 |
| | 2005 | 0.0042 | 98.8622 | 4.6164 | 96.7282 | 98.3038 | 44.9818 | 0.0226 | 0.0709 | 0.0364 | 98.5153 |
| | 2010 | 0.0042 | 98.8622 | 4.6133 | 96.7299 | 98.3050 | 44.8566 | 0.0226 | 0.0709 | 0.0364 | 98.5165 |
| | 2015 | 0.0043 | 98.8622 | 4.6169 | 96.7173 | 98.3035 | 45.9214 | 0.0226 | 0.0711 | 0.0365 | 98.5151 |
| | 2020 | 0.0044 | 98.8622 | 4.6231 | 96.4232 | 98.3010 | 52.9085 | 0.0226 | 0.0713 | 0.0398 | 98.5124 |
| 2–3 km | 1990 | 0.0061 | 98.5037 | 5.9790 | 95.7233 | 97.7515 | 46.2642 | 0.0297 | 0.0913 | 0.0469 | 97.9622 |
| | 1995 | 0.0061 | 98.5037 | 5.9805 | 95.7237 | 97.7509 | 46.2456 | 0.0297 | 0.0913 | 0.0469 | 97.9616 |
| | 2000 | 0.0061 | 98.5037 | 5.9836 | 95.7170 | 97.7496 | 46.6191 | 0.0297 | 0.0915 | 0.0470 | 97.9603 |
| | 2005 | 0.0060 | 98.5037 | 5.9692 | 95.7269 | 97.9555 | 46.0622 | 0.0297 | 0.0913 | 0.0469 | 97.9662 |
| | 2010 | 0.0060 | 98.5037 | 5.9636 | 95.7300 | 97.7578 | 45.8999 | 0.0297 | 0.0912 | 0.0469 | 97.9687 |
| | 2015 | 0.0060 | 98.5037 | 5.9692 | 95.7167 | 97.7555 | 46.7481 | 0.0297 | 0.0915 | 0.0470 | 97.9663 |
| | 2020 | 0.0059 | 98.5033 | 5.9677 | 95.7020 | 97.7562 | 47.5625 | 0.0297 | 0.0918 | 0.0472 | 97.9671 |

Analysis within the 1–2 km buffer zone of ethnic villages shows that LSI fluctuates steadily from 1990 to 2020; CONTAG shows an upward and then downward trend; DIVISION remains unchanged, and the AI decreases; SHDI fluctuates steadily and shows an increasing trend. The rest of the indicators are consistent with the 0–1 km buffer zone changes. The analysis of the 2–3 km buffer zone of ethnic villages shows that PD remains unchanged from 1990 to 2015 and decreases slightly from 2015 to 2020, indicating that patchiness and fragmentation remain unchanged. LPI and CONTAG show a decreasing trend, and LSI fluctuates and tends to decrease; DIVISION, AI, SHDI, and SHEI change in line with the 1–2 km buffer zone.

## 4 CONCLUSION

This paper presents a spatial analysis of the elevation, slope, and aspect of ethnic villages in Guizhou Province and establishes three different buffer zones around the villages to analyze the changes in land use and landscape patterns in the study area. The results show significant differences in the distribution of different ethnic villages in terms of elevation, slope, and aspect. Most of the ethnic villages are located at medium altitude and lower slope areas, the reason being that the lower slope is conducive to the construction of houses and reclamation of arable land, and there is less risk of mudslides and other disasters, which is beneficial to the villages' safety and future development and can also reflect the obvious ethnic characteristics. The land use pattern around ethnic villages at different scales was studied, and it was found that the land conversion of the villages was related to their distribution. The change of arable land in ethnic villages in Qiandongnan and Qiannan mainly showed the mutual conversion of arable land and forestland; the conversion of ethnic villages in Qianzhong from other land use types to building land was very significant, accompanied by a small amount of conversion from other land use types to water bodies, which was related to the policies of different regions. It was found from the landscape pattern of villages that the patch aggregation within the 0–1 km buffer zone of ethnic villages is developing toward fragmentation, with complex patch shapes and increased landscape diversity, while patch fragmentation is higher in the 1–2 km buffer zone and highest in the 2–3 km buffer zone.

Land use changes in ethnic villages are important in promoting the comprehensive development of ethnic villages. Economic development is inseparable from land, and while land provides resources for economic development, it is also affected. Once land use patterns change, it will lead to ecological problems, especially in these ethnic areas, where land use is simple and resilient, and ecological problems caused by economic development are particularly prominent. Therefore, research on land use dynamics and driving forces in ethnic areas can help understand land change patterns, make early warnings of risks and rational planning before ecological problems caused by land change arrive, and reduce ecological damage and economic losses.

## REFERENCES


Heqing, Z. & Yiping, L. (2018). *Tourism and China's ethnic rurality: A tale of two villages, Anthropological notebooks*, 24(1):23.

Li, X., Xie, C., Morrison, A.M. & Nguyen, T.H.H. (2021). *Experiences, Motivations, Perceptions, and Attitudes Regarding Ethnic Minority Village Tourism, Sustainability*, 13(4): 2364.

Ma, H. &Tong, Y. (2022). *Spatial differentiation of traditional villages using ArcGIS and GeoDa: A case study of Southwest China, Ecological informatics*, 68: 101416.

Mooney, H.A., Duraiappah, A. & Larigauderie, A. (2013). Evolution of natural and social science interactions in global change research programs, *Proceedings of the National Academy of Sciences - PNAS*, 110(1):3665–3672.

Sterling, S.M., Ducharne, A. & Polcher, J. (2013). The impact of global land-cover change on the terrestrial water cycle, *Nature climate change*, 3(4):385–390.

Xiao, Y., Zhao, J., Sun, S., Guo, L., Axmacher, J. & Sang, W. (2020). Sustainability Dynamics of Traditional Villages: A Case Study in Qiannan Prefecture, Guizhou, China, *Sustainability*, 12(1):314.

Yang, L. (2011). Ethnic tourism and cultural representation, *Annals of tourism research*, 38(2):561–585.

Yihong, Z., Tingdong, M., Bin, B. & Qing, Y. (2022). Spatial Distribution Characteristics and Influencing Factors of Traditional Villages in Guizhou Province, *Journal of landscape research*, 14(1): 37–46.

Zheng, G., Jiang, D., Luan, Y. & Yao, Y. (2022). GIS-based spatial differentiation of ethnic minority villages in Guizhou Province, China, *Journal of mountain science*, 19(4):987–1000.

*Advances in Urban Construction and Management Engineering – Cha (Ed)*


# Study on characteristics of coil and water loss under bamboo forest based on different configurations

Wei Wang*
*Yuhang District Forestry and Water Conservancy Bureau, Hangzhou, China*

Zhaowei Shen*
*Zhejiang Key Laboratory of Water Conservancy Disaster Prevention and Reduction, Zhejiang Institute of Hydraulics & Estuary, Hangzhou, China*

ABSTRACT: Bamboo is an important cash crop in southern China. In this paper, the soil and water loss and nutrient loss of pure bamboo forest, bamboo forest, and mixed forest of bamboo-paulownia-hemerocallis Hemerocallis with different rainfall-runoff were analyzed. The results showed that: (1) the runoff coefficient of the pure bamboo forest was the largest, and that of the mixed forest of Paulownia ARUNDINACEA and Hemerocallis Hemerocallis was the smallest. (2) the changing trend of runoff coefficient and nutrient loss is basically the same, and both show an obvious fluctuation trend.

## 1 INTRODUCTION

Rainfall is one of the important factors which cause the loss of nutrients, and the characteristics of rainfall intensity and rainfall intensity have a great influence on the surface runoff and nutrient loss. In addition, the intensity of rainfall and the degree of loss of nutrients and occurrence frequency, and so on are closely related (Li & Huang 2009; Li 2009; Wei et al. 2007; Wei 2007; Yan 2014; Yan et al. 2014). The research on the relationship between rainfall and soil nutrient loss is one of the important contents of soil and water conservation scientific research work, which can provide a scientific basis for the development of soil and water conservation projects. Therefore, the selection of pure bamboo forest, bamboo forest, and bamboo Tong Yang Yang Tong - Hemerocallis mixed forest as the research object, analysis of rainfall intensity, runoff, and nutrient loss as well as the relationship between rainfall and the latter, to explore the three types of forest soil erosion characteristics for forest type configuration and soil erosion provide on the basis of science.

## 2 RESEARCH METHOD

The study area sets three vegetation configuration modes: pure bamboo forest, bamboo forest, and bamboo Tong Yang Yang Tong - Hemerocallis mixed forest. Each vegetation configuration mode was repeated two times, that is, a total of 9 residential runoff layouts. The rainfall collection device was measured using a simple artificial rainfall level measuring device installed in the open space, built-in water storage equipment, and indoor space with rain.

When rainfall occurs, the rainfall time (date), rainwater storage device, and water storage are recorded. When the rain ceases (about 1H rain), we record the time, remove the water equipment, and record the water consumption amount of the cylinder. We determined the runoff and sediment

*Corresponding Authors: 2583273597@qq.com and 656375320@qq.com

 DOI 10.1201/9781003348023-23

yield by measuring the depth of water in the pool when rainfall exceeds 20mm. Then we determined the total nitrogen by alkaline potassium sulfate ultraviolet spectrophotometry (11894-89 GB) and the total phosphorus determination using the vanadium molybdenum phosphoric acid colorimetric method.

## 3 RAINFALL CHARACTERISTICS

As shown in Figure 1, statistical analysis of the 6-10 test areas has 25 rainfall events, and the main characteristics are as follows. From June to October, rainfall intensity and rainfall fluctuated, the minimum rainfall was 6.0 mm, the maximum rainfall was 106.0 mm, and the rainfall was mainly concentrated in 30mm. The intensity of rainfall is the same as that of rainfall, which is obviously fluctuating, but the fluctuation pattern is slightly different from that of rainfall, which is related to the duration of rainfall. According to Figure 1, the rainfall intensity is between 0.18-5.67mm/h. According to the rainfall intensity, the grade can be divided into: light rainfall intensity (<0.42mm/h), moderate rainfall intensity (0.42mm/h = <1.25mm/h), heavy rainfall intensity and storm 1.25mm/h = <2.5mm/h) ([13] = 2.5mm/h).

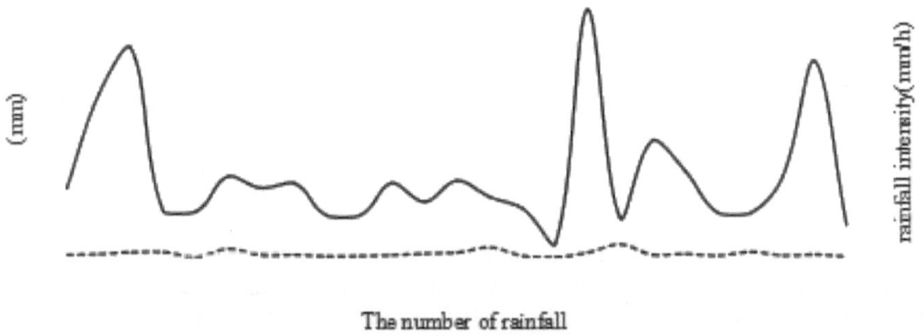


Figure 1. Distribution of rainfall intensity and rainfall intensity.

Due to the concentration of rainstorms and raindrops, it is more likely to cause soil loss than moderate rain and heavy rain. In this study, the soil loss of the three forests increased with the increase in rainfall intensity. However, there are differences in the amount of soil loss among the three forests. Like the runoff coefficient, the amount of soil loss of pure Phyllostachys pubescens forest is the largest, followed by the Phyllostachys pubescens poplar Paulownia mixed forest and the Phyllostachys pubescens poplar Paulownia Hemerocallis mixed forest. This is related to a large number of bamboo shoot digging activities in Anji County, and there are plant species such as Paulownia and Hemerocallis in the mixed forest, which can attach to the surface of the soil and reduce the amount of soil loss. Soil loss is accompanied by a certain amount of nutrient loss. Similarly, the loss of total nitrogen and total phosphorus increases with the increase in rainfall intensity. Among the three kinds of forests, the loss of pure Phyllostachys pubescens forest is also the largest, followed by the Phyllostachys pubescens poplar Paulownia mixed forest and the Phyllostachys pubescens poplar Paulownia Hemerocallis mixed forest.

## 4 SURFACE RUNOFF DYNAMIC

The surface runoff dynamics of three kinds of forest stands were analyzed, and the results are shown in Figure 2. From Figure 2, we can see that the three types of runoff coefficient change trends are basically the same. There was a significant fluctuation trend; the maximum value appeared on June 19th, reaching 10%, corresponding to heavy rain, and the minimum value appeared on July 13th, only 0.7%. The runoff coefficients in June and August are relatively large, in the collection of 11 rain, five heavy rain in June and August, and five of the three heavy rain in four fields are concentrated over more than two months. So the runoff coefficient in June and August is relatively large compared to the other two months. In addition, the pure bamboo forest runoff coefficient is always greater than the other two stands.

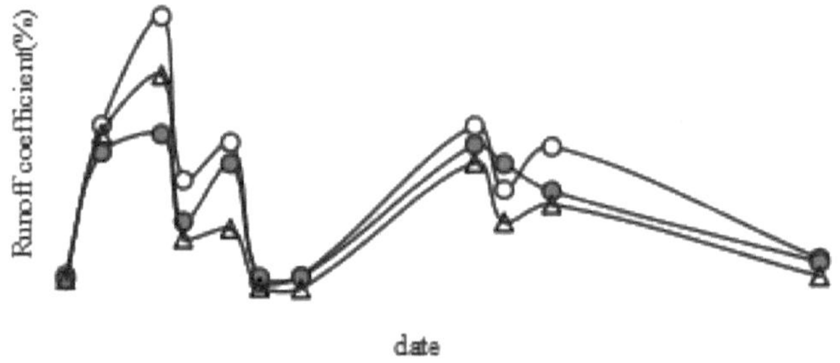


Figure 2. Dynamic variation of runoff coefficient.

## 5 DYNAMICS OF NUTRIENT LOSS

The research on soil nutrient loss was mainly focused on the loss of total nitrogen and total phosphorus (Figures 4 and 5). 6-10 analysis of 3-month loss of total nitrogen and total phosphorus content of the stand, the changing trend of the two and the amount of soil loss is basically the same, were the highest in June, reached 10.62 kg/ha and 0.15 kg/ha, the lowest in July, only 0.09 kg/ha and 0.003 kg/ha. This is due to soil nutrients (i.e., total nitrogen and total phosphorus) attached to soil particles. When there is runoff migration in soil particles, surface runoff, and nutrient migration cause nutrient loss. Similarly, compared with the other two kinds of pure bamboo forest, the nutrient loss is more severe.

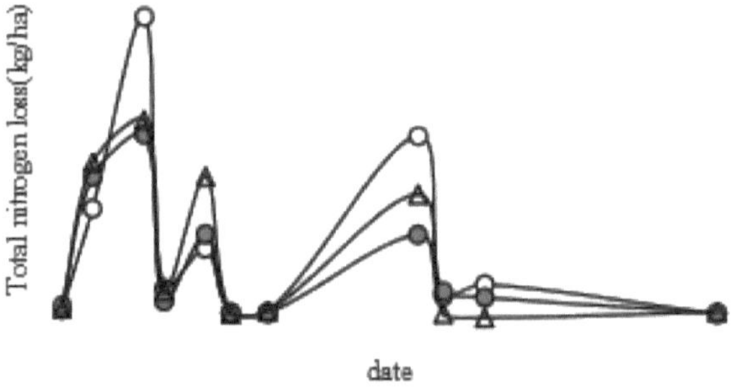


Figure 3. Dynamic changes in total nitrogen loss.

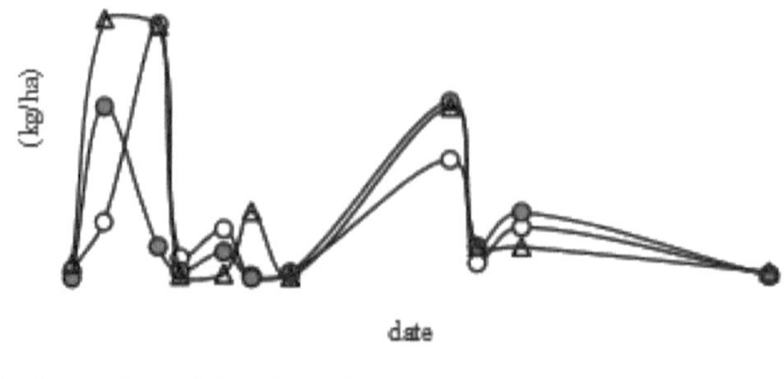


Figure 4. Dynamic changes in total phosphorus loss.

## 6 RELATIONSHIP BETWEEN RAINFALL AND SOIL EROSION

The Pearson correlation analysis of rainfall, runoff coefficient, soil loss, total nitrogen loss and total phosphorus loss was carried out by SPSS. It was found that there was a correlation between each factor, and R2 was between 0.650-0.995. Carry out regression analysis on rainfall and runoff coefficient and construct regression equation.

Application of Pearson on rainfall, runoff coefficient, soil loss, total nitrogen loss, and total phosphorus loss for SPSS correlation analysis found that there was a correlation between the various factors, R2 in 0.650-0.995. The rainfall and runoff coefficients were analyzed, and the regression equation was constructed. The results are shown in Table 1.

Table 1. Regression model of rainfall and surface runoff.

| Sample type | regression equation | $R^2$ |
|---|---|---|
| Pure bamboo | y=24.537x+4.599 | 0.726 |
| Bamboo - Yang Tong | y=20.127x+6.794 | 0.674 |
| Bamboo - Yang Tong - Lily | y=13.869x+1.978 | 0.820 |

The regression equation between rainfall and surface runoff and the regression equation between surface runoff and soil loss is linear regression equations, indicating that there is a linear relationship between rainfall and surface runoff, surface runoff, and soil loss. Among them, the runoff increases with the increase of rainfall. This is because the ground surface is easy to form muddy mud under the impact of raindrops. The mud blocks the soil pores in the infiltration process, resulting in the reduction of rainwater infiltration, which makes the surface runoff increase rapidly. Generally, the greater the rainfall intensity, the easier it is to form surface runoff. The amount of soil loss increases with the increase of runoff. Rainfall causes soil erosion through surface runoff. Rainfall peels the surface soil, forming raindrop splash erosion, and runoff is responsible for transporting the stripped soil particles away. Therefore, the amount of soil loss increases with the increase of runoff.

## 7 CONCLUSION

(1) In the experimental area, the rainfall amount of 6-10 obviously fluctuated, and the rainfall amount was 6.0-106.0 mm. According to the rainfall intensity grade 25, rainfall is divided into light rain, moderate rain, and heavy rain.
(2) The runoff coefficient in June and August is relatively large compared to the other two months. In addition, the pure bamboo forest runoff coefficient is always greater than the other two stands.
(3) The rainstorm runoff coefficient is relatively high, followed by heavy rain. The runoff coefficient of the pure bamboo forest is the largest, and that of the bamboo - Yang Tong - Hemerocallis mixed forest is minimum. The trend of runoff coefficient and nutrient loss was basically the same, all of which showed a trend of fluctuation.
(4) Application of Pearson on rainfall, runoff coefficient, soil loss, total nitrogen loss, and total phosphorus loss for SPSS correlation analysis found that there was a correlation between the various factors, R2 in 0.650-0.995. The rainfall and runoff coefficients were analyzed, and the regression equation was constructed.

## ACKNOWLEDGMENTS

The project was supported by the Natural Science Foundation of Zhejiang Province of China (Grant No. LY14E090005) and the Department of Science and Technology of Zhejiang Province (Grant No. 2014F50015).

## REFERENCES


Bi H X, Liu B, Wu J, et al.Effects of precipitation and Landuse on runoff during the past in a typical watershed 50 years in the loss plateau, China [J]. *International Journal of Sediment Research*, 2009, 24 (3): 352–364.

Fang N F, Shi Z H, Li L, et al. The effects of rainfall regimes and land use change on runoff and soil loss in a small mountainous watershed [J]. *Catena*, 2012(99): 1–8.

Li Guang, Huang Gaobao. Effects of rainfall intensity and land use on Soil and water loss in Loess Hilly Region [J]. *Journal of Agricultural Engineering*, 2009, 25 (11): 85–90.

Wei W, Chen L D, Fu B J, et al. The effect of land uses and rainfall regimes on runoff and soil erosion in the semi-arid loess hilly area, China [J]. *Journal of Hydrology*, 2007, 335 (3-4): 247–258.

Yan Qinghong, Yuan Cuiping, Lei Tingwu, et al. Types of rainfall and soil and water conservation effect of soil erosion of small watershed in loess area [J]. *Journal of Agricultural Machinery*, 2014, 45 (2): 169–175.

*Advances in Urban Construction and Management Engineering – Cha (Ed)*


# Research on the application of construction technology of prefabricated small box girder viaduct across railway

Aihong Qin, Liya Han & Pooya Saffari*
*Qingdao City College, Qingdao, Shandong, China*

Kejian Li
*Qingdao Geological Mineral Rock Engineering Co., Ltd., Qingdao, Shandong, China*

ABSTRACT: Prefabricated box girder is an important structural form in highway bridge construction. Taking the prefabricated box girder of Qilu Avenue Viaduct as an example, this paper analyzes the construction technology of the prefabricated small box girder spanning the third and fourth lines of the Beijing and Shanghai Railway. It can be used as a reference for the construction of similar bridges across railways.

## 1 INTRODUCTION

The explosive growth of expressways in China has led to the rapid development of railway viaducts. As the main structural form in the construction of highway bridges, prefabricated box girders are particularly common in railway viaducts (Yulong 2022). Before constructing a prefabricated small box girder across the railway, the girder erection plan, key points, and timeline should be determined in consultation with the railway department. During the construction process, the construction should be carried out orderly, and safety should be ensured. Therefore, it is beneficial to explore the construction and safety process of the railway prefabricated small box girders (Xiaohua 2017). This paper aims to study the construction technology of the prefabricated box girder of the Qilu Avenue Viaduct, which was constructed across the Beijing-Shanghai numbers 3 and 4 railway lines.

## 2 PROJECT OVERVIEW

Qilu Avenue is an important traffic trunk road in Jinan City. The main road of the K0+119.30-K1+306.14 section is in the form of a viaduct in a south-north direction. It crosses Qingyuan Road, Binnan Road, Xiaoqing River, South-to-North Water Diversion Container, and Binhe North Road and finally lands about 300 meters south of the Jiqi Road intersection. The viaduct crosses the Beijing-Shanghai numbers 3 and 4 railway lines. The Beijing-Shanghai Line is a busy trunk line with high traffic density. The construction of the upper bridge without affecting the traffic requires strong safety hazards and causes construction difficulties. In addition, the viaduct spans the Xiaoqing River, and the construction of the box girder shall not affect the normal flood flow of the river.

## 3 CONSTRUCTION SCHEME OF PREFABRICATED SMALL BOX GIRDER

The fifth section of the viaduct is assembled and erected by the prefabricated small box girder. It is planned to build a prefabricated component stationary yard on the east side of the road near the

*Corresponding Author: pooya_saffari@qdc.edu.cn

DOI 10.1201/9781003348023-24

lower bridge entrance of the Qilu Avenue viaduct for on-site prefabrication. During the erection, cranes are used to lift, and beam transport trolleys are transported to the site for beam feeding. JQIVW160t/40m stepping bridge span erecting machine is used for erection. Before erecting, a protective shed should be set up above the line. Additionally, all overhead catenary lines shall be cut off for 120 minutes at each blockade point, and two pieces of beams should be erected at each key point. After the beams and small box beams are all erected, the wet joints and crossbeams are placed, and finally, the bridge deck is paved.

## 4 CONSTRUCTION TECHNOLOGY OF PREFABRICATED SMALL BOX GIRDER

### 4.1 *Small box girder production*

In this project, only in the fifth section of the project, the 35m prefabricated small box girder is adopted. Ten sets of the box girder need to be prefabricated. It is planned to set up a small box girder prefabrication plant near the north side of the viaduct and on the east side of Qilu Avenue using a hoisting crane. The beam transport trolley transports the beam to the site, and the road beam erection machine carries out the beam erection. The layout plan of the prefabrication yard for the viaduct components in the north extension of Qilu Avenue is shown in Figure 1.

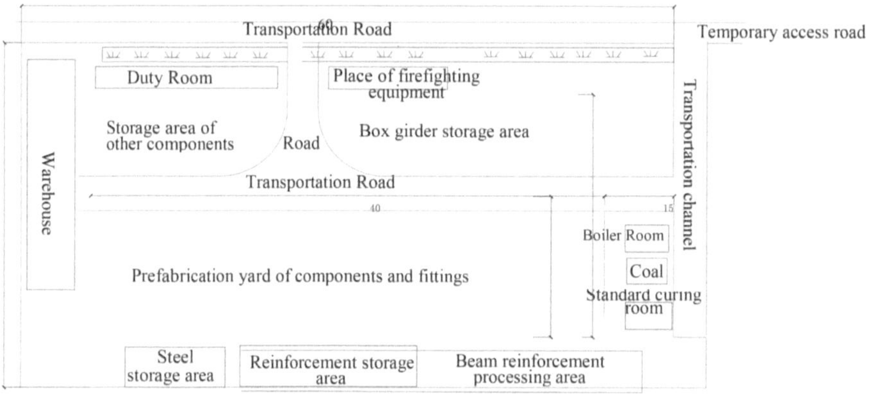


Figure 1. Layout plan of prefabrication yard for components and fittings of Viaduct on Qilu Avenue North Extension.

During box girder construction, box girder supports and box girder bellows should be embedded, the position of the drainage hole should be reserved in advance, and various embedded parts should be pre-buried.

#### 4.1.1 *Pedestal and bottom mold*

The base of the beam-making pedestal is subjected to concrete hardening treatment, and the pedestal is made of cast-in-place 50cm thick concrete. A steel plate is laid in the middle to be used as the bottom formwork of the box girder. Mounting holes or bolts for fixing the box girder core mold and side mold shall be set aside on the pedestal as required.

#### 4.1.2 *Side molds and end molds*

The side molds and end molds of the box girder are made of steel plates. The side molds on both sides can be connected and fixed with bolts from the bottom. In order to ensure that the concrete is well vibrated and compacted without the phenomenon of concrete honeycomb, an attached vibrator should be installed on the side mold. The positions of the pre-stressed rib holes placed on the end mold must meet the requirements of the relevant specifications.

4.1.3 *Preparation and hoisting of reinforcement framework*

Reinforcing steel bars must be strictly inspected for their factory certificate and the quality assurance certificate before entering the site. According to the storage requirements, all kinds of steel bars shall be identified separately, stored according to the category, and cannot be mixed. The steel skeleton is bounded on the welded mold in advance, and the various grooves on the mold are painted with different colors, and the corresponding steel bars are placed in the grooves and bound into a steel skeleton. This can ensure the position, geometric size, and quantity of various steel bars and make the finished skeleton firm, ensuring the quality of the steel skeleton and improving the production speed. After the steel frame of the box girder is completed, the gantry crane is used for the overall lifting. Embedded parts and bellows are inserted into the reinforced skeleton and bound stably. Various embedded parts and bellows should be firm and reliable to ensure they will not collapse or displace during subsequent concrete pouring. The connection of the bellows should be absolute to ensure the anti-seepage performance of the bellows (Zhiyong 2019).

4.1.4 *Concrete pouring of the beam body*

Concrete is prepared according to the design requirements. Cement, sand, stone, water, etc., must meet the design and specification requirements. The concrete pouring of the beam body adopts the inclined layer method, which is poured from both ends of the beam body to the mid-span and is closed at the mid-span. The vibration should be mainly carried out by the attached vibrator set on the side formwork, assisted by the inserted vibrator on the surface. Finally, the surface of the beam body is vibrated with a flat vibrator to ensure that the concrete of the box girder is well vibrated and compacted.

4.1.5 *Pre-tensioning of prestressed tendons*

The key process that affects the strength and quality of prefabricated beams is the pre-tensioning of pre-stressed steel bars. The prestressed tendons should be tensioned when the concrete age of the box girder is less than seven days, and the strength reaches 85% of the design strength. Before the pre-stressed steel bars are tensioned, the strength test on the concrete block is required.

The pre-stressed steel bar shall be tensioned according to the design requirements. In order to reduce the loss of prestressing, the pre-stressed steel bar must be over-tensioned or repeatedly tensioned. The steel hinge's wire should be stretched at both ends, and it should be stretched symmetrically according to the requirements.

4.1.6 *Tension check*

The elongation value of the prestressed tendons during tension is taken as the control stress check index.

4.1.7 *Duct grouting*

Duct grouting should be completed as soon as possible and no more than seven days at the latest. The mortar pump for grouting needs to have enough working pressure. The mortar must meet various indicators such as strength, fluidity, and bleeding. The grouting is generally injected from one end and flows to the other. In order to ensure that the mortar fills the pores, two sets of grouting are conducted. The two grouting directions are opposite, and the time interval is 30 minutes.

4.1.8 *Transportation and storage of beam segments*

The prefabricated beams should be transported after the pre-stressed tendons are tensioned and anchored. During the transportation and storing of the beam segments, the following points should be considered: the position of the supports in the beam body; the beam should be well supported during the sliding process of the beam to prevent the beam body from overturning; the beam should not be stored more than three months (Shilong & Xinli 2021).

### 4.2 *Line protective shed*

#### 4.2.1 *Line protective shed*

Before the erection of beam segments and construction of anti-collision walls and wet joints, it is necessary to isolate and protect the third and fourth lines of the Beijing and Shanghai railway. The protective shed adopts a C20 concrete single foundation in which the depth of the foundation is 120cm, and the length and width are both 80cm. The foundation is completely buried in the road shoulder. The top of the foundation is covered with a steel plate (500×500×10mm), and the bottom of the steel plate is welded with U-shaped Φ20 steel bars. The steel bars are anchored into the concrete at a depth of 40cm.

The columns are made of steel pipes with a diameter of 30cm. There are nine columns with a spacing of 6m outside the third and fourth lines, and the column height is 11m. No. 10 channel steel is used as diagonal bracing between the two columns, and the column is welded on the embedded steel plate on the top surface of the foundation (500×500×10mm).

Longitudinal beams are made of AH30 steel, and beams are made of 20# steel channels. Planks with a thickness of 5cm are set on the beams, and 2 mm thick steel plates are nailed to the top of the planks. Insulation panels are placed under the beams to prevent the normal operation of the line from being affected by the catenary discharge.

During the construction of the protective shed, it is necessary to reduce the speed limit of the railway lines to 45 km/h. during the construction of the foundation and columns of the protective shed, the blocking point shall be established.

#### 4.2.2 *Demolition of line protection shed*

Before demolishing the protective shed, all the workload of the viaduct above the railway should be completed.

Main construction process: wet joints, anti-collision wall, shutter installation, and vertical wall construction → manual removal of iron sheets → manual removal of wooden boards and insulating boards → longitudinally movement of beams to the outside of beams→ vertical lifting of the beams → removal of longitudinal beams → removal of columns → Subgrade cleaning and repairing.

For the removal of the protective shed, the upper and lower ends are removed at the same time, and the 100T crane is used for lifting.

The demolition of the protective shed shall be carried out once a day for each of the upper and lower protective sheds at the point of departure according to the key plan approved by the road bureau. The demolition of the protective shed must be completed within the blockade point. The beams are directly moved out of the line with a truck crane.

### 4.3 *Process and lifting method of prefabricated small box girder*

The fifth link of the viaduct on the north extension of Qilu Avenue spans the Beijing-Shanghai No. 3 and No. 4 lines. The superstructure adopts a prefabricated small box girder structure. While erecting the substructure and box girder adjacent to both sides of the railway, reducing the impact on the operation of the railway line and ensuring the safety of the line operation are the top two priorities of the construction.

#### 4.3.1 *Arrangement of key points for the erection of box beams on the third and fourth lines of the Beijing-Shanghai railway*

In order to ensure the safety of the Beijing-Shanghai railway, it is necessary to temporarily close the key points of the line for construction when crossing the line and erecting the beams. According to previous experience, it is necessary to block the points 5-10 times, each time 120 minutes, to erect 1-2 small box girders. The key point time of the closed line must be negotiated and confirmed by the railway department one month before the beam erection. During the construction of the beam, the erection shall be carried out in strict accordance with the key point time given by the railway, and the relevant personnel of the railway department shall be invited for safety supervision (Baoxiang 2021).

The erection steps at the key points are as follows:

(1) Erection preparation: The railway department must approve the construction plan before construction. Then the bridge erection machine needs to be set up, and the box girder should be transported in place with a beam-carrying flat car. Besides, various girder erection preparations have to be carried out.
(2) Beams erection: The erection of beams needs to carry out in 5 days in total, and each day the third and fourth lines of the Beijing-Shanghai Railway should be blocked for two hours. After the erection, the barriers are lifted when the box beams are fixed, and the third and fourth lines of the Beijing-Shanghai Railway will be opened. On each day, two beams will be erected.

#### 4.3.2 *Beam construction*

The box girder can be installed after its strength reaches the design requirements. A crane carries out the installation for lifting purposes, and the prefabricated box girder shall be transported to the bridge site by the beam transporting trolley. The JQIV160t/40m type stepping bridge erection frame is adopted to set up the box girders.

Before the beam's construction, inspections should be conducted. All dimensions should meet the design requirements. The beams of the same hole must have the same age and similar cambers. The end of the beam before beam erection shall be roughened for easy connection.

The highway-railway interchange bridge is set up with a double guide bridge girder erection machine hoisted by a truck crane and transported by a beam trolley. The assembled bridge erection machine is installed behind the P14 pier and erected from the P14 pier to the P13 pier.

### 4.4 *Construction of small box girder anti-collision walls and wet joints above the railway*

#### 4.4.1 *Construction of the crash barrier*

The formwork of the crash barrier adopts steel formwork, and the support and reinforcement of the steel formwork are shown in Figure 2.

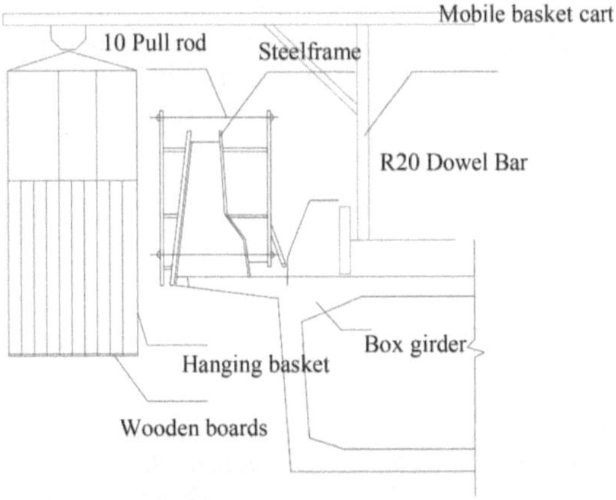


Figure 2. The schematic diagram of the crash barrier.

#### 4.4.2 *Diaphragm beams and wet joints for cross-railway construction*

This bridge is an overpass over the third and fourth lines of the Beijing-Shanghai Railway. Therefore, the construction of wet-joint concrete cannot affect the operation of the railway. Consequently, it is decided to set up a protective hoist net under the two prefabricated box girders for wet joint construction. This protection method is carried out to ensure that objects on the bridge will not fall under the bridge during the construction of the wet joints. The construction of wet joints can be arranged in the blockade points during the erection of the new piece of girder and after the two adjacent box girders are erected to ensure driving safety (Huihui 2021). See Figure 3 for the construction of the protective hoist net.

800m before and after the construction area and on the bridge, security guards shall be assigned. The security guards will use walkie-talkies to communicate with each other and monitor the traffic conditions. All construction operations on the bridge deck will be stopped when vehicles pass by, and the construction will be carried out at intervals between trains. Special personnel will be assigned under the bridge to check the lines to ensure operational safety.

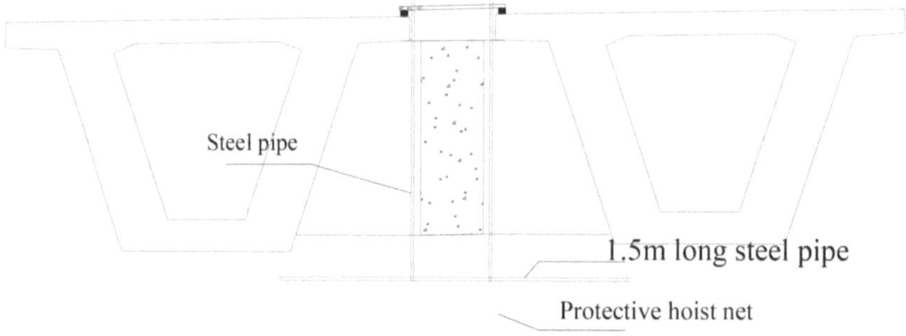


Figure 3. The schematic diagram of the protective hoist net.

## 5 CONCLUSION

The viaduct of Qilu Avenue spans the third and fourth lines of Beijing and Shanghai, and ten small box girders are erected. The small box girder is prefabricated on the construction site. When erecting small box beams, blockade points shall be constructed, and two beams shall be erected at each blockade point within 120 minutes. The successful application of this construction technology has accumulated certain experience for the same kind of railway box girder bridge project in China.

## REFERENCES

Baoxiang, L. (2021). Construction technology of cross-iron box girder frame. *Smart City*. 7(24):141–142.

Huihui, Z. (2021). Research on key technologies for the construction of a 35 m cast-in-place PCR box girder on the existing reconstruction railway on the upper span. *Construction Technology*. 50(07): 128–131.

Shilong, L. & Xinli, W. (2021). Development and application of construction technology and key equipment for high-speed railway box girder transportation. *Construction Machinery Technology and Management*. 34(05): 30–35.

Xiaohua, C. (2017). Research on the application of prefabricated box girder construction technology for cross-railway bridges. *Construction Technology Development*. 69–70.

Yulong, F. (2022). Construction technology for hoisting prefabricated small box girders across railway lines in small urban spaces. *Architectural Science and Technology*. 6(01):19–21.

Zhiyong, W. (2019). Analysis of the main construction technology of prestressed concrete box girder + steel box girder in the urban cross-railway station. *Qinghai Transportation Science and Technology*. (01): 116–119.

*Advances in Urban Construction and Management Engineering – Cha (Ed)*


# Study on-road performance of glass fiber asphalt macadam sealing layer

Jianxin Yang
*JSTI Group, Nanjing, Jiangsu, China*

Linlin Liu*
*Jiangsu Xiandai Road & Bridge Co., LTD., Nanjing, Jiangsu, China*

Jian Lu
*JSTI Group, Nanjing, Jiangsu, China*

ABSTRACT: In China, the basic highway base adopts a semi-rigid base and sub-base, which are mainly composed of cement-stabilized material and lime-ash stabilized material. In the process of use, the semi-rigid base material is prone to reflect cracks and lead to surface cracking. In addition, the bond between the pavement and the base layer is not tight enough, which will put the pavement structure in a discontinuous state and reduce the ability of the pavement to resist damage. To prevent cracking and water damage caused by semi-rigid base reflecting cracks on asphalt pavement, a glass fiber asphalt macadam sealing structure was proposed, and its interlayer bonding performance, anti-reflecting crack performance, and glass fiber distribution uniformity were experimentally studied. The results show that the order of the influence level on the bonding strength of glass fiber asphalt gravel sealing layer is glass fiber > modified emulsified asphalt > fiber length and gravel dosage. The influence level of crack resistance is modified by emulsified asphalt > glass fiber > fiber length and gravel dosage. The adhesive property of rubber asphalt fiber is the strongest, and the adhesive property of the thin slurry-sealing layer is the worst. Based on the bond performance test, anti-reflection crack performance test, and fiber distribution uniformity test, the optimal fiber length of glass fiber asphalt gravel sealing layer was found to be 6 cm.

## 1 INTRODUCTION

The basic base of highways in China adopts a semi-rigid base and sub-base, which are mainly composed of cement-stabilized material and lime-ash stabilized material. In the process of use, the semi-rigid base material is prone to reflect cracks and lead to surface cracking. In addition, the bond between the pavement and the base layer is not adequately tight, which will put the pavement structure in a discontinuous state and reduce the ability of the pavement to resist damage. It has been proved that the design of a sealing layer between the upper base layer and the lower layer can improve the bonding effect between the base layer and the surface layer and the waterproof and shear resistance between the layers. However, due to material reasons, the traditional slurry-sealing layer technology cannot provide a better waterproof and crack resistance effect (Liu et al. 2012).

To prevent the cracking and water damage caused by the reflection crack of semi-rigid base pavement, researchers at home and abroad have carried out many studies on the prevention of cracking of semi-rigid base itself in recent years. For example, the pavement is added, the thickness of the surface layer is appropriately thickened, the skeleton is made of more dense base, the amount

*Corresponding Author: liulinlin@jsxdlqyxzrgs.wecome.work

DOI 10.1201/9781003348023-25 

of cement used is reduced, and the gravel relief layer is spread at the crack. The above methods have a certain anti-crack effect but cannot completely solve the semi-rigid base reflective crack problem. Therefore, it is necessary to propose an economical and efficient functional, structural layer to prevent the occurrence of the disease completely. The glass fiber asphalt macadam seal is a continuous layered structure with high-strength fiber based on the traditional macadam seal. This paper studies the road performance of the glass fiber asphalt macadam sealing layer, including the interlayer bonding performance test, anti-reflection crack performance test, and glass fiber distribution uniformity test. The research results provide a reference for promoting the glass fiber asphalt macadam sealing layer (Lin 2019; Zhang 2020).

## 2 TEST

### 2.1 *Main raw materials*

Glass fiber, Shandong Taicheng Fiber Co., LTD., short cut glass fiber with a diameter of fewer than 10 μm and a length of 6 mm, the quality of glass fiber meets the requirements of "Glass fiber untwisted roving" (GB/T 18369) standard specification. SBR modified emulsified asphalt is prepared by Shell 70# asphalt from Royal Dutch Shell Oil Company and cationic fast-cracking emulsifier from Mervich Wick company in the USA. The quality of modified emulsified asphalt meets the requirements of the standard specification of Highway Asphalt Construction (JTG F40-2004). Limestone gravel, products of Xi'an Jelide Building Materials Co., LTD., particle size 4.75 mm–9.5 mm, the quality of limestone gravel meets the requirements of "Technical Specifications for Highway Asphalt Construction" (JTG F40-2004) standard specification (GB/T 18369-2008. (2008)./JTG F40-2004. (2004)).

### 2.2 *Main equipment and instruments*

XH-40 type multifunctional strength detector: manufactured by Beijing Heaven and Earth Spark Science and Technology Development Co., LTD, model: XH-40, with a measurement range of–40 KN. Automatic wheel grinding and cutting machine: model: TDCX-2, manufactured by Shanghai Shengshihui Test Equipment Co., LTD; model: HYCZ-6, manufactured by Shanghai Lei Yun Test Instrument Manufacturing Co., LTD.

### 2.3 *Specimen preparation*

#### 2.3.1 *Specimen for bond performance test*

For cement concrete specimen + fiber sealing layer + asphalt concrete specimen (100 mm×100 mm), the upper and lower surfaces of the specimen (asphalt concrete on the top and cement concrete on the bottom) should be clean and smooth. The fixture is firmly bonded to both sides of the specimen with a high-strength epoxy adhesive.

#### 2.3.2 *Specimens for anti-reflection performance test*

① Add a layer of 30 cm × 10 cm × 1 cm thick rubber plates at the bottom, two rubber plates spacing 10 cm ② The size of the bottom concrete slab is 30 cm × 30 cm × 4 cm, and the strength grade of C30 cement concrete is made according to the proportion requirements. After 28 d, the concrete slab is cut, and about 3 mm cracks were reserved in the middle. To ensure the crack width in the middle of the concrete slab, two pieces of wood chips with a width of 3 mm are pressed into the end when making the specimen ③ according to the prepared mix ratio parameters for the production of glass fiber asphalt sealing layer ④ The asphalt mixture surface layer is prepared according to the method for Making Asphalt Mixture Specimen (T0703-2011).

## 2.4 *Performance test*

### 2.4.1 *Bonding properties*

The specimen with a good bond performance test is placed in the standard test for 24 h static, the specimen is placed in the 50°C ± 2°C thermostat for 4 h, and then the specimen is taken out. The specimen bonded with the fixture is loaded into the tensile testing machine, the vertical tension is acted on the specimen surface vertically, and the tensile rate is adjusted to 10 mm/min. Then, we start the tensile testing machine to damage the bond and measure the tension required to break the seal/substrate adhesion. The orthogonal test scheme is shown in Table 1.

Table 1. Table of orthogonal test factors.

| Specimen number | Fiber dosage (g/m$^2$) | The length of the fiber (cm) | The amount of modified emulsified asphalt (kg/m$^2$) | Amount of crushed stone (kg/m$^2$) |
|---|---|---|---|---|
| 1 | 50 | 3 | 1.6 | 7 |
| 2 | 80 | 6 | 1.8 | 8 |
| 3 | 100 | 9 | 2 | 9 |
| 4 | 120 | 12 | 2.2 | 10 |

### 2.4.2 *Anti-reflection performance*

① The specimen is placed in the incubator, which has reached the test temperature, and the insulation is not less than 8 h ② The rut testing machine is adjusted to the test temperature, placed in the rut testing machine, and then place the test wheel in the middle of the asphalt mixture surface ③ The testing machine is started, the direction of the test wheel is perpendicular to the direction of the pre-crack in the test, and the insulation is not less than 1.5 h. At this time, the rut deformation recorder is closed, and the test wheel walks back and forth on the surface of the composite pavement structure. The rut recorder automatically records the round trip times of the test wheel and the ambient temperature ④ Then we observe the asphalt concrete surface layer and record the number of round trips of the rut test wheel when cracks appear on the surface of the asphalt layer and cracks run through the specimen surface. The orthogonal test scheme is shown in Table 2.

Table 2. Table of orthogonal test factors.

| Specimen number | Fiber dosage (g/m$^2$) | The length of the fiber (cm) | The amount of modified emulsified asphalt (kg/m$^2$) | Amount of crushed stone (kg/m$^2$) |
|---|---|---|---|---|
| 1 | 50 | 3 | 1.6 | 7 |
| 2 | 80 | 6 | 1.8 | 8 |
| 3 | 100 | 9 | 2 | 9 |
| 4 | 120 | 12 | 2.2 | 10 |

### 2.4.3 *Glass fiber uniformity*

The glass fiber length ranges from 3 to 12 cm, with 3 cm as the first gear, and the test is divided into four grades. The fiber dosage used in each grade is shown in Table 5.5. Fibers of different amounts are evenly distributed in an area of 900 cm$^2$, which is further divided into nine units of the same size. Then the net fiber weight on each unit is obtained by trial and error.

# 3 RESULTS AND DISCUSSION

## 3.1 *Bonding properties*

### 3.1.1 *Influence of different dosage ratios of the same material on bond strength*

The test results of adhesive strength of fiber asphalt seal with different mixing ratios are shown in Figure 1.

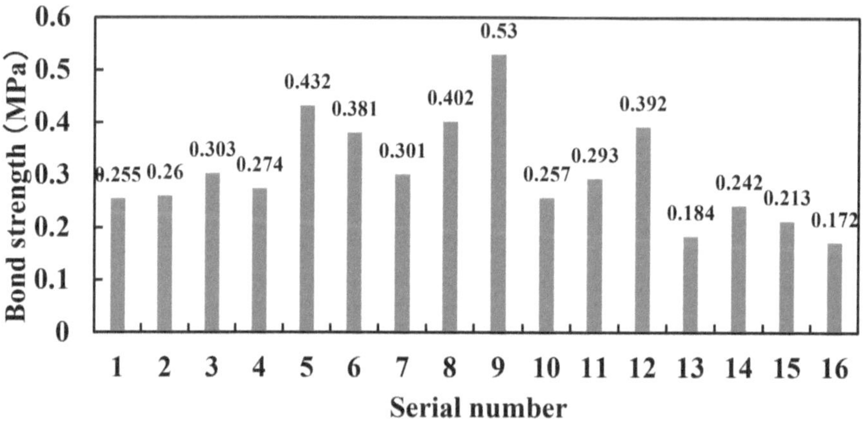


Figure 1. Bonding strength test results of different mixing ratios.

According to the test results in the table above, the order of the level of influence on the bonding strength of the glass fiber sealing layer (GFSL) is glass fiber > emulsified asphalt > fiber length and gravel dosage. GFSL bonding strength is the strongest dosage combination: glass fiber dosage of 100 $g/m^2$, modified emulsified asphalt dosage of 2.0 $kg/m^2$, fiber length of 3 cm, gravel dosage of 9.0 $kg/m^2$.

### 3.1.2 *Analysis of bonding properties of different sealing materials*

The bonding strength tests on GFSL, thin slurry sealing layer (TSSL), modified emulsified asphalt sealing layer (MEASL), and rubber asphalt glass fiber sealing layer (RSGFSL) are carried out. The test results are shown in Figure 2 below.

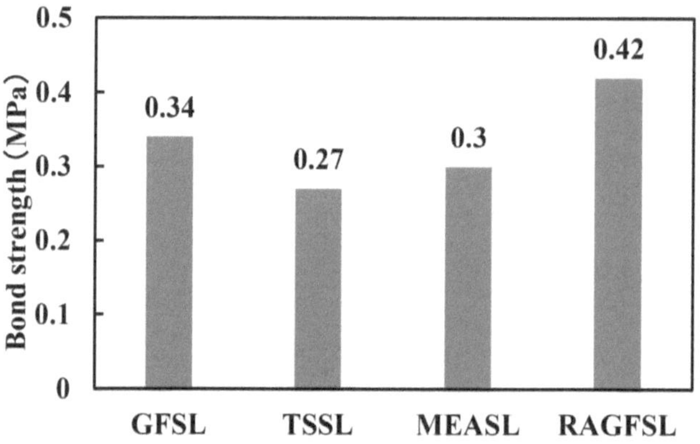


Figure 2. Test results of adhesive properties of different sealing materials.

It can be seen from the test results in the figure above that the order of sealing layer and pavement structure viscosity is: RSGFSL > GFSL > MEASL > TSSL.

## 3.2 *Anti-reflection crack performance*

According to the different mix ratios of fiber asphalt gravel sealing layer, the test piece was made, and some crack penetration tests were carried out. The time of anti-reflective crack penetration under different mixing ratios of glass fiber asphalt sealing layer was tested. The test results are shown in Figure 3 below.

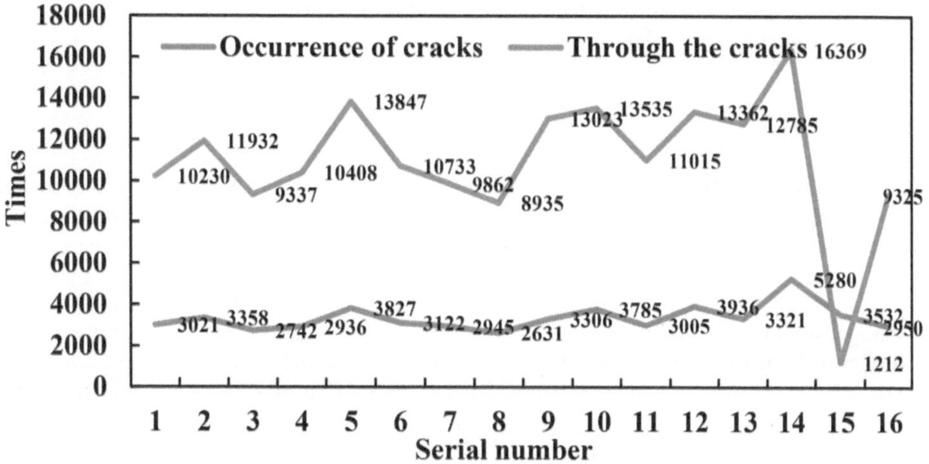


Figure 3. Test results of anti-reflection crack times with different mixing ratios.

According to the test results in the table above, the order of the level of influence on the crack resistance of glass fiber asphalt sealing layer is emulsified asphalt > glass fiber > fiber length and gravel dosage. Glass fiber sealing layer anti-reflection crack performance is the strongest dosage combination: modified emulsified asphalt dosage of 2.0 kg/m$^2$, glass fiber dosage of 120 g/m$^2$, fiber length of 6 cm, gravel dosage of 9.0 kg/m$^2$.

### 3.3 *Uniformity of glass fiber distribution*

The glass fiber length ranges from 3 cm to 12 cm, with 3 cm as first gear, which is divided into four grades for the test. Different amounts of fiber are evenly distributed in an area of 900 cm$^2$, which is further divided into nine units of the same size. Then, the net weight of fiber on each unit is obtained through repeated tests. The test results determined by the cell grid method are shown in Table 3 below.

Table 3. Test results of the cell grid method.

| The length of the fiber (cm) | Fiber mass on cell mesh (g) | | | | | | | | |
|---|---|---|---|---|---|---|---|---|---|
| | 4 | | | 10.2 | | | 15.3 | | |
| 3 | 0.405 | 0.482 | 0.494 | 1.506 | 1.053 | 1.002 | 1.901 | 1.723 | 1.608 |
| | 0.513 | 0.428 | 0.408 | 1.611 | 0.977 | 1.029 | 1.579 | 1.702 | 1.546 |
| | 0.407 | 0.426 | 0.437 | 1.034 | 0.959 | 1.029 | 1.653 | 1.956 | 1.632 |
| 6 | 0.467 | 0.449 | 0.406 | 1.527 | 1.086 | 0.934 | 1.708 | 1.664 | 1.858 |
| | 0.479 | 0.401 | 0.438 | 1.661 | 1.063 | 0.918 | 1.859 | 1.701 | 1.581 |
| | 0.496 | 0.457 | 0.407 | 1.053 | 1.064 | 0.894 | 1.585 | 1.885 | 1.459 |
| 9 | 0.489 | 0.458 | 0.461 | 0.864 | 0.909 | 1.078 | 1.856 | 1.754 | 1.688 |
| | 0.457 | 0.471 | 0.369 | 1.226 | 1.131 | 0.809 | 1.902 | 1.758 | 1.459 |
| | 0.473 | 0.448 | 0.374 | 1.96 | 1.107 | 1.116 | 1.978 | 1.497 | 1.408 |
| 12 | 0.451 | 0.489 | 0.512 | 1.254 | 1.202 | 0.759 | 1.602 | 1.295 | 1.901 |
| | 0.495 | 0.417 | 0.457 | 0.903 | 1.147 | 0.799 | 1.459 | 1.629 | 1.887 |
| | 0.459 | 0.379 | 0.341 | 1.201 | 1.126 | 1.829 | 1.968 | 2.151 | 1.408 |

According to the test results in Table 3, the shorter the fiber length is, the more uniform is the fiber distribution. On the contrary, the longer the fiber length is, the more uneven is the distribution, and the clumping phenomenon occurs. According to the bonding performance test, anti-reflection crack performance test, and uniform distribution test, the optimal fiber length of glass fiber asphalt gravel sealing layer is 6 cm.

## 4 CONCLUSION

(1) The order of the influence level on the bonding strength of glass fiber asphalt gravel sealing layer is as follows: glass fiber > modified emulsified asphalt > fiber length and gravel dosage. The order of influence level of crack resistance is modified emulsified asphalt > glass fiber > fiber length and gravel dosage.
(2) The strongest bonding performance of glass fiber asphalt gravel sealing layer is glass fiber dosage of 100 g/m$^2$, modified emulsified asphalt dosage of 2.0 kg/m$^2$, fiber length of 3 cm, and gravel dosage of 9.0 kg/m$^2$. The combination of materials with the strongest cracking resistance is 2.0 kg/m$^2$ of modified emulsified asphalt, 120 g/m$^2$ of glass fiber, 6 cm of fiber length, and 9.0 kg/m$^2$ of gravel.
(3) By comparing the bonding strength test of different sealing materials, it can be concluded that the rubber asphalt fiber has the strongest bonding performance, while the thin slurry-sealing layer has the worst bonding performance. Based on the bond performance test, anti-reflection crack performance test, and fiber distribution uniformity test, the optimal fiber length of glass fiber asphalt gravel sealing layer is 6 cm.
(4) In this paper, only the fiber sealing technology of modified emulsified asphalt is studied, and the fiber sealing technology of high viscosity and high elasticity composite modified asphalt and rubber asphalt needs to be studied further.

## REFERENCES


GB/T 18369-2008. (2008) *Glass fiber roving without twist* [S].

Huang X J. (2020) *Study on fiber sealing properties and finite element simulation based on stress absorption principle* [D]. Chongqing Jiaotong University.

JTG F40-2004. (2004) Technical Code for Construction of Highway Asphalt Pavement [S].

Lin B Y. (2019) *Study on mechanical properties and application of glass fiber asphalt gravel sealing layer* [D]. Chang'an University.

Liu Y Y, Ling T Q, Jiang K. (2012) Research on Crack Resistance of Emulsified Asphalt Macadam Sealing Layer Based on Fracture Energy Fiber [J]. *Highway*,(08):201–207.

Wang C. (2022) Maintenance technology of fiber Sealing layer of asphalt Pavement for municipal Roads [J]. *Urban Roads, Bridges, and Flood Control*,(03):32–34+11.

Zhang N. (2020) Experimental study on sealing properties of modified emulsified asphalt fiber [J]. *Transportation World*, (29):32–33.

*Advances in Urban Construction and Management Engineering – Cha (Ed)*


# Spatial performance of urban-rural interface based on LBS data—Take the Huangyan district of Taizhou as an example

Liyuan Fei*
*Shanghai Academy of Fine Arts, Shanghai University, Shanghai, China*

ABSTRACT: The location of the urban-rural interface makes it easier for this region to breed rich industrial clusters. Many people are attracted to the region, and the low land rent contributes to the high level of production efficiency. Therefore, improving the spatial performance of urban-rural interface space has an important impact on city development. This paper first attempts to evaluate the spatial performance indicators of urban intersection areas - traffic circulation and service popularity. Under the current new data environment, GPS-LBS (GPS embedded and location-based service) data are used to collect application signal and location information of different periods and different people. By comparing the daily commuting between the same streets in 2017 and 2020, it is found that with the development of the urban-rural interface, the link between streets and towns based on industry changes most significantly. Further comparing the distribution of working places and living places, it is found that the urban-rural interface can achieve a higher level of "jobs-housing balance". Subdivide people with different income levels and explore their different travel modes in space. It is found that the commuting between different districts in Taizhou is relatively independent, but the urban-rural transition area makes the connection between different districts closer. Lower-income groups have more flexibility in activities in urban-rural transition areas. Finally, through the case study of some communities or villages in the urban-rural junction area of Huangyan District, Taizhou, it is found that the spatial performance of Xifan village is better than that of other villages because of the coupling mechanism of its industrial cluster and residential space. Finally, some suggestions are put forward to improve the spatial performance of urban-rural transition areas in future urban planning.

## 1 INTRODUCTION

### 1.1 *Research background*

At present, China is in a period of rapid urbanization. The expansion of urban land, the reorganization of urban internal functional structure, and the change in the relationship between cities and suburban areas have a profound impact on the evolution of suburban settlements (Burgess 1925 & 1984; Li 2021, 2017). In the process of urban-rural integration, there are many social problems, governance difficulties, and development bottlenecks, which restrict the overall development of urban and rural areas, and most of these problems are concentrated at the junction of urban and rural areas (Chen 2021; Douglass 1992). It is of great significance for the overall development of urban and rural areas to solve the outstanding problems in the urban-rural fringe. The urban-rural fringe is a complex regional space, where most of them have rural living styles and urban living habits. The research on urban-rural transitional zones is worthy of our in-depth discussion and research.

*Corresponding Author: feiliyuan@shu.edu.cn

DOI 10.1201/9781003348023-26

### 1.2 *Research purpose and significance*

As an area with the strongest interaction between urban and rural areas, the urban-rural interface not only has the advantage of being close to urban areas but also has the advantage of relatively low land rent, labor, and other factors (Chen 2021). Therefore, it often breeds industrial clusters based on local resource endowments in suburban streets and towns, which plays an important role in driving the urban-rural development of surrounding areas. Therefore, the significance of this research on the comprehensive effect of urban space in the urban-rural interface is to investigate whether the area meets the comprehensive requirements of the development of various functions of the city, whether it can reasonably allocate resources, and whether it can make various elements develop in a balanced manner to effectively improve the comprehensive benefits of the city.

## 2 RESEARCH METHODS AND BIG DATA PLATFORM CONSTRUCTION

### 2.1 *Research methods*

This paper focuses on the spatial performance indicators of the urban-rural interface, including traffic circulation and service popularity. Traffic circulation is to compare the commuting data in 2017 and 2020 through the integration of spatiotemporal data. It is expected that the spatial mobility relationship of people will be obtained while the industrial development in the urban-rural transition areas in three years. In addition, it compares the balance between occupation and residence based on the spatial distribution of residence and employment. By dividing different income groups, we can judge whether each factor in the highland area can develop in a fair and balanced way. Then, we investigate the reasons for improving spatial performance through in-depth interviews and case studies. We are striving to develop a theoretical and practical comparison of the spatial performance of the villages in Huangyan District and to identify the differences in the distribution of population and facilities. This study reveals the time- and space-based change laws for hot spots of crowd activity within an urban environment.

### 2.2 *Data acquisition and portrait rules*

The quality of empirical data is of great significance to the proof/falsification of this study. In terms of big data, this study uses location-based service data based on GPS embedded. GPS-LBS data, like mobile signaling data and smart card swiping data, is a common data source for urban big data research (Chen 2020; Mondal 2021). Regional and urban spatial research conducted by scholars at home and abroad based on LBS data is often carried out in combination with social network services (SNS) (Li 2017), such as Twitter and microblogs that need event location (Zhang 2020). LBS data collection relies on the application that opens the location service to sign in and report location information. Compared with the current mobile signaling data, LBS has the advantage of more accurate and reliable positioning. Therefore, it can be used to analyze the regional and urban macrostructure, as well as the specific analysis of the space at the medium and micro levels. However, traditional LBS data also has obvious limitations when applied to specific research issues such as "job residence relationship". For example, only logging in to a specific app can retrieve data, a relatively single application leads to group deviation, data granularity accuracy is too large, and failure to identify device terminals (PC/Mac, pad, and mobile phones) leads to a large number of "SOHO users" without displacement, which leads to the deviation between the analysis results and the actual overall job residence space, Therefore, it is more necessary to carefully judge the representativeness and data quality of data samples.

To overcome the defects of traditional LBS data, this study screened the data only for mobile terminal users through the data outsourcing SDK platform and was able to effectively collect LBS information called by multiple mainstream mobile applications (Apps) within a certain period. LBS location service data is collected using the following mechanism periodically updating data, that is, when a user opens any of the above apps that request location information, the user reports the GPS location data of the user to the SDK platform at a particle accuracy of one hour/time. Whenever users log in, search, send, or receive information, pushes and other events occur within any of the above apps, real-time location data will also be derived.

The temporal and spatial range of data collection is Taizhou City, Zhejiang Province, the same month with an interval of three years between October 2017 and October 2020, and 20 consecutive working (school) weeks, during which the climate is mild, including the National Day golden week. The LBS basic data obtained includes user desensitization ID, timestamp, longitude and latitude GPS coordinates, and other information. A total of 179,752 independent mobile terminal users who entered the district of Huangyan City from October 1, 2017, to October 31, 2017, and from October 1, 2020, to October 31, 2020, were screened out. By tracing all the LBS trajectories of these users in the country in the past two months, the city with the highest frequency of users was identified as their place of belonging, and on this basis, the portrait analysis of users' residence and employment was carried out. The rule of the portrait is to count the most frequently occurring coordinates of a user during non-working hours (21:00-07:00) on weekdays and weekends as the user's residential address; According to the statistics of the user's working hours on weekdays (10:00-07:00), the coordinate with the most frequent occurrence is the user's working place. Based on this, the algorithm has been iteratively optimized. First, the LBS data of holidays are removed, and then 25% weight is given to the occurrence times of the predicted office address and residential address in the previous month as the input variable to participate in the prediction of the office address and residential address in the next month, with a cycle of one year. In this process, random sampling is conducted for manual verification to improve recognition accuracy.

## 3 EMPIRICAL ANALYSES ON SPATIAL PERFORMANCE OF TAIZHOU CITY

### 3.1 *Overview*

Zhejiang Province is located on the southeast coast of China. Taizhou city is a prefecture-level city in Zhejiang Province. It is located on the central coast of Zhejiang Province, adjacent to the East China Sea in the East, Shaoxing City and Ningbo City in the north, Wenzhou City in the south, Jinhua City and Lishui City in the West. Taizhou is inclined from west to East, mainly mountainous. It has authority over three municipal districts, three county-level cities, and three counties, Jiaojiang, Huangyan, and Luqiao. By the end of 2013, Taizhou had a permanent population of 6.038 million, accounting for 58.1% of the urban population.

Huangyan District of Taizhou city is connected to the Jiaojiang District of Taizhou City in the East, Luqiao District, Wenling City, and the Yueqing City of Wenzhou City in the south, Yongjia County and Xianju County in the West, and Linhai City in the north. The whole area is 989.89 square kilometers. At the level of planning practice, in response to the new normal of social and economic development, Zhejiang Province has been promoting the construction of "characteristic towns" since 2015, that is, to create a new development space platform of "small but refined" non-towns and non-districts", break the bottleneck of land resources, and promote the accumulation of productive services, innovative talents, and other high-end production factors, to realize the upgrading of "massive economy" (industrial clusters) or the agglomeration of emerging industries (Chen 2018). As the experience of Zhejiang characteristic towns is introduced to the whole country, the national characteristic towns approved with more than half a book have certain basic industrial clusters. This kind of "industry-based characteristic town" is usually located in the urban-rural fringe, and its industrial development has certain "rooted" or "localized" characteristics. The construction of characteristic towns is the first time that the discipline of urban and rural planning systematically deals with the practical problems of the interactive development of industrial areas and communities on the block scale, providing rich samples for the research on the interactive mechanism between industrial clusters and surrounding communities.

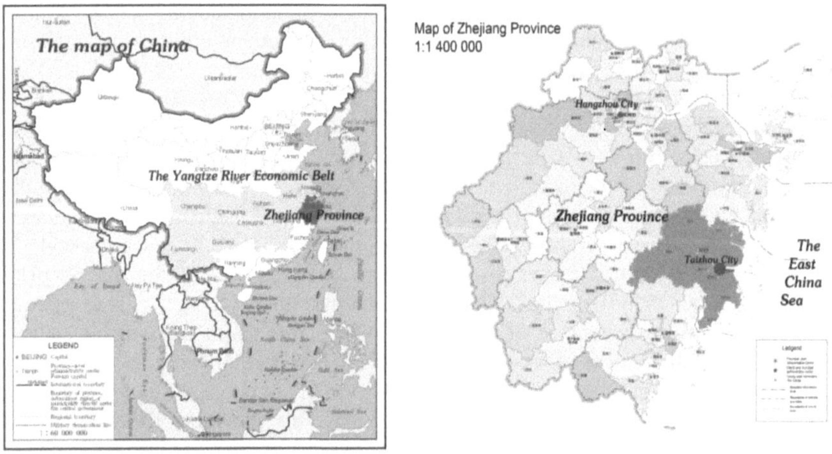


Figure 1. Location of the study area.

## 3.2 *Huangyan spatial performance analysis*

Through LBS data analysis, we found that each urban area of Taizhou is relatively independent, which is equivalent to the formation of the group structure. Taizhou implemented the integration of multiple counties and cities in the 1990s, so from the perspective of the development process, all regions are relatively fragmented. The main urban areas include Huangyan District, Luqiao District, and Jiaojiang District. The connection between districts is not very close. Due to the natural terrain and river segmentation, the pattern of cluster cities has become more obvious. From the OD (Od line means Origin-Destination line, which is the connecting line between the starting point and the ending point. Here means the daily commuting traffic flow between the centers of streets and towns) in streets of Taizhou in 2017 and the OD among streets of Taizhou in 2020, Taizhou has a high level of local job and living balance, and Huangyan District, Luqiao District, and Jiaojiang District have become independent clusters.

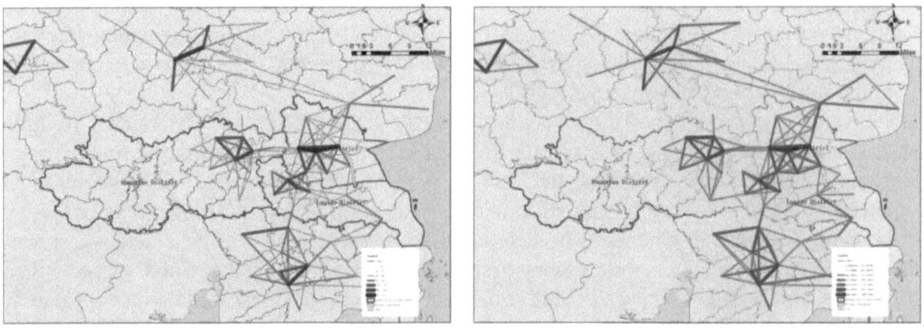

Figure 2. 2017 Taizhou inter Street commuting link (left) and 2020 Taizhou inter Street commuting link (right).

## 3.3 *Analysis of spatial activities of different populations*

By dividing the daily commuting rules of Taizhou People with different income levels in space. The income capacity level data includes three sources, namely, the tag that the developer gives the user and the user app list; push type, push frequency and other information, as well as the

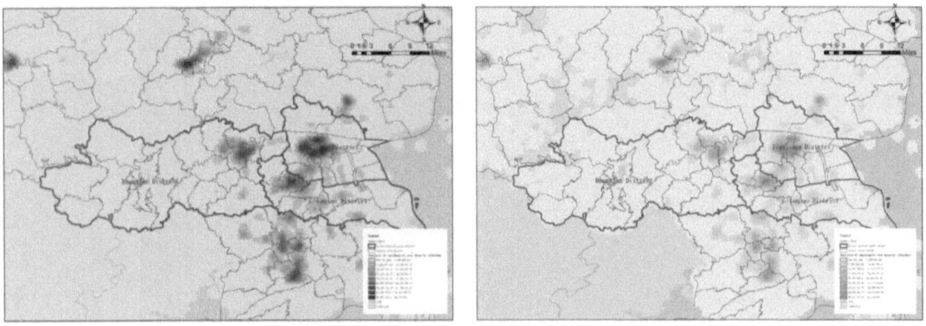

Figure 3. Analysis of Taizhou residential core density in 2020 (left) and employment core density in 2020 (right).

location information, parsed according to the longitude and latitude of LBS, identify the location information of the user's residential address and office address (hospital, school, business district, etc.). The AHP analytic hierarchy process is used to calculate the weighted sum of variables of different dimensions and convert them into 0 to 100 points. The income capacity is differentiated according to the size of the score. The income capacity score is normally distributed. For first-tier cities, 0 to 40 points (corresponding to a monthly income of less than 7,000 yuan) are designated as low-income capacity, and 41 to 50 points (corresponding to a monthly income of 7 k to 1.5 k) are designated as medium-income capacity, 51 to 100 points (corresponding to a monthly income of more than 1.5 k) are judged as high-income ability. For tier 2 and tier 3 cities: 0 to 40 points (corresponding to a monthly income of less than 2.5000 yuan) are designated as low-income ability, and 41 to 50 points (corresponding to a monthly income of 2.5 k to 5 k) are designated as medium-income ability, and 51 to 100 points (corresponding to a monthly income of more than 5 k) are designated as high-income ability. The number of people with high education levels identified in Taizhou is small and scattered. There are people with high education levels in the center of each district. The largest number of people with a middle education level are identified in Taizhou and followed the overall population travel trend. The total number of people with a low education level is less than that of people with a middle education level, but the difference is not much. The overall structure is similar to that of people with secondary education levels, but if it is enlarged to the district level, the difference will be obvious.

Compared with low-income groups, high-income groups are very few in magnitude. The center of Jiaojiang District forms an obvious flow center. High-income groups in Tiantai County, Xianju County, Wenling County, and Yuhuan city form relatively independent flow centers.

The total middle-income population is about three times more than the high-income population. The flow of middle-income people in Luqiao District, Huangyan District, and Jiaojiang District are closer, thus forming the main flow core. Wenling city is relatively independent and forms a secondary center.

The total number of low-income people is the largest. We can know that the overall income level of Taizhou residents is still low. From this map, it is obvious that the number of activities of the low-income group is greater in the south of Taizhou. Yuhuan city in the south of Taizhou is a village dominated by fishery and a market town with service functions, with less mold industry.

It can be drawn from the analysis chart that although the Chinese local government invests a lot in rural infrastructure to enhance urban-rural integrated development, the mobility data tells that the daily urban-rural ties are not so close. Therefore, it seems that small towns, not cities, play a more important role in the integrated development of urban and rural.

In the macro analysis, we observed several regions with high spatial performance, and Xifan village is one of them.

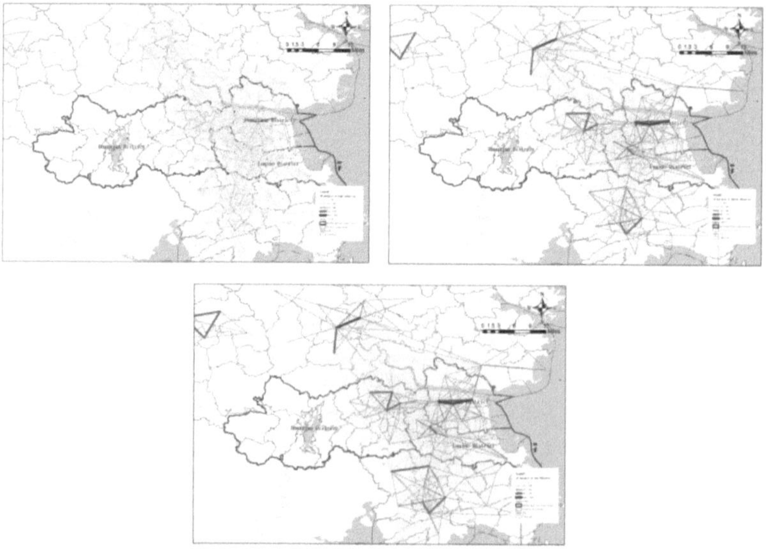

Figure 4. People's activity tracks at different levels (high, medium, and low).

## 4 EXCELLENT SPATIAL PERFORMANCE CASES IN HUANGYAN DISTRICT – XIFAN VILLAGE

### 4.1 *Xifan village, the urban-rural fringe of Taizhou City*

Xifan village is one of the villages with excellent spatial performance found in the above macro research. Xifan village and the surrounding mold factories selected in this paper are typical cases of "special effect towns based on local industrial clusters". Taizhou is known as the "degree of mold". Taizhou is the largest plastic mold production base in China, in which the plastic mold of TV shells covers 50% of TV manufacturers in China. Taizhou's plastic molds enjoy a high reputation at home and abroad. The industrial scale, production process, technical level, and quality management are in the leading position in the same industry in China. Taizhou is known as the "hometown of molds" and "Kingdom of plastic products". Taizhou has unique advantages in mold technology and talents. There are nearly 1,000 mold manufacturing enterprises and tens of thousands of plastic products manufacturing enterprises. Mold is the basic process equipment of industrial production. In the whole product production process, mold manufacturing is between product design and product manufacturing, and in the middle of the whole product value chain. Mold manufacturing is the preliminary preparation for product production, which is closely related to the production process. Therefore, mold manufacturing should be an important part of the production process. Xifan village in Taizhou and its surrounding villages have important production functions in mold manufacturing. At the same time, because its environment is located at the junction of urban and rural areas in the west of Huangyan District, it is taken as the research object by us.

In Xifan village, the local original residents continuously build illegal buildings to provide rental housing for the floating population in pursuit of profits, while the floating population from Xianju county continuously enter Xifan village by relying on their kinship and township relationship, and continuously expanding the production scale and establish a self-service living system by relying on mold production. All kinds of signs show that the nested society and production relations based on the relationship between relatives and townships play a vital role in shaping the space of Xifan village. In other words, this is a kind of "weak" space production formed by the migrant population, a vulnerable group, relying on their power. Through on-the-spot investigation and interview, we can conclude that the spatial production mechanism of Xifan village is that under the current

Figure 5. Typical urban and rural coexistence in Xifan Village.

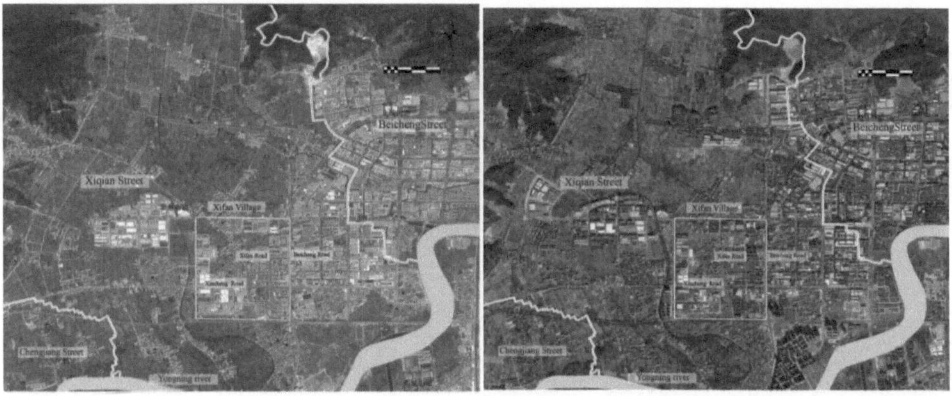


Figure 6. Spatiotemporal texture changes in the urban-rural interface.

macro institutional background, the floating population occupies the space through the flexible use of the spatial characteristics of the original villages in the city, and consolidates the space by playing games with other social actors based on social network relations to complete the process of reshaping the space.

When the floating population enters the city, they are mostly engaged in labor-intensive work because of their low cultural quality and lack of corresponding labor skills, and the mold production in Xifan village just meets this demand. Xifan village, as the junction of urban and rural areas, has relatively low rent for rental houses, which saves the living and production costs of the floating population to the greatest extent. Moreover, Xifan village has convenient transportation, and there are many subway and bus lines nearby for use. In the long-term development, another supply market for belt production materials has been formed in the east of Xifan village, and it is close to Zhongda Cloth Market in the north. It is very convenient to obtain orders, purchase raw materials, ship finished products, and commute to daily life. In Xifan village, the floating population of the Xianju nationality has taken full advantage of the spatial characteristics of the village, chose Xifan village for living and production purposes, and occupied and reconstructed its space.

First, analyze the "resident employment balance index" of the five functional areas. According to the LBS data, the working and living population data and resident population data of the five functional areas are counted. The calculation formula for calculating the "resident employment balance index" indicates that the "resident employment balance index" for Ao'an Village, Shuangfeng Village, and Mou Village is low, indicating that the area has many residents and relatively complete distribution of public facilities. It is a concentrated area for living, and the area has insufficient jobs or a single type of employment. Secondly, it analyzes the "employment balance index" of five functional areas. Through the mobile signaling data, the working and living population data and working population data of the five functional areas are counted. According to the calculation formula of the "employment balance index", it is analyzed that the "employment balance index" of Ao'an village, Shuangfeng Village, and Mou Village is low, indicating that there are more employment opportunities in this area, the regional public facilities are imperfect, and the livability is general. The author believes that the results of the separate analysis of the two indicators are uncertain and one-sided. To make the conclusion more objective and truer, it is suggested to combine the two indicators for comprehensive analysis. According to the analysis of the above two indexes, it can be inferred that if the "residents' employment balance index" is low and the "employees' housing balance index" is high, it indicates that the total resident population in the region is greater than the total employed population (jobs), which reflects that the region needs to increase more jobs compared with the high total resident population, such as the old urban area, the southern area of Sucheng new area and the central area of Suyu. If the "residents' employment balance index" is high and the "employees' housing balance index" is low, it indicates that the total resident population in the region is less than the total employed population (jobs), which reflects that the region needs to increase public facilities and improve livability relative to the high total employed population to attract more employed people to stay here, such as Susu Park. If both are low, it indicates that the employment and living in the region are extremely unbalanced, showing a trend of working here, but most of them do not live here, living here, and most of them do not work here. For example, the data for Shifu New Area and Suqian High-Tech Zone are the most obvious. There are many reasons for this imbalance, such as the general livability of the region and the single type of employment in the region. If both are relatively high, it means that there is a balance between jobs and living in this area, such as Xifan Village and Mou Village.

Table 1. Housing balance index and employment balance index.

## 4.2 *"Economic integration" and "social isolation" in Huangyan urban-rural interface*

The case shows that the mold production market and the surrounding communities are highly "integrated" in terms of economic functions, which is highlighted by the extension of market functions to the surrounding communities. However, the two sides are significantly "isolated" in social development, which is highlighted by the opposite social network structure from the above two perspectives. In fact, "economic integration" is not always a good thing for the development of the community. Although the functions of living, packaging, subcontracting, and storage provided by the community have reduced the factor cost of the entire mold trading industry cluster, they have paid a great social cost. Residents must endure the 24-hour continuous noise, light pollution, poor environment, and even potential safety hazards, The surrounding communities of the entire industrial cluster have also been in a state of "black and white reversed, disorderly piled, and also city and township". At the same time, "social isolation" is highlighted by the lack of communication and understanding between people in different circles in the social network structure and even the existence of deep "prejudice". In the structured interview with characteristic groups, the research group found that there was a lack of basic communication and understanding between mold makers and residents (usually property owners and local enterprise employees), the two largest social groups in Huangyan. Mold owners in Huangyan Area are usually immigrants, who have more prejudices against residents. Many mold dealers believe that local people develop a lazy and enterprising lifestyle by relying on the rent income. In contrast, the residents in the communities around the mold factory believe that although they do not earn as much money as the flower merchants, they have stable and decent jobs and can consider their families. Many well-known mold makers hold a disdainful attitude that they live a life of black and white.

This kind of industrial and social relationship of economic integration and social isolation makes the economic and social costs of the renovation of the surrounding areas of the Xifan village mold factory extremely high, which not only creates great obstacles to the benign interaction between the Dounan flower market and the surrounding communities but also locks the environmental quality and spatial style of the region in the state of "city and township" for a long time. This spontaneously formed contradiction cannot be automatically resolved through the role of the market mechanism. It needs to be solved through effective institutional supply and intervention mechanisms, and urban and rural planning should play a leading role in this kind of public intervention.

Xifan village has truly achieved the comprehensive utilization of land, realized the matching of the number of jobs and residential units, and realized the informal balance of work and housing. Because of the regional area, the smaller the city, the easier it is to maintain a certain degree of spatial balance. After investigation, 68.43% of couples work in the same place.

## 5 CONCLUSIONS AND PROSPECT

From the LBS analysis, we can see that the urban-rural interface has played a significant role in promoting the functional integration of various districts, and effectively formed the balance between working and living. Low-income people are more flexible in the urban-rural interface. In the future, we hope to increase the regional spatial performance by improving the following ways.

- Road construction, including roads and railways, is necessary to provide the country with convenient transportation links with the rest of the world.
- The introduction of the fund appears to be an easy method of allocating and raising funds for investments in this field through some actions taken by the government.
- The introduction of talent (labor) is aimed at promoting the return of personnel by providing some job positions.

Therefore, we are discussing ways to narrow the gap between urban and rural areas, which can be divided into several categories:

- The key to improving space performance lies in the industry. Industry determines whether people can stay. In the past, the way of building first and then attracting investment was inefficient or even ineffective.
- Industry selection should be based on local characteristics. It is not advisable to move the industries with urban functions directly to the countryside. It is also unsuccessful to attract investment through the establishment of micro-industrial parks. Only by extracting local cultural customs and integrating them into production can it be more successful.

Taking Xifan village and the surrounding villages as an example, the high-level efficiency of job and living balance and spatial performance is mainly reflected in the relevant mechanisms of industrial clusters and living space. The formation of spatial performance is generally rooted in the local natural endowment, cultural foundation, social relations, and institutional structure, which constitute the internal conditions for the formation and development of industrial clusters in a certain region. Therefore, spatial performance should be the core issue that must be considered in the development of the region where the urban-rural fringe is located. This study has a strong reference value for the development of these characteristic towns. With the economic development entering the new normal, the exogenous industrial development power is gradually declining, while the importance of local industrial development is becoming increasingly prominent. This endogenous industrial development power has strong regularity, but there is still a lack of planning research in this area. The development of "Characteristic Towns Based on Industrial Clusters" across the country provides an excellent opportunity. Through in-depth and solid industrial research and social research, planners should investigate new models for improving spatial performance development in urban-rural transition areas in the future by absorbing the "Zhejiang experience".

## REFERENCES


Burgess, E. (1925). *The Growth of the City: An Introduction to a Research Project.* In: Park, R.E. Burgers, E.W. and McKenzie, R.D., Eds., (1984). *The City*, University of Chicago Press, Chicago.

Chen C, You L, Zhu J, (2021). Interaction mechanism between industrial clusters in urban-rural fringe and surrounding communities – a case study of Dounan flower market in Kunming [j] *Southern architecture*, 2021 (05): 64–70.

Chen X, Li M, Wang Z, (2020). Exploration of job housing balance path based on big data analysis [j] *Urban and rural construction*, (20): 42–43.

Chen J, (2018). Research on upgrading of Huangyan mould industry cluster from the perspective of global value chain [j] *Economic and trade practice*, (13): 38–39.

Douglass, M. (1992). Global interdependence and urbanization: planning for Bangkok mega-urban regional. Paper presented at the international conference on managing mega-urban regional inASEAN countries: policy challenges and responses. *Asia Institute of Technology.Bangkok*, 30 Nov.–3 Dec.

Li F, Zhao M, Huang J, (2021). On the performance and development mode selection of the spatial structure of big cities [j] *Journal of urban planning*, (01): 18–27.

Li F, Zhao M, Wu M, Huang J, (2017). On the "spatial performance" mechanism of "multi center" spatial structure in big cities – a study based on LBS portrait data and conventional census data in Xiamen [j] *Journal of urban planning*, 2017, (05): 21–32.

Mondal, A. R., Chang, X. Y., Von Stroh, D., Lemp, J., & Viswanathan, K. (2021). *Using Location-Based Services Data for Climate Resilience and Emergency Planning* (No. TRBAM-21-03823).

Zhang R, (2020). *Research on spatial and temporal distribution of Hefei population based on big data and spatial syntax* [d] Hefei University of technology, 2020.

*Advances in Urban Construction and Management Engineering – Cha (Ed)*
*© 2023 Copyright the Author(s), ISBN 978-1-032-39018-5*

# Experimental study on mechanical properties of new assembled swinging column structure

Xiaomeng Zhang*
*China Architecture Design & Research Group, Beijing, China*

Bingzhen Zhao
*Future City Innovation Technology Co., Ltd., Shaanxi Construction Engineering Holding Group, Xi'an, China*

Xianghui Kong & Wenting Liu
*China Architecture Design & Research Group, Beijing, China*

Jiashu Hao
*Tianjin University, Beijing, China*

ABSTRACT: In this paper, a kind of swing column structure is proposed. Rubber pads are set up above and below the column to release the temperature and displacement under impact load. Through the structure test, the corresponding test scheme is set up. The test structure shows that the system has good deformation performance and mechanical performance, and can be effectively applied to practical engineering.

## 1 INTRODUCTION

In this paper, a kind of swing column structure is proposed. Rubber pads are set above and below the column to release the temperature and displacement under impact load. Many experts and scholars have carried out certain studies on the form of the swing structure. Palermo et al. (Palermo & Pampanin 2008; Palermo et al. 2007) proposed combining unbonded prestress and energy dissipation device to apply it to the structure of a swing bridge pier, and carried out quasi-static tests. Mander and Cheng (1997)applied swing structure to the seismic Design of Bridges and proposed a Damage Avoidance Design of self-resetting swing pier. The development history of the swinging and self-reset structure is reviewed, and the basic principle of the swinging and self-reset structure is briefly introduced (Lu et al. 2014, 2019). In this paper, the mechanical characteristics of two swinging columns are verified by quasi-static tests, and their seismic performance is analyzed.

## 2 EXPERIMENT DESIGN

### 2.1 *Specimen design*

Two specimens were designed and made for the test. The geometric dimensions and section reinforcement structures of the specimens are shown in Figure 1 and Table 1.

*Corresponding Author: 155203255@qq.com

DOI 10.1201/9781003348023-27 

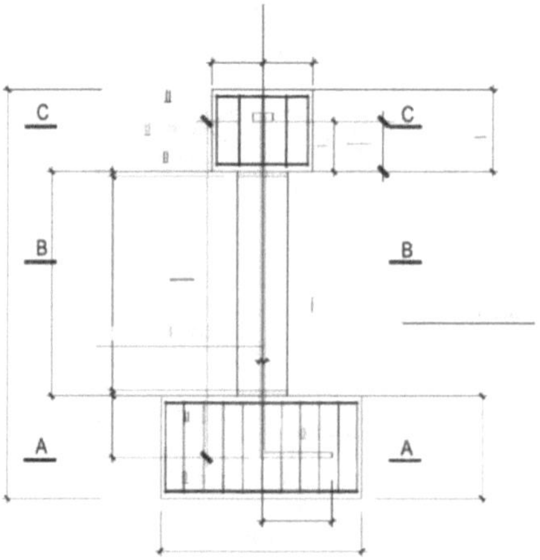


Figure 1. Structures of RC-1 and RC-2.

Table 1. Component size.

| Serial number | The steel bar diameter (mm) | The thickness of the rubber pad on top of the column (mm) | The thickness of the rubber pad at bottom of the column (mm) | Swing column aspect ratio | Axial pressure (kN) |
|---|---|---|---|---|---|
| RC-1 | 25 | 25 | 25 | 4.2 | 150 |
| RC-2 | 20 | 25 | 25 | 4.2 | 150 |

### 2.2 *Test equipment and loading equipment*

The test loading device is shown in Figure 2. The specimen floor beam is fixed to the rigid ground by a steel compression beam and anchor bolt. In the test, a predetermined vertical axial pressure was first applied on the top of the column and kept constant during the loading process, and then monotonically increasing horizontal load was applied on the top of the specimen through a 100-ton horizontal jack. According to *Concrete Structure Test Method* (Standard ID: GB/T50152-2012), for formal preload, a test load solution should be used to make sure the instrument is functioning correctly, and to ensure correct contact load calculation using hierarchical load. Once 2.5 mm of specimen damage has been observed per level, the load should not be continued.

## 3 TEST PROCESS AND FAILURE CHARACTERISTICS

### 3.1 *RC-1*

As there were rubber pads between the RC-1 column pier, the top beam, and the foundation, the load-displacement curve of the specimen developed linearly in the initial loading stage. When the RC-1 was loaded to 7.5 mm, the rubber pad at the bottom of the swing column was slightly bulging, and there was no gap between the rubber pad at the tensile side of the swing column and the foundation. When the specimen RC-1 was loaded to 10 mm, a gap appeared between the upper rubber pad on the tensile side of the swing column and the top beam, and a gap appeared between the lower rubber pad on the tensile side of the swing column and the foundation. With the increase of the horizontal displacement load of the top beam, the gap between the rubber pad on the tension side of the upper part of the swing column and the top beam does not change significantly, but

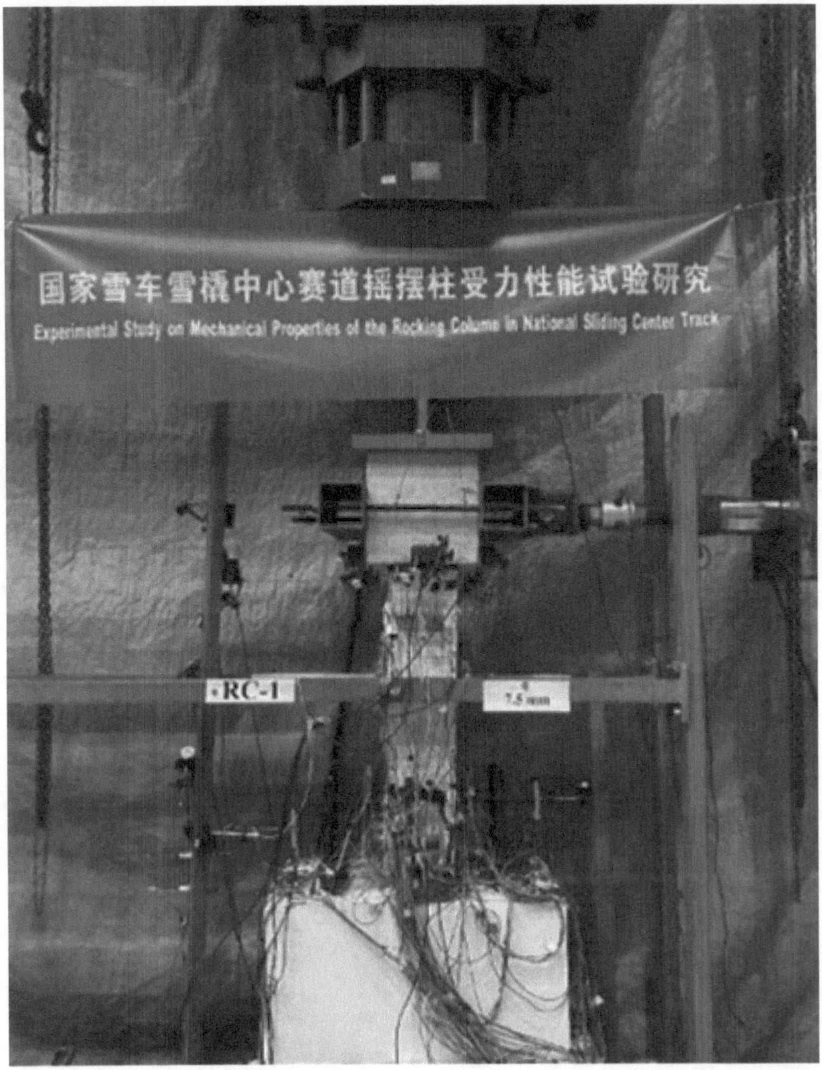


Figure 2. Schematic diagram of the loading device.

the rubber pad on the compression side of the lower part of the swing column pops out obviously, and the gap between the rubber pad on the tension side of the swing column and the foundation gradually increases. When the horizontal load is loaded to 60 mm, the maximum gap is 10 mm; when the specimen was loaded to 55 mm, the gap between the rubber pad on the tension side of the lower part of the swing column and the foundation remained stable, the rubber pad on the compression side of the lower part of the swing column popped out, and the rubber pad and the swing column were empty at the position of the steel rod. At this time, the load-displacement curve began to rebound, indicating that the steel rod began to be tensile deformation. As the horizontal displacement load continues to increase, the gap between the rubber pad on the tensile side of the lower part of the swing column and the foundation continues to increase, and the compression deformation of the rubber pad on the compression side of the lower part of the swing column has no obvious change. When the specimen RC-1 was loaded to 120 mm, the compression side of the swing column decreased by 10 mm, and the tension side was 15 mm. When specimen RC-1 was loaded to 130 mm, the load continued to rise. Because the section size of specimen RC-2 swing column was 250 mm × 250 mm, when the specimen was loaded to 130 mm, half of the column section size had been exceeded, and the loading stopped. The rubber pad at the bottom of SPECIMEN RC-1 was twisted after the test stopped.

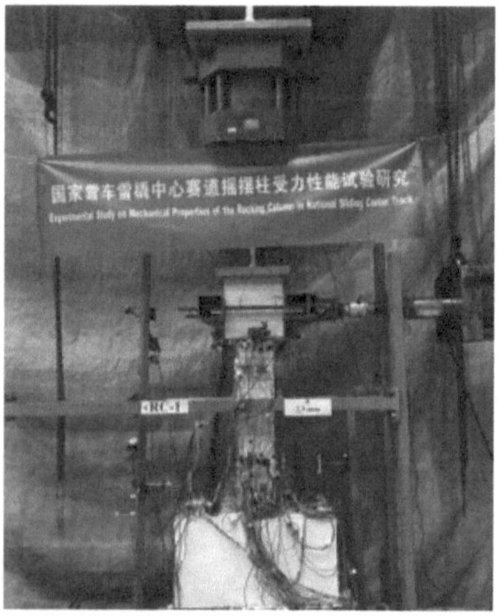


(a) Overall diagram of force on RC-1

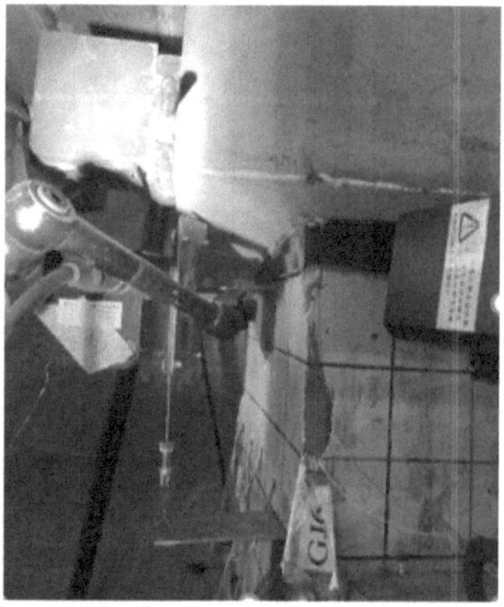

(b) Pressure on the upper rubber pad

(c) On RC-1, the overall force was drawn at 130 mm

(d) The lower rubber pad is pulled at 130 mm

Figure 3. Experiment phenomenon of RC-1.

## 3.2 *RC-2*

Due to the existence of a rubber cushion between the RC-2 column pier, the top beam, and the foundation, the load-displacement curve of the specimen develops linearly in the initial loading stage. When the specimen RC-2 was loaded to 7.5 mm, the pressure side of the rubber pad at the bottom of the swing column was slightly bulging, and there was no gap between the rubber pad at

(a) At 60 mm, the overall force diagram of RC-2 was obtained

(b) At 60 mm, the lower rubber pad is subjected to a force diagram

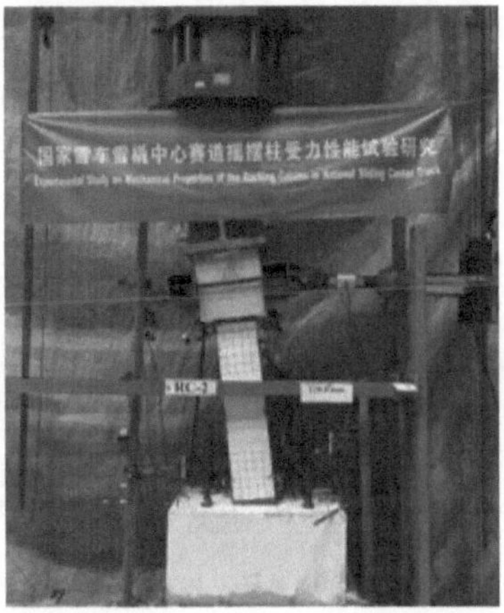


(c) At 120 mm, the overall force diagram of RC-2 was obtained

(d) At 120 mm, the lower rubber pad is subjected to a force diagram

Figure 4. Experiment phenomenon of RC-1.

the tension side of the swing column and the foundation. When the specimen RC-2 was loaded to 10 mm, a gap appeared between the upper rubber pad on the tensile side of the swing column and the top beam, and a gap appeared between the lower rubber pad on the tensile side of the swing column and the foundation. With the increase of the horizontal displacement load of the top beam, the gap between the rubber pad on the tension side of the upper part of the swing column and the top beam does not change significantly, but the rubber pad on the compression side of the lower part of the swing column pops out obviously, and the gap between the rubber pad on the tension side of the swing column and the foundation gradually increases. When the horizontal load is loaded to 60 mm, the maximum gap is 10 mm. When the specimen was loaded to 70 mm, the gap between the rubber pad on the tension side of the lower part of the swing column and the foundation remained stable, and the rubber pad on the compression side of the lower part of the swing column popped out, and the rubber pad and the swing column were empty at the position of the steel rod. At this time, the load-displacement curve began to rebound, indicating that the steel rod began to be tensile deformation. As the horizontal displacement load continues to increase, the gap between the rubber pad on the tensile side of the lower part of the swing column and the foundation continues to increase, and the compression deformation of the rubber pad on the compression side of the lower part of the swing column has no obvious change. When the specimen RC-2 was loaded to 120 mm, the compression side of the swing column decreased by 10 mm, and the tension side was 15 mm. When the specimen RC-2 was loaded to 130 mm, the load decreased to 5.2 kN. Because the section size of specimen RC-2 swing column was 250 mm × 250 mm, when the specimen was loaded to 130 mm, half of the column section size had been exceeded, and the loading stopped.

## 4 CONCLUSIONS

This chapter introduces the mechanical properties of the swing column structure. Through the mechanical experiment of the swing column, we can obtain the following conclusions.

(1) The swinging column structure has good mechanical performance and large deformation performance. It is a deformable structure.
(2) Compared with the general structural system, the swing column structure has greater deformation performance and can be quickly recovered after deformation. It can be effectively deformed and recovered under impact load and temperature load, which is a good structural form.

## ACKNOWLEDGMENTS

This paper is supported by the Advanced Innovation Center for Future Urban Design of the Beijing University of Civil Engineering and Architecture (Grant No. UDCGJ002).

## REFERENCES

Lu Xilin, Zhou Ying, Chen Cong. Research progress on innovative earthquake-resilient structural systems[J]. *Earthquake Engineering and Enginrring Dynamics*, 2014, 34 (4): 130–139.

Lu Xilin, Dayang W U, Zhou Y. State-of-the-art of earthquake resilient structures[J]. *Journal of Building Structures*, 2019.

Mander J B, Cheng C T. Seismic resistance of bridge piers based on damage avoidance design[J]. *Technical Report NCEER*, 1997, 97–0014.

Palermo A, Pampanin S. Enhanced Seismic Performance of Hybrid Bridge Systems: Comparison with Traditional Monolithic Solutions[J]. *Journal of Earthquake Engineering*, 2008, 12(8):1267–1295.

Palermo A, Pampanin S, Marriott D. Design, Modeling, and Experimental Response of Seismic Resistant Bridge Piers with Posttensioned Dissipating Connections[J]. *Journal of Structural Engineering*, 2007, 133(11):1648–1661.

*Advances in Urban Construction and Management Engineering – Cha (Ed)*


# Analysis of the application of the teaching of Engineering Geology Course based on case teaching method

Sai Zhang & Pooya Saffari*
*Qingdao City College, Qingdao, Shandong, China*

Wenzheng Jiang
*No. 19 Middle School of Qingdao, Qingdao, Shandong, China*

Reza Andasht Kazeroon
*Xi'an University of Architecture and Technology, Xi'an, Shaanxi, China*

ABSTRACT: Engineering geology is a course that emphasizes both theory and practice. Applying the case teaching method in the teaching of Engineering Geology can effectively realize the integration of theory and practice. Combining the nature of the course and the characteristics of the case teaching method, this paper takes the geological disaster module in the course as an example, studies the debris flow disaster in the Hailuo Valley gully on the Qinghai-Tibet Plateau, and analyzes the application of the case teaching method in the teaching of Engineering geology.

## 1 COURSE BACKGROUND AND OVERVIEW

### 1.1 *Course background*

Disasters and emergencies have been increasing all over the world. The 42nd United Nations General Assembly in 1987 designated the ten years at the end of the 20th century as the "International Decade for Natural Disaster Reduction." Later on, the World Summit on Sustainable Development in the early 21st century included "Natural Disaster Reduction" in the Johannesburg Plan of Implementation (Wapner 2003). In 2009, the United Nations International Agency for Disaster Reduction (ISDR) proposed education on natural disaster prevention to build a global platform for disaster risk reduction. In the same year, the "Global Disaster Risk Reduction Assessment Report: Risk and Poverty in Climate Change" (GAR) was published, which comprehensively discussed the location, causes, and ways of the occurrence of disaster risk. It can be seen that global disaster events and national disaster education are receiving more and more attention.

China is one of the countries with the most serious natural disasters in the world, especially geological disasters. Geological disasters refer to phenomena that cause direct or indirect losses to the safety of human life and property due to natural or man-made reasons, including landslides, collapses, debris flows, lagging collapses, ground fissures, and ground subsidence (Bin 2020). With the improvement of national infrastructure and mineral resources development capacities, the problem of geological disasters is becoming more and more serious. In order to meet China's needs for talent training in the field of geological disasters, the teaching of civil engineering courses that permeate disaster education is becoming increasingly important.

*Corresponding Author: pooya_saffari@qdc.edu.cn

DOI 10.1201/9781003348023-28

## 1.2 *Course overview*

Civil engineering major educates students to comprehensively analyze and solve relevant problems using qualitative and quantitative analysis tools and have the professional quality of engaging in civil engineering-related work. The course takes engineering practice as the core, pays attention to theoretical knowledge, practical ability, and professional quality, and cultivates innovative talents in civil engineering. Based on the national curriculum outline and the national gold curriculum standard of colleges and universities in China, the curriculum takes constructivism as the theoretical support. It creates a curriculum system for the cultivation of innovative talents through "information technology-assisted teaching," "geology engineering teaching case," "field investigation practice", and other modes.

Engineering geology is an important professional basic subject based on the engineering practice ability. It establishes the objectives of talent training with knowledge, ability, and literacy. The main goal is to train students to master the basic concepts and theories, main analysis methods, and main prevention measures of basic geology and engineering geology, so that they can be closely combined with subsequent professional courses. In addition, to develop students' ability to solve relevant geological problems in civil engineering design effectively, construction, and operation management to ensure the engineering buildings' safety, economy, and stability. It aims to explore internal potential, stimulate innovation consciousness, cultivate the core quality of students' craftsmanship spirit, and improve social adaptability.

# 2 APPLICATION OF TEACHING CASE METHOD IN GEOLOGY ENGINEERING COURSE

1. Overview of the teaching case method
In 1870, Christopher, Dean of the school of law at Harvard University, formally proposed the "Teaching Case Method." The teaching case method refers to that teachers set teaching objectives according to the curriculum system, combined with appropriate cases, and guide students to measure, investigate, think, calculate, analyze, and communicate around the case materials, engaging them in active exploration. In this method, students can deepen their understanding of basic knowledge and principles, improving their ability to face real cases and analyze and solve problems (Weiqiang 2020). The case teaching method is an interactive and open method that combines theoretical knowledge with practical skills. Based on typical real cases, the teaching process is organized with specific, clear, and real situational materials, with students as the main body and teachers as the auxiliary, to complete the teaching objectives in the case of inquiry.

2. The significance of teaching case method application
Engineering geology is an important professional basic course. According to the teaching of seven project points in the two modules of basic geology and engineering geology, and teaching case method is adopted. Each unit or module takes engineering examples as clues and adopts the independent exploration mode. In the process of solving engineering geological problems, it pays attention to the understanding of theoretical concepts, the explanation of the formation process of geological disasters, and the measures to solve geological problems. The summary and reflection modes are carried out to expand the breadth and depth of the profession.

## 2.1 *Student dimension*

The application of the teaching case method in the engineering geology course is the reproduction of typical engineering geological events. It allows students to experience the direct feelings of the real situation through analysis and research. In the process of inquiry, students' subjective initiative and interest in learning are stimulated, and student's ability to ask questions, study, and solve problems in practice is improved.

### 2.2 *Teacher dimension*

The case selection in the teaching case method tests teachers' professional knowledge reserve and classroom management skills. Teachers need to select the best cases from numerous materials according to the actual classroom teaching practice and teaching objectives and organize the relevant materials of the case. They should have both basic theoretical knowledge and practical application skills. In the process of case teaching, teachers play the role of facilitators, promoting the research and development of the whole case situation, which poses a high challenge to the professional quality of teachers.

### 2.3 *Classroom dimension*

The teaching content of engineering geology class includes not only the various geological structure types, but also the complex geological process, which challenges the students' abstract ability, generalization ability, and spatial ability. Conducting teaching in the case teaching mode can integrate obscure and difficult textbook knowledge into clear, practical cases, deepen students' theoretical knowledge and analytical ability, truly develop students' professional quality, and greatly improve the efficiency and effectiveness of classroom teaching.

## 3 CASE ANALYSIS - TAKING THE DEBRIS FLOW IN THE HAILUO VALLEY OF THE QINGHAI-TIBET PLATEAU AS AN EXAMPLE

The currently used textbook of "Civil Engineering Geology" (third edition), edited by Hu Houtian and Bai Zhiyong of the Higher Education Press, describes the debris flow in Chapter 5, which is titled "Unfavorable Geological Phenomenon and Prevention". It selects the Hailuo Valley gully in the Qinghai-Tibet Plateau as a case study. The material of the debris flow disaster in the Hailuo Valley of the Qinghai-Tibet Plateau is from the paper "study on the genesis of rainfall-glacier mixed type debris flow of Hailuo gully in northwest Yunnan on July 28, 2019" by Zhaoxin et al. (2020) and was adapted as the 18th question of the comprehensive question of geography in the 2021 Shandong Provincial Academic Level Examination. The debris flow disasters in the case have significant regional representation, which can reflect the causes and hazards of debris flow disasters under the specific background of the region. It fully reflects the nature of the physical and dynamic conditions of debris flow as a typical debris flow case.

Table 1. Material list of debris flow cases in the Hailuo Valley gully in the Qinghai-Tibet Plateau.

| | |
|---|---|
| Teaching Chapters | Civil Engineering Geology (Third Edition)<br>Chapter 5 "Unfavorable Geological Phenomenon and Prevention," Section 3 Debris Flow |
| Teaching objectives | 1. According to the specific case materials, students can analyze the causes of this debris flow disaster. On this basis, general conditions for the formation of debris flow disasters are summarized.<br>2. In the process of group cooperative learning, combined with specific case materials, the teacher explains the harm caused by this debris flow disaster and tries to put forward reasonable measures for disaster prevention and mitigation. |
| Case presentation | (1) Materials<br>The Hailuo Valley is located on the southeastern edge of the Qinghai-Tibet Plateau, in the middle of the Hengduan Mountains, with an area of 53.4 square kilometers. The main ditch originates from the west side of Haba Snow Mountain and merges into the Chang Jiang River, a first-class tributary of the Jinsha River, with 12.8 kilometers. From bottom to top, the basin is divided into a high mountain and cold zone peak ridge area. Wide valley area, temperate zone narrow valley area, and subtropical low valley area. The precipitation in the basin increases significantly with the elevation, and the annual precipitation in the peak ridge area above 4200 m above sea level exceeds 1100mm. On July 28, 2019, the 6-hour precipitation in the peak ridge area reached 60.4 mm, which stimulated the mixed debris flow of ultra-large-scale precipitation and glacial meltwater (Xin 2020). |

*(continued)*

Table 1. Continued

| | |
|---|---|
| | (2) Image material<br>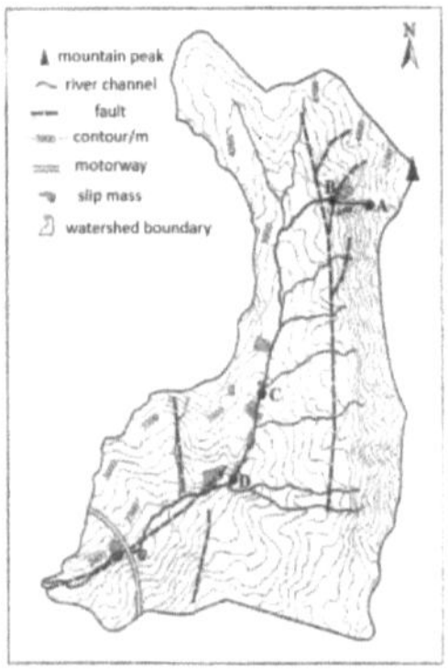<br> |
| Case Analysis | The case takes the 7.28 debris flow disaster in the Hailuo Valley gully of the Qinghai-Tibet Plateau as an example and presents the typical mixed debris flow of precipitation and glacial meltwater in this region. In combination with the requirements of the course syllabus and textbooks for the content of this part, and based on the case material information, two inquiry topics are planned to be designed as the starting point to promote students to carry out inquiry learning of the case.<br>(1) Topic design<br>Inquiry topic 1: State the source of the solid matter of the debris flow disaster, and analyze the effect of the topography of different sections of the main gully of the Hailuo Valley on the formation of debris flow.<br>Research topic 2: Analyze the reasons why the debris flow in the Hailuo alley poses a serious threat to local construction engineering and propose rational disaster prevention and mitigation measures.<br>(2) Design intent<br>This case is based on a rainfall-glacier meltwater mixed debris flow in the Hailuo Valley gully on the Qinghai-Tibet Plateau, which involves the source of debris flow and the relevant content of its formation process. It can improve the basic literacy level of students. Research topic 1 focuses on analyzing the two main causes of debris flow disasters: physical and dynamic conditions. Students are required to logically and clearly explain the causes of debris flow in combination with regional background and special illustrations. Research topic 2 focuses on the hazards and prevention of debris flows. Students are required to summarize the causes of serious disasters caused by regional debris flow disasters in combination with the region's geological conditions and construction engineering needs and put forward specific measures for disaster prevention and reduction.<br>(3) Topic assumption<br>Research topic 1: Material sources mainly include moraine, colluvial, landslide deposits, fault fragments, and weathered debris. The moraine is mainly formed by the widely developed alpine glaciers on the Qinghai-Tibet Plateau. The glaciers transported the deposits formed during the glaciation process; the landslide deposits are mainly produced and accumulated by the micro collapses and landslides in the local area. The rock mass is mainly derived from the special legend fault, the relative displacement near the fault, and the dislocation of the rock mass to form broken rock mass debris. There are differences in various topographical conditions of the main channel. The terrain in the high section is steep and prone to collapse, and the runoff confluence speed is fast, providing material and hydrodynamic conditions for forming debris flow. The terrain in the middle section is gentle and open, providing a place to accumulate many loose solid materials. The gully in the low section is narrow, straight, and steep, which speeds up the passage time of debris flow.<br>Research topic 2: This area has abundant solid material sources and hydrodynamic conditions, leading to frequent debris flows. The debris flow is fast, large in scale, and extremely destructive. In the peak and ridge areas of the basin, the precipitation is extremely large and superimposed with glacial meltwater, collecting a large amount of runoff. Therefore, the debris flow is mostly developed in the high-altitude area, while they are not easy to detect in the living and production areas in the downstream areas, and it is very easy to cause devastating damage to the project construction. Based on the regional background and actual needs, measures such as advanced monitoring and prediction mechanisms, improving disaster emergency response capabilities (Tao 2014), and strengthening self-rescue capabilities can be taken. |

## 4 CONCLUSION

Adopting the case teaching method in the classroom teaching of engineering geology can effectively improve the teaching quality and efficiency and make the classroom teaching content more abundant and vivid. In the practical application process, attention should be paid to the coordination and cooperation of subjects, objects, and case materials in the process of teaching implementation. Thus, students can truly feel the improvement in ability brought by case analysis. Meanwhile, teachers can guide the advancement process of classroom teaching to maximize the benefits of case teaching.

## REFERENCES


Bin, Z., Nengxiong, X. & Zhongjian, Z. (2020). Construction and application of a practical teaching base for geological disaster management, monitoring and early warning (in Chinese). *China Geological Education*, 29(01): 111–114.

Tao, X. (2014). A new water-sediment separation structure for debris flow defense and its model test (in Chinese). *Bulletin of engineering geology and the environment*, 73(4): 947–958.

Wapner, P. (2003). World Summit on Sustainable Development: toward post-Jo'burg environmentalism. *Global Environmental Politics*, 3(1): 1–10.

Weiqiang, Z, Shuyun, Z. & Liwen, C. (2020). The application of the case teaching method in the course of "Disaster Geology" (in Chinese). *Curriculum Education Research*, (17): 254–256.

Zhaoxin, Z., Haitai, Z. & Zhifang, Z. (2020). Study on the genesis of the "7·28" rainfall-glacier meltwater mixed debris flow in Haibaluo Valley, northwestern Yunnan (in Chinese). *Journal of Engineering Geology*, 28(06): 1339–1349.

*Urban management and optimization of public environmental governance*

*Advances in Urban Construction and Management Engineering – Cha (Ed)*
 *ISBN 978-1-032-39018-5*

# Analysis for the government governance of highway PPP projects based on CSF and KPI

Jie Li*, Lijuan Yao* & Yanmei Guo*
*Shandong Institute of Commerce & Technology, Jinan, Shandong Province, China*

ABSTRACT: In recent years, the Chinese government has put more attention to the introduction and popularizing of PPP projects, and relevant departments are actively improving the PPP regulation system. In addition, to ensure the legitimacy of the project, government governance should have the responsibility to promote the success of the project. On the basis of literature research, we extract the government governance elements from two dimensions of CSF and KPI, and compared them with the current situation of government governance in our country, propose how to improve the regulatory system in our country, and guarantee the effectiveness and comprehensive of the governance.

## 1 INSTRUCTION

PPP (Public-Private-Partnership) model was applied earlier in the field of expressways in my country and achieved good results, making great contributions to the construction of expressways in my country. The problem, which has been criticized by all walks of life, has affected the successful implementation of the project, reflecting that the government's supervision in such projects needs to be strengthened.

In order to ensure that the supervision of such projects is more reasonable and efficient, the government should not only supervise the legal and compliant behavior of all parties involved in the project but also scientifically and rationally select the focus of government supervision from the perspective of ensuring the success of the project. If there is too much intervention in the market operation of the project, it is necessary to grasp the key points to ensure the effectiveness and comprehensiveness of supervision and protect the public interest. CSF (critical success factor) and KPI (key performance indicator) are the key factors to ensure the success and benefit of the project, and a large number of experts and scholars have conducted in-depth and detailed research in this area, so this paper adopts the literature research method. This paper studies the Government Governance Factors (GGF) from the two dimensions of CSF and KPI, compares them with the current state of government supervision in expressway projects in my country, finds out the existing problems, and puts forward suggestions for further improving the government supervision system of expressway PPP projects in my country (Chan1 2010).

## 2 KEY FACTOR EXTRACTION

This paper searches the literature on CSF and KPI in PPP projects and expressway projects at home and abroad in recent years, selects 62 representative articles from them, and extracts the elements closely related to government supervision (Li 2005).

*Corresponding Authors: 425698086@qq.com, 786112657@qq.com and 425698086@qq.com

DOI 10.1201/9781003348023-29

### 2.1 *Extraction of key success factors related to government*

The research on the project CSF mainly adopts the methods of expert interview, literature research, and questionnaire survey. Among them, the key success factors summarized by Li Bing based on the questionnaire cover a wide range and are comprehensively classified. Therefore, this paper uses this as the basis for analysis and sorting. The frequency of 35 papers related to CSF (Yamout 2007) was calculated according to the following equation (1), and the frequency of occurrence of each CSF is shown in Table 1.

$$f = V/\lambda \tag{1}$$

F indicates frequency, V represents the number of samples containing this parameter, and λ represents the total number of samples.

Table 1. Frequency of occurrence of CSF in literature.

| CSFs | Frequency f | CSFs | Frequency f |
|---|---|---|---|
| 1. Stable macroeconomic conditions | 0.400 | 10. Technical feasibility of the project | 0.286 |
| 2. Good legal system | 0.400 | 11. Public sector and corporate rights sharing | 0.143 |
| 3. Sound economic policies | 0.228 | 12. Political support | 0.371 |
| 4. Available financial markets | 0.200 | 13. Social Support | 0.200 |
| 5. Multi-benefit goals | 0.343 | 14. Well-established organizations and public institutions | 0.143 |
| 6. Appropriate risk allocation and sharing | 0.571 | 15. Transparency of the procurement process | 0.343 |
| 7. Commitments and responsibilities of the public sector and businesses | 0.400 | 16. Competition in the procurement process | 0.314 |
| 8. Good social actors | 0.486 | 17. Government Assurance | 0.314 |
| 9. Good Administration | 0.257 | 18. Detailed and true assessment of costs and benefits | 0.257 |

From the above analysis, it can be seen that proper risk allocation and sharing is the most frequent occurrence; good private institutions come second; stable macroeconomic conditions, sound legal systems, and commitments and responsibilities of public and private institutions appear. The frequencies are all 0.4, which indicates that the government plays a very critical role in PPP projects. The franchise agreement clarifies the responsibilities and rights of the government and enterprises in the project, serving as the most important contract document in a PPP project. The government should strengthen its supervision and ensure that the process of determining risks and benefits for both parties is standardized and scientific. Some CSFs not covered in the above list have also appeared in the literature, such as information exchange and disclosure, capital structure, meeting public requirements, pricing mechanism, reward and punishment mechanism, compensation mechanism, public sector capacity, and professional knowledge capacity (Goran & Nevena 2013). According to the above analysis, combined with the characteristics of expressway PPP projects, there are 9 main CSFs of expressway PPP projects related to the government, as shown in Table 2.

#### 2.1.1 *Extraction of government-related KPIs*

Due to the different thinking angles of different scholars, the selection of performance indicators is quite different, and it is not suitable to use the frequency statistics method. Therefore, this paper selects 25 KPI-related literature Focus is placed on the indicators related to government regulation, which are redefined and reclassified (Zu 2014).

This paper extracts the KPIs of expressway PPP projects related to government supervision from the four aspects of the economy, technology, society, and ecological environment, as shown in

Table 2. CSFs of expressway PPP projects related to government regulation.

| Number | CSFs | Number | CSFs | Number | CSFs |
|---|---|---|---|---|---|
| 1 | Reasonable risk allocation and sharing | 4 | Reasonable allocation of commitments and interests between public and private | 7 | Government guarantee |
| 2 | Good social participant | 5 | Good purchasing process | 8 | Reasonable project plan |
| 3 | Sound legal system | 6 | Good administration | 9 | Information exchange and disclosure |

Table 3. The economic KPIs focus on the cost-effectiveness of the expressway project using the PPP model and whether it can achieve value for money (VFM); the technical KPIs focus on whether the project meets the quality requirements and whether the supporting facilities are complete; the social KPIs focus on the project, the social impact and social benefits of the surrounding area, as well as the user's (driver's) evaluation and satisfaction of the road conditions; the KPIs in the ecological environment mainly reflect the impact on the surrounding ecological environment during the construction and operation of the expressway (Liu 2014).

Table 3. Expressway PPP project KPIs related to government regulation.

| Economic aspect KPIs | Technical aspects KPIs | Social aspect KPIs | Ecological environment KPIs |
|---|---|---|---|
| 1. Can it achieve value for money? | 1. Highway detection IRI | 1. Public satisfaction | 1. Degree of pollution control |
| 2. Whether the project life cycle cost is reduced | 2. Road smoothness | 2. The degree of influence of regional economic development | 2. Degree of protection of natural resources utilization |
| 3. Cost-effectiveness of the project | 3. Completeness of transportation facilities | 3. Influence degree of the regional road network | 3. Degree in environmental management |
| 4. Project risk management capabilities | 4. Main engineering and traffic safety | 4. The degree of influence of social progress | 4. Degree of influence on ecological balance |
| | | 5. The degree of influence of external contacts | |

## 3 ANALYSIS OF GOVERNMENT SUPERVISION ELEMENTS

Combined with the whole life process of the expressway PPP project, as well as the extracted 9 CSFs and 17 KPIs, the key issues and key links that the government should pay attention to in the supervision process are further analyzed, so as to form the key supervision elements of the government, that is, the government supervision factors (GGFs).

In the project approval stage of the project, the government supervision department should review the feasibility of the project, review the VFM of the project, judge whether the project is suitable for the PPP financing model, and avoid blind adoption (Solomon 2012). Social benefits are also one of the focuses of expressway PPP projects. Although the review of the project starts from the project approval stage, its actual effect is more reflected after the project is put into operation, so it is placed in the operation in Figure 1.

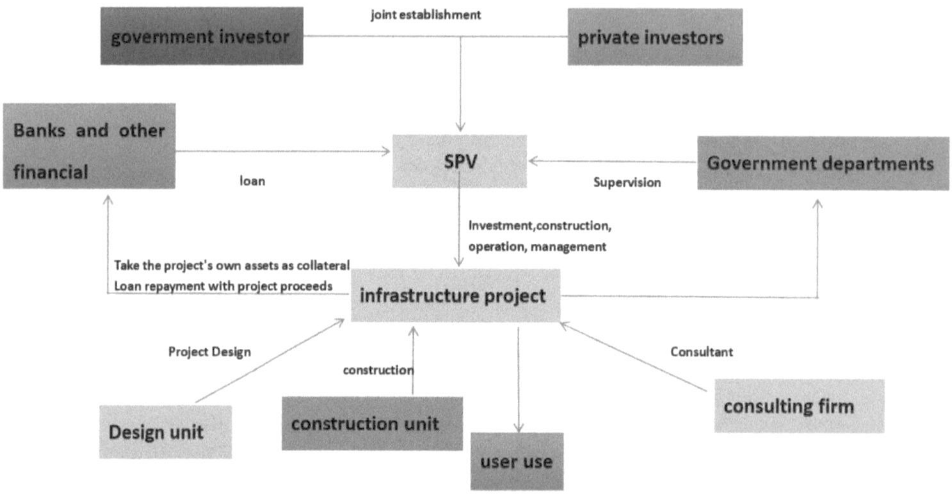


Figure 1. PPP architecture.

The bidding stage of a project has always been the focus of government supervision, and emphasis should be placed on the supervision of the transparency of the project procurement process and fair competition to ensure that capable and creditworthy social capital is selected to participate in the construction of infrastructure projects. The signing of the concession agreement is related to the success or failure of the project (Wang 2007). Attention should be paid to the reasonable division of interests, rights and responsibilities, and risk sharing between the public and private, as well as the reasonable effectiveness of government guarantees to ensure the smooth operation of PPP projects. The supervision of the project franchising stage is highly specialized and has a large workload. The government supervision department can fully cooperate with the industry, professional associations and third-party institutions, and ensure the participation of the media and the public in supervision through information disclosure and other means.

The construction stage of the project is an important stage in the formation of the project quality. In the expressway PPP project, it is mainly manifested in the smoothness of the road surface, the completeness of the supporting transportation facilities, the quality of the main project, and traffic safety. The system design of the PPP project makes the project company very concerned about the quality of the project, otherwise, it will affect its operation, the government only needs to focus on monitoring the public interest and environmental aspects at this stage.

During the operation of the project, the government supervision department should conduct supervision from the economic, social, and environmental aspects. They should measure the economic and social benefits of the project from the perspective of the whole life cycle, and assess whether the VFM of the project is achieved. The government's regulatory factors at this stage mainly relate to project fees, handover, maintenance, upkeep, and project company actions. The charging supervision of the project can be subject to social supervision through the information disclosure platform to ensure the public interest; in the handover stage of the project, the government can entrust a third-party agency to assist in the acceptance and the handover of the project.

In general, the elements of government supervision in expressway PPP projects mainly include the review of the project plan, the project VFM evaluation, the selection of social participants, the supervision of the procurement process, the supervision of the franchise agreement, the behavior of the public sector and social participants, the completion time of the project, the quality of the main project, the completeness of the supporting transportation facilities, the traffic safety, the smoothness of the road surface, the degree of environmental protection, the charges for the project, the handover of the project, the maintenance and maintenance of the project, and the behavior of the project company. In the process of supervising expressway projects, the government should actively play the role of various professions, industry associations, and professional third-party institutions, and make use of their advantages in talents and technology to improve the professionalism of supervision. The elements of government regulation and cooperation with third-party institutions are shown in Table 4.

Table 4. Analysis of government regulatory factors and regulatory responsibilities.

| Elements of government regulation | Government regulation | Thirdparty assistance | Government Governance Factors | Government regulation | Thirdparty assistance |
|---|---|---|---|---|---|
| Project plan review | | √ | traffic safety | √ | √ |
| Project VFM Evaluation | | √ | road smoothness | √ | √ |
| The choice of social actors | √ | √ | Degree of environmental protection | √ | √ |
| Procurement process | √ | | project fee | √ | |
| Franchise Agreement | √ | √ | handover of the project | √ | √ |
| Public Sector and Corporate Behaviour | √ | | user satisfaction | | √ |
| Project completion time | | √ | Maintenance and upkeep of the project | | √ |
| Main project quality | √ | √ | Other behaviors of social participants | √ | |
| Completeness of supporting transportation facilities | √ | √ | | | |

## 4 CONCLUSIONS

In order to better supervise expressway PPP projects, improve the efficiency of supervision, and ensure the effectiveness of supervision, the government should further improve the following aspects.

(1) They should strengthen the construction of laws, regulations, and systems, and establish corresponding supervision standards according to the particularity of PPP projects to ensure the standardization and effectiveness of government supervision of PPP projects.
(2) Commitment is made to actively play the role of various professions, industry associations, and professional third-party institutions. The government supervision of expressway PPP projects has strong professionalism, which can make full use of the talents and professional advantages of third-party institutions to improve supervision efficiency.
(3) Expressway PPP projects are closely related to public interests. An information platform should be established, the information disclosure system should be improved, and extensive supervision by the media and the public should be accepted to truly ensure the social welfare of the project.

With the in-depth study of expressway PPP projects by Chinese scholars and the advancement of PPP project practice, it is believed that my country's expressway PPP supervision system will be more perfect so that PPP will play a greater role in the expressway field.

## REFERENCES

AlbertP. C. Chan1. Critical Success Factors for PPPs in Infrastructure Developments: Chinese Perspective [J]. *JCEM*. 2010:484–494.

Bing Li. Critical success factors for PPP/PFI projects in the UK construction industry. *Construction Management and Economic*, 2005: 459–471.

Ghina Yamout. A critical assessment of a proposed public-private partnership (PPP) for the management of water services in Lebanon [J], *Water Resource Management*, 2007.
Goran and Nevena. Use of key performance indicators for PPP transport projects to meet stakeholders' performance objectives[J]. *Built Environment Project and Management*, 2013. 228–249.
Junxiao Liu. Life Cycle Critical Success Factors for Public-Private Partnership Infrastructure Projects[J]. *JME*. 2014.
Solomon. Critical success factors in public-private partnership (PPP) on infrastructure delivery in Nigeria [J]. *JFM*, 2012.
Wang Wen-Xiong. Critical Success Factors of Infrastructure Projects under PPP Model in China[J]. *International Conference on Wireless Communications, Networking and Mobile Computing*, 2007:4970–4974.
Zuhaili. Critical Success Factors for Value Management Workshops in Malaysia[J]. *JME*. 2014:1-9.

*Advances in Urban Construction and Management Engineering – Cha (Ed)*


# Spatial distribution and regional difference of carbon emissions in China: A spatial econometric analysis

Xiaotong Tang & Hong Mi*
*School of Public Affairs, Zhejiang University, Hangzhou, China*

ABSTRACT: Based on energy consumption data of each province in China from 2000 to 2019, this paper uses spatial autocorrelation analysis and Spatial Durbin Mode to explore the spatial distribution pattern of China's regional carbon emissions and the impact of population, economy, industrial structure, energy structure and urbanization level on carbon emissions. The results demonstrate that: (1) Carbon emissions from energy consumption of different provinces all increased from 2000 to 2019, while carbon emission intensity decreased year by year. (2) There is a spatial aggregation effect of regional carbon emissions in China, the spatial distribution of carbon emissions mainly shows a pattern of high in the East and low in the west. (3) Economic growth, population, industrial structure, energy structure, and urbanization rate are the main driving factors to carbon emissions. Different carbon control methods should be selected according to the influence mechanism of various driving factors in different regions on carbon emissions.

## 1 INSTRUCTIONS

The carbon game generated by the unbalanced development among countries in the world has become the knot of carbon reduction, and the difference in carbon emission has become the focus of solving the carbon emission problem (Tong 2015). Through the difference analysis of carbon emissions and carbon emission intensity, it can evaluate the emission reduction achievements and future emission reduction potential of each region, which can provide a theoretical basis for reducing carbon emissions and promoting the development of the low-carbon economy. As the country with the largest carbon emissions in the world, the spatial agglomeration of industry and population makes significant differences in carbon emission intensity in different regions of China (Jiang 2014). The spatial distribution of population, economy and carbon emission deviates significantly from the equilibrium state (Yue et al. 2010). Specifically, the spatial distribution of the population is relatively uniform, the degree of economic concentration is the highest, and the concentration of carbon emissions is between the two. The high level of concentration indicates that carbon emissions are primarily concentrated in some cities (Chuai et al. 2012; Cheng et al. 2013).

Therefore, scientifically refining regional carbon emission differences, analyzing regional carbon emission differences and their influencing factors, and seeking targeted emission reduction approaches have important theoretical value and practical significance for China to actively reduce carbon emissions and realize regional sustainable development. In this study, firstly, we calculate the carbon emissions of 30 provinces in China from 2000 to 2019 and compare the differences and change trends of carbon emissions and carbon emission intensity in different provinces. Then, aiming at the spatial correlation of carbon emissions, we empirically study the impact of different driving factors on carbon emissions in the region by using the Spatial Durbin Model, which identifies the spatial spillover of carbon emissions and its influencing factors. Finally, we put forward suggestions on how to reduce regional carbon emissions.

*Corresponding Author: spsswork@zju.edu.cn

DOI 10.1201/9781003348023-30

## 2 DATA AND METHODS

### 2.1 *Data sources*

This paper selects the panel data of 30 provinces from 2000 to 2019 (autonomous regions and municipalities). Due to the lack of relevant data in the Tibet Autonomous Region, Taiwan Province, Hong Kong and Macao Special Administrative Regions, all data sources and results in this paper did not include these areas. Among them, energy consumption data is from *China Energy Statistical Yearbook* (2000–2019). Other data such as regional population, per capita GDP, urbanization rate, the proportion of the tertiary industry, and the proportion of non-fossil energy are from the *China Energy Statistical Yearbook* and the *China Statistical Yearbook* (2000-2019).

### 2.2 *Estimation of carbon emissions*

At present, the main energy types include fossil energy, electricity, the solar, wind, and so on, and traditional energy represented by fossil energy contributes the most to carbon emissions. Due to the lack of direct monitoring data, existing studies generally use energy consumption to estimate carbon emissions. In this paper, the Intergovernmental Panel on Climate Change (IPCC) calculation method of carbon emissions was used. The carbon emissions are obtained by energy consumption, carbon emission coefficients, and carbon oxidation coefficients as the following equation:

$$C = \sum_{i=1}^{n} C_i = \sum_{i=1}^{n} M_i E_i V_i \tag{1}$$

where $C$ is the total amount of carbon emissions, $C_i$ is carbon emissions from energy type $i$, $M_i$ is the consuming amount of energy $i$, and $E_i$ is the carbon emission coefficient of energy $i$. $V_i$ is the carbon oxidation coefficient.

### 2.3 *Spatial econometric model*

The panel data econometric model of carbon emission standard without considering the spatial effect is:

$$\ln I_{it} = b \ln P_{it} + c \ln A_{it} + d \ln T_{it} + f \ln U_{it} + g \ln IN_{it} + h \ln RE_{it} + \mu_{it} + \upsilon_{it} + \varepsilon_{it} \tag{2}$$

where, $i$ represents the cross-section province, $t$ represents the period, $b$, $c$, $d$, $f$, $g$, and $h$ are the constant regression parameters to be estimated; $\varepsilon_{it}$ are independent and identically distributed random error term, $\mu_{it}$ represents the spatial effect, $\upsilon_{it}$ represents the period effect, so Equation (2) is a panel model with two effects of space and period.

At present, the widely used spatial models mainly include Spatial Lag Model (SLM), Spatial Error Model (SEM), and Spatial Durbin Model (SDM). The three econometric models represent different economic connotations, that is, what mechanism spatial interaction effects are generated through. SLM model believes that all explanatory variables in the model will directly act on the dependent variables through the spatial transmission mechanism. SEM assumes that the error term is the source of spatial interaction effects, and the spatial spillover effects formed by regions are caused by random shocks (Anselin et al. 2008). SDM model includes two types of assumptions at the same time. The standard panel econometric model ignores the problem of biased parameter estimates of spatial effects, so this paper includes the spatial effect of the provincial carbon emission function. When the explanatory variable of carbon emissions in the region is determined by the observed carbon emissions in its neighboring regions, and considering that the influence factors of spatial neighboring regions also have an impact on the carbon emissions of the province, it is necessary to use the Spatial Durbin Model (SDM):

$$\ln I_{it} = \rho\sum_{j=1}^{N}\omega_{ij}\ln I_{jt} + b\ln P_{it} + c\ln A_{it} + d\ln T_{it} + f\ln U_{it} + g\ln IN_{it} + h\ln RE_{it}$$

$$+\alpha\sum_{j=1}^{N}\omega_{ij}\ln P_{jt} + \beta\sum_{j=1}^{N}\omega_{ij}\ln A_{jt} + \chi\sum_{j=1}^{N}\omega_{ij}\ln T_{jt} + \delta\sum_{j=1}^{N}\omega_{ij}\ln U_{jt}$$

$$+\xi\sum_{j=1}^{N}\omega_{ij}\ln IN_{jt} + \zeta\sum_{j=1}^{N}\omega_{ij}\ln RE_{jt} + \mu_{it} + \upsilon_{it} + \varepsilon_{it} \quad (3)$$

Where, $\rho$ is the spatial lag (autoregressive) coefficient, and is the element of the spatial weight matrix W. The weight matrix is row-normalized, and the sum of the elements in each row is 1. The proximity matrix is used to set the weight matrix W in this paper. Wlnp, WlnA, WlnT, WlnU, WlnIN, and WlnRE represent the spatial lag variables of population growth, economic growth, technological progress, urbanization rate, industrial structure, and energy structure in neighboring provinces, $\alpha$, $\beta$, $\chi$, $\delta$, $\xi$ and $\zeta$ are the constant regression parameters to be estimated.

## 3 RESULTS AND DISCUSSION

### 3.1 *Temporospatial evolution pattern analysis of carbon emissions*

According to estimates, the total carbon emissions of China's 30 provinces increased from 4185.96Mt in 2000 to 14,025.34Mt in 2019, which means a total increase of 235% in the past 20 years.

From the perspective of time sequence, in 2000, Liaoning had the highest carbon emissions at 350.98Mt, followed by Shanxi, Hebei, and other provinces. Qinghai and Hainan had the lowest carbon emissions, neither exceeding 15Mt. In 2010, Shandong had the highest carbon emissions from energy consumption, reaching 1086.67Mt, followed by Hebei, Shanxi, Liaoning, and Jiangsu Provinces. In 2019, Shandong's carbon emissions still ranked first in the country, up to 1515.24Mt, followed by Shanxi and Inner Mongolia, both exceeding 1000Mt.

From the perspective of spatial distribution, China's carbon emissions generally show a distribution pattern of high in the east and low in the west, high in the north and low in the south, and cluster in large and medium-sized cities. The high-value areas of total carbon emissions are mainly concentrated in metropolitan areas, such as the Bohai Rim region centered in Beijing, the Yangtze River Delta region centered in Shanghai, the Pearl River Delta region centered in Guangdong, and the Chengdu-Chongqing metropolitan area. They all have obvious characteristics of high carbon emission accumulation, However, Tibet, Xinjiang, Gansu, and Qinghai in the western region show the spatial clustering effect of low-carbon emissions.

### 3.2 *Temporospatial evolution pattern analysis of carbon emission intensity*

From the perspective of time sequence, although the total carbon emissions of China's energy consumption continued to increase from 2000 to 2019, the economic growth rate was much higher than the growth rate of carbon emissions, and the carbon emission intensity of energy consumption generally showed a downward trend year by year. The national average carbon emission intensity dropped from 5.4 in 2000 to 2.06 in 2019, a decrease of 48.48%.

In 2000, Shanxi and Guizhou had the highest carbon emission intensity at 19.08 and 11.02 respectively. The carbon emission intensity of Xinjiang, Gansu, Inner Mongolia, Ningxia, Liaoning, Jilin, and other provinces in the northwest and northeast were also relatively high. Hainan, Zhejiang, Fujian, Guangdong, Guangxi, and other southern coastal provinces have low carbon emission

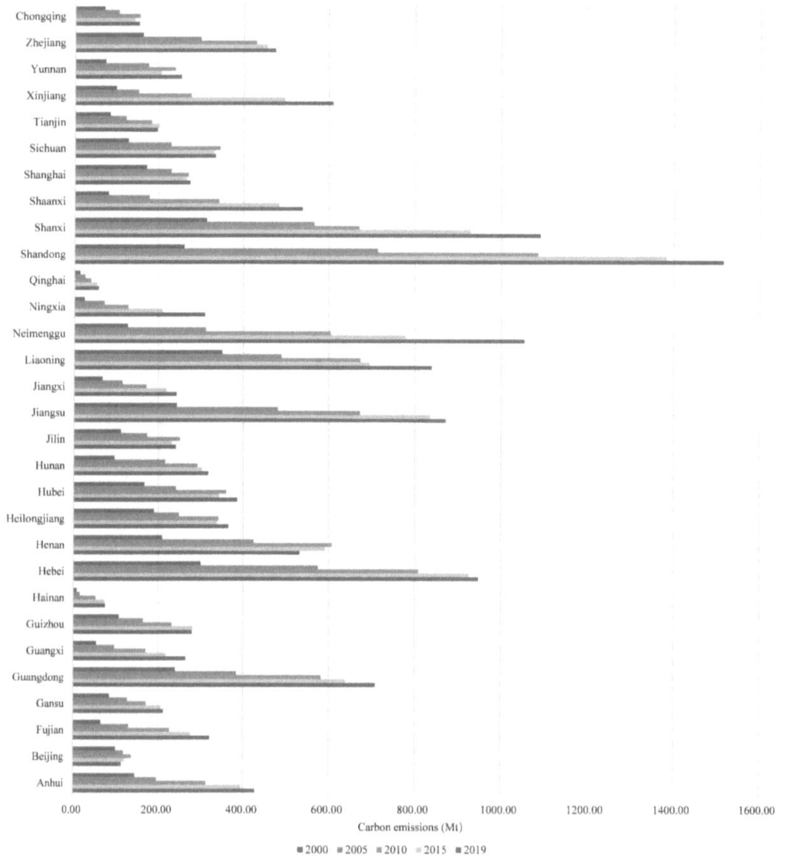


Figure 1. Provincial distribution of carbon emissions in 2000–2019.

intensity. In 2005, Shanxi and Ningxia had the highest carbon emission intensity at 13.39 and 12.03 respectively, followed by Guizhou, Gansu, and Inner Mongolia. Compared with 2000, Jiangsu and Hunan also joined the ranks of low carbon emission intensity. Overall, it still shows a trend of high in the north and low in the south. In 2010, Ningxia and Shanxi had the highest carbon emission intensity, followed by Inner Mongolia, Xinjiang, and Guizhou Provinces. Provinces with low carbon emission intensity increased more, including Jiangxi, Hubei, Sichuan, Chongqing, and other inland provinces. In 2019, Ningxia had the highest carbon emission intensity (8.28), followed by Shanxi and Inner Mongolia. The carbon emission intensity of Beijing, Shanghai, Sichuan, Guangdong, and Zhejiang Provinces were all below 1.

Analysis of the distribution of provinces at different levels shows that the proportion of provinces with a carbon emission intensity of less than 3 increased from 23% in 2000 to 63% in 2010 and 83% in 2019. The decrease in the number of provinces with high carbon emission intensity and the substantial increase in provinces with low carbon emission intensity further supports the fact that the carbon emission intensity of China continues to decline.

From the perspective of global spatial autocorrelation, the spatial distribution of China's carbon emission intensity shows an obvious regional characteristic of being high in the north and low in the south. The carbon intensity of Shanxi, Ningxia, and Guizhou Provinces has always been at a high level. The number of low-carbon-intensity provinces is also increasing. In addition to some coastal provinces such as Zhejiang, Guangdong, and Fujian, inland provinces such as Chongqing, Hubei, and Sichuan have also joined the ranks of low-carbon intensity.

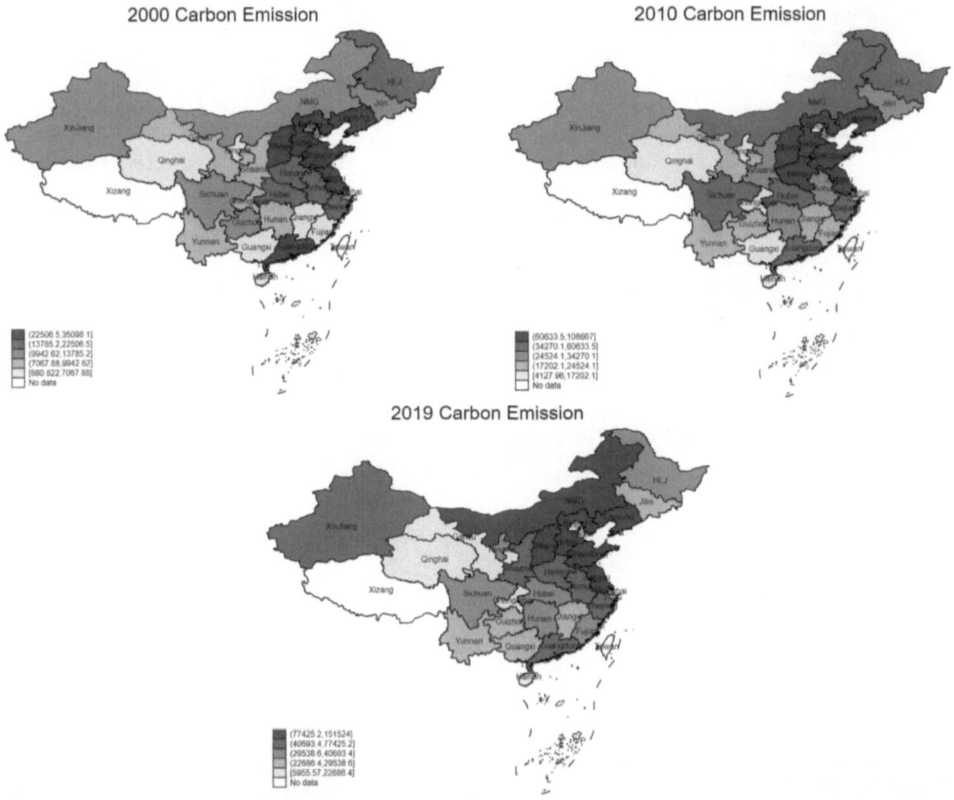


Figure 2. Spatial patterns of China's carbon emissions in 2000, 2010 and 2019.

### 3.3 *Spatial econometric analysis of carbon emission.*

Table 1. Estimation and test results of the SDM.

| Parameters and testing values | Regression coefficient | z-value | p-value |
|---|---|---|---|
| lnPGDP | –0.034 | –0.74 | 0.047 |
| lnPop | 7.627 | 7.69 | 0.000 |
| lnIS | 12,218.95 | 1.27 | 0.002 |
| lnES | 20,560.36 | 4.42 | 0.000 |
| lnUR | 89,743.65 | 5.81 | 0.000 |
| W*lnPGDP | 0.350 | 2.61 | 0.009 |
| W*lnPop | –13.139 | –1.91 | 0.000 |
| W*lnIS | 24,457.29 | 1.22 | 0.002 |
| W*lnES | –1166.177 | –0.05 | 0.043 |
| W*lnUR | –61,754.16 | –1.46 | 0.045 |
| rho | 0.450 | 4.56 | 0.000 |
| obs=600 | $R^2$=0.73 | LogL=–767.91 | |

Economic growth (GDP per capita, PGDP), population size (Pop), industrial structure (IS), energy structure (ES), and urbanization rate (UR) are the main factors for the increase in carbon emissions

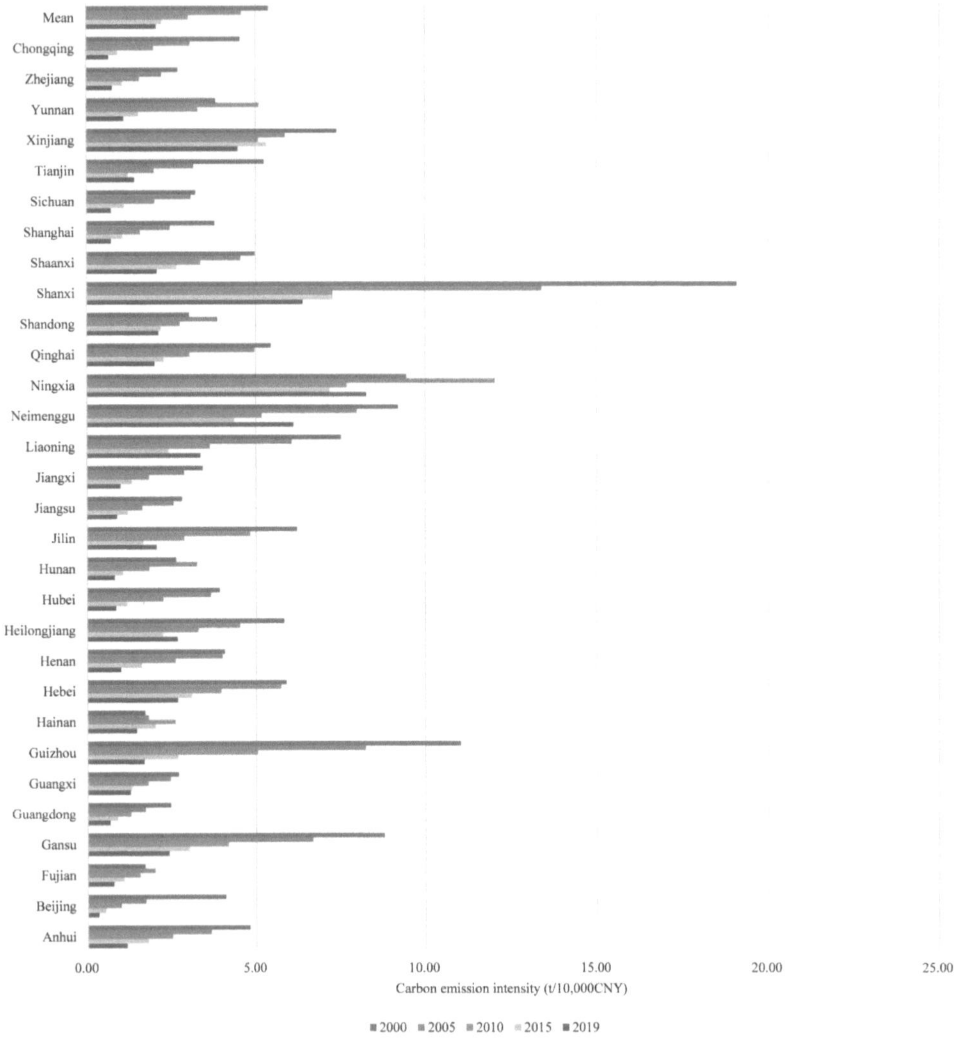


Figure 3. Provincial distribution of carbon emission intensity in 2000–2019.

(Dietz & Rosa 1994; Zou et al. 2009). In order to explore the influence mechanism of each variable on carbon emissions, this paper established a model to clarify the influence direction and coefficient of each variable.

In the SDM, the spatial autoregressive coefficient rho is significantly positive at the level of 1%, indicating that the spatial effect of carbon emissions is significant. Every 1% increase in carbon emissions in neighboring areas will result in a 0.45% growth in carbon emissions in the region. The estimation and test results of the spatial panel econometric model show that the population, urbanization rate, industrial structure, and energy structure are all significantly positive at the 1% level, and their coefficients are 7.627, 12,218.95, 20,560.36, and 89,743.65. It shows that the expansion of China's economic scale and population size means more energy consumption, and traditional energy is the main source, and their carbon emissions are much higher than new energy and clean energy, resulting in more carbon emissions. Besides, due to the heavy industry accounts for a high proportion of the secondary industry, which is a typical high-energy-consuming industry, there is a strong correlation between carbon emissions and energy structure and industrial structure,

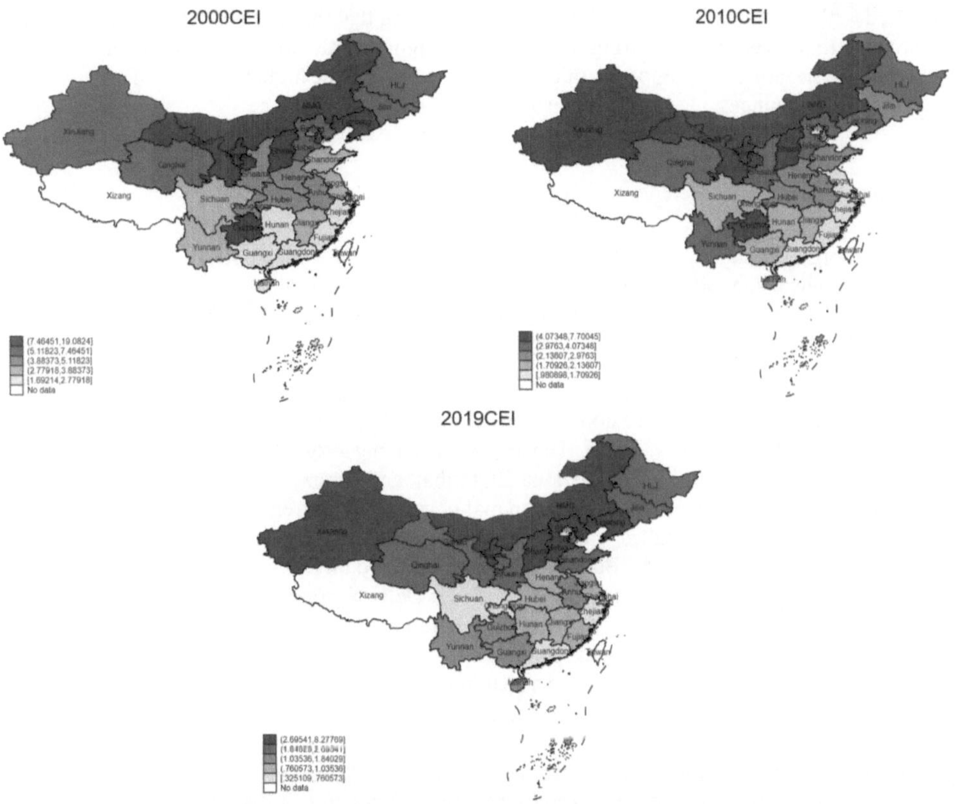


Figure 4. Spatial patterns of China's carbon emission intensity in 2000, 2010 and 2019.

the larger the proportion of the output value of the secondary industry and the proportion of coal consumption, the greater the carbon emissions. Moreover, the GDP per capita is significantly negative at the level of 5%, indicating that with the increase in GDP per capita, regional carbon emissions gradually decrease. With the transformation of the industrial structure, the dependence of economic development on carbon emissions will decrease, so GDP per capita and carbon emissions are negatively correlated.

In addition, variables such as GDP per capita, population size, industrial structure, energy structure, and urbanization level also have significant spatial autocorrelation and spatial spillover effects. The enhancement of these variables in a province will not only increase the carbon emissions of this province but also cause changes in the carbon emissions of neighboring provinces through a spatial conduction mechanism. Among the weighted independent variables, the coefficient of W*lnCO2 is significantly positive, indicating that the carbon emissions of the province are affected by the carbon emissions of neighboring provinces, and the two change in the same direction. The greater the carbon emissions of the neighboring provinces, the greater the carbon emissions of the province. Besides, W*lnPGDP and W*lnIS are significantly positively correlated with carbon emissions at the 1% significance level, indicating that the economic development level and industrial structure of neighboring provinces significantly affect the carbon emissions of the province. W*lnPOP is significantly negatively correlated with carbon emissions at the 1% level, indicating that the increase in the population size of adjacent regions will suppress carbon emissions in the region. This is due to differences in regional carbon emissions caused by population migration and agglomeration (Cole & Neumayer 2004). W*lnES and W*lnUR are significantly negatively correlated with carbon emissions at the 5% level, indicating that the energy consumption structure and urbanization level of neighboring provinces will significantly affect the carbon emissions of the province. The larger the proportion of coal consumption and the higher the level of urbanization in neighboring provinces, the lower the carbon emissions of the province.

## 4 CONCLUSIONS

By using energy consumption data of each province in China from 2000 to 2019 and making Stata16 as technical support, this study explores the amount and structure of carbon emissions and their interacted relationship with the population and economy respectively in different provinces in China. The results demonstrate that:

(1) Carbon emissions from energy consumption of different provinces all increased from 2000 to 2019, however, carbon emission intensity decreased year by year. Judging from the spatial pattern, Shandong, Shanxi, Hebei, Henan, and Liaoning Provinces always had high carbon emissions but carbon emissions in Hainan, Qinghai, and Ningxia Provinces were always low.
(2) China's regional carbon emissions have spatial positive autocorrelation in spatial distribution. The provinces with the highest carbon emissions are mostly located in economically developed coastal areas, such as the Bohai Rim region centered on Beijing, the Yangtze River Delta region centered on Shanghai, and the Pearl River Delta region centered on Guangdong, followed by areas with more developed secondary industries, such as Shandong, Shanxi, Hubei, etc. Besides, there is a spatial aggregation effect of regional carbon emissions in China. For example, the Bohai Rim region shows a spatial cluster effect of high carbon emissions, while Gansu, Qinghai, and Ningxia Provinces in the western region show a spatial cluster effect of low carbon emissions.
(3) Economic growth, population, industrial structure, energy structure, and urbanization rate are the main driving factors of carbon emissions. Population, urbanization rate, industrial structure, and energy structure have a significant positive effect on regional carbon emissions, while per capita GDP has a significant negative effect on regional carbon emissions.

## 5 POLICY RECOMMENDATIONS

With China already committing to peak carbon dioxide emissions before 2030 and achieving carbon neutrality before 2060, the Central Economic Work Conference urged quicker steps to come up with an action plan that enables the peaking of emissions.

To achieve the win-win goal of economic development and emission reduction, we need to fully consider the spatial correlation and spillover, incorporate the spatial dependence into the analysis of carbon emission intensity and the formulation of carbon emission reduction policies, accelerate the economic and technological cooperation between developed and underdeveloped region (Fu et al. 2015). On the basis of regional differentiation, we should consider the actual situation of all regions, and adopt differentiated policies and measures for key industries and regions while promoting the overall emission reduction of all provinces and industries. For the northeast and northwest regions with high energy dependence, reduce the proportion of high-energy consumption industries by accelerating the adjustment of industrial structure, improve energy efficiency by accelerating the pace of technological innovation, gradually change the energy consumption structure dominated by coal energy, and reduce carbon emission intensity while promoting the transformation of economic growth mode. For the eastern coastal areas with rapid economic development and low carbon emission intensity, we should continue to develop new energy, high-tech industries, and modern service industries, develop a circular economic growth mode, and take a low-carbon and energy-saving economic development route. In addition, all regions should not only continue to accelerate the pace of technological progress and improve energy efficiency, but also accelerate the upgrading of industrial structure, optimize the energy consumption structure, promote the use of clean energy and renewable energy, and improve environmental quality.

Moreover, the carbon emission trading market is the key market-oriented mechanism to achieve China's carbon peak and carbon neutralization goals. In the process of carbon trading, we can refer to the EU trading system and promote the coordinated and sustainable development of the regional population-economy-environment via carbon trading on the premise of controlling the decline of total carbon emissions.

## ACKNOWLEDGMENTS

This work was supported by The National Key Research and Development Program of China (No. 2020YFD1100202) and the Zhejiang Provincial Natural Science Foundation of China (No. LD21G030001).

## REFERENCES

Anselin, L., Gallo, J. L. & Jayet, H. 2008. Spatial panel econometrics. In Mátyás László & Sevestre Patrick (eds). Williams & H. Faure (eds), *The econometrics of panel data*: 625–660. Berlin: Springer.

Cheng, Y.Q., Wang, Z.Y., Zhang S.Z., Ye, X.Y. & Jiang, H.M. 2013. Spatial econometric analysis of carbon emission intensity and its driving factors from energy consumption in China. *Acta Geographica Sinica* 68(10): 1418–1431. (in Chinese)

Chuai, X., Huang, X., Wang, W., Wen, J., Chen, Q. & Peng, J. 2012. Spatial econometric analysis of carbon emissions from energy consumption in china. *Journal of Geographical Sciences* 22(4): 630–642.

Cole, M.A. & Neumayer, E. 2004. Examining the impact of demographic factors on air pollution. *Population and Environment* 26(1): 5–21.

Dietz, T. & Rosa, E.A. 1994. Rethinking the environmental impacts of population, affluence, and technology. *Human ecology review* 1(2): 277–300.

Fu, Y.P., Ma S.C. & Song, Q. 2015. Spatial Econometric Analysis of Regional Carbon Intensity. *Statistical Research* 32(6): 67–73. (in Chinese)

Jiang, H.M. 2014. Research on the effect of population on citizen's lifestyle-related carbon emissions in Chinese cities (2000-2010). *Fudan University*. (in Chinese)

Tong X. 2015. Research on regional carbon emissions differential analysis and reduction approaches in China. *Northeastern University*.

Yue, C., Hu, X.Y. & He, C.F. et al. 2010. Provincial carbon emissions and carbon intensity in China from 1995 to 2007: Carbon emissions and social development. *Acta Scientiarum Naturalium Universitatis Pekinensis* 46(4): 510–516. (in Chinese)

Zou, X.P., Chen, S.F. & Ning, M. 2009. Empirical research on the influence factor of carbon emission in Chinese provincial regions. *Ecological Economy* 3: 34–37. (in Chinese)

*Advances in Urban Construction and Management Engineering – Cha (Ed)*


# Passenger transport safety analysis based on 4R crisis management and passenger perception

Jianglian Luo*
*Southwest Jiaotong University Hope College, Chengdu, China*

Xiaojie Liu*
*Electric Engineering Co., Ltd of China Railway Seventh Group, Zhengzhou, China*

ABSTRACT: China's urban rail transit construction has developed rapidly in recent years. More and more cities have opened rail transit in China, and the total mileage of rail transit operations is also increasing. With the expansion of the rail transit network and the increasing trend of passenger flow year by year, the pressure on urban rail transit stations' safety management is increasing daily. How to adhere to people-oriented thought, put the safety of passengers' life and property in the first place, and constantly improve the level of safety management of urban rail transit stations is an urgent problem to be solved by urban rail transit enterprises. In this paper, from the perspectives of enterprise security management and passenger safety awareness, namely, based on the 4R crisis management theory and SERVQUAL model index system of building enterprise and passenger perception perspective, and then through the investigation and analysis, it is concluded that urban rail transit passenger station safety management in the safety management in different stages of the 4R crisis management mode exists a difference of perception, and adjust enterprise security management strategy according to perception difference.

## 1 INTRODUCTION

Urban rail transit, like a large capacity public transportation tool in the city, has the technical advantages of safety, comfort, environmental protection, fast, large volume, and so on, which has become the main choice for urban residents to travel. With the rapid development of urban rail transit in China, the number of cities and the total mileage of urban rail transit are increasing. As of January 1, 2020, 40 cities in Mainland China have opened urban rail, with a total operating mileage of 6,730.27 km and an annual passenger volume of more than 24 billion (Yu 2020). According to the 14th Five-Year Development Plan, the Chinese mainland will usher in a new round of rail transit construction. With the continuous expansion of the rail transit network and the increasing trend of passenger flow year by year, the pressure on safety management of urban rail transit stations is increasing day by day. Urban rail transit stations are divided into ground stations, underground stations, and elevated stations. As underground stations are relatively closed and narrow underground spaces, during daily operations, such as morning and evening rush hours, large passenger flows within a short period of time make waiting trains gather together. Once fire, explosion, and other emergencies occur, the closed space may cause huge casualties and property losses. It may even become the focus of social media with negative social effects. Safety is an eternal theme. The safety management of urban rail transit enterprises should always adhere to the idea of being "people-oriented", constantly improve the level of urban rail transit safety management, put the safety of passengers' lives and property in the first place, and implement it. Therefore, how to

*Corresponding Authors: 1020659792@qq.com and 275551004@qq.com


DOI 10.1201/9781003348023-31

prevent and respond to emergencies encountered in the operation of urban rail transit, eliminate or reduce the occurrence of accidents, and reduce the harm caused by accidents, is an important issue for urban rail transit enterprises (Xi 2015).

In the current research status, 4R crisis management theory is mostly used in the medical, government public security, and university crisis management fields. Passenger perception is mainly used to analyze service quality management, and one of its analysis methods, the SERVQUAL model, is to improve enterprise service quality by comparing passenger expectations and perception (Millayne2020). In the construction and implementation of the nursing safety and quality management system based on 4R crisis management theory, Wang Feifan, Xiang Kelan, et al established and improved the nursing safety and quality management framework based on 4R crisis management theory from the perspective of enterprises, and improved patient care safety and quality standards (Wang 2017). Meng Liang and Zhang Guanglei used the 4R crisis management theory to analyze the main causes of security problems and establish a sound university security prevention and control system from the perspective of university management in the construction of university security prevention and control system (Meng 2017). In terms of the safety management of urban rail transit enterprises, the 4R crisis management theory has also been applied. For example, Lu Wengang and Peng Jing studied the evaluation index system of the public emergency response capacity of the Guangzhou metro. From the perspective of enterprise safety management, the evaluation system of the Guangzhou metro public emergency response capacity is constructed based on the crisis objective management method, crisis management 4R model, and analytic hierarchy process (Lu 2012). The application of the SERVQUAL model in passenger perception, such as Lin Lin, Jiguang Port, and others used the SERVQUAL model in the evaluation study of urban rail transit operation service quality to build an evaluation model of urban rail transit operation service quality. Through perception analysis, countermeasures for improving the service quality of urban rail transit operations and suggestions for accelerating the service quality of urban rail transit operation enterprises are put forward (Lin 2020).

Based on the current application status of 4R crisis management in urban rail transit enterprises, this paper considers safety issues from two aspects of enterprise safety management and passenger perception from an innovative perspective. That is, the perfection of the enterprise safety management system guarantees passengers' travel, and passenger perception feedback on the deficiencies of enterprise safety management. The passenger transport safety analysis model of rail transit enterprises is constructed from the perspective of both enterprises and passengers, the safety emergency management ability of rail transit enterprises is improved, and the basic principle of "people-oriented, safety first" is earnestly implemented.

## 2 4R CRISIS MANAGEMENT THEORY

American Robert Heath proposes the 4R crisis management theory in His book Crisis Management, which integrates system management with staged operation steps. The main working steps are divided into four stages: reduction, readiness, response and recovery. It presents a continuous, dynamic and interactive cyclic process, including all aspects of management before, during and after the crisis (Lin 2020). The four stages of 4R crisis management theory are Reduction, Readiness, Response and Recovery respectively. The initial of each stage is referred to as 4R crisis management. 4R crisis management can effectively reduce the possibility of risk occurrence and improve the quality of safety management. Among them, reduction force is the core stage of crisis management, which is mainly to take a series of preventive measures to reduce the aggression and possibility of the crisis as much as possible, identify the source of danger, stifle the danger in the bud and stage, and reduce the possibility of the crisis from the root. The reduction force stage is carried out from the aspects of environment, structure, personnel and system. The readiness stage is to carry out crisis prevention work, form a crisis management team, formulate a crisis management plan, and carry out daily crisis management and other work, such as monitoring and management of hazard sources and management personnel training. The response stage refers to the response strategies adopted in time when a crisis occurs, such as launching emergency plans, on-site command and coordination, rescue and other work. Recovery is to quickly summarize experience after the occurrence of a crisis, minimize the negative impact of the crisis and provide an empirical basis for future safety management, such as out of the shadow of the crisis, and restoring the state of operation.

## 3 URBAN RAIL TRANSIT PASSENGER TRANSPORT SAFETY

Rail transit passenger safety incident refers to the sudden occurrence in rail transit station or train during the operation of rail transit, which causes certain negative impacts on society and passengers, property loss and personal safety threats. As urban rail transit is the backbone of the urban public transportation system, an important part of the urban comprehensive transportation system, and the preferred means of transportation for residents to travel, while most urban rail transit stations are underground stations, which are characterized by semi-closed structures, poor air circulation and concentrated passenger flow, etc., bringing more challenges and tests for the safety management of urban rail transit stations. China's urban rail transit has relatively perfect systems and emergency plans in terms of safety management and accident emergency rescue. However, in various emergencies and application practices, due to the characteristics of time urgency and multi-dimension of emergencies, it is necessary to pay attention to the harm degree of various safety accidents all the time. Taking a look at some historical accidents at home and abroad, the Paris subway fire in 1903 killed 84 passengers. Arson on the Daegu subway killed nearly 198 passengers in 2003; in 2006, 41 people were killed when a train derailed in an eastern Spanish city. in 2011, more than 260 people were injured in a Shanghai subway crash; in 2019, one staff member was killed due to the civil air defence door intrusion in Chongqing rail transit ring Line (Liu 2021).

It can be seen that once a safety crisis occurs in an urban rail transit station, it may pose a serious threat to the life safety of passengers. In order to maximize the people-oriented protection of passenger safety, and improve the safety management of urban rail transit enterprises, exploring the theory and method of safety management is endless, in order to more and more perfect safety management systems, and play a practical role in rail transit operation management.

## 4 INDEX ESTABLISHMENT AND ANALYSIS

### 4.1 *Based on 4R crisis management and passenger perception index establishment*

The reduction stage of passenger safety management of urban rail transit stations is the core part of 4R crisis management, which runs through the whole process of safety management, and mainly refers to the prevention process. In the reduction stage, enterprises should focus on reducing the risk of emergencies, which needs to be analyzed from the station environment, structure, system, personnel, and other aspects, so as to reduce the risk of crisis and improve the ability to cope with the crisis. The reduction stage, mainly includes environmental safety, hazard identification, equipment safety, personnel safety education, and so on. The readiness stage is mainly to do a good job in the early warning of emergencies and crisis prevention and control when the urban rail transit station once a crisis situation can be ready to respond to measures. It mainly includes the improvement of the safety system, safety publicity, and management, formulation and drill of the emergency plan, etc., and emergency plan should make clear the organization and responsibility, early warning and information report, and other comprehensive contents. The response stage refers to the emergency rescue when a crisis occurs, mainly referring to the emergency response process and emergency information disclosure, including emergency alarm, accident reporting, accident control, rescue, and other aspects. The recovery stage is the recovery stage after the crisis event of an urban rail transit station, also refers to as the recovery process. When the crisis situation is controlled and the crisis is resolved, it is necessary to immediately respond to the post-crisis recovery, in order to restore the operation indicators and personnel status to the normal state in advance. The recovery stage is not only a process of recovery, but also a process of summarizing experience, restoring normal operations, and reducing the social impact and personnel panic caused by disasters. This process includes accident investigation, operation restoration, summary evaluation, and public advocacy.

4R crisis management theory provides a mature theoretical framework for urban rail transit safety management. By sorting out the characteristics and action objectives of each stage, the enterprise safety managers can grasp and deal with the occurrence of emergencies in a timely and accurate manner, and reduce injury or loss to the minimum. 4R crisis management theory is to manage the crisis from the perspective of enterprises, eliminate security risks from the source, and achieve operational safety. The object of crisis management is operation safety, passengers are the main body of operation safety, so passengers, passengers themselves feel safe, which is also an important aspect of safety management. Passenger perception is a series of processes in which passengers perceive, feel, pay attention and perceive urban rail transit enterprises. The sensory information includes physiological state and mental activity as well as the existence of the external environment. The index system is established based on passenger perception. Through the perception indicators, the safety management of enterprises is investigated and analyzed from the perspectives of both passengers and enterprises. Thus, an indicator system based on 4R crisis management and passenger perception is established, as shown in Table 1.

### 4.2 *Passenger perception model survey analysis*

The passenger perception section analysis uses the SERVQUAL model, which is a tool for measuring the quality of service. Since safety is the most important part of the urban rail transit service quality system, this paper uses the SERVQUAL model for the safety management of rail transit operation enterprises, by measuring the passengers' perception of each stage of enterprise safety management, reflecting the difference between enterprise safety management and passenger perception from the perspective of passengers, and obtaining problems in station passenger safety management through comparative analysis of differences.

The SERVQUAL value for the individual indicator is calculated as follows:

$$SQ_1 = \sum_{i=1}^{n}(P_i - E_i) \tag{1}$$

The SERVQUAL value of the stage indicator is equal to the average of each individual indicator in this stage, which is calculated as follows:

$$SQ_2 = \frac{1}{m}\sum_{x=1}^{m} SQ_1 \tag{2}$$

The average SERVQUAL value of the passenger is equal to the weighted average of the indicators of each stage divided by the number of valid questionnaires, which is calculated by:

$$SQ_3 = \frac{1}{n}\sum_{j=1}^{m} \omega_j SQ_2 \tag{3}$$

Of the above, $E_i$ is the ith factor in terms of passenger expectations. $P_i$ is the i factor score in terms of passenger perception. $n$ is the number of valid questionnaires recovered. $m$ is the number of indicators in each phase. $X$ is the x index in each phase. $\omega_j$ is the weight of the j stage. $SQ$ is the passenger perception value.

According to Maslow's hierarchy of needs, since security is the most basic human need, it is a lack of sexual needs based on physiological needs. The need for security is the need of human

Table 1. Establish indicators based on 4R crisis management theory and passenger perception.

| 4R Crisis Management Phase ($R_i$) | Passenger Perception Index ($R_{ij}$) | Indicators of parsing |
|---|---|---|
| Reduction ($R_1$) | ($R_{11}$) Environmental safety | Familiar with the platform, station hall and other environments |
| | ($R_{12}$) Hazard identification | Is the setting of hazard identification clear and reasonable |
| | ($R_{13}$) Safety equipment | The safety and reliability of passenger facilities and equipment in the station |
| | ($R_{14}$) Safety education | The level of safety awareness of the station staff at work |
| Readiness ($R_2$) | ($R_{21}$) The management system | Understanding of station emergency management system |
| | ($R_{22}$) Safety propaganda | Satisfaction with station emergency safety propaganda |
| | ($R_{23}$) The emergency response plan | Understanding of station emergency plan |
| | ($R_{24}$) Emergency drill | Knowledge of station emergency drills |
| Response ($R_3$) | ($R_{31}$) The emergency alarm | Are you familiar with all kinds of emergency alarm terminal Settings at the station |
| | ($R_{32}$) The accident report | Understand the accident reporting process of station staff |
| | ($R_{33}$) The accident control | Satisfaction with station accident control |
| | ($R_{34}$) Self-help and mutual aid | Understanding of self-rescue and mutual rescue in station emergencies |
| Recovery ($R_4$) | ($R_{41}$) Accident investigation | Concern for the results of the investigation after the station emergency |
| | ($R_{42}$) Recovery operation | Level of knowledge of the conditions under which operations will resume after an emergency at the station |
| | ($R_{43}$) Summary evaluation | The level of concern for the post-emergency reflection summary at the station |
| | ($R_{44}$) Social propaganda | Satisfaction with the social communication channels after the station resumes operation. |

beings to ensure their own safety, to escape danger and invasion. From The hierarchical analysis of Maslow's needs, passengers' individual safety needs are pursued to maximize safety, so in this passenger perception model application, the passengers' expectations for safety are set to full scores. In the need for safety, human receptors are one of the tools for individuals to seek safety, so in the application of the SERVQUAL model, the passenger perception part is mainly investigated and analyzed, and the shortcomings of corporate safety management are feedback from the perspective of passengers.

In this survey, the stations of a rail transit operator were selected, a total of 300 questionnaires were distributed, and 262 valid questionnaires were recovered. The questionnaire set is divided into three parts: basic information about passengers, weight scoring of each stage, and setting of individual indicator questions. Regarding the proportion of weights, the four stages of reduction force, reserve force, reaction force and resilience are the tasks of different stages of enterprise crisis management, and each stage occupies irreplaceable importance in crisis management. According to the survey data, the respondents scored 29.1%, 23%, 28.5% and 19.4% on the weight of the four stages. Regarding the setting of the single indicator question, it is set to a 10-point system, and the respondents score one by one according to 16 individual indicators, with a scoring range of 0-10 points, 0 points indicate that the passenger perception value of the indicator is completely dissatisfied, and 10 points indicate that the passenger perception value of the indicator is highly satisfied.

Summarizing the survey results, the $SQ_1$ values of 16 indicators in the 4R crisis management stages and the $SQ_2$ values of each stage are calculated from the above formulas 1 to 2, and the $SQ_3$ values are derived from formula 3, that is, the average SERVQUAL value of passengers, as shown in Table 2:

Table 2. The calculation results.

| 4R Crisis Management Phase ($R_i$) | $\omega_j$ | Indicators | Expectations $\sum_{i=1}^{n} Ei$ | Perceived value $\sum_{i=1}^{n} Pi$ | $SQ_1$ | $SQ_2$ |
|---|---|---|---|---|---|---|
| $R_1$ | 0.291 | $R_{11}$ | 2620 | 2056 | –564 | –683 |
| | | $R_{12}$ | 2620 | 1786 | –834 | |
| | | $R_{13}$ | 2620 | 1940 | –680 | |
| | | $R_{14}$ | 2620 | 1966 | –654 | |
| $R_2$ | 0.23 | $R_{21}$ | 2620 | 1903 | –717 | –766.25 |
| | | $R_{22}$ | 2620 | 1930 | –690 | |
| | | $R_{23}$ | 2620 | 1849 | –771 | |
| | | $R_{24}$ | 2620 | 1733 | –887 | |
| $R_3$ | 0.285 | $R_{31}$ | 2620 | 1995 | –625 | –756.5 |
| | | $R_{32}$ | 2620 | 1894 | –726 | |
| | | $R_{33}$ | 2620 | 1747 | –873 | |
| | | $R_{34}$ | 2620 | 1796 | –802 | |
| $R_4$ | 0.194 | $R_{41}$ | 2620 | 1796 | –824 | –737.75 |
| | | $R_{42}$ | 2620 | 1968 | –652 | |
| | | $R_{43}$ | 2620 | 1839 | –781 | |
| | | $R_{44}$ | 2620 | 1926 | –694 | |
| | | Average passenger SERVQUAL value ($SQ_3$) | | | | –2.8 |

### 4.3 *Data conclusion analysis*

The average passenger SERVQUAL value $SQ_3$ is –2.8, which is the gap between passenger perception and passenger expectation. According to the final SQ score, there is still a certain gap between the safety management of this rail transit enterprise and the full expectation of passengers on safety. This gap needs to be further improved by the operating enterprises.

By dividing the $SQ_2$ value of the four stages of 4R crisis security management by the number of valid questionnaires, the average value of security perception of each indicator is calculated. The distribution of points in the four stages is shown in Figure 1. It can be seen from the figure that the perceived value of Reduction is the highest, followed by Recovery, Response, and Readiness. It can be seen that in the 4R crisis safety management mode, the adjustment of the station safety management strategy should focus on strengthening the management in the Recovery stage, Response stage, Readiness stage, and Reduction stage successively.

Among the four stages, the Readiness stage has the lowest score, especially in the management system, safety publicity, emergency plan, and emergency drill. It can be seen that passengers attach importance to the Readiness stage, emergency drills, and improvement of emergency plans. Enterprises should also establish and improve management systems and emergency plans from the perspective of passengers' perception, do a good job in safety publicity and regularly carry out emergency drills.

In the investigation and analysis of the four indicators in the Response stage, safety perception of accident control is the lowest value, and accident control is an important indicator in the Response stage, which is also the key to determining whether the crisis accident can be contacted quickly. It shows that when passengers face crisis accidents, the proper emergency treatment of rail transit

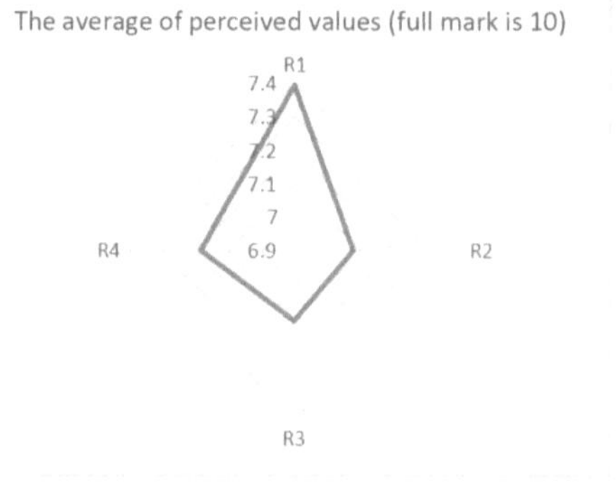


Figure 1. Survey results of passenger safety perception in 4R crisis safety management mode.

enterprises is the problem that passengers pay attention to. From the perspective of passengers' perception, enterprises find deficiencies in the stage of Response and strengthen their ability to control crises and accidents.

Among the four indicators of Recovery, the low score of accident investigation indicates that passengers pay more attention to the follow-up investigation after the accident and expect the truth of the crisis accident. In the Recovery stage, enterprises should not only quickly resume operations, strengthen social publicity, learn lessons and guide passengers to identify with the enterprise again, but also seek truth from facts, restore the truth of the crisis and accident, find the root cause and face the weakness, so as to establish a long-term foothold.

In terms of environmental safety, hazard identification, equipment safety, and safety education in the Reduction phase, the safety perception of hazard identification is the lowest, indicating that passengers not only pay attention to the safety of the overall environment but also pay more and more attention to the safety of small details, such as whether there are signs and labels for hazard sources in stations. At the same time, we also pay more attention to the safety education of equipment and staff. It is urgent to improve the safety management of equipment and personnel training. Enterprises should regularly carry out safety training and education for employees, standardize the management of facilities and equipment, record the safe use of the ledger, regularly carry out hazard identification and do a good job of marking, starting from small details, for passengers' environment safety atmosphere. The perceived average values of 16 indicators in 4 stages are summarized as shown in Figure 2.

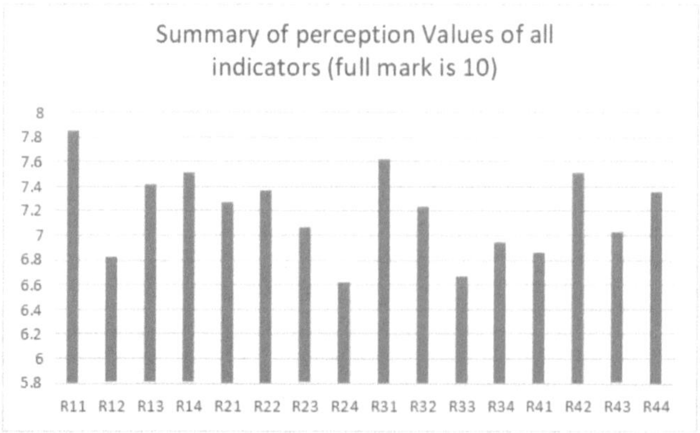


Figure 2. Average perception value of 16 indicators in 4 stages.

## 5 CONCLUSION

This paper combines the 4R crisis management theory with passenger perception and constructs an index system from the perspective of passenger psychology and enterprise safety management. By selecting the target station for investigation and analysis, the safety management of rail transit enterprises is not perfect from the perspective of passengers, and some targeted suggestions are given.

The main conclusion of this paper is to integrate the 4R crisis theory into the safety management of urban rail transit stations. By trying to apply it in the field of urban rail transit operation management, innovating the perspective of passenger perception, and integrating relevant knowledge of passenger psychology, the supervision and implementation of safety management can realize a dual-subject response mechanism, that is, the enterprise is the main body of safety implementation, and the passenger is the main body of safety function. At present, the safety management analysis of the combination of the 4R crisis management theory and passenger perception is still in the preliminary exploration stage, and there are still many shortcomings. For example, the study and application of the 4R crisis management theory are still in a relatively simple stage, and the analysis of passenger perception value is not deep enough. In the follow-up research, the close combination of the 4R crisis management theory and passenger perception will be further expanded, and new methods will be actively explored to analyze passenger safety in rail transit stations.

## REFERENCES


Lin Li, Ji Guang Gang, Tang Lin. Research on evaluation of urban rail Transit Operation Service Quality [J]. *Railway Transport and Economy*, 2020, 42(12):112–116.

Liu Peiyao. *Research on Coal mine emergency Management Based on "4R" Theory* [D]. Xi'an University of Science and Technology, 2021(01):10–11.

Lu Wengang, Peng Jing. Study on the Evaluation Index System of Guangzhou metro Public Emergency Response Capability[J]. *Urban Development Studies*, 2012, 19(04):118–124.

Meng Liang, Zhang Guanglei. Construction of University Security Prevention and Control System based on 4R Crisis Management Theory [J]. *Modern University Education*, 2017, (04):91–96.

Millayne Nogueira Medeiros, Anderson Tiago Peixoto Gonçalves, Luiz Joaquim Diniz da Silva, et al. Application of the SERVQUAL Model in the evaluation of the quality of the service provided by a security company in Higher Education Institution[J], *Revista Gestão da Produção Operações e Sistemas*,2020,Volume 15, No. 2.

Wang Fanfei, Xiang Kelan, QU Hong, et al. Construction and implementation of nursing safety and quality management system based on 4R theory of crisis Management [J]. *Journal of Nursing*, 2017, 32(05):52–55.

Xi He, Hanli Cai, Haipeng Li, et al. Subway Counter-terrorism Strategy Study Based on 4R Model [C], *Proceedings of 2015 International Conference on Education Research and Reform*(ER; R 2015 V8), 2015:319–324.

Yu Xinyuan. Research on public health Epidemic Prevention and control of urban rail Transit [D]. *China Academy of Railway Sciences*, 2020.

*Advances in Urban Construction and Management Engineering – Cha (Ed)*
*© 2023 Copyright the Author(s), ISBN 978-1-032-39018-5*

# Sensitivity analysis and optimization scheme of waterlogging control measures in a coastal plain of Wenzhou

Benjun Shi*
*Zhejiang Guangchuan Engineering Consulting Co., LTD, Hangzhou, China*
*Zhejiang Institute of Hydraulics & Estuary, Hangzhou, China*
*Zhejiang Provincial Key Laboratory of Hydraulic Disaster Prevention and Mitigation, Hangzhou, China*

Rui Zhang*
*Zhejiang Design Institute of Water Conservancy and Hydroelectric Power, Hangzhou, China*

ABSTRACT: Based on the optimization of flood control measures in a typical coastal plain area of Wenzhou, the sensitivity of river scale, sluices scale, and other factors on drainage effect and project investment are analyzed, and the drainage project scheme is optimized accordingly. On the basis of this, proposed measures for coastal plain drainage and evaluation criteria for project scale rationality are put forward, which can provide a reference for seeking the optimal drainage scheme with both benefits and investment.

## 1 INTRODUCTION

The main idea of flood control and drainage in coastal plain is to increase flood discharge capacity and flood storage and detention capacity of the plain. The main engineering measures include widening river channels, increasing lowland regulation and storage, expanding sluices, and adding or expanding strong drainage capacity. The above flood control ideas and engineering measures are generally applied in combination (Wang 2013). For specific hydraulic projects, the investment includes construction, land acquisition, and environment. Land acquisition and environment investment mainly include construction land acquisition and resettlement investment. In the preliminary demonstration of a project, technical and economic comparison is often required for different combination schemes of river channels, sludges, pumping stations, and other factors, so as to determine the most economical and reasonable drainage scheme. All of these factors are interwoven in the process of drawing up the drainage scheme, which makes it difficult to obtain the optimal drainage scheme (Liu 2012).

## 2 BRIEF INTRODUCTION OF DRAINAGE PROJECT OF A COASTAL PLAIN IN WENZHOU

The plain has a dense water system and flat terrain, and the basic water surface rate is about 9.14%. At present, the width of the main drainage channel is mainly concentrated in 9 ~ 30 m, with an average width of about 19 m. In addition, the existing residential houses are close to the river bank, and if the river is widened, the amount of land expropriation will be large. At present, the waterlogged in this area is discharged into the Aojiang River in the north or into the sea in the east through the existing sluice gate. Located on the east coast of southern Zhejiang province, this plain

*Corresponding Authors: 632666121@qq.com and 13738048683@163.com

 DOI 10.1201/9781003348023-32

is frequently affected by typhoons, often hit by heavy rain, and prone to flood disasters. The lack of backbone drainage channels, a large number of river bayonets, and uncoordinated scale of river sluices lead to the low drainage capacity of the plain (Zhejiang Guangchuan 2016).

## 3 SENSITIVITY ANALYSIS AND OPTIMIZATION SCHEME OF WATERLOGGING CONTROL MEASURES

Based on the current situation of planning water conservancy projects in this plain, this paper focuses on the sensitivity analysis of the XL sluice, PC sluice, and the river scale in front of the sluice. The 5km river in front of XL sluice and PC sluice was set as a regular river with uniform width. In this calculation, the one-dimensional unsteady flow dynamics MIKE11 model developed by the Danish Institute of Hydraulic Science (DHI) was used to establish the hydraulic calculation model. The calculated time step was 10s, and the channel roughness was 0.025-0.028. The calculation conditions mainly include adjusting the scale of the XL sluice and PC sluice and the river channel in front of the sluice, calculating the maximum flood level and submerging time of each characteristic point of the plain, and the displacement of the two sluices, etc. The changes in land area, construction investment, land acquisition, resettlement investment, total engineering investment, and other factors under each calculation condition were analyzed, and sensitivity analysis was conducted to further optimize the drainage scheme (Cai 2008). There are 12 calculation conditions, as shown in Table 1. There are 16 characteristic points.

Table 1. Calculation conditions (width of sluice and river) (Unit: m).

| Scheme number | XL sluice | XL sluice front river | PC sluice | PC sluice front river |
|---|---|---|---|---|
| Scheme 1 | 3 | 3 | 42 | 42 |
| Scheme 2 | 3 | 6 | 42 | 84 |
| Scheme 3 | 3 | 9 | 42 | 126 |
| Scheme 4 | 6 | 6 | 49 | 49 |
| Scheme 5 | 6 | 12 | 49 | 98 |
| Scheme 6 | 6 | 18 | 49 | 147 |
| Scheme 7 | 9 | 9 | 56 | 56 |
| Scheme 8 | 9 | 18 | 56 | 112 |
| Scheme 9 | 9 | 27 | 56 | 168 |
| Scheme 10 | 12 | 12 | 63 | 63 |
| Scheme 11 | 12 | 24 | 63 | 126 |
| Scheme 12 | 12 | 36 | 63 | 189 |

### 3.1 *Sensitivity analysis of drainage effect*

(1) Flood level

The scale of XL sluice and PC sluice is fixed, the sluice-river width ratio is adjusted to 1:1, 1:2, and 1:3, and the increments of flood water level in different places are compared, so as to explore the sensitivity of flood water level to the change of sluice-river width ratio. The calculation result of XL sluice width of 6 m and PC sluice width of 49 m is shown in Figure 1.

By comprehensive comparison of flood water level increment in different areas within the affected area, it can be found that, compared with the adjustment from 1:2 to 1:3, the flood level decreased more when the width ratio of sluice and river is adjusted from 1:1 to 1:2. Therefore, it can be concluded that under this calculation condition, when the sluice-river width ratio is in the range of 1:1~1:2, the flood level is more sensitive to the change of the river scale in front of the sluice. In terms of a single factor of flood level, comparing schemes 4- 5 and schemes 5-6, namely

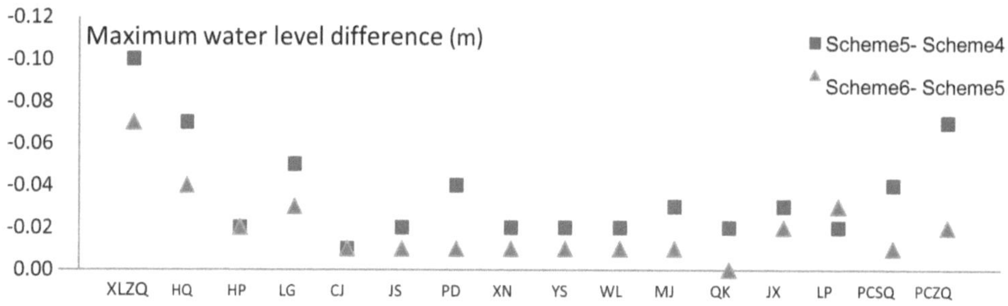


Figure 1. Maximum water level difference under different sluice-river width ratios (Sluice-river width ratio: Scheme 4 1:1, Scheme 5 1:2, Scheme 6 1:3).

the sluice-river width ratio of 1:2, is the most suitable. By comparing schemes 1, 2 and 3, schemes 7, 8 and 9, and schemes 10, 11 and 12, the above conclusion can also be drawn.

The width ratio of sluice and river was fixed as 1:2, and the width of sluice and river was adjusted at the same time to compare the increment of flood level, so as to explore the sensitivity of flood level to the change of sluice width and river width in front of the sluice. The calculation results are shown in Figure 2.

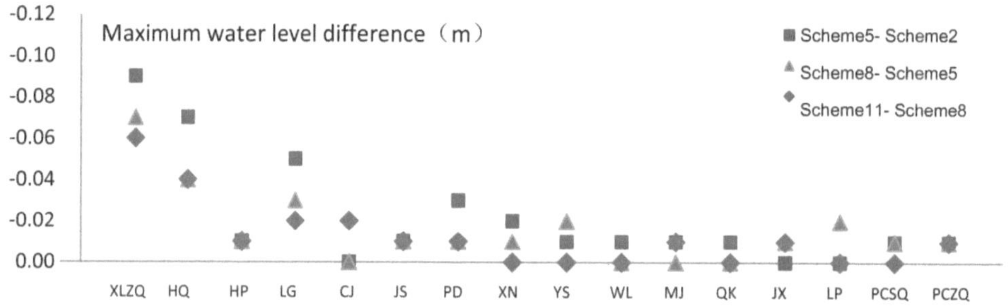


Figure 2. Maximum water level difference under different sluice widths and river width.

According to Figure 2, when the width of the XL sluice varies from 3 m to 6 m, the flood level is sensitive to the width of the sluice and river. Considering the single factor of flood level, it is suggested that the width of the XL sluice should be about 6 m, and the width of the river in front of the sluice should be 12 m. When the width of a PC sluice varies from 42 m to 56 m, the flood level is sensitive to the width of the sluice and river. Considering the single factor of flood level, it is suggested that the width of the PC sluice should be about 56 m, and the width of the corresponding river in front of the sluice should be 112 m.

(2) Submerging time

The tester fixed the scale of the XL sluice and PC sluice, adjusted the sluice-river width ratio to 1:1, 1:2, and 1:3, and compared the increment of submerging time of different places, so as to explore the sensitivity of submerging time to the change of sluice-river width ratio. The calculation result of XL sluice width of 6 m and PC sluice width of 49 m is shown in Figure 3.

It can be concluded that under this calculation condition when the sluice-river width ratio is in the range of 1:1～1:2, the submerging time is more sensitive to the change of the river scale in front of the sluice. In terms of the single factor of submerging time, comparing schemes 4-5 and schemes 5-6, namely the sluice-river width ratio of 1:2, is the most suitable. By comparing schemes 1, 2 and 3, schemes 7, 8 and 9, and schemes 10, 11 and 12, the above conclusion can also be drawn.

The width ratio of sluice and river was fixed as 1:2, and the width of sluice and river was adjusted at the same time to compare the decrement of submerging time, so as to explore the sensitivity of

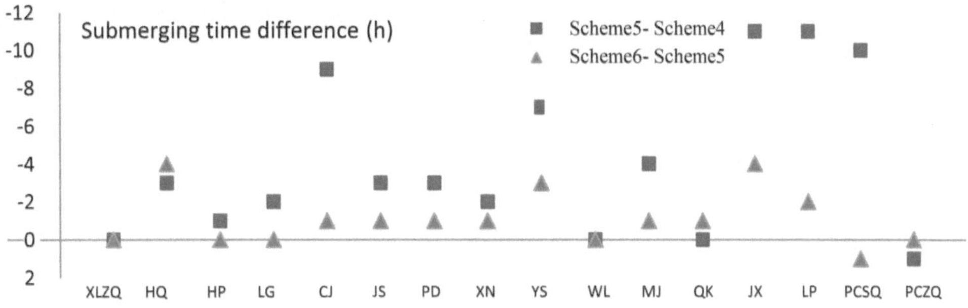


Figure 3. Submerging time difference under different sluice-river width ratios (Sluice-river width ratio: Scheme 4 1:1 Scheme 5 1:2 Scheme 6 1:3).

submerging time to the change of sluice width and river width in front of the sluice. The calculation results are shown in Figure 4.

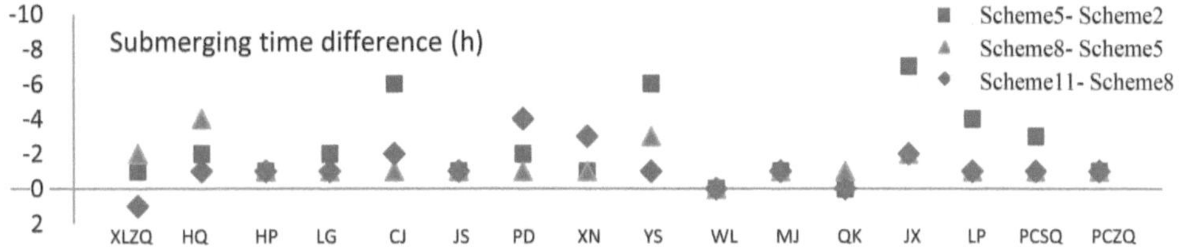


Figure 4. Submerging time difference under different sluice widths and river width.

According to Figure 4, when the width of the XL sluice varies from 3 m to 6 m, the submerging time is sensitive to the change in width of the sluice and river. Considering the single factor of submerging time, it is suggested that the width of the XL sluice should be about 6 m, and the width of the river in front of the sluice should be 12 m. When the width of the PC sluice varies from 42 m to 49 m, the submerging time is sensitive to the width of the sluice and river. Considering the single factor of submerging time, it is suggested that the width of the PC sluice should be about 49 m, and the width of the corresponding river in front of the sluice should be 98 m.

(3) Displacement

The calculation results of the displacement of XL and PC sluice under each scheme are shown in Figures 5 and 6.

According to Figure 5, when the sluice scale remains unchanged and the sluice-river width ratio is within the range of 1:1 to 1:2, the displacement is more sensitive to the change of the river scale in front of the sluice, compared with 1:2 to 1:3. Considering the single factor of displacement, the most suitable width ratio of sluice and river is 1:2.

By comparing schemes 2, 5, 8, and 11, it can be seen from Figure 6 that when the width ratio of sluice and river is fixed at 1:2, the sensitivity of displacement decreases with the increase of sluice and river scale, but the decrease is not significant. Therefore, considering the single factor of displacement, it is suggested that the width of the XL sluice can be selected in the range of 3~12 m, and the width of the PC sluice can be selected in the range of 42~56 m.

### 3.2 *Numerical simulation results of stress under water pressure component*

According to the study on the influence of sluice and river scale on waterlogging drainage effect in Section 3.1, it can be found that: the scale of XL sluice and front river mainly affects the flood level and submergence time of XL sluice to LG Town, and the affected length of the river in front of the sluice is about 2 km; the scale of PC sluice and front river mainly affects the flood level

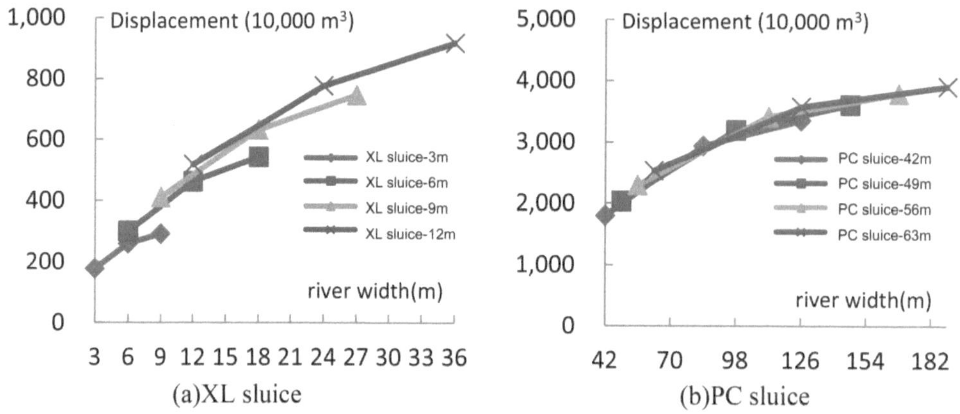


Figure 5. Displacement – River width (fixed sluice width).

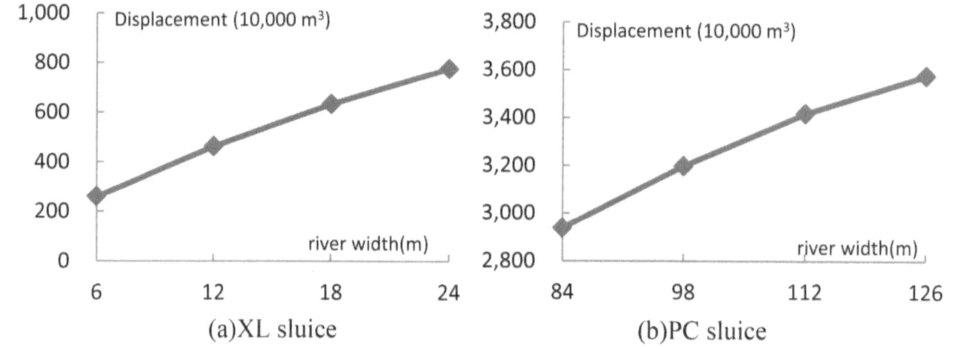


Figure 6. Displacement – River width (the width ratio of sluice and river is 1:2).

and submergence time of PC sluice to JX Town, and the affected length of the river in front of the sluice is about 5 km. According to this foundation, this paper calculates the area and total investment of the two sluice engineering. The investment in this study mainly considers the cost of land expropriation and demolition, the cost of river construction and installation, and the cost of sluice engineering construction and installation. In order to explore the sensitivity of unilateral displacement investment (total investment/displacement) to changes in sluice and river scale, the calculation results are sorted out in Figures 7 and 8.

It can be found from Figure 7(a) that when the scale of XL sluice is fixed and the sluice-river width ratio is in the range of 1:1~1:2, the unilateral displacement investment is more sensitive to the change of river scale in front of the sluice and decreases with the increase of river width. In terms of a single factor of investment of XL sluice, the sluice-river width ratio is the most suitable at 1:2. In the same way, it can be found from Figure 7(b) that in terms of a single factor of investment of PC sluice, the sluice-river width ratio is the most suitable at 1:1~1:2.

It can be found from Figure 8(a) that the unilateral displacement investment of XL sluice is sensitive to the change of sluice and river scale, and the sensitivity decreases with the increase of scale. As for the single factor of unilateral displacement investment, it is suggested that the width of the XL sluice can be selected within the range of 6~9m according to needs. It can be found from Figure 8(b) that the unilateral displacement investment of PC sluice is not sensitive to the change of sluice and river scale.

In addition, according to the data of the proportion of each item in the engineering investment, it can be found that: for the relatively small XL sluice, the river construction and installation cost

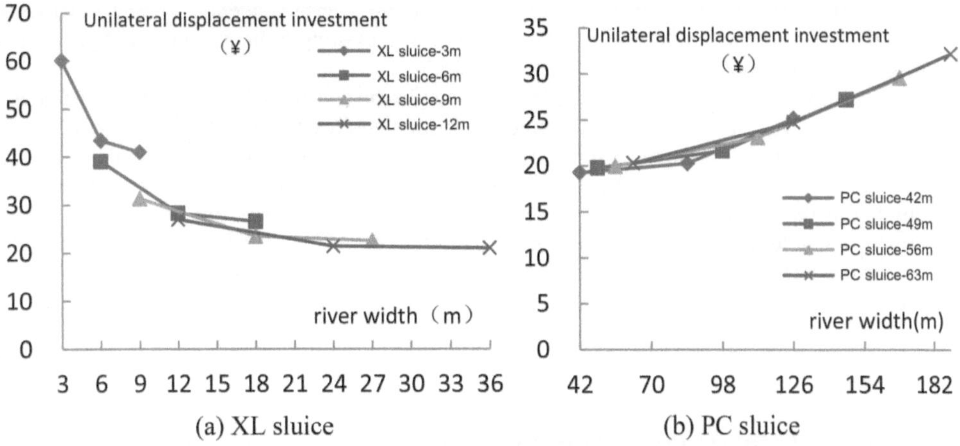


Figure 7. Unilateral displacement investment – River width ((fixed sluice width).

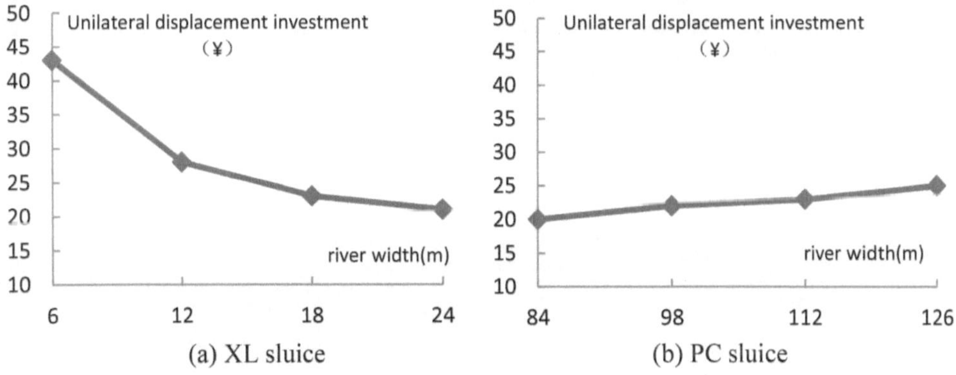


Figure 8. Unilateral displacement investment – River width (the width ratio of sluice and river is 1:2).

in front of the dam occupies an absolutely dominant position in the total investment, accounting for more than 75%. And for the larger ones, PC sluice, this ratio is about 50%, followed by the cost of land expropriation demolition, the same gate construction and installation cost accounted for the least proportion. Therefore, when optimizing the drainage scheme, on the premise of meeting flood control function and drainage effect, it should be considered to minimize the width of the river in front of the sluice and reduce the area of the project, so as to reduce the total investment of flood control and drainage project.

## 4 CONCLUSION

Based on the sensitivity analysis of drainage measures in this coastal plain of Wenzhou, the drainage scheme is optimized. On this basis, the following evaluation criteria for the rationality of drainage measures in the coastal plain of Zhejiang Province are proposed:

(1) The ratio of the total clear width of the sluice to the average width of the river in front of the sluice should be about 1:2, and should not be less than 1:2.
(2) For different regions and locations, the sensitivity of drainage effect factors and engineering investment to the scale of sluice and river is different, and there is no statistical law. Therefore,

in the process of determining the specific drainage engineering scheme, it is necessary to set different working conditions by referring to similar existing projects and making scheme comparisons to determine the most reasonable scale.

(3) The investment in land acquisition and resettlement for the construction of wide river channels is usually large, and the investment in land acquisition and resettlement for the current urban areas is even more than the investment in engineering and occupies a large amount of precious cultivated land, so the local government has huge pressure on the balance. Therefore, when optimizing the drainage scheme, on the premise of meeting flood control function and drainage effect, it should be considered to minimize the width of the river in front of the sluice and reduce the area of the project, so as to reduce the total investment of flood control and drainage project.

## REFERENCES


Cai Y, Xin Y, Hu D. Review of sensitivity analysis[J]. *Journal of Beijing Normal University* (Natural Science), 2008, 01:9–16.

Liu Y, Lin D.R, Wang B, Bao Z.J. An optimized calculation of drainage scale for first phase reclamation project of Oufei[J]. *Advances in Science and Technology of Water Resources*, 2012, 04:63–65+69.

*The feasibility study report of the Jiangnanyang Plain Backbone Drainage Project in Cangnan County*. Zhejiang Guangchuan Engineering Consulting Co. Ltd. 2016.

Wang J, Cao M.K, Yang C.X, Wang Y. Study on tidal river network in coastal plain and floodgate scale in lake area [J]. *Jiangsu Science and Technology Information*, 2013, 13:66–69.

*Advances in Urban Construction and Management Engineering – Cha (Ed)*


# Research on quantitative indicator system of transfer efficiency in airport general transportation center

Kun Zhang & Jian Zhang*
*China Airport Planning and Design Institute Co., Ltd., Beijing, China*

ABSTRACT: This paper aims to establish a comprehensive evaluation system for the airport general transportation center, which can take into account the co-interest of travelers, operational benefits of the enterprise, and social benefits simultaneously, and a quantitative method combined with scoring criteria was proposed as well. Specifically, this paper innovatively realizes the quantitative comprehensive evaluation of the airport general transportation center and also ascertains the evaluation standard and quantitative calculation method of the three-level indicator system of the general transportation center. The indicator system was quantitatively evaluated by taking Beijing Daxing International Airport's general transportation center as a case, and the result of the evaluation is observed to be excellent. The result proves that the method proposed in this paper has a significant guidance effect on future airport planning and design.

## 1 INTRODUCTION

The transportation hub is a huge system. The airport aims to build a seamless three-dimensional general transportation center (GTC) that integrates air transportation, urban public transportation, underground rail transportation, social vehicles, taxis, and long-distance buses. As a part of the airport hub, the airport's landside transportation facilities connect air traffic and urban traffic, providing service for the "last mile" of passengers arriving or leaving the airport. As a comprehensive transportation hub for aiding the operation of the terminals, the general transportation center is responsible for solving the transfer of traffic generators between various means of transportation. With the expansion of the airport area, the growth of passenger traffic, the shortage of land resources, and the diversification of transportation modes for passengers entering and leaving the airport, the internal and external resources of the general transportation center should be reasonably and effectively allocated to meet the needs of airport operation, expansion, and requirements for urban traffic development. Currently, in the context of the transformation of Chinese airports from the type of large-scale with high developing speed to the type of high operating quality with efficiency, the study of indicators of general transportation centers is of great significance.

The first and foremost factor of the airport general transportation center is based on the optimal allocation and layout of space resources, the second is to select the most satisfactory combination of various transportation modes according to certain standards within the center, and the last is to determine the collaborative relationship between various transportation modes. Therefore, using various transportation modes as basic variables to describe the corresponding organic integration methods will involve issues such as the overall planning of the airport field, the integrated layout of the airport, and the coordination of airport functions. It must be emphasized that the optimal allocation of spaces and the determination of the coordination relationship of the airport's general transportation center must consider the requirements of the decision-making level. The lower-level decision-making of the airport's general transportation center configuration should be subject to

*Corresponding Author: 243106763@qq.com

DOI 10.1201/9781003348023-33

the constraints and regulations of the higher-level decision-making, in other words, to satisfy the requirement of upper-level planning.

This article is recorded in the "Research and Application of Key Indicators of Airport Intelligent Comprehensive Transportation Center", which is the outstanding achievement of the project "China Civil Aviation High-quality Development Indicator Framework System (Trial) (Civil Aviation Administration of China 2021)." After numerous engineering practices and experience summarization, the project team proposed that the standard for the general transportation center has to be established upon several indicators of timeliness, comfort, turnover rate, safety, and travel structure balance. And at the same time, the indicators have to be quantified to establish a complete indicator system.

## 2 THE APPROACH OF ESTABLISHING THE QUANTIFIABLE INDICATOR SYSTEM

The establishment of the traffic layout and operation indicator system for a general transportation center within a large-scale airport can accurately reflect the airport's production and operation effects and operating costs, and at the same time provide support for the next stage of strategic decision-making. Therefore, the following principles should be followed while establishing the operation indicator system: scientific principles, systematic principles, hierarchical principles, comparability principles, principles in line with industry characteristics, and principles of combining qualitative and quantitative aspects.

The construction of the layout indicator system for the large-scale airport general transportation center comprehensively considers the service level for travelers, the economy of enterprise investment and operation, and social benefits. Through data calculation and construction of scoring criteria for the indicators, the layout of a large-scale airport general transportation center can be described more comprehensively, deeply, and accurately, thereby providing a more comprehensive and objective scientific basis for reflecting the pros and cons of the layout of the general transportation center, followed by an effective airport construction and operation management for public travel.

In the process of establishing the indicator system, firstly, the first-level indicators which include three sub-indicator systems, namely, traveler level of service indicators, enterprise economic indicators, and social indicators should be clearly defined and distinguished. Secondly, for the travel service and enterprise economic indicators, it is necessary to distinguish the travel demand of different modes of transportation (cars, taxis or airport buses, rail transit), determine the level of service on the landside of the airport, and measure the construction and operation costs of facilities (parking lots, lanes, rail stations) corresponding to different modes of transportation. Thirdly, for social benefits indicators, the indicator system takes into account the airport emergency guarantees level and travel structure optimization, and other aspects to subdivide and evaluate indicators in this aspect. The abovementioned are the construction ideas of the second-level indicators in the indicator system. Finally, the third-level indicators are used to describe specific indicators at a certain level (Wu 2016). The overall framework of the evaluation indicator system for the layout of the airport's general transportation center is shown in the following Figure 1:

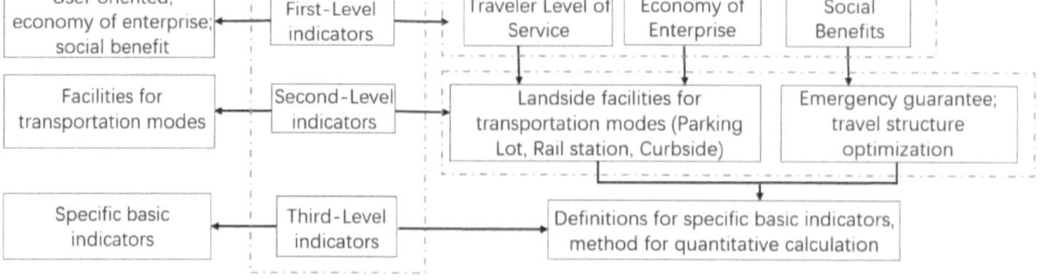


Figure 1. The overall framework of the evaluation indicator system for the layout of the general transportation center.

## 3 ORGANIZATION OF THE INDICATOR SYSTEM

The indicator system of the airport general transportation center is divided into three levels (Zhang 2012), of which six first-level indicators can be specifically divided into the following categories: passenger level of service indicators (timeliness, comfort, safety), business economics (facility availability), and social benefits (emergency guarantee level, travel structure optimization). The second-level indicators mainly consider the timeliness, comfort, and safety of landside traffic at the airport for travel by different transportation modes (cars, taxis or airport buses, rail transit), as well as the corresponding facilities (parking lots, curbside, rail station) utilization, and then subdivide the passenger level of service and enterprise economic indicators; At the same time, the indicators of social benefits are subdivided in terms of congestion relief, disaster relief, severe weather response, as well as transportation supply and guarantee. Finally, based on the first and second-level indicators, a specific three-level indicator system is formulated, and the specific meaning and quantitative calculation method of each third-level indicator are clarified. The three-level indicators are detailed in Level 3 of the third column of Table 3. It should be noted that the third-level indicators in Figure 2 are not complete, and only a sample of third-level indicators are listed accordingly to their upper-level indicators. A complete set of third-level indicators can be found in Table 3.

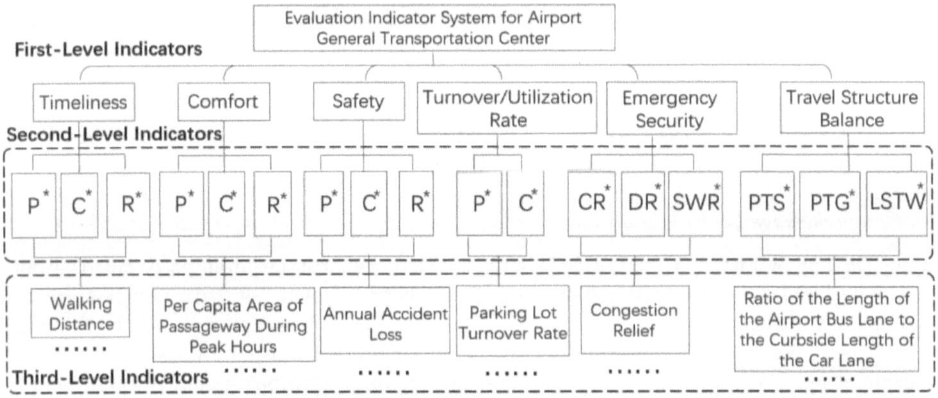


*P = Parking Lot; C = Curbside; R = Railway Station; CR = Congestion Relief; DR = Disaster Relief; SWR= Severe Weather Protection; PTS=Public Traffic Supply; PTG = Public Traffic Guarantee; LWTS = Low-Speed Traffic Supply.

Figure 2. Evaluation indicator system for airport general transportation center.

The value of each evaluation indicator is within the score range of 0-10. After normalization, the score range of the layout evaluation of large-scale airport transportation hub is still 0-10. The ratings of different scores are shown in Table 1 below:

Table 1. Evaluation score and rating table of large airport transportation hub layout.

| Score Range | Rating |
|---|---|
| [9,10] | Excellent |
| [8,9) | Good |
| [7,8) | Average |

## 4 QUANTITATIVE ANALYSIS AND EVALUATION METHOD

### 4.1 *Analytic hierarchy process method*

Analytic Hierarchy Process (AHP) is a decision-making method based on Multi-level weight analysis (Chen 2012). It is a multi-objective decision-making analysis method that combines qualitative and quantitative analysis. Analytic Hierarchy Process is to decompose the decision-making problem into different hierarchical structures in the order of the overall objective, sub-objectives, evaluation criteria, and specific investment plans, and then use the method of solving the eigenvectors of the judgment matrix to obtain the priority weight of each element of each level to an element at the previous level. Finally, the final weight of each alternative to the total objective is recursively merged by the weighted sum method, and the one with the largest final weight is the optimal plan.

### 4.2 *Fuzzy comprehensive evaluation method*

After the construction of the evaluation indicator system, the fuzzy comprehensive evaluation method is used to process the indicator system, and then the layout plan of the large airport general transportation center can be evaluated. The so-called fuzzy comprehensive evaluation method is to use fuzzy language to describe the qualitative things appropriately, quantifying and evaluating the qualitative descriptions. It quantifies the fuzzy indicators (i.e., determines the degree of membership) by constructing a hierarchical fuzzy subset, and then uses the corresponding fuzzy variable principle to comprehensively process each indicator.

## 5 CASE ANALYSIS – EVALUATION OF GENERAL TRANSPORTATION CENTER OF BEIJING DAXING INTERNATIONAL AIRPORT

### 5.1 *Comprehensive landside transportation layout of Beijing Daxing international airport*

Beijing Daxing International Airport is a large-capacity public transportation-oriented airport. It strives to build a comprehensive transportation backbone network of "five verticals and two horizontals," and organically integrates various transportation such as highways, urban rail transit, and intercity railways. At the same time, a comprehensive ground transportation system is formed that centered upon Beijing Daxing Airport with strong regional radiation capability, which includes the "Highway Radiation Circle" and "Railway Radiation Circle" (Liu 2015).

It is particularly important to note that several rail transit lines are running through and beneath the terminal building from north to south; at the same time, rail transit stations are set up on the second underground floor of the terminal building, and passengers can directly enter the terminal building through large-capacity elevators or escalators. "Seamless connection" and "zero-distance transfer" for air and landside traffic are built. These all together make Daxing Airport a truly integrated transportation hub.

The general transportation center of Beijing Daxing Airport contains a variety of core facilities and auxiliary facilities, and their layout relationship directly affects the operation efficiency and level of service of the hub. Through research, induction, analysis, and other means, the subject studies the composition, layout, advantages, and disadvantages of various facilities in the hub, and summarizes the applicable layout principles in combination with the specific characteristics of Beijing Daxing Airport to guide the layout planning and management planning of various transportation facilities in the hub.

The evaluation object indicator set is obtained from the large-scale airport general transportation center evaluation indicator system, then the judgment matrix needs to be constructed, and the corresponding weight vector needs to be calculated as well. At the same time, the consistency test of the constructed judgment matrix is carried out. According to the comparison scale of the analytic hierarchy process method, the expert scoring method is used to compare and score the selected indicators in pairs. The mode of the scores scored by the experts is taken as the score of

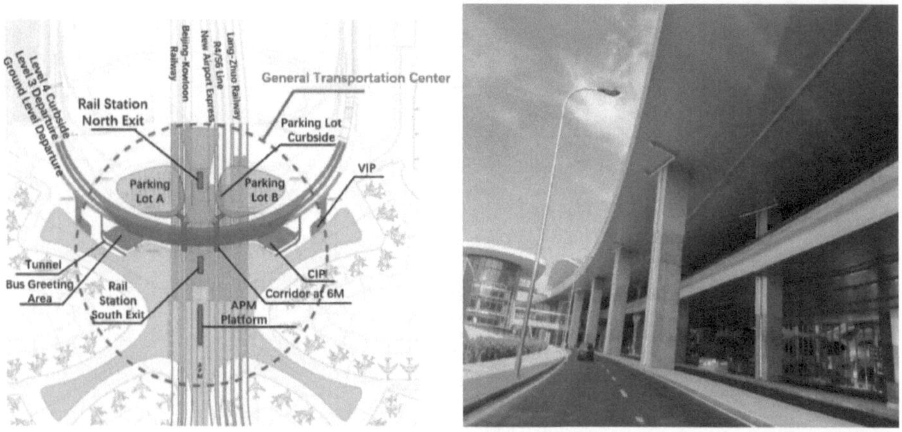


Figure 3. Beijing Daxing international airport layout (Left) & multilevel curbsides (Right).

the pairwise comparison of each indicator, and the priority vector of each element is calculated at the same time.

The first-level indicator set is A=(Timeliness, Comfort, Safety, Facility utilization rate, Emergency support level, Travel structure optimization)=(A1, A2, A3, A4, A5, A6), and its judgment matrix is as follows in Table 2. The meaning of each scale in the table is shown in Table 3 below.

Table 2. First-level indictor judgement matrix.

| First-Level Indicator | A1 | A2 | A3 | A4 | A5 | A6 | Priority Vector Score |
|---|---|---|---|---|---|---|---|
| Timeliness A1 | 1 | 3 | 1/5 | 4 | 1/3 | 2 | 0.13 |
| Comfort A2 | 1/3 | 1 | 1/5 | 3 | 1/5 | 2 | 0.08 |
| Safety A3 | 5 | 5 | 1 | 7 | 1 | 5 | 0.36 |
| Facility Utilization A4 | 1/4 | 1/3 | 1/7 | 1 | 1/5 | 1/3 | 0.04 |
| Emergency Support A5 | 3 | 5 | 1 | 5 | 1 | 5 | 0.31 |
| Travel structure optimization A6 | 1/2 | 1/2 | 1/5 | 3 | 1/5 | 1 | 0.08 |

AHP is a method of transforming the thinking process of an analyst into a mathematical form. Due to the large and wide variety of factors involved, the evaluators cannot give accurate comparative judgments for certain factors, which may lead to inconsistency in judgments. The inconsistency of this judgment can be reflected by the change of the characteristic root of the judgment matrix, and the consistency of the matrix needs to be checked.

The consistency test is carried out on the constructed first-level indicator judgment matrix, and the result is calculated as CR=0.0579<0.1, which proves that the established first-level indicator judgment matrix is considered to have good consistency and be acceptable.

Similarly, the overall priority vector results of the evaluation indicators of the airport general transportation center are obtained as shown in the following table.

## 6 EVALUATION RESULTS

According to the evaluation indicator priority vector established above and the evaluation set of Beijing Daxing Airport, the layout evaluation result can be obtained, and the calculation formula is as follows:

$$P = \sum_{i} \sum_{j} \sum_{l} p_{A_i} \cdot p_{B_j} \cdot p_{C_l} \cdot n_{ijl} \cdot \partial_{B_j}^{A_i} \cdot \partial_{C_l}^{B_i}$$

Table 3. Beijing Daxing airport layout evaluation set.

| Level 1 | Level 2 | Level 3 |
|---|---|---|
| Timeliness A1 | Private Vehicle B1 | Walking Distance C1 |
| | | Average Delay in Peak Hours C2 |
| | | Inducible Marker Condition C3 |
| | Ride-Hailing B2 | Walking Distance C1 |
| | | Average Delay in Peak Hours C2 |
| | | Inducible Marker Condition C3 |
| | Rental Car B3 | Walking Distance C1 |
| | | Average Delay in Peak Hours C2 |
| | | Inducible Marker Condition C3 |
| | Bus Stop B4 | Walking Distance C1 |
| | | Average Delay in Peak Hours C2 |
| | | Inducible Marker Condition C3 |
| | Reserved Parking Lot B5 | Access Distance C4 |
| | | Passage Specificity Status C5 |
| | Departure Level B6 | Walking Distance C1 |
| | | Average Delay in Peak Hours C2 |
| | | Inducible Marker Condition C3 |
| | Arrival Level B7 | Walking Distance C1 |
| | | Average Delay in Peak Hours C2 |
| | | Inducible Marker Condition C3 |
| | Railway Station B8 | Walking Distance C1 |
| | | Average Delay in Peak Hours C2 |
| | | Inducible Marker Condition C3 |
| | Urban Rail Station B9 | Walking Distance C1 |
| | | Average Delay in Peak Hours C2 |
| | | Inducible Marker Condition C3 |
| Comfort A2 | Private Vehicle B1 | Per Capita Area of Passageway During Peak Hours C6 |
| | | The proportion of Automated Escalators/Walkways to the Total Length of the Passage C7 |
| | | Parking Services C8 |
| | Ride-Hailing B2 | Per Capita Area of Passageway During Peak Hours C6 |
| | | The proportion of Automated Escalators/Walkways to the Total Length of the Passage C7 |
| | Rental Car B3 | Per Capita Area of Passageway During Peak Hours C6 |
| | | The proportion of Automated Escalators/Walkways to the Total Length of the Passage C7 |
| | Bus Stop B4 | Per Capita Area of Passageway During Peak Hours C6 |
| | | The proportion of Automated Escalators/Walkways to the Total Length of the Passage C7 |
| | Reserved Parking Lot B5 | Per Capita Area of Passageway During Peak Hours C6 |
| | | The proportion of Automated Escalators/Walkways to the Total Length of the Passage C7 |
| | Departure Level B6 | Per Capita Area of Curbside During Peak Hours C9 |
| | | Number of Driveways C10 |
| | | Separation of People and Vehicles C11 |
| | Arrival Level B7 | Per Capita Area of Curbside During Peak Hours C9 |
| | | Number of Driveways C10 |
| | | Separation of People and Vehicles C11 |

*(continued)*

Table 3. Continued

| Level 1 | Level 2 | Level 3 |
|---|---|---|
| | Railway station B8 | Per Capita Area of Passageway During Peak Hours C6 |
| | | The proportion of Automated Escalators/Walkways to the Total Length of the Passage C7 |
| | Urban Rail Station B9 | Per Capita Area of Passageway During Peak Hours C6 |
| Safety A3 | Private Vehicle B1 | Annual Accident Loss C12 |
| | | Weighted Number of Conflicted Points C13 |
| | | The proportion of Passengers Affected by the Conflicted Points C14 |
| | Ride-Hailing B2 | Annual Accident Loss C12 |
| | | Weighted Number of Conflicted Points C13 |
| | | The proportion of Passengers Affected by the Conflicted Points C14 |
| | Rental Car B3 | Annual Accident Loss C12 |
| | | Weighted Number of Conflicted Points C13 |
| | | The proportion of Passengers Affected by the Conflicted Points C14 |
| | Bus Stop B4 | Annual Accident Loss C12 |
| | | Weighted Number of Conflicted Points C13 |
| | | The proportion of Passengers Affected by the Conflicted Points C14 |
| | Reserved Parking Lot B5 | Annual Accident Loss C12 |
| | | Weighted Number of Conflicted Points C13 |
| | | The proportion of Passengers Affected by the Conflicted Points C14 |
| | Departure Level B6 | Annual Accident Loss C12 |
| | | Weighted Number of Conflicted Points C13 |
| | | The proportion of Passengers Affected by the Conflicted Points C14 |
| | Arrival Level B7 | Annual Accident Loss C12 |
| | | Weighted Number of Conflicted Points C13 |
| | | The proportion of Passengers Affected by the Conflicted Points C14 |
| Turnover Utilization A4 | Private Vehicle B1 | Parking Lot Turnover Rate C15 |
| | | Parking Lot Utilization Rate C16 |
| | Ride-Hailing B2 | Parking Lot Turnover Rate C15 |
| | | Parking Lot Utilization Rate C16 |
| | Rental Car B3 | Parking Lot Turnover Rate C15 |
| | | Parking Lot Utilization Rate C16 |
| | Bus Stop B4 | Parking Lot Turnover Rate C15 |
| | | Parking Lot Utilization Rate C16 |
| | Reserved Parking Lot B5 | Parking Lot Turnover Rate C15 |
| | | Parking Lot Utilization Rate C16 |
| | Departure Level B6 | Curbside Saturation Rate C17 |
| | Arrival Level B7 | Curbside Saturation Rate C17 |
| Emergency Guarantee A5 | Congestion Relief B8 | Response Time C18 |
| | | Effect of Plan Implementation C19 |
| | Disaster Relief B9 | Response Time C18 |
| | | Effect of Plan Implementation C19 |
| | Severe Weather Protection B10 | Response Time C18 |
| | | Effect of Plan Implementation C19 |
| Travel Structure Balance A6 | Public Traffic Supply B11 | Bus Supply Ratio C20 |
| | | Bus Supply and Demand Ratio C21 |
| | Public Traffic Guarantee B12 | The ratio of the Length of the Airport Bus Lane to the Curbside Length of the Car Lane C22 |
| | | The ratio of the Length of the Airport Bus Lane to the Curbside Length of the Car Lane C23 |
| | Low-Speed Transport Guarantee B13 | The ratio of the Area of Slow Traffic Facilities to the Area of Car Facilities C24 |
| | | Guarantees for Special Lanes for Slow Traffics C25 |

Where $P$ = obtained layout evaluation result; $P_{A_i}$ = priority vector set under the first-level indicators; $P_{B_j}$ = priority vector set under the second-level indicators; $P_{C_l}$ = priority vector set under the third-level indicators; $n_{ijl}$ = obtained evaluation value of each indicator; $\partial_{B_j}^{A_i}$ = whether there is a $B_j$ level index under the $A_i$ level indicator, it is 1 if it exists, and 0 if it does not exist; $\partial_{C_l}^{B_j}$ = whether there is a $B_j$ level indicator under the $C_l$ level index, it is 1 if it exists, and 0 if it does not exist.

After evaluation, the result shows that the layout rating of the general transportation center of Beijing Daxing Airport is excellent.

## 7 CONCLUSIONS

The airport general transportation center should consider the airport as the foothold, according to the goal of optimizing the allocation of resources, characteristics of social and economic development in the region, the positioning of the airport in the national comprehensive three-dimensional transportation network, the airport development goals, the short-term and long-term planning and various transportation elements and the scope of the airport's radiation area, giving priority to the development of green and intensive road transportation, promoting the goal of effective air-rail intermodal transportation, increasing the proportion of public transportation trips, and building a general transportation center in the combination of efficient transfer and close connection between multiple transportation modes and the airport.

This paper provides a solution for evaluating the operational efficiency of the airport's general transportation center through the way of establishing a comprehensive evaluation system of the layout of the airport's general transportation center which can consider the following three aspects or more into account, namely, the traveler's travel, enterprise operation, and social benefits. The establishing process is undertaken by considering the main factors such as transportation development strategy, urban transportation system configuration, comprehensive transportation travelers, transportation facilities, transportation distribution, etc., combined with the development status of the airport general transportation center. Moreover, this paper put forward a quantitative method and scoring criteria for the evaluation of the airport general transportation center, and innovatively proposed the three-level indicator evaluation standard of the general transportation center. The evaluation method formulated in this paper has been applied to Daxing Airport, and the evaluation result is observed to beexcellent. These project results can be applied to the other airports to guide the layout optimization and management planning of the general transportation center of the airport.


## ACKNOWLEDGMENTS

China Airport Planning and Design Institute Co., Ltd. 2020. Research and Application of Key Indicators of Airport Intelligent Comprehensive Transportation Center, *China Civil Aviation High-Quality Development Indicator Framework System (Trial).*

Thanks to *China Airport Planning & Design Institute Co., Ltd* for its generous funding of this research project.


## REFERENCES


Chen, Y.Y et al. 2012. Real-time Evaluation Method of Passenger Crowding in Integrated Transport Hub. *Journal of Highway and Transportation Research and Development* 29(S1): 75–80.

Civil Aviation Administration of China. 2021. *Research and Application of Key Technologies in the Construction of Intelligent Integrated Transportation hub for Beijing New Airport.*

Liu, W.J. 2015. *Integrated and Sustainable Planning for Multimodal Terminals*. Urban Transportation of China 13: 30–35.
Wu, P. & Sun, X.N.2016. *Research on functional matching and evaluation of comprehensive passenger transport hub*. China Communicators Press Co., Ltd.
Zhang, G.H et al. 2012. Study on Evaluation Method of Operational Efficiency for Intermodal Terminal. *Journal of Wuhan University of Technology* (Transportation Science & Engineering) 36: 25–28.

*Advances in Urban Construction and Management Engineering – Cha (Ed)*
*© 2023 Copyright the Author(s), ISBN 978-1-032-39018-5*

# Research on low impact landscape design in urban rainwater utilization technology environment

Jun Yan & Ruixin Zhang*
*Jingdezhen Ceramic University, Jiangxi, China*

ABSTRACT: In order to further explore how to design a low-impact landscape under the technical environment of urban rainwater utilization, this study takes the low-impact landscape design of a certain place as an actual case, makes a quantitative analysis of the actual situation of the area based on GIS and other technologies, then formulates a low-impact landscape design scheme according to the analysis results, and finally analyzes the overall effect of the scheme. The analysis results show that the design scheme has more obvious benefits under the condition of medium and small precipitation, which can provide some reference for future related work.

## 1 INTRODUCTION

In recent years, landscape construction in various regions has been continuously promoted. In order to avoid environmental pollution in the process of landscape project construction and further reduce the construction cost, low-impact landscape design has become the focus of the industry. In order to achieve the goal of low-impact landscape design, it is a feasible way to adopt urban rainwater utilization technology. Of course, due to the complexity of low-impact landscape design, in the actual design work, it is necessary to conduct deeper research on the low-impact landscape design work under the urban rainwater utilization technology environment based on both qualitative and quantitative aspects (Wang, 2022).

## 2 PROJECT OVERVIEW

A certain place plans to optimize the design of the municipal landscape of some blocks in the area. The total area of the target area is about 24 hectares, of which the water area is about 1.8 hectares. The water level is controlled by a sluice, and some rainwater wells in the surrounding roads and plots are connected to the water system of the area through municipal pipelines. At present, most of the blocks in the target area are residential areas and commercial areas. Since the previous block construction has experienced more artificial transformation, this area basically no longer has purely natural hydrological characteristics. It is still difficult to introduce the surface runoff and optimized water resources in this area into the landscape water environment by means of urban rainwater utilization technology (Sheng, 2021).

## 3 CORRELATION ANALYSIS OF LOW-IMPACT LANDSCAPE DESIGN IN THIS BLOCK

### 3.1 *Building of the block GIS model*

In this work, the construction of the GIS model is mainly based on the existing basic topographic map data, relevant data is imported into ArcGIS software, and the storage volume is calculated based on the hydrology module of this software. The calculation results are shown in Table 1.

*Corresponding Author: 003240@jcu.edu.cn

 DOI 10.1201/9781003348023-34

Table 1. The existing water storage capacity of this block.

| Design return period | Storage water level/m | Submerged area/$m^2$ | Water storage capacity regulation/$m^3$ |
|---|---|---|---|
| Once in 2 years | 12.90 | 15348 | 11985 |
| Once in 5 years | 13.69 | 20632 | 28166 |
| Once in 10 years | 14.58 | 24607 | 46223 |
| Once in 20 years | 15.32 | 30078 | 65199 |
| Once in 50 years | 16.23 | 41665 | 96992 |
| Once in 100 years | 17.02 | 57601 | 136593 |

According to the data in Table 1, at present, this block has a certain storage capacity. However, after further on-the-spot investigation, it is found that the runoff path in this block is chaotic, the distribution of storage space is uneven, some spaces are difficult to be effectively applied, and there is a lack of sufficient storage capacity for 20-year and higher-level precipitation, so it needs to be optimized. The calculation results are also used as the basis of subsequent basic data (Liu, 2021).

### 3.2 *Design of the block SWMM model*

In this study, the design of the SWMM model of the block is mainly based on remote sensing satellite images and CAD mapping drawings. After obtaining the model data, open the file with CAD, divide the sub-catchment areas in the figure, and export them to SWMM with the help of inps plug-in tools to complete the construction of the final model. This study is divided into six sub-catchment areas, numbered S1-S6 respectively.

After each partition is determined, the parameter values of each partition are determined. After referring to the relevant authoritative literature, the parameter values in this study are shown in Table 2.

Table 2. Parameter calibration table.

| Manning roughness coefficient | | Surface injection and storage volume coefficient | | |
|---|---|---|---|---|
| N-imperv | N-perv | Des-imperv/mm | Des-perv/mm | Zero-Imperv/% |
| 0.013 | 0.24 | 1.27 | 3.8 | 25 |
| Horton penetration model | | | | |
| Max.Infil(mm/h) | | Min.Infil(mm/h) | Decay Constant | Drying Time/d |
| 76.2 | | 2.16 | 4.14 | 7 |

After determining the above parameters, the characteristic width value of the sub-catchment is further determined and adjusted. The following formula is applied to this link:

$$Width = K \times \sqrt{A}$$

In this formula, K is the characteristic Width coefficient, which takes the value of 0.8, and A is the area of the sub-catchment. Based on this, the characteristic Width of the catchment can be obtained according to the above formula.

### 3.3 *Collection and analysis of rainfall data*

Rainfall data is an essential element of landscape design in the environment of urban rainwater utilization technology. In this study, the extraction and calculation of rainfall data are mainly based on the following steps (Fu, 2021).

First, the rainstorm intensity is calculated. The following formula is used in this link:

$$q = \frac{167A_1(1 + ClgP)}{(t + b)^n}$$

In this formula, $t$ represents rainfall duration, and the unit is min. $P$ is the design return period, and the unit is A. Other parameters are local parameters, which can be obtained by querying the local corresponding formula. In this study, $A_1 = 64.1718$, $C = 0.837$, $b = 32.900$, and $n = 1.011$.

The above data can be substituted to calculate the rainstorm intensity Q, with the unit of L/(s $hm^2$).

The second is to apply the Boltzmann formula to curve fitting of rainfall frequency data. The formula of this link is as follows:

$$y = 0.06089 + \frac{0.09339}{1 + e^{\frac{x-1047.3404}{117.0752}}}$$

Third, based on the above formula, the rain pattern used in the analysis is selected. In this study, the "Chicago Rain Pattern" is used for research, and the rain peak coefficient r=0.4 is determined. According to the selected "Chicago Rain Type", the long-duration rainfall data is used for simulation analysis to ensure the accuracy of the simulation analysis results. The data is selected from the local continuous daily rainfall detection data in the recent three years, and t=120min is selected for simulation. The simulation analysis results are shown in Table 3.

Table 3. Simulation results of basic characteristics of rainfall in different design return periods.

| Duration of rainfall | 120min | | | | | |
|---|---|---|---|---|---|---|
| Return period | P=1a | P=2a | P=3a | P=5a | P=10a | P=20a |
| Rainfall/mm | 47.53 | 59.50 | 66.51 | 75.33 | 87.30 | 99.28 |
| Average rainfall intensity(mm/min) | 0.792 | 0.992 | 1.108 | 1.255 | 1.455 | 1.656 |

All the above data can be used as a reference for low-impact landscape design.

### 3.4 *Preliminary design of the scheme*

Based on the above research data, it can be seen that it already has a certain urban rainwater utilization capacity and drainage capacity in advance for this research area. It is only necessary to properly optimize its runoff path, drainage and storage space, and other elements, without a lot of additional construction work. Based on this premise, in the design of this scheme, the designer applies the joint optimization scheme of front-end+middle-end+end of low-impact development, optimizes the existing runoff path, and builds a water storage body centered on Block 6, and the runoff from Block 1-5 is collected to the water storage body of Block 6. The main structure of the water storage body is an artificial lake with a reservoir structure, with a volume of 11450$m^3$. At the same time, in the design process, the elevation of the normal water level of the artificial lake is generally about 10 cm lower than the overflow, while the elevation of the artificial lake revetment is designed according to the normal water level to further improve the water storage capacity of the artificial lake (Wang, 2020).

## 4 TECHNICAL POINTS OF LANDSCAPE DESIGN BASED ON A LOW-IMPACT SCHEME

### 4.1 *Rainwater garden design*

Specifically, in the design of the "rainwater garden" adopted this time, the designers chose the design pattern of a multi-level structure and divided the rainwater garden into five parts: water storage layer, covering layer, preparation, planting soil layer, artificial filling layer, and gravel layer. As for the plant configuration of this "rainwater garden", while considering the integrity, the designers also actively take measures to avoid the repetition of plant types, and carry out research based on the overall perspective, taking into account the factors such as roads and green belts, and effectively integrating the involved factors. In order to further improve the rainwater interception capacity, the designers adopted the vertical design mode in the plant configuration, that is, arranged in the order of "Arbor-shrub-herb". Because these three types of plants are different in height, they will have the effect of interception layer by layer. On this basis, in order to further improve the design effect of rain gardens, and give consideration to practicality and aesthetics, the designers adopted the arrangement of interspersed herbs among trees and shrubs.

### 4.2 *Design of sunken green space*

In order to further improve the ability of rainwater collection, utilization, and discharge in this area, in this design, the designers have laid out several sunken green spaces in the landscape area. Before the formal design begins, the designers first determine the relevant parameters in the design work according to the following two steps.

The first is to determine the area of sunken green space. This link should be calculated by the following formula:

$$S = 60kJF_gT$$

In this formula, S is the rainfall infiltration amount, k is the steady infiltration rate of the soil, J is the hydraulic gradient, and T is the calculation time of infiltration, thus determining the area of sunken green space $F_g$.

The second is to determinc the depth of sunken green space, and its calculation formula is as follows:

$$U = 0.001HF_g$$

In this formula, U is the water storage capacity of the sunken green space. According to this formula, the depth H that the sunken green space should have can be determined by substituting the relevant data for calculation.

After the above parameters are determined, the design of the sunken green space is started, and the sunken green space is separated from other areas by kerbs. On this basis, in order to further improve the water storage capacity of the sunken green space, a gap with a width of 20-60cm is set on the kerb for drainage.

### 4.3 *Ground pavement design*

Considering various factors such as road traffic and rainwater collection and utilization in this area, in this design process, porous asphalt materials are used to design the pavement, while turf bricks and crushed stones are used to design the areas outside the pavement. On this basis, in order to ensure that the pavement has changeable features in a visual sense, a combination design mode of various materials should be adopted, such as using the differences in color and texture of various materials, drawing patterns by painting artistic techniques, and further enhancing the aesthetic significance of the pavement. At the same time, considering the wide road surface in this area, a sand-sinking well is added under the paved road, and it is combined with a reservoir to realize the purification, collection, and utilization of rainwater. In this design, the rapid filtration and flocculation mode of the filter screen is set in the sand sinking well, in which the filter screen aperture is set to 200μm, the polyaluminium chloride is used as flocculant, and the feeding density is 10mg/L (Yang, 2020).

## 4.4 *Overall design effect analysis*

In this link, the technical personnel uses the continuous daily rainfall monitoring data of the area in the recent three years to analyze the overall effect of this design scheme, and the analysis results are shown in Table 4.

Table 4. Analysis results of total annual runoff control.

| Index | Parameter |
|---|---|
| Total rainfall/mm | 3553.50 |
| Total holdup/mm | 3244.35 |
| Total external displacement/mm | 309.15 |
| Annual total runoff volume control rate/% | 91.3 |

From the data in Table 4, it is not difficult to see that through this design, the regulation and storage function of the landscape has been further improved, especially compared with the previous index of about 70%.

On the other hand, in order to further clarify the water environment regulation and storage capacity of this area in different return periods, taking one of the main discharge outlets as an example, based on the Chicago rain pattern, the 2-hour heavy rainfall under different conditions is simulated, and the rainfall-runoff data under this environment is analyzed. The analysis results are shown in Table 5.

It can be seen from the data in Table 5 that, considering the discharge port under the most unfavorable conditions, it still has a high rainwater discharge capacity under the condition of a small return period, which can effectively control the total amount of runoff and weaken the peak rainfall flow in heavy rainfall weather. This is also beneficial for building a "sponge city". Of course, in the case of a long return period, the pressure of rainwater discharge is still prominent, and there may be blocks with excessive rainwater. This shows that the low-impact landscape design based on urban rainwater utilization technology has more obvious benefits under moderate and small rainfall conditions, while the landscape design under special conditions still needs further research and optimization.

Table 5. Rainwater runoff at different return periods.

| Return period | P=1a | P=2a | P=3a | P=5a | P=10a | P=20a |
|---|---|---|---|---|---|---|
| Total discharge (106/tr) | 1.369 | 1.763 | 2.140 | 2.608 | 3.355 | 4.129 |

## 5 CONCLUSION

On the whole, in this study, following the logic of "asking questions-analyzing questions-solving problems", based on the principles of eco-environmental protection and high efficiency, and combined with practical cases, the low-impact landscape design in urban rainwater utilization technology environment was studied from both theoretical and practical aspects. The results show that the low-impact landscape of this design has certain practical value. Of course, there is still room for improvement in this study. In future work, it is still necessary to further introduce intelligent technology and other elements to further improve the quality of landscape design.


## ACKNOWLEDGMENTS

Project fund: 2021 Science and Technology Research Project of Jiangxi Education Department (Project Name: Research on Ecological Approaches and Design Strategies of Urban Landscape Based on Low Impact Development (LID) Technology) (Project No. GJJ211337).

## REFERENCES


Fu Manyi (2021).Ecological Landscape Reconstruction Design of Mine Abandoned Land Based on Low Impact Development— A Case Study of Zhanggou Village in Xinmi City.Journal of Green Science and Technology, 23(05):22–25.

Liu Bingying (2021). Study on Landscape Design of Mountain Ecological Residential Area under the Concept of Low Impact Development. Chongqing Jiaotong University.

Shen Huifang (2021). Analysis of Urban Waterfront Landscape Space Design Based on Low Impact Development. Xiandai Horticulture, 44(20):105–106.

Wang Lian (2020). Research on Campus Landscape Reconstruction Design Based on Low Impact Development (LID)—Taking Jiangsu Institute of Architecture and Technology as an example. Green Environmental Protection Building Materials, (11):77–78.

Wang Pei (2022). Study on Low Impact Design Strategy of Wetland Landscape in Western Sichuan Plateau—A Case Study of Zoigeka Hardjo Wetland Reserve. Art & Design, (02):142–144.

Yang Jin (2020). Study on Effect Evaluation of Low Impact Rainwater System Based on Landscape Ecology Method. Beijing University of Civil Engineering and Architecture.

*Advances in Urban Construction and Management Engineering – Cha (Ed)*
*© 2023 Copyright the Author(s), ISBN 978-1-032-39018-5*

# Research on forest disaster prevention system in Chengdu Longquanshan urban forest park

Kangcai Nie*
*College of Architecture, Southwest Minzu University, China*

Xuehong Zhou*
*Sichuan Institute of Territorial Spatial Planning, China*

ABSTRACT: With the concept of ecological civilization growing in China, it has become the consensus of the whole society to respect nature, conform to nature and protect nature. As one of the important spatial carriers in Sichuan Province to implement the concept of ecological civilization and the concept of park city, Chengdu Longquanshan Urban Forest Park has become a major place for Chengdu people to feel nature, integrate with nature and carry out suburban activities such as suburban leisure, scenery appreciation and fitness. With the development of all kinds of activities, the spatial density and intensity of development of people, facilities and places in the park have increased, and the risk of disasters such as forest fires, forest pests and diseases, flash floods, landslides and secondary disasters has also increased. In order to effectively prevent the occurrence of disasters and ensure the safety of excursions, this paper focuses on the prevention and control of fire in the Wanxing area of Chengdu Longquanshan Forest Park and is oriented toward the implementation of a disaster prevention system for urban forest parks that combines prevention, resistance, and rescue.

## 1 INTRODUCTION

The report of the 18th Party Congress points out that "in the face of the severe situation of tightening resource constraints, serious environmental pollution and degradation of the ecosystem, it is necessary to establish the concept of an ecological civilization that respects nature, responds to nature and protects nature". This is a scientific judgment based on a profound reflection on the relationship between man and nature and is of great significance as a guide to building a beautiful China and achieving the sustainable development of the Chinese nation. In the community of life that people depend on for their survival, forest resources are a general term for forest land and the forest organisms that grow on it, an invisible environmental resource and a potential "green energy source". It includes not only forest resources but also forest and understorey plants, wild animals, soil microorganisms and other natural environmental factors (Liu 2021). A forest park is an area that is managed on the basis of the protection of forest resources, making full use of the various ecological functions of the forest, providing tourism services for people and allowing for scientific and cultural activities (Liang 2012).

Chengdu, as the first place where China's "city fully embodies the new development concept" and the first place where "park city" is mentioned, has been vigorously carrying out the creation of park scenes since 2018, adding 38.85 million square meters of green space in the city. The city's "park city under the snowy mountains" is becoming increasingly beautiful. As Chengdu's urban "green heart", the ecologically beautiful Longquanshan Urban Forest Park has become the main place for

*Corresponding Authors: niekangcai@163.com and 313631774@qq.com

 DOI 10.1201/9781003348023-35

Chengdu people to experience nature, integrate with nature and carry out suburban activities such as leisure, scenery appreciation and fitness. However, various safety incidents, such as forest fires, forest pests and diseases, flash floods, landslides and secondary disasters, have occurred from time to time. On the basis of giving full play to ecological and landscape benefits, how to strengthen the planning and construction of forest disaster prevention and mitigation systems, build up the basic surface of the material carrier of the resilient safety defence of urban forest parks through the technical means of territorial spatial planning, enhance the "resilience" against disasters, ensure and implement various types of facilities at all levels required for disaster prevention and control in space, and improve the response to disasters. It is one of the key issues that need to be addressed in Longquanshan Urban Forest Park to improve the level of forest protection and ensure the safety of tourism by improving the prevention and control of high-risk disasters such as forest fires, forest pests, and geological disasters (Li 2021).

## 2 BASIC INFORMATION ABOUT CHENGDU LONGQUANSHAN FOREST PARK

Chengdu Longquanshan Urban Forest Park is located in the Chengdu section of the Longquan Mountain Range in Chengdu, Sichuan Province, stretching 90 kilometres north-south and 10-12 kilometres east-west, including the ecological area of Longquan Mountain with Longquan Mountain as the main body and Sanquan Lake, Longquan Lake and Cuiping Lake as representatives, involving 38 villages in 6 districts (cities) and counties, including Jintang County, Qingbaijiang District, Longquanyi District, Janyang City, East District of High-tech Zone, Tianfu New Area Directly Administered District, and 268 villages in 6 districts (cities) and counties, with a total area of about 1275 square kilometres. The planning is positioned as a "world-class quality urban green center and an international city living room", with a spatial structure of "one urban forest park, two functional zones, ten recreational units, three distinctive landscapes, three traffic rings and a whole area supported by greenways" (Figure 1).

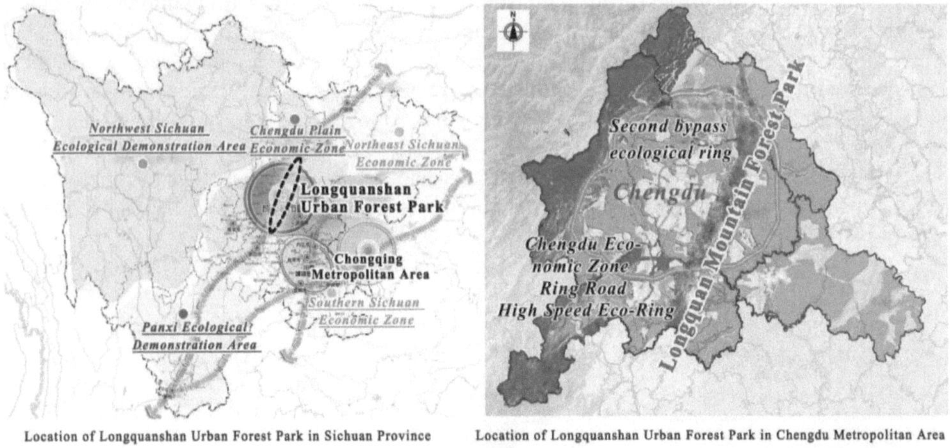


Figure 1. Location of Longquanshan Urban Forest Park (Source: Longquanshan Urban Forest Park Plan).

Wanxing Area is located in the middle section of Longquanshan Urban Forest Park, in the mountainous forest landscape area of Huahai Linlu, with forest recreation and mountain sports functions as the main attractions, one is Wanxing Forest Park and the other is cycling off-road, which is an important tourist area of Longquanshan Forest Park. With an area of 51 square kilometres and a population of 13,000, the district is a typical suburban traditional mountainous area of Chengdu. The terrain is deep hills and mountains (Figure 2), with an average altitude of 670 meters. The area is rich in forest resources, mostly natural secondary forests, artificially planted tree forests

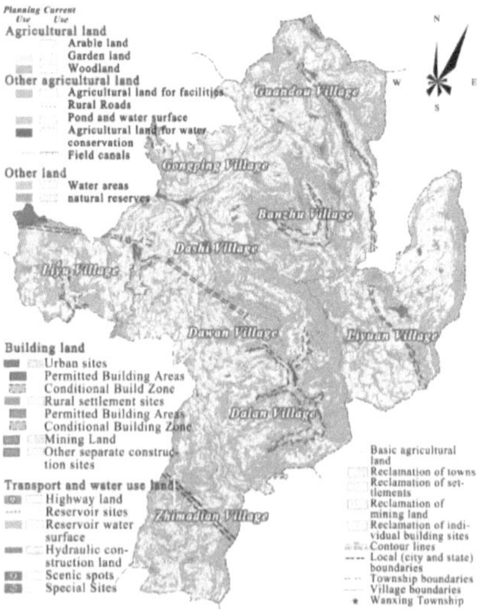


Figure 2. Land use for Wanxing area (Source: Self-drawn).

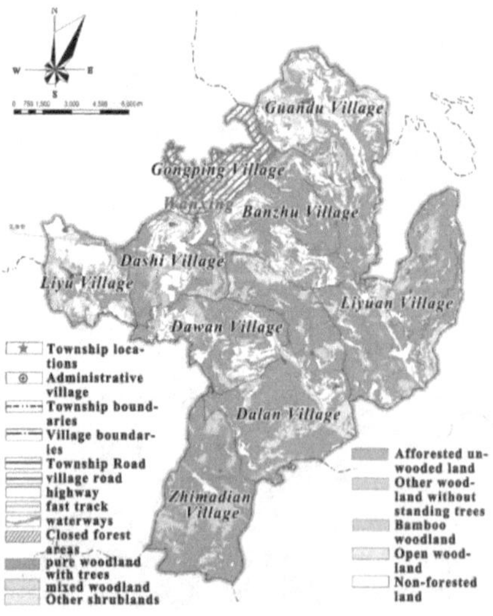


Figure 3. Distribution of woodland in Wanxing area (Source: Self-drawn).

and sparse grass slope vegetation, including cypress forests, alder forests and patches of spotted bamboo forests, etc. The forest coverage rate has reached 79.93% in 2019 (Figure 3).

Starting from 2017, in order to protect the fragile ecological environment of Longquan Mountain, Wanxing Area started to close the mountains for forestry and began to implement the ecological immigrant relocation project for the whole countryside and the project of returning farmland to forest, with 13,000 people from seven administrative villages gradually settling into the ecological immigrant concentration area of Quejia Village in Xihe Town at the end of 2017. After the aborigines moved out, the villages within the area gradually decayed, water plants, sewage treatment plants, cemeteries, Chengdu Chang'an Intravenous Industrial Park (partly) and Chengdu Renewable Energy, Environmental Protection Technology and Solid Waste Sanitary Treatment Site were scattered and distributed, and various types of pipelines were intertwined, increasing the risk of forest disasters. While the construction of Longquanshan Urban Forest Park was being promoted, the construction of the disaster prevention system for key areas of the forest park has become the focus of work.

## 3 CONSTRUCTION OF A FOREST DISASTER PREVENTION SYSTEM IN THE WANXING AREA

### 3.1 *Forest hazard prediction and risk analysis assessment*

There are many disaster risk factors affecting the safety of urban forest parks, such as forest fire, pest and disease risk, geological disaster risk, earthquake risk and flood risk. Risk assessment is an important basic task to protect the ecological environment of urban forest parks and ensure the safety of visitation by combining the physical and geographical background characteristics of urban forest parks and the results of dual evaluations, assessing the level of risk factors and constructing disaster prevention systems according to different levels, and proposing relevant planning and disaster prevention strategies (Zhang 2012). The risk assessment itself involves both inherent deterministic factors such as forest resources and uncertainties such as the movement of people, and its uncertainties include both random and fuzzy uncertainties, so the analysis of disaster sources is crucial (Guo 2014). According to the analysis of the natural geographical pattern of Wanxing

Area, the data of various disasters in the area in recent years were extracted, and the preliminary assessment concluded that the risk factors affecting the safety of Wanxing Area mainly include forest fire factors, pests and disease risk factors and geological disaster risk factors, and each major disaster factor was then evaluated by single factor analysis according to the environmental impact elements. On the basis of the single factor multi-factor evaluation, the results of the assessment of forest fires, forest pests and diseases, geological hazards and floods in the Wanxing area were superimposed and coupled, and the activity patterns of the population were integrated to divide the whole township into comprehensive disaster high-risk areas, comprehensive disaster medium-risk areas and comprehensive disaster low-risk areas. The high-risk areas are mainly located in the northern part of the commune, the central part and the eastern edge of the forest fire-prone areas (Figure 4).

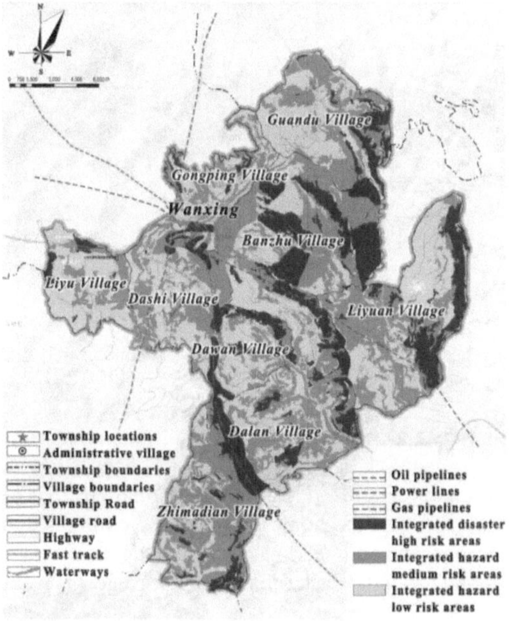


Figure 4. Results of the comprehensive assessment of hazard factors in the Wanxing area (Source: Self-drawn).

### 3.2 *Spatial zoning for disaster prevention and mitigation*

Based on the results of the comprehensive assessment of the disaster factors in the Wanxing area, and based on the distribution of landscape resources in the Longquanshan Urban Forest Park, combined with the tourism market demand, the possibility of resource development and the scientific nature of the park management, the principle of "forestry as appropriate, tourism as appropriate and industry as appropriate" is adhered to, and the Wanxing area is divided into three main functional areas, namely ecological restoration area. The park is divided into three main functional areas: the ecological restoration area, the production prevention and control area and the scenic tour area. It is a disaster prevention system that combines "prevention, resistance and rescue", with a series of points and lines, forming a combination of "points, lines and surfaces" and complementary human, physical, technical, and institutional prevention (Figure 5).

#### 3.2.1 *Ecological restoration areas*

The ecological restoration zone mainly includes areas at high risk of comprehensive disasters, as well as areas requiring ecological restoration and cultivation, such as sparse forest land, unstocked afforestation land, other woodlands without standing trees and low-quality forested land. These

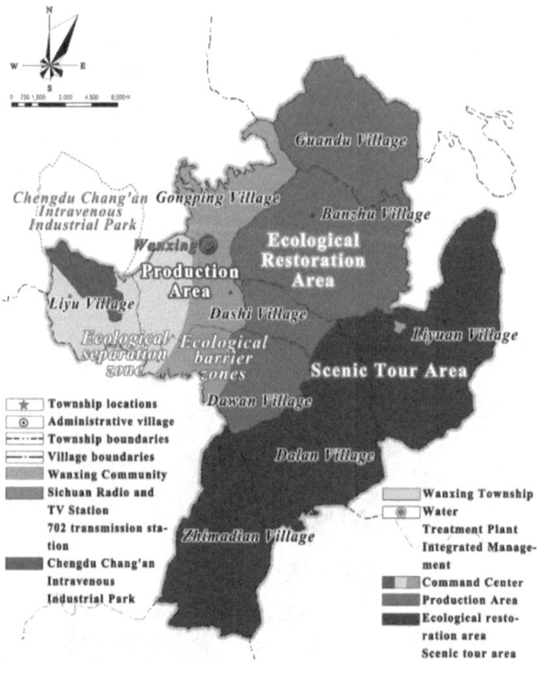


Figure 5. Spatial zoning for disaster prevention in Wanxing area (Source: Self-drawn).

areas are to be closed to forestry and are strictly prohibited from human activities. On the one hand, the self-regulating and self-organizing capacity of the forest ecosystem should be relied upon to allow it to evolve in an orderly direction and restore the original virtuous cycle of habitats. On the other hand, the existing disaster hazards should be gradually eliminated and secondary disasters avoided through certain physical and engineering measures, such as forest transformation projects, biological control projects and ground disaster management projects (Wang 2006).

### 3.2.2 *Production prevention and control area*

The production prevention and control zone mainly includes areas such as renewable energy, environmental protection technology companies and solid waste sanitary disposal sites, which are located in the Wanxing area. The zone focuses on preventing the occurrence of production safety accidents, building a good ecological buffer zone, improving the environment around the park, purifying the atmosphere, water system and soil pollution, effectively isolating the environmental impact of treatment and disposal projects in the park on other areas, and at the same time focusing on solving the problems of insufficient vegetation biodiversity and poor landscape effects (He 2016). Based on the topographic and geomorphological characteristics of Wanxing Township, the north-south ridge line on the western side of the township will be used as the boundary to set up an ecological buffer zone to further reduce the impact of production projects on other areas. Within the ecological buffer zone, the construction of tourism projects and public facilities will be restricted; ecological conservation will be carried out, and the current timber and protection forests will be restored by the vegetation system itself; the current economic forests will be mainly planted with trees such as eucalyptus and cypress, and the ecosystem will be gradually restored by planting other trees such as cycad, balsam fir, heather, giant eucalyptus and poplar, supplemented by shrubs and herbs; other areas with a low vegetation cover factor will be restored by. In other areas with low vegetation cover, ecosystem cultivation and protection forests are built to increase vegetation cover and ecosystem stability.

3.2.3 *Scenic excursion areas*

It mainly includes areas other than ecological restoration areas and production areas, where citizens and tourists can enter and participate in recreational tours. The comprehensive disaster risk in the scenic tour area is relatively low, and the main focus is on preventing man-made forest fires, strengthening forest fire monitoring and early warning, and planning and building supporting facilities such as forest fire isolation, forest fire water sources and firefighting facilities (He 2016). It is needed to refine the system for setting up places of refuge and evacuation routes and strictly control important public buildings and lifeline projects (Zhang 2012).

## 4 CONCLUSION

Forests are a precious part of nature and play an important role in the stable development of the ecological environment, and urban forest parks are one of the important ways to build an ecological civilization. In the context of the construction of ecological civilization, which insists on "ecological priority and green development", urban forest parks have become an important measure to improve the current ecological environment and a key point to creating a beautiful and livable living environment and promoting the sustainable development of human society. As an important part of the national spatial ecosystem, urban forest parks should, in the process of planning, construction and management, pay full attention to the assessment of risk factors such as forest fires, pests and geological disasters and disaster monitoring, and strengthen the construction of comprehensive disaster prevention and mitigation system and incorporate it into the national spatial planning system. On the basis of a full understanding of the park's natural geography and forest distribution, the scale of forest park tours, visitor capacity and potential disaster types should be carefully analysed. The forest park will be guided by planning, through human and physical prevention and modern technology, to ensure the normal functioning of all functions.

## REFERENCES


Guo Yourui. *Research on the safety of landscape garden planning and design* [D] Fujian Agriculture and Forestry University 2014.

He Guangliang. A brief discussion on the transformation of disaster prevention function of urban forest park - an example of disaster prevention transformation of Dafshan Forest Park in Guangzhou [J] *Urban Architecture* 2016.

Li Shuna. Analysis of urban forest park construction based on ecological civilization [J] *Environmental Protection and Energy Conservation* 2021 (10): 93–94.

Liang Dongcheng, Bian Junjing. Research on the construction of environmental protection system in forest parks [J] *Guangdong Forestry Science and Technology* 2012 (2).

Liu Xingtong. Cao F C. Wang Manqi. Research on urban forest park planning and design under the concept of ecological civilization [J] *Modern Horticulture* 2021(15):132–133.

Wang Meisong. Discussion on the construction of disaster prevention and mitigation system in forest parks [J] *Fujian Forestry Science and Technology* 2006 (6).

Zhang Xu. On the characteristics and measures of forest fire prevention planning in nature reserves [J] *Guangdong Forestry Science and Technology* 2012 (2).

Zhang Yunsheng. Research on risk assessment of forest fire occurrence [J] *Forest Fire Prevention* 2012 (12).

*Advances in Urban Construction and Management Engineering – Cha (Ed)*
*© 2023 Copyright the Author(s), ISBN 978-1-032-39018-5*

# Research on cost of underwater bored piles quality based on earned value method

Shu Zong*
*Department of Construction Management, Kunming University of Science and Technology Oxbridge College, Kunming, Yunnan, China*

Chao-wang Zhang
*Department of Water Engineering, Kunming University of Science and Technology Oxbridge College, Kunming, Yunnan, China*

ABSTRACT: At present, with the development of China's social economy, construction projects continue to flourish, and underwater bored piles are also increasingly widely adopted in these projects. Therefore, ensuring the quality and safety of this type of construction project is an urgent problem that needs to be solved. In this paper, the earned value method is used to study the quality problem that occurred during the construction process of underwater bored piles. Through the quality earned value method, the project is managed and controlled to minimize the risks that may arise in the actual construction process, reduce rework, and ensure the quality of the project. By improving the traditional earned value method, quality is introduced into the evaluation system of the earned value method, so that the quality earned value method can be applied to practical projects and ensure their successful completion.

## 1 INTRODUCTION

In the traditional earned value method, the cost index is used to reflect the value of each project task. At present, the consideration of quality cost is often ignored in the evaluation of the traditional earned value method. The control of quality factors is limited to some agreements, without considering the impact of quality on cost. Therefore, if the factors affecting the quality are more complex, the cost of construction will not be well controlled by the traditional control method, and it is easy to cause adverse effects on the completion of the project. The quality earned value method is based on the earned value method but considers project quality factors. The basic theory is the same as the earned value method. With the consideration of quality factors, project evaluation can be more practical, and factors that may affect the project can be discovered as early as possible in the project process, and deviations can be corrected in time to ensure the smooth completion of the project. The quality earned value method refines the basic parameters based on the traditional earned value method. As they are the quality functions, the basic parameter analysis is used to judge the completion of construction quality and progress.

## 2 QUALITY COST ANALYSIS OF UNDERWATER BORED PILE

At present, in diverse types of port projects, underwater bored piles are the most frequently used technology, and most large water conservancy projects have used underwater bored piles. In such

*Corresponding Author: 928288652@qq.com

 DOI 10.1201/9781003348023-36

circumstances, there will be long-span concrete bridge construction, so impact drilling construction is indispensable. Due to some of the terrain steep and geological report factors, impact drilling in highway engineering construction also made repeated achievements. As more long-span Bridges and super-tall buildings appear, bored piles of pile foundation have become the most important foundation forms of a building structure, which have been widely used in practical construction and have made great contributions.

The actual main quality costs of bored pile construction are as follows: (1) Prevention cost; (2) Appraisal cost; (3) Internal loss cost; (4) External loss cost (Babu 2007). Prevention costs refer to the cost of ensuring that the product quality of the underwater bored cast-in-place pile reaches the expected target agreed in the contract. It includes the cost of quality improvement, quality assurance, and quality training for personnel. In the construction process of underwater bored piles, appraisal cost refers to the cost generated by the construction process whether it conforms to the requirements to identify the quality of underwater bored piles and all kinds of materials used. It includes the cost of quality inspection of all kinds of raw materials used in the construction process of underwater bored piles, the cost of maintenance of testing equipment, and the cost of evaluating the construction quality of bored piles. This item refers to the loss caused by quality problems after the project is delivered to the owner and the costs incurred to make up for these quality problems, including warranty costs and legal costs caused by quality problems.

## 3 QUALITY EARNED VALUE COST ANALYSIS

### 3.1 *Determine the influencing factors of quality*

During the construction of an underwater bored pile, there will be many problems that will affect the quality and cost of the project. Therefore, we need to distinguish the impact of these factors on the project, find out the primary ones, and focus on the factors that affect the quality cost, to improve control efficiency and achieve reasonable control of the project.

The Pareto method can be used to analyze the cost of quality loss of underwater bored piles. Firstly, ABC analysis was conducted on the quality loss, which affects the quality cost of the underwater bored pile. The net cost of various quality losses was calculated and ranked from largest to smallest, and then the ratio of each item to the cost of quality losses was also calculated. According to the classification of accumulative proportion, the next work that needed to be focused on was known. The analysis results of the quality loss cost of the underwater bored pile are shown in Figure 1.

As can be seen from Figure 1, internal repair loss and internal rework loss are the main factors affecting quality. The cumulative frequency is between 0~80%, which is a factor in A. Therefore, to ensure quality and reduce loss, we should pay attention to these two factors and improve them at the same time. According to the above analysis, we can find that reducing internal repair and rework loss is the main way to reduce quality costs. According to the above analysis, we can find that reducing internal repair and rework loss is the main solution to reduce quality costs. Most of the problems in the actual construction can be avoided by strengthening the quality prevention and guarantee measures in the early stage (GB/T 13339-1991).

### 3.2 *Construction of quality evaluation system*

To analyze the quality cost of a project through the quality earned value method, a quality evaluation system and the weight of each evaluation index should be established first. The evaluation indexes of the traditional earned value method are the cost performance index (CPI) and schedule performance index (SPI) (Juran 1999). In the quality earned value method, by referring to the relevant data of a bridge project, the partial project with the greatest impact on the project cost is selected for quality cost analysis. The partial earthwork has the greatest influence on the engineering quality of a project, so the hierarchy diagram of the construction engineering quality evaluation system is established and shown in Figure 2.

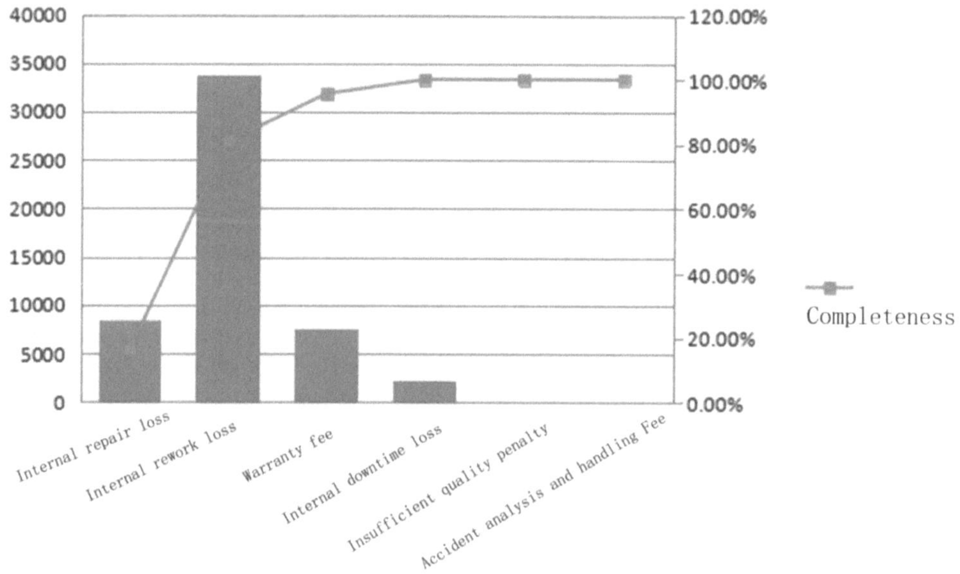


Figure 1. Proportion of quality loss cost of underwater bored pile.

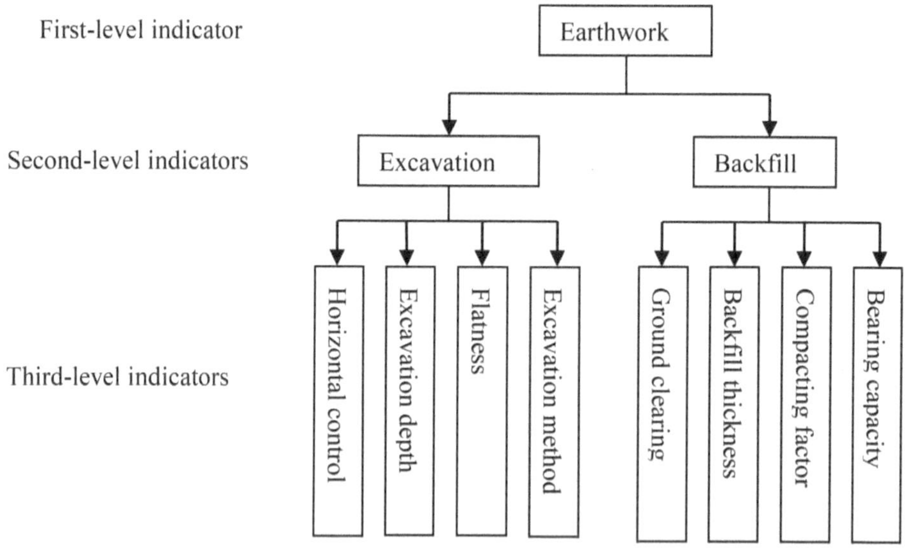


Figure 2. Hierarchy of quality evaluation system.

Where the weight set of Second-level indicators is $b = \{0.3, 0.7\}$; the index weight set of the three layers is $C_1 = \{0.267, 0.381, 0.157, 0.195\}$ and $C_2 = \{0.02, 0.262, 0.357, 0.361\}$.

The evaluation set $S = \{5, 4, 3, 2, 1\}$ and the corresponding evaluation are excellent, good, qualified, unqualified, and poor. The weights are determined by the expert scoring method, and the completed projects are scored. The summary results of the expert scoring survey are shown in Table 1.

Table 1. The summary results of the expert scoring survey.

| | C11 | C12 | C13 | C14 | C21 | C22 | C23 | C24 |
|---|---|---|---|---|---|---|---|---|
| Excellent | 1 | 2 | 3 | 0 | 0 | 8 | 7 | 8 |
| Good | 3 | 8 | 3 | 2 | 4 | 2 | 2 | 1 |
| Qualified | 6 | 0 | 4 | 8 | 6 | 0 | 1 | 1 |
| Unqualified | 0 | 0 | 0 | 1 | 0 | 0 | 0 | 0 |
| Poor | 0 | 0 | 0 | 1 | 0 | 0 | 0 | 0 |
| Σ | | | | 10 | | | | |
| rij | r11 | r12 | r13 | r14 | r21 | r22 | r23 | r24 |

According to the above table, the fuzzy evaluation matrix of indicators can be obtained:

$$R=\begin{pmatrix} 0.1 & 0.3 & 0.6 & 0 & 0 \\ 0.2 & 0.8 & 0 & 0 & 0 \\ 0.3 & 0.3 & 0.4 & 0 & 0 \\ 0 & 0.2 & 0.8 & 0.1 & 0.1 \end{pmatrix}$$

Through the fuzzy transformation of this matrix, the fuzzy evaluation vector of the second-level index U1 can be calculated:

$$y_1=C_1R_1=\{0.15, 0.471, 0.379, 0.0195, 0.0195\}$$

Similarly, the fuzzy evaluation vector of U2 can be obtained; thereby, the fuzzy evaluation matrix R can be constructed.

$$R=\begin{pmatrix} 0.15 & 0.471 & 0.379 & 0.0195 & 0.0195 \\ 0.7483 & 0.1679 & 0.0838 & 0 & 0 \end{pmatrix}$$

Fuzzy evaluation is conducted on the indicators, and the formula is

$$y=bR=\{0.5688\ 0.2588\ 0.1724\ 0.0059\ 0.0059\}$$

Therefore, the grade of earthwork in this project is as follows:

$$P_A=y\cdot s=4.4141$$

The calculation process of PB, such as the planned quality level score of the PA project, is

$$P_B=y\cdot s=4.224$$

Therefore, in this project, the quality compliance coefficient of earthwork is

$$Q=\frac{P_A}{P_B}=1.045$$

In the same way, the quality compliance coefficient of each sub-project in pile foundation construction can be obtained.

### 3.3 *Analysis of quality deviation by earned value method*

The comprehensive analysis of project cost data found that the construction cost in the construction process is too high. The reason for the high construction cost is that there is material waste in the construction process. In the construction process of the project, the cost of materials accounts for about 60% of the total costs, and the waste of materials can be reduced through reasonable management and control (Li 2008). Such management and control are effective ways to reduce project costs. Therefore, in the construction process of the project, it is necessary to realize fine management, strictly control each link of construction, and reduce the waste of materials to obtain higher income.

By introducing the quality earned value method into the management control of the project, we can find that the introduction of the quality earned value method has a positive effect on the cost quality control of the project.

Therefore, by introducing the quality earned value method into the construction management process of the project, there are obvious advantages. The cumulative cost deviation analysis curve was drawn, as shown in Figure 3.

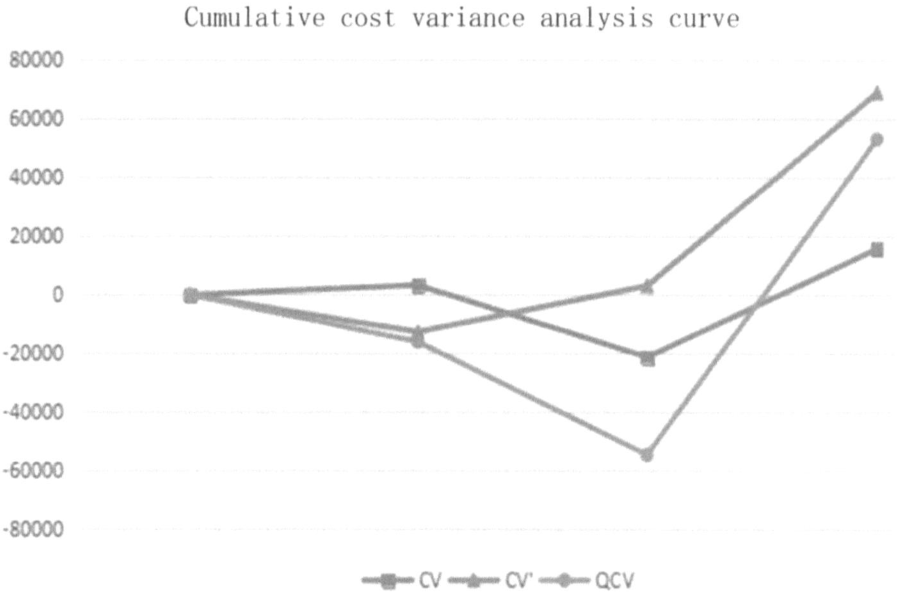


Figure 3. Cumulative cost deviation analysis.

It can be found that in the third control node, according to the traditional earned value method, the cost is reduced, but according to the actual cost deviation, after offsetting the impact of quality on the cost, we can find that the actual cost is increased. Through this analysis, we can also find that the traditional earned value method has false positives or omissions in the evaluation of this node. If this phenomenon frequently occurs in the management and control of the whole project, it is extremely detrimental to the smooth completion of the whole project. Compared with traditional methods, the evaluation system of quality earned value method can find problems existing in cost control in the early stage and take measures in advance to control the project cost back to the normal level.

At the same time, by introducing the factor of quality into the earned value method, the overall project management and control ability of construction enterprises can be better reflected in the evaluation indicators (Wang 2010). For this project, although the project completed the pre-project target in the construction process, due to the lax control of quality factors, rework was caused, thus causing a certain cost loss. However, in the later stage of the project, combined with the construction conditions and the construction level of the construction enterprise, the construction method of the project has been optimized to some extent, and the profitability of the project has been improved. Compared with the traditional earned value method, the introduction of the quality earned value method can make the project's profitability better evaluated.

## 4 CONCLUSIONS

In this paper, the quality earned value method is used to control the quality cost in the construction stage of the concrete bored pile, and the following conclusions are drawn: The introduction of the quality earned value method in the construction process of bored piles can find the problems in the construction process earlier and solve these problems. At the same time, the introduction of the quality earned value method can help the construction contractors better control the quality, project duration, and cost, which can also make the construction contractors better control the overall project.

## ACKNOWLEDGEMENTS

This paper is supported by the Program for Innovative Research Team (in Science and Technology) at the University of Yunnan Province (IRTSTYN) and is one of the phased achievements of the scientific research fund program of the Yunnan Education Department.

## REFERENCES

Babu A J G, Suresh N. Project management with time, cost, and quality considerations[J]. *European Journal of Operational Research*, 2007, 88(2):320–327.

GB/T 13339-1991, *Quality Cost Management Guidelines* [S] (In Chinese).

Juran J M. Juran's Quality Handbook[J]. *Juran's quality handbook*, 5th ed. McGraw-Hill, 1999.

Li T. *Application Research of Engineering Project Cost Control Based on Earned Value Theory* [D]. Tianjin University, 2008. (In Chinese).

Wang T. *Application research of EARNED Value Management Based on CM Management Model* [D]. Chongqing University, 2010. (In Chinese).

*Advances in Urban Construction and Management Engineering – Cha (Ed)*


# Research on rural air pollution control under the background of "beautiful countryside"

Ying Sun*
*Institute of Governance, Party School of Liaoning Provincial Party Committee, Shenyang, P.R. China*

ABSTRACT: In recent years, with the acceleration of China's modernization process, while the urban environment is improving day by day, the pollution problem in rural areas is becoming more and more prominent. Especially in the developed areas with a high degree of industrialization and urbanization, the decline of rural environmental quality has formed a strong contrast with economic and social development. The serious problem of rural air pollution is actually caused by many reasons. In addition to automobile exhaust, illegal emissions from some factories and dust caused by construction waste in rural areas are the culprits affecting rural air. Compared with urban environmental protection and industrial pollution, rural environmental protection started late and has a weak foundation. It is difficult to control due to the limitations of manpower, capital, farmers' apos, production, and lifestyle. Rural environmental problems involve many aspects and departments. Improving the rural environment requires not only the government to increase investment, but also the coordination of governments at all levels and departments. We need to take the strategy of overall urban and rural development, strengthen the management of the rural ecological environment, establish and improve relevant laws and regulations on the prevention and control of air pollution, and advocate for farmers to gradually apply organic fertilizers to reduce pollution, with a focus on industrial pollution control. Although some work has been done on rural environmental protection in recent years, there is still a long way to explore.

## 1 GENERAL INSTRUCTIONS

Agricultural pollution is the source of air pollutants, but it is not the only one. There are many other sources, such as industrial pollution and automobile exhaust emissions. Agricultural pollution mainly comes from the application of a large amount of chemical fertilizer and organic fertilizer when planting wheat, corn, vegetables, cotton, peanuts, rice, soybeans, and other crops, resulting in a large number of rural aquacultures. The implementation of various agricultural activities has a certain impact on the ambient air quality and induces some heavily polluted weather.

Crop straw is the main by-product of agricultural planting. A large amount of straw is produced after crop harvest, and straw incineration produces flue gas, which discharges a large amount of particulate matter and gaseous pollutants (such as VOCs and $NO_x$) into the atmosphere, resulting in the rapid increase of pollutant concentration in a short time. Large area straw burning will also cause regional haze pollution and photochemical pollution. In addition, with the development of agricultural technology and the improvement of living standards, the amount of agricultural waste (such as plastic film, greenhouse plastic cloth, fertilizer packaging bags, and agricultural product residues) is huge, and the incineration of rural waste will also aggravate air pollution (Yuan & Lu 2021). Farmland fertilization is an important means to increase soil nutrients and improve the yield of agricultural products. However, only part of the fertilizer applied to farmland enters the soil

*Corresponding Author: 110161.12017366@qq.com


DOI 10.1201/9781003348023-37

and is absorbed by crops, and a considerable part of the fertilizer volatilizes into the atmospheric environment. The application of nitrogen fertilizer (such as ammonium bicarbonate, ammonium chloride, ammonium nitrate, and ammonium sulfate) in chemical fertilizer will release a large amount of $NH_3$ into the atmosphere (especially under dry and high-temperature conditions), significantly promote the formation of secondary inorganic aerosols, and easily lead to ammonia poisoning of fertilizer applicators. The use of chlorinated fertilizer will also produce active chlorinated compounds to enhance photochemical reaction and atmospheric oxidation. In addition, the direct use of organic fertilizer (farm fertilizer) will also release pollutants such as $NH_3$, $H_2S$, and VOCs, which will have a certain impact on air quality. Rural breeding is one of the important sources of income for many farmers. Retail or small-scale breeding accounts for a certain proportion of the breeding industry, especially pigs, cattle, sheep, and chickens. A large amount of $CO_2$, $CH_4$, $NH_3$, $H_2S$, and other gases will be produced in the process of raising poultry and livestock. $CO_2$ and CH4 are important greenhouse gases, and their large release will aggravate the greenhouse effect and climate change. The emission of polluting gases such as $NH_3$ and $H_2S$ will produce a peculiar smell and promote the formation of secondary inorganic aerosols, aggravating air pollution to a certain extent.

## 2 SOURCES OF RURAL AIR POLLUTION

### 2.1 *Pollution of industrial wastewater, waste, and waste gas*

With the development of the rural economy, the rural industrial structure has also undergone corresponding changes. Township and village-run enterprises have mushroomed, followed by the pollution of industrial waste in rural areas. For example, the production of building materials enterprises such as brick and tile factories, lime factories, cement factories, and granite plate factories has increased the dust concentration in the rural atmosphere. During the production process, harmful gases such as smoke, sulfur dioxide, carbon monoxide, and carbon dioxide can also be produced to pollute the atmosphere, especially in the rural areas around the city. Wastewater from various small coal mines, mercury mines, manganese mines, gold mines, and chromium-containing wastewater from small electroplating enterprises can pollute water sources, and some even pollute crops and remain in all kinds of food. There are more and more small-scale processing plants around the village. There are many steel smelters, sand turning plants, paper mills, breeding bases, and residential areas (Huang 2019). Some factories have imperfect sewage treatment facilities, most large-scale livestock and poultry farms lack necessary sewage treatment facilities, and organic wastewater is directly discharged without effective treatment, resulting in a large number of pollutants entering the water body with surface runoff.

### 2.2 *Air pollution caused by burning straw in rural areas*

Lignin, cellulose, hemicellulose, and other flammable substances in straw are partially transformed into carbonaceous particles during combustion. A large number of particles are suspended in the air, which is conducive to the formation of haze weather and provides rich condensation nodules for the formation of fog droplets. When burning straw, the three pollution indexes of sulfur dioxide, nitrogen dioxide, and inhalable particulate matter in the atmosphere reach a high peak, in which the concentration of sulfur dioxide is 1 times higher than usual, and the concentration of nitrogen dioxide and inhalable particulate matter is 3 times higher than usual, which is equivalent to the fifth level of daily average concentration. When the concentration of inhalable particulate matter reaches a certain level, it will cause great irritation to the part containing the mucous membrane of the human eyes, nose, and throat. At least, it will cause cough, chest tightness, and tears. In serious cases, it may lead to bronchitis. At the same time, the smoke formed by burning straw will reduce the air visibility and visible range, directly affect the normal operation of civil aviation, railway, and expressway, easily lead to traffic accidents and affect personal safety.

### 2.3 *Atmospheric pollution caused by chemical fertilizers and pesticides*

The harm of chemical fertilizer to agricultural production is mainly manifested in the excessive application of chemical fertilizer, the unscientific application proportion of nitrogen, phosphorus, and potassium fertilizer, and the uneven use of chemical fertilizer in different regions, resulting in soil hardening and decline of productivity. The low utilization rate of chemical fertilizer makes the nutrient components of chemical fertilizer greatly lost through leakage and surface runoff, causing water pollution and eutrophication, and a large number of greenhouse gases discharged by ammonia nitrogen nitrification pollute the air. Nitrous oxide will be formed after nitrogen volatilization in chemical fertilizer. It is a greenhouse gas and will destroy the atmospheric ozone layer. At present, nitrous oxide in the atmosphere increases at a rate of $0.2\% \sim 0.3\%$ every year, which will accelerate global warming, destroy biodiversity and threaten the balance of the ecosystem. Due to the change in ecological environment, the natural enemies of some agricultural pests have decreased and agricultural pests have increased, resulting in the massive application of pesticides, which not only pollutes crops and soil but also pollutes air and water sources.

### 2.4 *Pollution of domestic waste*

Domestic wastes mainly include cinders, empty bottles of sanitary pesticides, washing powder, plastic bags, plastic films, and other domestic wastes. These pollutants contain a wide variety of harmful substances and pose a great threat to human health. In addition to domestic sewage, domestic waste is becoming increasingly messy, and the pollution components such as plastic bags, batteries, and household appliances are very complex. In the next few years, electronic products, electronic and electrical products are at the peak of scrapping. If they are not handled properly, they will do great harm. It contains a large number of harmful substances: lead, cadmium, mercury, plastics, etc. once it enters the soil, it will seriously pollute the water source, endanger plants and microorganisms, and cause great harm to the human body.

### 2.5 *Pollution of human and animal feces*

Zoonosis has been a worrying problem since ancient times, but this phenomenon still exists, and the improper use and treatment of livestock and poultry manure is the main reason. There are a large number of pathogenic bacteria and parasite eggs in untreated livestock and poultry fecal sewage. If directly used, it will pollute land resources. If directly discharged, it will cause serious pollution to the water body and rural environment, resulting in deterioration of water quality, breeding of mosquitoes and flies, the proliferation of bacteria and induced diseases, seriously affecting the quality of the rural living environment, endangering human physical and mental health and causing human and animal diseases, human and animal feces in some rural areas can

Table 1. Classification of agricultural wastes.

| Agricultural Waste | | | | | | |
|---|---|---|---|---|---|---|
| Crop straw | | | | livestock manure | | |
| Rice straw, wheat straw, rape straw, etc | Cotton stalk, soybean stalk, etc | Peanut vine, sweet potato vine, etc | Corn straw, etc | Pig manure, cow manure, and horse manure | Chicken dung, pigeon dung, quail dung | Sheep dung and rabbit dung |
| Perishable and low nutrient | Not easy to rot and low nutrient | Perishable and low nutrient | It is easy to rot and low in nutrients | It is easy to rot and low in nutrients | It is easy to rot and has high nutrients | Extremely perishable, medium nutrient |

not be treated harmlessly (Zhang 2019). Pathogenic microorganisms pollute the atmosphere, water, soil and crops, spread diseases, and endanger people's health.

## 3 MAIN PROBLEMS OF RURAL AIR POLLUTION

### 3.1 *Weak awareness of pollution control*

Most people think that the rural ecological environment is generally good and lacks sensitivity to air pollution prevention and control. Individual localities and departments have not really established a scientific outlook on development and a correct outlook on political achievements, pursued rapid economic growth, and ignored the improvement of the ecological environment. To a certain extent, the problems of low illegal costs and high law-abiding costs still exist. Farmers' traditional lifestyle is difficult to change. A large number of chemical fertilizers and pesticides are used for planting plants in rural areas, and herbicides and other chemicals are used for weeding. The land planted in rural areas needs a lot of fertilization, and pesticides are chemicals for killing insects and weeds. Pesticides are often used in rural areas. Many farmers do not know that these fertilizers and pesticides contain a large amount of nitrate nitrogen and other drugs that pollute the environment, and are difficult to volatilize and degrade. In addition, unreasonable use leads to the reduction of soil organic matter and damage to the land and air.

### 3.2 *Wide coverage of pollution sources*

Cement, gravel, and other materials on the construction site are stacked in the open air, resulting in serious construction dust and residue. There are many domestic coal-fired boilers, so it is difficult to supervise and eliminate them. Garbage is stacked in the open air, the cleaning and transportation are not timely, and the cooking fume of the catering industry is directly discharged. Some enterprises do not have a strong subject awareness of energy conservation and emission reduction, and there are often illegal phenomena such as facility shutdown and excessive emissions. Farmers' economic income level has increased year by year, private cars have increased geometrically, and automobile exhaust pollution has increased sharply. Farmers can not get benefits from the comprehensive utilization of straw, and it is difficult to prohibit open-air incineration (Ao 2016). Some factories have moved to the suburbs or even rural areas. These factories have sewage that seriously pollutes the environment. Moreover, these factories are not allowed to be established in places with many buildings, so they can only move to places with few people, which has a great impact on the rural environment. Rural farming is mainly open-air free-range farming and small-scale captive farming. The breeding places are scattered, and it is extremely difficult to collect and treat waste gas. There is a lack of efficient waste gas recovery and treatment devices, and the feces are not cleaned and treated in time, resulting in the direct discharge of harmful gases produced by poultry and livestock farming into the ambient air.

### 3.3 *The linkage control mechanism is not smooth*

Air pollution prevention and control is a comprehensive and complex system engineering, involving many departments. At present, there are overlapping regulatory functions and multi-head management in some fields, the subject of law enforcement is unknown, the responsibility is unclear, and there are repeated law enforcement and mutual prevarication. The joint prevention and control mechanism is not perfect enough. Departments work independently and fight alone, which can not form an overall joint force. There are some vacuum and loopholes in the prevention and control of air pollution. Grassroots law enforcement personnel are poorly equipped, lacking ability, backward methods, and single means. The most rural economy is in the stage of extensive economic development. The task of technological transformation and elimination of backward production capacity is arduous, the demand for funds is large, and the investment is generally insufficient. Relying too much on administrative means to promote, the relevant laws are lagging behind, the illegal cost is set too low, the provisions on construction dust, volatile organic pollutants, motor vehicle, and ship pollution, regional air pollution prevention and control are not detailed, and the indicators are unclear, so it is difficult to grasp and implement.

## 4 RURAL AIR POLLUTION CONTROL PATH

### 4.1 *Implement the strategy of overall urban and rural development*

Air pollution in rural areas has the characteristics of large areas, scattered point sources, random sewage discharge, hidden pollution, and difficult monitoring. Relying on weak grass-roots environmental protection institutions is difficult to effectively meet the work needs, and the current situation of relying solely on administrative forces needs to be changed. Therefore, we should strengthen environmental publicity, improve people's environmental awareness, mobilize the participation of the whole people, encourage people's supervision through the establishment of reward reporting and other systems, and form a working pattern of everyone participation, with joint prevention and control of air pollution. To solve the problem of waste pollution in rural areas, we should first advocate a civilized lifestyle and improve the ecological environment awareness of rural residents. People should be guided to realize the harmfulness of garbage pollution, establish a scientific concept of ecological civilization, and resolutely change and correct the wrong habits and lifestyles such as littering and sewage discharge. We should incorporate the civilized lifestyle into the content of township regulations and civil conventions, gradually change the uncivilized ideas, advocate the civilized lifestyle, ideologically eliminate the source of "garbage" and solve the problem of rural garbage pollution. Environmental protection departments at all levels shall make use of newspapers, radio, television, Internet, and other media as well as posters, blackboard newspapers, and other publicity methods to widely publicize the harm caused by air pollution to mankind, publicize the aspects of air pollution sources, and publicize how to prevent and control air pollution, so as to make every citizen consciously enhance their awareness of preventing and controlling air pollution and know how to prevent and control air pollution. It is attempted to publicize the social concept of green, environmental protection, and sustainable development, and promote every farmer to actively participate in the construction and practice of green agriculture. In view of the large number of air pollutants discharged from rural areas, it is necessary to conduct a sufficient investigation and extensive demonstration, and study and issue relevant policies and legal provisions on this basis. It is important to exert the role of grass-root cadres (Liu et al. 2015). In accordance with the principle of responsibility within the jurisdiction and territorial management, townships and towns shall implement cadre village contracting, grid management, point-to-household, round-the-clock inspection, without leaving a dead corner, implement corresponding prevention and control measures and solve practical problems. It is needed to give full play to the advantages of village cadres' familiarity with the situation, timely information, and flexible patrol, encouraging them to do a good job in publicity, mobilization, inspection, and patrol. At the same time, we should also strictly enforce work discipline and strictly investigate the responsibilities of cadres who do not work seriously, do not implement in place, and have prominent problems, so as to strictly enforce discipline and ensure the implementation of work. The environmental governance model of paying equal attention to rural environmental governance and urban environmental protection is gradually transferred to rural areas. At the same time, we will implement the strategy of coordinated urban and rural development, guide the coordinated urban and rural development of rural environmental protection, make a strategic layout of the overall ecological environment of rural society, and accelerate the formation of an integrated urban-rural development pattern.

### 4.2 *Strengthen the management of the rural ecological environment*

According to the requirements of various indicators of farmland irrigation water quality standards, the water quality of sewage irrigation water shall be tested to eliminate the use of substandard sewage for farmland irrigation. If it is necessary to use it, it shall be treated innocuously. The approval of large-scale breeding must comply with the overall land use plan of towns and villages, and land use agreements shall be signed with local rural collective economic organizations. The government shall strengthen the supervision and management of the approved large-scale breeding farms and use them in strict accordance with the approved purposes. Establish environmental protection institutions at the township government level, allocate special environmental protection monitoring, investigation, and implementation personnel, organize agriculture, water conservancy, and other departments involved in the prevention and control of rural environmental pollution, establish township and rural environmental joint law enforcement agencies, share resources among multiple departments, and improve law enforcement efficiency. Relevant rural departments in all localities should send cadres to go deep among the masses and educate farmers not to burn weeds and straws, so as to reduce air pollution. At the same time, accelerate the research and development of crop straw recycling technology and related equipment to turn straw into high-value-added products such as feed, fuel, chemical, and commodity raw materials. Efforts should be made to develop stable and efficient chemical fertilizer, improve fertilization technology, bury the chemical fertilizer deeply and maintain sufficient water, and reduce the volatilization and release of chemical fertilizer. Commitment is made to research and develop the comprehensive utilization and treatment technology of farm waste gas, and to turn the waste gas into gas and chemical raw materials after collection. It is significant to strengthen the supervision of straw burning, implement the financial reward and subsidy policy, guide farmers to comprehensively promote the mechanized crushing and returning of straw to the field and the comprehensive utilization of feed and energy, and gradually realize the comprehensive prohibition of straw burning. We should strengthen publicity, education, supervision, and law enforcement, and prohibit the open burning of domestic waste, cleaning waste, construction waste, asphalt, and other substances that produce smoke and dust. We ought to increase project support, promote the development of a circular economy, support enterprises to comprehensively utilize straw and domestic waste, and rural waste resources, vigorously develop bioenergy and produce renewable resources, and promote the transformation from traditional agriculture to ecological agriculture. In the adjustment of rural economic structure, we should actively guide farmers to develop ecological agriculture and organic agriculture, establish organic food and green food bases, and reduce the use of pesticides and chemical fertilizers. Farmers should be actively guided to develop large-scale breeding communities and promote the professional and large-scale development of animal husbandry and aquaculture. By strengthening the publicity of organic food and green food, urban residents and consumers should be guided to implement quality supervision over agricultural products, so as to form a virtuous production and consumption cycle.

### 4.3 *Establish and improve laws and regulations related to the prevention and control of air pollution*

Although China has issued relevant laws, there are no corresponding implementation measures at the local grass-roots level, especially for rural air pollution prevention and control, the measures are not in place, and the responsible person is not clear, which has become a blind area of air pollution prevention and control. The provincial, municipal, and county levels shall issue corresponding implementation rules or local regulations in accordance with national laws. We need to establish sound management and supervision system, take Township and village cadres as the main body of managers, provide technical guidance, supervision and evaluation by agricultural departments and environmental protection departments, strictly implement relevant policies and laws, and promote the promotion and application of green agricultural technologies and products. It is essential to establish a government-led, multi-party linkage, clear rights and responsibilities, sound network supervision, and a governance system to eliminate blind spots and loopholes and form a joint force (Yu & Gao 2016). We will increase financial budget investment at the same level and transfer payments from higher levels, and encourage private capital and social funds to participate in air pollution control. We will speed up the revision and improvement of laws, and ensure that grass-roots units have laws to abide by. We will increase human and financial investment in pollution prevention and ecological protection, establish sanitation stations in villages and towns, allocate full-time sanitation personnel to perform cleaning and environmental health supervision duties, strengthen the construction of rural environmental protection infrastructure, and provide rural areas with garbage collection vehicles and garbage cans, so as to achieve unified collection and classified treatment of garbage. Further, it is required to strengthen the reward and compensation for biogas construction and establish an agricultural ecological compensation mechanism. It is planned to promote the construction of waste harmless treatment facilities according to the plan, step by step and batch by batch, and finally achieve full coverage, so as to ensure that the problem of waste pollution in rural areas is comprehensively and effectively solved.

Table 2. Flow chart of straw conversion into Bioenergy.

| | | | | | | | |
|---|---|---|---|---|---|---|---|
| | Drying treatment | | Fuel tank | Gas emission up to standard | Fly ash collection | Oxygen supply system | |
| Straw collection | Straw collection tank | Straw crushing | Supply oil pump | Gas treatment system | Control system | | |
| | | | Mixing and stirring | Flue gas monitoring system | Radiant heating system | Ash treatment system | Resource reuse |
| | | | | Feeding system | Incineration system | | |
| garbage collection | Garbage pool | Garbage crushing | | Flue gas monitoring system | Boiler turbine | | |
| | Percolation treatment | | | | Generator Set | | |
| | | | | | Power access | | |

Table 3. Statistical table of the utilization rate of a village after environmental protection and energy conversion.

| Administrative village | Area (MU) | Straw yield per mu (ton) | The total output of straw (ton) | Straw utilization (ton) | Utilization% |
|---|---|---|---|---|---|
| 1 | 824 | 0.85 | 700.4 | 680.8 | 97 |
| 2 | 770 | 0.85 | 654.5 | 641.6 | 81 |
| 3 | 950 | 0.85 | 807.5 | 790.8 | 98 |
| 4 | 566 | 0.85 | 481.1 | 462.8 | 96 |
| 5 | 422 | 0.85 | 358.7 | 340.8 | 95 |
| 6 | 120 | 0.85 | 102.0 | 95.8 | 94 |
| total | 3652 | 0.85 | 3104.2 | 3012.6 | 97 |

### 4.4 *Encouraged farmers to gradually apply an organic fertilizer to reduce pollution*

Facing the increasingly serious situation of rural environmental pollution, taking the development road of ecological agriculture is the only way to improve the current situation. We should increase the popularization and application of advanced agricultural production technology, advocate the use of new pesticides with high efficiency, low toxicity, and low residue, and promote the integrated development of intensive livestock and poultry breeding and ecological agriculture and animal husbandry. We will continue to strengthen the construction of biogas, and strive to rationalize the agricultural industrial structure, make the production technology ecological, clean the production process and make the production products harmless. The state has issued some incentive policies for the application of organic fertilizer, such as reducing the production of chemical fertilizers and pesticides and properly subsidizing farmers. It is suggested that chemical fertilizer and pesticide manufacturers should transform and upgrade, adjust the product structure, reduce the production of inorganic fertilizer and toxic pesticides, and reduce the market supply. Local manufacturers should actively cooperate with breeding farms and large breeding households to collect the feces of pigs, cattle, sheep, chickens, ducks, rabbits, and other livestock, make granular organic fertilizer through scientific fermentation, put it on the market, and sell it to farmers. We should widely promote biological pest control technology, light pest control equipment, duck and fish farming in rice fields, and minimize the use of chemical pesticides. Governments at all levels should increase investment and encourage manufacturers and farmers to produce and use organic environmental protection and low-carbon fertilizers. To strengthen the rural financial guarantee mechanism, financial departments at all levels should include rural environmental protection in their budgets and gradually increase investment. The government should reduce direct administrative intervention in rural areas, and use more economic means and incentive mechanisms to control the input of natural resources and the output of pollutants, so as to make the behavior of producers and consumers more environmentally friendly. We will increase subsidies for structural adjustment and ecological environment protection, vigorously support the development of ecological agriculture, change taxes and subsidies that are not conducive to environmental and social sustainable development, and shift to a non-point source pollution tax.

### 4.5 *Give prominence to industrial pollution control*

There is an exigent need to accelerate the transformation of development mode and industrial structure adjustment, adhere to the road of new industrialization, and actively develop high-tech industries, strategic emerging industries, and low-carbon smoke-free industries. We will strengthen scientific and technological innovation, transform and upgrade traditional industries, resolutely eliminate backward production capacity, and shut down and transform enterprises with serious pollution and backward production capacity. Enterprises should be encouraged and guided to actively develop and utilize clean energy such as natural gas, electricity, solar energy, and biomass energy, and gradually reduce coal consumption. We will strengthen the control of particulate matter emissions from industrial coal-fired boilers and limit the construction of new coal-fired boilers with a steam capacity of 10 tons or less per hour. High-efficiency dust removal transformation shall be carried out for those coal-fired industrial boilers whose smoke and dust can not be stably discharged up to standard, and bag dust removal devices shall be installed for fluidized bed boilers and pulverized coal boilers, thereby vigorously promoting the cleaner production process of industrial enterprises, eliminate or reduce pollutants in the production process, so as to realize less waste and no waste discharge at the end of production.

We need to strengthen the control of automobile exhaust and catering oil fume pollution, regulate dust management on construction sites, bring dust pollution prevention and control into the scope of project supervision, implement measures such as "enclosure, cover, sprinkling and washing" in construction, and promote the transformation of construction and construction methods. It is critical to strengthen the administrative law enforcement of urban management, carry out the dust control of residue transportation, concrete mixing plant and material storage yard, urge the implementation of measures such as sealing, enclosure, covering, spraying, greening, and setting up wind and dust suppression nets in accordance with the requirements of wind and dust prevention. We are committed to vigorously promoting the forest growth and quality improvement project, greening and hardening the bare land, and strengthening the ecological restoration of mines. We will comprehensively implement the management of environmental protection signs for motor vehicles, strengthen the construction of environmental inspection institutions and testing lines for rural motor vehicles, promote the regional restriction of "yellow standard vehicles", complete the elimination task of "yellow standard vehicles", and expand the scope of demonstration and promotion of new energy vehicles in the field of public services. We should double down on efforts to strictly comply with the environmental protection requirements of new catering service business projects, prohibit the construction of catering business projects emitting oil fume in unapproved non-commercial buildings, strengthen policy guidance, promote the use of purified household range hoods, ban outdoor smoking barbecue and promote indoor carbon free barbecue.

## 5 CONCLUSION

At present, rural financial, material and human resources and technology are difficult to solve the rural garbage problem independently. The government must take the lead in establishing corresponding management institutions, innovate and implement scientific management mechanisms, bring rural drinking water safety, ancient tree protection, road hardening and greening into the rural ecological civilization planning and work agenda, and mobilize the combination of the masses and experts. We will work twice to comprehensively improve the level of rural environmental improvement, thoroughly improve the current situation of rural ecological civilization and environment, and provide a clean and hygienic ecological environment in rural areas.

## REFERENCES


Ao Xiang. (2016) Rural ecological environment protection and construction countermeasures. *Henan science and technology*. 146–147.

Huang Yanjiong. (2019) Study on improvement of rural point pollution control measures. *Journal of Hubei University of Economics* (HUMANITIES AND SOCIAL SCIENCES). 56–58 + 75.

Liu Jianqin, Wan dajuan, Wang Tingshi, Wang Ding, Li Mei. (2015) Research progress of rural air pollution at home and abroad [J]*Environmental protection and circular economy*. 9–11.

Yu Junhong, Gao Guilin. (2016) Legal Thoughts on improving the prevention and control of rural air pollution in China. *Environmental protection*. 51–53.

Yuan Xiangling, Lu Wei. (2021) Study on rural land pollution and soil environmental protection methods. *Resource conservation and environmental protection*. 35–36.

Zhang Hangang. (2019) Causes and Countermeasures of rural straw incineration from the perspective of environmental sociology. *Green technology*. 113–116.

*Advances in Urban Construction and Management Engineering – Cha (Ed)*


# Study on the design of improving the performance of the damping system of railway vehicles

Zhidan Wu & Ying Huang*
*Liuzhou Railway Vocational Technical College, Liuzhou Guangxi, Guang Xi, China*

ABSTRACT: With the continuous improvement of China's economic growth, China's railway is also gradually speeding up with the development of the times. However, the increasing problem of train vibration has seriously affected the passengers' experience and the safety of the driver during the operation. At present, the dynamic performance of railway vehicle damping systems is mostly simulated by experimental equipment, and some pantograph and catenary tests need to be carried out on-site, and some indoor tests cannot be truly simulated to test and improve the shock absorption effect of vehicles. Based on this, this paper finds that the computer simulation experiment can effectively combine the mathematical model and the damping system to form a new hybrid simulation test system. Therefore, this paper analyzes the relevant principles of the shock absorption system of the railway vehicles by using the software and hardware detection technology in the computer system and puts forward the scheme design for the improvement and promotion of the shock absorption system of the railway vehicles in the future, hoping to provide relevant reference and reference for the subsequent improvement of the shock absorption system of the vehicles.

## 1 INTRODUCTION

### 1.1 *Literature review*

Method of parameter modeling is used to establish the hydraulic control model and simulation model of railway lateral shock absorber under the environment of EASY5, then verified the accuracy of the two models, selected the data and parameters of the required vehicle shock absorber, and used the model to establish the digital experiment to comprehensively analyze the influence of parameters on vehicle damping characteristics. The results show that the damping structure parameters and the joint stiffness of the bottom valve aperture have a great influence on the damping characteristics of the shock absorber. When the parameters are higher than the value range, the damping characteristics will be distorted (Xiao et al. 2014). Lei Rufeng found that when the suspension system of the vehicle was blocked, it seriously affected the normal driving of the vehicle and even could not reach the designated area safely. To achieve a good effect of the damper and ensure the normal operation of the vehicle suspension system, it is necessary to study the performance of the damper, put forward a reasonable design scheme, and carry out tests to select the best damper to use with the suspension and achieve the best damping effect (Huo 2015). Lei Jianfeng and Guo Xin pointed out that simulating the actual running vibration condition of the vehicle is the most critical link for the research of the test-bed and the performance test of the vehicle before leaving the factory. The dynamic analysis and simulation design are carried out by adding spring to the excitation system, and a new type of shock absorber in the design of the rolling vibration table is discussed (Lei & Guo 2014). Ma et al. used ANSYS to simulate the electromagnetic part of the system from the perspective of the mechanical and electromagnetic properties of the system, analyzed and elaborated the design scheme and simulation results of the system, and found that the

*Corresponding Author: huangying800816@163.com

DOI 10.1201/9781003348023-38

hybrid intelligent damping system is superior to the general cylinder hydraulic damping system in damping performance and automatic flexibility function, which can ensure the smooth driving of the vehicle and ensure that the battery can fully meet the actual use needs (Ma et al. 2016).

### 1.2 *Research purposes*

With the rapid economic growth in China, the construction and improvement of transportation and its infrastructure are also accelerating and innovating. With the development of the economy, the technology of high-speed trains in our country has been continuously improved and the operation speed has been improved. However, when the train meets people's daily traffic, the comfort needs to be improved. For example, the vibration and noise of the train in the process of operation is very serious, which seriously affects the passengers' travel experience and the driver's driving safety. To ensure the weakening of anti-seismic and noise problems in the process of vehicle operation, the optimization and improvement of shock absorbers have become one of the key areas of vehicle damping research and one of the indicators to evaluate vehicle quality. The shock absorber is the key component of all kinds of motor vehicles. The performance of the shock absorber is directly related to the safety performance and driving comfort performance of the whole vehicle. Based on this, this paper summarizes the structure and principle of the microcomputer monitoring system of railway vehicle shock absorbers and makes optimization and improvement plans for some problems of the current shock absorber to continuously improve the performance and innovation of railway vehicle shock absorbers system.

## 2 STRUCTURE AND PRINCIPLE OF MICROCOMPUTER TESTING SYSTEM FOR RAILWAY VEHICLE SHOCK ABSORBER

Based on the mechanical shock performance test-bed, a microcomputer-based main measurement, and control technology is used to detect the indicator function and speed characteristics of the shock absorbers of railway vehicles.

### 2.1 *Hardware*

At present, the performance of the shock absorber of railway vehicles is realized in the computer testing system, and then the shock absorber is clamped by the mechanical part according to the shock absorber testing standard to achieve vehicle shock absorption. The upper shock can be used to test the function and speed characteristics of the shock absorber of railway vehicles, and the lower shock can be used to test the durability of the shock absorber. Because most railway vehicles use electromagnetic speed regulating motor, it cannot only effectively detect the indicator function characteristics and speed characteristics of the shock absorber under 1.67 Hz excitation, but also understand the change frequency of indicator power and speed characteristics and the change of part of the performance of the shock absorber under different frequencies (Chen et al. 2014).

The 8,0286 microcomputers with low prices and powerful functions are selected as the main control computer of the system. The main measurement and control machine includes the host, keyboard, mouse, hard disk, driver, printer, and VGA display screen, as well as ps-2112 16 channel 8-bit A/D converter adc0816 as the shock absorber of railway vehicle travel, as well as the resistance stress 2 signal analog-to-digital converter. The ADC0816 conversion board is matched with a 286 microcomputer, the display screen resolution is 12 bits, the time setting is about 25s, the relevant data conversion is carried out for the signal between $-5$ and $+5$ V, and the error is controlled at 0.1%, to ensure the detection accuracy.

Strain type pull (press) force sensor can be used to detect the blocked pressure of the shock absorber. According to the measuring range of railway vehicles, different pull (press) force sensors and strain gauges can be continuously converted to preamplifiers. The output voltage signal ranges from $-5$ to $+5$ V, and the accuracy of detection is less than 0.5%. For the shock absorption of railway vehicle travel, the differential transformer type displacement sensor can be used to change the measurable maximum mileage through different displacement sensors of the vehicle, and the output voltage signal is −5 to +5 V, with an accuracy of 0.5%. The frequency regulator can adjust the excitation frequency and speed continuously within the measurable range of the vehicle, or it can use the flange position of the vibration table to change the vehicle excitation formation and adjust the excitation speed within the range of 0 to 1.2 m/s.

### 2.2 *Software*

The software used in the computer detection system is compiled by C language and language assembly. Using Chinese character screen display, human-machine dialogue, and other ways, the operating system is more convenient and intuitive, and the software is set as modular. To accurately detect the accuracy of sampling and the processing speed of computer operation, it is necessary to use 286 program microcomputer assembly programming language for sampling and preprocessing subprogram operation of sampling, as shown in Figure 1. A sampling point is controlled to 300 points by the function of the subroutine; the interference signals that may exist in the sampling can

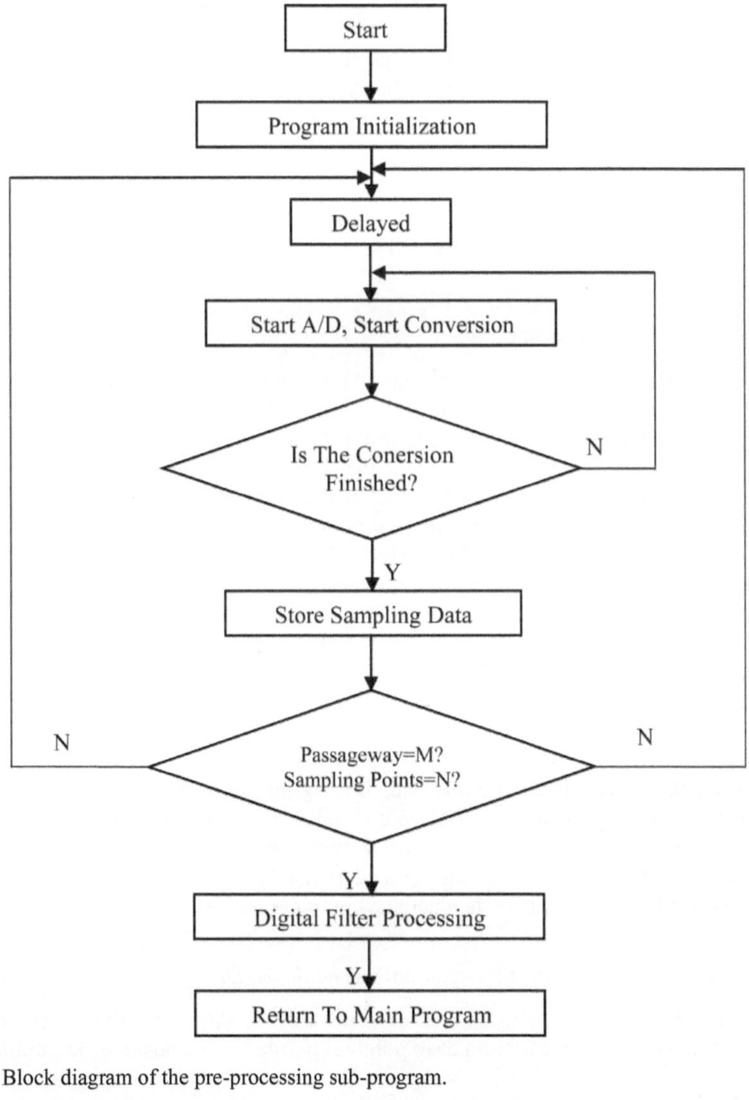


Figure 1. Block diagram of the pre-processing sub-program.

be filtered by a digital filtering function; the sampling frequency can be arbitrarily changed. In addition, the main program used in the vehicle damping test is compiled in C language, which is also designed as a modular mechanism. From top to bottom, it is designed into six major functional modules: the main control module, the system preparation module, the data test module, the data output module, the system standard module, and the exit system module. The structure is shown in Figure 2.

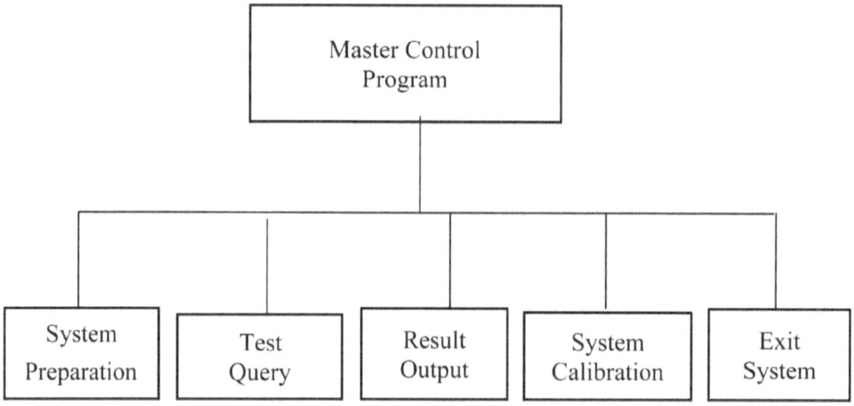


Figure 2. Main program composition architecture.

## 3 IMPROVEMENT SCHEME OF PERFORMANCE OF DAMPING SYSTEM FOR RAILWAY VEHICLES

### 3.1 *Key points of software system design*

The system software design of computer testing the performance of shock absorber includes collecting data, processing data, and the output of the results. The software uses the most advanced graphic interface technology, through the Chinese pull-down menu and full-screen text editing interface, to effectively help users operate. Since the software is simple to use, users do not need to have the professional computer knowledge to calculate and view the process through on-screen prompts. The video interface of the software is programmed by Turbo C++, and data collection and calculation are carried out by combining Turbo C++ and language compilation. In addition, the software also has the following functional modules.

The experimental module of recording sample data: the editor of the module is a full-screen field, which can use the cursor arbitrarily within the effective range of the full screen, and input the test date, test sample data, personnel, and other data. Moreover, it is convenient to record and modify the data content that needs to be changed arbitrarily. Sensor and system detection standard template: including A/D acquisition conversion and PCL-711s drive detection module. According to the different mechanics modules used in the vehicle shock absorber test, the measured data of the sample is analyzed, including the numerical filtering displayed by the software. Output data results, storage, and query module: the conclusion module includes the overall data displayed on the computer screen, printing and drawing, and analysis data conclusion. The memory module saves all kinds of test data. The query module can query the data content of all data, materials, etc., which is convenient for users to know the data and computer usage.

### 3.2 *Construction of the improved mathematical mechanic model*

For the signal requiring displacement, the top dead point of the shock absorber is regarded as the negative maximum value, and the bottom dead point is regarded as the positive maximum value, in which the speed and acceleration are set as the positive values. The time interval of a/D sampling is ∇T, the displacement is s, the force is f, and the data acquisition is n, that is, $S_i$ (i = 0, 1, 2, …, N-1), and fi (i = 0, 1, 2, 3, …, N-1).

Because the clamp is located under the sensor, its weight can eliminate the balance and withering effect brought by the weight, but when the shock absorber is made into variable-speed movement, the inertial force generated by the clamp can only be reflected in the force test signal of the tension pressure sensor, so it can be found that when the force signal processes the sample data in the mathematical mechanics model, the oil carried by the clamp must be removed previously Inertial force.

Taking one of the clamps in the shock absorber as the research object, the force signal acquired by PCL-711s is the force F of the tension and compression sensor, and the force signal used in the shock absorber is f′, using Newton's second law to study the clamp.

$$F' - F = ma \tag{1}$$

where $m$ is the fixture mass and $a$ is the speed in fixture detection.

$$F' = F + ma \tag{2}$$

$$\begin{aligned} V &= \frac{ds}{dt} \\ V_i &= \frac{ds_i}{dt_i} = \frac{S_{i+1} - S_i}{\Delta_t} \, (i = 0, 1, 2 \cdots\cdots N-1) \end{aligned} \tag{3}$$

$$a_i = \frac{dvi}{dti} = \frac{V_{i+1} - V_i}{\Delta_t} \, (i = 0, 1, 2 \cdots\cdots N-1) \tag{4}$$

At present, the domestic shock absorber test-bed often does not consider the elimination of the inertial force effect brought by the fixture, which leads to the overall level of the indicator function characteristics and speed characteristics in the test results moving up, so most people think that the resistance of the shock absorber is far greater than the recovery resistance of the vehicle (Zhang 2014). Through the above calculation results, it can be found that the compression resistance of the shock absorber with superior performance is equal to the basic resistance of its recovery, sometimes greater than the compression resistance.

Through the determination and test of the above calculation method, it can be found that the system used for vehicle damping is accurate and reliable in data detection, and has strong universality, which is suitable for any vehicle damping system detection. At the same time, to further improve the accuracy of the test and some characteristics of the response, we can also start from the angle of the machine part and elimination of the inertial force of the fixture. As the excitation frequency of the shock absorption system needs to be improved, it is necessary to control the frequency at f = 1.67 Hz during the structural optimization test of the shock absorber to ensure the operation stability of the mechanical equipment and the supervision of the system software. In addition, to eliminate the inertial force of the fixture, the acceleration of the fixture and its internal structure can be directly optimized and improved by adding an acceleration sensor to avoid the deviation of the acceleration data obtained by the displacement second derivative algorithm, and the inertia force of the fixture cannot be effectively eliminated by using mechanics (Yan et al. 2015).

### 3.3 *Elimination of errors and interferences*

Because of the bad environment, the noise and vibration are very serious problems at the test site. In addition to the error caused by the system hardware and software itself, it is also easy to be affected by various kinds of uncontrolled external interference. If not eliminated in time, it will seriously affect the test results, and even lead to the failure of the test. The main errors of the test system come from sensors, preamplifiers, and A/D. Therefore, to better eliminate the error caused by the test system, it is necessary to select a higher level of the precision monitor according to the actual test requirements to meet the accuracy required by the test. In this regard, the following measures are taken to eliminate the interference during the test.

All data lines using signal transmission shall be replaced with shielded data lines and the shielding layer shall be grounded. The AC power supply and UPS with high-performance voltage stabilization and purification are used to prevent 50 Hz and its harmonic components and the instantaneous interference generated by the AC power grid. Strictly separate and clarify all kinds of bottom lines to avoid interference between lines. For common mode interference, all bottom lines, casings, etc. are connected to the ground pile according to different functions. The median filter can be used in the system software to suppress the impact of random interference data and improve detection accuracy. In addition, between the interface board of the host and the preamplifier signal, the resistance-capacitance filtering with different values is used to prevent high and low-frequency interference. Through the above measures, it can be found that the computer detection system of the vehicle shock absorber can better reduce error and interference (Peng et al. 2013).

Generally, in many experiments and researches, the results sometimes appear as random error problems according to data, measurement value, and other problems, but the standard of error control range is widely used in experiments. The error and interference in the vehicle damping system need to be brought into the evaluation repeatedly through calculation and detection value, and then the error and interference value will be reduced continuously, close to the real test data. In addition, in the engineering test, the number of observations is far more than the data from the external test. Therefore, it is necessary to find out the average value of observation times, and then get the approximate true value, which can also be called the best value to improve the performance of the vehicle damping system, optimize the lifting scheme, and reduce the relevant errors and interference.

## 4 CONCLUSION

The research field of railway vehicle damping mostly focuses on damping control, including the theoretical method of damping systems and the adoption of new technology. Moreover, in the past, the overall mass of the train is small and the speed is low. The shock absorption is to press the shock absorber through the train driver's operation and control mechanism to meet the needs of vehicle shock absorption. However, with the continuous improvement of the quality of railway vehicles, the optimization and improvement of the damping system have become very important for vehicle damping. By testing the indicator function and speed characteristics of the vehicle shock absorber, the vibration frequency of the shock absorber can be accurately detected to regulate the motion speed and sampling speed of the shock absorber. The microcomputer monitoring system is very helpful for the data display of vehicle damping performance and the optimization and improvement of the following vehicle damping system. Moreover, the microcomputer detection system of the shock absorber is based on the mechanical excitation performance test-bed, through the microcomputer main measurement and control detection technology, to carry on the intuitive data monitoring to the indicated power function characteristic and the speed characteristic of the vehicle shock absorber performance. In this paper, a microcomputer testing system for vehicle shock absorber performance is developed to improve the speed and accuracy of testing and data accuracy based on traditional shock absorption testing. Through the improvement of shock absorption performance, the identification ability of indicator function characteristics and speed characteristics of the shock absorber is improved. In addition, the vibration damping microcomputer detection system has the advantages of strong function, convenient operation, and other stable advantages, and then improve the vehicle damping performance vibration damping microcomputer system.

## ACKNOWLEDGMENTS

This research was supported by the Guangxi Department of Education Scientific Research Project (Research on Fault Information Extraction and Intelligent Diagnosis Algorithm of

Gear Transmission System in Multi-source Strong Noise Environment of Metro) (Grant No. 2022KY1401).

## REFERENCES

Chen Z.H., Yang Q., Ning X.L., et al. (2014). *Magnetic Revenant Damper Design, Test and Its Vibration Reduction Performance Study, Seismic Engineering and Engineering Vibration*, 34 (01), 204–210.

Huo R.F. (2015). *Vehicle Suspension Damping System Damping Design Ideas Industry and Technology Forum*, 14 (10), 56–57.

Lei J.F., Guo X. (2014). *Preliminary Analysis of the Vibration System of the Rolling Vibration Test Bench of Locomotive Vehicles*. Science and Technology Innovation Guide, 16 (28), 99–100.

Ma C., Li G.L., Song J., Li X.Y. (2016). The Design and Simulation of Hybrid Intelligent Shock-Absorbing System. *Journal of Southwestern University* (Natural Science Edition), 38 (01), 188–196.

Peng H.M., Li S.Q. (2013). Development of Lateral Shock Absorbers for Seismic Softos of Railway Vehicles. *Rail Transportation Equipment and Technology*, 27 (3): 52.

Xiao Q., Li Q.H., Wang C.G., et al. (2014). Study on the Impact of Hydraulic Shock Absorber Structural Parameters on Their Damping Characteristics of Railway Vehicles. *Journal of Railways*, 41 (10), 21–27.

Yan L.X., Zhou H., Peng C., et al. (2015). *Adaptive Active Resonance Shock absorber design and vibration reduction performance research Experimental Mechanics*, 30 (5): 556–566.

Zhang B. (2014). The Overall Design of the Railway Vehicle Acceptance Technology Information Platform. *Rail Transportation Equipment and Technology*, 27 (03): 45–48.

*Advances in Urban Construction and Management Engineering – Cha (Ed)*
*© 2023 Copyright the Author(s), ISBN 978-1-032-39018-5*

# Construction technology of large tree-like steel support ETFE sky curtain membrane structure

Xiaoyan Wang*
*Guangzhou City Construction College, Guangzhou, China*
*Tun Abdul Razak University, Malaysia*

Jinhong Lao*
*Guangzhou City Construction College, Guangzhou, China*

ABSTRACT: In consideration of the technical problems of scaffold erection, steel support hoisting, and membrane structure assembly in the construction of large-scale tree-like steel-supported membrane structures, this paper develops a product of self-locking clips for building scaffolding, which is applied to the process of regionally differentiated scaffolding, making the scaffolding each the rods are quickly and accurately connected, which improves the overall stability of the scaffold; based on BIM technology, the large-scale tree-shaped steel support rod construction technology and the self-expanding ETFE sky curtain membrane structure construction technology have been developed, enhancing the hoisting efficiency and installation accuracy of the steel support membrane structure and achieving first-time excellent installation quality.

## 1 INTRODUCTION

Currently, supporting membrane structures resembling sun umbrellas are frequently installed in open squares surrounding large buildings to shield individuals from wind and rain during leisure and recreation. Membrane structures have been gradually used in large-scale stadiums, exhibition centers, cultural entertainment, and other public buildings, as well as in public green spaces and parks (Zhang 2009). Membrane structure building is the most representative new architectural form in the 21st century and has become one of the main forms of large-span space architecture (Chen & Yan 2009). Membrane structure has the characteristics of novel shape and simple force transmission. As a typical form of flexible tension structure (Wu et al. 2014; Yang et al. 2019), membrane structure has many differences in mechanical performance from traditional rigid structures. Lin (2015) used a project to illustrate the application of steel frame membrane structure installation technology in stadium construction. Li (2020) conducted theoretical and simulation research on the shape and load analysis of membrane structures based on current codes and theories using the steel structure design software 3D3S. Li et al. (2021) investigated and innovated the construction technology of a large-area reverse-mounted membrane structure by using a hoist and pulley block to hoist the keel and double-control graded pensioning of the membrane surface in conjunction with engineering practice. Xue et al. (2010) studied the construction technology of the main steel pipe truss structure, membrane structure, LED screen light bar, and sound installation during the construction of the canopy membrane structure. Lain Hui analysand the installation and construction technology of ETFE membrane curtain wall in a cultural and athletic facility. Lian (2020) investigated the installation and construction of an ETFE membrane curtain wall in a cultural and sports center. By using winch traction, we can improve the overall lifting method to

*Corresponding Authors: 530702153@qq.com and 1278065386@qq.com

 DOI 10.1201/9781003348023-39

ensure construction quality and safety, as well as compress the construction period, reduce costs, and increase efficiency. Huang (2013) designed the ETFE air pillow node on the lighting roof of Guangzhou South Railway Station and combined local characteristics with a good drainage system.

## 2 PROJECT OVERVIEW

The outdoor steel support membrane structure project of an international shopping center in Hair requires an investment of nearly 350 million yuan. The membrane structure's projected area is 17,000 $m^2$, its highest point is 33 meters above the ground, and it is divided into nine high and low areas. Steel consumption is approximately 3,200 tons. The ETFE film has a thickness of 0.25 mm, a transmittance of 89%, the ultimate tensile stress of 50 MPa, standard width of 1.6 m, and a light blue color. The membrane plate and the fixing parts are made of professional aluminum alloy components with an anodized surface, the oxidation thickness is not less than 10 $\mu$m, and the screws are 316 stainless steel bolts.

Figure 1. Aerial view of the tree-like self-expanding ETFE canopy membrane structure.

The project site is divided into nine construction areas, of which the 1−2#, 3–4#, and 5–8# areas are constructed simultaneously in parallel, and the 9# and membrane structure are constructed lastly. Actual distinctions are made. This project employs steel support resembling a tree, with a complex structure and distinctive form, for the installation of the canopy membrane. It is composed of numerous rods with a variety of angles, elevations, and thicknesses. The precision of the production and processing of the steel support is high, and the installation process is challenging. This project's membrane structure covers a vast area, and its construction is susceptible to quality issues such as wrinkles, stains, color differences, and stagnant water.

## 3 REGIONAL DIFFERENTIATED SCAFFOLDING CONSTRUCTION TECHNOLOGY WITH SELF-LOCKING FUNCTION

### 3.1 *Research and develop scaffold node self-locking technology*

In the nine areas of this project, there are many scaffolds erected, the erection height is high, the construction work is cumbersome, the workload is heavy, the operation is difficult, and there are many potential safety hazards. In response to the above problems, the author developed "a self-locking section of scaffolding for construction", which is used to connect scaffold components quickly, conveniently, and accurately. The self-locking clip section includes a vertical rod jacket, a vertical rod clip is arranged in the middle hole of the vertical rod jacket, and the outer surfaces of the four sides of the vertical rod jacket are provided with a horizontal rod clip comprising an adjustment bin and a horizontal rod clip. The crossbar clamp is locked by a self-locking structure. The adjustment bin is equipped with two racks connected by gears, with the first rack located in the crossbar clamp and the second rack located on the side of the crossbar clamp's hairpin end. On the side of the hairpin connection end of the second gear rack, an ejector sleeve is sleeved to automatically lock the clamp of the ejector crossbar. The crossbar is fixedly connected by the crossbar clamp on the outer surface of the vertical bar jacket's four sides and the crossbar clamp's self-locking structure.

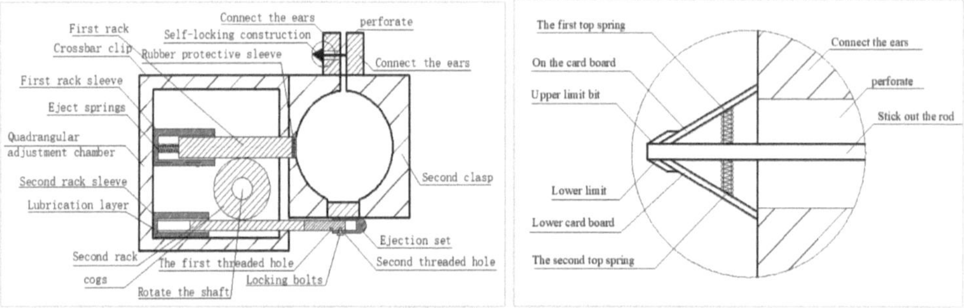


Figure 2. Depicts its sectional view, self-locking structure large-scale drawing, and three-dimensional drawing.

### 3.2 *Regional differentiated scaffolding construction technology*

#### 3.2.1 *The overall installation idea of scaffolding*

In this project, nine areas are designated for the installation of the canopy membrane structure. Due to the various buildings surrounding each area, the height of the membrane structure varies in each area. Consequently, during the process of scaffolding erection, scaffolding is erected according to the surrounding environment of various areas. General scaffolding and portal scaffolding are the two types of scaffolding.

The scaffolding construction plan for the installation of membrane structures on-site should follow these principles. For the installation of membrane structures, scaffolding is usually erected along the outer eaves of the surrounding floors, with 1–3# area scaffolding constructed simultaneously with 4–5# area scaffolding. The scaffolding is constructed from the ground up, and the 6–9# area uses two rows and four layers of mobile portal frames on wheels. The scaffolding is positioned 0.5 m from the lower surface of the steel structure and extends 2 m inwards and 3 m outwards to facilitate the installation of the membrane structure.

#### 3.2.2 *Regional differential scaffolding scheme*

This project is separated into nine sections. Due to the varying plane positions and scopes of the membrane structure, the scope and form of the scaffolding are likewise distinct, as are the form and height of the buildings in each region. The manner and height of the scaffolding must also be modified extensively. The scaffolding in the 1–3# area, for example, is erected from the building floor. On the levels of Buildings 1, 2, and 6, Buildings 3 and 5, and Building 6, three distinct regions are separated. As shown in Figure 3, a row of scaffolding should be erected along the exterior border of the building. The scaffolding in the 4–5# area is built from the ground, and should leave a passageway of 5 × 4 m for cranes and personnel, and should be built away from the highway. Since the overall height of the 6–9# area is less than 10 m, a group of gantry frames are installed at around 10 m intervals around the outer perimeter of the steel structure. The installation of the membrane structure is carried out using a moving cycle.

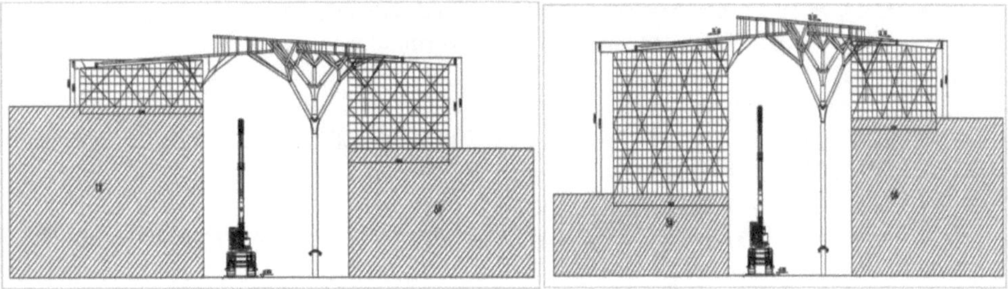

Figure 3. Schematic of the plan and elevation of the scaffolding plan.

### 3.3 *Self-locking scaffold erection process and technical points*

Line drawing and lofting → set the full-length scaffolding board and base → vertical sweeping rod → vertical pole → self-locking clip fixing → horizontal sweeping rod → large and small cross rods (frame board) → scissors braces → connecting wall pieces → laying scaffolding boards → protection Railing → hang safety net.

(1) During the process of line drawing and staking, the distance between the inner and outer vertical poles from the wall at the four corners of the structure must be measured and marked according to the structural requirements. Straighten it with a steel tape measure, mark the pole's position with a little piece of bamboo, and then mark the pole.
(2) When the backing plate and the base are erected, the backing plate must be positioned flat on the positioning line, not hung.
(3) Sweeping poles and vertical poles must be installed with skids, and sweeping poles in vertical and horizontal directions must be installed, attached to the standing poles, and approximately 20 cm from the base.
(4) When securing the self-locking segment, the developed self-locking segment must be secured to the vertical rod, and then the upper vertical rod must be inserted to secure it. Utilize the self-locking concept to firmly secure the horizontal rod when inserting it.
(5) When establishing large cross bars and tiny cross bars, the distance between the large cross bars in the height direction of the scaffold must be 1.8 m for the vertical net to be suspended. The outer frame is fitted with a small crossbar at the intersection of the vertical bar and the huge horizontal bar, and both ends are attached to the vertical bar to create the space structure's overall force.
(6) A pair of scissor braces shall be provided at each end of the outside facade of the scaffold and shall be continuously placed from the bottom to the top, with a maximum net distance of 15 m between each pair of scissor braces in the middle.
(7) The connecting wall components are set between the scaffolding and the building at intervals of 4.5 m horizontally and 3.6 m vertically, and a cable-stayed wire rope is put every 6 m horizontally. The pull node is encrypted within the corner range and at the top, and pull nodes are placed every 3.6 m in the vertical direction and within 1 m of the corner.
(8) When laying scaffolding boards, boards with a board width of 200 mm must be placed between the inner rows of vertical poles and the structural layer, and the inner and outer poles must be covered with scaffolding boards without probing boards. A full-layer scaffold should be positioned horizontally and perpendicular to the wall and must cover the entire surface without voids. It is essential to adopt sufficient protection measures when it is not possible to cover the full coverage.

(9) During the installation of protective railings, 1.2-meter-tall protective railings and 30-centimeter-tall skirting poles must be placed on the side scaffolding, and there must be at least two 0.9- and 1.3-meter-tall protective railings in the top row.
(10) An approved green dense mesh safety net is attached to the inner side of the outer pole of the scaffold, which covers the outside of the scaffold. The safety net is suspended by 18 lead wires, and the requirements are stringent and uniform.

## 4 CONSTRUCTION TECHNOLOGY OF LARGE TREE-LIKE STEEL SUPPORT ROD BASED ON BIM TECHNOLOGY

### 4.1 *Application of BIM technology to the manufacturing and processing of tree-shaped steel support rods*

This project's steel tree-like support structure is comprised of columns, trunks, branches, crowns, gutters, steel castings, and other components. In the process of constructing steel support, the connection method of rods is complex, and it is necessary to improve the assembly efficiency and ensure the installation's overall appearance. Consequently, with the aid of BIM technology, the BIM model diagram of the steel support is created, and the processing list scale is automatically counted through the model to precisely process the components. Utilizing 6-axis plasma intersecting line CNC cutting technology, the factory processes many rods rapidly into finished goods. All rods use 6-axis plasma intersecting line CNC cutting technology in the factory to quickly process many rods into finished products. The on-site assembly utilizes full penetration welding, which greatly improves construction speed and quality, and performs finite element analysis on key structural nodes to ensure the steel structure's safety and stability.

### 4.2 *Installation and construction technology of large tree-shaped steel support rods*

#### 4.2.1 *Hoisting preparation*

**(1) Simulation of the construction site using BIM technology.** Construction simulation is performed based on the original building structure, including the installation position and rotation range of the tower crane on the construction site. This ensures that hoisting covers all construction areas, which facilitates the selection of tower crane equipment and the development of special tower crane construction plans.
**(2) The size and positioning of the steel support rod are reviewed.** Before installation, use a GPS total station and manual to conduct a comprehensive and detailed review of the size, elevation, and angle of the steel support rods that have been prepared on-site, and use BIM technology to carry out scientific and reasonable cutting, cutting, and processing, and then transport them to the site, overall hoisting to ensure the size and accuracy of the installation.
**(3) The steel support rod is processed and formed.** Through the accurate BIM model, the inventory and scale of the steel structure canopy's components are automatically counted and linked to the CNC cutting machine to precisely cut and mass-produce the rods.
**(4) The components are checked and accepted for storage.** After the components have been delivered to the installation site, they must be carefully inspected for their marks, quantity, and appearance quality, and a record of inspection and acceptance must be kept. The components should be stored in a specialized facility and organized by category.
**(5) Secondary component transport.** When transferring the components from the stacking site to the assembly site for the second time, a reasonable lifting point should be chosen and the necessary precautions should be taken to prevent component damage.

#### 4.2.2 *Positioning and paying out*

The GIS technology is used in the project, and the Trimbletx55 3D laser high-speed scanner scans obtain 3D coordinate data of the surface of the measured object with a large area and high resolution, collects spatial point information, and quickly establishes the object's 3D point cloud data model. The measurement efficiency has been drastically enhanced. Based on the 3D point cloud data model, a BIM model is developed. The steel structure canopy support column is aligned with the original building structure column, combined with 3D3S for structural safety analysis and calculation, repeated simulation adjustment, and optimization, to maximize the steel structure's rationalization. Through optimization analysis and calculation, the design value of each steel column's maximum bearing reaction force under each effect combination is determined. As depicted in Figures 4 and 5, a secondary reinforcement design will be conducted for the column that cannot meet the structural requirements based on the original building structure calculation.

Figure 4. 3D point cloud data model.

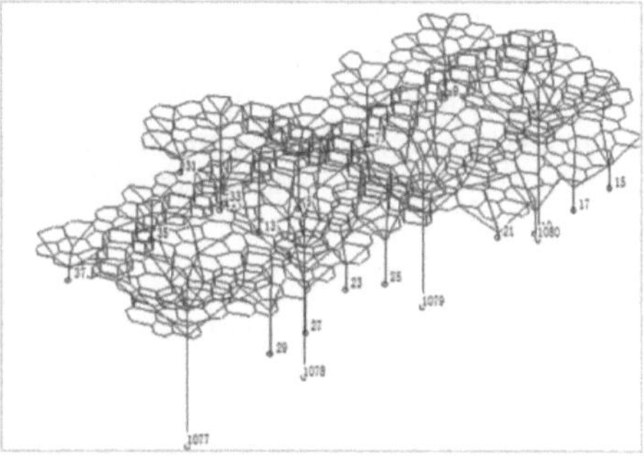

Figure 5. Calculation diagram of the large tree-like steel support rod.

#### 4.2.3 *Hoisting steel support rods*

(1) On-site hoisting sequence of steel support rods: typically install the 1–3# area with a large area first, followed by the corresponding adjacent areas based on the on-site delivery conditions. The hoisting of the steel support rod adheres to the principle "first the middle, then the outer side, from high to low, first the steel tree pole, then the steel tree branch and the steel branch, first down and then up," and five construction teams are arranged to work simultaneously.
(2) Selection and arrangement of lifting equipment: According to the site plan, a total of 9 construction areas need to install steel support rods. Combined with the analysis of its lifting performance and working conditions, 12 NK250 50t truck cranes are deployed on-site.
(3) Steel support rod hoisting process: positioning review, site acceptance → installation of steel tree poles → installation of steel branches → installation of steel tree branches → installation of steel tree crowns → installation of the next steel tree pole → and so on → installation Completion, structural acceptance.
(4) Main points of steel support rod hoisting
    a. The steel-supported assembly platform is constructed on-site, and the components are assembled and welded on the platform bench to ensure the components' precise dimensions.
    b. All steel components should be assembled and welded in a sequence that minimizes the generation of internal welding stresses.
    c. Sandblasting or mechanical rust removal is required for all steel components in the factory. The rust removal grade is 2.5 Sa. The surface is sprayed with two coats of epoxy zinc-rich primer and two coats of polyurethane topcoat with a total thickness of not less than 180 $\mu$m.
    d. The lifting process for four on-site assemblies and processing is performed by a rented 50 t truck crane or crawler crane. After assembling the steel pipe, it is hoisted using a 25-ton truck crane, and the process is coordinated with the erection of local scaffolding.

## 5 CONSTRUCTION TECHNOLOGY OF SELF-EXPANDING ETFE SKY CURTAIN MEMBRANE STRUCTURE

The area with the largest surface area should be installed first, followed by the adjacent area; installation should begin within the center and proceed outward to the sides, from high to low, depending on the site's delivery situation and the steel structure's work surface. On either side of the fire escape are two 50-ton construction cranes. A secondary transfer platform consists of a temporary steel structure and wooden frame of 10 × 10 m used to place the membrane structure, auxiliary materials, accessories, and construction equipment and tools required for the construction project.

### 5.1 *Self-expanding ETFE sky curtain membrane structure hoisting process*

Size review of steel support rods → erection of operating platform → incoming acceptance of membrane materials → installation of the integrated structure of film and canopy membrane sandwich water tank → ETFE canopy membrane material hoisting, secondary transfer in place → ETFE canopy membrane pensioning and fixing → canopy membrane installation of the electromagnet and related equipment → acceptance of the canopy membrane structure → commissioning, test inflation, and acceptance.

### 5.2 *Installation preparation of self-expanding ETFE sky curtain membrane structure*

(1) **Simulating a construction site using BIM technology.** Combined with the original building structure, the construction simulation of the tower crane position and cantilever on the construction site is conducted to ensure that the self-expanding ETFE sky curtain membrane structure is hoisted to cover the entire construction area. It is also convenient for the selection of tower crane equipment and the preparation of special construction plans for tower cranes.
(2) **Positioning via GPS or total station.** Use a GPS or total station to determine the horizontal position, size, and elevation of each node before installation, and check and adjust the site construction drawings and plans promptly to avoid membrane tension, wrinkling, and failure to blow out the set shape.
(3) **Docking and disclosure of technical information.** Provide excellent service in docking and verifying steel membrane structure drawings, processing, transportation, hoisting, cross-construction, overall hoisting, etc. Various technical disclosures, safety education, safety records, safety protection measures, as well as emergency planning are available.
(4) **The air film equipment installation platform is reserved.** Relevant professional units must reserve platforms and interfaces for the installation of air-film equipment in advance, arrange equipment lines and lightning protection installations in advance, and manage critical construction schedule nodes in strict accordance with the predetermined construction procedure.
(5) **Conform to the membrane's size structure.** Strictly adhere to the on-site main building structure and steel structure membrane structure model, carefully review the chord length and angle of each segmented area, and mark and label each cut piece and related auxiliary materials and accessories to prevent transportation and on-site errors, confusion, etc.
(6) **Examine the load of the operating platform.** A working platform that is convenient for the placement of construction materials and the installation of the membrane structure is erected following the design specifications, and the load of the scaffolding and platform erection is estimated to ensure that the platform remains within its safe load range.

### 5.3 *The installation method and main points of the self-expanding ETFE sky curtain membrane structure*

(1) First, transport the factory-processed membrane materials to the on-site warehouse for acceptance, and use the honeycomb steel structure as the fundamental unit for sorting, unfolding,

and reversing the membrane surface. From the center to the two sides, the order of hoisting is outward.

(2) The ETFE sky curtain film uses a honeycomb monomer as the processing base for the membrane structure and completes the production of the overall air pillow without inflating it. The inflatable and power pipelines and lines are connected and inflated multiple times after being transported to the site in batches to achieve the desired result.

(3) Compare the descriptions and illustrations of the on-site installation drawings of the membrane materials with the distribution of the diaphragms on the construction drawings, and mark the front and back, diagonal vertical, and membrane material numbers of each upper membrane material.

(4) According to the construction abortionist plan, hoist the membrane material from the lower yard to the operation platform, or stack it nearby in a location that facilitates secondary transfer. Within the scope of the operation surface of each construction area, it is planned to use two 50 t truck cranes to hoist the site's upper membrane structure using scaffolding and an operation platform. The location of the crane is determined on the basis that it can cover the membrane structure's working surface.

(5) The membrane structure is also installed from the center to the sides and from high to low. Install the upper ETFE canopy membrane structure and related equipment; the membrane structure's after-installation effect diagram is depicted in Figure 6.

Figure 6. Installation process and effect diagram of tree-like self-expanding ETFE sky curtain film.

## 6 CONCLUSIONS

The research, development, and application of the above-mentioned construction technology demonstrate that this technology can effectively solve the problems associated with the production and installation of large ETFE membrane structures supported by steel trees, optimism the process flow and construction technology, and achieve precise construction and installation. After completion, the effect of the decoration is satisfactory, the construction is straightforward, and the environment is unpolluted. The application of this technology improves the installation and construction efficiency of tree-like steel support rods and canopy membrane structures, shortens the construction period, reduces the waste of steel and membrane materials, saves energy, protects the environment, and has obvious benefits without increasing the membrane structure's cost or compromising its quality.


## ACKNOWLEDGEMENTS

This research was supported by the Sheen Zhukov Construction Decoration Engineering Co., Ltd., Chief Engineer Li Hui for providing engineering cases and suggestions during the writing process and the Guangdong Province General Colleges and Universities Young Innovative Talents Project (Project Title: Research on Green Construction and Virtual Simulation Technology Application of Industrialized PC Buildings in Complex Environment; Grant No: 2020KQNCX252).

## REFERENCES

Chen, Z. H., & Yan, X.Y. (2007). Two commonly used membrane materials PTFE and ETFE. *J. Architectural Knowledge*, (04): 64.

Huang, B. (2013). Application of ETFE membrane structure in Guangzhou South Railway Station J. *Urban Architecture*, (12): 56–57. DOI:10.3969/j.issn.1673-0232.2013.12.047.

Li, F. C., Wang, S., Han, X., Zhang, X. Y., & Li, Q. (2021). Construction technology of special-shaped high-altitude reverse-mounted skeleton-supported membrane structure. *J. Building Construction*, 2021, 43(10): 2073–2075. DOI:10.14144/j. cnki.jzsg.2021.10.030.

Li, N. (2020). *Design and analysis of skeleton-supported membrane structure*. D. Hebei University of Engineering. DOI: 10.27104/d.cnki.ghbjy.2020.000534.

Lian, K. (2020). *ETFE membrane structure curtain wall installation technology*. C//. Proceedings of the 26th East China Six Provinces and One City Civil Engineering Construction Technology Exchange Conference (Volume 2)., 2020: 282–285. DOI:10.26914 /c.cnkihy.2020.021319.

Lin, Q.M. (2015). Talking about the application of steel frame membrane structure installation technology in stadiums. *J. Doors and Windows*, (06):53–56.

Wu, Y., Yang, Q. S., & Shen, S. Z. (2014). Current Situation and Prospect of Membrane Structure Analysis Theory Research. *J. Engineering Mechanics*, 31(2): 1–14. doi:10.6052/j.issn.1000-4750 .2013.05.ST03 .

Xue, F., & Hou, Y. Q. (2010). Construction technology of the canopy project of Yuanrong Times Square. *J. Construction Technology*, 39(07): 44–47.

Yang, T., Ding, X., Yang, X. D., & Li, Q. (2019). Development Status and Trend of Membrane Structure Materials for Buildings. *J. Textile Herald*, (S1): 95–97. DOI:10.16481/j.cnki.ctl.2019 .s1.041.

Zhang, Q.L. (2009). Application and development of membrane structure system [J]. *World Architecture*, (10): 36–39. DOI:10.16414/j.wa.2009.10.012.

*Advances in Urban Construction and Management Engineering – Cha (Ed)*
*© 2023 Copyright the Author(s), ISBN 978-1-032-39018-5*

# Evaluation of basic public service level in Xining city based on entropy weight TOPSIS method

Yabo Wang* & Haifa Jia*
*College of Civil Engineering, Qinghai University, Xining, Qinghai, China*

ABSTRACT: To comprehensively measure the development level of basic public services in Xining City, Qinghai Province, and reveal its changing trend, the evaluation system for basic public services was constructed from these five aspects: basic education, public health, public culture, social security, and infrastructure. TOPSIS was used to comprehensively evaluate the level of basic public services in Xining City from 2012 to 2020. The results show that from 2012 to 2020, the level of basic public services in Xining City has seen an obvious upward trend, infrastructure and social security have increased significantly, basic education has been on a declining trend, and public cultural level and public health service level shows an upward trend. As a typical quantitative evaluation method, the TOPSIS entropy method can provide a reference for the evaluation of basic public services in other regions.

## 1 INTRODUCTION

For a long time, basic public service has always been an important issue in academic research. Adam Smith first proposed basic public service abroad, he pointed out that the state has the responsibility to ensure the fairness of public service (Collection of Circular Economy and Sustainable Development of Coal Industry 2006). On formulating *the 11th Five-year Plan for National Economic and Social Development*, the report of the Fifth Plenary Session of the 16th Central Committee of the Communist Party of China pointed out that "to actively promote social harmony, enrich the people's spiritual and civilized life, accelerate the improvement of the social security system". *The Twelfth Five-Year Plan*, adopted by the Fourth Session of the Eleventh National People's Congress in 2011, clearly pointed out that to establish and improve the basic public service system (Xiong et al. 2018). With the continuous growth of China's economy, the level of basic public services has been continuously improved, but the gap between regional basic public services has gradually widened, showing a trend of high in the east and low in the west. Han Xiaowei and Yin Luanyu (Han & Yin 2010) said that basic public services should include two meanings: basic and public services. Ma Guoxian (Ma 2007) pointed out that China's basic national conditions can meet the lowest public service standards. Wang Xiaobin and Shao Yanfei et al. used the coefficient of variation to analyze the current supply and demand situation and the existing problems of five basic public services in different regions of Jiangmen City and put forward policy suggestions to make the basic public services of Jiangmen City more reasonable. Lu Xiaojun and Zhang Xinyu (Lu & Zhang 2017) used the entropy method and spatial autocorrelation analysis to measure and analyze the regional differences and spatial patterns of basic public services in small and medium-sized cities from 2008 to 2013. It is concluded that in small and medium-sized cities in China, the overall level of basic public services is low, and there is still an obvious distribution pattern of high in the east and low in the west. Wang Xiaoling (2013) constructed a comprehensive evaluation index system of basic public services, including basic education service, basic medical service, social security service, public cultural service, ecological environment service, and public infrastructure service, to measure the basic public service level of all provinces and regions in China in 2011 by using

*Corresponding Authors: 312371353@qq.com and 153608568@qq.com

DOI 10.1201/9781003348023-40

entropy method. Wang Luyun and Wang Chongju constructed a comprehensive evaluation index system of basic public services from six aspects: education and cultural services, basic medical services, social security services, ecological environment services, public infrastructure services, and information services. The level and difference of basic public services in 12 provincial regions in the western region from 2005 to 2014 were measured, and the spatial pattern was analyzed. According to the existing research on the public service level, a conclusion can be drawn that most researchers choose education, health care, culture, and social security systems as the research factors. From the perspective of the research area, as the capital city of Qinghai Province, the research on the basic public service level of Xining City still has great deficiencies. Therefore, with a comprehensive evaluation index system constructed, this paper measures and analyzes the basic public service level of Xining City, which has important practical significance for the improvement of the basic public service level of Xining City and even for Qinghai Province.

## 2 OVERVIEW OF THE STUDY AREA

Xining, as the capital city of Qinghai Province, has a permanent population of 2.468 million by 2020, governing 4 districts (Chengdong District, Chengxi District, Chengzhong district, and Chengbei District) and 3 counties (Huangzhong County, Datong County, and Huangyuan county). The administrative area covers an area of 7606.75 square kilometers, with terrain high in the southwest and low in the northeast. Its average altitude is 2434.2 m, and its annual average temperature is 6.1°C.

## 3 RESEARCH METHODS AND DATA SOURCES

### 3.1 *TOPSIS entropy method*

The entropy method is an objective weighting method. When the dispersion degree of each evaluation index is large, the entropy value is small, indicating that the greater the impact of the index on the comprehensive evaluation is, the greater the weight is, and vice versa. Due to the wide application of the entropy weight TOPSIS method, this paper no longer elaborates on the formula in detail. The specific formula can be seen in Xiong Xing and Yu Xinghou's research (Xiong et al. 2018) on the measurement and influencing factors of regional equalization of basic public services in China.

### 3.2 *Data source*

The data for this paper comes from *Xining Statistical Yearbook* and *Qinghai Statistical Yearbook*; the study period is 2012–2020.

## 4 MEASUREMENT AND EVALUATION OF BASIC PUBLIC SERVICES IN XINING

### 4.1 *Build the index system of basic public service level*

The paper selects 23 indicators from five aspects based on the existing research and *the 12th Five-Year Plan for National Economic and Social Development of the People's Republic of China*: basic education, public culture, public health, social security, and infrastructure (Chen 2017). See Table 1 for more information.

### 4.2 *Determine the index weight*

The entropy method is used to calculate the weight of each index (generally retaining four decimals) as shown in Table 1.

Table 1. Indicator system of basic public services in Xining.

| Criterion layer | Indicator layer | Unit | Index types | First level index weight | Secondary index weight |
|---|---|---|---|---|---|
| Public health | Number of health institutions per 10,000 people | Number | P | 0.0350 | 0.0310 |
| | Number of beds per 10,000 people in hospitals and health centers | Sheet | P | | 0.0362 |
| | The number of health technicians per ten thousand people | Person | P | | 0.0347 |
| | The proportion of health expenditure in public budget expenditure | % | P | | 0.0379 |
| Elementary education | The number of students borne by full-time teachers in ordinary colleges and universities | Person | N | 0.0366 | 0.0507 |
| | Number of students borne by full-time teachers in general secondary schools | Person | N | | 0.0369 |
| | Number of students borne by full-time teachers in ordinary primary schools | Person | N | | 0.0179 |
| | The proportion of education expenditure to public budget expenditure | % | N | | 0.0161 |
| | The number of schools per ten thousand ordinary higher students | All/10,000 people | P | | 0.0295 |
| | Number of schools per 10,000 middle school students | All/10,000 people | P | | 0.0387 |
| | The number of schools per ten thousand pupils | All/10,000 people | P | | 0.0663 |
| Public culture | Number of books in public libraries per 10,000 people | Books/10,000 people | P | 0.0463 | 0.0768 |
| | The proportion of Cultural Sports and Media Expenditure in General Public Budget Expenditure | % | P | | 0.0382 |
| | Broadcasting population coverage | % | P | | 0.0241 |
| Social security | The proportion of the population participating in basic pension insurance | % | P | 0.0564 | 0.0748 |
| | The proportion of the population attending basic medical care | % | P | | 0.0856 |
| | The proportion of social security and employment expenditure in public budget expenditure | % | P | | 0.0329 |
| | Domestic building area per person | $m^2$/person | P | | 0.0323 |
| Infrastructure | Public green space area per capita | $m^2$/person | P | 0.0478 | 0.0209 |
| | Per capita urban road area | $m^2$ | P | | 0.0780 |
| | Water supply | tons | P | | 0.0145 |
| | Urban gas penetration rate | % | P | | 0.0425 |
| | Water use per capita living | Liters per day | N | | 0.0832 |

The weight calculation of the first-level indicators uses the arithmetic mean to average the weights of the second-level indicators (Xiong et al. 2019). It can be seen from Table 1 that the weight of social security at the level of basic public services is the largest, followed by infrastructure, public culture, basic education, and public health. This shows that social security is the basis of basic public services. With the improvement of people's living standards, people's requirements for infrastructure, public culture, and public health are getting higher and higher. However, due to the government's low investment in public culture and low per capita level, the weight of public culture is small. From the secondary indicators, the proportion of health expenditure in public budget expenditure is relatively large, and the weight of the number of health institutions per ten thousand people is small. In elementary education, the number of schools per ten thousand pupils accounts for a larger weight, and the proportion of education expenditure to public budget expenditure is relatively small. In public culture, the weight of the number of public library collections per ten thousand people is larger, and the weight of the coverage rate of the broadcast population is relatively small. In social security, the proportion of people who participate in basic medical care accounts for a large proportion of the population, and the proportion of per capita residential areas is relatively small. The weight of per capita domestic water consumption in infrastructure is larger, and the weight of water supply is relatively small.

## 5 COMPREHENSIVE EVALUATION OF BASIC PUBLIC SERVICES IN XINING CITY

According to the weight of the index system of basic public services in Table 1, the TOPSIS model is used to calculate the comprehensive score of the first-level index and the comprehensive score of basic public services. The calculation results are shown in Figure 1 and Figure 2.

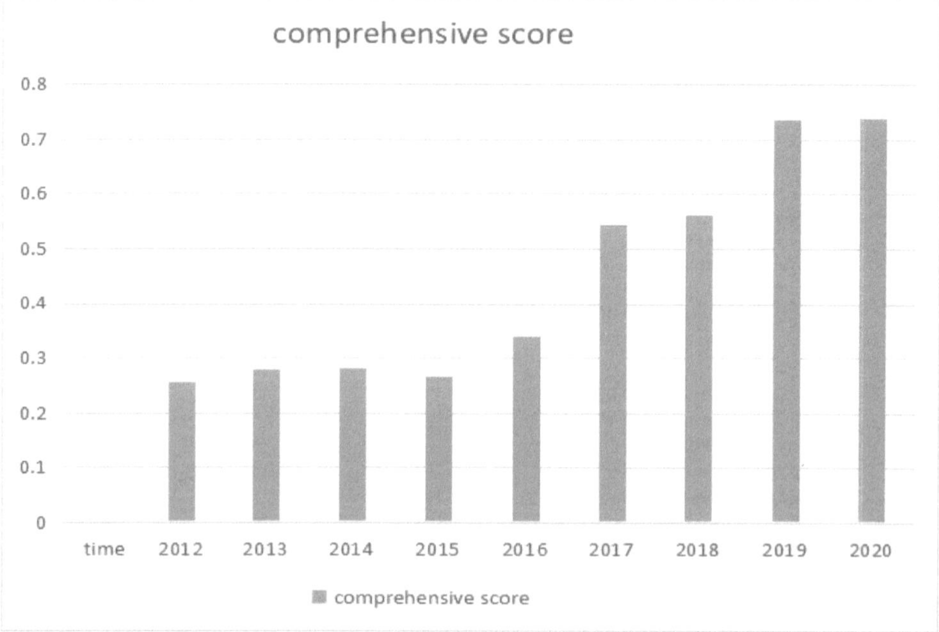


Figure 1. The comprehensive score of first-level indicators.

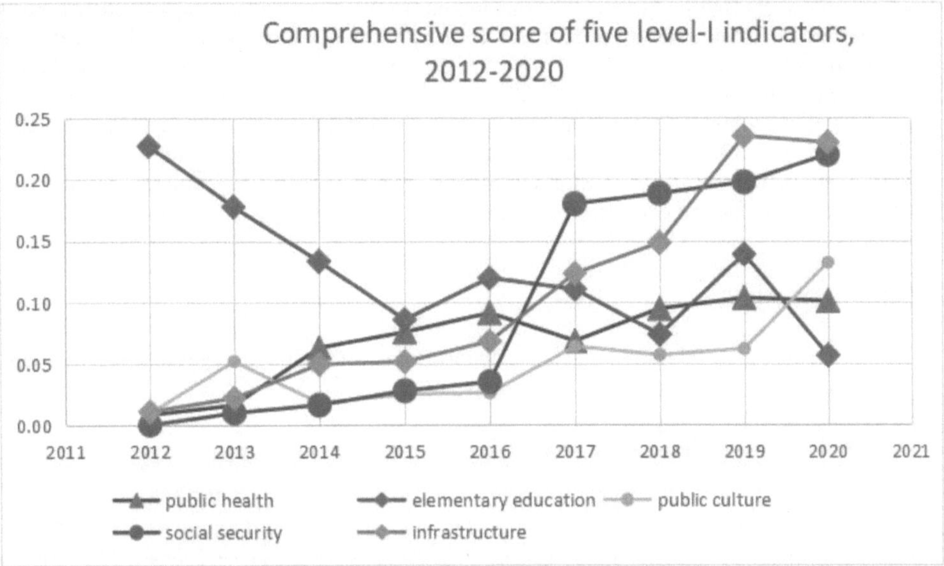


Figure 2. The comprehensive score of second-level indicators (2012–2020).

It can be seen from Figure 1 that the overall level of basic public services in Xining City has an upward trend from 2012 to 2020, and this trend is obvious. The comprehensive score has risen from 0.26 in 2012 to 0.74 in 2020. As seen in Figure 2, the composite score of public health increased from 0.01 in 2012 to 0.10 in 2020, due to the increase in the share of health expenditure in public budget expenditure from 0.07 in 2012 to 0.08 in 2020. The overall score of elementary education dropped from 0.23 to 0.06, showing an overall downward trend, but there was a turning point in 2015. The reason was that the number of students burdened by full-time teachers in ordinary primary schools dropped from 28.1 in 2012 to 26.05 in 2020 or 10,000 primary school students. The average number of schools dropped from 17.00 in 2012 to 8.63 in 2020. The comprehensive score of public culture rose from 0.01 to 0.13, and the upward trend was obvious because the number of books per 10,000 people rose from 49,612.89 to 92,504.05. The comprehensive score of social security increased from 0.00 to 0.22. The reason is that the proportion of people participating in basic medical care in the population increased from 0.11 to 0.73, and the proportion of social security and employment expenditure in public budget expenditure increased from 0.10 in 2012 to 2020. 0.16. The comprehensive score of infrastructure increased from 0.01 to 0.23, because the per capita domestic water consumption dropped from 193.12 L per day to 120.79 L per day. The level of public service in Xining is constantly improving.

## 6 CONCLUSIONS

Based on the TOPSIS entropy method, this paper selects 23 secondary indicators from five aspects of public health, basic education, public culture, social security, and infrastructure, and then a comprehensive evaluation index system modeled to comprehensively measure the basic public service level of Xining City from 2012 to 2020. The results show that the level of basic public services in Xining City has an overall upward trend and an obvious upward trend from 2012 to 2015, indicating that the government of Xining City has conscientiously implemented the Twelfth

Five-Year Plan, increased investment in pension, employment, medical care, and other aspects so that the level of basic public services has been significantly improved. Social security and infrastructure indicators score increased significantly; the score of basic education is declining, and the share of education expenditure in fiscal expenditure decreased from 0.23 in 2012 to 0.16 in 2020; the scores of public health and public cultural indicators are on the rise. It can be concluded that the overall level of basic public services in Xining is constantly improving, which indicates that the government has made great efforts to promote public services. In future construction, we should pay more attention to education and increase investment in education funds.

## ACKNOWLEDGMENT

The authors are very grateful for the financial support from the General Fund Project of Philosophy and Social Sciences of Qinghai Province (Grant No. 21005).

## REFERENCES

Adam Smith, Tang Risong, Yang Zhaoyu. *The Wealth of Nations* [M]. 1. Huaxia Publishing House: 430.

Chen Juan. *Research on the Equalization of Regional Basic Public Services and the Reform of Fiscal System* [D]. Jilin University, 2017.

Han Xiaowei, Yin Luanyu. Differentiation and Analysis of Basic Public Service Concepts[J]. *Jianghan Forum*, 2010(09): 42–44.

Lu Xiaojun, Zhang Xinyu. Regional Differences in Basic Public Service Levels in Small and Medium Cities in my country [J]. *Journal of Dalian University of Technology* (Social Science Edition), 2017, 38(01): 139–143.

Ma Guoxian. Research on Public Finance Policy of Equalization of Basic Public Services[J]. *Fiscal Research*, 2007(10): 74–77.

Suggestions of the Central Committee of the Communist Party of China on Formulating the Eleventh Five-Year Plan for National Economic and Social Development (Abstract) [C]. *Collection of Circular Economy and Sustainable Development of Coal Industry*, 2006.

Wang Luyun, Wang Chongju, Deng Lin. Research on the Level and Spatial Pattern of Basic Public Services in Western China [J]. *Western Forum*, 2016, 26(05): 51–60.

Wang Xiaobin, Shao Yanfei, Huang Minyi, et al. Research on the Evaluation of the Equalization Level of Basic Public Services in Jiangmen City [J]. *Journal of Wuyi University* (Social Science Edition), 2015, 17(04): 58–62+91-92.

Wang Xiaoling. The Basic Public Service Level and Regional Difference Analysis of Provinces and Regions in my country [J]. *Journal of Zhongnan University of Economics and Law*, 2013(03): 23–29+158-159.

Xiong Xing, Yu Xinghou, Pu Kunming. Comprehensive Evaluation and Spatial Analysis of Basic Public Services in the Yangtze River Economic Belt [J]. *East China Economic Management*, 2019, 33(01): 51–61.

Xiong Xing, Yu Xinghou, Wang Yuxin. Measurement and Influencing Factors of Equalization Level of Regional Basic Public Services in my country [J]. *Journal of Southwest University for Nationalities* (Humanities and Social Sciences Edition), 2018, 39(03): 108–116.

*Advances in Urban Construction and Management Engineering – Cha (Ed)*
 *ISBN 978-1-032-39018-5*

# Evaluation of the effects of street-stall economic governance from the perspective of good governance: Taking Hangzhou as an example

Qing Xu*, Jianbo Qiu*, Jia Zhou*, Yuefeng Li*, Jiayi Wu* & Zhenglan Lu*
*Department of Urban and Rural Planning, Zhejiang Gongshang University, Hangzhou, China*

ABSTRACT: Based on the concept of good governance, from the three dimensions of vendors, consumers, and governors, the paper summarizes ten elements, categorized as different players, of good governance to refine the index system for evaluating the effects of the street-stall economic governance. The evaluation of the effects of the street-stall economic governance in Hangzhou has led to the following conclusions: Vendors are the most important players in the street-stall economy, followed by consumers, and then governors; flexible enforcement represents a significant concern; consumers prioritize convenience over stability. Hangzhou has made headway in the street-stall economic governance, which initially leads to good governance.

## 1 INTRODUCTION

With the three advantages of low threshold, low risk, and low commodity prices (Peng 2021), the street-stall economy quickly drove employment in 2020 when the epidemic was stably controlled, ensuring the basic needs of people at the community level. Therefore, it is in line with the requirements of expanding employment proposed in the "14th Five-Year Plan". However, the current model for street-stall economic governance is too outdated to give full play to its function of ensuring people's livelihood and stimulating domestic demands. Therefore, how to improve the governance model of the street-stall economy has attracted scholars' attention (Cui 2019; Sun 2021). As a highly inclusive informal economy with a wide range of transactions as well as commodities and random trading venues (Xiao 2021), the street-stall economy can help many migrant workers and local people at the bottom to find jobs, broaden channels for their income, and make life easier for surrounding residents, thus promoting social stability (Maneepong 2013; Sun 2021; Wang 2019, 2018). At present, research on the street-stall economy is mainly focused on its public relations, and economic functions, including its ability to coordinate with the formal economic sector, welfare organizations, and street-stall operators (Martínez 2017); vendors' ability to propel the development of street-stall economy (Steel 2012), economic role in the fields of culture, tourism, supply, etc. (Li 2019; Sekhani 2019). However, China's vast territory, regional differences, and complex population composition have led to the complexity of the research on the street-stall economy (Meng 2021). Finding general laws from the complexity is of great significance for promoting the development of the community market and ensuring people's basic needs.

Influenced by governance ideas in recent years (Yu 2001), the management mode of the stalled economy has changed from "rule" to "governance". The current research focuses on the dimension of power, the identity and social status of personnel working on the ground, and the participation of non-governmental organizations, games, and systems (Fang 2013; Langford 2019; Liu 2021; Lyons 2013; Ye 2016), etc. Many research emphasizes the transformation of community market management from rule to governance, but the research on the shift from governance to good governance is lacking. At present, research on good governance revolves around public management (Yu 2002). As little research focuses on community market governance, they mainly stress theories

*Corresponding Authors: xuqing6886@163.com, q15858802093@163.com, 1397458487@qq.com, 1827122651@qq.com, 1651050633@qq.com and zlluhz@mail.zjgsu.edu.cn

DOI 10.1201/9781003348023-41

rather than applications. Therefore, this paper intends to integrate methods for the evaluation index system drawn by other researchers (Shen et al. 2011; Yu 2021, 2020). Ten elements of good governance are categorized as second-level indicators including legitimacy, transparency, accountability, responsiveness, effectiveness, rule of law, engagement, stability, and justice (Yu 2001). The paper aims to build an index system for evaluating the street-stall economic governance from the perspective of good governance and explore the development direction of the street-stall economy through the empirical research on the street-stall economy in downtown Hangzhou, thus providing the theoretical guidance of good governance for the development of the street-stall economy, as well as the case reference for the current research on the application of good governance.

## 2 OVERVIEW OF HANGZHOU AND RESEARCH METHODOLOGY

### 2.1 *Overview of Hangzhou*

Hangzhou, a key city of the Yangtze River Delta, is one of the important economic centers in eastern China. Guided by the Notice on the Building of Pilot Cities of Provincial Nighttime Economy, the street-stall economy in Hangzhou has developed rapidly, and night markets featured by the Wushan Night Market for neighboring residents and the Wulin Night Market mainly for foreign tourists have become symbols of the street-stall economy in the downtown area, with new night markets emerging including the West Lake Night Market and the Kerry Center Night Market. The basic data of this study were collected from the questionnaires and interviews of street-stall vendors, management agencies, and customers in Hangzhou from July to December 2021, and the index system for evaluating the effects of the street-stall economy governance has been constructed based on the data by the Delphi method and hierarchical analysis to evaluate the street-stall economic governance in Hangzhou.

### 2.2 *Construction of the evaluation index system*

Based on the research of Yu Keping et al., the ten management elements of good governance are analyzed and categorized as different players from the three dimensions of governors, consumers, and vendors in the street-stall economy, and an index system for evaluating the effects of street-stall economy governance is constructed (Table 1).

Table 1. Index system for evaluating the effects of the street-stall economy governance.

| First-level indicators | Second-level indicators | Third-level indicator | Definitions |
|---|---|---|---|
| Vendors (V) | Vendors' engagement (v1) | Stall cost | Whether vendors are satisfied with the explicit costs (mainly capital) required to operate the stalls |
| | | Income from stalls | Whether vendors are satisfied with the income from the stalls |
| | Qualification legitimacy (v2) | Stall time | Whether vendors are satisfied with the specified time of opening stalls |
| | | Stall design | Whether vendors are satisfied with the stall design |
| | | Access requirements | Whether vendors are satisfied with the access requirements for entering the gathering area of street stalls |
| | | Information registration | Whether vendors are satisfied with the convenience of information registration of opening the street stalls |
| | Policy justice (v3) | Public measures | Whether vendors are satisfied with the governor's public measures for managing the street stall |
| | | Rigid regulations | Whether vendors are satisfied with the extent to which the governor follows the management system for the street stall |
| | | Flexible enforcement | Whether vendors are satisfied with the way the governor enforces the law |

(*continued*)

Table 1. Continued.

| First-level indicators | Second-level indicators | Third-level indicator | Definitions |
|---|---|---|---|
| | Convenience (c1) | Location planning | Whether consumers are satisfied with the plan of street stalls |
| Consumers (C) | | Consumption distance | Whether consumers are satisfied with the distance from the consumers' houses to street stalls |
| | | Transportation convenience | Whether consumers are satisfied with the convenience of transportation from their houses to street stalls |
| Validity (c2) | Commodity quality | Whether consumers are satisfied with the quality of goods in street stalls | |
| | | Commodity prices | Whether consumers are satisfied with the price of goods on street stalls |
| | | Commodity categories | Whether consumers are satisfied with the variety of goods on street stalls |
| | | Environment satisfaction | Whether consumers are satisfied with the consumer environment of street stalls |
| | | Cultural features | Whether consumers are satisfied with the characteristic cultural atmosphere of street stalls |
| | Stability (g1) | Market order | Whether governors are satisfied with the measures to maintain the order of the street stalls |
| | | Environment measures | Whether governors are satisfied with the measures to improve the environment around the street stalls |
| Governors (G) | | Vendors' quality | Whether governors are satisfied with the qualities of the vendors in the business process |
| | Responsiveness (g2) | Association's work | Whether governors are satisfied with the work of the Street Stall Autonomous Association |
| | | Consultation mechanism | Whether governors are satisfied with the communication mechanism to resolve disputes in the street-stall economy |
| | | Technology | Whether governors are satisfied with the technology in the management when enforcing laws |
| | Accountability (g3) | Financial support | Whether governors are satisfied with the loan support and public goods provided to vendors |
| | | Input cost | Whether governors are satisfied with the input cost when managing street stalls |

Vendors (V): Vendors refer to those who do not have a fixed store but sell products at approved locations and within specified hours, while vendors' engagement (V1) serves as a prerequisite for evaluating the degree of good governance of the street-stall economy. The more satisfied street-stall vendors are with the number of costs spent on operation and the income received, the more likely they are to engage in the street-stall economy which holds greater appeal to them. The qualification legitimacy (V2) reflects the legal cost of street stalls. The more satisfied vendors are with the access requirements and the convenience, the more likely they are to accept the legal cost of street stalls. The justice of the policy (V3) indicates vendors' satisfaction with governors, which includes public policy, the regularity of law enforcement, and the specific forms of law enforcement. The more satisfied vendors are, the more likely they are to accept the governors.

Consumers (C): Consumers refer to individuals who purchase products and services for personal consumption or to users of final products, whose opinions are important feedback on the governance of the street-stall economy. Convenience (C1) is the main factor that affects consumers' purchases.

The closer the street stalls are to consumers' houses and the more convenient the transportation is, the more likely consumers are to make purchases, which also indicates how reasonable the location planning of the street stall is. Validity (C2), another important factor that determines whether consumers will purchase products again, is mainly influenced by five factors: commodity prices, commodity quality, commodity categories, consumer environment, and environmental and cultural characteristics. The more satisfied consumers are, the more willing they are to purchase products again.

Governors (G): Governors refer to those who, with their position and knowledge, shoulder the responsibility of the street stalls, and are capable of substantially influencing vendors' operations and achieving effectiveness. Considering the policy changes during the epidemic, the indicator of stability (G1) is divided into the market order and environmental measures, and the more satisfied governors are, the more effective the measures are in this environment. Responsiveness (G2) is the governors' evaluation indicator of the existing civil governance and individual organizations, including vendors' quality, associations' work, and consultation mechanism. The more-educated vendors are, the more satisfied governors are with the associations' work and consultation mechanism. Technology input is the evaluation of the new management tools under the background of "Internet plus"; the accountability (G3) is the governors' evaluation indicator of the higher government's expenditure on the management of the street-stall economy, mainly divided into financial input and input of public service facilities. The more satisfied governors are, the more responsible they will be.

### 2.3 *Weight computing of the evaluation index system and testing methods*

This paper, based on the analysis of hierarchical weight computing conducted through the Analytic Hierarchy Process (AHP), aims to propose an index system for evaluating the effects of street-stall economy governance. The first-level indicators are related to targets evaluated from the perspective of vendors, consumers, and governors; the second-level indicators are revolved around criteria including decisions or evaluation indicators; the third-level indicators are concerned with measures including objects to be evaluated which in this paper refer to the street-stall economy in downtown Hangzhou (including Shangcheng District, Xiacheng District, and Gongshu District).

After the evaluation system of analytic hierarchy is built, the fuzzy quantification method of qualitative indicators is used to measure the relative importance of each indicator to the corresponding indicator of the previous level to set the weights of a single ranking and calculate the total weights and ranking of each indicator in the whole hierarchical evaluation system. The method of solving the eigenvector of judgment matrix is used to calculate the importance or priority weight of each indicator of each level to the indicator of the previous level to determine the rank and weight according to its importance. The judgment matrix proposed in this paper is based on Table 2 to determine the importance and assigned values of each indicator.

Table 2. Criteria of importance in pairwise comparisons.

| Importance | Definition |
|---|---|
| 1 | Both compared factors are equally important |
| 3 | The former factor is slightly more important than the latter |
| 5 | The former is more important than the latter |
| 7 | The former is very important to the latter when the two factors are compared |
| 9 | When the two factors are compared, the former is the most important to the latter |
| 2, 4, 6, and 8 | These factors are in the intermediate level of importance compared to the above judgments |
| Reciprocal | If the ratio of the importance of factor *i* to factor *j* is *a*, then the ratio of the importance of factor *j* to factor *i* is 1/*a*. |

After the previously described matrix is created, it is also necessary to determine whether it is a consistent matrix, so a consistency test is needed to obtain the CI and CR which are calculated as follows.

$$CI = \frac{\lambda_{max} - n}{n - 1} \tag{1}$$

The larger the value of CI is obtained from the above equation, the more inconsistent it is. If the value is close to 0, it means better consistency; if the value is equal to 0, there is complete consistency. The more consistent it is, the more credible the weight judgment is. In addition, RI, a random consistency indicator, is a necessary one for calculating CR, and the standard values of the average RI are shown in the following table.

Table 3. Standard value of average random consistency indicator *RI*.

| n | 1 | 2 | 3 | 4 | 5 | 6 | 7 | 8 | 9 | 10 | 11 | 12 | 13 | 14 |
|---|---|---|---|---|---|---|---|---|---|---|---|---|---|---|
| RI | 0 | 0 | 0.52 | 0.89 | 1.12 | 1.24 | 1.36 | 1.41 | 1.46 | 1.49 | 1.52 | 1.54 | 1.56 | 1.58 |

When the consistency is calculated, the final value of CR prevails, and the calculation of CR is based on CI and RI, which are calculated as follows.

$$CR = \frac{CI}{RI} \tag{2}$$

According to the calculation result of CR, with 0.10 as the standard, if the value of CR is less than 0.10, it indicates that the consistency of the judgment matrix is reasonable and thus passes the consistency test. After the judgment matrix and weight calculation method are used, in this paper, the weights of indicators of each tier are obtained from the calculation of EXPERT CHOICE software.

## 3 EVALUATION RESULTS AND ANALYSIS

### 3.1 *Analysis and results of the weights of the evaluation index system*

In this paper, based on Yu Keping and other scholars' insights on good governance, an index system for evaluating the effects of street-stall economy governance has been constructed, with a total of 36 indicators in three levels. In the following section, weights are assigned to the 36 indicators according to their relative importance by the Delphi method, and their single ranking of importance at that level as well as the total ranking for the target level are determined according to the weights.

Based on the constructed evaluation index system, in this paper, the importance of 36 indicators is evaluated by 16 experts and practitioners in related fields. To consider all perspectives and operations, these experts fall into the categories of street-stall consumers, vendors, economic researchers, and governors, with four experts from each category. With the help of the Delphi method, a questionnaire for experts was created based on the index system mentioned above. Sixteen experts compared the importance of each indicator in the evaluation system in pairs according to the judgment criteria in Table 2 and the results were averaged to produce the final judgment matrix. Based on the calculation results of the Delphi method, in this paper, EXPERT CHOICE software is used to calculate the importance values of each hierarchical indicator, i.e., the weight of each indicator. The following table shows the results of weight calculation and the results of the consistency test.

The results reveal that the importance scores of vendors, consumers, and governors, relative to the target level in the evaluation model of the innovation management of the street-stall economy, are 0.512, 0.308, and 0.18, respectively. As the value of the CR index in the consistency test is

0.0049 which is less than 0.1, it passes the consistency test. This indicates that concerning the evaluation of the innovative management model of the street-stall economy, vendors are the most important players, followed by the consumers and governors. In addition, it means that in the street-stall industry, the top priority is that vendors provide better services. Consumers' demands drive the development of the industry, while governors provide guarantees for maintaining sound order.

Table 4. Importance (weights) results of the first-level evaluation indicators relative to the target level.

| First-level indicators | Weights | Consistency test |
|---|---|---|
| Vendors | 0.512 | CR = 0.0049 |
| Consumers | 0.308 | CR < 0.1 |
| Governors | 0.180 | |

Table 5 shows the importance of the second-level evaluation indicators relative to the first-level indicators to which they belong, and the weight values are 0.139, 0.305, and 0.556. For the value of CR is 0.00075, which is less than 0.1, this means that vendors and justice of policy management are the most important indicators followed by qualification legitimacy, and finally the vendors' engagement. Among vendors, consumers, and governors, as the vendors are very important, specifically their engagement, they are encouraged to engage in the street-stall economy industry. However, compared to vendors' engagement, the justice of policy management and the requirements of qualification legitimacy for them are more important. Regarding consumers as the first-level indicator, the convenience of consumption which promotes the effectiveness of consumption gains more importance. For the governors, the stable performance of the street-stall economy is the most important indicator followed by the vendors' responsiveness, and then their responsibility.

Table 5. Importance (weights) results of second-level evaluation indicators relative to the first-level indicators.

| First-level indicators | Second-level indicators | Importance (weights) | Consistency test |
|---|---|---|---|
| Vendors | Engagement | 0.139 | CR = 0.00075 |
| | Qualification legitimacy | 0.305 | CR < 0.1 |
| | Justice of policy management | 0.556 | |
| Consumers | Convenience of consumption | 0.628 | CR = 0 |
| | Validity of consumption | 0.372 | CR < 0.1 |
| Governors | Stability of street stalls | 0.538 | CR = 0.00543 |
| | Vendors' responsiveness | 0.293 | CR < 0.1 |
| | Responsibility | 0.168 | |

In the results of the importance (weights) of the three-level evaluation indicators relative to the second-level indicators to which they belong, all the CR values are less than 0.1, which means all of them have passed the consistency test. In terms of weights, compared with engagement as the second-level indicator, the importance of the third-level indicators is in order of income, cost, design, and time. This means that the fundamental factors affecting the vendors' engagement are income, cost, and then design as well as time. For the qualification legitimacy, experts reach a consensus that access requirements are the most important one, followed by the convenience of information registration. Concerning the justice of policy management, the most important indicator is flexible enforcement, while the less important ones are public measures and rigid policies, which meet the requirements for "good governance".

Concerning the convenience of consumption, the order of importance is respectively consumption distance, location norms, and transportation convenience. For the validity of consumption, experts believe that consumers take the price of goods and quality as the most important indicators followed by the types of goods and cultural characteristics. The reason behind this is that although street-stall consumption involves cultural components, it is targeted at the low-end consumer market, so consumers have lower expectations.

For the stability of the street-stall economy, governors feel that measures to maintain market order are the most important indicator followed by environmental measures such as hygiene measures. The most important indicator in terms of vendors' responsiveness is the consultation mechanism followed by vendors' quality, the work of industry associations, and the input of technological tools in management. In terms of governors' responsibility, their input in management is very important. However, because governors are not on the input side of the street-stall economy market, the economic benefits brought by the development of the street-stall economy have less impact on them; thus, their financial support for the development of the street-stall economy is less important.

Table 6. Results of the importance (weights) of the third-level evaluation indicators relative to the second-level indicators.

| Second-level indicators | Third-level indicators | Importance (weights) | Consistency test |
|---|---|---|---|
| Vendors' engagement | Stall cost<br>Income from stalls<br>Stall time<br>Stall design | 0.282<br>0.415<br>0.114<br>0.188 | CR = 0.02<br>CR < 0.1 |
| Vendors' qualification legitimacy | Access requirements<br>Information registration | 0.692<br>0.308 | CR = 0<br>CR < 0.1 |
| Policy justice | Public measures<br>Rigid policy<br>Flexible enforcement | 0.288<br>0.178<br>0.534 | CR = 0.02<br>CR < 0.1 |
| Consumption convenience | Location planning<br>Consumption distance<br>Transportation convenience | 0.275<br>0.566<br>0.159 | CR = 0.01<br>CR < 0.1 |
| Consumption validity | Commodity quality<br>Commodity price<br>Commodity categories<br>Environment satisfaction<br>Cultural features | 0.25<br>0.36<br>0.184<br>0.081<br>0.124 | CR = 0.02<br>CR < 0.1 |
| Stability of street stalls | Market order measures<br>Environmental measures | 0.682<br>0.318 | CR = 0<br>CR < 0.1 |
| Vendors' responsiveness | Vendors' quality<br>Association's work<br>Consultation mechanism<br>Technology | 0.256<br>0.177<br>0.453<br>0.115 | CR = 0.01<br>CR < 0.1 |
| Responsibility | Financial support<br>Input cost | 0.315<br>0.685 | CR = 0<br>CR < 0.1 |

Table 6 shows the importance weighting of the third-level indicators relative to the target level. As all CR values are less than 0.1, all pass the consistency test. For the evaluation of the innovative management model of the street-stall economy, the most important indicator is the flexible enforcement with an importance weight of 0.15, followed by consumption distance (0.566), access requirements (0.692), public measures (0.081), commodity prices (0.061), and market order measures (0.051). These indicators reveal that consumers care more about such factors as distance and price, while vendors focus on flexible enforcement, access requirements, and public measures, and governors emphasize measures to maintain market order.

Table 7. Results of importance weights of third-level evaluation indicators relative to the target level.

| Serial number | Variables (no order of importance) | Importance (weights) | Variables (in order of importance) | Importance (weights) |
|---|---|---|---|---|
| 1 | Stall cost | 0.026 | Enforcement flexibility | 0.15 |
| 2 | Income from stalls | 0.038 | Consumption distance | 0.102 |
| 3 | Stall time | 0.01 | Access requirements | 0.083 |
| 4 | Stall design | 0.017 | Public measures | 0.081 |
| 5 | Access requirements | 0.083 | Commodity price | 0.061 |
| 6 | Information registration | 0.037 | Market order measures | 0.051 |
| 7 | Public measures | 0.081 | Rigid policy | 0.05 |
| 8 | Rigid policy | 0.05 | Location planning | 0.05 |
| 9 | Flexible enforcement | 0.15 | Commodity quality | 0.042 |
| 10 | Location planning | 0.05 | Income from stalls | 0.038 |
| 11 | Consumption distance | 0.102 | Information registration | 0.037 |
| 12 | Transportation convenience | 0.029 | Commodity categories | 0.031 |
| 13 | Commodity quality | 0.042 | Transportation convenience | 0.029 |
| 14 | Commodity price | 0.061 | Consultation mechanism | 0.028 |
| 15 | Commodity categories | 0.031 | Stall cost | 0.026 |
| 16 | Environment satisfaction | 0.014 | Environmental measures | 0.024 |
| 17 | Cultural features | 0.021 | Cultural features | 0.021 |
| 18 | Market order measures | 0.051 | Stall design | 0.017 |
| 19 | Environmental measures | 0.024 | Vendors' quality | 0.016 |
| 20 | Vendors' quality | 0.016 | Input expenses | 0.016 |
| 21 | Association's work | 0.011 | Environment satisfaction | 0.014 |
| 22 | Consultation mechanism | 0.028 | Association's work | 0.011 |
| 23 | Technology | 0.007 | Stall time | 0.01 |
| 24 | Financial support | 0.007 | Technology | 0.007 |
| 25 | Input cost | 0.016 | Financial support | 0.007 |

### 3.2 *Results of the empirical study on the street-stall economy in Hangzhou*

In this study, based on the evaluation system and weights proposed in the previous section, three types of questionnaires are respectively targeted at vendors, consumers, and governors, and an empirical study was conducted which lasted five months from July to December 2021.

Among the questionnaires, concerning vendors' engagement, qualification legitimacy, and public policy, nine questions were designed to include such indicators as stall cost, income, time of stall appearance, stall design, access requirements, the convenience of information registration, public policy, rigid regulations, and flexible enforcement. With a total of 563 questionnaires distributed to the street-stall vendors in Hangzhou, 156 questionnaires were available for analysis, accounting for 27.71% of the total questionnaires. Of these 156 analyzable questionnaires, 89 (57%) were male and 67 (43%) were female vendors.

According to the vendors' questionnaire results, the responses to all indicators are assigned mean values, with the most important indicator being five and the least important one being one. As seen from the mean value results, vendors are content with the stall design in Hangzhou, and care less about stall time, income from stalls, access requirements, the convenience of information registration, rigid policy, and flexible enforcement, but they are dissatisfied with the operating costs of stalls and the public measures.

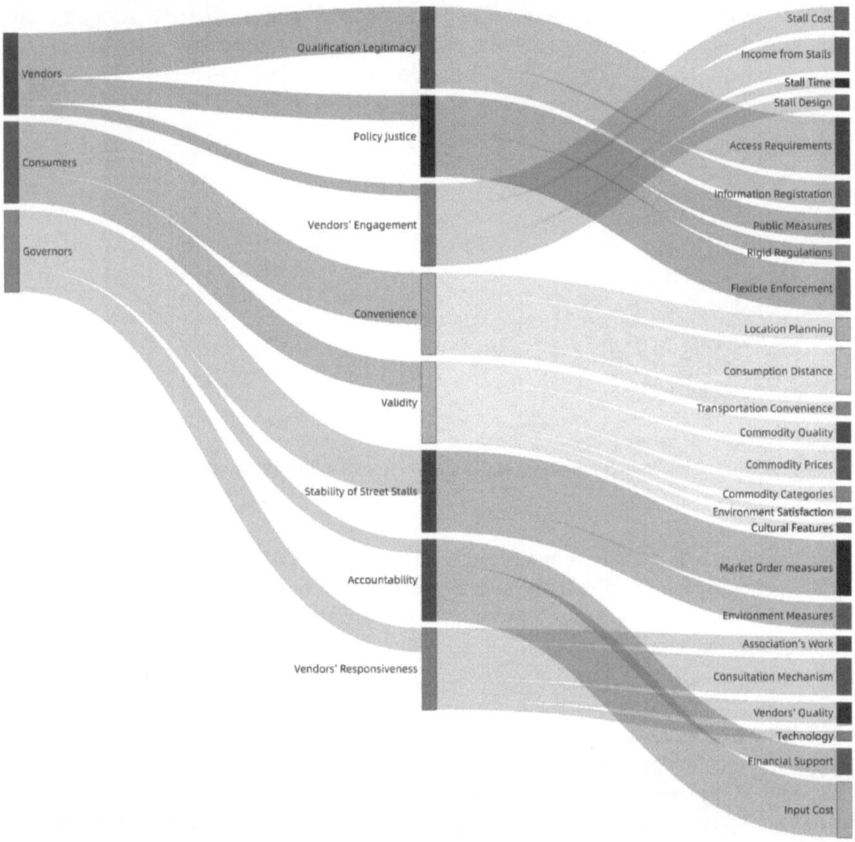


Figure 1. Map of the importance of the evaluation indicators of the effect of the street-stall economic governance.

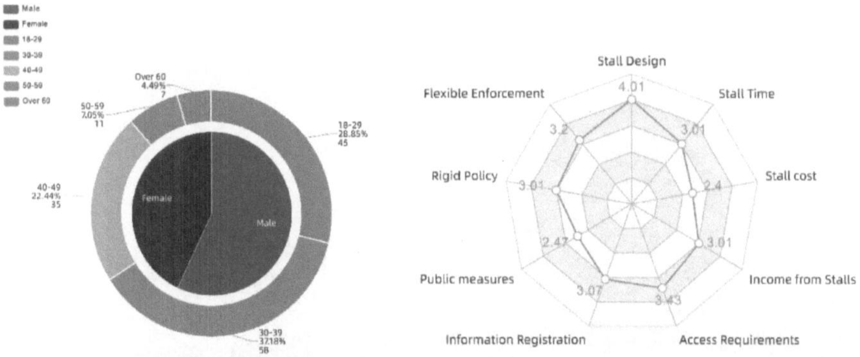


Figure 2. Population portrait of street-stall vendors and the mean value of their scores on each indicator.

For the questionnaire of the street-stall consumers, the convenience and validity of consumption, eight questionnaire questions were designed to target eight indicators including location planning, consumption distance, transportation convenience, commodity quality, commodity prices, commodity categories, environment satisfaction, and cultural features. With 1,937 questionnaires distributed in Hangzhou and aimed at the consumers of street stalls, a total of 921 analyzable questionnaires were obtained, with a questionnaire recovery rate of 47.86%. Among them, 516 consumers were female, accounting for 56%, and 405 were male, accounting for 44%.

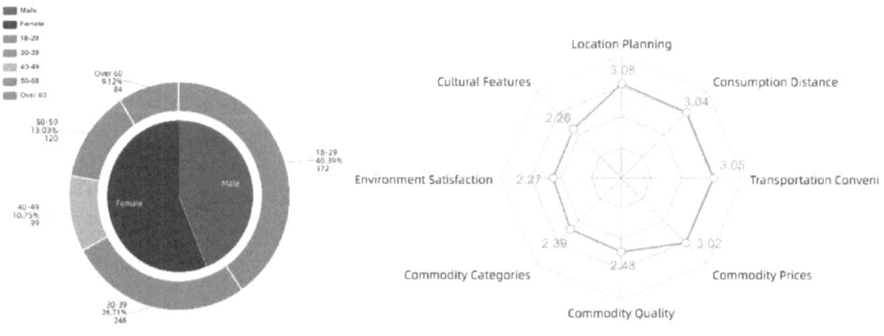


Figure 3. Portrait of the consumers of street stalls and the mean value of the scores for each indicator.

The overall results of the questionnaire targeted at street-stall consumers in Figure 3 show that they are generally satisfied with the location planning, consumption distance, transportation convenience, and commodity prices of street stalls in Hangzhou, but they are dissatisfied with the commodity quality, categories, environment, and cultural characteristics.

Finally, a questionnaire was aimed at governors of street stalls. Concerning stability, responsiveness, and accountability, eight questionnaire questions were designed to include eight indicators: market order measures, environmental measures, vendors' quality, association's work, consultation mechanism, technology, financial support, and input cost. Because of the limited number of governors, only 112 questionnaires were distributed in the whole city of Hangzhou, and a total of 32 questionnaires were available for analysis, accounting for 29.09% of the total number. Of these 32 analyzable questionnaires, 8 governors were female, accounting for 25%, and 24 were male, accounting for 75%.

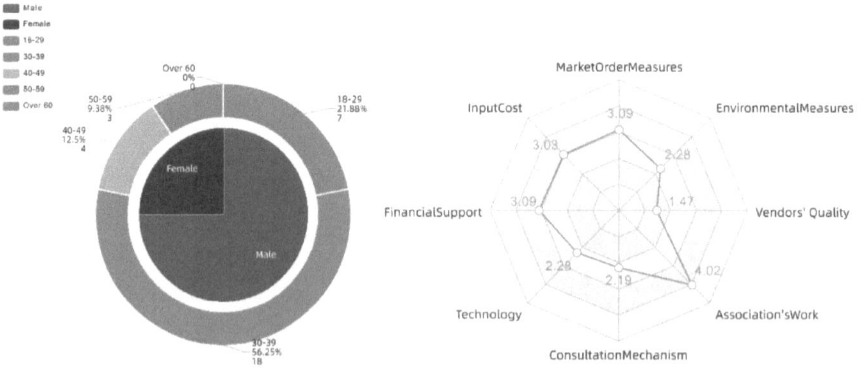


Figure 4. Population portrait of street-stall governors and the mean value of scores on each indicator.

From the results of the questionnaire for governors in Figure 4, they are relatively satisfied with the work of the street-stall association, believing that the vendors' quality needs to be improved, but they are less satisfied with the measures of environmental management, consultation mechanism, and investment in scientific and technological means.

### 3.3 *Evaluation of the effects of street-stall economic governance in Hangzhou*

Table 8 shows the total score of the innovative management model for evaluating the street-stall economy in Hangzhou and the individual scores of each indicator on three levels. As seen from the table, the total weighted score of the innovative management model for evaluating the street-stall economy in Hangzhou is 2.89, which means that the street-stall economic governance in Hangzhou is in the middle to the upper level, and there is still some room for improvement.

As can be seen in the individual scores, the flexible enforcement scores are the highest, which shows that people are satisfied with this indicator. Following flexible enforcement are indicators of consumption distance, access requirements, and public measures. However, the scores of other indicators reveal that to develop the street-stall economy in Hangzhou, investment in science and technology and financial support must be increased, while certain training must be provided to vendors to improve the way they conduct business.

Table 8. Total evaluation score of governance effects in Hangzhou.

| Indicators | Evaluation score | Weights | The total score of indicators | Ranking of indicators | The total score of indicators |
|---|---|---|---|---|---|
| Stall design | 4.00641 | 0.017 | 0.068109 | Flexible enforcement | 0.452885 |
| Stall time | 3.00641 | 0.01 | 0.030064 | Consumption distance | 0.309987 |
| Stall cost | 2.403846 | 0.026 | 0.0625 | Access requirements | 0.284647 |
| Income from stalls | 3.00641 | 0.038 | 0.114244 | Public measures | 0.200423 |
| Access requirements | 3.429487 | 0.083 | 0.284647 | Commodity prices | 0.183993 |
| Information registration | 3.00641 | 0.037 | 0.111237 | Market order measures | 0.157781 |
| Public measures | 2.474359 | 0.081 | 0.200423 | Location planning | 0.154072 |
| Rigid policy | 3.012821 | 0.05 | 0.150641 | Rigid policy | 0.150641 |
| Flexible enforcement | 3.019231 | 0.15 | 0.452885 | Income from stalls | 0.114244 |
| Location planning | 3.081433 | 0.05 | 0.154072 | Information registration | 0.111237 |
| Consumption distance | 3.039088 | 0.102 | 0.309987 | Commodity quality | 0.101922 |
| Transportation convenience | 3.045603 | 0.029 | 0.088322 | Transportation convenience | 0.088322 |
| Commodity prices | 3.016287 | 0.061 | 0.183993 | Commodity categories | 0.074117 |
| Commodity quality | 2.42671 | 0.042 | 0.101922 | Stall design | 0.068109 |
| Commodity categories | 2.390879 | 0.031 | 0.074117 | Stall cost | 0.0625 |
| Environment satisfaction | 2.270358 | 0.014 | 0.031785 | Consultation mechanism | 0.06125 |
| Cultural features | 2.259501 | 0.021 | 0.04745 | Environmental measures | 0.05475 |
| Market order measures | 3.09375 | 0.051 | 0.157781 | Input cost | 0.0485 |
| Environmental measures | 2.28125 | 0.024 | 0.05475 | Cultural features | 0.04745 |
| Vendors' quality | 1.46875 | 0.016 | 0.0235 | Association's work | 0.044344 |
| Association's work | 4.03125 | 0.011 | 0.044344 | Environment satisfaction | 0.031785 |
| Consultation mechanism | 2.1875 | 0.028 | 0.06125 | Stall time | 0.030064 |
| Technology | 2.28125 | 0.007 | 0.015969 | Vendors' quality | 0.0235 |
| Financial support | 3.09375 | 0.007 | 0.021656 | Financial support | 0.021656 |
| Input cost | 3.03125 | 0.016 | 0.0485 | Technology | 0.015969 |
| The total score of Hangzhou | | | 2.894148 | | 2.894148 |

## 4 DISCUSSION

Compared with more studies that focus on the street-stall economy from the perspectives of public management and economics, this study introduces the concept of good governance, aiming to improve the governance of the street-stall economy from the top-level design and realize the development of the street-stall economy from governance to good governance at an early stage. By introducing the method of evaluation indicators, the paper builds an index system for evaluating the effects of street-stall economic governance with the characteristics of good governance and the street-stall economy coupled. Also, the paper intends to provide new ideas for the development of the street-stall economy and offer a reference for the study of good governance, but there are still shortcomings in the study.

The street-stall economy is complicated. Although in this study a general index system is proposed for evaluating the effect of street-stall economic governance effectiveness, it is limited as it is specific to Hangzhou and fails to apply to other regions. Having said that, it provides a reference for the development of street-stall economy in other regions when the specific effect is becoming universally applicable.

The analysis needs to take into consideration of the special and temporal characteristics of the development of the street-stall economic governance. This paper only assesses the effects of street-stall economic governance in Hangzhou, and an analysis is not carried out of spatial and temporal characteristics, which means there are some shortcomings in terms of the completeness of the study. Therefore, more research on both spatial and temporal dimensions will be conducted in the future to ensure sound study.

## 5 CONCLUSIONS

In this paper, based on ten management elements of good governance, eight second-level indicators are selected and categorized as the first-level indicators including vendors, consumers, and governors, and are further classified into 25 third-level indicators, thus building an index system for evaluating the effects of street-stall economic governance. In addition, the weights are calculated for each indicator using the judgment matrix and consistency test in the analytic hierarchy process, and the index system is used to evaluate the effects of street-stall economic governance. The following conclusions are drawn.

In the performance of the street-stall economy, the most important factor is the vendors, followed by consumers and again governors. In terms of street-stall economic governance, vendors or sellers play an indispensable role as they are governed, making them the bridge between governors and consumers. Compared with the governors' work and consumers' purchases, vendors play an active part in promoting the development of the street-stall economy. As for governors, the effect brought about by governance to a large extent can also be reflected through the consumers' purchases. Governors help to maintain sound order, and when shifting to good governance, they need to provide guidance. As governors bring fewer effects on governance than consumers and vendors do, the most important players are vendors followed by consumers and governors. In the refined third-level indicators, flexible enforcement is of the greatest concern. Governors emphasize the stability of street stalls while consumers value the convenience of consumption.

In the empirical study of the street-stall economy in Hangzhou, based on the previous evaluation index system, the total weighted score of the effect of street-stall economic governance in Hangzhou is 2.89, which is more than half of the total score. This indicates that the effect of governance is moderately favorable and is moving towards good governance. Among them, indicators that meet the requirements of good governance are flexible enforcement, consumption distance, and access requirements.

The results of the survey for vendors in this study reflect that they are not satisfied with the operating costs and public measures for stalls, while consumers believe that the quality, variety, environment, and cultural characteristics of goods need to be improved. Governors are not satisfied with the vendors' quality, measures for environmental management, consultation mechanism, and technological tools. This reveals the direction of the future governance of the street-stall economy in which the Internet and technologies of the new era can be applied to achieve transparent and efficient management.

## ACKNOWLEDGMENTS

This work was financially supported by the Philosophy and Social Science Planning Project of Zhejiang Province (Grant No. 20NDJC098YB); the Graduate Project of Zhejiang Gongshang

University (Grant No. 3100JYN4118002G-546) in 2021; and the National College Students' innovation and entrepreneurship training program (Grant No. 202110353038).

## REFERENCES


Cui Z F, Wu H Y. "Stall" Economy for 70 Years: The Logic of "Embedded" Flexible Governance in Urban Governance [J]. *Economic Issues*, 2019(06): 1–10.

Edipally S. Street vending in urban 'informal' markets: Reflections from case-studies of street vendors in Delhi (India) and Phnom Penh City (Cambodia)[J]. *Cities*, 2019, 89(6):120–129.

Fang L. The Return of Government Power and the Direction of Local Governance[J]. *Theoretical Investigation*, 2013(01): 27–31.

Langford N J. The Governance of Social Standards in Emerging Markets: An exploration of actors and interests shaping Trustea as a Southern multi-stakeholder initiative[J]. *Geoforum*, 2019, 104(8):81–91.

Ll A, Pw A, Sr B. A marriage of convenience: Street vendors' everyday accommodation of power in Dhaka, Bangladesh—ScienceDirect[J]. *Cities*, 2019, 84(11):143–150.

Lyons, Michal. Pro-Poor Business Law? On MKURABITA and the Legal Empowerment of Tanzania's Street Vendors[J]. *Hague Journal on the Rule of Law*, 2013, 5 (1): 74–95.

Maneepong C, Walsh J C. A new generation of Bangkok Street vendors: Economic crisis as opportunity and threat[J]. *Cities*, 2013, 34(10): 37–43.

Martínez L, Short J R, Estrada D. The urban informal economy: street vendors in Cali, Colombia[J]. *Cities*, 2017, 66(6): 34–43.

Meng, Y Z, Yao, M. The theoretical Logic of the "Street Vendor Economy" and China's Path Choice[J]. *Journal of Central China Normal University* (Humanities and Social Sciences), 2021, 60(01): 45–52.

Peng J, Peng X B, Zhang J J. Identification of the Key Influencing Factors of the Integration of Street-stall Economy Development into Urban Governance: Empirical Research Based on DEMATEL Method[J]. *Modern Urban Research*, 2021(02):8-13.

Sekhani R, Mohan D, MLiu M. Politics and Governance in Emerging Market Countries[J]. *Global Journal of Emerging Market Economies*, 2021, 13(2): 265–286.

Shen L Y, Ochoa J J, Shah M N, et al. The application of urban sustainability indicators—A comparison between various practices[J]. *Habitat International*, 2011, 35(1):17–29.

Steel G. Local encounters with globetrotters: Tourism's Potential for Street Vendors in Cusco, Peru[J]. *Annals of Tourism Research*, 2012, 39(2):601–619.

Sun Y L. The Empowerment Effect of Street-stall Economy on Small and Medium-sized Enterprises under the New Development Pattern[J]. *Social Science Journal*, 2021(04):122–130.

Wang J G. Inclusiveness and Sharing, Explicit and Implicit Mutual Learning, Livability Predicted: Historical Prospect and Contemporary Creation of Urban Vitality[J]. *Urban Planning*, 2019, 43(12): 9–16.

Wang Y. How Urban Refined Management Accommodates the Stall economy—Based on an Empirical Investigation and Research in the Central Urban area of Shanghai [J]. *China Development Observation*, 2018(13):47–51.

Xiao X, Zhang X J. On Internal Logic and Realistic Path of the Current Development of "the Stall Economy" in China: From the Perspective of Marxism Political Economics [J]. *Journal of Hunan University of Science & Technology* (Social Science Edition), 2021, 24(02):143–148.

Ye M. The Governance of Street Vendors in Megacities of China and India: Administrative Elimination and Political Acceptation [J]. *Journal of East China University of Science and Technology* (Social Science Edition), 2016, 31(03): 10–19.

Yu J X, Ren J. Social Governance Community and Its Implementation Institutions Design [J]. *Journal of Political Science*, 2020(01): 45–56+125–126.

Yu K P, Li J P, Mao S L, Gao X P, Peng X Y, Yang X D, Dong L S. How far is China from "Good Governance" – An Academic Discussion on "Governance and Good Governance" [J]. *China Administrative Management*, 2001 (09):15–21.

Yu K P. Governance and Good Governance: A New Framework for Political Analysis[J]. *Social Sciences in Nanjing*, 2001(09): 40–44.

Yu K P. Important issues on the Innovation of Chinese Urban Governance Based upon Case Studies of Megacities [J]. *Wuhan University Journal* (Philosophy & Social Science), 2021, 74(03): 88–99.

Yu K P. Introduction to Global Governance[J]. *Marxism and Reality*, 2002(01): 20–32.

*Advances in Urban Construction and Management Engineering – Cha (Ed)*
*© 2023 Copyright the Author(s), ISBN 978-1-032-39018-5*

# Application of plant landscaping in landscape design

Shan Gao*
*The Changchun University of Architecture and Civil Engineering, China*

ABSTRACT: In this paper, the author examines the application of plant landscaping in landscape design based on relevant work experience. Then the author presents the corresponding design measures, hoping that landscape designers will be able to use them as a reference, and the line of landscape design.

## 1 INTRODUCTION

In garden landscape design, plants as landscape elements should exert their characteristics in plant landscaping and meet the needs of the design scheme. Therefore, in the design process, the method of plant landscaping needs to be selected according to local conditions, and different design measures and configuration methods of plant landscaping should be used reasonably to improve the artistic effect of the garden landscape.

## 2 ANALYSIS OF THE ROLE OF PLANT LANDSCAPING

### 2.1 *Garden plants can reflect the landscape timing*

In garden landscape design, plants as landscape elements should exert their characteristics in plant landscaping and meet the needs of the design scheme. Therefore, in the design process, the method of plant landscaping needs to be selected according to local conditions, and different landscaping techniques and design measures should be used reasonably to improve the artistic effect of garden landscapes.

### 2.2 *Garden plants can form spatial changes*

Like buildings, plants will form a specific spatial structure and have various spatial functional characteristics. Therefore, through the change of the spatial structure in the plant landscaping, people can increase the spatial change of the landscape by changing the viewpoint and line of sight. Through the method of hierarchical control, the scenery of plants can be formed into different levels of spatial structure.

### 2.3 *Garden plants can be grown in different ways*

Garden plants have unique postures and colors. Plant landscaping involves selecting specific planting methods based on the characteristics of different plants so that their characteristics can be reflected not only in the plants themselves but also in the plant combinations. Common planting methods include solitary planting, pair planting, column planting, cluster planting, and patch planting (Chai 2013). In practical application, solitary planting can fully show the beauty of the plant's

*Corresponding Author: 1035285449@qq.com

 DOI 10.1201/9781003348023-42

personality. Planting a tree with a beautiful shape in an open area can play a role in spotting the scene and guiding the line of sight; the opposite planting is a planting that forms a symmetrical and balanced relationship according to a certain axis. The method emphasizes the sequence of the landscape space; the planting is neatly arranged, with a strong sense of rhythm; the cluster planting can reflect the group beauty of the trees through the combination of tree species at different levels; the patch planting can be applied to the scenic area or woodland.

## 3 PROBLEMS OF PLANT LANDSCAPING IN GARDEN LANDSCAPE DESIGN

### 3.1 *Irrational configuration*

In some landscape design works, the overall landscape ornamental is poor, with single plant species, unreasonable plant configuration, and lack of coordination among plants. Guo Hua et al. found through their research that the landscape effect shown by many landscapes has a poor sense of hierarchy; Ma Chunping et al. discussed the problems of the hierarchy of garden landscaping, plant species, plant growth environment, and concluded that the selection of plants needs to consider its coordination with the surrounding architecture, emphasizing the need to have its characteristics in the landscape design of the garden.

### 3.2 *Neglected ecological protection work*

In the specific design process of landscaping, many designers are prone to ignore the design core in landscaping and tend to pursue exotic visual feelings. Zhu Qiwen points out that blindly introducing expensive and rare plants in the design as a way of planting landscape design is an act that does not follow the law of natural development. Song Xiaobin's research shows that reducing the overall functionality of plant landscaping cannot achieve the true purpose of landscape design. Some garden designers believe that trees should be as tall as possible, and many local landscape gardens have experienced many tree deaths due to the failure to select suitable plants according to local climatic characteristics, resulting in serious economic losses, which also directly reduces the ornamental nature of landscape gardens due to a major discount in the functionality of tree species.

### 3.3 *Less variety of plants*

As the variety of plants available in a region is very limited, and with the continuous expansion of urban land areas, human demand for plant landscaping is also growing year by year. Feng Jifu's research shows that urban landscaping needs to increase the quality of landscape plants, but due to geographical restrictions, many places have fewer plant species to choose from, thus affecting the overall effect of landscape gardening.

For example, in the northern garden design, tall trees and shrubs from the south are chosen to show the uniqueness of the garden, but in the actual gardening process, due to the low production of such plants, it is easy to have a shortage of garden landscape design. The situation is that the supply of landscape design exceeds the demand.

### 3.4 *Less cultural taste*

The design work of the garden landscape should be in line with the actual local situation. Du Jinze Yu pointed out that understanding the local customs and natural conditions is the premise of making the plant landscape consistent with the overall style of the garden, the garden landscape has the characteristics of history and culture, in the design of the landscape needs to have innovative ideas, to bring a certain economic value to the people around to carry forward the traditional Chinese culture. Some garden landscapes do not have the aura of artistic products, and the lack of their cultural connotation in plant landscaping seriously reduces the overall effect of landscaping.

## 4 MEASURES TO IMPROVE THE ARTISTIC EFFECT OF GARDEN PLANT LANDSCAPING

### 4.1 *Pay attention to seasonal changes in plants*

In the design process, it is necessary to understand the changing characteristics of plants in four seasons, and at the same time, make comprehensive matching according to the changing characteristics of plants. In the landscaping stage, it is not only necessary to reflect the landscape characteristics of one season, but also to consider the changing needs of other seasons. For example, in spring, people can use more plants with earlier flowering periods, such as Yingchun and Forsythia, to give people a feeling of recovery and vitality; in summer, they can use more green-leafed plants and cool-colored flowers to give people a cool and refreshing look; in autumn, more autumn leaf species can be used to reflect its rich color changes; in winter, more evergreen species can be used to add landscape color. In short, in the stage of plant landscaping, the need for the four seasons changes in the garden scenery need to be considered, and the seasonal changes of different plants can be constructed through the rational use of different plant combinations (Zhou 2015).

Table 1. Flowers and plants of different colors in different seasons in northern China.

| Season\color | Red series | Yellow series |
|---|---|---|
| Spring (March to May) | Arbor: Magnolia<br>Shrubs: red rambutan, snow hills, and white lilac | Arbor: Bauhinia,<br>Shrubs: Forsythia, Yingchun, Begonia, Peony |
| Summer (June to August) | Shrubs: wood hydrangea, Bella anna, bellflower, columbine, and spiraea | Arbor: Ash,<br>Shrubs: Tianmu Qionghua, elderberry<br>Fujimoto: yellow wood, windmill jasmine |
| Autumn (September to November) | Shrubs: Cone Hydrangea,<br>Ground Covers: Chrysanthemum, Hosta | Ground cover: Hemerocallis, calamus, chrysanthemum |
| Winter (November to February) | Shrubs: poinsettia | Trees: Lamei, Tiger-legged plum |
| Season\color | Red series | Yellow series |
| Spring (March to May) | Arbor: Magnolia<br>Shrubs: red rambutan, snow hills, white lilac | Arbor: Bauhinia,<br>Shrubs: Forsythia, Yingchun, Begonia, Peony |

The seasonal phase of plants is undoubtedly one of the most temporally changing elements in spatial color. Plant landscaping is difficult because the seasons are evolving and alternating, showing different scenes throughout the year, each season has its characteristics, and needs to be comprehensively considered according to the ecological requirements of plants, as well as in terms of shape and color. The variety of plant colors is very rich, which is displayed through different parts of the plant, such as leaves, inflorescences, peels, branches, and so on. The color of its leaves is the most important landscaping element, and it plays a leading role in the landscaping process (Ai 2020). As shown in the figure below, it is the seasonal change of maple leaves throughout the year.

Figure 1. Seasonal changes of maple leaves throughout the year.

### 4.2 *Plant landscaping should be done in an orderly manner*

Plant landscaping should be used to ensure that the configuration of plants meets the needs of density, suitable for trees, and minimizes artificial traces. In the design of the need to carry out a comprehensive layout from the whole, and then on the details of the part of the consideration, the effective vision, close together. From the overall level of analysis, if the design area is relatively large, it is necessary to do a good job in some small areas of tree species coordination collocation to make it can set off each other, to meet the needs of garden art effect, to achieve the design goal of coordination, unity, and integrity.

### 4.3 *Use different landscaping techniques*

Since plants also have a certain sense of space, different landscaping techniques can be used in the design process to meet changing needs such as open space and semi-open space. Chinese classical Chinese classics such as framed view, obstructed view, suppressed view, borrowed view, and opposite view can be used. To achieve the effect of changing the scenery at every step and seeing the big in the small, the gardening method uses modern landscape elements to create a rich and varied landscape space (Yang 2014).

### 4.4 *Application of enclosed space*

In the process of plant landscaping, there are many types of plant enclosed space, including open space, semi-open space, and vertical enclosed space. Therefore, in the specific design, the type of enclosed space for plants should be selected according to the terrain conditions. It may be possible, for example, to create enclosed spaces on both sides of the road in a residential area, which will keep dust and noise outside and create a peaceful and harmonious environment for living.

### 4.5 *Enrich the visual experience in plant landscaping*

The intuitive feeling of the garden landscape is transmitted through vision, so in the design stage, the designer needs to conduct a comprehensive analysis of the visual effect. Combining the visual perspective relationship with the far and near views is applied to the design scheme. For example, a large tree with an isolated value has a unique visual sense. Therefore, when designing a plant landscape, it is necessary to understand the planting method of plants. Through solitary planting, for planting, column value, and cluster planting, the expression form of the plant is constructed to improve the viewing effect and enrich the visual experience.

### 4.6 *Design a plant landscape that incorporates nature*

The specific design link needs to take the natural landscape as the basis, and fully integrate the landscape design with the natural landscape. In the design, it is necessary to consider the needs of the ecosystem, adapt the trees to the right place, and choose the plants reasonably according to the local conditions (Zhao 2013). In addition, different plants should be selected according to the functional characteristics of different spaces, so that people can feel the changes in different spaces.

## 5 CONCLUSION

In the process of landscape design, plant landscaping is the key factor to improve the quality of landscape design. Strengthening the application of plant landscaping in the process of landscape design can not only improve the artistic effect of landscape, improve people's living environment to meet people's spiritual needs, but also promote the development of landscape engineering to a certain extent, which is the purpose and significance of this study.

## REFERENCES


Ai Jiahui. *Research and application of plant landscaping in public spaces in northern residential areas* [D]. Tianjin University of Technology, 2020.

Chai Xiaoxian. *Investigation and Research on the Application of Ornamental Aromatic Plants in Quzhou City*[D].Fujian Agriculture and Forestry University, 2013.

Yang Chen, Yang Na. A brief analysis of the application of new Chinese gardens in the human settlement environment [J]. *Architectural Engineering Technology and Design*, 2014(10):949.

Zhao Jie. Discussion on the rationality of the arrangement of green plants in the square[J].*Modern Horticulture*, 2013(4):113.

Zhou Changmei. The Problem of Plant Configuration in Landscape Architecture Design[J]. *Theoretical Research on Urban Construction*, 2015(15):1532–1533.

*Advances in Urban Construction and Management Engineering – Cha (Ed)*


# Research on the promotion path of green total factor productivity of China's logistics industry based on fsQCA method

Xiaoyu Qu* & Yaqi Cao*
*Dalian University of Technology, Dalian, China*

ABSTRACT: In the transformation of China's economy from high-speed development to high-quality development, the green development of the logistics industry is one of the driving forces to promote the high-quality development of China's economy. This paper first uses the DEA Malmquist model to measure the green total factor productivity of the logistics industry in 30 provinces and cities in China, and then uses fuzzy set qualitative comparative analysis (fsQCA) to explore the influencing factors and promotion path of green total factor productivity of logistics industry in China. Finally, it concludes that logistics energy intensity and urbanization rate are the core factors to improve the green total factor productivity of the logistics industry in China. It also puts forward four paths that can provide a reference for improving the green total factor productivity of China's logistics industry and accelerating the green development of China's logistics industry.

## 1 INTRODUCTION

With the steady growth of the national economy and the accelerating process of urbanization, the logistics industry of comprehensive services is also developing rapidly and becoming stronger. At the same time, people's awareness of environmental protection is also increasing, and the concept of green logistics is gradually rooted in the hearts of the people. China is at the intersection of the "14th five-year plan" and the realization of China's "two centenaries" goals. The state issued the "opinions on promoting the high-quality development of logistics and promoting the formation of a strong domestic market", in which it proposed to accelerate the development of green logistics, take green logistics as a breakthrough, and drive upstream and downstream enterprises to develop green supply chains.

## 2 LITERATURE REVIEW

At present, the analysis of the influencing factors of green total factor productivity in the logistics industry mostly focuses on infrastructure, economic level, technical level, and environmental regulation. In terms of infrastructure construction, scholars' research conclusions are relatively unified, and they believe that having a good infrastructure can promote the green development of the logistics industry (Gao 2013; Guo 2016; Kherbash 2015; Mangla 2016; Wang 2014). In terms of economic level and technological innovation, most scholars have the same conclusion that technological progress and good economic development level can promote the green total factor productivity of the logistics industry (Dang 2018). In terms of environmental regulation, scholars' research results have not reached a unified conclusion. Some scholars believe that environmental regulation will increase the cost of enterprises (Gu 2019; Walley 1994; Zhao 2006), while others

*Corresponding Authors: 670047115@qq.com and 1358325569@qq.com

DOI 10.1201/9781003348023-43

hold the opposite view. They believe that to make up for the losses caused by environmental regulation and obtain new profit space, enterprises need to carry out technological innovation to reduce emission costs, improve output efficiency, and then promote the growth of green total factor productivity (Porter 1991). With the deepening of research, some scholars believe that the impact of environmental regulation on green total factor productivity is not a simple linear relationship (Li 2012). In addition, there is also a correlation between various factors. Some scholars have found that there is a correlation between the technological level and environmental regulation. In central and eastern China, there is an inverted U-shaped relationship between technological progress and environmental regulation, but affected by the form of regulation, this relationship is not significant in Western China (Zhang 2011). Some scholars have also found that when the infrastructure level is at a low or low level, the urbanization rate has an inhibitory effect on the green total factor productivity of the logistics industry; when the infrastructure level is at a high and high level, the urbanization rate can effectively promote the growth of green total factor productivity in the logistics industry (Liu 2020).

## 3 PATH ANALYSIS OF GREEN TOTAL FACTOR PRODUCTIVITY IMPROVEMENT IN THE LOGISTICS INDUSTRY

### 3.1 *Calibration variables*

Table 1. Name and meaning description of variables.

| Variable type | Primary index | Variable name | Meaning |
|---|---|---|---|
| Result variable | | Green total factor productivity of logistics industry | Calculated from the above |
| Conditional variable | Infrastructure | Logistics energy intensity | Logistics energy consumption/logistics added value |
| | | Transportation intensity | Comprehensive logistics turnover/regional GDP |
| | Economic development | Urbanization rate | Urban population/total population of the region |
| | | Industrial structure | The added value of logistics industry/regional GDP |
| | | Per capita GDP | Real GDP/population |
| | Scientific and technological level | Scientific and technological innovation ability | Regional patent authorization |
| | External constraints | Environmental regulation | $CO_2$ emission of logistics industry/added value of logistics industry |

### 3.2 *Data source*

It is calculated according to the relevant data in the National Bureau of statistics and the average value of relevant indicators in the *2017-2019 China Energy Statistics Yearbook.*

### 3.3 *Variable calibration*

When performing fsQCA analysis, the result variables and condition variables should be calibrated first.

## 3.4 *Necessity test*

This paper analyzes the configuration of the following seven conditional variables to explore the configuration that has an important impact on the green total factor productivity of the logistics industry.

Table 2. Necessity test results.

| Variable | Consistency | Coverage | Variable | Consistency | Coverage |
|---|---|---|---|---|---|
| EFF | 0.594184 | 0.633099 | PER | 0.498348 | 0.492167 |
| ~ EFF | 0.465962 | 0.446203 | ~ PER | 0.608063 | 0.626703 |
| TTI | 0.443490 | 0.476224 | INO | 0.525446 | 0.544521 |
| ~ TTI | 0.645737 | 0.614079 | ~ INO | 0.545274 | 0.535714 |
| URB | 0.555188 | 0.567951 | ER | 0.615334 | 0.621495 |
| ~ URB | 0.513549 | 0.510848 | ~ER | 0.465301 | 0.468708 |
| IND | 0.500330 | 0.505003 | | | |
| ~ IND | 0.561137 | 0.565623 | | | |

## 3.5 *Path analysis*

Table 3. Empirical results of complex solutions.

| | Complex solution | | |
|---|---|---|---|
| Configuration | Raw coverage | Unique coverage | Consistency |
| EFF*~TTI*~URB*~IND*INO*ER | 0.1269 | 0.0713814 | 0.936585 |
| ~ EFF*~TTI*URB*IND*~PER*~INO*~ER | 0.0799735 | 0.054858 | 0.864286 |
| EFF*~TTI*URB*~IND*~PER*~INO*ER | 0.105089 | 0.0608064 | 0.946429 |
| EFF*~TTI*URB*~IND*PER*INO*~ER | 0.117647 | 0.064772 | 0.824047 |
| EFF*TTI*URB*IND*PER*~INO*ER | 0.0568407 | 0.0356907 | 0.886598 |
| Solution coverage | | 0.355585 | |
| Solution consistency | | 0.896667 | |

Table 4. Empirical results of reduced solution.

| | Parsimonious solution | | |
|---|---|---|---|
| Configuration | Raw coverage | Unique coverage | Consistency |
| ~ TTI*URB*~PER | 0.179775 | 0.0508923 | 0.928328 |
| EFF*URB*~INO | 0.161269 | 0.0337079 | 0.769716 |
| EFF*~ER | 0.197621 | 0.0198282 | 0.814714 |
| EFF*~URB*INO | 0.132849 | 0 | 0.939252 |
| ~ URB*INO*ER | 0.167217 | 0.011236 | 0.951128 |
| ~ TTI*URB*~IND*~ER | 0.198282 | 0 | 0.852273 |
| ~ TTI*~IND*INO*~ER | 0.220753 | 0 | 0.778555 |
| ~ TTI*~URB*~IND*INO | 0.163913 | 0.00396565 | 0.763077 |
| Solution coverage | | 0.481163 | |
| Solution consistency | | 0.81523 | |

Table 5. Empirical results of intermediate solutions.

| Configuration | Intermediate solution | | |
|---|---|---|---|
| | Raw coverage | Unique coverage | Consistency |
| EFF*~TTI*~URB*~IND*INO*ER | 0.1269 | 0.0713814 | 0.936585 |
| ~ EFF*~TTI*URB*IND*~PER*~INO*~ER | 0.0799735 | 0.054858 | 0.864286 |
| EFF*~TTI*URB*~IND*~PER*~INO*ER | 0.105089 | 0.0608064 | 0.946429 |
| EFF*TTI*URB*IND*PER*~INO*ER | 0.0568407 | 0.0356907 | 0.886598 |
| EFF*~TTI*URB*~IND*PER*INO*~ER | 0.117647 | 0.064772 | 0.824074 |
| Solution coverage | | 0.355585 | |
| Solution consistency | | 0.896667 | |

## 4 ANALYSIS OF RESEARCH RESULTS

The specific empirical results are shown in Table 6.

Table 6. Configuration analysis results of conditional variables.

| Conditional variable | Configuration | | | | |
|---|---|---|---|---|---|
| | 1 | 2 | 3a | 3b | 4 |
| Logistics energy intensity | ● | ⊗ | ● | ● | ● |
| Transportation intensity | ⊗ | | | • | |
| Urbanization rate | ⊗ | ● | ● | ● | ● |
| industrial structure | | • | | • | |
| Per capita GDP | | | | • | • |
| Scientific and technological innovation ability | ● | ⊗ | ⊗ | ⊗ | ⊗ |
| Environmental regulation | ● | ⊗ | ● | ● | ⊗ |
| Configuration consistency | 0.937 | 0.864 | 0.946 | 0.887 | 0.824 |
| Configuration coverage | 0.127 | 0.080 | 0.105 | 0.057 | 0.118 |
| Unique coverage | 0.071 | 0.055 | 0.061 | 0.036 | 0.065 |
| Overall consistency | | | 0.897 | | |
| Overall coverage | | | 0.356 | | |

Note: "●" indicates the occurrence of the cause variable, and this cause variable is the core condition; "•" indicates the occurrence of the cause variable, but this variable is a non-core condition; "⊗" indicates that the cause variable does not appear, and this variable is a core condition or non-core variable; blanks indicate that the occurrence of the cause variable has little impact on the production of the result.

Path 1: the "facility + innovation + regulation" logistics industry green total factor productivity improvement path, including configuration 1. In this path, the core conditions are logistics energy intensity, scientific and technological innovation ability, and environmental regulation, which shows that in the absence of economic development, taking infrastructure factors, scientific and technological level factors and environmental regulation factors as growth conditions is the key to promoting the green total factor productivity of the logistics industry. The regions applicable to this route include Sichuan and Hunan.

Path 2: "economic" logistics industry green total factor productivity improvement path, including configuration 2. In this path, the core condition is the urbanization rate, and the marginal condition is the industrial structure, which shows that the economic development factor is a necessary condition for some regions to improve the green total factor productivity of the logistics industry. The region applicable to this route is Jilin.

Path 3: "facility + economy + regulation" logistics industry green total factor productivity improvement path, including configurations 3a and 3b. On this path, the core conditions are logistics energy intensity, urbanization rate, and environmental regulation. In configuration 3a, under the environmental regulation of the government, we should not only improve the green total factor productivity of the logistics industry but also pay attention to the construction of infrastructure and the strength of regional economic development. The region suitable for configuration 3a is Heilongjiang. The marginal conditions of configuration 3b are transportation intensity, industrial structure, and per capita GDP, which indicates that in the case of insufficient scientific and technological innovation ability if we want to improve the green total factor productivity of the logistics industry, we must take the road of building infrastructure, improving economic strength, and complying with regulations. The region suitable for configuration 3b is Liaoning.

Path 4: "facility + economy + innovation" logistics industry green total factor productivity improvement path, including configuration 4. On this path, the core conditions are logistics energy intensity, urbanization rate, and scientific and technological innovation ability, and the marginal conditions are per capita GDP, indicating that based on perfect infrastructure construction and good economic development level, we should continue to improve the level of scientific and technological innovation to provide a broad development space for improving the green total factor productivity of the logistics industry. The region applicable to this route is Jiangsu.

## 5 CONCLUSIONS AND SUGGESTIONS

This paper explores the green total factor productivity improvement path of the logistics industry in different regions of China based on the fsQCA method, and finds the following conclusions:

(1) A single conditional variable does not constitute a necessary condition for the improvement of green total factor productivity in the logistics industry. The five configurations that produce high green total factor productivity in the logistics industry constitute sufficient conditions for the result variables.
(2) There are four different paths to improve the green total factor productivity of the logistics industry, and these paths apply to different regions.
(3) Logistics energy intensity and urbanization rate appear in most paths, and are the core conditions.

Based on the above conclusions, the following suggestions are put forward.

(1) First, we should strengthen infrastructure construction, reduce logistics energy intensity, and lay the foundation for improving the green total factor productivity of the logistics industry. The government should act as a guide and supporter.
(2) Second, we should also improve the regional economic level and accelerate the pace of urbanization. By strengthening the regional economic level, improving the urbanization rate, attracting talents and technology, and then improving the green total factor productivity of the logistics industry.
(3) Finally, we should enhance the ability of scientific and technological innovation and promote the sustainable and green development of the logistics industry.

## REFERENCES


Dang VL, Yeo GT. Weighing the Key Factors to Improve Vietnam's Logistics System [J]. *The Asian Journal of Shipping and Logistics*, 2018, 34(4): 308–316.

Gao Xiuli, Meng feirong Analysis of total factor productivity and its influencing factors in China's logistics industry [J]. *Technical Economy*, 2013, 32(2): 51–58.

Gu W, Wang C, Dai SF, Wei LR, Chiang IR. Optimal strategies for reverse logistics network construction: A multi-criteria decision method for Chinese iron and steel industry [J]. *Resources Policy*, 2019, 2(2): 1–12.

Guo caihuan Empirical analysis of logistics industry on Industrial Structure Optimization – Taking Hebei Province as an example [J]. *Business Economics*, 2016, 7(7): 60–61+108.
Kherbash O, Mocan ML. A Review of Logistics and Transport Sector as a Factor of Globalization [J]. *Procedia Economics and Finance*, 2015, 27(27): 42–47.
Li Ling, Tao Feng The choice of optimal environmental regulation intensity in China's manufacturing industry – from the perspective of green total factor productivity [J]. *China Industrial Economy*, 2012, 28(5): 70–82.
Liu Haohua, Chen Xiuling, Wang Xuefeng Research on the threshold effect of urbanization and logistics green total factor productivity [J]. *Economic and Management Review*, 2020, 36(02): 123–13.
Mangla SK, Govindan K, Luthra S. Critical success factors for reverse logistics in Indian industries: a structural model [J]. *Journal of Cleaner Production*, 2016, 129(129): 608–621.
Porter MA. America's green strategy [J]. *Scientific American*, 1991, 264(4): 168.
Walley N, Whitehead B. It's not easy being green [J]. *Harvard Business Review*, 1994, 72(3): 46–51.
Wang Jian, fan Yuejiao Test and analysis of the dynamic effect of China's logistics service industry agglomeration on regional economic growth – An Empirical Study Based on dynamic panel data model [J]. *China Circulation Economy*, 2014, 28(7): 39–46.
Zhang Cheng, Lu Min, Guo Lu, Yu Tongshen Environmental regulation intensity and production technology progress [J]. *Economic Research*, 2011, 46(2): 113–124.
Zhao Xikang *Guide Green Innovation* [M]. Beijing: Economic Science Press, 2006

*Advances in Urban Construction and Management Engineering – Cha (Ed)*
*© 2023 Copyright the Author(s), ISBN 978-1-032-39018-5*

# Fruits and vegetables distribution route optimization considering fuzzy demand and customer satisfaction

Fachao Li*, Shaojing Li* & Chenxia Jin*
*School of Economics and Management, Hebei University of Science and Technology, Hebei, China*

ABSTRACT: For the vehicle routing problem with fuzzy demand (VRPFD), this paper introduces the expected value of fuzzy numbers to deal with customer demand satisfaction, builds a multi-objective programming model with minimum cost and maximum satisfaction, and uses a genetic algorithm to solve it. Taking the fruit and vegetable distribution of the *M* supermarket chain as an example, considering the changes in the distribution scheme under different parameters, and comparing it with the previous methods, the results show that the optimized distribution route has lower cost and higher satisfaction. This paper provides a reference for the fruit and vegetable distribution path planning and provided a solution to the VRPFD.

## 1 INTRODUCTION

Vehicle Routing Problem (VRP) is a classic problem in logistics and distribution systems, often with deterministic assumptions, but there is often certain deterministic information that cannot be known in advance, such as a customer demand of about 50 kg; vehicle from point A, travel time to point B is about 10 to 15 minutes.

With the objective of fresh produce logistics distribution path optimization under uncertain demand, a demand uncertainty model was constructed, and an ant colony algorithm and particle swarm algorithm were selected to solve the problem (Chen 2018; Li 2021). A single-objective fuzzy chance constraint model was established based on fuzzy plausibility measure theory and the problem was solved using an improved hybrid difference algorithm and hybrid genetic ant colony optimization algorithm (Cao 2009; Kuo 2017). The research combined characteristics of supermarket logistics distribution optimized its distribution path and solved it by using a simulated annealing algorithm (Wang 2019). These studies consider fuzzy shipping volume and customer satisfaction, a fuzzy expectation value model, and a fuzzy opportunity constraint planning model were constructed, and the vehicle path problem was solved by the heuristic algorithm and sensitivity analysis method (Fan 2020; Hu 2019; Yu 2018).

In summary, the research considering the existence of customer demand fuzziness in the actual distribution process is closer to reality, however, the vehicle routing problem with fuzzy demand (VRPFD) still has problems, so this paper, based on the existing VRP research, provides a multi-objective planning model with minimum cost and maximum satisfaction, and solves it with genetic algorithm to get the optimal path.

## 2 PROBLEM DESCRIPTION AND CONDITION AND CONDITIONAL ASSUMPTIONS

This paper selects the *M* chain supermarket in Shijiazhuang City as the research object. The supermarket has 32 stores and only one fruit and vegetable distribution center. It belongs to the

*Corresponding Authors: lifachao@tsinghua.org.cn, 1277747887@qq.com and jinchenxia2005@126.com

DOI 10.1201/9781003348023-44

problem of multiple vehicles departing from the same distribution center and providing logistics and distribution services to multiple stores. The following conditions are assumed.

(1) The locations of the distribution center and all stores of the $M$ supermarket chain are known; (2) there is only one distribution center and there are enough vehicles to be called, and the demand of each customer point cannot exceed the capacity of each vehicle; (3) the distribution vehicles depart from the distribution center with a full load and return to the distribution center after completing the distribution; (4) each distribution vehicle can serve more than one store; (5) the demand of chain stores is ambiguous. and the distribution center can meet the demand of all chain stores without shortage; (6) the distribution process is under constant temperature, and the influence of temperature on product deterioration is not considered; (7) the vehicle distribution process travel speed is constant.

The symbols and variables are explained as follows.

$D$: distribution center, represented by natural number 0;

$N$: number of customer points, $i,j \in \{0, 1, 2...N\}$;

$K$: the number of vehicles owned by the distribution center, $k \in \{1, 2...K\}$;

$d_{ij}$: distance from customer $i$ to customer $j$;

$Q$: maximum loaded weight of distribution vehicles;

$C_0$: fixed cost of a single distribution vehicle;

$F$: travel cost of a vehicle per unit distance;

$\theta$: minimum satisfaction;

$q_i$: demand of customer point $i$;

$x_i$: the actual distribution quantity of $i$;

When customer point $i$ to $j$ is dispatched by vehicle $k$, $x_{ijk} = 1$; otherwise $x_{ijk} = 0$. When customer point $i$ is dispatched by vehicle $k$, $y_{ik} = 1$; otherwise $y_{ik} = 0$.

## 3 MODEL CONSTRUCTION

### 3.1 *Distribution cost analysis*

Fixed costs mainly include the maintenance fee, and the cost of hiring drivers for delivery vehicles, which are related to the number of vehicles.

$$C_1 = C_0 \sum_{k=1}^{K} \sum_{j=1}^{N} x_{0jk} \tag{1}$$

The cost of transport is proportional to the distance traveled, regardless of the weight of the vehicle, which is denoted as

$$C_2 = F \sum_{i=0}^{N} \sum_{j=0}^{N} \sum_{k=0}^{K} \left(d_{i0} + d_{ij} + d_{j0}\right) x_{ijk} \tag{2}$$

### 3.2 *Fuzzy demand analysis*

When it comes to actual distribution, store demand for goods is often fuzzy and is usually within a certain range. In this paper, we use triangular fuzzy numbers to represent each store's demand for goods. Set the fuzzy demand of stores $q_i = (q_{1,i}, q_{2,i}, q_{3,i})$, where $q_{1,i}$ is the minimum demand; $q_{2,i}$ is the most likely demand; $q_{3,i}$ is the maximum demand.

To derive the demand at the point of maximum customer satisfaction. The expected value of the fuzzy number $E(q_i) = \frac{q_{1,i}+2q_{2,i}+q_{3,i}}{4}$ is introduced.

Satisfaction is 1 when $x_i = E(q_i)$; when $q_{1,i} \leq x_i < E(q_i)$ or $E(q_i) < x_i \leq q_{3,i}$ the customer accepts the service, but satisfaction tends to increase or decrease linearly with the demand volume; satisfaction is lowest when the upper and lower limits of the customer's fuzzy demand volume are exceeded. The specific function diagram and the customer's demand satisfaction function $U(q_i)$ are expressed as follows.

$$U(q_i)=\begin{cases} 0, x_i < q_{1,i} \\ \frac{x_i - q_{1,i}}{E(q_i) - q_{1,i}}(1-\theta)+\theta, q_{1,i} \leq x_i < E(q_i) \\ 1, x_i = E(q_i) \\ \frac{q_{3,i} - x_i}{q_{3,i} - E(q_i)}(1-\theta)+\theta, E(q_i) < x_i \leq q_{3,i} \\ 0, x_i > q_{3,i} \end{cases} \tag{3}$$

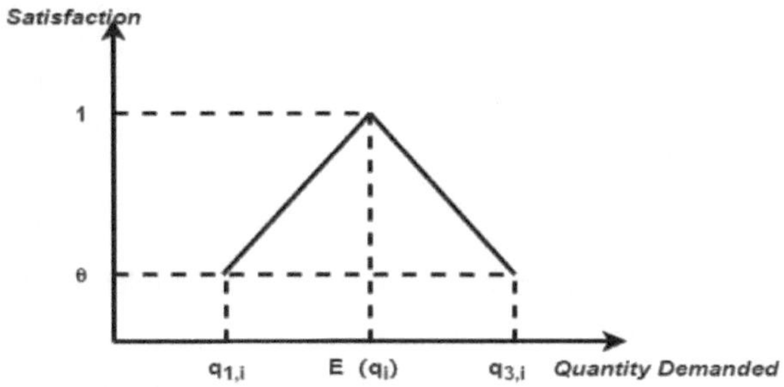


Figure 1. Diagram of the customer demand satisfaction function.

### 3.3 *Model construction*

Based on the above analysis, the model developed is

$$\min f_1 = C_1 + C_2 = C_0 \sum_{k=1}^{K} \sum_{j=1}^{N} x_{0jk} + F \sum_{i=0}^{N} \sum_{j=0}^{N} \sum_{k=0}^{K} (d_{i0} + d_{ij} + d_{j0}) x_{ijk} \tag{4}$$

$$\max f_2 = \frac{1}{N} \sum_{i=0}^{N} U(q_i) \tag{5}$$

To facilitate the solution, converted into a minimization problem with a single objective function and then solved. As the calculation criteria for $f_1 f_2$ are different, convert $f_1$ to $\frac{C_1+C_2}{C}$, where $C$ is the maximum cost per store for individual distribution ($C = C_0 * N + F \sum_{i=0}^{N} (d_{0i} + d_{i0})$) and $\lambda_1 \lambda_2$ is the weighting factor to satisfy $\lambda_1 + \lambda_2 = 1$

$$\min f = \lambda_1 f_1 + \lambda_2 (1 - f_2) = \lambda_1 \frac{C_1 + C_2}{C} + \lambda_2 \left[ 1 - \frac{1}{N} \sum_{i=1}^{N} U(q_i) \right] \tag{6}$$

The constraints are shown below:

$$U(q_i) \geq \theta, i \in \{1, 2 \ldots N\} \tag{7}$$

$$\sum_{k=1}^{K} y_{ik} = 1, k \in \{1, 2 \ldots K\} \tag{8}$$

$$\sum_{i=0}^{N} x_{0ik} = \sum_{i=0}^{N} x_{i0k} = 1, k \in \{1, 2 \ldots K\} \tag{9}$$

$$\sum_{i=0}^{N} x_{ijk} - \sum_{i=0}^{N} x_{jik} = 0, j \in \{1, 2...N\} \tag{10}$$

$$\sum_{i \in S} \sum_{j \in S} x_{ijk} \leq |S| - 1, S \in J \tag{11}$$

$$x_{ijk} \in \{0, 1\}, i \in \{1, 2...N\}, j \in \{1, 2...N\} \tag{12}$$

$$y_{ik} \in \{0, 1\}, i \in \{1, 2...N\} \tag{13}$$

Equation (6) is the objective function required to be solved in this paper; Equation (7) indicates that the customer demand satisfaction is not lower than the minimum value $\theta$; Equation (8) indicates that each customer can only be served by one vehicle; Equation (9) indicates that each vehicle starts at the distribution center and returns to the distribution center when the distribution is completed; Equation (10) represents the in-out balance constraint; Equation (11) is the elimination of the subloop constraint; Equations (12) and (13) represent the value constraints of the variables.

### 3.4 *Algorithm design*

In this paper, a genetic algorithm is used to solve the constructed model, and the specific steps are as follows.

(1) Coding. According to the characteristics of fruit and vegetable distribution center distribution path optimization and the generation method of encoding, this paper adopts the natural number encoding method to construct chromosomes.
(2) Population initialization. In selecting the population size, the value is generally taken between 20 and 200. To avoid falling into local optimum and ensure global convergence, the population size is set to 80 in this paper.
(3) Adaptation function. The fitness function $f' = {}^{1}/_{\min f} = {}^{1}/_{\lambda_1 f_1 + \lambda_2(1 - f_2)}$ is determined by setting the parameters of the algorithm, and by selecting the next generation of individuals with a large fitness genetically based on the specific objective function values.
(4) Selection. In this paper, the roulette wheel method is chosen to select the parents of the crosses. An elite retention strategy is also used, where individuals with high fitness are retained unconditionally to ensure that the optimal chromosome is inherited to the next generation.
(5) Crossover. Selected the crossover point, the gene fragments of the two crossover points are exchanged to produce a new chromosome. When performing the crossover operation, it is necessary to avoid generating duplicate gene points.
(6) Mutation. To preserve the diversity of individuals in the population to mutate the chromosomes of the offspring, this paper chooses inverse order mutation as the mutation process of the improved genetic algorithm.

## 4 CASE AND RESULTS ANALYSIS

### 4.1 *Case study*

*M* Supermarket is a chain of supermarkets opened by Northland People's Hundred Group Limited Company in Shijiazhuang, Hebei Province. Its main business scope includes daily necessities textiles, sporting goods, foodstuffs, cereals, and oil, fruit and vegetable products, home appliances, clothing, etc. To date, there are 32 chain stores, a general merchandise distribution center, and a fruit and vegetable distribution center. This paper researched the distribution path problem from the fruit and vegetable distribution center of the *M* supermarket chain to each outlet. The distribution locations are shown in Figure 2, where the dots represent the distribution customers and the triangles represent the distribution center. Distance data for distribution centers to stores and stores to stores $d_{ij}$ are shown in Table 1.

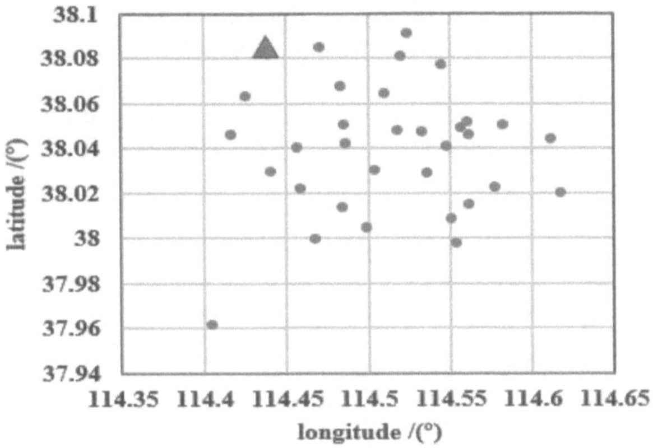


Figure 2. Location point map of distribution centers and individual stores.

Table 1. Distribution center to store and store to store distances.

| Location code | 0 | 1 | 2 | … | 21 | … | 31 | 32 |
|---|---|---|---|---|---|---|---|---|
| 0 | 0 | 12 | 13 | ... | 13 | … | 15 | 9 |
| 1 | 12 | 0 | 2.2 | … | 13 | … | 15 | 9 |
| … | … | … | … | … | … | … | … | … |
| 32 | 9 | 9 | 8.9 | … | 3.6 | … | 12 | 0 |

The store demand is expressed as a triangular fuzzy number and the data on the demand for fruit and vegetables in each store is shown in Table 2. Each reference value is shown in Table 3.

Table 2. Fruit and vegetable demand in stores.

| Location code | Fuzzy demand | Expectations |
|---|---|---|
| 1 | (384, 425, 486) | 430 |
| 2 | (435, 480, 540) | 484 |
| … | … | … |
| 32 | (346, 386, 436) | 389 |

Table 3. Model-related parameter value settings.

| Parameters | Parameters values |
|---|---|
| $Q$ | 2000 kg |
| $C$ | 4904 yuan |
| $C_0$ | 100 yuan |
| $F$ | 2 yuan/km |

### 4.2 *Analysis of results*

In this paper, the MATLAB programming language was used to solve the genetic algorithm, setting the population size of the genetic algorithm to 80, the probability of crossover to 0.9, the probability of variation to 0.1, and the maximum number of iterations to 500. Firstly, the effect of $\theta$ on the distribution cost and satisfaction when taking different values on (0, 1) was analyzed, and the initial weighting coefficient $\lambda_1 = \lambda_2 = 0.5$ was set, and the results are shown in Table 4.

Table 4. Comparison of solution results for different $\theta$.

| $\theta$ | 0.1 | 0.2 | 0.3 | 0.4 | 0.5 | 0.6 | 0.7 | 0.8 | 0.9 |
|---|---|---|---|---|---|---|---|---|---|
| Objective function value | 0.187 | 0.195 | 0.193 | 0.180 | 0.183 | 0.159 | 0.160 | 0.144 | 0.163 |
| Cost | 1301 | 1299 | 1291 | 1272 | 1263 | 1274 | 1270 | 1269 | 1290 |
| Average satisfaction | 0.891 | 0.875 | 0.876 | 0.900 | 0.891 | 0.941 | 0.939 | 0.971 | 0.977 |

The results show that when the cost and satisfaction two factors are considered, the minimum demand satisfaction $\theta$ is 0.8. Currently, the distribution cost is 1,269 yuan and the average satisfaction is 97.1%. The effect of changing the weight coefficients on the objective function is considered for (6), $\lambda_1$ and $\lambda_2$ are chosen between [0, 1], and the solution results are shown in Table 5.

Table 5. Comparison of results under different weighting factors.

| $\lambda_1$ | 0 | 0.1 | 0.2 | 0.3 | 0.4 | 0.5 | 0.6 | 0.7 | 0.8 | 0.9 | 1 |
|---|---|---|---|---|---|---|---|---|---|---|---|
| $\lambda_2$ | 1 | 0.9 | 0.8 | 0.7 | 0.6 | 0.5 | 0.4 | 0.3 | 0.2 | 0.1 | 0 |
| Objective function value | 0.017 | 0.047 | 0.07 | 0.097 | 0.131 | 0.144 | 0.172 | 0.192 | 0.213 | 0.237 | 0.249 |
| Cost | 1766 | 1405 | 1304 | 1259 | 1310 | 1269 | 1280 | 1273 | 1257 | 1256 | 1221 |
| Average satisfaction | 0.983 | 0.98 | 0.978 | 0.972 | 0.96 | 0.971 | 0.961 | 0.965 | 0.961 | 0.939 | 0.887 |

The results show that the cost is 1,766 yuan when considering only satisfaction, and 0.887 when considering only cost. The values selected for $\lambda_1$ and $\lambda_2$ are 0.3 and 0.7 when considering cost and satisfaction together. The optimized vehicle distribution scheme is shown in Table 6. Table 7 shows the comparison of the data before and after the optimization of the distribution scheme.

Table 6. Optimized vehicle distribution scheme.

| Number | Distribution routes | $x_i$ | Distance | Cost | Satisfaction |
|---|---|---|---|---|---|
| 1 | 0-27-18-9-26-8-0 | 0-332-407-446-401-395-0 | 43.3 | 187 | 0.978 |
| 2 | 0-23-10-14-15-0 | 0-441-451-444-450-0 | 23.6 | 147 | 0.978 |
| 3 | 0-2-22-31-29-0 | 0-492-437-428-410-0 | 32.7 | 165 | 0.961 |
| 4 | 0-19-25-13-11-0 | 0-373-502-392-452-0 | 29.7 | 159 | 0.938 |
| 5 | 0-3-4-1-17-32-0 | 0-441-324-426-424-384-0 | 30.4 | 161 | 0.987 |
| 6 | 0-21-30-16-20-7-0 | 0-372-342-307-447-447-0 | 47.7 | 195 | 0.973 |
| 7 | 0-12-28-5-6-24-0 | 0-340-333-420-435-467-0 | 72.2 | 244 | 0.980 |

Table 7. Comparison of data before and after optimization.

| Program | Number | Distance | Cost | Average satisfaction |
|---|---|---|---|---|
| Before optimization | 10 | 476.1 | 1829 | 0.759 |
| After optimization | 7 | 279.6 | 1258 | 0.971 |
| Comparative analysis | 3 | 196.5 | 571 | 0.212 |

It can be seen from the table that a total of seven vehicles are required to complete the distribution of 32 stores with a full load from the distribution center, the cost is 1,258 yuan and the satisfaction is 97.1%, which meets the minimum satisfaction requirement. The number of vehicles used after optimization was reduced by three, the distance was reduced by 196.5 km, the cost was reduced by 571 yuan and the satisfaction was increased by 21.2%, indicating that the optimized delivery solution improved the delivery efficiency and had strong rationality. Figure 3 shows the iteration diagram of the genetic algorithm, and Figure 4 shows the optimized vehicle distribution route.

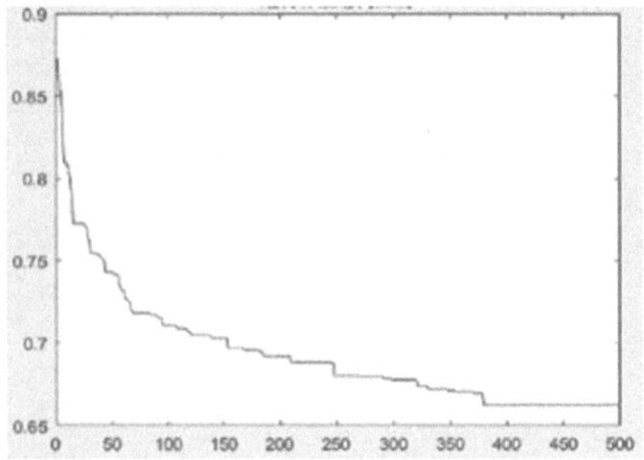


Figure 3. Optimization process diagram.

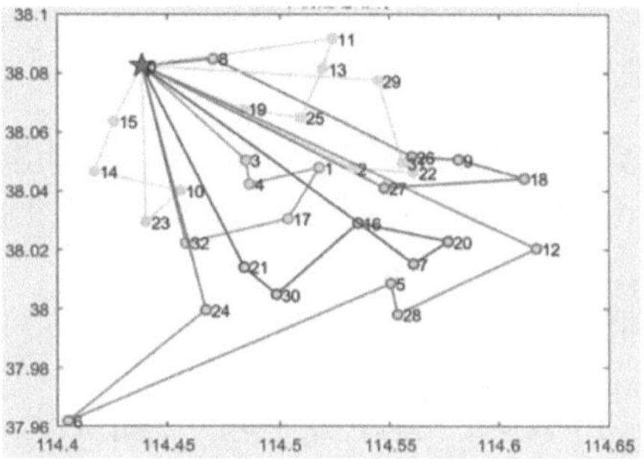


Figure 4. Distribution solution roadmap.

## 5 CONCLUSION

In this paper, a multi-objective programming model is constructed with the objectives of minimizing distribution costs and maximizing customer demand satisfaction for the vehicle path problem in a fuzzy demand environment. By comparing with the distribution path before optimization, we know that the fuzzy demand at the customer point can reduce the number of vehicles used and shorten the distance, thus reducing the distribution cost, while ensuring customer satisfaction. This paper investigates the VRPFD. In the future, we will investigate the fuzzy travel time and fuzzy time window that exist, and make further improvements to the solution algorithm.

## ACKNOWLEDGEMENTS

This work was supported by the National Natural Science Foundation of China (Grant No. 72101082); the National Natural Science Foundation of Hebei Province (Grant No. F2021208011); the Youth Top Talent Project of Research Project of Humanities and Social Sciences in Colleges and Universities of Hebei Province (Grant No. BJ2021088); and the 2022 Hebei Province Human Resources and Society Guarantee Scientific Research Cooperation Project Record (Grant No. JRSHZ-2022-01079).

## REFERENCES

Cao Erbao, Lai Mingyong. (2009). Fuzzy demand vehicle path problem based on hybrid differential evolutionary algorithm. *J. Systems Engineering Theory and Practice*, (02)106–113.

Chen Jiumei, Zhou Nan, WangYong. (2018). Path optimization of multi-compartment cold chain distribution vehicles for fresh agricultural products. *J. Systems Engineering*, 36(08):106–113.

Fan Houming. (2020). Vehicle path problem with fuzzy demand and time window and hybrid genetic algorithm for solving it. *J. Journal of Systems Management*, 29(01):107–118.

Hu Zuoan, Jia·Yezi, Li·Bowei, LIU·Lu. (2019). A study on vehicle path optimization considering customer satisfaction. *J. Industrial Engineering*, 22(01):100–107.

Kuo R J, Zulvia F E. (2017). Hybrid genetic ant colony optimization algorithm for capacitated vehicle routing problem with fuzzy demand —A case study on garbage collection system. C. 4th International Conference on Industrial Engineering and Applications. *IEEE*, 244–248.

Li Yufeng. (2021). Research on the optimization of fresh produce logistics distribution path under uncertain demand. *J. Science and Technology Innovation*. 177–179.

Wang Qiang. (2019). Research on supermarket logistics distribution path optimization based on simulated annealing algorithm. *J. Journal of Guangzhou Maritime Academy*, 27(03).

Yu Xueqing. (2018). Multimodal transport path optimization considering fuzzy demand. *J. Jounal of Beijing Jiaotong University*, 42(03):23–29+36.

*Advances in Urban Construction and Management Engineering – Cha (Ed)*
*© 2023 Copyright the Author(s), ISBN 978-1-032-39018-5*

# The effect of safety training and safety incentives on Chinese construction workers' risk-taking behavior based on the structural equation model

Chunpeng Liu
*School of Marxism, Yancheng Teachers University, Yancheng, P.R. China*

Yan Wang*
*Business School, Yancheng Teachers University, Yancheng, P.R. China*

ABSTRACT: The construction industry is notorious for its high accident and work injury rates. Behavior-based safety is an important application of behavioral science to solve safety problems in the construction sector. This study aims to examine the effects of safety training, safety incentives, and risk tolerance on construction workers' risk-taking behavior by quantitative methods. A cross-sectional survey of 1,026 construction workers at 53 construction sites in China was conducted. Confirmatory factor analysis of structural equation modeling was used to examine the relationship between these variables. Results show that safety training has a significant direct negative impact on risk-taking behavior, while safety incentives have an indirect negative impact on risk-taking behavior through risk tolerance. Therefore, safety training and safety incentive are different ways to reduce the risk-taking behavior of construction workers and provide different intervention ideas for the behavior intervention of construction workers.

## 1 INTRODUCTION

The development of the construction industry has been plagued by frequent accidents or injuries. Around 60,000 people worldwide are reported to be killed by construction accidents every year, which equates to one accident every nine minutes (Wang et al. 2016). Previous studies have found that more than 80 percent of accidents are caused by workers' risk-taking behavior (Haslam et al. 2005; He et al. 2020). Safety training and safety incentives are commonly used safety management methods on construction sites (Sukamani et al. 2021), but no research has quantitatively analyzed their effects on workers' intrinsic risk tolerance and extrinsic risk-taking behavior. This study aims to explore the effects of safety training and safety incentives on risk tolerance and risk-taking behavior and to compare their different mechanisms.

Safety training has always been regarded by managers of high-risk industries as an important means to ensure safe production. Many studies have shown that employees can improve their safe operation behavior after effective safety training. Williams et al. (2010) proved that the use of personal protective equipment by workers has significantly improved after safety training. Man et al. (2021) found that safety training has a significant negative impact on risk-taking behavior through a study of 536 construction workers in Hong Kong. Accordingly, we hypothesized:

H1: Safety training negatively influences risk-taking behavior.

Safety incentive refers to the use of reward or punishment measures for people in the system to ensure the safety of individuals, enterprises, and the environment to prevent safety accidents, which is regarded as effective safety management means. Zulkefli et al. (2014) found that both monetary and non-monetary incentives will affect workers' safety-related behaviors. Based on the social exchange theory, when individuals perceive that they have obtained economic, rights, status,

*Corresponding Author: wangyan1818daisy@163.com

 DOI 10.1201/9781003348023-45

and other resources from the organization, they will consciously and actively repay the organization based on the purpose of exchange (Blau 1968). That is, the behavioral state of employees is the structure of the enterprise leadership's investment in employee management. When employees perceive higher safety incentives, they tend to show lower risk-taking behaviors. Therefore, we hypothesized:

H2: Safety incentives significantly negatively affect risk-taking behavior.

In addition, the external environment not only affects the individual's behavior but also affects the individual's internal cognition, such as risk tolerance (Wang et al. 2016). Risk tolerance refers to an individual's perception and willingness to accept existing risks. Harvey et al. (2001) conducted a 16-month intervention study of employees in the nuclear industry and found that the safety attitudes of employees who had received safety training changed significantly. Scott (2014) found that safety training and education significantly negatively affected pedestrians' risk-taking attitudes. In addition, safety incentives can also affect workers' inherent risk tolerance, and for construction workers, leaders' safety incentives can affect their risk tolerance. Thus, we hypothesized:

H3: Safety training negatively influences risk tolerance.

H4: Safety incentives negatively affect risk tolerance.

According to social cognitive theory, individual behavior is influenced by the body's self-cognition (Tolman 1952). Man et al. (2021) proposed that there is a significant positive correlation between construction workers' attitudes towards risk-taking and risk-taking behaviors and that risk-taking attitudes have a significant mediating effect between safety training and risk-taking behaviors. Bhandari & Hallowell (2022) showed that work-related risk tolerance significantly affects individuals' willingness to violate safety rules, and risk tolerance has a significant mediating effect between safety climate and safety rule violations. Therefore, we hypothesized:

H5: Risk tolerance has a significant positive effect on construction workers' risk-taking behavior.

H6: Risk tolerance has a mediating effect between safety training and risk-taking behavior.

H7: Risk tolerance has a mediating effect between safety incentives and risk-taking behavior.

Building on previous studies, the conceptual model of the present study is shown in Figure 1. First, we explored the impacts of safety training (H1) and safety incentives (H2) on risk-taking behavior. Second, we explored the impacts of safety training (H3) and safety incentives (H4) on risk tolerance. Then, we explored the impacts of risk tolerance on risk-taking behavior (H5). Finally, we explored the mediating effect of risk tolerance on the relationships of safety training (H6) and safety incentives (H7) with risk-taking behavior.

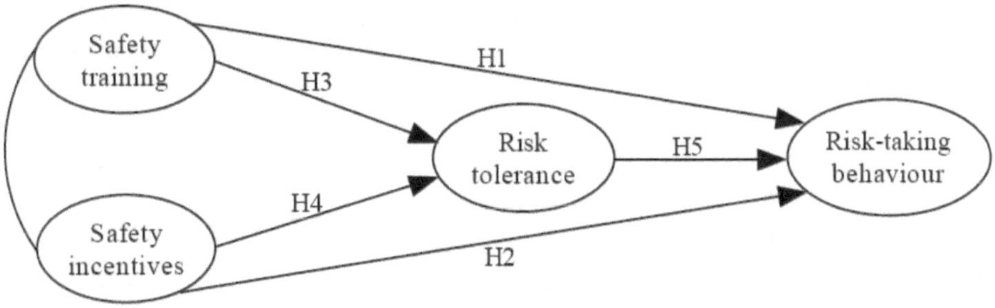


Figure 1. Conceptual model and hypotheses of the study.

## 2 MATERIALS AND METHODS

### 2.1 *Procedures and participants*

From September 7, 2021, to January 20, 2022, a total of 1,102 questionnaires were distributed at 53 construction projects in China, and all questionnaires were collected. Questionnaires that were not seriously answered were omitted. A total of 1,026 valid questionnaires were collected from front-line construction workers, with an effective recovery rate of 93.10%.

## 2.2 *Measures*

### 2.2.1 *Safety training*

Safety training was measured by using the six-item scale of Vinodkumar and Bhasi (2010). Each item was rated using a 5-point Likert scale ranging from 1 (strongly disagree) to 5 (strongly agree), with higher scores indicating greater safety training. Sukamani et al. (2021) demonstrated that the scale has good reliability and validity for the measurement of construction worker safety training. In the current study, Cronbach's alpha was 0.901.

### 2.2.2 *Safety incentives*

Safety incentives were measured by five items from Fernández-Muñiz (2017). Each item was rated using a 5-point Likert scale ranging from 1 (strongly disagree) to 5 (strongly agree), with higher scores indicating greater safety incentives. Internal consistency, as measured by Cronbach's alpha was 0.877.

### 2.2.3 *Risk tolerance*

Risk tolerance was measured by the scale of Wang et al. (2016), and the scale has a total of 8 items. Each item has five options to evaluate the acceptability of risk status, ranging from 1 (very small) to 5 (very large), and the higher the score, the greater the risk tolerance. In this study, Cronbach's alpha was 0.950.

### 2.2.4 *Risk-taking behavior*

Risk-taking behavior was assessed using 5-items by modifying the scale of Low et al. (2019). All items were rated on a 7-point Likert scale ranging from 1 (strongly disagree) to 7 (strongly agree). It had been proved to have good reliability and validity. In the current study, Cronbach's alpha was 0.953.

## 2.3 *Statistical analysis*

We used SPSS 26 software for internal consistency, descriptive statistics, and correlation analysis. Next, we used Mplus 8.7 software, conducting the structural equation model (Figure 1), with maximum likelihood estimation (ML). Then, we used the Delta method to calculate the standard error of the mediation effect with 1,000 nonparametric bootstrap replicates.

# 3 RESULTS

## 3.1 *Descriptive findings*

Scale descriptive statistics (mean, SD), and correlation between primary variables are displayed in Table 1. All variables were significantly intercorrelated.

## 3.2 *Model testing*

The conceptual model (Figure 1) showed reasonable fit based on all the indices: ML, $\chi^2/df=3.542$, RMSEA=0.050, CFI=0.969, TLI=0.965, SRMR=0.028. As shown in Figure 2, relationships among the variables were further analyzed using a structural model with standardized parameter estimates. We found that safety training had a significant negative association with risk-taking behavior ($\beta = 0.238$; $p < 0.001$), while safety incentives had no significant influence on risk-taking behavior ($\beta = 0.054$; $p = 0.388$), leading to the support of H1 and rejection of H2. However, we found that

Table 1. Means, standard deviations and correlations among the primary study variables.

| | M | SD | ST | SI | RT | RTB |
|---|---|---|---|---|---|---|
| ST | 3.94 | 0.745 | 1 | | | |
| SI | 3.78 | 0.785 | 0.706** | 1 | | |
| RT | 2.42 | 1.046 | -0.334** | -0.360** | 1 | |
| RTB | 2.52 | 1.491 | -0.252** | -0.201** | 0.251** | 1 |

Notes: ST refers to safety training; SI refers to safety incentives; RT refers to risk tolerance; RTB refers to risk-taking behavior; M refers to means; SD refers to standard deviations; "**" means that the value of p is smaller than 0.01.

safety training has no significant influence on risk tolerance ($\beta = -0.115$; $p = 0.054$), while safety incentives negatively significant effect on risk tolerance ($\beta = -0.300$; $p < 0.001$), leading to the rejection of H3, and supporting of H4. In addition, we also found that risk tolerance positively influenced risk-taking behavior ($\beta = 0.203$; $p < 0.001$), supporting H5.

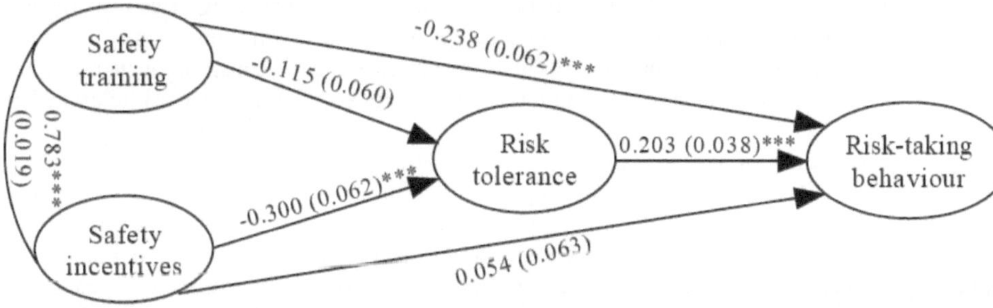


Figure 2. Structural model of safety training and safety incentives predicting risk tolerance, and risk tolerance predicting risk-taking behavior, with standardized path coefficients and SEs (in parentheses). "***" means that the value of p is smaller than 0.001.

### 3.3 *Mediation results*

Table 2 demonstrates mediation testing results. Risk tolerance mediated relations between safety incentives and risk-taking behavior, $\beta = -0.061$, $SE = 0.017$, $z = -3.471$, and $p = 0.001$ (supporting H7), but not between safety training and risk-taking behavior, $\beta = -0.023$, $SE = 0.013$, $z = -1.815$, and $p = 0.070$ (rejecting H6).

Table 2. Results of the mediation with a standardized estimate.

| | $\beta$ | S.E. | Z | P |
|---|---|---|---|---|
| ST -> RT -> RTB | -0.023 | 0.013 | -1.815 | 0.070 |
| SI -> RT -> RTB | -0.061 | 0.017 | -3.471 | 0.001 |

Notes: ST refers to safety training, SI refers to safety incentives, RT refers to risk tolerance, and RTB refers to risk-taking behavior.

## 4 DISCUSSIONS

Safety training and safety incentives, as commonly used safety management methods on construction sites, have been discussed by many scholars for their role in promoting safety behaviors

(Fernández-Muñiz et al. 2017; Sukamani et al. 2021). However, no studies have explored the effects of safety training and safety incentives on construction workers' risk-taking behavior. Through the study of 1,026 construction workers in 53 construction sites in China, this study found that safety training has a direct negative impact on construction workers' risk-taking behavior, while safety incentives have an indirect negative impact on construction workers through risk tolerance.

In the structural equation model path analysis, it is found that safety training has a significant negative effect on the risk-taking behavior of construction workers, but safety training has no significant effect on the inherent risk tolerance of construction workers. This suggests that safety training can reduce construction workers' risk-taking behavior, but it does not change construction workers' inherent risk tolerance. In addition, safety incentives have no significant effect on construction workers' risk-taking behavior, but safety incentives have a significant negative effect on construction workers' inherent risk tolerance. This indicates that safety incentives cannot directly affect construction workers' risk-taking behavior, but safety incentives can significantly negatively affect workers' inherent risk tolerance. Therefore, it can be seen from the above conclusions that safety training negatively affects the risk-taking behavior of construction workers, and safety incentives negatively affect the risk attitude of construction workers.

In the mediation effect analysis, it is found that risk tolerance has no significant mediating effect between safety training and risk-taking behavior, but risk tolerance has a significant mediating effect between safety incentives and risk-taking behavior. That is, safety training cannot affect the change of risk-taking behavior by changing the inherent risk tolerance of construction workers, but safety incentives can reduce their risk-taking behavior by acting on the inherent risk tolerance of construction workers.

Integrating path analysis and mediation effects concluded that both safety training and safety incentives have significant effects on inhibiting construction workers' risk-taking behaviors, but their mechanisms are different. Safety training directly inhibits the occurrence of risk-taking behaviors of construction workers, but safety incentives inhibit the occurrence of risk-taking behaviors by affecting workers' inherent risk tolerance. Construction site managers can focus on choosing two safety management methods, safety training, and safety incentives, according to different mechanisms (Bhandari & Hallowell 2022; Low et al. 2019).

## 5 CONCLUSION

The current study constructed a structural equation model to explore the predictive powers of safety training and safety incentives on risk-taking behavior. We found the negative effect of safety training on risk-taking behavior and safety incentives on risk tolerance. Furthermore, we found that risk tolerance mediated the relationship between safety incentives and risk-taking behavior, but not the relationship between safety training and risk-taking behavior. It is reasonable to infer that improving both safety training and safety incentives can hinder risk-taking behavior, but their effect mechanisms are different. Safety training directly reduces risk-taking behavior, and safety incentives indirectly reduce risk-taking behavior by affecting intrinsic risk tolerance.

## REFERENCES

Bhandari, S., & Hallowell, M. R. (2022). Influence of safety climate on risk tolerance and risk-taking behavior: A cross-cultural examination. *Safety science*, 146, 105559.

Blau, P. M. (1968). Social exchange. *International encyclopedia of the social sciences*, 7(4), 452–457.

Fernández-Muñiz, B., Montes-Peón, J. M., & Vázquez-Ordás, C. J. (2017). The role of safety leadership and working conditions in safety performance in process industries. *Journal of Loss Prevention in the Process Industries*, 50, 403–415.

Harvey, J., Bolam, H., Gregory, D., & Erdos, G. (2001). The effectiveness of training to change safety culture and attitudes within a highly regulated environment. *Personnel Review*, 1–12.

Haslam, S. A., O'Brien, A., Jetten, J., Vormedal, K., & Penna, S. (2005). Taking the strain: Social identity, social support, and the experience of stress. *British journal of social psychology*, 44(3), 355–370.

He, C., McCabe, B., Jia, G., & Sun, J. (2020). Effects of safety climate and safety behavior on safety outcomes between supervisors and construction workers. *Journal of construction engineering and management*, 146(1), 04019092.

Low, B. K. L., Man, S. S., Chan, A. H. S., & Alabdulkarim, S. (2019). Construction worker risk-taking behavior model with individual and organizational factors. *International journal of environmental research and public health*, 16(8), 1335.

Man, S. S., Chan, A. H. S., Alabdulkarim, S., & Zhang, T. (2021). The effect of personal and organizational factors on the risk-taking behavior of Hong Kong construction workers. *Safety science*, 136, 105155.

Sukamani, D., Wang, J., & Kusi, M. (2021). Impact of safety worker behavior and safety climate as mediator and safety training as moderator on safety performance in construction firms in Nepal. *KSCE Journal of Civil Engineering*, 25(5), 1555–1567.

Scott, D. (2014). *A Multimedia Pedestrian Safety Program and School Infrastructure: Finding The Connection To Pedestrian Risk-taking Attitudes And Perceptions Of Pedestrian Behavior*. 12, 1–15.


Sukamani, D., Wang, J., & Kusi, M. (2021). Impact of safety worker behavior and safety climate as mediator and safety training as moderator on safety performance in construction firms in Nepal. *KSCE Journal of Civil Engineering*, 25(5), 1555–1567.


Tolman, E. C. (1952). A cognition motivation model. Psychological review, 59(5), 389.

Vinodkumar, M. N., & Bhasi, M. (2010). Safety management practices and safety behaviour: Assessing the mediating role of safety knowledge and motivation. *Accident Analysis & Prevention*, 42(6), 2082–2093.

Wang, J., Zou, P. X., & Li, P. P. (2016). Critical factors and paths influencing construction workers' safety risk tolerances. *Accident analysis & prevention*, 93, 267–279.

Williams Jr, Q., Ochsner, M., Marshall, E., Kimmel, L., & Martino, C. (2010). The impact of a peer-led participatory health and safety training program for Latino day laborers in construction. *Journal of safety research*, 41(3), 253–261.


Wang, J., Zou, P. X., & Li, P. P. (2016). Critical factors and paths influencing construction workers' safety risk tolerances. *Accident analysis & prevention*, 93, 267–279.


Zulkefli, F. A., Ulang, N. M., Baharum, F. (2014). Construction health and safety: effectiveness of safety incentive programme. *SHS Web of conferences*, 11:01012

*Advances in Urban Construction and Management Engineering – Cha (Ed)*


# Analysis of the impact of network public opinion on urban public security and strategy research

Xitao Liu* & Xinyao Zhou*
*College of Public Finance and Administration, Harbin University of Commerce, Harbin, China*

ABSTRACT: With the advent of the information age, netizens are more willing to express their wishes on the Internet. However, while narrowing the distance between the public and the government, there are often problems of uncontrolled information dissemination. Network public opinion emerges as the times require a combination of traditional public opinion and modern communication media. The uncertain factors and security risks of the virtual environment also bring great challenges to urban public safety to some extent, and the multiplier effect of public opinion also has a great impact on it. Therefore, from the perspective of analyzing the impact of network public opinion on public security, this paper proposes relevant mechanisms and strategies for city managers in terms of big data, opinion leaders, and laws and regulations. It is more necessary to improve network security governance in public security.

## 1 INTRODUCTION

According to the 49th *Statistical Report on China's Internet Development Status* released by the China Internet Network Information Center (CNN-IC) in Beijing on February 25, 2022, as of December 2021, the number of Internet users in my country reached 1. 032 billion, which is higher than that in 2021. In December, it increased by 42. 96 million, and the Internet penetration rate reached 73. 0%, an increase of 2.6 percentage points from December 2021 (China Internet Network Information Center, 2022). Against the background of such data, increased netizens are willing to use the Internet to participate in political life, but there are also situations where the age of netizens tends to be younger and their quality tends to be lower. In a certain sense, this also shows that online public opinion will bring certain challenges to urban public security. This paper analyzes the impact of network public opinion on public security and proposes corresponding strategies for the negative impact.

## 2 THE INFLUENCE OF NETWORK PUBLIC OPINION ON URBAN PUBLIC SECURITY

### 2.1 *The positive impact of network public opinion on urban public safety*

#### 2.1.1 *The enthusiasm for public political participation has increased significantly*

With the change in media technology, various media platforms such as Weibo, WeChat, and short videos continue to emerge, and there are increased platforms where the public can express their views (Bi et al. 2019). If they use the Internet, they can participate in political life anytime, anywhere (Bi et al. 2019). In addition, the online world is relatively free and low-cost, especially with the advent of the 5G era and the popularization of the Internet, the public's enthusiasm for participating

*Corresponding Authors: liu_xitao@126.com and 1902761618@qq.com


DOI 10.1201/9781003348023-46

in political life through the Internet has been continuously improved, enabling the public security system to effectively respond to public opinion events.

### 2.1.2 *Promote the construction of urban public safety*

Network public opinion usually prompts a certain event or topic to arouse high attention on the Internet, which may attract the attention of government decision-makers, and this problem can easily be transformed from a single social problem to a public problem. At this time, heated public opinion played a driving role in urban public security, effectively improving the efficiency of public security governance. Such as Shandong fake vaccine, Changchun Changsheng fake vaccine, and other similar incidents, the state quickly promulgated the "Vaccine Administration Law of the People's Republic of China (Draft for Comment)" and officially implemented it on December 1, 2019, which to a certain extent promoted the Urban public safety construction.

### 2.1.3 *Promote the openness of urban governance services*

The openness and equality of the Internet provide a great bridge for the openness of urban governance services, which can easily and quickly contact the public and city managers, greatly stimulate the public's sense of ownership, and network public opinion is more convenient for the government. Effectively collecting opinions from the masses make the government pay more attention to people's livelihood.

## 2.2 *The negative influence of network public opinion on urban public security*

### 2.2.1 *Infectious public opinion can easily harm public safety*

SIR is a classic model of the spread of infectious diseases. When the main body of public opinion is brought into this model, the public participating in public opinion can be divided into susceptible (S), infective (I), and removal (R). Therefore, the public who issued or forwarded the public opinion event can be assumed to be infective (I), the public who has a neutral or agreeable opinion on this event and spread it can be assumed to be susceptible (S), and the public who do not comment on this event or the public objecting to the dissemination of the opinion is assumed to be removal (R). Assume that the number of people participating in the discussion of this public opinion event is N, the proportional coefficient between the number of infectives (I) and susceptible (S) in time t is $\beta$, and the proportionality coefficient between the number of removals (R) and the number of infectives (I) is $\gamma$, then we can infer that the infection model of the public participating in public opinion events is:

$$N = I + R + S$$

$$S + I \xrightarrow{\beta} I_{(t)} + I$$

$$I_t \xrightarrow{\gamma} R_t$$

According to the model, it can be known that the public has a great influence on the dissemination of public opinion events, so in the process of information dissemination, public opinions that endanger public safety may be formed.

### 2.2.2 *Hidden security risks are a threat to public security governance*

Network information security is an important guarantee to ensure the accuracy of public security governance, but its hidden dangers mainly lie in the loss of authenticity and integrity. Therefore, the security risks of network public opinion will threaten the realization of public security governance to a certain extent. In the process of spreading online public opinion, it may be intercepted and illegally altered by some criminals, resulting in the loss of authenticity and integrity of the information. From this point of view, urban public security is facing unprecedented challenges, which is a security threat to the process of public security governance.

2.2.3 *The outbreak of public opinion crisis affects the efficiency of public security governance*
When the crisis of public opinion broke out, the city, as the carrier of crisis, risk, and development coexisted, and the cost of governance became higher and higher. Therefore, the relevant indicators of public security governance capability are calculated according to the analytic hierarchy process, and the results are shown in Tables 1 to 4. Among them, CIU, CIU1 represent consistency indicators, RIU, RIU1 represent average random consistency indicators, CRU, CRU1 represent consistency ratio; i = 1, ..., 6.

Table 1. Judgment matrix and analysis results of public security governance.

| $U$ | $U_1$ | $U_2$ | $U_3$ | $U_4$ | $U_5$ | $U_6$ | $\alpha_1$ |
|---|---|---|---|---|---|---|---|
| $U_1$ | 1 | 1/3 | 1/3 | 1/2 | 1/3 | 2 | 0. 0831 |
| $U_2$ | 3 | 1 | 1 | 2 | 1 | 5 | 0. 2506 |
| $U_3$ | 3 | 1 | 1 | 2 | 1 | 5 | 0. 2506 |
| $U_4$ | 2 | 1/2 | 1/2 | 1 | 1/2 | 2 | 0. 1276 |
| $U_5$ | 3 | 1 | 1 | 2 | 1 | 3 | 0. 2321 |
| $U_6$ | 1/2 | 1/5 | 1/5 | 1/2 | 1/3 | 1 | 0. 0560 |
| $\lambda_{max}^{U}$ =6.0556, $CI^U$ = 0. 111 $RI^U$ =1. 24 and $CR^U$ = 0. 0090 | | | | | | | |

Table 2. Pre-warning judgment matrix and analysis results.

| $U_1$ | $V_{11}$ | $V_{12}$ | $V_{13}$ | $V_{14}$ | $V_{15}$ | $B_{1j}$ |
|---|---|---|---|---|---|---|
| $V_{11}$ | 1 | 2 | 2 | 3 | 2 | 0. 3298 |
| $V_{12}$ | 1/2 | 1 | 3 | 4 | 3 | 0. 3086 |
| $V_{13}$ | 1/2 | 1/3 | 1 | 2 | 2 | 0. 1577 |
| $V_{14}$ | 1/3 | 1/4 | 1/2 | 1 | 1/3 | 0. 0723 |
| $V_{15}$ | 1/2 | 1/3 | 1/2 | 3 | 1 | 0. 1316 |
| $\lambda_{max}^{U1}$ =5. 2836, $CI^{U1}$ =0. 0709, $RI^{U1}$ =1. 12, and $CR^{U1}$ = 0. 0633 | | | | | | |

Table 3. Judgment matrix and analysis results of emergency management.

| $U_2$ | $V_{21}$ | $V_{22}$ | $V_{23}$ | $V_{24}$ | $B_{2j}$ |
|---|---|---|---|---|---|
| $V_{21}$ | 1 | 1 | 1/4 | 1/5 | 0. 0889 |
| $V_{22}$ | 1 | 1 | 1/4 | 1/5 | 0. 0889 |
| $V_{23}$ | 4 | 4 | 1 | 1/2 | 0. 3182 |
| $V_{24}$ | 5 | 5 | 2 | 1 | 0. 5039 |
| $\lambda_{max}^{U2}$ =4.0277, $CI^{U2}$ = 0. 0092, $RI^{U2}$ =.9, and $CR^{U2}$ = 0. 0102 | | | | | |

We calculate the overall consistency of the measure layer to the target layer and obtain the following results: CI = 0.0211, RI = 0.858; CR = CI/RI = 0.0246 < 0.1. The results meet the overall consistency requirement. Through the above indicators, it can be determined that the public security governance capacity needs to be strengthened, and the outbreak of the public opinion crisis affects the efficiency of public security governance.

Table 4. Judgment matrix and analysis results of aftercare and recovery.

| $U_3$ | $V_{31}$ | $V_{32}$ | $V_{33}$ | $V_{34}$ | $B_{3j}$ |
|---|---|---|---|---|---|
| $V_{31}$ | 1 | 2 | 1/2 | 1/3 | 0. 1539 |
| $V_{32}$ | 1/2 | 1 | 1/4 | 1/5 | 0. 0809 |
| $V_{33}$ | 2 | 4 | 1 | 1/2 | 0. 2880 |
| $V_{34}$ | 3 | 5 | 2 | 1 | 0. 4773 |
| $\lambda_{max}^{U3}$=4.0211, $CI^{U3}$ = 0. 0070, $RI^{U3}$ =.9, and $CR^{U3}$ = 0. 0078 | | | | | |

## 3 CONSTRUCTION STRATEGY OF NETWORK PUBLIC OPINION RESPONSE MECHANISM UNDER URBAN PUBLIC SECURITY

With the advent of the 5G era and the increase in network penetration, the Internet will continue to change the structure of traditional network public opinion, and the public is more willing to use the Internet as a virtual venue to express and maintain themselves. The Fifth Plenary Session of the 19th Central Committee of the Communist Party of China clearly stated, "Like the real world, cyberspace should not only promote freedom but also maintain order. We must not only respect the rights of netizens to exchange ideas and express their will, but also build a good network order following the law, which will help protect the legitimate rights and interests of many netizens." Under such circumstances, it is necessary to actively respond to urban crisis events, strengthen the awareness of online public opinion, strengthen early warning monitoring and research and judgment, provide the ability to respond to public security crises, and improve the level of crisis governance.

### 3.1 *Improve the big data screening and judgment mechanism*

In the current era, this phenomenon of the public being in the whirlpool of public opinion is more prominent because it has itself become an important part of the source of big data. Then it is necessary to use the big data system to filter such information, intervene in the event itself in advance, especially when the urban crisis event occurs, and pay more attention to the appearance of such information.

In addition to the vast amount of data resources, big data involves a fundamental shift in the way it is processed, laying the foundation for public safety governance. Therefore, it is necessary for government departments to conduct comprehensive research and judgment on the current public opinion through big data technology, to ensure that there is a more scientific and reliable early warning mechanism in urban crisis management, and to accurately judge the current situation of the public opinion field and the public's cognition and expression of public opinion events.

At the same time, consideration should be given to using advanced technologies such as big data to push relevant content of public opinion guidance on new media platforms, providing the public with the truth, disseminating relevant knowledge about dealing with crises, and reducing the public's sense of panic (Yi et al. 2018).

### 3.2 *Constructing a guiding mechanism for online opinion leaders*

In the current environment of two-level communication, opinion leaders, as an important role, play a particularly important role in the outbreak of public opinion and are an important factor in transforming the huge and complex public opinion until the incident subsides. It also plays an indispensable role in stabilizing the public sentiment and correctly guides public opinion by gaining the public's trust.

Actively play the role of opinion leaders, use life-like words to get closer to the public, maintain close contact, promote the public's rational thinking on public opinion events, and avoid the phenomenon of group emotional extremes. The public who accepts the guidance will have a correct understanding of the occurrence of urban public security crises, and at the same time can try to avoid the occurrence of new crises. The greater spread of positive energy is helpful to the effective governance of urban public safety, and it can actively transform users' emotions and prevent the phenomenon of "information cocoons".

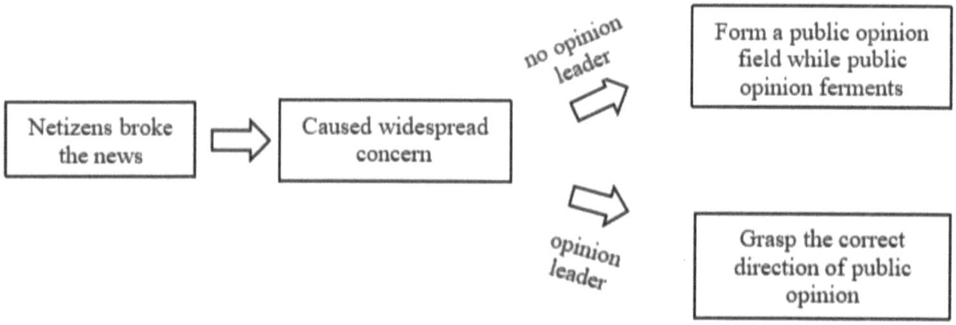


Figure 1. Whether there are opinion leaders and the development process of public opinion.

### 3.3 *Improve the mechanism for setting relevant laws and regulations*

In response to online public opinion, corresponding laws and regulations should also be improved, and existing relevant laws and regulations should be checked and filled. The current system of laws and regulations is lacking. Compared with the rapid update and iteration of the Internet, there is a certain lag in legal updates. At the same time, there are problems of insufficient practicality and binding force, which often lead to unsatisfactory public security governance effects.

Therefore, relevant laws and regulations should be actively improved to fundamentally reduce the frequency of online public opinion crises. Given the current network development trend, formulate laws and regulations that are more in line with the status quo, and improve the legalization and modernization of public security governance. The online world should not be allowed to become an extrajudicial place to protect public rights and the effective implementation of government crisis management.

## 4 CONCLUSION

With the strong rise of emerging media, although online public opinion can improve the public's participation and enthusiasm in political life to a certain extent, it also inevitably has some negative effects and has its limitations. Therefore, we should correctly grasp the laws of public security governance and find scientific methods to guide network public opinion.

Based on expounding on the impact of network public opinion on urban public security, this paper analyzes the current situation and countermeasures of network public opinion with the help of the SIR model and AHP, and proposes to improve the big data screening and judgment mechanism, constructing a guiding mechanism for online opinion leaders and improve the mechanism for setting relevant laws and regulations. Only in this way can the urban crisis be better resolved rather than worsened, and a reasonable and correct resolution of public opinion will inevitably become the mainstream way of public security governance in the future.

## REFERENCES


Bi xiang, Tang cunchen, Xiao qiao. (2019) Research on Short Video Social Media Public Opinion Monitoring and Crisis Prevention Mechanism. *Intelligence Theory and Practice*, 42(10): 102–108.

Chen shaowei, Bao jing, Jia kai. (2020) Research on the Governance of Intelligent Push Business: Policy Implications from the Perspective of International Comparison. *Chinese Administration*, (09): 149–154.

China Internet Network Information Center. (2022) The 49th *Statistical Report on China's Internet Development Status*. http://www.cnnic.net.cn/hlwfzyj/hlwxzbg/hlwtjbg/202202/t20220225_71727.htm

Yi Jing, Liu Peiyu, Tang Xiaobing. (2018) Improved SIR Advertising Spreading Model and Its Effectiveness in Social Network. *Procedia Computer Science*, 129: 215–218.

*Advances in Urban Construction and Management Engineering – Cha (Ed)*


# Research on comprehensive integration based on urban intelligent traffic management system

Can Chen*
*Chongqing Energy College, Chongqing, China*

Xi Ni*
*China Railway Southwest Research Institute Co., Ltd., Chengdu, China*

ABSTRACT: To discuss the current situation of urban intelligent traffic management system integration and promote the development of urban intelligent traffic, based on analyzing issues with integration of urban intelligent traffic management system, this paper builds a perfect intelligent urban traffic integrated management system, proposes to safeguard measures for an integrated intelligent traffic management system, implements the integrated development of traffic management in smart city, and promotes the progress of urban intelligence.

## 1 INTRODUCTION

With the continuous development of urban roads, urban traffic problems are becoming increasingly obvious, and the construction of an intelligent traffic management system is becoming more important. Traffic plays an important role in urban development. Through the integration of urban traffic intelligent management and the use of information technology and intelligent technology, we can realize the intelligent development of urban traffic and effectively alleviate the problem of urban traffic congestion, and ensure the safe operation of urban traffic (Brazil 2021).

## 2 THE PROBLEMS AND CURRENT SITUATION OF INTELLIGENT DEVELOPMENT OF URBAN TRAFFIC IN CHINA

The intelligent construction of urban road traffic was put forward by Chinese experts at the end of the last century, mainly to alleviate the phenomenon of traffic congestion, alleviate the pressure of traffic congestion, comply with the development needs of the times, and move forward, and build a smart city. However, because leaders blindly pursue economic development and ignore these sub infrastructures, road traffic jams occur every year. At present, the fundamental reasons for the lack of functionality, low security of information resources, backward infrastructure and technology, and low overall quality of management personnel in road traffic intelligent construction in China are as follows: lack of unified planning and management standards, lack of management information technology, lack of financial security, and incomplete related laws and policies. These factors are also the key problems to be solved in the construction and application of intelligent transportation systems in the current environment.

*Corresponding Authors: xiexiaohong@cug.edu.cn and gzmetro11@126.com

 DOI 10.1201/9781003348023-47

### 2.1 *Problems existing in the development of urban traffic intelligence*

#### 2.1.1 *Policy issues*

Without sound laws and policies, the construction of intelligent transportation cannot be properly restricted, and the construction risk will be further increased. Fundamentally, the government's efforts are inseparable from the effective improvement of intelligent transportation management regulations. The government can also make necessary adjustments to the current policies in combination with its reality, but the actual effect is very limited and still needs the government's attention. At present, more policies in China focus on coping with downward pressure and pay insufficient attention to intelligent transportation systems. An intelligent transportation system is the core of the city's intelligent transportation construction and application policy (Yu 2020).

#### 2.1.2 *Funding issues*

The construction of intelligent transportation requires a lot of humans, material, and financial resources. Therefore, the government and traffic management departments increase and contact the financial pressure. In recent years, the municipal government has begun to appropriately increase investment in traffic construction and pay attention to the establishment and optimization of the intelligent traffic management system, but the results are not very big, and the gap between the nation's first and second-tier cities is still large. Given the great downward pressure on China's economy, this financial pressure is difficult to solve in the short term, there is a shortage of funds for intelligent transportation construction, and many construction projects are difficult to carry out in an all-around way. Therefore, the core of the capital is also an important obstacle to the city's intelligent transportation construction.

#### 2.1.3 *Technical problems*

Economic development has greatly promoted the innovation and upgrading of information technology and intelligent transportation construction is also undergoing dynamic changes. To achieve the best effect of intelligent transportation construction, we must keep up with the pace of information development, quickly introduce advanced information management technology, build a perfect technical system, and speed up the process of intelligent transportation construction. However, due to the lack of information technology insight, lack of funds, and other factors, the information management technology of intelligent transportation construction is still in a backward state and cannot be fully adjusted in the short term (Zhang 2020). Technical points are also the main problems faced by intelligent transportation construction in country-level cities.

#### 2.1.4 *Mass consciousness problems*

Most county-level city citizens are transformed by farmers who enter the city. They usually do not have a good education and are highly random. It is common to bring rural travel habits to cities, violate traffic rules, drive without a license and drive after drinking among the people who caused traffic accidents, citizens from rural to urban accounted for a large proportion. The causes of most traffic accidents are simple and clear. In many cases, interests are intertwined, resulting in traffic accidents. No matter who is right or wrong, they will put forward requirements beyond expectations (Zhang 2019). To some extent, they accuse the government of negligence and so on, which is also true. On the other hand, some prefecture managers, especially local middle-level managers, live comfortably, are less impacted by the outside world, have conservative thinking, and have ambiguous attitudes in the face of innovation and change. If a few county-level cities are changed, this group will be subject to resistance Many design concepts in road construction still follow the old practice in stages. A few years ago, when designing road intersections, a city still followed the existing roundabout design, which could not solve the contradiction between roundabout vehicles and vehicles leaving the roundabout area. The construction of green space in the middle of the road only considers the beauty and does not consider the contradiction between left turning and equidistant reversing on the bifurcated roads on both sides of the road. It is considered to cause many traffic hazards.

## 2.2 *Development status of urban traffic intelligence*

From 2016 to 2021, China's intelligent transportation industry has been increasing year by year. In 2016, the market size of China's intelligent transportation industry was 97.3 billion yuan, and by the end of 2021, the market size increased to 191.7 billion yuan. It is estimated that the market size of the intelligent transportation industry in 2022 is 213.3 billion yuan, as shown in Figure 1.

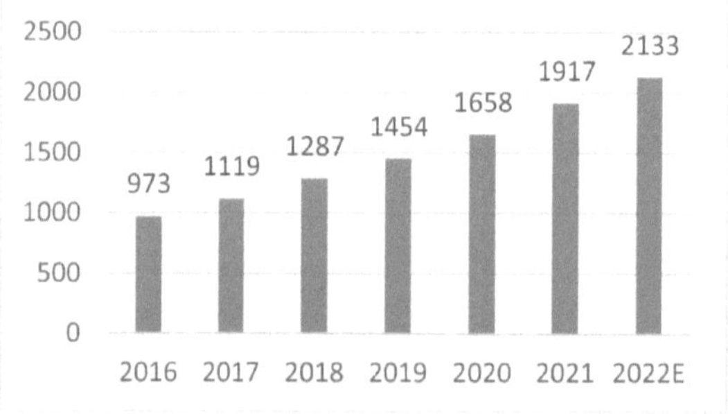


Figure 1. Development scale of China's Intelligent Transportation Industry from 2016 to 2021. (Unit: 100 million Yuan) (Data source: China Competition Information)

The enterprise with the largest market share in China's urban smart transportation industry is China Transinfo Technology Co., Ltd., which accounts for 17.1% of the market share, followed by Enjoyor Technology Co., Ltd., which accounts for 14%, mainly highlighting urban brain operation; And Hisense Transtech accounts for 9.9% of the market share, which mainly use cloud computing, big data, and artificial intelligence to jointly build urban traffic management cloud brain. Meanwhile, other enterprises account for 42.2%, as shown in Figure 2.

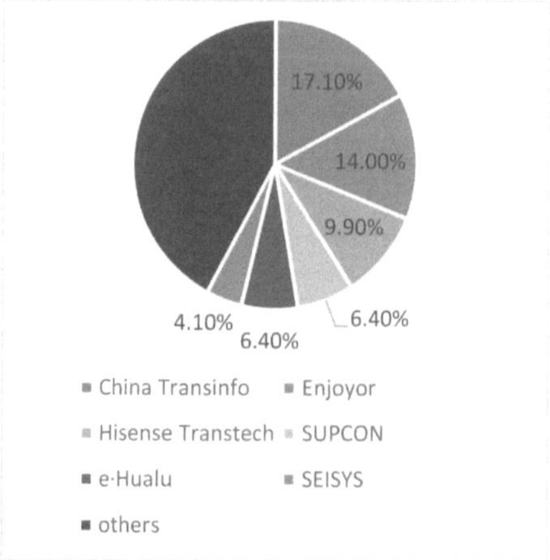


Figure 2. Market share of China's urban smart transportation enterprises in 2020. (Data source: China Competition Information)

# 3 STRUCTURAL COMPOSITION OF URBAN INTELLIGENT TRAFFIC MANAGEMENT SYSTEM

## 3.1 *Structure of urban intelligent transportation system*

The structure of an urban intelligent transportation system is mainly composed of four levels. Each level oversees different service contents and management. The first level is the information application service layer, which mainly realizes the management of transportation business, transportation information release service, transportation safety management, operation related, urban transportation electronic payment, and other information services. The second layer is the decision optimization layer, which mainly realizes the optimization decision of traffic management through information integration, information mining, scheme optimization, emergency plan, and other contents to support the formation of urban traffic. The third layer is the traffic information management layer, which is mainly to realize the storage of information, data screening, data manipulation, and data interaction, sort out and implement data, and update and store traffic data in time to ensure the good operation of traffic. The fourth layer is the traffic information collection layer, which can also be said to be the department coordination layer. It mainly collects data information including traffic police brigade of urban public security bureau, traffic bureau, planning bureau, urban management bureau, etc. The traffic police brigade of the urban public security bureau also needs to monitor traffic dynamic information and static information to better ensure the stable operation of urban traffic and avoid urban congestion (Figure 3).

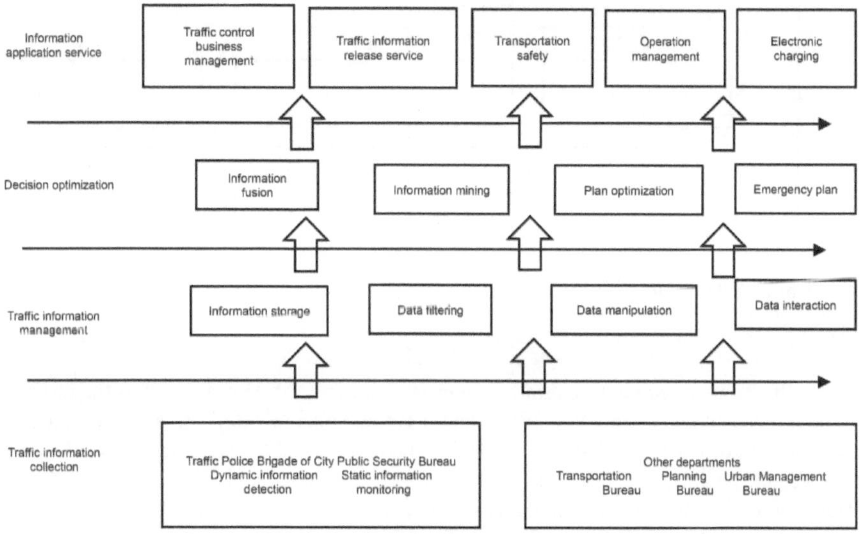


Figure 3. Structure integration diagram of the urban intelligent transportation system.

### 3.2 *Traffic information collection system*

The intelligent traffic management system needs to first locate and analyze the vehicle, plan the vehicle driving route, timely solve the congestion problem of important sections and intersections, and provide real-time road traffic information (An 2019). The composition of the traffic guidance system is shown in Figure 4. The traffic information control center can collect information, including road status, and flow, and then process the information to guide and dredge the traffic flow.

## 4 SUPPORTING MEASURES OF URBAN INTELLIGENT TRAFFIC MANAGEMENT SYSTEM

### 4.1 *Coordinate with various departments to develop an urban intelligent transportation system*

Building an intelligent system of "Internet + city traffic command" requires multi-department coordination. The labor force stands at the forefront of solving problems to ensure the implementation of the functions of the urban intelligent transportation system.

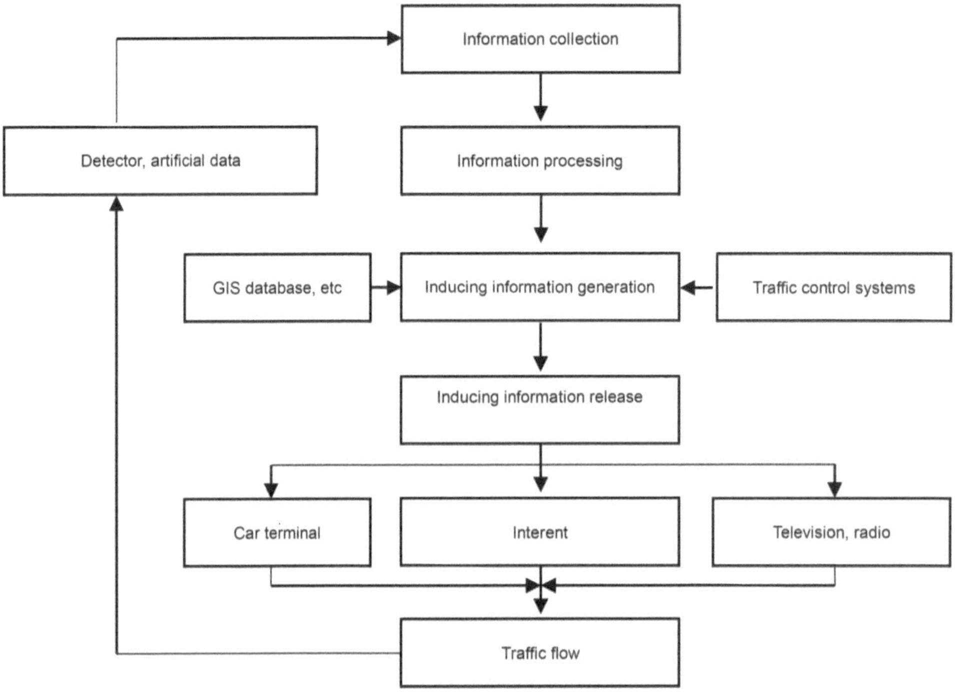


Figure 4. Traffic system guidance diagram.

(1) The intelligent management system of "Internet + urban traffic command" is the traffic management department of the whole city. The traffic management department of public security has multiple powers in road safety, traffic management, and traffic rulemaking. It can operate, safeguard rights, and build an intelligent traffic system, which can greatly improve the efficiency and level of traffic management and is also an important embodiment of "Internet + government affairs" (Dong 2019). The public security traffic management department is responsible for project application, planning, and promotion, while government departments carry out financing, departmental coordination, and other work. The public security traffic management department and government departments jointly lead the effort to create intelligent transportation.
(2) Research and design personnel are required to participate in the "Internet + city traffic command" intelligent system. The implementation and promotion of the project need scientific argumentation and careful decision. Therefore, it needs the participation of relevant professional institutions such as urban planning process institutions, technology research, and design institutions, and urban planning institutions to better understand the urban traffic planning blueprint, traffic governance policies, and the frontier dynamics and progress of intelligent technology. From the perspective of scientific urban development, the planning and implementation can act as a "think tank" for the whole process of project promotion, and timely study and demonstrate the scientific characteristics of the project implementation process and the application of electronic sensing, network communication, big data, cloud computing, and other technologies, build an intelligent traffic management system, fully absorb the advanced experience at home and abroad, and combine urban traffic construction conditions to formulate feasible technical solutions for improvement and innovation.
(3) The construction of the intelligent traffic management system is also inseparable from the cooperation of the masses, the municipal traffic management department should be open

to the masses about the construction progress and pilot programs, and the public exchange of progress, significance, and ideas. It puts forward the vision of intelligent transportation and calls on the whole people to actively support and participate in the construction and management of an intelligent transportation system (Yu 2020).

### 4.2 *Development and utilization of urban traffic intelligent terminal*

Terminal application, as the entrance of the "Internet + city traffic command system", is a key product related to management efficiency and user experience, and an important product at the management level and user level. In the construction of an intelligent system for "Internet + city traffic command", we should pay attention to the development of terminal products, constantly promote intelligent terminals, and enhance the overall experience of traffic control departments and terminal product users.

(1) It is necessary to actively use mobile Internet, mobile positioning, human-computer interaction, and other technologies to develop an intelligent transportation App to display route query, route navigation, map positioning, intelligent guidance, voice response, and other contents of the App. Traffic management departments and users can interact with each other through mobile applications to realize traffic information queries and route optimization and plan more convenient traffic routes for the user (Liu 2017).
(2) In addition to mobile terminal applications, various network communication terminals shall be deployed in various urban areas to facilitate instant communication between urban traffic management departments. It should be emphasized that in special areas such as tunnels, high-speed railways, and subways, the signal coverage of telecom service providers should be strengthened to protect the communication signals of users in these sections so that they can receive relevant communication signals and timely information.
(3) It is also necessary to promote ground Wi-Fi terminal coverage so that the public can access the network through mobile phones, tablets, and other terminals to receive intelligent traffic command information more conveniently.
(4) In the context of the continuous development of the mobile Internet, WeChat has become an information dissemination tool with many functions. Intelligent traffic management also actively uses the advantages of WeChat in independent query and applet applications to promote the openness of urban traffic WeChat. You can take the initiative to open, and you can increase user activity by pushing traffic information promptly, interacting with users through relevant interfaces, and using WeChat mini-programs to provide users with faster information queries. With the WeChat official account as the core, the continuous construction of an intelligent transportation media matrix can provide users with efficient and convenient personalized information services, and improve the content and service ecology of the "Internet + Urban Traffic Command" intelligent system.

## 5 CONCLUSION

To sum up, there are four main problems in the development of urban traffic intelligence in China. The first is policy issues. With the proposal of the 14th five-year plan, the development of urban traffic intelligence is imperative, but the supporting policies and measures are not perfect. The second is the problem of capital. To develop urban traffic intelligence, it needs to spend a lot of money, but the funds in some areas are not in place, resulting in the slow development of urban traffic. The third is technical problems. In the process of intelligent development of urban transportation, intelligent technology and information technology cannot be perfectly integrated with urban transportation, and need to be further improved. The fourth is the awareness of the masses. The masses know little about traffic rules in traffic. In case of traffic accidents, they will think it is a problem of traffic design, and the awareness of abiding by traffic rules needs to be improved. By analyzing the composition of urban traffic intelligent management architecture, we can analyze the composition of the traffic information data induction system, improve the level of urban traffic intelligence, and ensure the stable implementation of the intelligent system by cooperating with various departments of the city. At the same time, urban traffic uses the advantages of "Internet +" to develop urban traffic intelligent terminals to meet the operational requirements of urban traffic.

## REFERENCES


An, S., (2019). Prospect of Xining intelligent traffic management system construction. *J. Transport World*, (23), 15–17.

Dong, W., (2019). Discussion on intelligent transportation technology based on wireless communication. *J. Computer Products and Circulation*, (07), 60.

Liu, Y., (2017). Urban Transportation and Road System Planning. *J. Private Technology*, 04: 238.

St.Paul, B., 2021. Uses Intelligent Means to Manage Urban Traffic. *J. Sensor World*, 27(06): 35.

Yu, H., 2020. Analysis and countermeasures of urban traffic congestion – A case study of Shenyang city. *J. Urban Roads Bridges and Flood Control*, (04), 11–13, 9.

Zhang, H., (2020). Effective improvement and analysis of traffic command and supervision under intelligent background. *J. Legal Expo*, (09), 143–144.

Zhang, Y., (2019). Research on urban road traffic congestion in the context of big data. *J. Mechanical and Electrical Information*, (29), 162–163.

*Advances in Urban Construction and Management Engineering – Cha (Ed)*


# Causes and stability evaluation of K86 landslide on Dayangyun expressway

Chunsheng Li*
*Broadvision Engineering Consultants Co. Ltd., Kunming, Yunnan, China*

Mei Yan
*Urban & Rural Planning & Design Institute of Yunnan Province, Kunming, China*

ABSTRACT: It is of great practical significance to research the causes and prevention of landslide geological disasters. Taking the landslide on the right side of the K86 section of Dayangyun Expressway as the research object, through the analysis of the topography, lithology, and hydrogeology, this paper explored the reasons for the formation of the landslide. It was found that the occurrence of the landslide was affected by a combination of factors. The stability analysis of the landslide was carried out, and the residual sliding force of each section under normal conditions, abnormal conditions I and abnormal conditions II was obtained. Its stable state was determined. Finally, according to the analysis, targeted treatment suggestions were put forward to provide a feasible plan for the prevention and control of landslides.

## 1 INTRODUCTION

China is one of the countries with the most frequent and severe geological disasters in the world. Geological disasters have the characteristics of wide distribution, concentrated time periods, and serious losses. Landslide disasters are one of the important types of geological disasters (Liu 2022). According to statistics, a total of 7,840 geological disasters occurred in China in 2020, including 4,810 landslides, accounting for 61.4% of the total disasters, causing 139 deaths, 58 injuries, and direct economic losses of 5.02 billion yuan (Zhao 2022). It can be seen that landslides seriously threaten the safety of human life and property and bring huge losses to the country and the people (Qi 2022). Therefore, it is particularly necessary to research the causes and prevention of landslide geological disasters. Based on this, this paper explores the reasons for the formation of the landslide and the treatment measures based on the landslide in the Dayangyun Expressway.

## 2 PROJECT OVERVIEW

### 2.1 *Topography*

The landslide on the right side of section K86+660~K86+995 of Dayangyun Expressway is located in Fangmeng, Guanping Township, Yunlong County. It belongs to the tectonic erosion middle mountain landform area. The landslide body is located on a slope, with an aspect of about 237°C, a slope angle of 19°C to 35°C, and a local steep sill. The thickness of the covering layer is 3.5-14.7m. The terrain is generally stepped, and two small gullies are in the middle of the landslide, presenting a ditch-shaped landform. The site photos are shown below.

*Corresponding Author: 331044068@qq.com

DOI 10.1201/9781003348023-48

Figure 1. The landscape of the landslide.

## 2.2 *Stratigraphic lithology*

According to the geological survey and drilling findings, the landslide area is overlaid with the fourth collapsed slope product ($Q_4^{c+dl}$) gravel soil, and the underlying bedrock is strongly-to-moderately weathered by the upper member of the Middle Jurassic Huakaizuo Formation ($J_2h^2$). The composition is mainly silty mudstone, argillaceous siltstone, and marl.

## 2.3 *Hydrogeology*

Surface water: The landslide area is located in the tectonic erosion Zhongshan landform unit, and the Shunbi River is under the secondary road below the landslide, and the flow rate during the investigation period is about 10-30L/s. Many gullies are formed by cutting surface water within the landslide range. The surface water is mainly influenced by the atmosphere, collects along the slope to the foot of the slope, seeps into the soil, and finally collects into the Shunbi River. Groundwater: According to the characteristics of groundwater stored in the rock storage space, the groundwater in the landslide area is mainly loose rock pore water and bedrock fissure water. Groundwater is mainly replenished by atmospheric precipitation and infiltration of surrounding surface water and is collected and discharged in low-lying valleys in the form of underground runoff.

## 2.4 *Landslide spatial form*

According to the strength of the deformation characteristics of the landslide, the landslide can be divided into the sliding area, the strong deformation area, and the weak deformation area, which is described as follows. Section 5 is the sliding area, and the structures affected are mainly the subgrade section. Sections 3 and 4 are strong deformation areas. The structures affected by it are mainly excavated subgrade sections and built drinking water pipelines. Sections 1 and 2 are weak deformation zone, they pass through the front edge of the weak deformation area, and the structures affected by it are mainly the excavated roadbed section and the ZK86+523 bridge.

According to the information provided by the construction party, this section has been excavated to the designed road elevation, estimated to be about $1067 \times 103m^3$ according to the drawn cross-section and geological survey, and the landslide is a large-scale landslide.

# 3 LANDSLIDE ORIGIN AND STABILITY EVALUATION

## 3.1 *Cause analysis of landslide*

Based on a comprehensive analysis of ground investigation and drilling exposure, the sliding deformation caused by landslides is controlled by geological structure, material composition, and topographic conditions (Xu 2022).

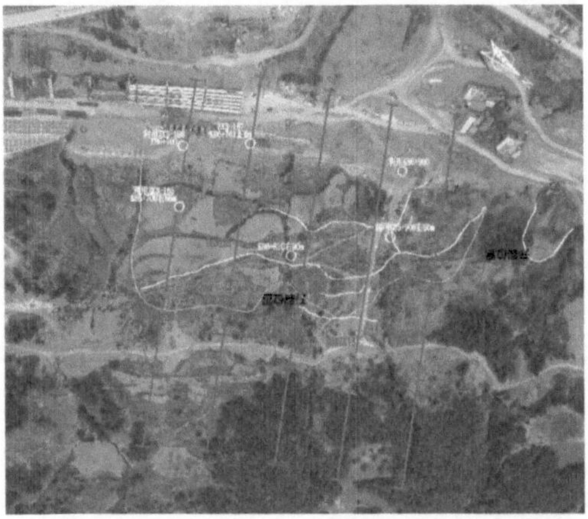

Figure 2. Schematic diagram of the landslide.

(1) Internal conditions: ① Formation lithology. The overlying soil layer of the landslide is mainly gravel soil and block rock soil, and the underlying layer is strong and moderately weathered silty mudstone and marl. ② Topography. The original slope is congruent, the slope angle is similar to the dip angle of the rock formation, and the rock surface is straight, which provides favorable conditions for the formation of landslides.
(2) Inducing factors: ① Seismic activity. The basic earthquake intensity of the site is 7 degrees. In 2021, the Yangbi area will experience frequent earthquakes, and the survey area will have obvious tremors. The main shock on May 21 reached a magnitude of 6.4, exceeding 6 degrees. Under the influence of the horizontal force of the earthquake, the formation and loss of landslides were aggravated. ② The role of atmospheric precipitation and groundwater. The surface water infiltration increases the weight of the slope body while reducing the strength of the sliding zone and the sliding body and promotes the formation and instability of the landslide. ③ Human engineering activities. With the excavation of the cut side slope, bridge abutment construction access road, and highway platform under construction, the slope forms a free surface, and the slope rate is too steep. Only temporary support measures are taken, which destroys the original stress state of the mountain. The formation of landslides played a certain leading role.

### 3.2 *Landslide stability evaluation*

#### 3.2.1 *Selection of working conditions and calculation methods*

Because the earthquake in the area where the landslide is located is grouped into the third group, the earthquake influence is considered in the calculation. According to the "Highway Subgrade Design Specifications" (JTG D30-2015), the arc sliding method (Liu 2020) was selected, and the landslide stability was checked in the natural condition I, heavy rain or continuous rainfall condition II, respectively. For the sliding zone and weak layer exposed by deep drilling, the sliding surface is a broken line, and the broken line sliding method is used for inverse calculation and calculation. The rock-soil interface is a broken line, and the broken line method is used for calculation.

#### 3.2.2 *Comprehensive value*

According to the results of the indoor test and back calculation, combined with the actual site, the comprehensive value of the shear strength of this sliding belt (surface) is:

Table 1. Comprehensive value of shear strength index of sliding zone soil.

| | | Natural | | Saturation | | |
|---|---|---|---|---|---|---|
| Location | Weights | C(KPa) | Φ(°C) | C(KPa) | Φ(°C) | Remark |
| Section 5 | 0.1 | 12.77 | 10.08 | 9.18 | 8.27 | Test value |
| | 0.9 | 10.22 | 19.60 | | | Inverse value |
| | 1.0 | 10.47 | 18.65 | 7.53 | 15.30 | Comprehensive value |
| Sections 3 and 4 | | 12.97 | 19.45 | 9.61 | 16.92 | Comprehensive value |
| Sections 1 and 2 | | 15.53 | 23.56 | 11.72 | 17.54 | Comprehensive value |

### 3.2.3 *Landslide stability check*

For calculation, select 1-1′, 2-2′, 3-3′, 4-4′, 5-5′ sections for stability checking. Because the 5-5′ section is the subgrade section, the line must be further excavated (press design excavation and fill elevation excavation). The calculation results are shown in the Table 2 below.

Table 2. Summary of stability calculation results.

| Section number | Working conditions | Stable state | Residual gliding force (kN/m) |
|---|---|---|---|
| Section 1-1′ | Normal working conditions | Stable | 0 |
| | Abnormal working conditions I | Unstable | 1802.57 |
| | Abnormal working conditions II | Basically stable | 288.11 |
| Section 2-2′' | Normal working conditions | Basically stable | 963.97 |
| | Abnormal working conditions I | Unstable | 1282.09 |
| | Abnormal working conditions II | Unstable | 1869.85 |
| Section 3-3′ | Normal working conditions | Basically stable | 1220.06 |
| | Abnormal working conditions I | Unstable | 2046.70 |
| | Abnormal working condition II | Less stable | 2172.97 |
| Section 4-4′ | Normal working conditions | Basically stable | 1014.14 |
| | Abnormal working conditions I | Unstable | 1749.08 |
| | Abnormal working conditions II | Less stable | 1833.26 |
| Section 5-5′ | Normal working conditions | Basically stable | 636.15 |
| | Abnormal working conditions I | Unstable | 1630.89 |
| | Abnormal working conditions II | Basically stable | 585.87 |
| Section 5-5′ after excavation | Normal working conditions | Basically stable | 691.15 |
| | Abnormal working conditions I | Unstable | 1951.88 |
| | Abnormal working conditions II | Less stable | 1261.28 |

## 3.3 *Landslide engineering geological evaluation*

Due to the support of the existing anti-slide piles (not completely ineffective) near sections 1 and 2 of the landslide, the landslide has not yet slipped and is in a basically stable to the under-stable state under normal working conditions (natural); under abnormal working conditions I (heavy rain or continuous rainfall) is in an unstable state; abnormal conditions II is in an unstable state. They are weak deformation areas.

Sections 3 and 4 of the landslide did not produce large slips, but many deformation cracks have occurred, which are basically stable in normal working conditions (natural); unstable in abnormal working conditions I (heavy rain or continuous rainfall). Normal condition II is in an under-stable state. They are strong deformation areas.

The 5-5' section of the landslide has slipped under normal working conditions (natural), which is a sliding area. After sliding, it is in a basically stable state under normal working conditions (natural); in abnormal working conditions I (heavy rain or continuous rainfall), it is in a steady state; in abnormal conditions II, it is in a basically stable state.

After excavation, Section 5 of the normal working condition (natural) is basically in a stable state. Section 5 of the abnormal working conditions I (heavy rain or continuous rainfall) is in an unstable state. Section 5 of the abnormal working conditions II (natural and earthquake) is in an under-stabilized state.

### 3.4 *Treatment suggestion*

(1) Suggested treatment plan 1: set up anchor cables + anti-slide piles + slope leveling + intercepting and drainage ditch above the route K86+665~K86+985 at the front edge of the landslide. Comparison plan 2: use local clearing + anchor rod frame beam + intercepting drainage ditch for support.
(2) An effective interception and drainage system shall be set up on the trailing edge of the landslide, and at the top of the slope, and at the same time, the cracks formed on the surface shall be sealed with a cement slurry to avoid adverse effects caused by the infiltration of surface water into the slope.
(3) The soil and rock composition of the landslide is relatively complex, the soil and rock are fragmented and loose, and the slope cracks are extremely developed. If the protection is not appropriate, the heavy rainfall in the current rainy season will infiltrate into the landslide body in large quantities, and there is a potential for sliding again. Disaster prevention and control should be done well.
(4) After the end of the support project, it is recommended to continue the deformation monitoring work for a period of time to test the disposal effect, so as to achieve the purpose of comprehensive remediation, safety, reliability, economy, and rationality.

## 4 CONCLUSIONS

In this paper, the reasons for the landslide formation and treatment measures are explored through the research on the landslide in the Dayangyun Expressway. The study found that the landslide formation was mainly caused by the combined influence of stratum lithology, topography, seismic activity, atmospheric precipitation, and groundwater and human activities. The stability analysis of the landslide was carried out. The residual sliding force of each section under normal conditions (natural), abnormal conditions I (heavy rain or continuous rainfall), and abnormal conditions II (natural and earthquake) were obtained to judge its stable state. Finally, according to the analysis, the targeted treatment suggestions are put forward to provide a feasible plan for the prevention and control of landslides.

## REFERENCES


Liu, Jie & Zhang, Jianhui (2022). Research on landslide susceptibility evaluation based on long-term scales: Taking Yongjing County, Gansu Province, as an example. *Chinese Journal of Earthquake Engineering*, 1–13.

Liu, Yongqiang. (2020). Causes and prevention of geological disasters of mine landslides. *World Nonferrous Metals* (04), 132–134.

Qi, Shenghui & Liu, Ran (2022). Stability analysis of highway subgrade high slope under the condition of water level change in the reservoir area. *Highway* (05), 106–111.

Xu, Qisong & Pan, Shijia (2022). Analysis of the genesis mechanism of typical landslides in northwestern Guizhou Province. *China Water Transport* (Second Half Month) (04), 98–100.

Zhao, Xin & Zhao, Yi (2022). *Application Research of BIM in Landslide Intelligent Monitoring. Engineering Survey*.

*Advances in Urban Construction and Management Engineering – Cha (Ed)*
 *ISBN 978-1-032-39018-5*

# Current situation and countermeasures of prevention and control of water pollutants in Inland River ships under the background of green shipping

Chaojie Zhou*, Xiaozhen Zhu* & Yanqi Zhao*
*Zhejiang Scientific Research Institute of Transport, Hangzhou, China*

ABSTRACT: Based on the concept and guidance documents of green shipping development, this paper takes Zhejiang Province as an example, and combines the development status of inland water pollutant reception, transshipment, disposal capacity building, and on-site management, analyzes the existing problems, and puts forward suggestions and strategies. This paper provides a reference for relevant regulatory authorities and shipping enterprises to promote the development of green shipping and help the construction of a strong transportation country.

## 1 INTRODUCTION

Water transportation has many advantages, such as low cost, large transportation volume, high efficiency, and low energy consumption (Yan 2021). With the continuous opening of the inland shipping market and the continuous growth of the number of ships, ships will produce various water pollutants during the navigation process.

The local shipping management departments have formulated relevant policies, rules and regulations, and action plans for the prevention and control of pollution from ships and ports, and actively carry out environmental governance and ecological protection actions in inland waters. Therefore, it is particularly important to analyze the new requirements for the development of green shipping under the new situation and study the current situation and existing problems of water pollutant prevention and control in inland waterway ships. Based on this, this paper researches and suggests corresponding strategies to promote the construction of inland waterway shipping ecological civilization.

## 2 NEW REQUIREMENTS FOR GREEN SHIPPING DEVELOPMENT

### 2.1 *New concept of ecological civilization construction*

The new concept of ecological civilization construction is to strengthen the reform of the ecological civilization system and promote the three-dimensional transportation of ecological priority and green development (Li 2021). To ensure the idea of both economic development and environmental protection, Zhejiang Province, as the first batch of "green transportation" pilot demonstration provinces, has planned eight major transportation constructions such as digital transportation and green transportation, implemented seven major projects such as "ecological environmental protection project," "green transportation industry cultivation project," and "pollution prevention and control key projects," and promoted green port navigation and green port creation, and constructed a sustainable development path of the green transportation system.

*Corresponding Authors: zcjcosima@163.com, startlife321@163.com and 994966926@qq.com

 DOI 10.1201/9781003348023-49

### 2.2 *Ship water pollution prevention guidance document*

To fight the battle of pollution prevention and control, China's various departments have clearly defined work priorities, objectives, and tasks and deployed the pollution prevention and control work of ships and ports. In April 2015, it was proposed to "strengthen the control of pollutants in ships and ports," accelerate the construction of garbage reception, transportation, and disposal facilities, improve the reception and disposal capacity of oily wastewater and chemical cabin washing water, and improve the emergency response capacity of pollution accidents. The management system of "three waters," namely, water pollution prevention, water ecological protection, and water resources management, was systematically promoted (Xu 2019). At the same time, it is required to promote the green development of the shipping industry and strengthen the prevention and control of pollution in ships and ports In August 2015, it was proposed to promote the construction of the receiving capacity of water pollutants from ships such as ports, docks, and shipyards, improve the receiving and disposal capacity of pollutants, and meet the requirements of receiving and disposal of pollutants from ships arriving at ports (Xu 2020). In January 2018, the requirements for the emission control of oily sewage, domestic sewage, sewage containing toxic liquid substances, and ship garbage from ships to environmental water bodies, as well as the implementation and supervision of the standards, were stipulated, and the emission control requirements for different pollutants from ships in different water areas and tonnages were further refined (Chen 2019). In May 2020, it was proposed to strengthen the prevention and control of port pollution, improve the emission control standards, accelerate the infrastructure construction of port ship pollutants receiving and transporting, chemical cleaning stations along the main channel of the Yangtze River, and other key waterways, and commit to environmentally friendly green shipping development mode (Xu 2021).

In addition, "Water Pollution Prevention Law of the People's Republic of China," "Regulations of the People's Republic of China on the Environmental Management of Inland Waters Contaminated by Ships," and "Action Plan for the Protection and Restoration of the Yangtze River," "Remediation Plan for Ships and Ports Contamination in the Yangtze River Economic Belt" and other laws and regulations are common objectives for the construction of green ports, the creation of green waterways, the promotion of ecological civilization construction, and the improvement of pollution prevention and control capabilities. Specific requirements are put forward for the pollution of the water transport industry caused by a large number of ports, dense waterways, large ship flow, ecological imbalance, and prevention and control pressure.

## 3 CURRENT SITUATION OF WATER POLLUTANTS CONTROL OF INLAND RIVER SHIPS IN ZHEJIANG PROVINCE

Zhejiang Province is rich in inland waterway resources and huge freight volume. The dense transport channel network promotes the rapid development of the shipping industry in Zhejiang Province. In recent years, Zhejiang has accelerated the process of promoting green ports, green channels, and ecological construction with the support of key green port and navigation projects, and has made remarkable contributions to the fight against pollution.

### 3.1 *Collection, reception, and disposal of water pollutants*

At present, Zhejiang Province has fully realized the comprehensive coverage of pollutant receiving facilities for inland ships, the full reception of waste from inland ships, the zero discharge of oily wastewater, and the substantial increase in the amount of domestic sewage received. All districts and cities have implemented the policy of free reception. According to the survey, the main processes of collecting, reception, and disposing of water pollutants from ships such as wharves and integrated water service areas in Hangzhou, Jiaxing, Huzhou, and Shaoxing, Zhejiang Province are as follows:

3.1.1 *Ship's domestic waste*

Ship's domestic waste is mainly received by the third-party receiving unit. The receiving methods include mobile receiving ship receiving and the ship on its shore. The third-party receiving unit transfers the domestic waste received by the mobile receiving ship or the shore to the nearby transfer stations in the city, and finally, it is transported to the garbage incineration or landfill treatment. The main garbage collection, reception, and disposal processes are shown in Figure 1:

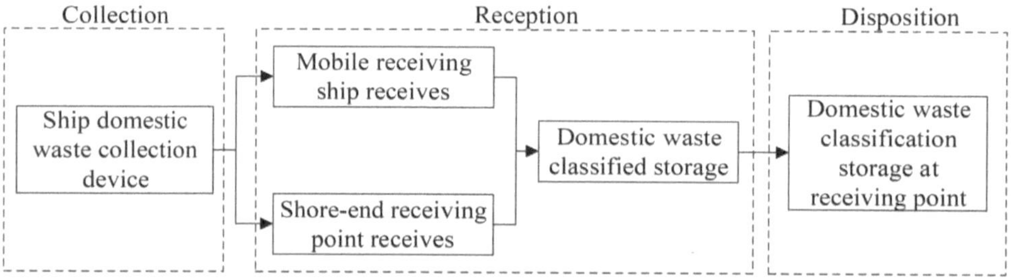


Figure 1. Flow chart of collection, reception, and disposal of domestic waste.

The mobile receiving ship and the shore end are equipped with four types of garbage bins (boxes) for classification reception. The crew put garbage into the garbage bins (boxes) set at the port terminals and the water service areas where the ship is convenient. Moreover, the receiving party establishes a standard paper ledger or information to record the amount of garbage, and both the ship and the receiving party have records and archives after each receipt.

3.1.2 *Ship's domestic sewage*

At present, Hangzhou, Jiaxing, and other places are equipped with mobile receiving ships equipped with domestic sewage storage facilities and actively receive domestic sewage from ships in docks and anchorage areas. The ship uses sewage lifting pumps to pump domestic sewage out, and the receiving ship extracts domestic sewage through self-priming pumps. Finally, the receiving vessel was sent to the shore receiving facility. After biochemical pretreatment, the receiving pipe entered the municipal sewage pipe network or the shore end for reuse. In Huzhou, Shaoxing, and other places, the ship actively docks to the designated domestic sewage receiving point through the shore receiving facilities to discharge domestic sewage onto the shore, and finally, the third party unit is transported to the urban sewage treatment plant for disposal. Domestic sewage collection, reception, and disposal process as shown in Figure 2:

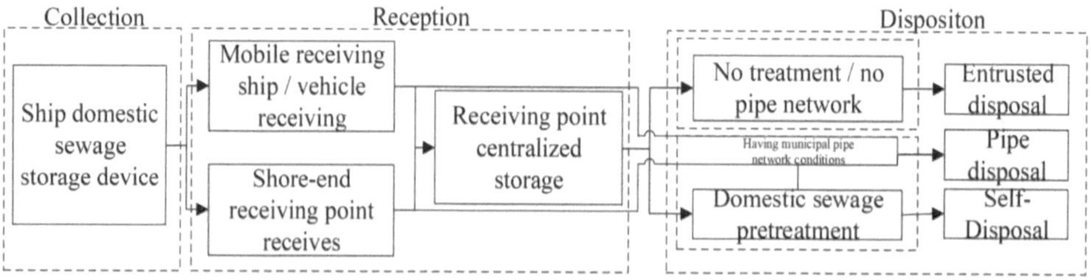


Figure 2. Flow chart of collection, reception, and disposal of domestic sewage.

3.1.3 *Ship's oily sewage*

The oily sewage from ships is pumped and transported to the shore-end oil-water separation device for treatment through the mobile receiving vessel or shore-end receiving facility and is divided into wastewater, waste oil and waste residue. The wastewater enters the urban pipe network system after being tested and up to the standard, and the waste oil, and waste residue are disposed of in the next step by a qualified third-party relevant unit. Due to the lack of oil-water separation devices in some terminals, third parties are required to transfer them to units with waste oil disposal qualifications for disposal. Oily sewage collection, reception, and disposal processes are shown in Figure 3:

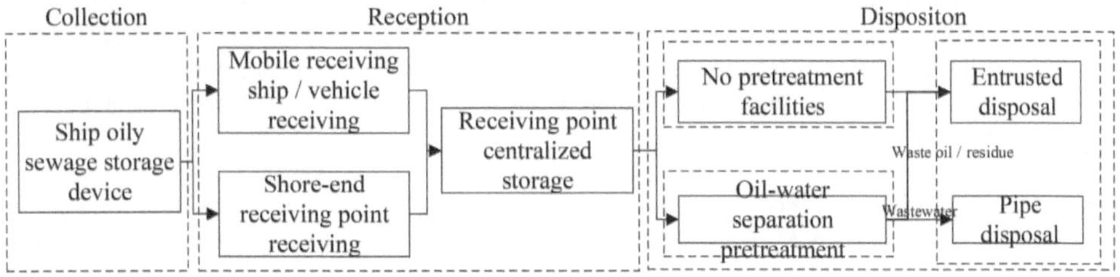


Figure 3. Flow chart of collection, reception, and disposal of oily sewage.

## 3.2 *Relevant standards and regulatory systems*

### 3.2.1 *Specifications for water pollution collection, reception, and disposal of ships*

All the districts and cities of Zhejiang province have compiled and released the Construction Plan of Port and Ship Pollutant Receiving, Trans-shipment and Disposal Facilities by the end of 2017, requiring that by the end of 2020. The transportation (port) administrative departments of ports and ship repair plants shall, together with industry and information technology, environmental protection, housing and construction, maritime affairs, and other departments, complete the assessment of the capacity of receiving, transporting, and disposing of pollutants from ships in the region, and complete the construction of relevant hardware facilities for receiving, transporting and disposing of pollutants. Huzhou issued the first Code for Transfer and Disposal of Domestic Sewage from Inland River Vessels in Zhejiang Province (*DB3305T/128-2019*). The Yangtze River Delta region has issued a uniform "Code for The Configuration of Inland River Receiving Facilities for Ship Water Pollutants" (*DB33/T 310001-2020*). However, the above standard code only provides for the transfer and disposal of water pollutants and the configuration of receiving facilities, and in fact, there is a lack of clear labels and provisions on the operation standard and management process.

### 3.2.2 *Supervision of water pollution prevention and control*

The means of document inspection without normalized emission monitoring has problems such as insufficient monitoring capacity and incomplete elimination of smuggling. The joint system of ship pollutant transfer is being gradually established, which is transformed from the one-way electronic joint form of a paper joint. That is, when the ship delivers water pollutants to the receiving party, it fills in relevant information by scanning the two-dimensional code and other means. The receiving party, the transshipment disposal party, and the maritime administration department can view the source and final destination of water pollutants through this platform. Due to the need for different departments to supervise the receiving, transshipment, disposal, and other environments, the degree of information resource sharing is limited, and it is urgent to establish a joint management and closed-loop mechanism.

### 3.2.3 *Supervision for operation and maintenance of receiving and disposal equipment*

The unreasonable layout and configuration of some port terminals led to inconvenient reception of water pollutants, low utilization of receiving stations, idle equipment, and waste of resources. At the same time, due to the successive construction of receiving facilities and the differences in facilities, it is impossible to formulate a unified facility management specification and operation process, so the normal operation and maintenance of facilities cannot be guaranteed.

## 4 SUGGESTION AND STRATEGY

### 4.1 *Speed up infrastructure construction and improve water pollutant collection, reception, and disposal capacity*

Firstly, for new port terminals, water service areas, and other receiving points, we should fully study their berthing capacity, ship tonnage and throughput, formulate and improve the evaluation mechanism standard of ship water pollutant receiving capacity, set up reasonable location and number of receiving facilities, and ensure that the actual receiving capacity and theoretical receiving capacity match.

Secondly, it is necessary to issue proof of the compliance of receiving capacity for the established receiving points that meet the requirements. For the receiving points that do not meet the requirements, it is necessary to study the actual situation of water pollutant receiving fully, timely reform the insufficient or idle receiving facilities, and comprehensively consider the regional water pollutant receiving methods to ensure the full utilization of resources.

Thirdly, according to the actual reception of domestic sewage, oily sewage pretreatment facilities to reduce transport and disposal costs; implement the third party responsibility subject, unified procurement, support the third party to receive water pollutants, and improve transport and disposal capacity.

### 4.2 *Improve regulatory standards and strengthen supervision*

Firstly, various management standards on the whole process of receiving, transferring, and disposing of water pollutants from ships and coordinating the implementation of multiple entities need to be further clarified, and the process of water pollutant prevention and control should be unified to regulate the construction, maintenance, and repair of receiving facilities as well as the connection of all aspects of transfer and disposal.

Secondly, the departments need to strengthen coordination and joint supervision further, while establishing a joint supervision mechanism of maritime, transportation, and environmental protection, implementing a closed-loop management system for the whole process of water pollutants from ships, establishing a joint order system for receiving, transferring and disposing of water pollutants and a service information system, clarifying the supervisory responsibilities of all parties and improving anti-pollution emergency response capabilities.

Thirdly, the departments need to carry out special actions to prevent and control water pollutants, strictly rectify the deficiencies of anti-pollution facilities and equipment in ports and ships, conduct centralized inspections of infrastructure equipment in the supervisory area, and use information technology tools such as drone inspections and ship pollutant alarm systems to increase monitoring and enforcement efforts.

### 4.3 *Enriching shipping elements and innovating operation mode*

Firstly, a blockchain circulation system for receiving, transporting, and disposing of ship water pollutants needs to be established. All processes are recorded in the whole process, which can be traced back and cannot be changed. The processing situation of various departments in each stage is reflected truthfully, and the shortcomings of insufficient supervision and information in maritime, transportation, and environmental protection departments are compensated.

Secondly, the evaluation mechanism is formulated for the service of marine water pollutant receiving units and disposal units, and the benign competition of each unit is formed to improve the service quality of port terminals and comprehensive water services.

Thirdly, building and operating new energy electric ships equipped with oily sewage, domestic sewage disposal, and ballast water treatment needs to be further accelerated to reduce the production of water pollutants at the source.

### 4.4 *Strengthen publicity and enhance the awareness of green environmental protection*

Firstly, the maritime, transportation, environmental protection, and other management departments regularly discuss and train the shipping enterprises and related units in the jurisdiction, formulate feasible supervision mechanisms and improve the business level of the management team.

Secondly, the assessment index of ship pollution prevention and control needs to be established to deepen the awareness of pollution prevention, comprehensively improve pollution prevention and control management, and implement environmental protection and ecological construction.

Thirdly, the management of shipping companies and crews needs to be further strengthened while raising the barriers to entry into the industry. Shipping companies must strengthen environmental awareness cultivation and policy advocacy and conduct mandatory training and assessment on relevant laws and standards.

## 5 CONCLUSION

This paper analyzes the new requirements for the development of green shipping. As an important part of the implementation of the concept of green development, shipping needs to run through the whole process of infrastructure planning, construction, operation, and maintenance while maintaining its high-quality and rapid high-level development. This paper discusses the current situation and existing problems of ship water pollution prevention and control and puts forward corresponding countermeasures and suggestions from four aspects of infrastructure construction, regulations and standards, industrial factors, and environmental protection consciousness, so as to provide a certain reference for relevant management departments and shipping enterprises.

## REFERENCES

Chen Xu, Wen Yanqing, Ye Ying, Cao Ying. Layout Planning of Inland River Ship Water Pollutant Receiving Stations in Hubei Province [J]. *Water Transport Management*, 2021, 43(06): 13–16.

Chen Zhaohu, Feng Jing, Chen Yuwen, Bai Yang. Interpretation of the newly promulgated ship water pollutant discharge control standards [J]. *Resource Conservation and Environmental Protection*, 2019(10):81–82.

Lei Li, Cheng Jinxiang. Key areas of green development of water transport in the Yangtze River Economic Belt [J]. *Water Transport Management*, 2021, 43(03): 7–10.

Xu Honglei, Han Zhaoxing. The implementation progress of the "Special Action Plan for Pollution Prevention and Control of Ships and Ports (2015-2020)" and the recommendations for the promotion of key tasks [J]. *Port Science and Technology*, 2016(06): 33–36+39.

Xu Min, Zhang Tao, Wang Dong, Zhao Yue, Xie Yangcun, Ma Lekuan. Review and Prospect of Water Pollution Prevention and Control in China for 40 Years [J]. *China Environmental Management*, 2019,11(03):65–71.

Yan Xinping. *Implementing self-reliance and self-improvement of technological innovation to open up a new situation of high-quality development of Yangtze River shipping* [N]. China Water Transport News, 2021-08-25(001).

*Advances in Urban Construction and Management Engineering – Cha (Ed)*


# Research on emergency rescue disposal from the perspective of urban public safety emergencies—A case study of a hotel collapse in Quanzhou, China in 2020

Yang Wang, Shuai Zhang*, Benshuai Liu, Xuan Liu & Yu Zhang
*National Earthquake Response Support Service, Beijing, China*

ABSTRACT: With the rapid development of urbanization, the public safety situation in China is generally stable. However, due to the large population, unbalanced economic development, and the coexistence of various city risks and hazards, once a sudden safety event occurs, its great influence and destruction will be inestimable. In the face of the current situation, it is of major importance to scientifically deal with public safety incidents and build effective emergency response mechanisms. Taking the collapse accident in Quanzhou, Fujian Province, as an example, this paper analyzes the emergency response, rescue decision-making, and disposal efficiency from the perspective of life-saving, summarizes the experience and inspirations of emergency rescue and onsite disposal for all regions, departments and rescue forces, which provides decision-making reference for urban public safety disposal in China.

## 1 INTRODUCTION

In recent years, China has been in the process of industrialization, informatization, and modernization while the field of public safety management is still facing challenges (Chen 2019; Liu 2020; Huang 2017; Zhu 2022). Ever since the further promotion of urbanization last century, a large number of high-rise buildings have been built densely, accompanied by illegal reconstruction, old and dilapidated houses, self-built rural houses, etc., leaving huge potential safety hazards (Li & Peng 2021; Liang 2021; Wang 2021; Yu & Lv 2014).

Accidents of building collapse have occurred from time to time. In 2021, a high-rise apartment in Florida collapsed, causing at least 97 deaths so far; the sudden collapse of a 5-story building during demolition in Gwangju, a southern city in South Korea, resulted in the burial of a bus and ten onsite staff; in 2022, a 10-story building collapsed in Abadan, a city in southwest Iran, leaving at least 80 people trapped under the ruins. In China, a hotel in Quanzhou collapsed in 2020, and a self-built house in Changsha, Hunan Province, collapsed in 2022. These building collapse events, which are not caused by natural disasters and fires, have the characteristics of proruption, serious casualties, difficult rescue operations, difficult coordination, and easy to cause secondary disasters (Liang 2021; Sun & Sun 2004), which have led to in-depth thinking on the emergency response and rescue disposal of collapsed buildings in the field of urban public safety management (Animesh & Aminur 2015; Yin 2022). In this paper, we take a hotel collapse accident in Quanzhou, Fujian Province as an example, focusing on the measures, strategies, and experience extracted in the disposal process of public safety emergency to provide references for urban emergency management.

*Corresponding Author: echo_zhang7@163.com


DOI 10.1201/9781003348023-50

## 2 CASE STUDY

At 19:14 on 7 March 2020, a seven-storey hotel in Quanzhou, Fujian Province, collapsed within 2 seconds. A total of 71 people were trapped in the building at the time of the incident, most of whom were people from other places to Quanzhou for centralized quarantine and health observation. After the intense rescue, 42 people survived.

### 2.1 *Emergency response*

#### 2.1.1 *Governmental response*

After the accident, the Ministry of Emergency Management immediately launched an emergency response and dispatched the Emergency Management Department as well as the Fire Rescue Corps of Fujian via a video connection to the disaster scene. Rescue forces of different levels are required to organize rescue operations and treat the injured, take scientific rescue measures to prevent secondary accidents, ensure rescue safety, investigate the case, and locate those responsible according to the law. At the same time, together with the Ministry of Housing and Urban-Rural Development, Health Commission, and other departments, an expert group was sent to the affected area overnight to guide and assist in the emergency disposal work. Quanzhou Municipal Government immediately activated the contingency plan and set up an onsite emergency search and rescue leading group with working groups of rescue, medical treatment, publicity, aftermath, and accident investigation to spare no effort to carry out rescue and accident disposal work.

#### 2.1.2 *Rescue forces' deployment*

Fujian Province and Quanzhou Municipal Government quickly organized forces to carry out the rescue operation. Fire Rescue Corps of Fujian immediately established ten heavy and light rescue teams with 1086 members in an organic and modular manner, carrying life detectors, rescue dogs, and special rescue equipment to the scene for disposal. A total of 118 teams and 5176 rescuers participated in this mission, including national comprehensive fire rescue teams, national production safety rescue teams, local professional teams, social rescue forces, etc.

### 2.2 *Rescue decision-making*

#### 2.2.1 *Onsite rescue environment*

Through combing the collected data, it is found that the site environment of the collapse is extremely complex, which mainly has three difficulties:

- Complex building structure. The collapsed buildings on the ground generated a high accumulation of construction waste, occupying a large floor area, and the broken building components with steel bars are crisscrossed in ruins. There are not only I-steel beams and columns but also reinforced concrete floors and brick concrete floors, which greatly increases the difficulty of manual breaking and breaching. Each steel plate layer is made of concrete floors, and the rooms are separated by solid walls. After the collapse, various components and items are stacked together. Breaking through each floor requires breaking down hard materials such as steel plates, concrete, and brick walls, as well as soft materials such as furniture decoration, which is extremely challenging.
- Collapsed and torsion staircase. On the one hand, the second to seventh floors of the building collapsed as a whole, tilting at a large angle to one side, brittle fracture and deformation and distortion occurred to local steel structures, and large displacement occurred to trapped people and objects. The depth of the buried in distress was different, and thus difficult to ascertain the specific location. On the other hand, multiple floors collapsed and superimposed, and the solid brick walls, beds, cabinets, sofas, and other furniture in the building filled much more space than living space for buried people to be rescued.
- Cramped surroundings. The accident site is too crowded, with only one side of the road for parking, living confined space for the arrived teams to conduct their rescue operations.

### 2.2.2 *Search and rescue operations*

After the rescue forces arrived at the scene, they adhered to the scientific and safe rescue. With the strict control of the scene, the implementation of the epidemic prevention and control measures, and the role of professional rescue personnel under the guidance of the national expert group. The rescue plan was continuously adjusted according to the situation, the support like communication, electricity, and medical treatment was coordinated and guaranteed, and the safety protection of rescuers was strengthened to avoid secondary disasters. Adhering to the principle of scientific and accurate treatment of "one person, one case," the wounded were treated carefully with the best resources and the unimpeded rescue channel. After 112 hours of arduous search and rescue, 42 of 71 buried people were rescued (see Appendix 1). The whole rescue process is mainly divided into the following four phases (see Figure 1):

Phase 1: Rescue of surface buried people. During the first three hours after the accident, the advance forces conducted a rapid search and rescue operation. Four groups were organized for safety observation, search and locating, and breaking and breaching, and three-dimensional search and rescue were carried out in different areas in the order of "from the outside to the inside, from easy to difficult," with 23 people buried on the surface rescued.

Phase 2: Rescue of shallowly buried people. The disaster site headquarters was quickly set up as soon as the reinforcement teams arrived. According to the site plane figure, the deployment map of rescue forces, and the pre-judgment distribution map of buried people, the key areas are gradually narrowed and located. Removing the way of demolition layer by layer and opening the observation window, rescuers could gradually go deep into the ruins' interior to save lives to the greatest extent. As of 24 hours after the incident, 26 trapped people had been rescued, and 18 survived.

Phase 3: Rescue of deeply buried people. After a comprehensive analysis, special hook machines and other equipment were dispatched to the site to stabilize the overall structure of the building and prevent large-scale secondary collapse by hoisting the steel beams of the building through steel bars. The narrow gap between the floors can be lifted, relieving the pressure on the trapped people in ruins and facilitating the rescue work. The heavy team and the light teams cooperated to "pair up" rescue work and took turns carrying out the tasks of heavy component demolition and rescue channel cleaning. During the 72-hour golden rescue time, 13 trapped people were rescued, of which three survived.

Phase 4: Closing rescue work. After repeated searching and confirming no signs of life, the search and rescue team took the demolition and stripping of construction machinery in cooperation with manual search and rescue and found another eight trapped victims. At 11:04 on 12 March, the last buried victim was rescued, announcing the end of the onsite rescue operation.

According to the analysis of the figure below, 44 survivors were rescued from 71 trapped people in the rescue operation, of which two died in hospital for treatment, with a survival rate of 59.15%. It is also miraculous to rescue three survivors 48 hours after the accident. The high number of people rescued, the survival rate, and the survival of deeply buried people all illustrate the success of this emergency response operation.

## 2.3 *Summary*

Compared with previous collapse accidents, the more favorable living environment in this operation has created favorable conditions for the buried under the ruins to cooperate with the rescue operation. From the perspective of collapse space, the ruins in the collapsing direction are stacked layer by layer, which causes the inverted space to be relatively complete. Furthermore, the location of the incident is a hotel, with beds, sofas, and other large furniture in the room, which created a certain living space under it. From the perspective of living conditions, the temperature is suitable for survival. During the rescue operation, the average temperature in Quanzhou remained between 13°C and 23°C. As it is used for isolation of COVID-19, there are food, water, and clothes in the room.

Combined with the above advantages, three survivors were successfully rescued 52 and 69 hours after the accident.

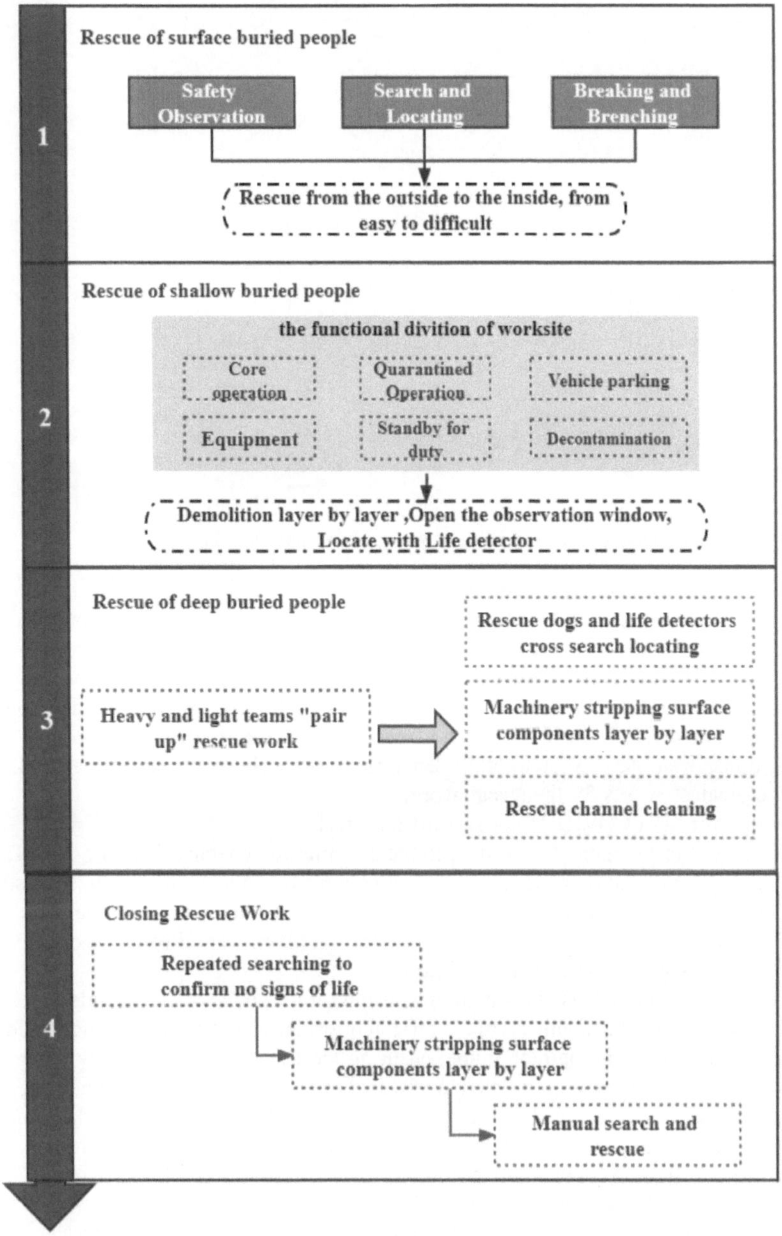


Figure 1. Flowchart for each stage of the hotel rescue operation in Quanzhou Fujian.

## 3 CONCLUSION AND INSPIRATIONS

As a complex public safety incident in the context of the global outbreak of COVID-19, the collapse has accumulated a series of experiences in emergency response, rescue decision-making, and epidemic control, which could sum up several key inspirations:

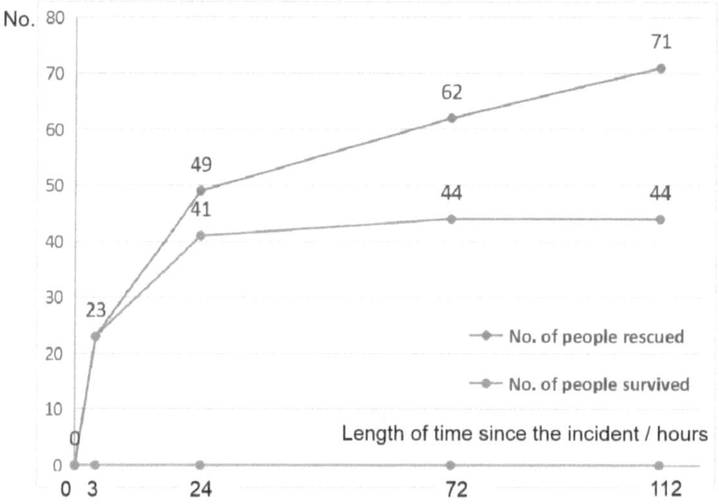


Figure 2. Efficiency analysis of the hotel rescue operation in Quanzhou, Fujian.

- Multi-sectoral linkage to rapid response
  The Ministry of Emergency Management and governments at all levels responded quickly to dispatch rescue forces to the scene. The first batch of rescue forces and reinforcements arrived at the scene 19 minutes and 3 hours after the accident. The expert group was sent to the scene to support the rescue decision-making. Governments at all levels, public security, housing and construction, medical and health, safety supervision, civil affairs, electric power, telecommunications, transportation, news, and other departments, as well as people from all walks of life, made the greatest efforts for the rescue work.
- Control the spread of COVID-19 for scientific control
  In response to this emergency, according to the transmission characteristics of COVID-19, sufficient epidemic prevention materials and disinfection equipment were mobilized to strengthen the protection of rescue team members and rescue dogs. At the rescue site, the prevention and control areas were scientifically divided with strict implementation of camp disinfection and sterilization management and health monitoring.
- Unimpeded channel to release information distribution for public opinion guide
  During the operation, the timeliness of the official information release was strengthened to grasp the initiative of the public opinion guide, ensuring the true and efficient dissemination of the accident progress. Through multiple channels, including Weibo, Tiktok, and news reports, the information onsite, like rescue disposal, casualties, and preliminary causes of the accident, was released to the whole society, so that the public widely supported the operation for the first time.
- Lessons learned from the accident to eliminate potential safety hazards
  Through the in-depth investigation and rectification of illegal land occupation, illegal construction, old and dilapidated houses, self-built rural houses, and "housing to business" buildings, the government would improve and implement the responsibility chain to "fundamentally eliminate potential accidents" to curb major public safety accidents from the source and protect the safety of people's lives and property.

ACKNOWLEDGEMENT

This research is supported by funding from the National Key Research and Development Plan [Grant Number 2020YFC1512100] and the National Natural Science Foundation of China [Grant Number 42061073].

## REFERENCES


Animesh, B. & R. Aminur (2015). Rescue and emergency management of a man-made disaster: lesson learned from a collapse factory building, Bangladesh. [J]. *The Scientific World Journal*, 2015.

Chen, S. Y. (2019). Research on the Current Situation and Countermeasures of Beijing Urban Public Security under the New Situation. *J. Intelligent City*. 5 (08), 50–51. (in Chinese)

Huang, Y. H. (2017). *Report on the development of urban public safety in China* (2016–2017). M. Beijing: Social Science Academic Press (China). 01–20. ( in Chinese)

Li, J. & Z. Peng (2021). Research on Decision Chain of Building Collapse Accident Based on Feature Extraction. *J. Journal of Engineering Management*. 35 (06), 125–130. (in Chinese)

Liang, S. T. (2021). Analysis of rescue countermeasures for building collapse accidents. *J. Housing and Real Estate*. (04):253–254. (in Chinese)

Liu, T. M. (2020). Bui2017lding a new era of national emergency management system. *J. China Fire*. 03, 14–17. (in Chinese)

Sun, L. M. & D. Sun (2004). The countermeasures for rescuing to the collapse of building. *J. Journal of The Armed Police Academy* (04):26–28. ( in Chinese)

Wang, F. (2021). Case statistics and cause analysis of existing urban building collapse accidents. *Shanghai construction science & Technology*. (02), 11–14. ( in Chinese)

Yin, L. C J. Li (2022). Thoughts on the development of emergency management and rescue from the typical building static collapse accidents. *J. The Scientific World Journal*. 41(03):425–427 (in Chinese)

Yu, X. H.& D. Lv (2014). Seismic sideway collapse fragility analysis based on structural typical failure modes. *J. Journal of Building Structures*. 35(08):8–14. (in Chinese)

Zhu, Z. P. (2022). Exploration of the Idea and Practice of the Management of Dynamic Hidden Dangers in Urban Public Security. *J.Shanghai Urban Management*. 31 (03), 47–53. (in Chinese)


## APPENDIX

Table 1. Timeline of the hotel rescue operation in Quanzhou Fujian.

| Date | Time | Events |
|---|---|---|
| 7 March | 19:14 | The whole collapse of the hotel occurred. |
| | 19:33 | The first rescue force arrived. (Netease News) |
| | 21:00 | Quanzhou high-speed traffic police detachment issued a traffic control notice. (Quanzhou high-speed traffic police official microblog) |
| | 21:00 | 17 victims have been rescued from the scene. (China Fire official microblog) |
| | 21:44 | Rescued security personnel from the first floor (China Fire official microblog) |
| | 21:50 | 23 victims have been rescued from the scene. (People's Daily News) |
| | 21:55 | Up to now, 25 victims have been rescued from the scene. (Press conference of District People's Government) |
| | 22:14 | The heavy earthquake rescue team of the Putian fire rescue detachment has arrived at the scene. (People's Daily News) |
| | 22:44 | Up to now 26 victims have been rescued from the scene. (China Fire official microblog) |
| | 22:45 | Quanzhou fire brigade rescued a man from the 5th floor with normal movement; 27 victims have been rescued so far. (China Fire official microblog) |
| | 22:56 | Quanzhou fire brigade rescued a man with leg injuries from the fifth floor; 28 victims have been rescued so far (China Fire official microblog) |
| | 23:00 | Fuzhou City and seconded Longyan, Putian, Xiamen, and Zhangzhou, a total of 7 fire squadrons,11 heavy and light rescue teams, 800 commanders, and 8 search and rescue dogs to participate in search and rescue. |
| | 23:06 | Quanzhou fire rescue detachment rescued a stranded woman with non-life-threatening injuries; 29 victims have been rescued so far (China Fire official microblog) |
| | 23:49 | Another trapped person was found. (China Fire official microblog) |
| | 23:59 | 31 victims have been rescued by now. (Quanzhou Evening News) |

(*continued*)

Table 1. Continued

| Date | Time | Events |
|---|---|---|
| 8 March | 00:06 | Jinjiang fire brigade found a trapped person under intense rescue operation. (China Fire official microblog) |
| | 00:16 | Xiamen Fire detachment rescued a 2-year-old boy and blindfolded him with a mask. So far, 32 victims have been rescued. (China Fire official microblog) |
| | 00:21 & 00:23 | The boy's parents were rescued. 34 victims are out of danger, and another 9 people escape from self-help. (China Fire official microblog) |
| | 00:55 | The Jinjiang fire brigade found a trapped person and conducted a breaching rescue operation. (China Fire official microblog) |
| | 02:48 | Quanzhou fire brigade rescued a male injured person with stable overall vital signs. At present, 35 people have been rescued from the scene. (Emergency Management Department) |
| | 03:15 | 36 victims have been rescued at the scene of the accident (CCTV News) |
| | 05:09 | 37 victims have been rescued. (Fujian Traffic Radio) |
| | 05:30 | After 6 hours of struggle to breach through three walls, the Quanzhou fire brigade successfully rescued a trapped person from the bathroom, the 38th trapped victim. (China Fire official microblog) |
| | 05:40 | The 38th person trapped in the collapse of the Hotel in Quanzhou was rescued, a male in his 50s, conscious. (Fujian Traffic Radio) |
| | 08:13 | Firefighters rescued one more trapped man, totaling 39 rescued victims (China Fire official microblog) |
| | 08:20 | By now, 51 victims have been rescued, two of whom have no vital signs. (Fujian Traffic Radio) |
| | 08:59 | More than 800 commanders and fighters conducted cross-regional reinforcement disposal. (Fujian Traffic Radio) |
| | 09:00 | Quanzhou fire brigade found a 12-year-old child who is conscious and can talk. (China Fire official microblog) |
| | 10:24 | After 3 hours of breaching and obstacle clearance, a life passage was finally opened up. (China Fire official microblog) |
| | 10:30 | 42 people have been rescued at the scene, including 4 dead, 1 critically ill and 4 seriously injured. (Emergency Management Department) |
| | 11:24 | After 16 hours, the boy was successfully rescued. At present, 43 trapped people have been searched and rescued. (China Fire official microblog) |
| | 11:30 | The scene searched and rescued 43 people who were trapped, and there are still 28 people trapped under the ruins. (Government Press Conference) |
| | 12:14 | A male survivor was successfully rescued from the scene. (China Fire official microblog) |
| | 13:30 | A trapped mother was successfully rescued. On the other side, firefighters were rescuing her trapped boy. (China News) |
| | Around 14:00 | The trapped boy was rescued. (China News) |
| | 16:00 | 48 victims have been rescued from the scene, of whom 10 died, and 38 were sent to the hospital for treatment. There are still 23 people being searched and rescued. (People's Daily) |
| | 16:36 | 49 people have been rescued from the scene, and there were still 22 people trapped. (China Fire official microblog) |
| 9 March | 02:43 | The site has been removed from the floor, colored steel tiles, debris, and all kinds of debris 45 cubic meters; cleanup work is continuing. (China Fire official microblog) |
| | 08:00 | The scene search and rescue of 49 people trapped, of which 8 people rescued when no vital signs. There are still 22 people being searched and rescued. (Quanzhou emergency rescue work leading group) |
| | Around 10:00 | A trapped person was found at the scene. |
| | 13:40 | Fujian firefighters rescued a male victim, and firefighters collectively observed a moment of silence for the victim. (People's Daily) |

(*continued*)

Table 1. Continued

| Date | Time | Events |
|---|---|---|
| | 14:00 | By now, 50 people have been rescued from the scene, 11 of them died, and 21 people are still being searched and rescued. (People's Daily) |
| | 19:44 | Firefighters rescued one more trapped person who had no vital signs at the time of rescue; 51 people have been rescued. |
| | 20:00 | The accident site search and rescue of trapped people 52 people, 13 people died, and 19 people are being searched and rescued now. (Quanzhou emergency rescue work leading group) |
| | 20:18 | A trapped mother and child were found under the ruins (China Fire official microblog) |
| | 23:05 & 23:08 | The trapped mother and child were rescued and sent to the hospital for treatment, at which time the mother and child had been trapped for nearly 52 hours. 54 people have been rescued so far (China Fire official microblog) |
| | 23:20 | The scene search and rescue of 55 people trapped, 14 people died, and there are still 16 people being searched and rescued. (Fujian Daily News) |
| 10 March | 00:24 | This was the 56th person rescued from the accident site, and 15 people were still trapped. (People's Daily) |
| | 03:44 | There was a drizzle on the scene. A female was found deeply buried, and the space was narrow. Firefighters drilled into the ruins and tried to open the rescue path. (China Fire official microblog) |
| | 04:41 | The woman was successfully rescued, and 57 people were rescued at the scene. (China Fire official microblog) |
| | 04:46 | The 58th victim has been rescued with no vital signs. (Southeast Morning Post) |
| | 06:20 | A boy was found under the ruins. (China Fire official microblog) |
| | 06:35 | The 59th victim was rescued at the worksite of Sanming, Quanzhou, and the Putian fire rescue detachment. (China Fire official microblog) |
| | 06:50 | 59 trapped people were rescued on the scene, 18 people have been killed so far, and 12 people are still waiting for rescue. (Strait Herald) |
| | 07:00 | Another child was found, and the two children were under intense rescue. (China Fire official microblog) |
| | 08:27 & 08:38 | Firefighters rescued a little boy; At 8:38, firefighters rescued a little girl. So far, 61 people have been rescued. (China Fire official microblog) |
| | 09:00 | Currently, 61 people have been rescued at the scene of the accident, of which 20 were killed, 41 were injured, and 10 are still trapped. (Government Press Conference) |
| | 16:38 | After 69 hours, the survivor who used the remote control to knock was finally rescued. |
| | 16:50 | Currently, 62 people have been rescued at the scene of the accident, of which 20 were killed, and 9 are still trapped. |
| 11 March | 4:43& 4:45& 4:50& 4:53& 5:22 | One trapped person was searched and rescued, respectively, and there were no vital signs at the rescue time. Currently, 67 people have been rescued. (People's daily) |
| | 05:50 | 67 trapped people were rescued on the scene, and 25 died, of which 23 had no vital signs when rescued, and 2 died after being sent to the hospital. |
| | 06:30 | One trapped person was rescued with no vital signs. |
| | 06:40 | As of now, 68 people have been rescued from the scene, 26 were killed, and 3 are being rescued. |
| | 08:43 | Firefighters rescue a male victim with no vital signs. (China Fire official microblog) |
| | 11:00 | As of now, 69 people have been rescued from the scene, 27 were killed, and 2 are being rescued. |
| | 12:45 | At present,70 people were rescued from the scene, 278 were killed, and the last trapped victim was being rescued. |
| 12 March | 11:04 | About 112 hours after the collapse, the last person trapped was found and was lifeless when rescued. All 71 trapped people have been rescued by now. |

*Advances in Urban Construction and Management Engineering – Cha (Ed)*


# Practice and exploration of BIM technology in engineering project management under the background of new infrastructure construction

Rui Zhao*, Xinyu Wang*, Lin Yuan* & Qiuyang Zhang*
*Sichuan Technology and Business University, China*

ABSTRACT: Under the background of new infrastructure construction, BIM technology is used to realize the digital, intelligent, and high-speed development of engineering project management, which provides convenience for the construction industry and effective guidance for the implementation of engineering projects. The construction period of a middle-school teaching complex building project in Sichuan is short, while the task is heavy and the site space is limited. In the bidding stage, a three-dimensional model is used to calculate the bidding quantity quickly, and an intuitive analysis of the construction model can quickly locate the construction difficulties. Zebra Menglong software is used to prepare the construction schedule and generate the animation of the construction schedule. In the construction stage, the construction deviation value is combined with the BIM model, and BIM 5D is used to simulate construction, pre-control risk, and dynamic control construction cost. We optimize the delivery process of BIM application results according to the progress of the project, use the visualization model of BIM technology to find the problems in each link of the project in advance, deal with them in advance, and timely deliver the corresponding results for the reference of all parties.

## 1 INTRODUCTION

Since the 14th Five-year Plan, it has been clearly proposed to promote infrastructure construction as a whole, build a complete modern infrastructure system that is highly efficient, practical, intelligent, green, safe, and reliable, and promote the coordinated development of intelligent construction, green building materials and building industrialization (CENTRAL Committee of the Communist Party of China 2020, Li 2020). New infrastructure has gradually become the development focus of the construction industry. The fundamental method for the construction industry and construction-related enterprises to achieve transformation is to make use of scientific and technological progress and management innovation. Therefore, the construction industry is constantly improving, optimizing, and standardizing under the influence of the trend of the times. BIM technology can be applied to all stages of engineering construction. It can make design, production, construction, management, and other links more informatized and intelligent and can optimize the management of the project. Engineering project management uses BIM technology to a large extent to provide convenience for the construction industry, and intelligent construction is leading a new round of construction industry revolution.

This paper introduces the application of BIM technology in engineering project management by taking a middle school teaching complex building in Sichuan province as an example. This project applies the Glodon cloud pricing platform to create a 3D model and uses Glodon bidding document compilation tool software to complete electronic bidding document compilation. Zebra Menglong

*Corresponding Authors: daishisara@qq.com, 1425789228@qq.com, 2180733031@qq.com and 192829785@qq.com

 DOI 10.1201/9781003348023-51

software was used for network planning, and BIM construction site 3D layout software was used for reasonable optimization of the construction site layout, combined with all kinds of information in the model to guide the site construction, providing effective guidance for the implementation of engineering projects.

## 2 PROJECT OVERVIEW

The project is a middle school teaching complex building (Figure 1), with a total construction area of $10079.34m^2$ and a construction area of $2002.29m^2$. The building is a frame structure with six floors on the ground and one floor underground. The span of the building is 63.5 meters, and the eaves are 23.70 meters high. The total investment is 16 million yuan, and the project duration is 150 days. The foundation of the project is the independent foundation under the column and the partial raft foundation. Although the excavation depth of the foundation pit is not more than 3m, the geological conditions, surrounding environment, and underground pipelines are complex, so special plans need to be prepared for construction.

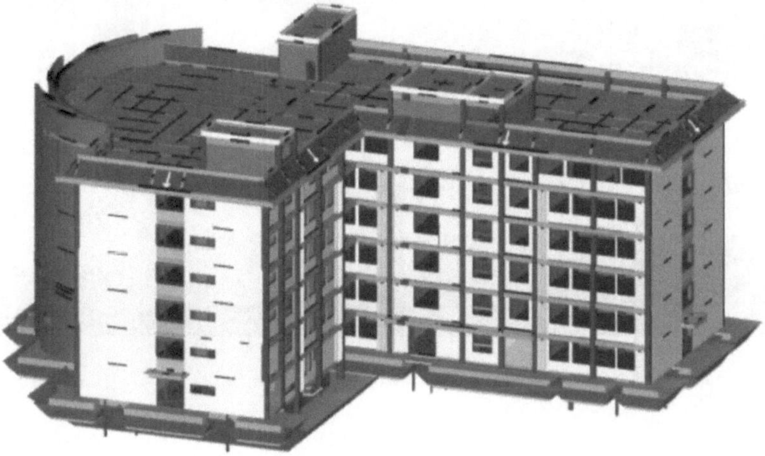

Figure 1. 3D model of teaching complex building.

## 3 APPLICATION OF BIM TECHNOLOGY IN ENGINEERING PROJECT MANAGEMENT

BIM technology includes the whole process management of construction projects from design, bidding, and construction (Zhang, 2022). By establishing the project information model, the delivery process of BIM application results is optimized according to the project schedule. The visualization model of BIM technology is used to find the problems in each link of the project in advance, deal with them in advance, and deliver the corresponding results in time for the reference of all parties.

### 3.1 *Application of BIM technology in the bidding stage*

According to the requirements of the State Council, bidding and tendering venues across the country are being comprehensively integrated, and the public resources and construction project trading center are transforming from a traditional supervision service mode to an informatized and electronic trading service platform (National Development and Reform Commission of the People's Republic of China 2016). BIM visualization of bidding document display and evaluation will also become the mainstream development direction in the future.

### 3.1.1 *Bidding stage*

We make a bill of quantities and bidding control price statements using the Glodon cloud pricing platform and complete the compilation of bidding documents through the electronic bidding document compilation system. There is a project basic information template in the software for compiling bidding documents, which only needs to be filled in according to the specific information, greatly shortening the time for compiling the basic information of the project. When compiling a bidding list, it is necessary to import the list and report form derived from civil construction modeling to obtain a bidding control price. In this process, only a part of the drawings was shown when importing PDF architectural drawings, so the drawings were modified and re-exported in PDF format in CAD and then re-imported into the bidding software, which successfully solved the incomplete problem after importing drawings (Figure 2).

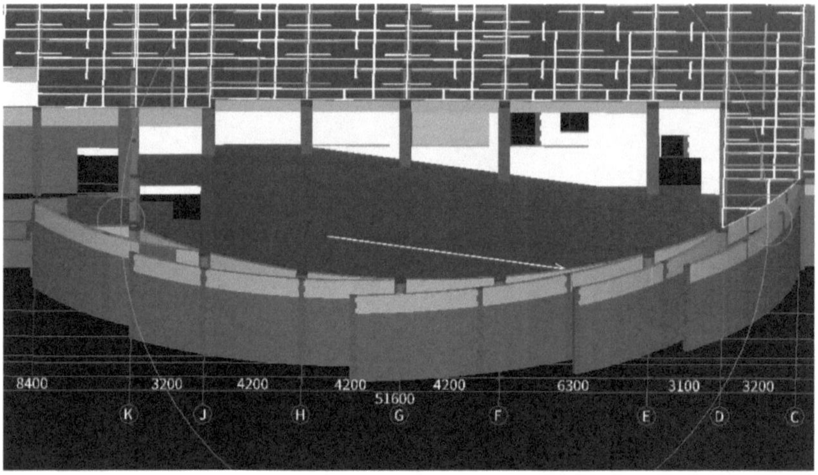


Figure 2. 3D graphic display of Glodon reinforcement calculation.

### 3.1.2 *Tendering stage*

In the bidding stage, in order to prepare a more accurate bidding document, the bidder establishes a 3D visual model according to the drawings and other information provided by the tenderee. After establishing the model, the bidder compares the bill of quantities of the bidding document and finds that there are obvious errors in the number of steel bars, so it is fed back to the tenderee in writing. The tenderee immediately clarified the relevant problems and corrected some data errors. After the correction, the bidder immediately made corresponding adjustments to the bidding documents by using BIM-related technologies. For example, they appropriately lower part of the demand for more material prices, adjust part of the demand for lower material prices according to the enterprise's situation and adjust the construction organization design content, to be able to use the right amount of human material machine to complete each part of the work, improve the enterprise's bid rate and profit rate.

Using Zebra Menglong software to prepare the construction progress plan and importing the progress plan and model into Glodon BIM bidding software, the animation demonstration of construction progress can be generated (Figure 3), and the whole process of the implementation of the construction organization plan can be seen intuitively, and the problems in the construction can be found more easily. When preparing the construction schedule, we should pay attention to the following points. First, we should comprehensively evaluate the construction site environment. Second, we should consider the order of each work, overlap relationship, and reasonable arrangement of the work schedule according to the level of the enterprise and the content of the work.

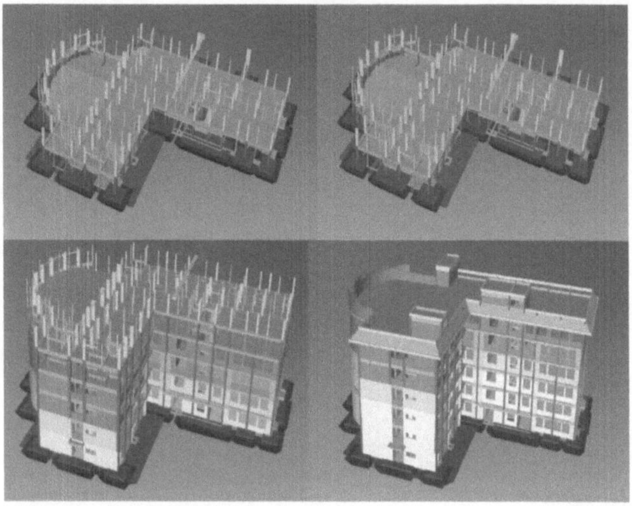

Figure 3. Construction progress demonstration.

### 3.2 *BIM technology is applied in the construction stage*

The successful realization of the dynamic management task of the BIM-5D series software of Glodon ensures project resource management and organization management (Dong 2021). Cycle combined with the actual project construction stage and points to the actual changes in the construction process of detailed control and inspection. This behavior is actually used to build a 3D model, adjust the BIM implementation, and finally, through the accumulation of a large amount of data, the formation of the actual cost database construction stage BIM-5D, risk pre-control, and dynamic control of construction cost. Pay attention to the importance of resource optimization management, and use Glodon BIM-5D software to ensure that the needs at all stages can be met.

#### 3.2.1 *Site layout*

Visionbank of BIM 3D site layout software can simulate the road around the construction site in the virtual space, the condition of the simulation model of the surrounding facilities, and materials piled up. According to the requirements of each stage machinery, materials, and engineering division, the optimal construction site layout plan can be determined before the construction (Figure 4), effectively reducing the number of secondary material handling and costs and improving work efficiency. By establishing the cost model, we can get a money progress curve so that the entire project funds of annual, quarterly, and monthly costs can be better reflected through the analysis of the planned cost and actual cost. We can know the expenditure of a job in advance; thus, the cost of the whole process can be under effective management.

#### 3.2.2 *Engineering change*

There is always a certain gap between the scheme design and construction. Therefore, an engineering change is an important link to correct the design and construction deviation. Changes will lead to changes in the quantity of the project, the construction schedule will be different, and then the actual cost of the project will bring fluctuations, and the phenomenon of exceeding the budget often appears. In this project, the change of the beam affects itself and the related plate based on the change information of the designing institute. The BIM team is responsible for modifying the data information of the BIM model and carrying on the operation of the time dimension and space dimension to the automatic operation of the corresponding quantity. The impact of engineering changes will be displayed, and the economic indicators of the project will be reflected.

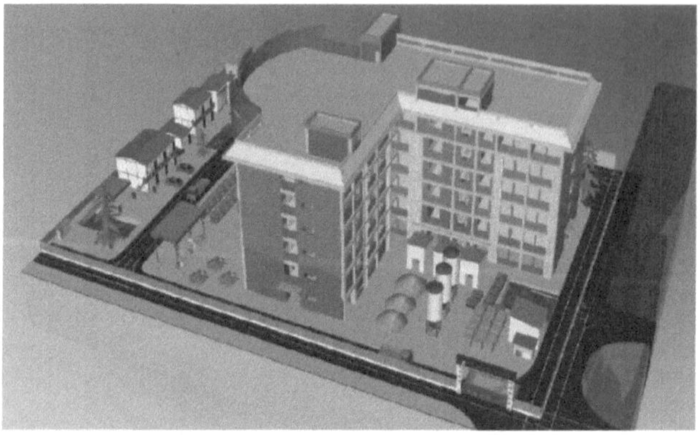

Figure 4. The layout.

### 3.2.3 *Schedule management*

In order to realize the steady implementation of the construction project, the site plane visualization model is established with the help of the BIM-5D view of Guangda so that the project managers can check and adjust the project nodes more intuitively. In the construction, the progress plan is updated in real-time according to the actual progress of the project, and BIM technology is used for real-time tracking management until the completion of the project (Figure 5). By simulating the construction process with a dynamic model, the possible problems in construction can be quickly located, dealt with in advance, and prevented from stopping in subsequent construction. Problems can also be dealt with quickly, and feasible solutions can be put forward in a short time, reflecting the maintenance of the effectiveness of construction schedule control. The project schedule delays two days in the basement structure, which is the key work. It affects the total time limit for a project to be completed on schedule. So we must adjust the construction plan as soon as possible by using BIM software to adjust the construction progress plan, solve the problem of their relatively slow, error-rone, quickly come up with solutions, and make engineering on the rails. Taking the case project as an example, the planned construction period is 180 days. In order to prevent the change of uncertain factors of the project, the construction period is optimized to 150 days according to the actual progress of the project and the real-time optimization management of BIM technology. BIM is reasonably planned to realize the reasonable connection and crossover between teams.

## 4 BIM DELIVERABLES

The demand for BIM technology and the required results in the bidding and construction stages of construction projects are different for each layer (Zhang 2022). The demand for BIM results in the bidding stage is to take the BIM information model as the basic reference basis for the compilation of bidding documents and select the construction enterprises with the most reasonable offer and the most advanced technology for cooperation. The construction scheme design in the construction stage is implemented throughout the project construction, and the scheme will be constantly adjusted according to the site situation. Therefore, the BIM application results in this stage are delivered according to the actual needs of the project (Table 1).

## 5 CONCLUSION

BIM technology has played a certain role in project planning, project component material control, bidding preparation, project construction schedule, site layout, cost management and other aspects.

Table 1. BIM Deliverables.

| No. | Implementation stage | Submissions | Submission file format |
|---|---|---|---|
| 1 | Bidding stage | Project drawings Bidding project background | .dwg/pdf/zip/rar |
| 2 | | Bill of quantities, bidding control price report | .xls/.xlsx/.PDF |
| 3 | | Bidding documents source documents | .GZ7 |
| 4 | Tendering stage | Double code time scale logical network schedule plan diagram | .JPG |
| 5 | | Three-dimensional layout of the construction site | .GBCB |
| 6 | | Bid construction organization design | .PDF |
| 7 | | Electronic tender document | .GT7 |
| 8 | | Capital progress curve | .JPG/PNG |
| 9 | Construction stage | Construction simulation animation video | .FLV/MP4 |
| 10 | | Multi-professional model integration picture | .JPG/PNG |
| 11 | | Flow section division table | .XLSX/XLS/CSV |
| 12 | | Total curve analysis chart of capital progress | .XLSL/XLS/CSV |
| 13 | | Total resource progress curve analysis chart | .XLSX/XLS/CSV |

The construction units should actively apply new technology, and understand the role of BIM, so that BIM technology is really applied to the project, rather than treating BIM as a burden on enterprises. With the rapid development of BIM technology, relevant units should do a good job in talent training and technical reserve in advance to get the top spot in the future technological and digital construction market and walk at the forefront of the development of the construction industry.

## ACKNOWLEDGEMENT

The Sichuan Association for Non-Government Education 2021 Research Project General Project: Practice and exploration of BIM teaching of construction engineering management courses under the background of new infrastructure construction (Project Number MBXH21YB226).

## REFERENCES

CENTRAL Committee of the Communist Party of China, Proposal of the CENTRAL Committee of the Communist Party of China on formulating the 14th Five-Year Plan for National Economic and Social Development and the Long-term Goals for 2035[S]. 2020.

Dong Wei *Research on the application of BIM Technology in housing construction cost management* [D] Liaoning University of Technology, 2021.

Li Xiaohua. “New infrastructure” for smart society and its policy orientation [J]. *Reform*, 2020 (5): 34-48.

National Development and Reform Commission of the People’s Republic of China, on carrying out the pilot work of National Electronic Tendering and Bidding in 2016 [S]. 2016.

Zhang Guanglei, Ji Wenbin, Zeng Zhiming, Qiao Guanxiang. Research on BIM technology in Engineering Project Application: A case study of Hezhengfang Zhourunyuan Project [J]. *Project Management Technology*, 2022, 20(03):106–110.

*Advances in Urban Construction and Management Engineering – Cha (Ed)*


# Visualized monitoring method for temperature target of running part of subway vehicle

Shenghua Wang, Xiaodong Xi & Sheng Lv
*Shanghai Metro Maintenance Guarantee Co., Ltd. Shanghai, China*

Ao Liu*
*China Railway Fourth Survey and Design Institute Group Co., Ltd., Hubei, Wuhan, China*

ABSTRACT: With the development and application of artificial intelligence technology, the research on target monitoring based on deep learning has made significant progress. However, most deep learning applications are based on visible light conditions, and few applications are related to infrared thermal imaging temperature scenarios. This paper considers using the Yolov3 deep learning algorithm framework to dynamically identify and monitor the temperature of key infrared targets under subway vehicles. The deep convolutional neural network is used to extract various abstract features through training and learning on a large amount of data. At the same time, the obtained original temperature video stream data is analyzed, the single-frame infrared heat map is filtered to improve the clarity of the target in the picture, and the error compensation of the key target temperature is performed by polynomial fitting. All these tests have proved that the network has high accuracy for target location and recognition, and the target temperature of different axle positions of the same train has a good consistency.

## 1 INTRODUCTION

The urban rail transit train is a very complex system, and the safety and stability of the vehicle operation are crucial. The temperature video stream data in this paper comes from the temperature detection system of the running part of the vehicle. The system is located on both sides and in the middle of the track, and uses industrial infrared thermal imagers to achieve dynamic temperature monitoring of key heating components, such as train bottom axle boxes, gearboxes, and motors. In practical application scenarios, dynamic monitoring of key targets under the vehicle is often accompanied by scene occlusion, object distance changes, and some environmental factors. At the same time, the infrared heat map is imaged by sensing the difference between the target object and the surrounding temperature, and the objects are accompanied by the interaction of energy simultaneously. These physical characteristics will cause the contour edge of the infrared image to be blurred, and the contrast will ultimately decrease (Li 2018; Lu 2008; Qiao 2013).

The detection of key targets by traditional image processing algorithms based on non-deep learning mainly includes frame difference method, matching method, optical flow method, feature background modeling method, threshold segmentation method, and others. (Feng 2018). The deep learning algorithms can be widely used in various scenarios under complex conditions. With the support of data-driven modeling and accelerated computing platforms, it has been widely used in face recognition, smart home, and other fields.

In this paper, the infrared scene is detected based on a deep learning algorithm, and it is optimized and improved by building a convolutional neural network, and strives to identify the target accurately. In addition, the limitation of frame rate and response time of industrial infrared thermal

*Corresponding Author: 18935931331@qq.com

 DOI 10.1201/9781003348023-52

imager creates technical challenges in the imaging of the target object during vehicle motion, such as smearing and noise, and these problems will cause errors in temperature monitoring. Therefore, the dynamic temperature error range needs to be considered, and the temperature measurement error of the target object during the moving process needs to be reduced to ensure the consistency of temperature data at different end positions when passing vehicles.

## 2 YOLOV3 ALGORITHM PRINCIPLE

This paper selects the Yolo algorithm in the deep learning algorithm framework to locate and identify the key heating targets under the vehicle. The algorithm can maximize the use of GPU for real-time target detection, and the detection effect and operation speed are high. At the same time, it is also the best-performing target detection algorithm in the Yolo series. The Faster R-CNN algorithm was mentioned earlier. Its core converts the target recognition problem into target classification, and it realizes multi-target localization detection in different scenes according to target frame extraction and category determination. Compared with R-CNN and Faster R-CNN, the latter has a faster detection rate; however, from the perspective of the search method, the calculation time is still long. The specific process is: first, obtain the feature map; second, obtain the feature information of the marked frame; finally, use the classifier to classify and combine the regressor to optimize the position of the marked frame further.

During the operation of Yolov3, the target classification is converted into data training, and the output layer in the entire network returns to the target position and output category, which improves the operation efficiency significantly. The specific process is as follows: first, use the classifier and locator to perform the detection process; second, combine the model and the image position scale to calculate the score; finally, define the target area with the best score as the visualization result (Dai 2018; Zhu 2018).

Compared with the previous version, Yolov3 replaced 19 with Darknet-53 and removed the pooling layer in terms of network structure. It is mainly used to extract network features. Its structure is 0–74 layers. The Yolov3 network does not have a fully connected layer; it has 53 convolutional layers, and the rest are residual networks. The Yolov3 network is responsible for calculating the loss function as the Yolo layer and outputs the target box information.

The detection accuracy of the entire network is closely related to the number of network layers. The more the layers, the better is the representation effect. Yolov3 adopts a residual network structure, ensuring that the training results are in the direction of convergence in the training process on a large amount of data. At the same time, in the deep network environment, the multi-channel convolutional layers are not correlated with each other. This way of linking the layers can greatly improve the efficiency of target prediction (Xu 2019).

Figure 1 shows the darknet53 algorithm framework in Yolov3. The advantage lies in the scale assignment between the target frame and the feature map. If a larger part of the feature map is distributed in the shallow network, it is assigned to the small object; otherwise, the feature map in the deep network is assigned to the large object. This operation mode can effectively improve the regional positioning accuracy of small targets.

## 3 DATASET PREPARATION AND MODEL TRAINING

### 3.1 *Training samples*

As deep learning models require a large amount of data support, most opening source databases are based on visible light images. The data set used in this paper is to analyze the temperature video stream of the train running part saved in real time by the trackside infrared thermal imager. The detection points are tentatively set as the axle box on the side of the vehicle, the motor on the bottom of the vehicle, the gearbox, and the coupling, as shown in Figures 2–4.

| | Type | Filters | Size | Output |
|---|---|---|---|---|
| | Convolutional | 32 | 3 × 3 | 256 × 256 |
| | Convolutional | 64 | 3 × 3 / 2 | 128 × 128 |
| 1× | Convolutional | 32 | 1 × 1 | |
| | Convolutional | 64 | 3 × 3 | |
| | Residual | | | 128 × 128 |
| | Convolutional | 128 | 3 × 3 / 2 | 64 × 64 |
| 2× | Convolutional | 64 | 1 × 1 | |
| | Convolutional | 128 | 3 × 3 | |
| | Residual | | | 64 × 64 |
| | Convolutional | 256 | 3 × 3 / 2 | 32 × 32 |
| 8× | Convolutional | 128 | 1 × 1 | |
| | Convolutional | 256 | 3 × 3 | |
| | Residual | | | 32 × 32 |
| | Convolutional | 512 | 3 × 3 / 2 | 16 × 16 |
| 8× | Convolutional | 256 | 1 × 1 | |
| | Convolutional | 512 | 3 × 3 | |
| | Residual | | | 16 × 16 |
| | Convolutional | 1024 | 3 × 3 / 2 | 8 × 8 |
| 4× | Convolutional | 512 | 1 × 1 | |
| | Convolutional | 1024 | 3 × 3 | |
| | Residual | | | 8 × 8 |
| | Avgpool | | Global | |
| | Connected | | 1000 | |
| | Softmax | | | |

Figure 1. Darknet-53 structure.

Figure 2. Axle box.

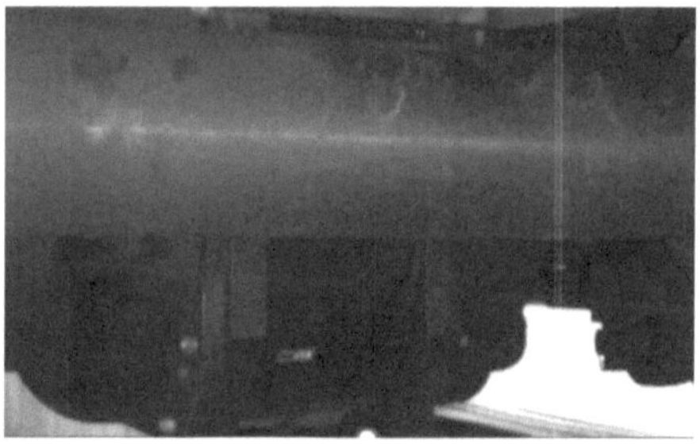

Figure 3. Motor.

Figure 4. Gear box.

Considering the generalization ability of the neural network model and the robustness of later feature extraction, other negative sample data without targets in the tunnel are selected to provide a richer and more comprehensive dataset for data training.

### 3.2 *Data processing*

The video collected by the infrared thermal imaging axial temperature detection system involved in this paper is analyzed to obtain an infrared heat map. Due to the large difference between infrared images and visible light images, infrared images have blurred target boundaries and low contrast between the background and the target due to their special physical characteristics, which increases the difficulty of model training and feature extraction. Therefore, it is necessary to filter and denoise the original image data.

The basic principle of image filtering is carrying out in the time domain or frequency domain through filtering operations to emphasize some features or remove unwanted parts of the image. Filtering is a domain operator that uses the pixel values around a given pixel to determine the final output value of this pixel. Common filtering methods include mean filtering, median filtering, Gaussian filtering, and bilateral filtering. In this paper, the bilateral filtering algorithm is selected. The purpose is to take advantage of the weight distribution advantage of the grayscale information in the operator; that is, in the field, the closer the grayscale value is to the center point, the greater is the weight.

In addition, bilateral filtering requires grayscale information of each center point area to determine its coefficients, so the speed is slower than that of general filtering, and the calculation amount increases at the square of the kernel. Figures 5 and 6 show the temperature heat map of key components under the vehicle before and after filtering:

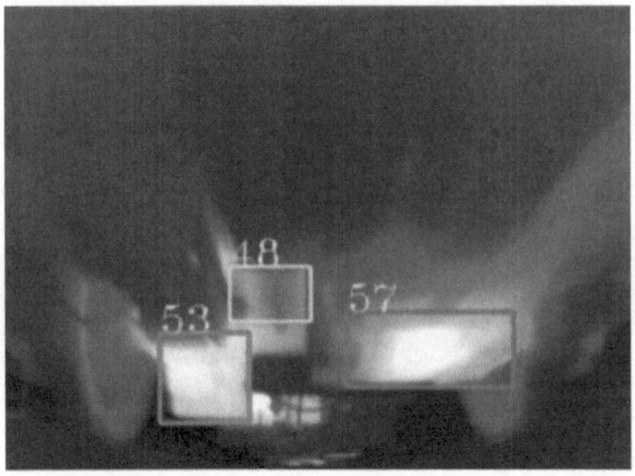

Figure 5. Original result display.

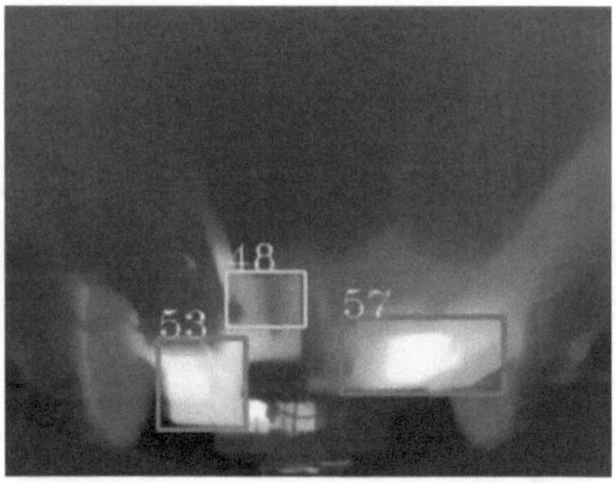

Figure 6. Processing data after filtering.

### 3.3 *Model training*

With the help of the image annotation tool, a certain number of sample data with target objects are labeled, including classification and target frame position. The process of training data includes: first, configure the parameters of the training network framework (batchsize, subdivisions, channels, learning rate, step size, etc.); second, modify the relevant parameters of the Yolo layer (the number of filters, anchors, and categories, etc.); finally, continuously adjust the above configuration parameters to optimize the model.

The data set used in this paper is composed of tens of thousands of positive samples containing the target and negative samples not containing the target, and the two are stored in equal proportions. The size of the original input image is 384*288, the number of inputs in a batch is 64, and the initial learning rate is 0.001. After the training starts, the loss value of the dataset shows a significant downward trend when the model iterates to 5000 times. As the number of training iterations increases, the loss value gradually flattens out.

The basis for the end of training is: when there are many error metrics in the training log, it is necessary to stop training when the average index no longer decreases, and the lower the average loss value, the better the training effect. Ultimately, the entire training metric should be such that the loss value ends up being distributed between 0.05 and 3.0.

## 4 TEMPERATURE CALIBRATION

### 4.1 *IR imaging calibration*

One of the main technical difficulties in the temperature detection of the key components of the train underbody running part described in this paper is to ensure the accuracy of the temperature detection. At high speed, the image smearing of the target and the increase in the proportion of the target size compared to the thermal imager's field of view will affect the temperature measurement accuracy of the infrared thermal imager. In addition, if the vehicle speed is very fast, the temperature of the target surface will spread faster, and the radiation energy received by the thermal imager will decrease. The fitting curve function is helpful for correcting the temperature of the target object at different speeds to ensure the accuracy of the data.

First, a device with the properties of a standard radiant heat source (black body: an ideal heat source with an emissivity of 1) is needed to help determine the deviation of the actual measured temperature from the actual temperature brought about by the movement. The specific idea is: that a black body heat source with a known set temperature passes through the front end of the infrared thermal imager at different speeds, captures the black body on the screen, marks its temperature value, and subtracts the theoretical temperature from the measured temperature at different speeds, that is A set of error data can be obtained as the object of the fitting. Figures 7 and 8 show the experimentally obtained black body dynamic temperature measurement results and their deviation from the true value:

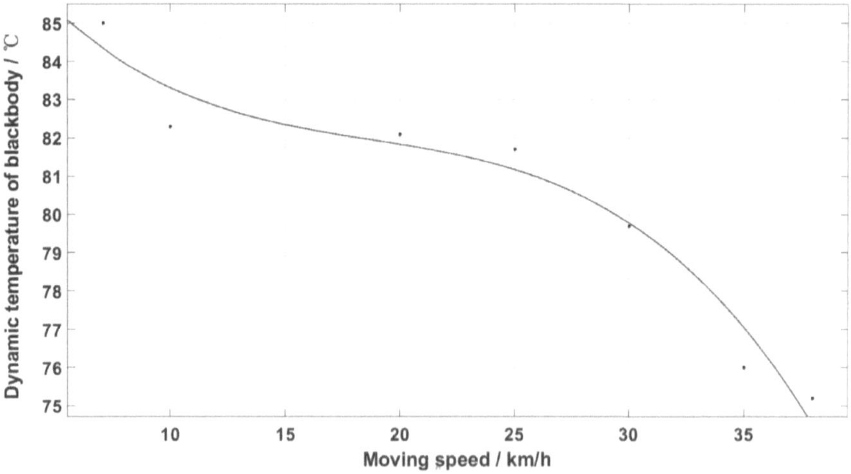


Figure 7. Measured temperature data of black body (theoretical value 85°C).

### 4.2 *Dynamic compensation*

On the basis of Section 3.1, this paper chooses a universal data fitting algorithm to achieve high-precision fitting with limited data volume. In practical applications, it is difficult to directly determine the correlation between values based on some isolated test data. The algorithm involved in the present invention can solve and improve the fitting effect of the limited amount of data, and the essence is to use the method of function approximation.

(1) Determining the Error Fit Function

First, the measured value of the black body temperature obtained from the dynamic black body calibration experiment is subtracted from the theoretical set temperature value and used as the input of the model. At the same time, according to different test motion speeds, they are sequentially stored as a one-dimensional array: y = [y1 y2 y3 y4...yi](i = 1~n).

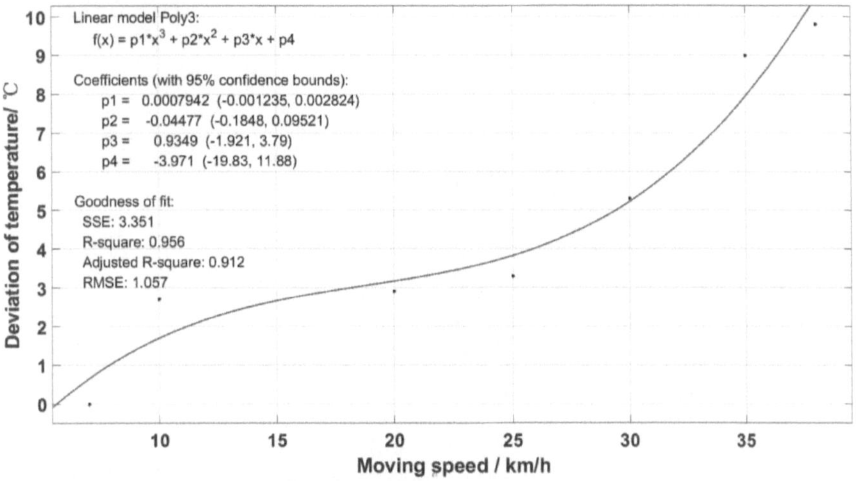


Figure 8. Polynomial fitting function of dynamic temperature measurement error.

Second, record and store the different experimental motion speeds into another array x = [x1 x2 x3 x4...xi](i = 1~n).

Then, determine the fitting accuracy *e* of the above input data and the initial order as 1.

Using the ployfit function we get the fitted array p = [p1 p2 p3 p4 ...pn], and then using the calculated velocity array x which is related to p to get the value y′ = [y′1 y′2 y′3...y′n], next we can calculate the square of the sum of the absolute values of the corresponding term errorsin E both in the y and y′, $E = \sum_{i=1}^{N} |yi' - yi|^2$

Finally, complete multiple loop iterations until E < e (if not satisfied, the order is incremented by 1, n=n+1), and convert the polynomials that meet the precision requirements into a higher order function by calling the poly2str function.

(2) Temperature compensation

The higher order error fitting function calculated in (1) will be used in the running temperature monitoring system, and the specific steps to be realized according to the trackside sensor signal are shown below.

1) All axles of the train pass through the two wheel sensors arranged on the track in turn and keep a fixed distance of L. When the wheels pass the No. 1 sensor, the time t1 will be recorded, and time t2 will be recorded after passing the No. 2 sensor, t2 > t1
2) After the fixed distance between the two sensors is known, the passing speed of each axle will be calculated in turn: v = L/(t2-t1)
3) Call the error fitting function in (1), and calculate the error compensation value of each axle in turn and compensate it to the measured temperature value.

Figure 9 reflects the temperature consistency of the key points of the same vehicle with different axle positions after the above temperature compensation processing.

## 5 SUMMARY AND OUTLOOK

This paper first introduces the commonly used algorithm models in deep learning region positioning, focusing on the Yolov3 network and its advantages. For complex scenes, such as infrared heat maps, it first can quickly capture and calculate the position of key targets. Second, this paper carries out relevant black body dynamic temperature detection and error analysis to evaluate the accuracy and precision of infrared thermal imaging temperature measurement of moving targets.

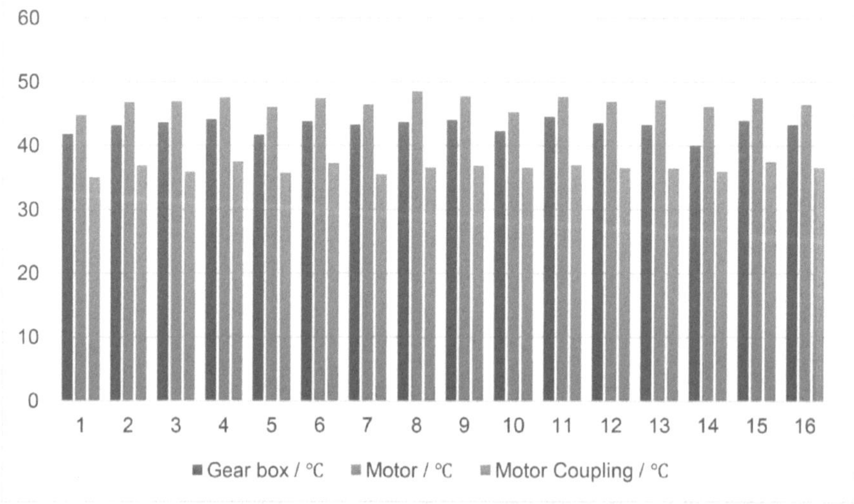


Figure 9. Corrected temperature of key components in different axle positions of the same vehicle. (Horizontal axis: axle position; Vertical axis: Temperature °C).

Finally, the paper also selects the polynomial function approximation to analyze data. The method fits the speed-temperature error function, and it finds the correlation and performs temperature compensation of the real target, to achieve better consistency of the temperature measurement data of the same vehicle target.

## REFERENCES


Chenyang Zhu, Hutian Feng, Yi Ou (2018). Research on automatic face tracking camera robot system based on YOLOv3 [J]. *TV Technology*. 42(9):64–69, 91.

Liyong Qiao, Lixin Xu, Min Gao (2013). Influence of Infrared Image Complexity on Target Detection Performance[J]. *Infrared and Laser Engineering*. (S1):253–261.

Shan Lu (2008). *Research on Infrared Target Detection and Recognition Technology* [D]. Master Thesis of Changchun University of Science and Technology.

Weicong Dai, Longxu Jin, Guoning Li, et al (2018). Improved YOLOv3 real-time detection algorithm for aircraft in remote sensing images [J]. *Optoelectronic Engineering*. 45(12):84–92.

Xiaoting Xu (2019.09). *Research on Infrared Scene Target Detection Technology Based on Deep Learning*. North Central University.

Xiaoyu Feng, Wei Mei, Dashuai Hu (2018). Aerial target detection based on improved Faster R-CNN [J]. *Journal of Optics*. 38(6):250–258.

Yi Li (2018). *Research on infrared target search of intelligent ammunition*[D]. Master's Thesis of Xi'an University of Technology.

*Advances in Urban Construction and Management Engineering – Cha (Ed)*


# A study on joint inventory decision of fresh produce considering service constraint level constraint and controllable lead time

Lei Zhou*, Xiaoli Ma* & Fachao Li*
*School of Economics and Management, Hebei University of Science and Technology, China*

ABSTRACT: The traditional inventory model takes minimizing the cost of a single inventory as the goal. However, there will be problems such as rising inventory costs and the bullwhip effect. To address these problems, this paper studies a fresh product inventory model from the perspective of joint inventory decisions of distribution centers and retailers. Under the constraints of random and normal distribution of retailer fresh product market demand in lead time, controllable lead time of distribution center, and corresponding service level, the joint inventory model is constructed with the goal of minimizing joint inventory costs and the order lot, lead time, reorder point and supply lot as decision variables. Finally, the reasonableness of the joint inventory model and the feasibility of the algorithm are verified by arithmetic examples.

## 1 INTRODUCTION

Inventory control is an important issue for modern supply chain management. The traditional supply chain inventory management model is aimed at minimizing the cost of supply chain members. Joint inventory management can weaken the demand amplification phenomenon caused by the independent operation of each node of the supply chain, and at the same time, it can reduce the business risks of each node of the supply chain.

Regarding the joint inventory problem, as early as 1976, Goyal proposed the joint economic lot model, which treats buyers and sellers as an integrated system and investigates how to achieve inventory collaboration to minimize the total system cost. Since then, many domestic and foreign scholars have promoted the model from different aspects and applied it to the joint inventory management in the secondary supply chain. Qunxia Li studied a two-level supply chain inventory problem consisting of a single producer and a single customer and proved that the joint objective function has the minimum joint supply chain inventory cost at both the optimal order quantity and the optimal number of orders (Li 2019). Wenjie Bi et al. proposed a joint inventory control and dynamic pricing method for fresh produce based on deep reinforcement learning methods (Bi 2022). Qing Zhang et al. studied the problem of inventory control strategy for a system consisting of traditional and online channels in the presence of demand perturbations (Zhang 2016). M. Vijayashree1 and R. Uthayakumar studied the relationship between ordering cost reduction and lead time to analyze the total inventory cost for single-supplier and single-retailer systems (Vijayashree 2017). By studying inventory management in the supply chain environment, Deng Hui focuses on the joint inventory management model and its implementation strategy in the supply chain environment (Deng 2016). Tiwari and Kumar et al. minimized the total cost of the system in the case where the lead time obeyed both normal and unknown distributions, but both ignored the consideration of reorder points (Tiwari 2018).

Therefore, this paper takes the secondary joint inventory decision consisting of distribution centers and retailers as the research content, the order lot, lead time, reorder point, and supply

*Corresponding Authors: zhoulei19@126.com, 1766764653@qq.com and lifachao@tsinghua.org.cn

DOI 10.1201/9781003348023-53

lot as the decision variables, constructs the joint inventory model, and makes the optimal strategy selection from the perspective of cost minimization.

## 2 JOINT INVENTORY DECISION MODEL CONSTRUCTION

### 2.1 *Symbol description*

To facilitate model construction, the following notation is defined.

Table 1. Symbols and meaning.

| | | | |
|---|---|---|---|
| $D$ | Retailers' average demand for fresh produce per unit time | $\theta_d$ | Distribution center unit fresh product spoilage rate |
| $X$ | The demand within the lead time period L, obeys a normal distribution | $L$ | Lead time, $L_1 \leq L \leq L_2$ |
| $F(x)$ | Distribution function of random demand for fresh produce | $R_d$ | Cost of spoilage of fresh products in distribution center units |
| $f(x)$ | Probability density function of random demand for fresh produce | $C_r$ | Retailers' order cost per order |
| $k$ | Mean standard deviation values corresponding to specific service levels | $C_d$ | Distribution center preparation costs per lot supplied |
| $Q$ | Retailer order volume per fresh produce order | $h_r$ | Retailers' unit inventory holding costs per unit of time |
| $nQ$ | The supply lot of the distribution center, where $n$ is a positive integer | $h_d$ | Distribution center in unit time unit inventory holding cost |
| $R_r$ | Retailer unit spoilage cost of fresh produce | $\beta$ | Percentage of backorders allowed during out-of-stock periods |
| $r$ | Reorder points | $\pi_1$ | Loss of profit per unit of product delivered late |
| $\theta_r$ | Retailer unit fresh produce spoilage rate | $\pi_2$ | Marginal profit per unit of goods |

### 2.2 *Model assumptions*

To ensure the applicability of the study, the following assumptions are presented in this paper.

1) The goods of each retail supermarket chain are supplied directly from the distribution center.
2) The reorder point r is the demand within the lead time plus the safety stock quantity.
3) The system uses a continuous inventory method, and whenever its inventory level drops to reorder point $r$, it issues an order with order quantity $Q$ to the distribution center. When the distribution center receives the retailer's demand, it immediately processes the order with operational efficiency $p$ and delivers it to the retailer immediately after completion.
4) The service level is denoted as $1 - \alpha$, which means that the demand is met, and $\alpha$ is denoted as the ratio of the number of out-of-stocks to the quantity ordered in a cycle.

### 2.3 *Model construction*

First, the expected total cost per unit of time is analyzed from the perspective of the distribution center, which mainly includes commodity preparation cost, inventory holding cost, and spoilage cost Thus, the total expected cost function per unit time of the distribution center can be obtained as follows.

$$TC_d = C_d\frac{D}{nQ} + \frac{h_dQ}{2}\left[n\left(1-\frac{D}{p}\right) - 1 + \frac{2D}{p}\right] + R_d \cdot \theta_d \cdot \frac{Q}{2}\left[n\left(1-\frac{D}{p}\right) - 1 + \frac{2D}{p}\right] \quad (1)$$

The second is the retailer's total expected cost per unit of time, which mainly includes order costs, inventory holding costs, spoilage costs, out-of-stock costs, and the costs associated with shorter lead times.

(1) Retailers' expected ordering cost per unit of time.

$$C_1 = C_r \frac{D}{Q} \tag{2}$$

(2) Retailers' expected inventory holding costs per unit of time.

Let the demand for each retailer's goods in lead time $L$ be $x$. When the demand is $x \leq r$, no stockout will occur in this ordering cycle, and vice versa, a stockout of $x - r$ units of goods will occur. Denoting the retailer's expected out-of-stock quantity in an ordering cycle by $G(r)$, we have $G(r) = E(x - r)^+$. The retailer's expected out-of-stock quantity in an ordering cycle can be obtained.

$$\begin{aligned} G(r) &= \int_r^\infty (x - r) f(x)\, dx \\ &= \int_{\mu L + k\sigma\sqrt{L}}^\infty \left(x - \mu L - k\sigma\sqrt{L}\right) \cdot \frac{1}{\sqrt{2\pi} \cdot \sigma\sqrt{L}} \exp\left[-\frac{1}{2}\left(\frac{x - \mu L}{\sigma\sqrt{L}}\right)^2\right] dx \end{aligned} \tag{3}$$

Let $k = x - \mu L / \sigma\sqrt{L}$, then $G(r) = \sigma\sqrt{L} f(k) - k\sigma\sqrt{L}[1 - F(k)]$. Let $M(k) = f(k) - k[1 - F(k)]$, the expected net inventory at the end of each cycle can be obtained as $k\sigma\sqrt{L} + (1 - \beta)G(r)$. Then the retailer's expected inventory holding cost per unit time can be obtained as:

$$C_2 = h_r \left(\frac{Q}{2} + k\sigma\sqrt{L} + (1 - \beta)\, G(r)\right) \tag{4}$$

(3) Retailers' spoilage costs per unit time of fresh goods.

$$C_3 = R_r \cdot \theta_r \left[\frac{Q}{2} + k\sigma\sqrt{L} + (1 - \beta)\, G(r)\right] \tag{5}$$

(4) Retailers' expected out-of-stock costs per unit of time.

The number of backorders allowed in each out-of-stock period is $\beta G(r)$. When retailers experience backorders, lost profits and lost sales occur.

Therefore, the retailer's expected out-of-stock cost per unit time can be obtained as

$$C_4 = \frac{D}{Q}[\pi_1 \beta + \pi_2 (1 - \beta)] G(r) \tag{6}$$

(5) Cost per unit time of lead time reduction for retailers.

The cost of lead time reduction in each cycle is $C(L) = \alpha(L_2 - L)$, and the cost of lead time reduction per unit of time is

$$C_5 = \frac{D}{Q} a (L_2 - L) \tag{7}$$

Thus, the retailer's expected total cost per unit of time can be expressed as

$$\begin{aligned} TC_r = {} & C_r \frac{D}{Q} + h_r \left(\frac{Q}{2} + k\sigma\sqrt{L} + (1 - \beta)\, G(r)\right) + R_r \cdot \theta_r \cdot \left[\frac{Q}{2} + k\sigma\sqrt{L} + (1 - \beta)\, G(r)\right] \\ & + \frac{D}{Q}[\pi_1 \beta + \pi_2 (1 - \beta)] G(r) + \frac{D}{Q} a (L_2 - L) \end{aligned} \tag{8}$$

and satisfies the constraint $G(r)/Q \leq \alpha$.

Based on the above analysis, the total cost of the joint inventory is expected to be

$$TC = C_d\frac{D}{nQ} + \frac{h_dQ}{2}\left[n\left(1-\frac{D}{p}\right)-1+\frac{2D}{p}\right] + R_d\cdot\theta_d\cdot\frac{Q}{2}\left[n\left(1-\frac{D}{p}\right)-1+\frac{2D}{p}\right]$$
$$+C_r\frac{D}{Q} + h_r\left(\frac{Q}{2}+k\sigma\sqrt{L}+(1-\beta)G(r)\right) + R_r\cdot\theta_r\cdot\left[\frac{Q}{2}+k\sigma\sqrt{L}+(1-\beta)G(r)\right]$$
$$+\frac{D}{Q}[\pi_1\beta+\pi_2(1-\beta)]G(r)+\frac{D}{Q}a(L_2-L) \tag{9}$$

### 2.4 *Model solving*

The optimal decision variables are solved for the established objective function.

(1) First, making first-order partial derivatives and second-order partial derivatives of the joint total cost function with respect to L yields.

$$\frac{\partial^2 TC}{\partial L^2} = -L^{-\frac{3}{2}}\cdot\sigma\cdot M(k)\cdot\frac{D}{4Q}[\pi_1\beta+\pi_2(1-\beta)] - L^{-\frac{3}{2}}\cdot\frac{\sigma(h_r+R_r\cdot\theta_r)}{4}[k+(1-\beta)M(k)] \tag{10}$$

In $\partial^2 TC/\partial^2 L < 0$, $TC$ is a concave function of $L$. Therefore, for any combination $(n, Q)$, the expected total cost minimization point must lie at the endpoint of the interval $(L_1, L_2)$.

(2) Then, calculate the first-order partial derivatives and second-order partial derivatives of the joint total cost function with respect to Q yields.

$$\frac{\partial^2 TC}{\partial Q^2} = \frac{2D}{Q^3}\left\{\frac{C_r}{n}+C_r+[\pi_1\beta+\pi_2(1-\beta)]G(r)+a(L_2-L)\right\} \tag{11}$$

From this, the retailer's optimal order quantity for a given $n$ and $L$ can be derived from the first-order condition, i.e., from $\partial TC/\partial Q = 0$.

$$Q_i = \sqrt{\frac{\left(2D+\left\{\frac{C_d}{n}+C_r+\left[\pi_1\beta+\pi_2(1-\beta)\sigma\sqrt{L_i}M(k)+a(L_2-L_i)\right]\right\}\right)}{(h_d+R_d\cdot\theta_d)\left[n\left(1-\frac{D}{p}\right)-1+\frac{2D}{p}\right]+(h_r+R_r\cdot\theta_r)}} \tag{12}$$

The above is the minimum point of the total cost function under the condition of no service level limitation if $\sigma\sqrt{L_i}M(k)/\alpha \le Q_i, Q_i$, is the minimum point of the total cost function; when $\sigma\sqrt{L_i}M(k)/\alpha > Q_i, \sigma\sqrt{L_i}M(k)$, is the minimum point of the total cost function. Therefore, the optimal joint order lot is $Q = max\{Q_i, \sigma\sqrt{L_i}M(k)/\alpha\}$.

## 3 EXAMPLE ANALYSIS

This paper takes the Y supermarket chain distribution center as the research background. The demand of each supermarket retailer obeys normal distribution. The operational efficiency of the distribution center $p$= 2000kg/day, the preparation cost per supplied batch of product $C_d$ = \$550/lot, the inventory holding cost per unit time unit of fresh produce $h_d$ = \$2.5/kg, the average demand of retailers $D$= 250kg/day, the order cost per order of retailers $C_r$ = \$200, the inventory holding cost per unit time unit of retailers $h_r$ = \$3/kg, and the lead time $L_1$ = 2 days and $L_2$ = 7 days.

According to the above solution steps, the optimal order quantity, optimal supply lot size, and optimal lead time of the model are discussed when the backorder ratio is 0.8, and the service level is 0.005, 0.01, 0.05, and 0.1, respectively. The specific solution procedure is shown in Table 2.

From the data in the table, it can be seen that for a given backorder ratio, for a given production lot ($n$) and lead time ($L$), the corresponding service level achieves the minimum value of joint

Table 2. Optimal solution solving process under different service level constraints.

| | n | Li | Qi | $\sigma\sqrt{L_i}M(k)\alpha$ | Q | r | TC |
|---|---|---|---|---|---|---|---|
| 0.005 | 1 | 2 | 141 | 236 | 236 | 54 | 1314 |
| | 1 | 7 | 137 | 441 | 441 | 166 | 1296 |
| | 2 | 2 | 325 | 326 | 325 | 54 | 1363 |
| | 2 | 7 | 309 | 441 | 441 | 166 | 1622 |
| | 3 | 2 | 404 | 236 | 236 | 54 | 1883 |
| | 3 | 7 | 381 | 441 | 441 | 166 | 2052 |
| 0.01 | 1 | 2 | 142 | 118 | 142 | 54 | 1743 |
| | 1 | 7 | 137 | 220 | 220 | 166 | 1415 |
| | 2 | 2 | 325 | 118 | 325 | 54 | 1363 |
| | 2 | 7 | 310 | 220 | 310 | 166 | 1400 |
| | 3 | 2 | 404 | 117 | 404 | 54 | 1866 |
| | 3 | 7 | 381 | 220 | 381 | 166 | 1866 |
| 0.05 | 1 | 2 | 140 | 24 | 140 | 54 | 1761 |
| | 1 | 7 | 135 | 44 | 135 | 166 | 1885 |
| | 2 | 2 | 323 | 24 | 323 | 54 | 1361 |
| | 2 | 7 | 308 | 44 | 308 | 166 | 1398 |
| | 3 | 2 | 402 | 24 | 402 | 54 | 1877 |
| | 3 | 7 | 379 | 44 | 379 | 166 | 1860 |
| 0.1 | 1 | 2 | 139 | 12 | 139 | 54 | 1770 |
| | 1 | 7 | 134 | 22 | 134 | 166 | 1895 |
| | 2 | 2 | 322 | 12 | 322 | 54 | 1359 |
| | 2 | 7 | 307 | 22 | 307 | 166 | 1397 |
| | 3 | 2 | 401 | 11 | 401 | 54 | 1874 |
| | 3 | 7 | 378 | 22 | 378 | 166 | 1857 |

inventory cost at $n$=2 and a shorter lead time ($L$=2). Since the lead time takes a shorter value, the corresponding reorder point is also smaller, while the joint inventory cost decreases as the service level decreases.

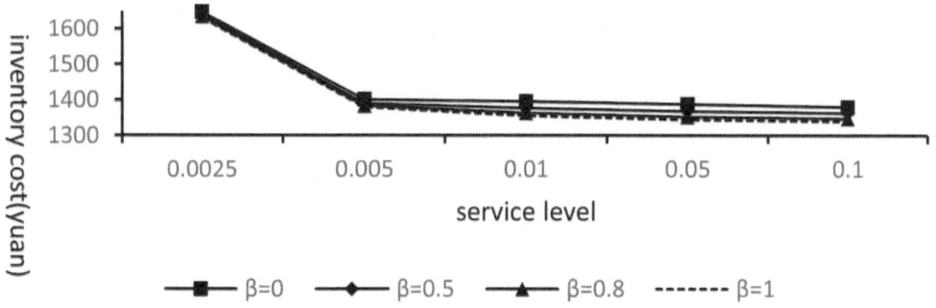


Figure 1. Impact of late delivery rate and service level on inventory optimal cost.

As can be seen from the figure, when the service level is higher, the cost is higher, especially when the requested service level is increased to a certain level, and the joint inventory cost rises steeply. This indicates that when the service level reaches a certain level, the benefits of continuing to pursue higher service levels are much smaller than the costs required. Also, it can be seen that the joint inventory cost decreases less with increasing backorder rate at the same service level, i.e., $Q*$ and $n*$ are not as sensitive to parameter $\beta$. But regardless of the value, the optimal inventory cost is obtained at a shorter lead time.

## 4 CONCLUSION

In the joint inventory decision problem, service level and lead time are two factors that affect and constrain each other. Shortening the lead time reduces the out-of-stock level to a certain extent, and the reduction of the out-of-stock level means an increase in customer satisfaction. In this paper, we consider two parts under different service level constraints and controllable lead time, discuss the optimal order quantity, lead time, reorder point, and supply lot size for distribution centers and retailers under random demand, and establish a two-level inventory model for a single distribution center and retailer. Our research can provide a reference for joint inventory decisions of enterprises in practical applications.


## ACKNOWLEDGEMENTS

This work was supported by the National Natural Science Foundation of China (72101082). National Natural Science Foundation of Hebei Province (F2021208011), the Youth Top Talent Project of Research Project of Humanities and Social Sciences in Colleges and Universities of Hebei Province (BJ2021088), the Research Project on the Development of Social Sciences in Hebei Province (20210201325), and the 2022 Hebei Province Human Resources and Society Guarantee Scientific Research Cooperation Project Record (JRSHZ-2022-01079).


## REFERENCES


Bi WJ, Zhou YB. (2022). Research on joint inventory control and dynamic pricing of fresh products based on deep reinforcement learning [J/OL]. *Computer Application Research*:1–6.

Deng Hui. (2016). Research on joint inventory management in supply chain environment[J]. *Modern Business Industry*, 37(31):44–47.

Li Qunxia. (2019). Supply chain synchronized inventory model optimization for supply and demand under joint decision making[J]. *Logistics Technology*, 38(06):109–115.

Sunil Tiwari, Shib Sankar Sana, Sumon Sarkar. (2018). Joint economic lot sizing model with stochastic demand and controllable lead-time by reducing ordering cost and setup cost[J]. *Revista de la Real Academia de Ciencias Exactas, Físicasy Naturales. Serie A. Matemáticas*, 112 (4).

Zhang Q, Zhang ZC, Zhao WF, Liu Z. (2016). A joint inventory control strategy for dual channels under demand perturbation[J]. *Statistics and Decision Making*, (15):44–48.

*Advances in Urban Construction and Management Engineering – Cha (Ed)*


# A tentative investigation of complex project management based on the system thinking

Zhenquan Zhou & Deprizon Syamsunur*
*Faculty of Engineering, Technology, and Built Environment, UCSI University, Kuala Lumpur, Malaysia*

ABSTRACT: During the execution of the vast majority of conventional projects, several complicated obstacles may be overcome using linear approaches and cutting-edge technologies. In the last decades, however, many professional managers have acknowledged the complexity of systems. Most projects across multiple industries failed because system complexity was ignored. It is difficult to predict the development of a complex system, so the project's urgency may increase the likelihood of failure. This paper aims to explain the characteristics of a complicated project and the associated system thinking by doing a literature review and analyzing an assumed case. The study process indicates that both hard system methodology and soft system methodology are essential for addressing linear and complex projects. The relevant research findings can help the new entrants to get in touch with constructions and complex project management disciplines to be familiar with management science and system thinking. Further research focusing on first-hand data collection and analysis is needed to validate current qualitative research findings and make the current conclusions accurate and up-to-date.

## 1 INTRODUCTION

According to the Project Management Institute (PMI 2017), a project is an initial endeavor to produce unique items. A capable project manager can organize a project efficiently and successfully, accomplishing tasks within budget, schedule, and quality constraints. However, many complex projects have failed over the last five years due to the inability of project teams to manage the complexity. Consequently, many management experts recognize the need to manage complex projects efficiently. In this paper, the second section explains what makes a project complex, and the third part discusses the significant distinctions between complex and complicated projects. After that, the fourth section considers traditional management methods and flexible governance approaches. Furthermore, the last main section demonstrates how to design a complicated but supposed project based on the appropriate methodology selections.

The research team has conducted online questionnaire surveys regarding the comparison between complicated and complex projects. The interviewees are classified into two groups: the university students studying engineering management degrees in their third year and the construction professionals who have worked for around three years. Besides, most of these interviewees anticipate or have been construction management professionals. Thus, the investigation results are professional and believable to a large extent. The investigation releases that the student and worker groups still need to enhance their understanding of dealing with complex projects. Based on these two surveys, there is a total of 35 effective responses. The analytical results are illustrated that less than 15% of interviewees understand the importance of HSM while dealing with a complex project. Thus, this article can help readers with a similar education background, bettering their understanding

*Corresponding Author: Deprizon@ucsiuniversity.edu.my

DOI 10.1201/9781003348023-54

of distinguishing between a complicated project and a complex project. Further research should refine and involve more experienced engineering management professionals to update and validate the current research direction. Moreover, a face-to-face interview could collect reputable and professional viewpoints from experts.

## 2 POTENTIAL ELEMENTS MAKE A PROJECT COMPLEX

Complex projects are more likely to go over budget, fall behind time, and provide substandard products after the project, consequently failing to fulfill the promised benefits to stakeholder groups (Locatelli et al. 2014). A panel of management experts assessed the most significant factors contributing to project complexity. Numerous stakeholder groups, unpredictability in project features, and interference from the external environment are typical contributors to project complexity. Human behavior, project uncertainty, and dynamic interactions among projects and subsystem components may significantly enhance the project complexity (Amaral & Ottino 2004; Locatelli et al. 2014). In other words, a complex project could contain multiple and interdependent systems, an unknown project environment, and a variety of human behaviors. Its increasing complexity will likely influence a project's predicable performance, leading to potential failures.

### 2.1 *Systems behaviors*

A complex project involves the random interaction of various system components with other internal system components and the external environment. These subsystems are interconnected with other system components, forming unique and unanticipated system behaviors. In other words, no established mechanism for imitating the subsequent system behaviors may significantly complicate the project management process. Component A, for instance, may interact with component B to produce Element C. Notwithstanding, A may interact with D to produce a new state of D or E. In the context of a simple and linear project, the management patterns generated by the reductionism method may successfully aid managers in forecasting system behaviors and managing project performance. Due to the inability to create an appropriate model for controlling disordered system behaviors, it is difficult to predict the future performance of a complex project precisely and comprehensively manage it (Amaral & Ottino 2004). As an alternative to a traditional top-down management style, managers must, for instance, continuously adapt their management techniques to the current project environment. Due to the various systems and their unpredictable behaviors, a project will likely be challenging under its complex situations.

### 2.2 *Humanistic concerns*

Organizing and synthesizing diverse human perspectives may increase the complexity of a project. This situation is also why the PMBOK adds the section on stakeholder management, containing the contemporary management discipline. The number of multicultural projects incorporating many stakeholder groups has increased recently. Diverse stakeholder groups could have divergent opinions and responses to the same endeavor and may act more hastily and randomly. It is unlikely that the regulating mechanism would absorb their thoughts and manage human behavior, increasing project governance's complexity (Gorod et al. 2008). For example, the stock market, a complex ecology system, is driven not just by the economic environment but also by dynamic investor confidence. In other words, measuring the degree of confidence of individual investors with diverse educational backgrounds and life experiences is relatively challenging owing to the diversity of human cognition. In addition, the independence of market participants will probably impact stock performance, putting investors and businesses in a bind. Consequently, human-related concerns may severely hamper the practical execution of a project.

## 2.3 *Ambiguity and unexpected events*

The ambiguity of a project scenario might result in several risks and uncertainties, raising the corresponding complexity. Most of the uncertainty comes from the dynamics and interactions of project components or internal systems, which may expose the project to "unknown unknowns" such as potential crises. Managing a complex issue, for instance, prevents managers from predicting the type of project risks and the frequency of crises. As seen in Figure 1, this circumstance will likely increase the risk level and may trigger the cascading risks associated with businesses worldwide (Helbing 2013). Similarly, the ambiguous state of the project will possibly make it more complex to manage project risks and crises, resulting in a more complicated and riskier project environment. Consequently, project ambiguity, uncertainty, and the ensuing crises may have detrimental effects on management performance, thus decreasing the chance of project success.

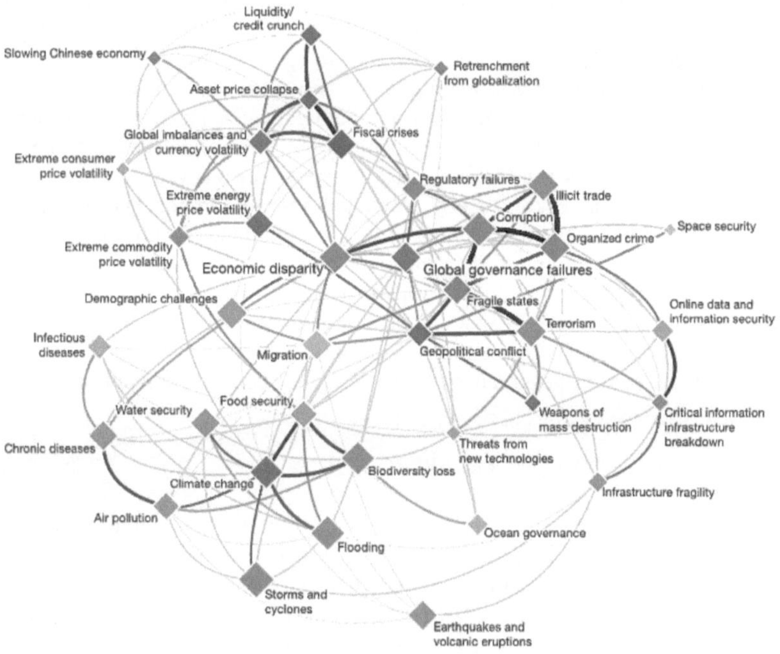


Figure 1. Systemic hazards with interconnected relationships (Helbing 2013, p. 52).

# 3 COMPARISONS BETWEEN TWO PROJECT TYPES

It is widely accepted that complicated endeavors are more likely to fail than straightforward and linear ones. In most situations, complicated projects may be categorized as "known unknowns," whereas complex projects are categorized as "unknown unknowns." It implies that people can recognize complicated endeavors but are unclear about the specific outcomes of such activities. However, a complex problem with unknown unknowns is a riskier project circumstance than a complicated environment because most people cannot predict what will occur and respond effectively to its potential disasters (Checkland & Poulter 2006). Therefore, complicated projects are amenable to foresee, but complex projects are unanticipated to a large extent.

Even though complicated tasks are often realizable, they are difficult to forecast precisely. Most of the time, managers have blueprints for handling complicated projects, but in the face of complex situations, they can only depend on timely input to adjust project performance. Many complicated projects are linear and inflexible and generally have evident cause-and-effect relationships between project components and throughout their lifecycle. Since project managers can simulate project performance and optimize management tactics, even relatively complicated and large-scale projects can also be handled effectively (Gorod Sauser & Boardman 2008). Based on these situations, engineering technologies are often associated with the most complicated challenges under challenging endeavors, so employing professionals and specialists helps mitigate this. In contrast, complex projects are often nonlinear and dynamic, making it challenging to set up a mathematical model for managing complexity. Due to a lack of applicable models for predicting project behavior, experts cannot successfully simulate and anticipate complex project outcomes (Keating & Katina 2019). Moreover, it is much more difficult to react effectively to project crises, lowering the likelihood of project success.

On the other hand, these two project types are conducted and addressed differently. Complex projects have more requirements and variation than complicated projects. According to Checkland, hard system methodology (HSM) is an appropriate technique for managing complicated linear situations, such as conventional engineering and technology-related projects (Checkland 1981). However, it has been shown that HSM is insufficient for managing large and complex projects since it cannot address problems resulting from system autonomy, adaptive behaviors, and interdependent component interactions. Thus, soft system methodology (SSM) is often used in complex projects to address several uncertainties and unanticipated disasters. For analyzing massive projects, the reductionism method, which seeks to study individual system components by simplifying the structure, is ineffective (Checkland & Poulter 2006). In addition, complicated projects usually utilize a top-down approach to controlling and regulating project development, while complex projects adopt bottom-up governance techniques. Most project management solutions are effective for complicated projects but ineffective for complex ones as system behaviors are adaptable, ever-changing, and spontaneous in a complex project (Locatelli et al. 2014). Under this potential issue, adaptive leadership and flexible governance could better guide complex projects on the proper path.

## 4 TRADITIONAL AND SYSTEMATIC APPROACHES

Conventional project management comprises five significant actions: initiation, planning, execution, monitoring and control, and closure. In many cases, standard project management uses a sequential linear approach. Conventional project management is based on lessons from comparable projects and recurrent patterns, such as setting up mathematical models for predicting project behaviors. Given that their project lifecycles are comparable, it may serve as a template for many other similar projects. Nevertheless, for complex projects, standardized project management processes are ineffective. Frequently, complex projects and systems lack a linear management paradigm. Emergencies arise in significant projects, and conventional managerial practices may fail if the linear approach is not followed (Amaral & Ottino 2004). Traditional project managers may use the notion to organize construction or industrial projects, but complex projects include several uncertainties and potential crises. Thus, there are no suitable and specific models for managing complex projects in most situations.

On the other side, traditional management often adopts a paradigm in which every project task should adhere to the needs and instructions of supervisors. This pattern of "receive and perform" is heavily regulated. On the other hand, adaptive techniques are more suitable for dealing with complex projects with possible contingencies. To effectively governance complex projects, it is crucial to modify the strategies depending on the up-to-date project's performance feedback. In addition, a suitable leadership style is required to lead a complex project effectively. It illustrates that appropriate governance is favored over management for complex projects. Leadership characteristics, as opposed to technical management strategies, are essential for completing a complex project (Gorod et al. 2008). Figure 2 describes agile project management for handling large projects. Agile project management is an iterative method for modifying the performance of a project throughout its lifecycle (Kissflow 2021). Besides, PDCA is also an iterative method for improving the precision of project development. Due to its agility, agile project management is more equipped to handle complexity in a complex project than traditional management systems.

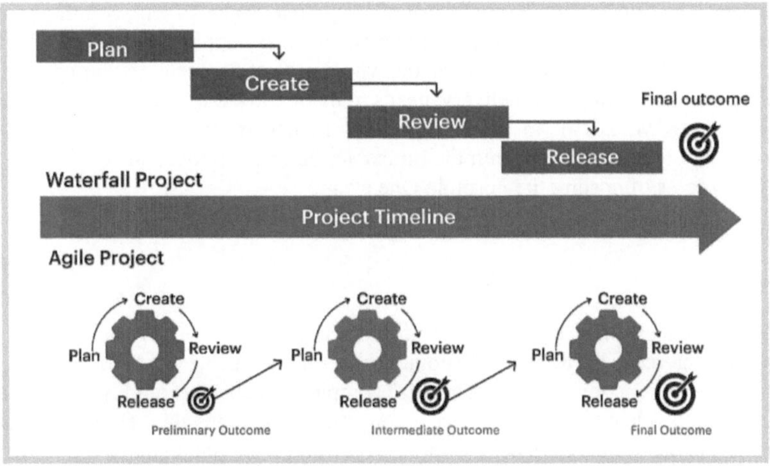


Figure 2. Conventional management method versus agile project governance (Kissflow 2021).

## 5 SUPPOSED CASE ANALYSIS

Hosting the Formula One World Championship should be regarded as a complex system since its primary components are interdependent and interact with their surroundings. Complex systems are often composed of interconnected and interacting components that may spontaneously develop subsystems. Due to system components' irregular growth, it is challenging to construct a suitable model for forecasting and simulating system behavior. Therefore, Formula One World Championship preparations need to involve a variety of approaches, resources, and strategies.

HSM was first developed for the engineering duties connected with this challenging project. HSM is extensively relevant to linear project execution, such as infrastructure, construction, and communications engineering projects. This technique is advantageous for tackling engineering-related concerns in this complex event. For example, someplace will host Formula One for the first time, yet there is no certified and suitable racing track for Formula One-related athletic activities. Consequently, the traditional competition land needs to be built before official acceptance (Bakhshi et al. 2016). The HSM will be tasked with ensuring the practical organization of this infrastructure project so that it is completed on time and under the requisite quality standards.

SSM typically consists of seven stages for systematically discovering, altering, and addressing complex people-related problems. Most people-related problems might be handled by SSM, which can identify, analyze, and recreate the system network of intricate issues more thoroughly. SSM is also essential, which tries to discover unconstrained concerns, explore operational feasibility, and effectively alter and handle the problem. This world-class event incorporates many stakeholder groups, such as sponsors, participants, the competition organizers, the broadcast party, spectaculars, security systems, medical sectors, and service companies. HSM can address the nonlinear challenges of engaging with the stakeholders, such as the varied and often competing interests of the several stakeholder groups involved in this event (Checkland & Poulter 2006).

Using HSM and SSM, relevant issues may be detected more rapidly, and both complicated and complex problem scenarios can be modified and enhanced. Moreover, given that a project is often characterized as a temporary endeavor to attain specific goals, hosting the Formula One event needs to be considered a sophisticated and substantial project. Many specific subprojects related to finances, security, tourism, energy supply, social services, and the health department should be handled and completed successfully during the event. Because efficient project management is essential to the success of a project, the PMBOK is highly suggested for the organization of this sports event. The categories are scope, cost, quality, time, stakeholders, communication, risk, procurement, human resource, and integrated management. By applying the PMBOK-illustrated project management knowledge and abilities, the whole project lifecycle, including project initiation, planning, execution, and final delivery, needs to be organized and managed more accurately and systematically (PMI 2017). In other words, this management knowledge system can help the organizers complete the event within the timetable, budget, safety, quality, and other criteria, enhancing the success of hosting the Formula One event.

## 6 CONCLUSIONS

Complex project management has a greater failure rate than complex projects in that complex systems are characterized by nonlinear, spontaneous, interconnected, adaptive, emergent, and stochastic qualities. Reductionism cannot analyze complex systems thoroughly as its system components are interdependent with other internal constituents. Besides this, effective strategies are required to analyze the project complexity, address relevant issues and manage the whole complex system. Complex obstacles will likely generate systematic issues and cascading dangers, worsening project cost, productivity, security, and other intangible aspects. The typical project management paradigm is unsuitable for managing complex projects because it cannot effectively handle the complexity, such as contingencies, uncertainties, and spontaneous behaviors that occur from complex projects. Based on the characteristics of complex projects, agile project management is one of the reasonable approaches because its flexible governing methods may more efficiently complete a complex project successfully. By discussing a supposed case, the appropriate adoption of HSM, SSM, and other practical tools play a significant role in hosting a potential worldwide sports event. More accurate quantitative investigations that the interviewees should be more experienced management professionals plan to be conducted in further research, which can present more up-to-date data and findings convincingly.

## REFERENCES


A guide to the project management body of knowledge (PMBOK guide). 2017, Sixth edition., Project Management Institute, Newtown Square, Pennsylvania.

Amaral, L & Ottino, J 2004, 'Complex networks,' *The European Physical Journal B*, vol. 38, no. 2, pp. 147–162.

Bakhshi, J, Ireland, V & Gorod, A 2016, 'Clarifying the project complexity construct: Past, present, and future,' *International Journal of Project Management*, vol. 34, no. 7, pp. 1199–1213.

Checkland, P & Poulter, J 2006, *Learning for action: a short definitive account of soft systems methodology and its use for practitioner, teachers, and students*, John Wiley & Sons, Chichester, West Sussex.

Checkland, P 1981, *Systems Thinking*, Systems Practice, Wiley, Chichester.

Dirk Helbing 2013, 'Globally networked risks and how to respond,' *Nature*, vol. 497, no. 7447, pp. 51–9.

Gorod, A, Sauser, B & Boardman, J 2008, 'System-of-Systems Engineering Management: A Review of Modern History and a Path Forward,' *IEEE Systems Journal*, vol. 2, no. 4, pp. 484–499.

Keating, CB & Katina, PF 2019, 'Complex system governance: Concept, utility, and challenges,' *Systems Research and Behavioral Science*, vol. 36, no. 5, pp. 687–705.

Kissflow 2021, *Traditional vs. Agile Project Management – Which One Fits You?* Kissflow, viewed May 10, 2022, <https://kissflow.com/project/agile/traditional-vs-agile-project-management/>.

Locatelli, G, Mancini, M & Romano, E 2014, 'Systems Engineering to improve the governance in complex project environments,' *International Journal of Project Management*, vol. 32, no. 8, pp. 1395–1410.

*Advances in Urban Construction and Management Engineering – Cha (Ed)*


# Government's behavioral decision in the prevention of urban geological disasters

Xianying Huang* & Chunfang Wang*
*Chinese Academy of Natural Resources Economics, Beijing, China*

ABSTRACT: In the urban construction that characterizes the progress of civilization, human beings can seize natural resources and avoid natural disasters with their powerful strength. Both natural forces and artificial activities can cause secondary disasters, including geological disasters such as collapses, landslides, debris flows, land subsidence, ground settlement, and ground fissures. In recent years, geological disasters have occurred frequently in China, and extreme climate change has continuously increased the threat to people's safety. To safeguard public interests, government departments have invested a lot of money and personnel to reduce the losses of geological disasters. The prospect theory proposed by Daniel Kahneman and Amos Tversky provides theoretical guidance for government departments to make decisions on the prevention of urban geological disasters. In different technical stages of urban geological disaster prevention, government departments' decision-making satisfaction is different. The geological disaster investigation can greatly reduce the disaster risk, and the government's decision-making shows a deterministic effect, and the results are very satisfactory. The construction of the monitoring and early warning system can smoothly promote urban infrastructure projects, and government departments will choose to avoid risks. Still, when it conflicts with urban development planning, government departments will choose risk bias, which is manifested as a reflection effect because the decision-making has changed significantly due to different risks. Prevention and control projects and popular science education apply to individual disaster sites and groups, and government behavior decisions will be determined by reference points, showing an isolation effect.

## 1 INTRODUCTION

In the process of interaction and evolution between humans and nature, the relationship is constantly changing. Scientific inventions and technological progress have greatly enhanced the ability of human beings to transform nature. Environmental absolutists believe that human beings and all other creatures are the products of the environment, and human activities are strictly restricted by the geographical environment. Environmental probabilists believe that humans have freedom of choice for the natural environment. Ecologicalists believe that humans seek to live in harmony with nature and even nurture the environment. Humans can show great strength in the face of nature but cannot change their fragile biological nature. Gradually, human beings began to learn to prevent and control the occurrence of natural disasters in an organized manner and no longer regard nature as the object of conquest and enslavement. As a public administration, the government is responsible for maintaining public safety and protecting people's property from damage. Governments need to act in predictable natural disasters (Du 2017).

*Corresponding Authors: xyhuang@canre.org.cn and cfwang@canre.org.cn

DOI 10.1201/9781003348023-55

There are various natural disasters, such as volcanic eruptions, earthquakes, floods, forest fires, and ocean waves. Humans cannot resist primitive natural forces, but the government is responsible for providing emergency rescue after disasters (Zhang & Okada 2006). Some secondary disasters, such as landslides, mudslides caused by heavy rain, and earthquakes, are sudden and destructive, but government departments can organize preventive work with the support of modern information technology to minimize losses. Some natural disasters are caused by human activities, such as land subsidence or ground fissures caused by overexploitation of groundwater. This type of disaster seriously threatens the safety of urban underground pipe networks, rail transit, and ground buildings. Human beings should respect the lithosphere and biosphere to avoid the collapse of the economic system and physical destruction (Capra & Spretnak 1988). When formulating urban development plans, government departments should consider promoting the coordinated development of human beings and the environment, coordinating the relationship between human beings and land, and protecting people from the threat of natural disasters. Human-land relationship research also needs theoretical support from psychology and behavioral sciences (WANG, Miao). Preventing and controlling disasters is the least costly method.

In the past research, the subject of human and geological disasters mainly focused on the overall relationship between human and natural disasters, emergency rescue after natural disasters, post-disaster engineering management of geological disasters, etc., and the research focus was biased towards engineering technology. Based on the "prospect theory," this paper innovatively explains the psychological expectations behind the government's decision-making and provides guidance for further geological disaster prevention work.

## 2 MAIN TYPES, CHARACTERISTICS, AND IMPACTS OF URBAN GEOLOGICAL DISASTERS

Just as the contradictory relationship between human beings and nature in the development process gives rise to ecology (Zhang 1993), geological disasters are caused by mistakes in the choice of human-land relationships. People cannot prevent geological disasters caused by internal dynamic geological processes such as earthquakes and volcanoes and mainly take measures to avoid them. Regarding the geological disasters induced by external dynamic geological action and human activities, such as collapse, landslides, debris flows, ground settlement, ground fissures, land subsidence, etc., people's prediction level is reliable (Sun 1991). In preventing urban geological disasters, the behavioral decision-making of government departments focuses on the latter type.

### 2.1 *Collapse*

Surface-exposed rocks experience partial loss of internal mineral components due to weathering, leading to loosening of the internal structure. Mountains with higher slopes are prone to the collapse of rock fragments under the induction of rainfall or earthquakes. The collapse will cause great harm to pedestrians on the road under the mountain, and it is easy to cause death.

### 2.2 *Landslide*

Landslide is the displacement of rock mass on the slope along the weak zone between rock layers under the action of gravity. Heavy rainfall and artificial changes in the overall structure of the mountain are the main causes of landslides, accounting for 90% and 9%, respectively (Li et al. 2004), this type of landslides are mainly small landslides. Earthquakes are the main factor that induces giant or large landslides. The early signs of landslides are not obvious, but in the stage of accelerated deformation, the speed changes prominently, increasing by 50 times within 24 hours (Bai et al. 2019), and the displacement of the landslide will continue to increase until it stops due to insufficient power.

## 2.3 *Debris flow*

Debris flow is the downward flow of a large amount of rubble, sediment, and other loose materials entrained by rainwater under the action of gravity. Long-term high temperatures and drought can easily lead to loose rocks and poor shear resistance, and debris flows are prone to occur in the event of sudden rainstorms. Unlike landslides, which are more inclined to move the entire mountain mass in which potential energy is converted into kinetic energy, debris flows are closer to the fluid state, and the movement encountered is less resistant and more harmful. For example, in 2010, a large debris flow in Zhouqu, Gansu, razed the impacted area to the ground, destroyed a large number of residential houses, and killed and disappeared more than 1,800 people (Xie 2010).

## 2.4 *Ground settlement, ground fissures, and land subsidence*

Land subsidence mainly occurs in karst areas and underground goaves, but the common inducing factor of urban land subsidence is the loss of underground supports caused by leakage of pipeline water, the degree or risk of collapse is related to the flow of seepage water, erosion Time, underground space structure and material composition, etc. Land subsidence often leads to urban safety accidents, which have the characteristics of concealment and suddenness. Ground fissures are mainly distributed along the fault zone and are associated with ground settlement. Over-exploitation of groundwater will cause ground fissures and ground settlement, and a large increase in ground loads will increase the degree of ground settlement. The occurrence scale of ground fissures in large and medium-sized cities is larger than that in surrounding towns (Qiao et al. 2020), which has a greater impact on the layout of buildings on the ground. The deformation of subgrades and bridges caused by ground settlement threatens the operation safety of high-speed railways (Liu et al. 2016).

Table 1. Annual change of the number of common geological disasters in China (unit: times).

| Index | Total | Landslide | Collapse | Debris Flow | Land Subsidence | Others |
|---|---|---|---|---|---|---|
| 2020 | 7840 | 4810 | 1797 | 899 | 183 | 151 |
| 2019 | 6181 | 4220 | 1238 | 599 | 121 | 3 |
| 2018 | 2966 | 1631 | 858 | 339 | 122 | 16 |
| 2017 | 7521 | 5524 | 1356 | 387 | 206 | 48 |
| 2016 | 10997 | 8194 | 1905 | 652 | 225 | 21 |
| 2015 | 8355 | 5668 | 1870 | 483 | 292 | 42 |
| 2014 | 10937 | 8149 | 1860 | 554 | 307 | 67 |
| 2013 | 15374 | 9832 | 3288 | 1547 | 385 | 322 |
| 2012 | 14675 | 11112 | 2152 | 952 | 364 | 95 |
| 2011 | 15804 | 11504 | 2445 | 1356 | 386 | 113 |

Data source: National Bureau of Statistics of China.

# 3 PROSPECT THEORY

The research on behavioral decision-making under uncertainty has had a positive and far-reaching impact on financial management, economics, and law. The focal research types of decision-making under uncertainty (Zhang 2004) include actual preference decision-making based on imperfect rationality, intuitive judgment based on reference values, decision-making style based on thinking style and decision style, decision-making strategy based on satisfaction principle, etc. People can objectively grasp the basic characteristics, occurrence mechanism, and potential impact of urban geological disasters, but how to prevent geological disasters, people can subjectively choose different behavioral decision-making theories and methods. The basic goal of government behavior decision-making is to seek public satisfaction, the reference point is the actual hazard of geological disasters, and government departments ultimately decide how to act according to the probability of disaster occurrence. The prospect theory proposed by Daniel Kahneman and Amos Tversky in 1979 provides theoretical guidance for government departments in the prevention of urban geological disasters.

### 3.1 *Fundamental contents*

Based on psychological experiments, prospect theory corrects the insufficiency of the hypothesis of a "rational economic man." It is a behavioral decision theory that integrates individual psychological perception and general population behavioral decision-making. The basic content of prospect theory is that facing the same environment, decision-makers will have different behavioral responses when they are in a state of income or loss. The experimental results show that: individuals in economic behavior are more averse to the amount of loss than they are satisfied with the same amount of income. When faced with certain gains, most people choose risk aversion; when faced with certain losses, most people choose risk preference.

### 3.2 *Theoretical model*

In prospect theory, people's behavioral decisions are based on Overall Value, which is described by Value Function and Weighting Function. The value function V(x) does not reflect the value of wealth or the economy, but a change relative to a certain reference point, describing the prospect of events and the state of human satisfaction. The weight function W(p) is not the probability of the event but describes the psychological probability of the decision maker under the expectation of the event. For different foreground events x and y, the psychological probabilities are p and q, respectively, then the overall value can be expressed as

$$F(x, p; y, q) = V(x)*W(p) + V(y)*W(q)$$

Among them: $W(0) = 0, W(1) = 1, F(0) = 0$

### 3.3 *Function feature*

The reference point in the value function is derived by the decision maker based on the individual's subjective impression. The value function has the characteristics of an S shape. It is convex when it is profitable and concave when it is losing. The slope of the concave curve at the time of loss is greater than the slope of the convex curve when it is profitable (see Figure 1). The asymmetry of the value function curve shows that under the same overall value conditions, people's behavioral decisions are loss aversion; that is, the absolute value of the loss is greater than the absolute value of the gain. The weight function exhibits nonlinear characteristics. When people make actual decisions, they often overestimate small probabilities and underestimate large probabilities. When the probability is near the extreme probability values of 0 and 1, people's cognition will change significantly.

### 3.4 *Decision effect*

Due to intuition bias, people will show three typical effects when making actual decisions. One is the certainty effect; that is, in the process of risk decision-making, the deterministic outcome is overestimated, and the possible outcome is underestimated. The second is the isolation effect, that is, when analyzing various prospects, ignoring the same part of the previous prospects, when the description of the prospects changes, it will lead to changes in decision-making. The third is the reflection effect, that is, the pursuit of risk in the face of losses and the avoidance of risks in the face of gains.

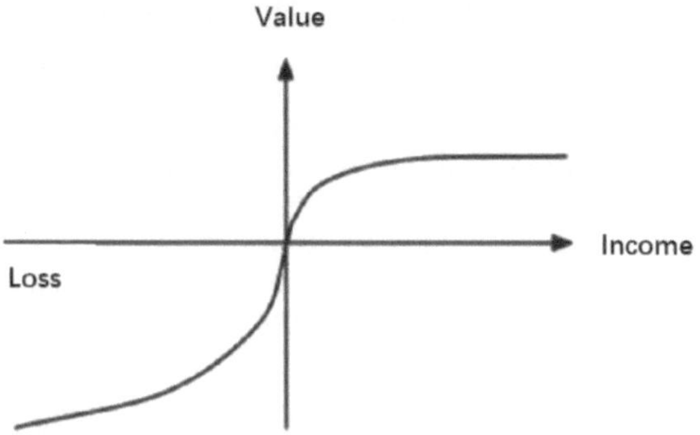


Figure 1. Value function curve in prospect theory.

## 4 GOVERNMENT'S BEHAVIORAL DECISIONS

Prospect theory holds that decisions depend on reference points. When government departments decide to prevent urban geological disasters, the reference point is the hazard of geological disasters or the safety of citizens' lives and wealth. As a part of natural disasters, urban geological disasters put people's activities under uncertainty. Preventing possible geological disasters is an important way for the government to protect people from natural disasters, and preventive decision-making has positive benefits (Yao et al. 2021). The main technical means and stages of geological disaster prevention include geological disaster investigation, construction of monitoring and early warning systems, geological disaster prevention engineering governance, and popular science education. In the process of geological disaster prevention in different technical means and stages, the behavioral decision-making of government departments will be continuously adjusted according to the changes in the overall value.

### 4.1 *Decision-making under certainty effects*

Human beings are bound to face risks when they live in nature, and absolute safety has only symbolic meaning. Theoretically, in any human activity area involving land, mountains, and bedrock, geological disasters are possible. The frequency of disasters can be reduced as long as people carry out some investigation work. China's laws related to geological disasters stipulate the responsibilities of the government. For example, the "Flood Control Law of the People's Republic of China" stipulates that the people's governments at or above the county level shall organize relevant departments to investigate the hidden dangers of landslides, collapses, and debris flows and take preventive measures. "The Law of the People's Republic of China on Earthquake Prevention and Disaster Reduction" stipulates that the people's governments at or above the county level shall incorporate earthquake prevention and disaster reduction work into their national economic and social development plans.

Based on the above considerations, the government will choose to avoid risks and actively investigate geological disasters to obtain safety benefits for the public. As far as its goals are concerned, the government must fulfill its legal responsibilities, avoid geological disasters in urban planning and construction, and investigate potential hazards in crowded areas. As far as its means are concerned, regional investigation of potential geological disasters in the city is a basic livelihood project. The government can complete the city-wide disaster investigation by setting up a geological disaster investigation project with a small number of financial expenses. Under the theoretical and legal framework, when government departments organize geological disaster investigations, the safety of life and property of all citizens and the satisfaction of the public are the reference points and goals for the government's decision-making. The purpose of decision-making is to reduce the risk of disaster to near zero; that is, the probability of no disaster is 1. The high degree of certainty of reducing the occurrence of disasters has prompted the extensive conduct of geological disaster surveys in various cities.

The truth is that government behavioral decisions significantly overestimate this deterministic outcome. Due to the wide range of investigations and weak technical means, government departments cannot control all disaster information, let alone accurately predict future emergencies. Geological disaster investigation is only a part of the prospect of the government's prevention of geological disasters, but not the whole. The results of a geological hazard investigation are not directly used to assess how much damage has been avoided. The overall value it brings is to earn the government's reputation or avoid legal sanctions for not properly performing its duties. The expected benefits of a geological hazard investigation are assessed by the relocation and resettlement costs of the threatened population.

### 4.2 *Decision-making under reflection effects*

The construction of geological disaster monitoring and early warning system is an inevitable requirement of smart city emergency management, and it has also become a basic responsibility of the government to maintain public safety. After organizing a geological disaster investigation, the government will delineate disaster spots within the city. In principle, government departments cannot plan and construct major projects at disaster sites but should deploy monitoring sites on time to prevent disasters. The construction of key municipal projects such as oil and gas pipelines, water and electricity pipelines, and transportation tracks must consider the threat of geological disasters. At some key points, the slight displacement of the ground may cause the pipeline to deform and break, resulting in oil and gas leakage, hydropower outage, and even fire and explosion, causing casualties and other accidents.

The state of infrastructure construction is an important label of a city and a major achievement of urban economic development. Government departments determine that infrastructure construction can bring huge benefits to urban development. Their behavioral decisions will choose risk avoidance, that is, to establish a geological disaster monitoring and early warning system to reduce engineering risks. In the construction of some major urban infrastructure, such as the construction of an urban, comprehensive pipe gallery, the government department will entrust survey units with professional qualifications for geological disaster prevention and control to carry out special evaluation and monitoring, which can minimize the risks caused by geological disasters and share the reputational risk borne by the government.

When geological disaster prevention conflicts with urban development planning, government departments sometimes choose to pursue risks. The perennial monitoring data in the Tongzhou area of Beijing shows that the ground settlement is obvious, and the impact of large-scale construction on land subsidence has been more prominent in recent years(Wang et al. 2018). The construction of the new city is to meet higher-level urban development needs such as industrial layout, population evacuation, and traffic congestion relief. The government's management work will be meaningless if the development needs are not met. To reduce losses, government departments will ignore the potential, slow and local risks brought about by ground settlement and tend to pursue risks and continue the construction of new towns.

The number and timing of the geological disaster monitoring and early warning system location settings depend on the urban development plan formulated by the government, the level of urban economic development, and financial status. Establishing a monitoring and early warning system is costly (Guan 2016) and requires data collection, transmission, imaging, and other equipment. The application of professional equipment has realized high-precision, real-time monitoring, and early warning, reducing the risks and deficiencies of on-site manual surveys. Still, it has also created huge financial pressures for small and medium-sized city government departments. To save funds for the development of other industries, government departments have shown a tendency to pursue risks, that is, to reduce the layout of geological disaster monitoring points.

### 4.3 *Decision-making under isolation effects*

When geological disaster prevention requires engineering governance, the government's decision-making will change due to the location of the disaster, the cause of the disaster, and the affected population. The large-scale land subsidence and ground settlement in urban areas caused by mining is a specific price in the industrial development of resource-based cities, and it is the result of government departments' behavioral decision-making ignoring preliminary investigation and evaluation, pursuing risks, and obtaining economic development benefits. However, the construction unit will be responsible for managing land subsidence and ground settlement caused by improper project construction, and the government department is only responsible for supervision. In different scenarios, government departments show different behavioral decisions. When the land transferred by the government is found to have active fault zones underground, it is not suitable to build high-rise buildings on the ground. Due to the deviation in geological safety before and after the land transfer, the government department will have to adjust the decision.

For the hidden dangers of geological disasters such as collapses, landslides, and debris flows that endanger public safety, government departments must invest funds and organize engineering governance. Disaster hidden danger points are distributed in different areas, and the affected people and buildings are also different. The comparability between prevention and control projects is poor and usually treated as special projects. The governance of each hidden danger point of disaster requires a decision. The value reference point of decision-making is the potential loss caused by the disaster, and the basis for decision-making is the evaluation results of professional technical institutions. After treatment, the hidden dangers of disasters will be eliminated. The hidden dangers that do not need to be treated may occur, and the loss caused by the decision of the risk preference is usually small.

Popular science education and training exercises in geological disaster-prone areas are low-cost and high-yield disaster prevention methods. In China, the organizational capacity of social groups is relatively weak, and government departments can mobilize all forces to popularize the basic knowledge of geological disaster prevention to the masses. "Group prevention and monitoring" is one of the main ways of preventing geological disasters in China, which helps reduce disaster risks. The masses are usually at the scene of disasters and have the opportunity to discover disasters. If the masses have weak awareness of disasters, it is easy to cause unnecessary losses. In the Touzhaigou landslide-detrital flow disaster in Shaotong City, Yunnan Province, in 1991, the local farmers had noticed a sharp increase in ground fissures the day before the incident. Still, they did not report the danger, resulting in heavy casualties and property losses (Zhong 1999). Government departments can use some actual case materials to organize popular science education, so the effect will be multiplied by half the effort.

### 4.4 *Case analysis*

At 17:00 on August 27, 2021, geological disaster investigators in Lianghe Community, Baishi Town, Zhongxian County, Chongqing City found that the nearby mountain was slightly deformed, so they immediately instructed the local government and monitoring personnel to conduct encrypted monitoring of the mountain. In the early morning of the next day, the early warning equipment issued an alarm message. After verification on the spot, the government department immediately organized personnel to evacuate 40 residents in an orderly manner. At 8:30, the mountain collapsed and slipped, with a sliding volume of 150,000 cubic meters, causing damage to 8 houses and a 200-meter road in the landslide. Thanks to the professional technical investigation, monitoring, and early warning system organized by government departments, there were no casualties in the disaster area.

On the afternoon of August 18, 2021, the geological disaster monitoring and early warning system organized by the Shenzhen municipal government showed the data on the north slope of Xiantong Sports Park, Liantang Street, Luohu District, had a sudden change, triggering a red warning of geological disasters. The local emergency department quickly evacuated dozens of tourists from the scene. Twenty minutes later, two small local collapses occurred on the north slope of the park. Facts have shown that the construction of geological disaster monitoring and early warning systems in key areas can most effectively ensure the safety of people's lives and property.

## 5 CONCLUSION AND SUGGESTION

Geological disasters will cause heavy losses to people's lives and properties. Still, ordinary people do not have the professional skills to discover the hidden dangers of geological disasters and even know little about the basic knowledge of geological disasters. Their ability to avoid risks is weak. As a public organization, the government has the power and funds to mobilize social forces to prevent the occurrence of geological disasters. In different technical stages of geological disaster prevention, the relative satisfaction degree of government departments to gains and losses varies significantly. In terms of gains, the geological disaster investigation has identified the risk points in the whole city, and the government has the highest satisfaction. The second is the geological disaster prevention project management, with which the benefits can be evaluated according to the value that threatens people's lives and properties, and the effect is obvious. Monitoring and early warning system construction is often included in urban infrastructure construction. It cannot be independently evaluated and accounted for, so the government's satisfaction is slightly lower. The lowest satisfaction is science education, which often has little effect due to limited scope of influence and lack of on-the-spot observations. In terms of losses, the construction of the monitoring and early warning system involves a wide range and costs a lot, and the government departments are the most disgusted. The second is the prevention and control of geological disasters, which is mainly to ensure the safety of specific people and assets and has little impact on the public and society. Compared with geological disaster investigations, government departments dislike popular science education the least because it can be used as a part of cultural education and will not cause losses.

Table 2. The government's relative satisfaction with different technical stages in urban geological disaster prevention.

| Different Technical Stages | Reference Point | Earnings Expectations | Loss Expectations |
|---|---|---|---|
| Geological Hazard Investigation | Life and property of urban residents | ++++ | - - |
| Construction of Monitoring and Early Warning System | Infrastructure Engineering | ++ | - - - - |
| Prevention Engineering Management | Specific Disaster Point | +++ | - - - |
| Science Education | Citizens of Disaster-Prone Areas | + | - |

Strengthening the prevention of urban geological disasters can not only avoid or reduce the loss of life and property caused by disasters, but also has great significance for the safe construction of national infrastructures such as water conservancy and hydropower projects and underground space utilization. The formulation of urban development plans and the effective development of the city's mineral resources, land resources, water resources, etc., also need to carry out geological disaster prevention work. In the decision-making process of geological disaster prevention, government departments should do the following. First, apply the investigation information of geological disasters to urban development planning to avoid the threat of geological disasters from the source. Second, conduct geological disaster investigation periodically, once every five years, and update survey data promptly. Third, a reasonable assessment of the risk level of geological disasters should not hinder "big projects" because of "small disasters." Fourth, the use of financial funds must be targeted, and disasters in key population areas must be eliminated in time. Fifth, disaster information should be comprehensively analyzed, and information from departments such as meteorology, water conservancy, and natural resources should be interconnected.

## AUTHORS

About the first author: Huang Xianying (1986-), male, Han nationality, from Juancheng, Shandong, associate researcher, majors in resource management and natural disaster research, E-mail: xyhuang@canre.org.cn.

About the corresponding author: Wang Chunfang (1988-), female, Han nationality, from Pingding, Shanxi, assistant researcher, mainly engaged in resource economy and policy research, E-mail: cfwang@canre.org.cn

## REFERENCES

Bai Jie, Ju Nengpan, Zhang Chengqiang, He Chaoyang, Liu Xiuwei. Study on the Characteristics and Successful Early Warning Of XingYi Landslide In Guizhou Province [J]. *Journal of Engineering Geology*, 2020, 28(06): 1246–1258. DOI:10.13544/j.cnki.jeg.2019-360.

Capra F, Spretnak C. *Green Politics: Global Hope*. Beijing: Dongfang Press. 1988 (57).

Du Yifang. The foreseeable possibility of the government in responding to natural disasters: the perspective of Japan's national responsibility [J]. *Global Law Review*, 2017, 39(01):96–109.

Guan Fengjun. Prevention and Control of Geological Hazards and Construction of "Four Systems" of Geological Environment Protection[J]. *Natural Resource Economics of China*, 2016, 29(10):4–7.

Li Yuan, Meng Hu, Dong Ying, Hu Shu-e. Main Types and characteristics of geo-hazard in China—Based on the results of geo-hazard survey in 290 counties[J]. The Chinese *Journal of Geological Hazard and Control*, 2004(02):32–37.

Liu Huanhuan, Zhang Youquan, Wang Rong, et al. Monitoring and Analysis of Land Subsidence Along the Beijing-Tianjin High-speed Railway (Beijing Section) [J]. *Chinese Journal of Geophysics*, 2016, 59(07):2424–2432.

Qiao Jianwei, Peng Jianbing, Zheng Jianguo, et al. Development rules and movement characteristics of earth fissures in China [J]. *Journal of Engineering Geology*, 2020, 28(05): 1016–1027. DOI:10.13544/j.cnki.jeg.2020-266.

Sun Ye. The Classification of Geological Hazard and The Strategical Analysis Of Disaster Reduction Countermeasure In China [J]. *The Chinese Journal Of Geological Hazard And Control*, 1991(04):12–19.

WANG Ai-min, Miao Leilei. Review On Theory Of Man-Nature Relationship[J]. *Advance In Earth Sciences*.

Wang Cong, Wang Yanbing, Zhou Chaodong. The Influence of Urban Expansion of Tongzhou on Land Subsidence[J]. *Journal of Capital Normal University* (Natural Science Edition), 2018, 39(04): 68–74. DOI:10.19789/j.1004-9398.2018.04.013.

Xie Yingxia. Introspection on Planning of Disaster Prevention and Mitigation and Planning of Post-disaster Reconstruction—Something Learned from the Zhouqu Debris-Flow Disaster [J]. *Journal Of Catastrophology*, 2010, 25(S1):16–19.

Yao Lin, Gao Xinghe, Ma Chunguang. Value Evaluation and Decision Support of Geological Disaster Prevention and Control Products. *Natural Resource Economics of China*, 2021, 34(05): 10–14. DOI:10.19676/j.cnki.1672-6995.000612.

Zhang Jiquan, Okada Norio, Tatano Hirokazu. Integrated natural disaster risk management: comprehensive and integrated model and Chinese strategy choice[J]. *Journal of Natural Disasters*, 2006(01):29–37.

Zhang Mo-nan. A Comprehensive Review of Research on Behavioral Decision-Making in Uncertain Circumstances [J]. *Modern Management Science*, 2004(11):37–40.

Zhang Yibing. The Deep Logic of Contemporary Ecological Horizon and Scientific View of History: Philosophical Identification of the Relationship between Human and Nature, Technology and Social Development [J]. *Philosophical Research*, 1993(08):11–20.

Zhong Lixun. Case Study on Significant Geo-hazards in China[J]. *The Chinese Journal of Geological Hazard and Control*, 1999(03):2–7+11.

*Advances in Urban Construction and Management Engineering – Cha (Ed)*
*© 2023 Copyright the Author(s), ISBN 978-1-032-39018-5*

# Research on seismic stability of highway tunnels in high-intensity areas based on numerical analysis

Xin Zhang
*China Road and Bridge Corporation, Beijing, China*

Fujie Zhou*
*School of Civil Engineering, Xian University of Architecture and Technology, Xi'an, Shaanxi, China*
*Shaanxi Key Laboratory of Geotechnical and Underground Space Engineering, Xi'an, Shaanxi, China*

Changwei Li & Bin Zhi
*China Road and Bridge Engineering Co., Ltd., Beijing, China*

ABSTRACT: With the Tajikistan-China border highway reconstruction and expansion project of the No. 2 tunnel project in the Karaihom-Wanqi section as the research background, the finite element numerical analysis method is used to study the seismic stability of the tunnel entrance and exit section. The results show that during the earthquake, the tunnel entrance section has the largest principal stress at the arch too far away from the slope and the smallest principal stress at the arch toe near the slope. During the seismic process of the exit section, the maximum principal stress is distributed at the arch foot of the lining on one side of the tunnel line, and the minimum principal stress is distributed along the arch foot. At the same time, based on the analysis results, measures are proposed to increase the stability of earthquake resistance at the entrance and exit of the tunnel, which provides a reference for the seismic design and construction of highway tunnels.

## 1 INTRODUCTION

In the process of infrastructure construction, it is inevitable to encounter the construction of tunnels in high-intensity seismic areas, which is of great significance to the study of seismic stability of highway tunnels in high-intensity seismic areas (Lin et al. 2016; Zhao et al. 2020). Based on this, many scholars have conducted a series of studies on tunnel seismic in high-intensity areas Taking a mountainous highway tunnel under construction in Yunnan Province as the background, Zhu Rongqing et al. (2018) sorted out and analyzed the field monitoring data and summarized the laws of surrounding rock deformation and lining stress of mountainous highway tunnels under complex geological conditions in high. Zhang Jiahua (2019) made a reliability analysis of the seismic stability of the tunnel face. Taking the highway tunnel crossing the Jiudu District as the research object, Wang Qiuyi (2019) summarized the actual seismic effect, advantages, and disadvantages of four seismic measures. Yu Ji et al. (2020) used the finite element numerical analysis method to analyze the seismic performance of each measuring point of back pressure backfill tunnel lining structure. Through the above research, it can be found that the seismic stability of the entrance and exit section of the tunnel in the high-intensity earthquake area should be paid attention to, and further research and exploration are needed for the safety and seismic stability of the lining structure of the highway tunnel crossing the high-intensity earthquake area.

*Corresponding Author: 18339182573@163.com

 DOI 10.1201/9781003348023-56

In this study, the import and export section of No.2 tunnel of Kalaihongm-Wanqi section of Tajikistan- China border highway reconstruction and expansion project (hereinafter referred to as No.2 tunnel of Tazhong highway phase II) is taken as the research background. Based on the finite element numerical analysis method, a three-dimensional finite element calculation model is established to study the safety and seismic stability of highway tunnel lining structures in high-intensity areas.

## 2 ENGINEERING SITUATIONS

The landform in the area of No. 2 Tunnel of Tazhong Highway Phase II belongs to the alpine landform. The mountain body where the tunnel is located crosses the ridge. The peak acceleration of the tunnel seismic fortification ground motion is 0.4g, and the seismic fortification intensity is IX degrees. In the area where the tunnel is located, there are Karakum Plate (II), Tarim Plate (III), and Central Iran-Gangdise Plate (IV). Therefore, affected by the above plate tectonics, there are three faults in the stratum of tunnel No. 2 of Tazhong Highway Phase II. In addition, the surrounding rock mass of the No. 2 Tunnel of Tazhong Highway Phase II is relatively broken, and multiple sets of joint fissures are developed. The valleys in the tunnel site area are strongly cut, and the slopes are steep. Most of the slopes are slope deposits. The stratum lithology in the tunnel site area is single, mainly composed of biotite and biotite plagioclase gneiss. The upper rock mass is a strongly-weakly weathered rock mass, and the joint fissures are developed with good permeability.

## 3 NUMERICAL MODELING

The entrance and exit section of No. 2 Tunnel in Phase II of Tazhong Highway is selected as the research object. The geological surrounding rock conditions of the entrance and exit section of the tunnel are all V-grade surrounding rock. The length of the defense range of the inner section of the tunnel is 25m, and the minimum defense length of the entrance section is 35m. Therefore, a finite element model of 40m × 40m × 55m is established according to the relationship between the surface elevation at the entrance and exit of the No. 2 tunnel and the position of the tunnel entrance. The general situation of the model is shown in Figure 1. The mountain is modeled by a 3D solid element, and the model type is Mohr-Coulomb. The material property is V-grade surrounding rock, and the specific parameters are referred to the Code for Design of Highway Tunnels (JTG D70-2-2014).

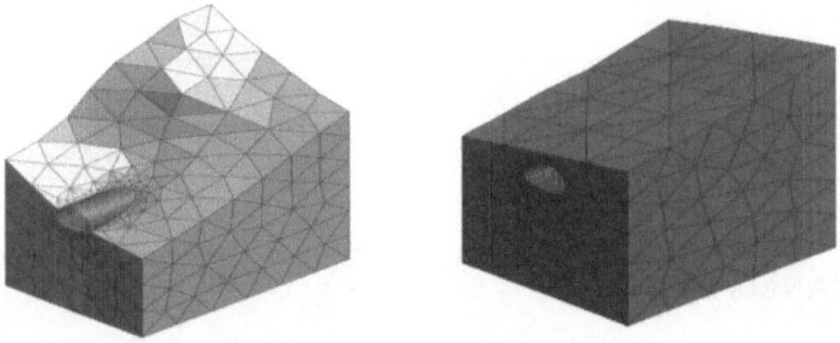

(a) Tunnel entrance (b) Exit zone

Figure 1. Overview of modeling of tunnel No. 2.

The tunnel lining is modeled by the plate-shell element. The material is C30 concrete. The lining thickness is 0.65m, according to the design data. The material property parameters are shown in Table 1.

Table 1. Material property parameters.

| Materials Name | Unit Weight ($kN/m^3$) | elastic Modulus ($GPa$) | Poisson's Ratio $\mu$ | Cohesion $c$ ($MPa$) | Internal Friction Angle $\varphi$ |
|---|---|---|---|---|---|
| V-Class Coffer | 20 | 2 | 0.35 | 0.2 | 27 |
| c30 concrete | 24 | 30 | 0.2 | — | — |

According to the Code for Seismic Design of Urban Rail Transit Structures (GB 50909-2014), the characteristic period Tg of site design seismic acceleration response spectrum should be adopted according to the site category, and the characteristic period analysis of seismic response spectrum in national standards should be adopted according to the Code for Design of Highway Tunnels (JTG D70-2-2014), as shown in Table 2.

Table 2. Design characteristic period *Tg* of seismic acceleration response spectrum.

| Characteristic Period Division of Response Spectrum | Site Class | | | | |
|---|---|---|---|---|---|
| | $I_0$ | $I_1$ | II | III | IV |
| 0.35*S* | 0.2 | 0.25 | 0.35 | 0.45 | 0.65 |
| 0.40*S* | 0.25 | 0.30 | 0.40 | 0.55 | 0.75 |
| 0.45*S* | 0.30 | 0.35 | 0.45 | 0.65 | 0.90 |

According to the Code for Seismic Design of Railway Engineering (GB50111-2006), the peak acceleration of seismic fortification ground motion of tunnel No. 2 of Tazhong Highway Phase II is 0.4g, and the envelope function parameters of the design response spectrum are shown in Table 3.

Table 3. Envelope function parameter of acceleration response spectrum of 0.4*g* ground motion peak.

| Exceedance Probability | Rise time (s) | Horizontal Time (s) | Cumulative Time (s) | Maximum Acceleration (g) |
|---|---|---|---|---|
| strong earthquake | 5 | 13 | 40 | 0.4 |

## 4 NUMERICAL ANALYSIS

### 4.1 *Tunnel entrance*

#### 4.1.1 *Structural safety analysis*

The stress nephogram of tunnel lining in the entrance section of the Tunnel of Tazhong Highway is shown in Figure 2. It can be seen from the figure that the compressive stress of the lining of the shallow buried section at the entrance of the No. 2 tunnel occurs in the arch waist and arch foot. T maximum compressive stress at the left arch waist is 2.72 MPa, and the maximum compressive stress at the right arch waist is 2.00 MPa. The compressive stress at the arch foot is about 1.3 MPa;

the maximum tensile stress of tunnel lining mainly occurs at the inverted arch of the tunnel, and the maximum value is 0.61 MPa.

The shear nephogram of the tunnel lining in the entrance section of No. 2 tunnel of Tazhong highway phase II is shown in Figure 3. It can be seen from the figure that the shear force on the vault, arch waist, and inverted arch of the entrance section of No. 2 tunnel is small, and the shear stress on the arch foot is large. The maximum shear stress at the left arch foot of the entrance section of the No. 2 tunnel is 0.59 MPa, and the maximum shear stress at the right arch foot is 0.45 MPa.

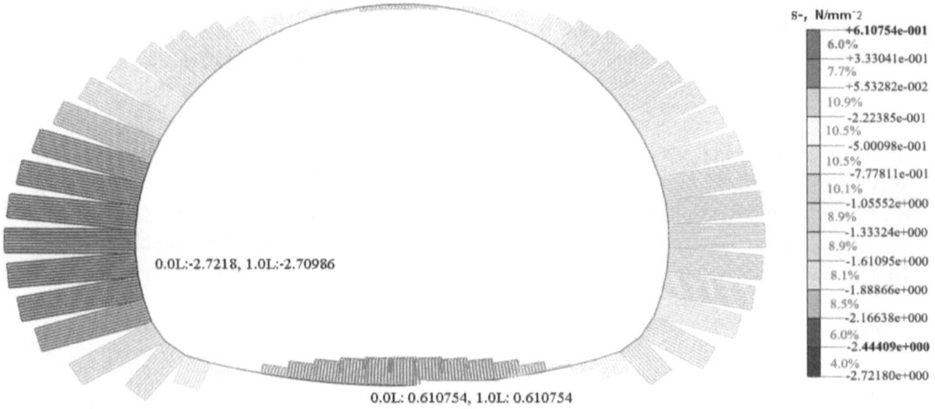


Figure 2. Lining stress cloud diagram at the entrance of No. 2 tunnel.

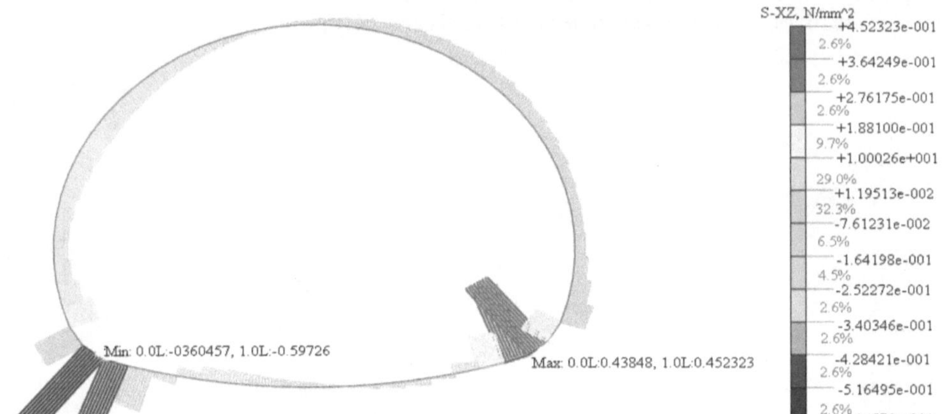


Figure 3. Shear stress cloud diagram at the entrance of No. 2 tunnel.

### 4.1.2 *Seismic stability analysis of tunnel*

The maximum principal stress nephogram of tunnel lining in the entrance section of No.2 Tunnel of Tazhong Highway Phase II is shown in Figure 4. According to the figure, the maximum principal stress of the No.2 Tunnel occurs at the arch foot of the line side during the earthquake. Combined with the topographic map, it is known that the maximum principal stress of No.2 tunnel lining is 1.98 MPa away from the side of the slope. The vault and the arch waist bear small compressive stress.

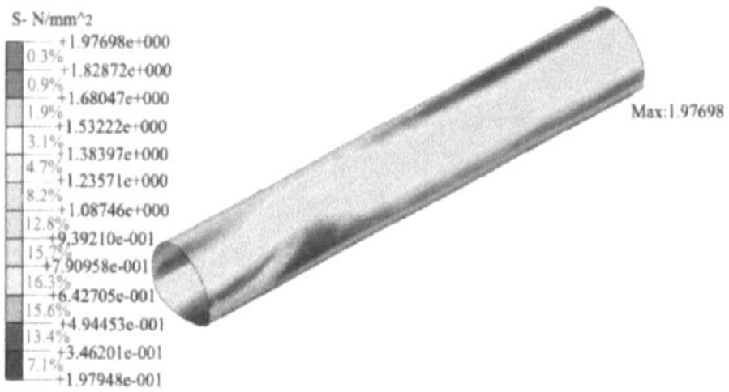


Figure 4. Maximum principal stress cloud diagram at the entrance of No. 2 tunnel.

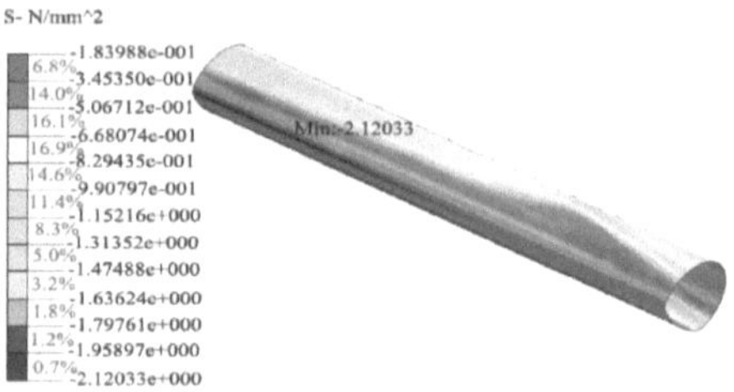


Figure 5. Cloud diagram of the minimum principal stress of the lining at the entrance of No. 2 tunnel.

The minimum principal stress nephogram of the entrance section of the No. 2 tunnel of Tazhong highway phase II is shown in Figure 5. It can be seen from the figure that the minimum principal stress of the entrance section of the No. 2 tunnel occurs at the arch foot near the side of the slope of the tunnel. The arch foot along the line is the main position of the distribution of the minimum principal stress of the tunnel lining. The minimum principal stress of the No. 2 tunnel is 2.12 MPa.

## 4.2 *Tunnel exit zone*

### 4.2.1 *Structural safety analysis*

The stress nephogram of the tunnel lining in the exit section of the No.2 tunnel of Tazhong Highway Phase II is shown in Figure 6. The figure shows that the tunnel lining in the exit section of the No.2 tunnel is less affected by the bias effect. The compressive stress mainly occurs at the arch waist and arch foot. The maximum compressive stress of the left arch waist is 1.84 MPa, and the maximum compressive stress occurs at the right arch waist position, about 2.06 MPa. The tunnel invert bears large tensile stress, and the maximum tensile stress is about 0.68 MPa.

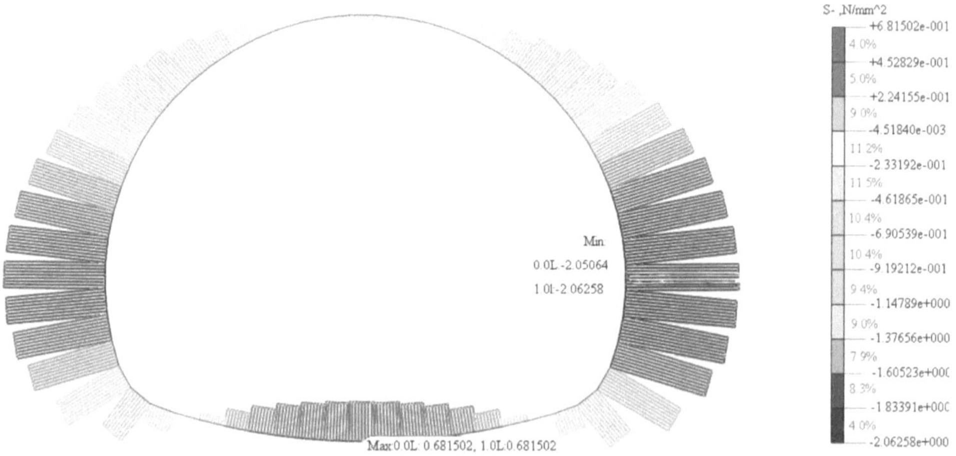


Figure 6. Lining stress cloud diagram of the exit section of No. 2 tunnel.

The shear nephogram of the tunnel lining in the entrance section of No. 2 tunnel of Tazhong highway phase II is shown in Figure 7. It can be seen from the figure that the shear force on the vault, arch waist, and inverted arch of the entrance section of No. 2 tunnel is small, and the shear stress on the arch foot is large. The maximum shear stress at the left arch foot of the exit section of the No. 2 tunnel is 0.44 MPa, and the maximum shear stress at the right arch foot is 0.49 MPa.

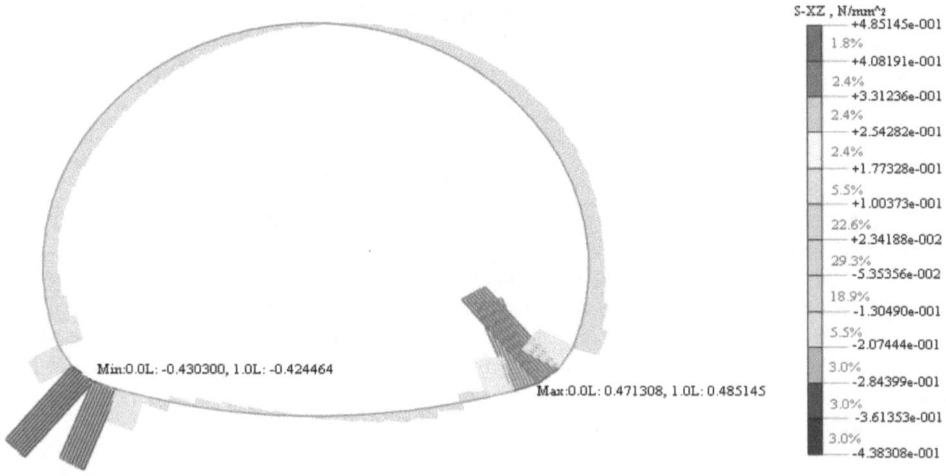


Figure 7. Shear stress cloud diagram of the exit section of No. 2 tunnel.

### 4.2.2 *Seismic stability analysis of tunnel*

The maximum principal stress nephogram of the tunnel lining in the exit section of the No.2 tunnel of Tazhong Highway Phase II is shown in Figure 8. According to the figure, the maximum principal stress of the tunnel lining in the exit section of No.2 tunnel is distributed at the arch foot of the tunnel line towards one side of the lining, and the area along the arch foot is the maximum principal stress distribution. The maximum compressive stress of the tunnel lining in the exit section of the No.2 tunnel is 1.79 MPa.

The minimum principal stress of tunnel lining in the exit section of No.2 tunnel of Tazhong Highway Phase II is shown in Figure 9. The minimum principal stress of tunnel lining in the exit section of the No.2 tunnel is distributed along one side of the arch foot in the earthquake. The minimum principal stress of tunnel lining in the exit section of the No.2 tunnel is about 1.75 MPa.

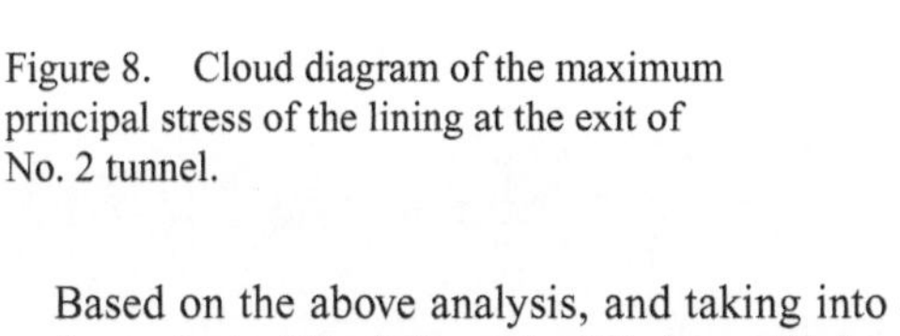


Figure 8. Cloud diagram of the maximum principal stress of the lining at the exit of No. 2 tunnel.

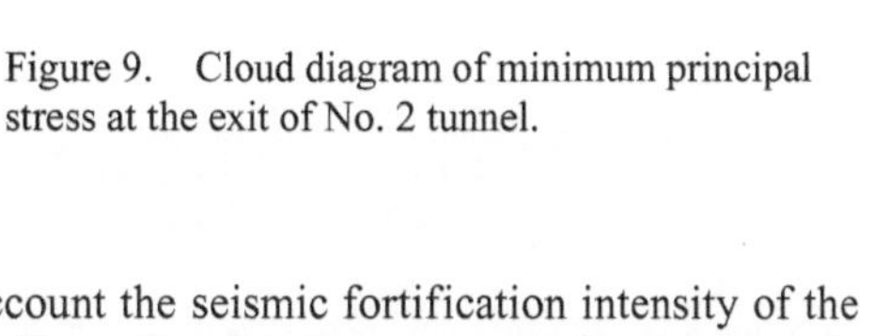


Figure 9. Cloud diagram of minimum principal stress at the exit of No. 2 tunnel.

Based on the above analysis, and taking into account the seismic fortification intensity of the main project of No.2 Tunnel of Tazhong Highway Phase II as IX degrees, the following seismic stabilization measures are recommended in tunnel engineering to ensure the safety of the project in construction and operation: (1) The key reinforcement should be carried out before the construction of the portal section, and the excavation height of the elevation slope of the portal section should be reduced as far as possible. (2) The portal wall should adopt the cast-in-situ reinforced concrete structure, and the open tunnel should adopt the reinforced concrete structure with an inverted arch to improve the ductility and stiffness of the structure. (3) A deformation joint is set every 20m in the range of 200 m in the tunnel entrance, and a deformation joint is set every 50m in the tunnel section.

## 5 CONCLUSION

By using the finite element analysis, the dynamic calculation and analysis of the No. 2 Tunnel of Tazhong Highway Phase II are carried out. It is concluded that during the earthquake, the maximum principal stress in the entrance section of the tunnel is distributed at the arch foot far from the slope, and the minimum principal stress is distributed at the arch foot near the slope. The maximum principal stress of the exit section is distributed at the arch foot of the lining on one side of the tunnel line, and the minimum principal stress is distributed along the arch foot. Based on this conclusion, the seismic stability enhancement measures of tunnels in high-intensity earthquake areas are proposed in order to provide a reference for similar projects.

## REFERENCES


Lin Zhi, Chen Xiang. Study on performance-based seismic fortification standards for highway tunnels[J]. *Modern Tunnel Technology*, 2016, 53(04): 9–15.

Wang Qiuyi, Yang Kui, Mao Jinlong, et al. Study on Comprehensive Seismic Measures of Secondary Lining Structure of Highway Tunnel in Nine Degree Earthquake Area[J]. *Modern Tunnel Technology*, 2019, 56 (05): 42–49+66.

Yu Ji, Yuan Song, Wang Xibao, et al. Study on Seismic Performance of Backfill Cave in High-Intensity Seismic Area[J]. *Chin J Undergr Space Eng*, 2020, 16(S1): 303–308.

Zhang Jiahua, Zhang Biao. Reliability analysis for seismic stability of tunnel faces in soft rock masses based on a 3D stochastic collapse model[J]. *Journal of Central South University*, 2019, 26(07): 1706–1718.

Zhao Fengbing, Luo Jianqiang, Liu Tao, et al. Setting effect of seismic joint in entrance section of super large section highway tunnel[J]. *Modern Tunnel Technology*, 2020, 57(S1): 203–210.

Zhu Rongqing, Ding Wenqi, Wang Qiuyi, et al. Study on Monitoring Law and Numerical Simulation of Highway Tunnel in High-Intensity Mountainous Area of Yunnan Province[J]. *Modern Tunnel Technology*, 2018, 55(S2): 1278–1287.

*Advances in Urban Construction and Management Engineering – Cha (Ed)*


# Study of the quality of the dark sky environment of urban parks in Fuzhou based on field measurements

Yuqing Lei
*College of Landscape Architecture and Art, Fujian Agriculture and Forestry University, China*

Ming Zhao*
*College of Economics and Management, Minjiang University, China*

Yuanli Xing & Yuanhui Chen
*College of Landscape Architecture and Art, Fujian Agriculture and Forestry University, China*

ABSTRACT: Taking Xihu Park, Zuohai Park, Jinjishan Park, and Jinniushan Park in Fuzhou as the research objects, the dark sky environment of Fuzhou urban parks was investigated by using an Unihedron Sky Quality Meter for field measurements and collect data. The following results were obtained. (1) The dark sky environment of Fuzhou urban parks was poor and worsened with time. (2) The magnitude of the brightness of Fuzhou urban parks decreased with decreasing measurement angle, with the highest brightness of zenith being associated with the best dark sky environment quality. (3) The zenith-equivalent brightness value of urban parks in Fuzhou was greatly affected by the light environment of the night sky layer, the range of 15°–45° was greatly affected by the light environment near the urban park canopy, and the internal differentiation was not significant. (4) Artificial lighting in parks causes light pollution in urban park areas. Elements such as artificial lighting, building facades, and large trees in the areas surrounding the parks also affected the quality of the dark sky environment of urban parks. Finally, corresponding recommendations are proposed to guide the planning and design of the nighttime light environment in urban parks and to provide a reference for a system of collecting field measurements and evaluating the dark sky environment of urban parks.

## 1 INTRODUCTION

Light is an important environmental factor that regulates many processes and functions in organisms. For millions of years, most organisms on Earth have evolved under the light-dark cycle of day and night, but artificial lighting has disrupted this day-night cycle (Singhal 2021). The term "light pollution," which refers to artificial light that adversely affects living things, was proposed by Verheijen in 1985 (Falchi 2016). The most superficial impact of light pollution on humans is the inability to directly see the stars in the sky. However, the impact of light pollution is far greater. Light pollution impacts organisms ranging from single-celled organisms to higher organisms (Bruening 2015). It can affect the diversity of ecosystems by regulating the abundance, number, and aggregation of organisms from the level of individuals to various levels of biological communities (Bennie 2016). By disrupting interspecific and intraspecific interactions and the process of landscape formation, light pollution interrupts population, genetics, and landscape structures (Botha 2017). The generation of greenhouse gases through luminescence exacerbates global warming (Chase 2020).

*Corresponding Author: 3314097390@qq.com

Due to the increasing impact of light pollution, the concept of the dark sky emerged. Subsequently, scholars from various countries and major international dark sky protection organizations have gradually carried out relevant research and promotion. The International Dark Sky Association is the most influential of these organizations, and its main form of dark sky protection is its certification of dark sky areas on a global scale. As of early 2022, the International Dark Sky Association certified 111 International Dark Sky Parks, the most common and important certification form. Additionally, this certification effectively protects natural habitats and promotes environmental education, thus receiving extensive attention and recognition throughout the international community.

Light pollution is an important issue in urban development. As an important place for urban residents' daily leisure activities, the dark sky environment of urban parks has rarely been investigated. In this paper, based on the experience of scholars from various countries and major dark sky protection organizations, we collected dark sky data in representative Fuzhou urban parks on the Light Pollution Map website. We measured the stars in the sky of representative urban parks in Fuzhou using the Unihedron Sky Quality Meter, an instrument used to measure the dark sky environment approved by the International Dark Sky Association. The characteristics of the current dark sky environment in Fuzhou urban parks were explored.

## 2 RESEARCH SUBJECTS AND METHODS

### 2.1 *Overview of the research parks*

By 2020, a total of 111 urban parks and green spaces were planned in Fuzhou (Liu 2022). To ensure that the survey results are generalizable, this study selected four typical, representative open parks in the main urban area of Fuzhou: Xihu Park, Zuohai Park, Jinjishan Park, and Jinniushan Park. All these parks have convenient transportation and are visited by many tourists.

### 2.2 *Data sources*

The data for this study were collected from the Light Pollution Map (LMP) website. The LMP is a mapping application that can display light pollution-related content on the base layer of Microsoft Bing. The data collected in this study mainly included the 2015 new world map of artificial night sky brightness published by Falchi and colleagues, which is currently the most influential world atlas of artificial brightness, and from which the SQM values in the study area in 2015 were obtained.

### 2.3 *Measurement instruments and background conditions*

The main instruments used in this study were the Unihedron Sky Quality Meter, the Spirit level, and the compass. The SQM measuring instrument is used to measure the brightness of the stars, and it is one of the methods for evaluating the dark sky environment officially recognized by the International Dark Sky Association. It is also the most widely used equipment for scientific measurement of the quality of sky brightness. The level meter was used to measure the inclination angle, and the compass was used to measure the direction.

Meteorological, atmospheric, celestial, and other conditions impact the dark sky environment more. Additionally, the SQM measuring instrument officially recommends that measurements be performed in the evening (at least one hour after sunset) with no cloud or a small amount of cloud cover and no moon or a crescent moon. The specific measurement background conditions for this study are shown in Table 1.

### 2.4 *Measurement indicators and levels*

The primary measurement index is the magnitude of sky brightness, and the unit is mag/arcsec2. The higher the number is, the darker the sky is (Liu 2007).

Table 1. Measurement background conditions.

| Area | Date | Time | Sunset | Moon phase | Temperature | Humidity | Cloud cover |
|---|---|---|---|---|---|---|---|
| Xihu Park | 22/05/03 | 21:10 | 18:35 | crescent | 20°C | 20% | 13% |
| Zuohai Park | 22/05/03 | 20:30 | 18:35 | crescent | 21°C | 23% | 13% |
| Jinjishan Park | 22/04/28 | 21:00 | 18:32 | crescent | 21°C | 92% | 11% |
| Jinniushan Park | 22/04/28 | 20:15 | 18:32 | crescent | 22°C | 81% | 11% |

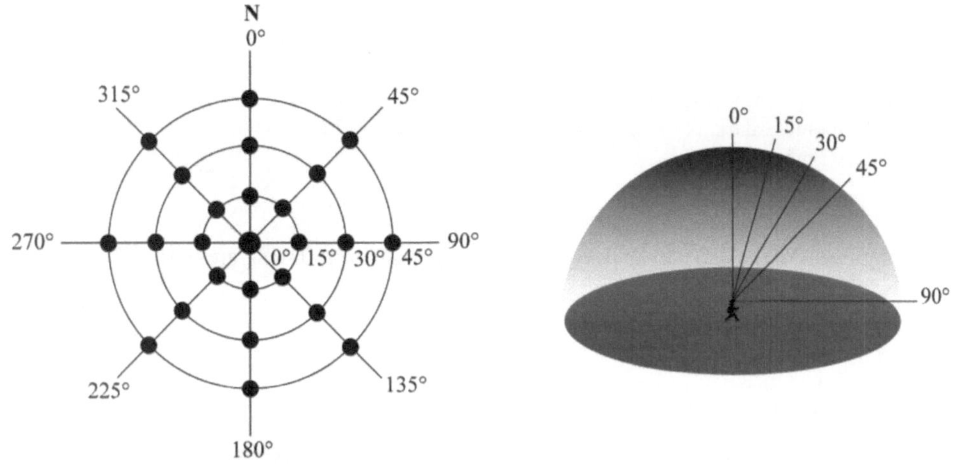


Figure 1. Measurement level, angle diagram.

In the analysis of the light environment of the whole sky, the measurement instrument cannot directly divide the brightness because sky brightness is much lower than ground surface brightness. Therefore, a 360° circular measurement of the space above the canopy was performed. To avoid interference caused by the projection of the researcher performing the measurements in the night-time light environment, the four angles of each point, namely, 0°, 15°, 30°, and 45°, as well as north (0°) and northeast (45°) were measured. There were a total of 25 points of magnitude in the eight directions, north (0°), northeast (45°), east (90°), southeast (135°), south (180°), southwest (225°), west (270°), and northwest (325°). Figure 1 shows a schematic diagram of the measurement levels and angles (Liu 2020).

## 3 MEASUREMENT RESULTS AND DATA ANALYSIS

### 3.1 *Selection of representative locations for the measurement studies*

Specific factors, such as the shadow of buildings and plants and irregular artificial lighting, impact the measurement results. Therefore, the measurement points in the parks need to be representative. All the measurement points in the four parks were located in relatively open and flat areas with tourists coming and going. There were no significantly strong light sources, no shadows of buildings or plants, and no irregular, artificial lighting.

Table 2. Typical measurement locations.

| Number | Area | Geographic information | Altitude (m) |
|---|---|---|---|
| Point 1 | Xihu Park | 119°17′24″E, 26°6′0″N | 11.5 |
| Point 2 | Zuohai Park | 119°16′48″E, 26°6′0″N | 9.4 |
| Point 3 | Jinjishan Park | 119°19′12″E, 26°6′0″N | 52.7 |
| Point 4 | Jinniushan Park | 119°15′36″E, 26°4′48″N | 19.1 |

3.2 *Measurement results and data analysis of the zenith magnitude at each measuring point*

The brightness of the zenith is one of the most important measurements used to evaluate the severity of light pollution (Liu et al. 2012). It is also one of the most important indicators used by the International Dark Sky Association to measure the quality of the dark sky environment. The International Astronomical Union defined that in the absence of light pollution, the brightness of the natural moonless dark sky is $2.1\times10^{-4}$ cd/m$^2$ (Yang et al. 2012). The conversion formula of magnitude to brightness is:

$$[\text{value in cd/m}^2] = 10.8 \times 10^4 \times 10^{[-0.4\times(\text{value in mag/arcsec}^2)]} \tag{1}$$

The brightness of the dark sky is converted to a magnitude of 21.6 mag/arcsec$^2$.

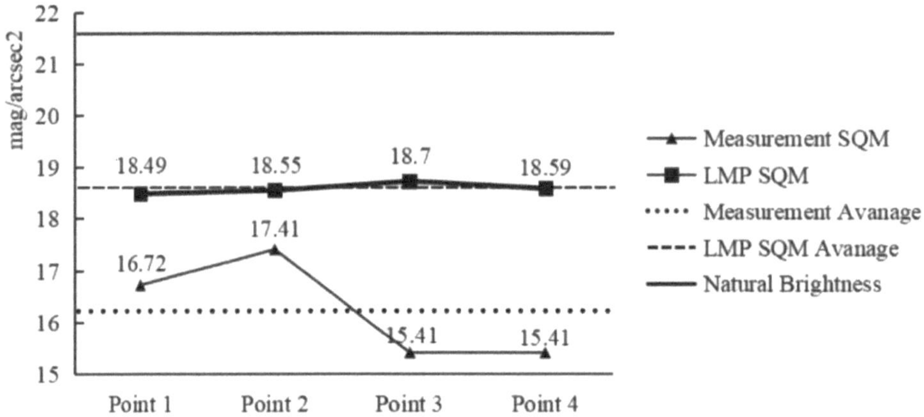


Figure 2. Line chart of measured zenith brightness data and LMP data at each measurement point.

Figure 2 shows the zenith brightness of urban parks in Fuzhou in 2015 and 2022 is far lower than that of the naturally dark sky. The dark sky environment of urban parks is poor, and a certain degree of light pollution is present. The value in 2022 is significantly lower than that in 2015, indicating that since 2015, with the city development of Fuzhou, the brightness of nighttime artificial lighting in urban parks has increased, and the dark sky environment in urban parks has worsened.

In 2015, the average SQM of the four parks was 18.58. However, the difference in the brightness values of the four parks was small. In 2022, the average SQM of the four parks decreased significantly to 16.24, and the magnitude of the difference among the four parks increased. The brightness of Jinniushan Park and Jinjishan Park was relatively high, while the brightness of Xihu Park and Zuohai Park was relatively low. These observed differences may be caused by the following reasons. On the one hand, the LMP data are calculated using high-resolution satellite data and an accurate sky brightness meter. The actual measurements, meteorological, atmospheric, and celestial conditions all impact the measurement results. On the other hand, as Xihu Park and Zuohai Park have existed in Fuzhou for a long time, their areas are smaller; thus, sites for selection are relatively limited. Additionally, the trees have grown for a long time, the crown is larger, artificial lighting produces larger projection at night, and lighting is hindered from all directions. Thus, Xihu Park and Zuohai Park have higher brightness values than the other two parks.

### 3.3 *The azimuth magnitude measurement results and data analysis of each measuring point*

According to the measurement results (Figure 3), there is a certain correlation between the magnitude of brightness at each measuring point and the angle of the measuring point. When the zenith brightness value of each measuring point was highest, the dark sky quality was best. Except for the individual tilt angle of the measuring point, the tilt angle gradually decreased from 0° to 45° at each level and each measuring point. The degree of change in the brightness magnitude was also greater. These data indicate that the canopy of urban parks in Fuzhou is greatly affected by light pollution and that closer to the ground surface, there is more severe light pollution and lower environmental quality of the dark sky.

#### 3.3.1 *Analysis of the average magnitude brightness of each measuring point azimuth*

The zenith magnitude at each measuring point is significantly higher than the brightness at different azimuth magnitudes. Therefore, to fully understand the current conditions of the dark sky

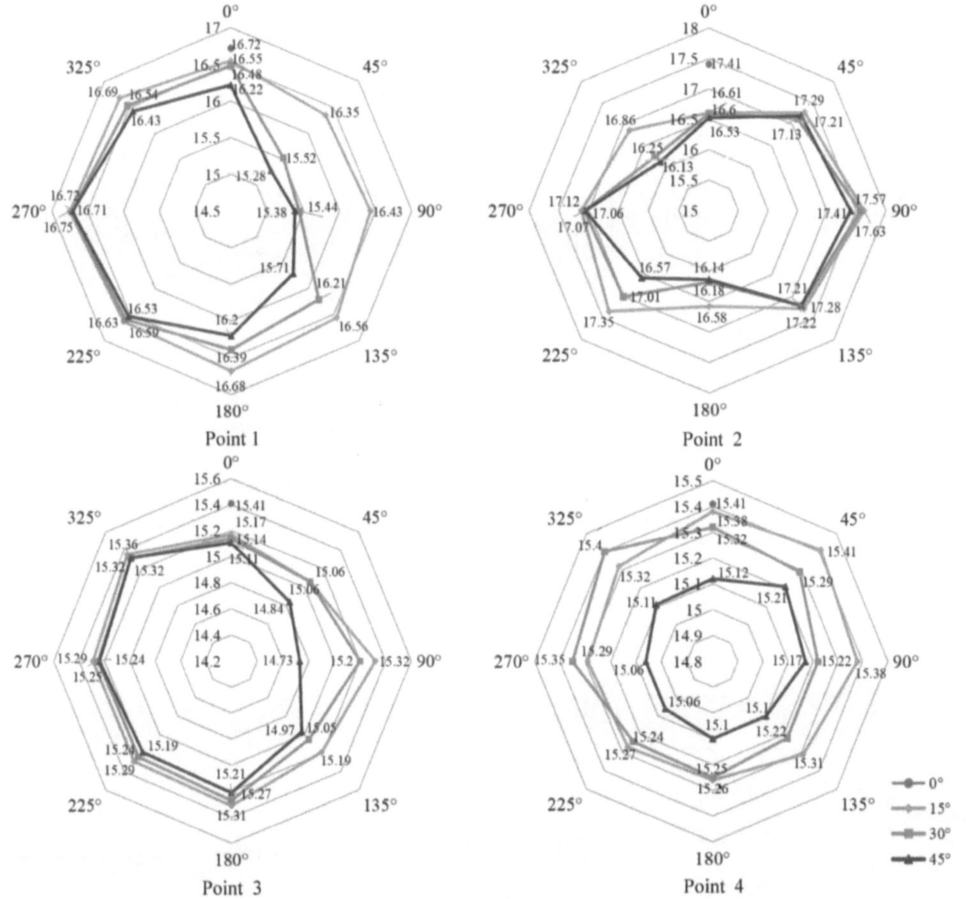


Figure 3. Radar map of the distribution of the magnitude of brightness at each measuring point in the urban parks of Fuzhou.

environment at each position of the park, the average values of 15°, 30°, and 45° at each measuring point were used as a reference to plot the city of Fuzhou. A radar map of the distribution of average magnitudes of brightness at various locations in the park was generated (Figure 4).

As shown in Figure 4, the brightness values of the satellites on the east and northeast sides of measurement point 1, the north and south sides of measuring point 2, the north and east sides of measuring point 3, and the south side of measuring point 4 are relatively low, and the light pollution is relatively serious. The brightness of the other azimuths is relatively high, and the light pollution is relatively small. Based on the field survey and measurement data, the main reasons that cause the magnitude of brightness to be lower at some angles than at other angles are as follows. First, the phenomenon of light intrusion generated by the commercial and residential areas around the city parks affects the local area of the city park. Second, a light source can have multiple refractions and reflections from the water surface of a park at night, affecting the urban park's local brightness. Third, a large tree canopy inside the park will have a greater impact on the nighttime illumination. For a night scene construction of the individual landscape nodes in the park, the brightness, color temperature, illuminance, and lighting fixtures that are different from the conventional lighting equipment are used, affecting the park's local area. The brightness has a certain impact.

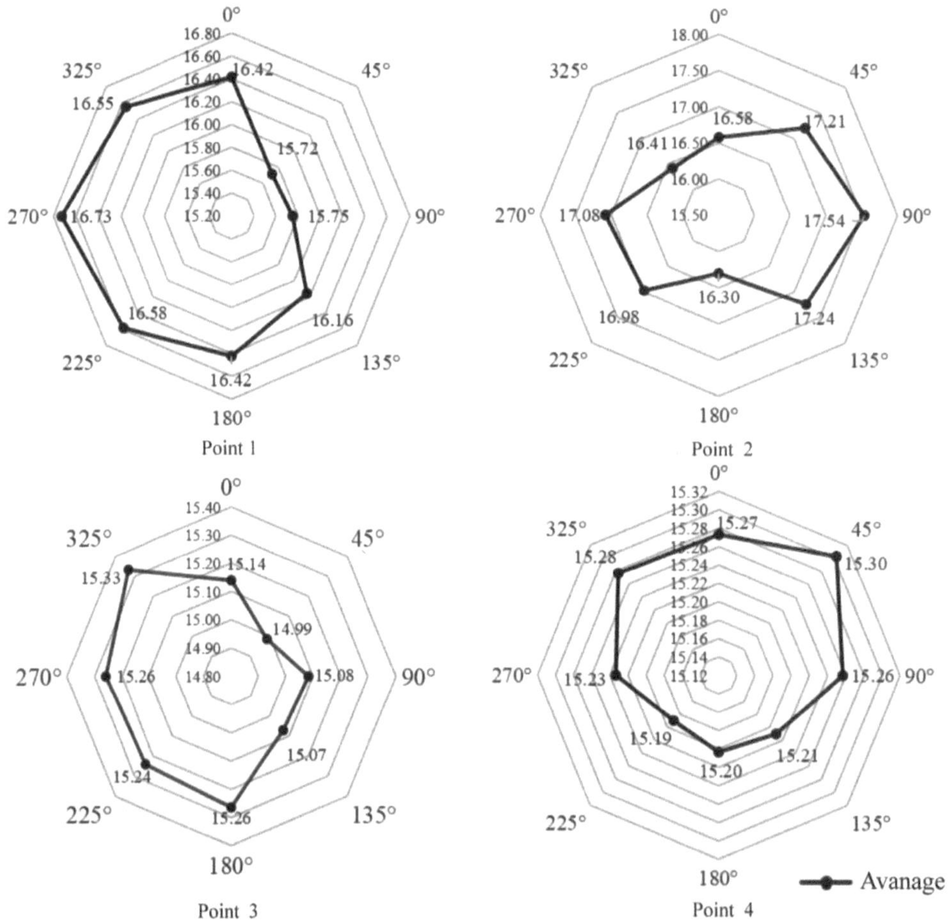


Figure 4. Radar map of the distribution of average magnitude brightness at various measuring points in Fuzhou urban park.

3.3.2 *Analysis of the average magnitude of the brightness of each measuring point at different inclination angles*

To understand the current situation of the dark environment at all levels of Fuzhou Urban Park, the average measurement results of 0°, 15°, 30°, and 45° were taken as the reference to draw a broken line map of the average magnitude of brightness at different incline angles for each measurement point in Fuzhou Urban Park.

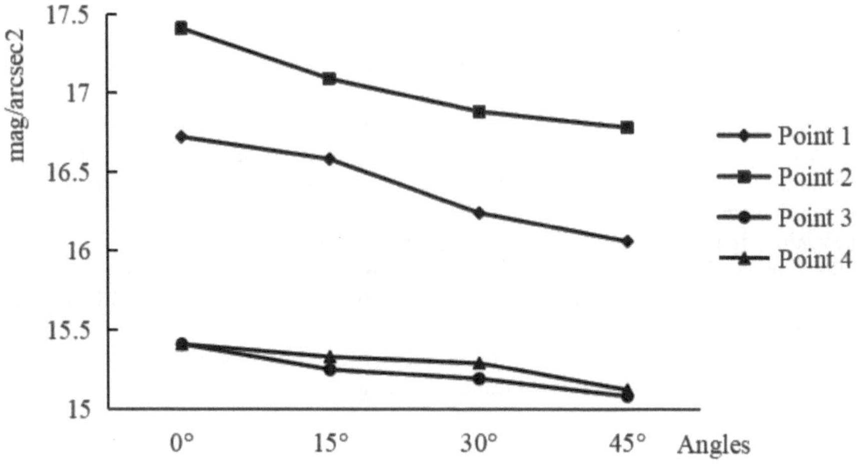


Figure 5. Line chart of average magnitude brightness at different obliquity angles of each measuring point in Fuzhou urban park.

As seen in Figure 5, except for point 1, the downward trend of the magnitude of brightness and the measurement angle of other measurement points is nearly similar. The slope of the broken line in the interval of 0°~15° is the largest and corresponds to the magnitude of brightness changing significantly. The slope of the broken line in the intervals of 15°~30° and 30°~45° gradually decreases, and the variation in the magnitude of brightness is gradually decreased. This indicates that the zenith-equivalent brightness value is greatly affected by the light environment of the night sky layer. While the 15°~45° interval is greatly affected by the light environment near the urban park canopy, the interval differentiation is not more significant than that of the 0°~15° interval. The light environment of the night sky layer is the collective result of the effect of the lighting of the entire park and even the entire urban space. In contrast, the light environment of the urban park canopy is mainly the result of the internal lighting of the urban park. When the inclination angles are 15°, 30°, and 45°, the magnitude of the brightness of each measuring point is mainly affected by the lighting inside the urban park, and the main influencing factors are the same. It seems that the lower the magnitude of brightness is, the lower the quality of the dark sky environment. The lighting inside an urban park directly affects the brightness of the zenith magnitude of the measurement point, and this is more strongly affected by the overall lighting of the entire park or even the entire urban space. Therefore, the brightness of the zenith is significantly greater than that of other measurement angles.

## 4 DISCUSSION

Light pollution endangers human health, affects the living environment of animals and plants, and exacerbates global warming. It is imperative to control light pollution and protect the dark sky environment. In this study, we sought to understand the current conditions of the dark sky environment in Fuzhou urban parks through field measurements and to analyze the current dark sky environmental damage in Fuzhou urban parks. The results of this study can guide the planning and design of nighttime lighting environments in urban parks and future field measurements of the dark sky in urban parks. The evaluation system provides a reference for future studies.

## 5 CONCLUSIONS

Through the measurement and analysis of the dark sky environment in Fuzhou urban parks, the following conclusions are drawn:

(1) Due to the impact of light pollution on urban parks in Fuzhou, the dark sky environment was poor and worsened over time.
(2) The magnitude of brightness at each different measurement point in the urban parks of Fuzhou shows a certain correlation with the inclination angle of the measuring points. The magnitude of brightness at the measuring point decreases with the decrease in the measuring angle. When the zenith magnitude of brightness is highest, the quality of the dark sky environment is best.
(3) The zenith brightness value of urban parks in Fuzhou is greatly affected by the light environment of the night sky layer. While the 15°~45° interval is greatly affected by the light environment near the urban park canopy, and the internal differentiation is greater than 0°~15°. The interval is not significant.
(4) Artificial lighting in parks is one of the sources of light pollution in urban areas. Elements such as artificial lighting, building facades, and large trees in the areas surrounding parks all affect the dark sky environment of urban parks.

## REFERENCES


Bennie, J. & Da Vies, T.W. & Cruse, D.& Gaston, K.J. (2016): Ecological effects of artificial light at night on wild plants. – *Journal of Ecology*, 104(3).

Botha, L.M. & Jones, T.M. & Hopkins, G.R. (2017): Effects of lifetime exposure to artificial light at night on cricket (Teleogryllus Commodus) courtship and mating behavior. – *Animal Behaviour*, 129:181–188.

Bruening, A. & Hoelker, F. & Franke, S.& Preuer, T. & Kloas, W. (2015): Spotlight on fish: Light pollution affects circadian rhythms of European perch but does not cause stress. – *Science of the Total Environment*, 511(apr. 1):516–522.

Chase, J.M. & Blowes, S.A.& Knight, T.M. & Gerstner, K. & May, F. (2020): Ecosystem decay exacerbates biodiversity loss with habitat loss. – *Nature*, 584(7820):238–243.

Falchi, F. & Cinzano, P. & Duriscoe, D. & Kyba, C. C. & Elvidge, C.D.& Baugh, K. & Portnov, B.A. & Rybnikova, N.A. & Furgoni, R. (2016): The new world atlas of artificial night sky brightness. – *Sci Adv*, 2(6):e1600377.

Liu, F. & Fu, W.C. & Wang, M.H. & DONG, J.W. (2022): Impact of Crowding Perception on Tourist' Restorative Experience: A Case Study of Fuzhou City Parks. – *Landscape Architecture*, 29(03):98–104.

Liu, M. & Yang, X.X. & Liu, Y.C. (2020): Comparison and Analysis of the Light Pollution Effect at Night in the Typical Commercial Areas of Milan and Dalian. – *China Illuminating Engineering Journal*, 31(06):8.

Liu, M. & Zhang, B.G. & Pan, X.H. & Yuan, J. (2012): Study on Evaluation Indexes and Methods of Light Pollution in Urban Lighting Planning. – *China Illuminating Engineering Journal*, 23(04):22–27.

Ma, J. & Zhang, B.G. & Zhang, M.Y. (2007): Measurement and Evaluation of the Night Sky Brightness-Taking the Night Sky in Tianjin as an Example. – *China Illuminating Engineering Journal*, (03):6–11.

Singhal, R.K. & Chauhan, J. & Jatav, H.S. & Rajput, V.D. & Singh, G.S. & Bose, B. (2021): Artificial night light alters ecosystem services provided by biotic components. – *Biologia Futura* 72(2):169–185.

Yang, X.X. (2020): Comparative analysis of light pollution at night in typical urban areas – Take the residential and commercial areas in Milan and Dalian as an example. – M.A. Thesis, Dalian University of Technology.

*Advances in Urban Construction and Management Engineering – Cha (Ed)*


# Productivity of rotary drilling rig construction in the Pearl River delta water resources allocation project

Zhiding Chen*
*College of Hydraulic & Environmental Engineering, China Three Gorges University, Yichang, China*
*Hubei Key Laboratory of Construction and Management in Hydropower Engineering, China*
*Three Gorges University, Yichang, Hubei, China*

Mei Zhang
*Guangdong Yue Hai PRD Water Supply Co., Ltd, Guangzhou, Guangdong, China*

Ran Song
*College of Hydraulic & Environmental Engineering, China Three Gorges University, Yichang, China*
*Hubei Key Laboratory of Construction and Management in Hydropower Engineering, China*
*Three Gorges University, Yichang, Hubei, China*

Shangge Li
*Guangdong Hydropower Planning & Design Institute, Guangzhou, Guangdong, China*

Yang Yang & Xuelian Zheng
*Guangdong Yue Hai PRD Water Supply Co., Ltd, Guangzhou, Guangdong, China*

ABSTRACT: In order to quickly and quantitatively calculate the construction platform time consumption of the rotary drilling rig according to the actual construction situation, based on the grey correlation analysis of influencing factors, the RBF neural network model was used to predict the time of the unit length of rotary drilling rig leading hole, and it was tested with the data in the PRD water resources allocation project. The results reveal significant discrepancies in pilot hole efficiency in the PRD water resources allocation project. The efficiency of pilot hole construction is greatly influenced by the trench specifications, staffing, and rock conditions. It is verified by the neural network that 27 groups of samples can stably predict the efficiency of a rotary drilling rig in the PRD water resources allocation project. The overall stability of the results is better than that of the BP neural network model based on the factors affecting the efficiency of rotary excavators.

## 1 INTRODUCTION

The Pearl River Delta (PRD) Water Resources Allocation Project diverts water from the Xijiang River system in Guangdong Province to the eastern part of the PRD to solve the water shortage problem in the Nansha District of Guangzhou, Shenzhen City, and Dongguan City, with a total length of 113.1 km. The project uses shield construction, with 34 shield intervals, including 36 working wells with an average depth of about 52.65m. The underground diaphragm wall construction of the working shaft was carried out by jump trench construction, which involved using a hydraulic grab trench wall machine in the soil layer, rotary drilling for the rock entry portion, and double wheel milling for the trench. The cost of diaphragm wall building projects includes a significant number of machinery table hours. However, accurate and reliable estimation of machinery consumption is

*Corresponding Author: zhidingchen@163.com

DOI 10.1201/9781003348023-58

still a challenge. In engineering practice, the table time consumption of construction machinery is estimated using quotas, which do not fully cover various actual construction conditions, and there is no method in the quotas to quantitatively calculate the consumption of rotary drilling rigs in trenching construction based on actual construction conditions.

The traditional construction time consumption is estimated based on mathematical statistics, using statistical analysis, technical determination, empirical estimation, and comparative analogy. In order to meet the generality of the quota, such methods require a large amount of engineering data accumulation, and the data processing work is complicated and more difficult to update and recode. IS (Information system) project (Lee & Kim 2001). To a certain extent, it solves the problem of complicated data processing, but the measured level of quotas is still affected by the abnormal factors encountered in the past construction. In order to solve the data noise problem, IAHP gray clustering method (Zheng 2011) and fuzzy integrated evaluation method (Yuan et al. 2019) were proposed to perform cost estimation (Jumas et al. 2018), which is less difficult to update the quotas compared to the traditional method. The superiority of Neural Networks (NNs) over Multiple Regression Analysis (MRA) and Case-Based Reasoning (CBR) has been demonstrated (Sonmez & Rowings 1998). The application of the BP neural network is promising (Chen 2013). However, the BP model approximates the minimum error by the gradient descent method, which is prone to the problems of slow convergence and local optimization (Fragkakis et al. 2010).

BRF neural network has the same strong nonlinear mapping and generalization ability as BP neural network, and the excitation function is the Gaussian function. The structural parameters can be separated for learning, which can fundamentally solve the problem of local optimization. As a result, in this study, the RBF neural network is used to compute the consumption of slot milling machine table hours, and its applicability is demonstrated using the example of water consumption. The findings of this study offer a novel approach for preparing water resources quotas, which can enable building construction businesses to adjust to cost reform, conduct construction work efficiency analysis, and produce an enterprise cost database.

## 2 DESCRIPTION OF PROJECT

The PRD water resources allocation project has 36 operational wells. The underground diaphragm wall is used to construct the working well enclosure, and it is separated into three stages: slotting, cage lifting, and concrete pouring. The slotting construction of an underground diaphragm wall uses jump slots, the soil layer uses a hydraulic grab slot wall machine, and the slot is milled with two wheels into the rock rotary drilling lead hole. The specific workflow is shown in Figure 1. Before slotting each slot section, the rotary excavator is used to lead the hole, then the cut-off sand bucket is used to drill into the soil layer, and then the barrel drill is used to extend the rotary

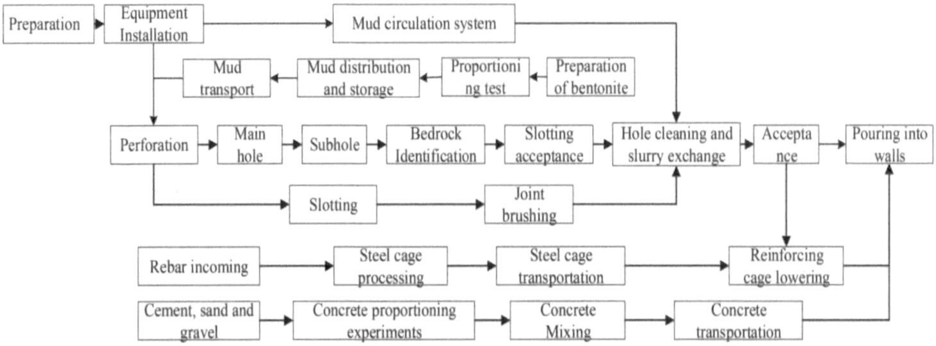


Figure 1. Underground diaphragm wall construction process.

excavator, and then the cut-off bit is used for fishing the sand, and the process is repeated until the bottom of the design wall is reached. Four holes are formed in the I sequence slot, which is located on both sides of the 1st and 2nd mill. The particular model parameters come primarily from Jintai, XCMG, and Sany Group.

## 3 METHODOLOGY

### 3.1 *Variable selection and screening*

According to *the budgetary estimate rules for Hydropower Construction Engineering (2007 Edition)*, the preliminary selected influencing factors are (1) Slot sequence (x1), distinguishing with I and II sequences according to the construction process. (2) Slot specifications: slot length (x2), slot width (x3) and slot thickness (x4). (3) Rock conditions (x5) are set according to the specific rock conditions, and the harder the rock, the more unfavorable the rotary drilling rig is to make holes and the longer the time required. (4) Workers, with the number of main workers (x6) as the main construction body and the number of auxiliary workers(x7) instead of their organization and management coordination.

After the initial qualitative analysis to find the influencing factors, the grey correlation method was used to screen the factors affecting construction efficiency quantitatively:

$$\zeta_t(i) = \frac{\min_t \min_i \left| Y(i) - X_t'(i) \right| + \eta \times \max_t \max_i \left| Y(i) - X_t'(i) \right|}{X_t'(i) + \eta \times \max_t \max_i \left| Y(i) - X_t'(i) \right|} \tag{1}$$

$$r_t = \frac{1}{n} \sum_{i=1}^{n} \zeta_t(i) \tag{2}$$

In equation (1), $Y(i)$ is the i-th sample of construction efficiency and $\eta$ is the resolution factor, which is taken as 0.5 here. It is generally considered that when $0.65 \leq r_t \leq 1$, the correlation is strong, which are selected to build the neural network model.

### 3.2 *Establishing the model*

Step 1: Create a neural network with six inputs and one output. Set the number of nodes in the hidden layer to r. The outputs of each unit in the hidden layer are

$$\mathrm{R_i} = \phi\,(\|\mathrm{X} - \mathrm{c}_i\|) \quad (i = 1, 2, \cdots, r) \tag{3}$$

$$\phi\,(v) = e^{\frac{-v^2}{2\sigma^2}} \tag{4}$$

In Eqs. (3) and (4), X is the factor affecting the construction efficiency; $c_i$, $\sigma_i$ are the unknown data, which are the center and width of the i-th unit of the hidden layer, respectively.

Step 2 Output layer.

$$y_i(X) = \sum_{i=1}^{r} w_i R_i \quad (i = 1, 2, \cdots, r) \tag{5}$$

In equation (5), $y_i(X)$ is the mechanical construction time; $w_i$ is the connection weight of the ith hidden layer to the output layer; $R_i$ is the output of each cell of the hidden layer, which is obtained from the above equation. To avoid the bias caused by the random selection of the center, the K-clustering algorithm (K-means) is used to cluster the samples to get the center.

Step 2: In actual computational engineering, the above model obtained through training is often unknown but can be infinitely approximated to obtain the predicted mechanical construction efficiency when the RBF network is to be approximated by the mapping relationship of

$$y = f(X) = w_0 + \sum_{i=1}^{r} w_i \phi(\|X - \mathrm{c}_i\|) \tag{6}$$

In Eq. (6), X is the input vector $[x_1,x_2,\cdots,x_6]$, i.e., the factors affecting the mechanical construction efficiency; y is the mechanical construction efficiency predicted by the simulation.

## 4 RESULTS

The construction process of the diaphragm wall rotary drilling rig hole formation in the Pearl River Delta Water Resources Allocation Project was recorded by the side stations, and the data is summarized as shown in Figure 2. Knife sequence is divided into I sequence, and II sequence, of which I sequence is more and II sequence is less. According to the working well design of the PRD water allocation project, the underground diaphragm wall slot specifications are $2.8\times1.2$, $5\times1$, $6.25\times1$, and $7\times1$. There are nine rock conditions in the sample data. The average number of workers is 4.6, and the main and auxiliary workers are similar in trend. The two need to cooperate during the actual construction process, which represents the construction organization level to a certain extent. As shown in Figure 3, the factors highly correlated with the construction efficiency of rotary drilling rigs are slot sequence, slot width, slot thickness, rock conditions, main workers, and auxiliary workers.

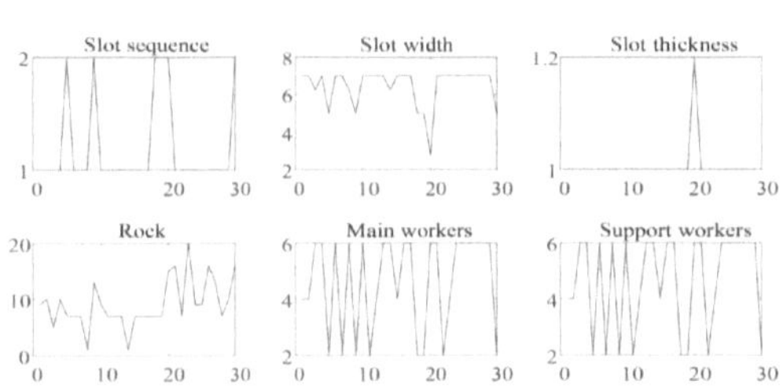


Figure 2. Analysis of original data.

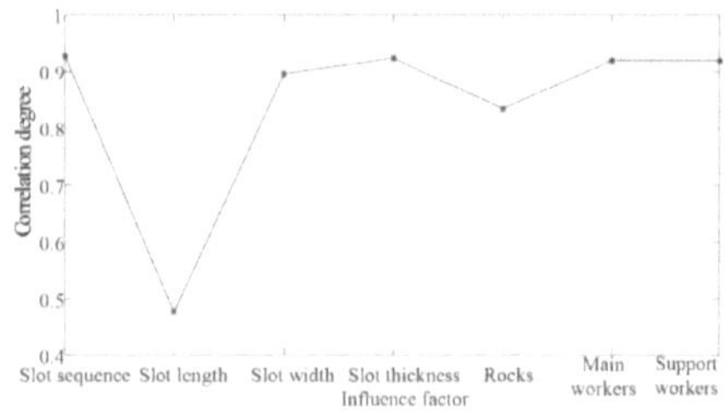


Figure 3. Correlation.

The sample data were divided into two parts, the first 30 groups of samples were used as training samples, and the last five groups were used as test samples. MATLAB was used to program the design. The target was reached after 20109 training times with an R of 0.99972, and Table 2 shows the actual and simulated efficiencies of the slot milling machine under different conditions:

Table 1. Actual efficiency and simulation efficiency.

| No. | Slot sequence | Slot width | Slot thickness | Main workers | Support workers | Actual efficiency | Simulation efficiency | Relation error |
|---|---|---|---|---|---|---|---|---|
| 31 | 1 | 7 | 1 | 6 | 6 | 0.65 | 0.63 | 3.01% |
| 32 | 2 | 5 | 1 | 2 | 2 | 0.78 | 0.78 | 0.72% |
| 33 | 1 | 7 | 1 | 6 | 6 | 0.75 | 0.73 | 2.80% |
| 34 | 1 | 7 | 1 | 6 | 6 | 0.74 | 0.75 | 0.64% |
| 35 | 1 | 7 | 1 | 6 | 6 | 0.83 | 0.90 | 8.31% |

The rock types in the table are Weathered soil; Weathered soil; Colluvial soil, weathered soil; Artificial filling soil, colluvial soil; Artificial soil fill, in that order. (1) Under other similar conditions, the rotary drilling rig construction efficiency is the lowest when the human worker fills the soil layer, and the actual efficiency is 0.830h/m. The rotary drilling rig construction efficiency is the highest when the soil is fully weathered, and the average actual efficiency is 0.715h/m. And the simulation efficiency and actual efficiency rules are roughly the same. (2) The relative errors between the actual efficiency and the simulated efficiency are small, 3.01%, 0.72%, 2.80%, and

0.64% for fully weathered soil, colluvial soil, fully weathered soil, artificial fill, and colluvial soil. (3) The relative error between the actual and simulated efficiency is large when artificially filling the soil layer, and the relative error is 8.31%.

## 5 DISCUSSION

### 5.1 *Influencing factors*

The large error between the actual efficiency and the Founder's efficiency when artificial filling may be because the training samples have fewer samples of artificial fill layers (only one group) or that the geotechnical properties of artificial fill layers are unstable and have high uncertainty. However, the training data analysis shows that the time required for the rotary drilling rig to feed 1m does not all increase with the enhancement of rock strength under other consistent conditions, which indicates that the rock condition as a natural environment cannot be changed. Still, the construction unit can improve its efficiency by improving other technical and management conditions. The efficiency decreases with the increase of slot specification under the same slot sequence. And the time required for the rotary drilling rig to reach 1m decreases and then increases as the number of workers increases. For a specific geological condition, as the number of workers increases, the time required for the rotary drilling rig to reach 1m generally follows the law of first decreasing and then increasing. This demonstrates that construction efficiency cannot be increased by increasing the number of workers but by improving the coordination of multiple parties.

### 5.2 *Sample size*

As shown in Figure 4, the samples of 29,27, 25, 22, and 20 groups were randomly selected, the save influencing factors were still taken, and training simulations were performed with the same parameters. (1) The mean error of the results decreases with the increase in the number of samples from 20 to 29 groups. (2) The training network is unstable when the number of samples is 20 and 22 groups, the average relative error increases rapidly, and the simulation results for new samples are also scattered, with the largest relative errors as high as 32.66% and 32.60%, which are 5-10 times the minimum relative errors. (3) The average relative error obtained from training is less than 10% when the number of samples is 25 groups, but there are still some training with relative errors greater than or equal to 10%. (4) While the average mean error and the known mean error of the network training results are less than 10% when the number of samples is 27 and 29 groups, which are more stable. (5) The overall average relative error decreases with the increase of sample size and stabilizes at about 7% by 27 sets of samples.

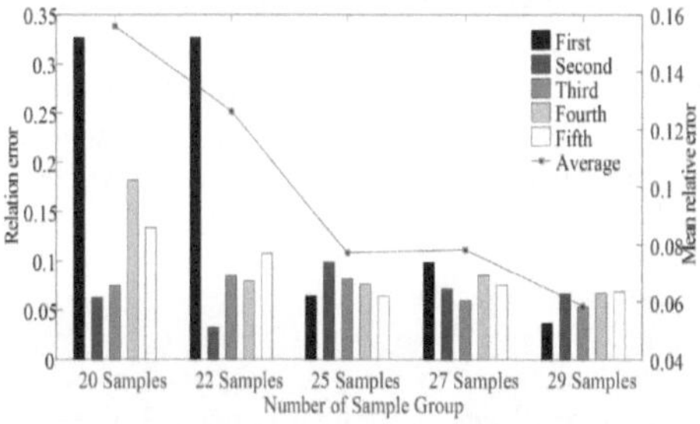

Figure 4. Sample size and relative error.

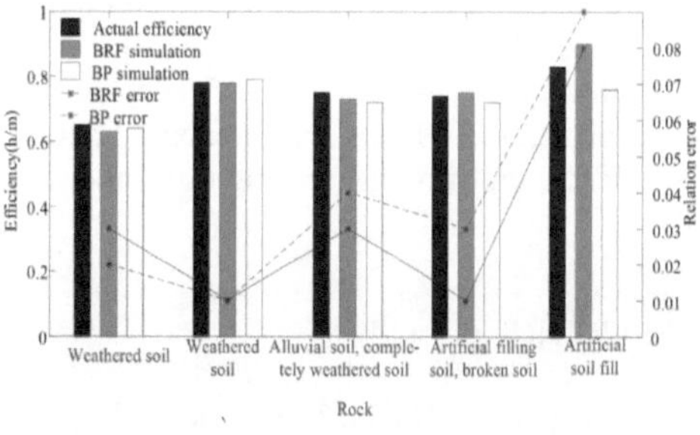

Figure 5. Comparison of simulation results.

### 5.3 *Compared with BP neural network*

Figure 5 presents the distinction between the result of the BP neural network and the BRF neural network, both of which can predict the efficiency of the slot milling machine.

(1) The results of BP neural network are 0.64h/m, 0.79h/m, 0.72h/m, 0.72h/m, 0.76h/m, the average relative error is 4%, the calculation results are still in line with the above-mentioned artificial fill layer. (2) The relative errors in the BP neural network calculation results are 3.60% for colluvial soil and weathered soil, 2.50% for artificial fill and colluvial soil, and 8.82% for artificial fill, all of which are greater than the BRF neural network model calculation results. (3) The BRF neural network model results are closer to the actual value in all cases except in the fully weathered soil1. Averaging the two cases of weakly weathered muddy sandy gravels, the results of the BRF neural network model are 0.709h/m, and those of the BP neural network model are 0.712h/m. The results of the BP model are closer to the actual value of 0.716h/m. The possible reason is that more samples of fully weathered soils are in the data, and the BP model has better fault tolerance for such cases. (4) Although the BP neural network model shows a strong fault tolerance in very individual data, the overall accuracy of the BRF neural network model pair is higher. Consistency of style is essential. Note the spacing, punctuation, and caps in all the examples below.

## 6 CONCLUSION

The present research set out to integrate the fragmented theory on construction quota into a model that captures six essential factors. The efficiency of the rotary drilling rig was simulated using neural networks, which were validated in terms of both relative error and sample size.

(1) The efficiency of rotary drilling rigs in the PRD water allocation project varies slightly overall, depending on the sequence of slots, slot specifications, rock conditions, employees, and other factors. Although the natural environment of the rocks cannot be modified, the building unit can increase efficiency by collaborating with different parties to determine the most efficient resource allocation technique.

(2) The efficiency of a rotary drilling rig can be measured using a BRF neural network with a small amount of data in a stable manner. The construction company does not need to record the entire construction process in the actual measurement process, but only selects some of the slots for measurement. Even if some of the data cannot be measured during the building time of some of the slots, other slots can be used to complement the measurement and fulfill the goal of ergonomics analysis of the construction machines.

(3) The overall stability of BRF neural network results is better than the results of the BP neural network model. In construction cost estimation, the superiority of BRF neural network compared with neural network (NN) has been proven. This paper nearly proves the applicability and overall stability of BRF neural network in the construction field, which provides theoretical support for construction and building enterprises to conduct ergonomics analysis and establish engineering cost database.

(4) In further studies, the model's adaptability is expected to be tested under different engineering conditions. While six influencing factors have been selected for generating the construction efficiency of rotary drilling rigs, other parameters can also be explored for different projects.

## REFERENCES


Chen Ming. (2013) *MATLAB neural network principle and example refinement.* Beijing: Tsinghua University Press.

Dwifitra Jumas, Faizul Azli Mohd-Rahim, Nurshuhada Zainon & Wayudi P. Utama. (2018). Improving the accuracy of conceptual cost estimation using MRA and ANFIS in Indonesian building projects. *Built Environment Project and Asset Management* (4).

Fragkakis Nikolaos, Lambropoulos Sergios & Pantouvakis John Paris. (2010).A cost estimate method for bridge superstructures using regression analysis and bootstrap. *Organization, Technology & Management in Construction: An International Journal*;(2).

Jin Woo Lee & Soung Hie Kim. (2001). An integrated approach for interdependent information system project selection. *International Journal of Project Management*(2).
National Cost Engineer Examination Textbook Editorial Committee. (2006) *Engineering Costing and Control*. Beijing: China Planning Press, 56–61.
Rifat Sonmez & James E. Rowings. (1998). Construction Labor Productivity Modeling with Neural Networks. *Journal of Construction Engineering and Management*(6).
The General Office of the Ministry of Housing and Urban-Rural Development of the People's Republic of China.(2020) Notice of the General Office of the Ministry of Housing and Urban-Rural Development on Printing and Distributing the Work Plan for Project Cost Reform. *Jianbanbiao* [2020] No.38.
Yuan Zhijie, Li hua, Han Jie... & Jin Ruiwen. (2019). Application of Fuzzy analytical hierarchy method and BIM in quota design of EPC project. *IOP Conference Series: Earth and Environmental Science* (1).
Zheng Xiuzhen. (2011) *Research on the application of the IAHP grey clustering method in the formulation of highway engineering construction quotas*. Changsha: Changsha University of Science and Technology.

*Advances in Urban Construction and Management Engineering – Cha (Ed)*
 *ISBN 978-1-032-39018-5*

# Similarity matching of railway business lines construction safety supervision clauses based on Word2vec model

Ya-feng Mei*
*Beijing Jiaotong University, Beijing, China*

ABSTRACT: To ensure safety during the construction of railway business lines, relevant departments have promulgated a series of relevant laws, regulations, rules, and standards. After the government-enterprise reform, the newly established Chinese railway company urgently needs to establish a supervision system that adapts to the development of the industry in the construction safety management of railway business lines. Currently, relying on manual identification of terms is inefficient, and regulatory requirements are constantly evolving. Therefore, we use the deep learning model Word2vec to achieve intelligent matching of similar terms in different texts, thereby supporting the establishment of a safety supervision model system for railway business line construction. After text data collection, preprocessing and model parameter adjustment, the similarity matching ability of the trained model in the test set reaches about 70%, which partially achieves our goal and promotes intelligent research on railway engineering safety.

## 1 INTRODUCTION

The key to the problem of similar text matching is the vectorized representation of words and sentences. Compared with traditional statistical models, deep learning neural network models can better reflect the overall semantic information of short texts. It is widely used in the field of NLP. Specifically, Bengio et al. (2000) built a classic neural network training language model earlier. Based on the language model N-Gram, a three-layer neural network model was constructed to train to obtain word vectors that are relatively generalized and can be used to calculate similarity, but there is a problem that the computational complexity is too large and is limited by the number of N. By learning from the former, Mikolov's team proposed the Word2vec model based on two prediction architectures: CBOW and Skip-Gram, high-quality word vector representations can be learned from large-scale unstructured datasets in a relatively short time (Le & Mikolov 2014; Mikolov et al. 2013; Rong 2014). Then, on the premise of massive data and massive computation, the Transformer framework model (Vaswani et al. 2017) and the Bert (Devlin et al. 2018) gradually emerged. The application of deep learning methods in the problem of similar retrieval of legal texts is being explored, and scholars continue to improve the effect and performance of the model (Viji & Revathy 2022). More and more combinatorial methods are used to compute text semantic similarities, such as the hybrid method based on weighted fine-tuned BERT feature extraction of the Siamese Bi-LSTM model (Mandal et al. 2021) and the Siamese network based on the BERT model (Wang et al. 2021) and so on.

Based on the document *Measures for the Safety Management of Railway Business Lines Construction*, this research establishes a similarity search based on the current safety supervision requirements of railway business lines construction and solves the existing clauses match problem by combining deep learning technology (Li et al. 2017) and similarity calculation method (Chen & Jiang 2017).

*Corresponding Author: m19800370975@163.com

 DOI 10.1201/9781003348023-59

## 2 FRAMEWORK AND MODEL

### 2.1 *Framework*

According to the matching requirements of the construction safety supervision clauses of business lines, relevant texts are collected, sorted out, preprocessed, vectorized, and matched by similarity calculation. Finally, matching targets are retrieved. The logical framework of the automatic similarity matching system is shown in Figure 1.

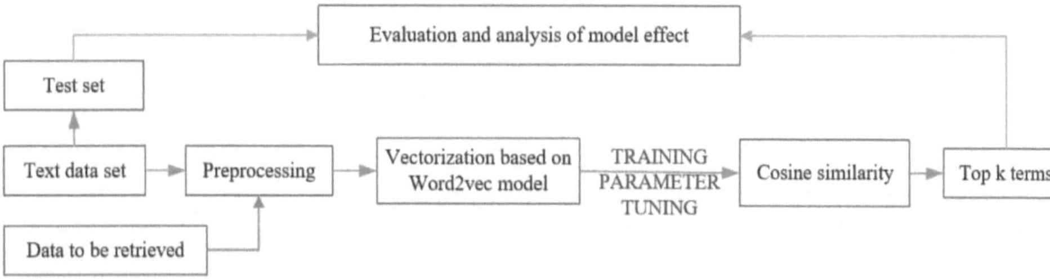


Figure 1. Process framework.

### 2.2 *Model*

In NLP tasks, Word2vec is a fast and effective method for obtaining distributed word vectors in deep learning. It is based on the Huffman tree and mainly adopts CBOW and Skip-Gram models. The goal of CBOW is to predict the current word according to the context. The goal of Skip-Gram is to predict the probability of the context based on the current word (Li et al. 2017). Compared with the statistical model, it can not only reduce the vector dimension but also extract the semantic information of the word. The structure is shown in Figure 2.

Because the accuracy of Skip-Gram is higher than that of CBOW, we choose the Skip-Gram model to train the task. While training the Skip-Gram model, firstly, input a one-hot vector $X_k(1 * V)$ of a central word into the input layer, and initialize the weight vector $W(V * N)$ by setting the dimension N of the output word vectors; Then calculate the hidden layer value $h_i(1 * N)$ by weighting, and initialize the weight vector $W'(N * V)$; weight C times, and finally obtain the output layer value through softmax, that is, $C$ vectors in $(1 * V)$ dimensions, representing $C$ words, establish a

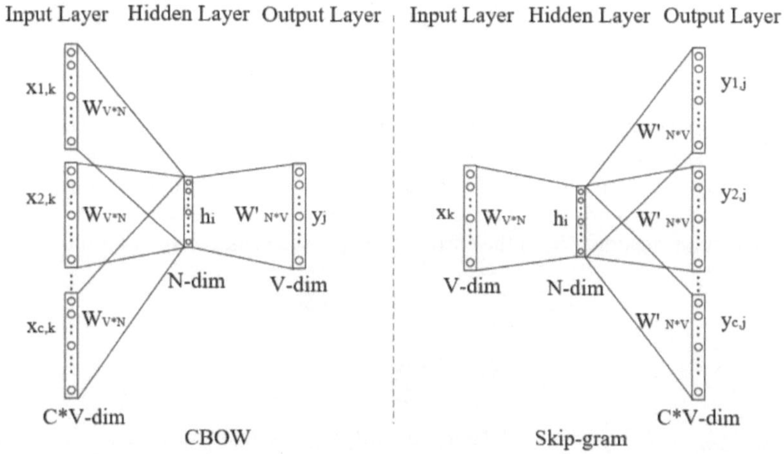


Figure 2. Word2vec model: CBOW (left) and Skip-Gram (right).

loss function based on the principle of maximizing the probability of the actual central word and obtaining the required word vector $W$ after the neural network is trained to convergence. The mathematical expression of the model is as Formula (1):

$$P(W_{t-k}, W_{t-k+1}, \ldots, W_{t+k-1}, W_{t+k} | W_t) \tag{1}$$

where $W_t$ is a word in the corpus; $K$ is the window size.

## 3 EXPERIMENTS

### 3.1 *Data sets and environments*

The text data used in this paper are the relevant rules for the construction safety supervision of railway business lines, collected and sorted out by the research group, as shown in Table 1. The experimental environment is also shown in Table 1.

Table 1. Introduction to the dataset and experimental environment.

| Data | Laws/regulations/regulations/standards |
|---|---|
| Text data subset (examples) | *Measures for the Construction Safety Management of Railway Business Lines*<br>*Work Safety Law of the People's Republic of China*<br>*Railway Safety Management Regulations*<br>*Construction Engineering Safety Production Safety Management Regulations*<br>... |
| Number of pairs of the test set (clauses) | 174 pairs |
| Operating system | Windows 10 |
| CPU | Intel (R) Core (TM) i5-7200U |
| Software | Python 3.9.0 |

### 3.2 *Evaluation index*

The purpose of this research is to train a model that can automatically match the correlation of the safety supervision clauses in the construction of railway business lines. The threshold of the effective similarity is 0.8, as shown in Formula (2), that is, in the training results, those with a similarity greater than 0.8 are valid clause matching pairs. Then the most important evaluation index is the matching accuracy $P_a$ in the corresponding test set, as shown in Formula (3).

$$S_{effective-threshold} > 0.80 \tag{2}$$

$$P_a = \frac{n_{m-train}}{n_{test}} * 100\% \tag{3}$$

where $S_{effective-threshold}$ is the critical effective similarity, $n_{test}$ is the number of similar clauses pairs in the test set, and $n_{m-train}$ is the number of similar clause pairs corresponding to the test set in the model training result within the critical effective similarity.

## 3.3 *Parameters*

Use the Gensim module in Python 3.9.0 to train the Word2vec model. The description of the main parameters and the values after tuning are shown in Table 2.

Table 2. Main parameters list.

| Main parameters | Description | Values |
|---|---|---|
| Sentences | Corpus to be analyzed | Sentences after preprocessing |
| Vector_size | Dimensions of the word vectors to be generated | 100 |
| Window | The maximum distance of the word vector context | 3 |
| Sg | 0: CBOW (default) or 1: Skip-Gram | 1 |
| Hs | 0: Negative Sampling (default) or 1: Hierarchical Softmax | 0 |
| Negative | Number of negative samples (default: 5) | 5 |
| Min_count | The minimum frequency of words that need to calculate the word vector | 5 |
| Max_vocab_size | Sets the RAM limit during word vector construction | 2,000 |
| Epochs | Maximum iterations in stochastic gradient descent | 500 |
| Alpha | Initial step size for iteration in stochastic gradient descent | 0.1 |
| Min_alpha | Minimum iteration step | 0.05 |
| Workers | Parallel number of the training | 1 |

## 3.4 *Results*

Write the text data to Excel, and the screenshots of part of the data are shown in Figure 3. The left side (sheet: output 1) is the pair of clauses with a similarity greater than 0.8 filtered out of the model output, and the right side (sheet: test set) is the test set of the clauses.

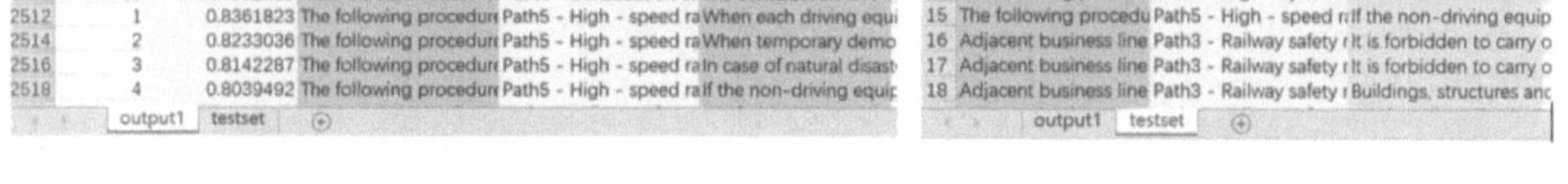

| | A | B | C | D | E |
|---|---|---|---|---|---|
| 1 | Order | Similarity | Path1-Regulation | Other-path | Similar-Regulation |
| 1846 | 1 | 0.84328103 | After the construction sc | Path4 - Construction er | Units undertaking const |
| 1848 | 2 | 0.83517605 | After the construction sc | Path4 - Construction er | The construction units s |
| 1850 | 3 | 0.834588 | After the construction sc | Path4 - Construction er | The principal person-in |
| 1852 | 4 | 0.82891196 | After the construction sc | Path4 - Construction er | Before the construction |
| 1854 | 5 | 0.8255449 | After the construction sc | Path4 - Construction er | The construction unit sh |
| 1856 | 6 | 0.816836 | After the construction sc | Path4 - Construction er | Where a construction p |
| 1858 | 7 | 0.8123592 | After the construction sc | Path4 - Construction er | Construction units shall |
| 1860 | 8 | 0.81152916 | After the construction sc | Path4 - Construction er | The departments of the |
| 1862 | 9 | 0.8110225 | After the construction sc | Path3 - Railway safety n | On the railway line prot |
| 1864 | 10 | 0.8088659 | After the construction sc | Path4 - Construction er | The competent adminis |
| 1866 | 11 | 0.80638325 | After the construction sc | Path5 - High - speed ra | These measures are for |
| 1868 | 12 | 0.80347973 | After the construction sc | Path2 - safe production | Where a production an |
| 1870 | 13 | 0.80222905 | After the construction sc | Path4 - Construction er | The construction unit th |
| 2512 | 1 | 0.8361823 | The following procedur | Path5 - High - speed ra | When each driving equi |
| 2514 | 2 | 0.8233036 | The following procedur | Path5 - High - speed ra | When temporary demo |
| 2516 | 3 | 0.8142287 | The following procedur | Path5 - High - speed ra | In case of natural disast |
| 2518 | 4 | 0.8039492 | The following procedur | Path5 - High - speed ra | If the non-driving equip |

output1 testset

| | A | B | C |
|---|---|---|---|
| 1 | Path1-Regulation | Other-path | Similar-Regulation |
| 2 | After the construction | Path4 - Construction e | Units undertaking const |
| 3 | After the construction | Path4 - Construction e | The construction units sl |
| 4 | After the construction | Path4 - Construction e | The principal person-in- |
| 5 | After the construction | Path4 - Construction e | Before the construction |
| 6 | After the construction | Path4 - Construction e | The construction unit sh |
| 7 | After the construction | Path4 - Construction e | Where a construction pr |
| 8 | After the construction | Path4 - Construction e | Construction units shall |
| 9 | After the construction | Path4 - Construction e | The departments of the |
| 10 | After the construction | Path3 - Railway safety r | On the railway line prot |
| 11 | After the construction | Path4 - Construction e | The competent adminis |
| 12 | The following procedu | Path5 - High - speed r | When each driving equi |
| 13 | The following procedu | Path5 - High - speed r | When temporary demol |
| 14 | The following procedu | Path5 - High - speed r | In case of natural disaste |
| 15 | The following procedu | Path5 - High - speed r | If the non-driving equip |
| 16 | Adjacent business line | Path3 - Railway safety r | It is forbidden to carry o |
| 17 | Adjacent business line | Path3 - Railway safety r | It is forbidden to carry o |
| 18 | Adjacent business line | Path3 - Railway safety r | Buildings, structures anc |

output1 testset

Figure 3. The screenshots of the results.

From the results in Figure 3, "Path1 - regulation" represents the clauses involved in the benchmark text file, "Other-path" represents the names of other corresponding text files, and Similar-Regulation represents the clauses similar to the benchmark. The effect of the Word2vec model is good (left), and the highest similarity reaches 0.87. In the results, the pairs of clauses with a similarity greater than 0.80 have been screened out, but not all of them match the test set (right) and need to be compared with the collated test set. After that, according to Formulas (2) and (3), we can calculate the value of $P_a$ is about 70%. The model achieves a similar matching goal to a certain extent, and preliminarily realizes the retrieval and matching application of the deep neural network method in the construction safety text of railway business lines.

## 4 CONCLUSIONS

This paper researches the application of neural network methods in the railway construction industry, specifically, it explores the use of NLP models to solve the similar or related matching problems of railway business line construction safety supervision clauses, and achieves intelligent similarity search matching to a certain extent, and works out the difficult and high-cost problems of manual sorting in the industry before. By combining the Word2vec model and the cosine similarity algorithm, after parameter training and adjustment, we obtain a model, whose accuracy index of the training results (in the data pairs whose similarity index is higher than 80%) reaches 70%, corresponding to the test set.

This can contribute to further intelligent matching and other research. In the following research, we will further try to apply better methods in this field and strive to come up with a method with a higher level of intelligent matching.

## REFERENCES


Bengio, Y. et al (2000). *A Neural Probabilistic Language Model. C. Advances in Neural Information Processing Systems 13*, Papers from Neural Information Processing Systems (NIPS).

Chen, E. J. & E. B. Jiang (2017). A review of text similarity calculation methods. *J. Data analysis and knowledge discovery*, 1(06), 1–11.

Devlin, J. et al (2018). BERT: Pre-training of Deep Bidirectional Transformers for Language Understanding. J.

Le, Q. V. & T. Mikolov (2014). Distributed Representations of Sentences and Documents. C. *International Conference on International Conference on Machine Learning-volume.*

Li, X. et al (2017). Research on semantic similarity calculation of sentences based on Word2vec. *J. Computer science*, 44(09), 256–260.

Mandal, A. et al (2021). Unsupervised approaches for measuring textual similarity between legal court case reports. *J. Artificial Intelligence and Law.*

Mikolov, T. et al (2013). Efficient Estimation of Word Representations in Vector Space. *J. Computer Science.*

Rong, X (2014). word2vec Parameter Learning Explained. *J. Computer Science.*

Vaswani, A. et al (2017). Attention Is All You Need. J.

Viji, D. & S. Revathy (2022). A hybrid approach of Weighted Fine-Tuned BERT extraction with deep Siamese Bi-LSTM model for semantic text similarity identification. *J. Multimedia Tools and Applications.*

Wang, K. et al (2021). *Comparison between Calculation Methods for Semantic Text Similarity based on Siamese Networks*. J.

*Advances in Urban Construction and Management Engineering – Cha (Ed)*


# Construction of multi-task corpus for safety production

Yi Zhou
*Beijing Academy of Emergency Management Science and Technology, Beijing, China*

Yanjun Guo*
*Beijing University of Artificial Intelligence, Beijing, China*

Nan Mei & Peng Zhang
*Beijing Academy of Emergency Management Science and Technology, Beijing, China*

Mingyuan Pan
*Beijing University of Posts and Telecommunications, Beijing, China*

Yifan Shi & Bingbing Liu
*Beijing Academy of Emergency Management Science and Technology, Beijing, China*

Xinbo Ai
*Beijing University of Artificial Intelligence, Beijing, China*

ABSTRACT: Under the background of the artificial intelligence era, the construction of corpus in the field of safety production is in a blank stage. This paper proposes a labeling system that adapts to the unstructured Chinese safety hazard inspection record. We use semi-automatic labeling technology to label six elements of information including structural knowledge information, hazard sources, hazard levels, accident types, hazard countermeasures, and character-level hazard description labels. Finally, we have formed the first safety production corpus, which is suitable for the analysis of various tasks in the field of natural language processing. We experimentally verify the high quality of the corpus.

## 1 INTRODUCTION

With the improvement of the information level of safety production, various industries have established information management systems such as hazard inspection systems, accident inspection systems, and major hazard source management systems. As these information management systems are widely promoted and used, a large amount of recorded data relating to safety production hazards has been accumulated (Guo 2019). Due to the lack of a safety production corpus, the records in the information management system are inconsistent and non-standard, making it difficult to carry out statistical analysis. The construction of the production safety corpus will also help the people to fully grasp the hazard inspections knowledge such as "what hazard to check, how to check for hazard, and how to deal with hazard."

The construction of corpus is an important part of natural language processing (Stubbs 2015). At present, there are three approaches to building a corpus of specialized domains: manual construction (Dolan 2005), semi-automatic construction (Li 2008; Luo 2012; Sheng 2019), and automatic construction (Vincze 2008; Wang 2016; Zhou 2017). There is a lot of unstructured and semi-structured text information in the field of safety production. Zhang (2020) used the crowdsourced

*Corresponding Author: gyj811@bupt.edu.cn

DOI 10.1201/9781003348023-60

database OSM (OpenStreetMap) and the collected geographic entities as basic place names to construct a dataset of four main entity types in the field of coal mine safety. Ai (2021) designed a sequence labeling method suitable for safety hazard description texts and extracted 7,627 valid triples of knowledge information semi-automatically. Although the above corpus construction methods can achieve good performance, they have the following problems.

(1) In the field of safety production, the cost of manually constructed corpora is high and the construction period is long. And this method is difficult to meet the timeliness requirements of corpus update. However, the automatic corpus labeling process lacks quantitative evaluation and is prone to generate a lot of noise, which affects the quality of the safety production corpus.
(2) Due to the distinctive language characteristics of safety production texts, the labeling system and method in the open field cannot be adapted to the construction of corpora in the field of safety production.

The object of this study is unstructured hazard description texts. We will not only annotate the hazard sources, risk levels, types of accidents, the possibility of causing accidents, and hazard countermeasures but also annotate the structured knowledge information in the text. Finally, we form a labeling system and corpus in the field of safety production.

## 2 THE LABELING SYSTEM OF SAFETY PRODUCTION CORPUS

Drawing on text classification task, sequence labeling task, and named entity recognition task in the field of natural language processing, we formulate a labeling system for safety production corpora. Figure 1 is a schematic diagram of our labeled features.

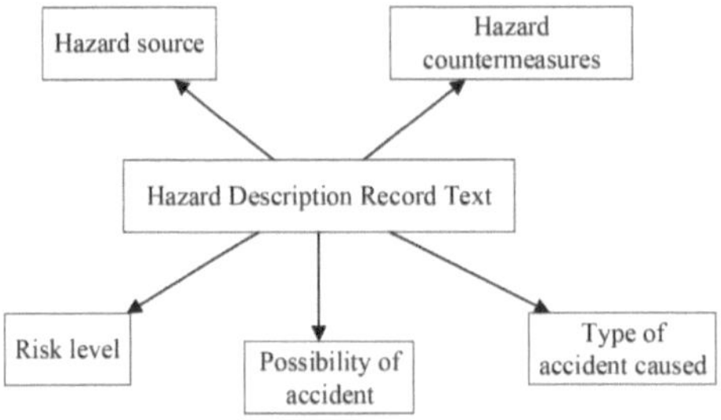


Figure 1. Labeling system elements.

**Hazard sources.** Referring to the named entity recognition task, we annotate the hazard source of the safety hazard description text. The entity types of hazard sources are mainly four categories: equipment, workers, working environment, and regulations. As shown in Table 1, in the safety hazard record text of "water entering the distribution box may cause a short circuit fire accident", the "distribution box" is labeled as a hazard source by the recording personnel.

**Hazard descriptions.** Referring to the sequence labeling task of natural language processing, we perform character-level labeling on the hazard description text (Ai 2021). We annotate the unsafe state of things in the data, the unsafe behavior of people, the unsafe factors of the environment, and the defects of the management system. The goal is to be able to use the corpus to analyze and mine equipment, workers, work environments, regulations, and their relationships with hazards.

**Risk levels.** Drawing on the text classification task, we quantify the risk level corresponding to the hazard description text. The labeling basis of the risk level is based on the Urban Safety Risk Assessment Pilot Work Manual. The risk level is assessed by the risk matrix method (Arimbi 2019). The risk level from high to low is divided into four colors: red, orange, yellow, and blue. The higher the risk, the more difficult it is to rectify the hazard.

Table 1. Annotation example of a corpus.

| Safety hazard description texts | Hazard source | Risk level | Possibility of accident | Types of accidents | Hazard countermeasures |
|---|---|---|---|---|---|
| Water entering the distribution box may cause a short circuit fire accident. | Distribution box | Blue | Level 1 | Electric shock; fire | The staff checks and records every day to find problems in time and reports them for solutions. |
| The boiler outlet valve is blocked, and the scale is not removed in time. | Boiler | Red | Level 5 | Fire; boiler explosion | Enterprises should set up alarms and protection devices on boilers, and regularly maintain and repair boilers. |
| Inflammable and explosive chemicals in the metering workplace encounter static sparks, electrical sparks, open flames, etc. | Metering workplace | Blue | Level 1 | Fire; explosion | Take safety precautions when measuring. |

**Types of accidents caused.** Incident types are defined as production incidents that can result from hazards. Drawing on the text classification task, the Enterprise Employee Casualty Accident Classification Standard is divided into 20 types according to the cause of the accident.

**Possibility of accidents.** Accident probability is defined as how likely a hazard is to cause an accident to occur. Considering the text classification task, we classify the probability of security incidents into 5 levels. The higher the level, the greater the probability of escalating from a hazard to an accident.

**Hazard countermeasures.** Hazard responses are defined as the response and corrective actions that a company should take when that hazard occurs. The content is labeled by professional safety technicians according to the situation on-site, labeling the emergency preparation and management measures that the enterprise should take.

## 3 CONSTRUCTION METHODS OF SAFETY PRODUCTION CORPUS

In the process of labeling the safety production corpus, we hand over five labeling tasks including hazard sources, risk levels, types of accidents caused, the possibility of an accident, and hazard countermeasures to professional safety technicians.

For the character-level labeling task of hazard description record texts, the labeling work will be divided into four stages. In the first stage, based on relevant laws and regulations, we constructed the first draft of the labeling specification. An expert in the field of safety production was invited to participate in the formulation and improvement of the specification. We adapted the open-source annotation tool (Yang 2018). In the second stage, we hired two graduate students in the safety field to participate in the task of the character-level labeling task. Other labeling tasks are calibrated by professional safety technicians. We conduct experiments on annotation consistency in Section 4.1 to ensure the quality of the annotation work. In the third stage, we select 100 texts from 1,500 hazard description texts for consistency evaluation.

To reduce the cost of manual annotation and improve the accuracy of the corpus, we use the corpus of 1,500 formal annotations when performing the task of character-level annotation, and use the automatic annotation method in the fourth stage. As shown in Figure 3, we adopt the related algorithm of sequence labeling.

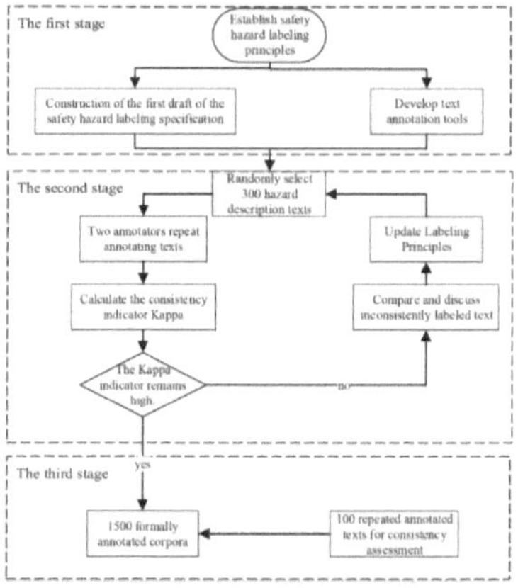


Figure 2. Multitasking safety production corpus labeling process.

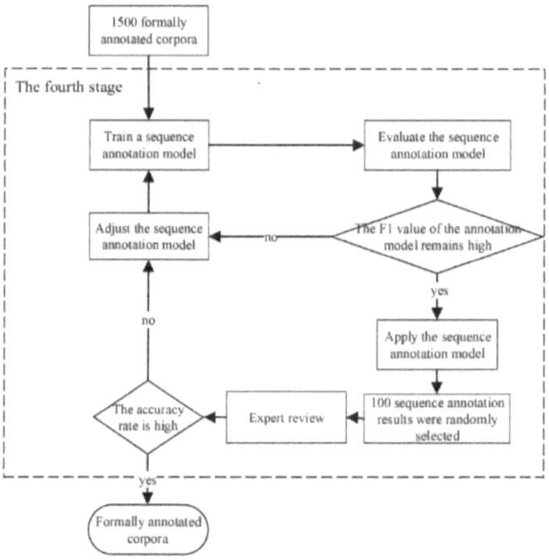


Figure 3. Character-level labeling process for safety production.

## 4 SAFETY PRODUCTION CORPUS ANALYSIS AND STATISTICS

### 4.1 *Annotation consistency check*

Although the labelers are professional safety production personnel and have a corresponding labeling system, due to personal subjectivity, there will be inconsistent labeling of the same text. Because some tasks in this corpus are labeled grade data, we use linear weighted kappa to test the consistency of labeling. The labeled data regarding the probability of an accident, for example, show "class = 1" as possible, while "class = 5" as very likely. Assuming that the hazard is likely to cause production accidents, the impact of predicting "class=1" as "class = 5" should be greater than that of predicting "class = 1" as "class = 2". The linear weighted Kappa calculation formula is

$$\text{Linear weighted kappa} = \frac{P_o - P_e}{1 - P_e} = \frac{\sum_{i=1}^{m}\sum_{j=1}^{m} w_{ij}p_{ij} - \sum_{i=1}^{m}\sum_{j=1}^{m} w_{ij}p_{i+}p_{+j}}{1 - \sum_{i=1}^{m}\sum_{j=1}^{m} w_{ij}p_{i+}p_{+j}} \quad (1)$$

where $P_o$ represents the observed agreement rate; $P_e$ represents the accidental agreement rate; $m$ represents the number of categories; $w_{ij}$ represents the weight of the true prediction of $i$ as $j$. We calculate the weights as follows. Let the score order be $c_1 < c_2 < \cdots < c_{m-1}$, the value is $c_i$, then we can obtain the calculation for the weight, as shown in Formula (2).

$$w_{ij} = 1 - \frac{|i-j|}{m-1} \tag{2}$$

It is generally considered that the weighted Kappa value greater than 0.8 indicates strong consistency, and less than 0.2 indicates poor consistency (Zheng 2006). We selected and calculated two annotators A and B, independently annotated the same 300 hazard description texts, and then conducted a consistency test. Through calculation, the linearly weighted Kappa values of the two labeling tasks of accident probability and risk level are 0.85 and 0.81, respectively. This shows that the annotation results for these two tasks are credible. For the character-level labeling task of hazard description texts, we adopt the same experimental approach as Ai (2021). We selected the best-performing ERNIE-CRF model for text character-level annotation. Other tasks are labeled by professional safety technicians according to relevant technical specifications, which ensures the accuracy of the labeling results.

## 4.2 *Safety production corpus statistics*

We finally labeled 30,417 hazard description texts. Then, we counted the frequency of occurrence of the same hazard in the safety production corpus and obtained 15 high-frequency hazard equipment and places respectively. As shown in Figure 4, electrical equipment, fuel distributor, flammable material, etc. belong to high-hazard equipment. As shown in Figure 5, the refueling area, unloading area, workshop, and storage tank area belong to high-hazard sites. Enterprises and governments

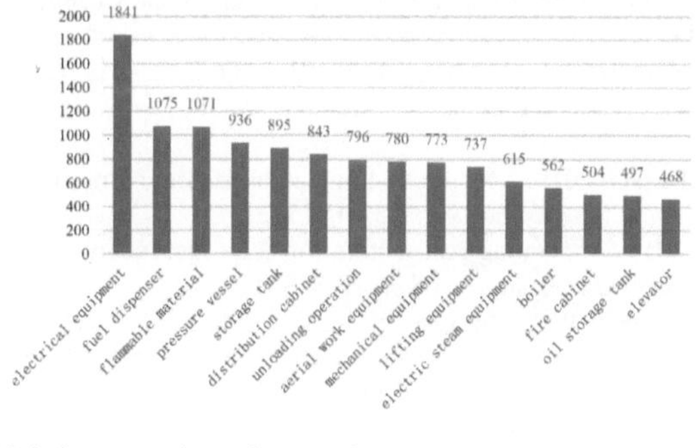


Figure 4. Statistics of high-frequency hazardous equipment.

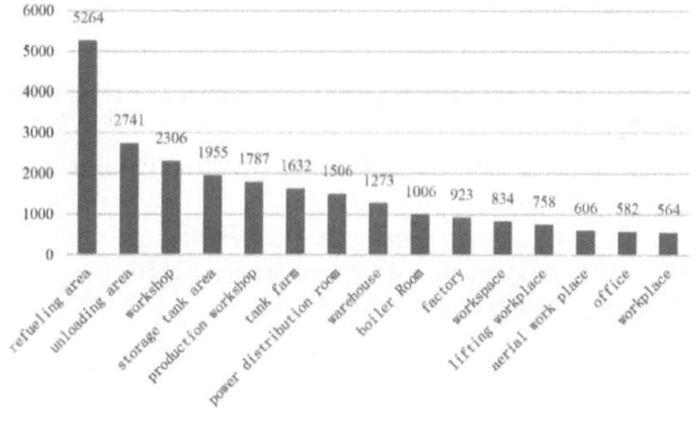


Figure 5. Statistics of high-frequency hazardous sites.

should pay more attention to hazardous sites and equipment with high frequency when conducting hazard self-inspection work. Therefore, this corpus will be helpful for hazard inspection work in the field.

## 5 CONCLUSIONS

According to the data characteristics of the security government system, we propose a labeling system for Chinese security hazard description texts. Then, we propose the first character-level annotation specification suitable for safety production. Finally, we semi-automatically construct a multi-task annotation corpus in the field of safety production through four stages. Through linearly weighted kappa computation, we verify the validity of the corpus. The corpus can effectively help the government and enterprises to conduct on-site refined hazard inspections. In the future, we will construct a knowledge graph in the field of safety products for a variety of industries to help people quickly grasp the relevant knowledge of safety production.

## REFERENCES


Ai, X.B. & Guo, Y.J. & Xie, Y.H. & Chen, C. 2021. Structural Analysis of Hidden Danger Description Text Based on ERNIE-CRF-ESL. *Journal of Beijing University of Posts Telecommunications*, 44, 107–113.

Arimbi, H.B. & Puspasari, M.A. & Syaifullah, D.H. 2019. *Hazard identification, risk assessment and risk control in a woodworking company*. 1ST INTERNATIONAL CONFERENCE ON INDUSTRIAL AND MANUFACTURING ENGINEERING505.

Dolan, B. & Brockett, C. 2005. *Automatically constructing a corpus of sentential paraphrases*. Proceedings of the 3rd International Workshop on Paraphrasing (IWP2005).

Guo, W. 2016. Research on intelligent search engine system against potential work safety hazard. *China Journal of Safety Science and Technology*, 12(11):171–175.

Li, P. & Zhu, Q. & Qian P. 2008. Construction Approach of Large-scale Corpus Based on Web. *Computer Engineering*, (07):41–43.

Luo S. & Liu Y. & Feng Y. & Han, L. & Chen, G. & Wang, Q. 2012. Method of Building BFS-CTC: a Chinese Tagged Corpus of Sentential Semantic Structure. *Journal of Beijing University of Technology*, 32(03):311–315.

Sheng, C. & Kong, F. & Zhou, G. 2019. Construction of a zero-element corpus of Chinese chapters. *Journal of Peking University* (Natural Science Edition), 55(01):15–21.

Stubbs, A. & Uzuner, O. 2015. Annotating risk factors for heart disease in clinical narratives for diabetic patients. *Journal of Biomedical Informatics*, 58(S): S78–S91.

Vincze, V. & Szarvas, G. & Farkas, R. & Móra, G. & Csirik, J. 2008. The BioScope corpus: biomedical texts annotated for uncertainty, negation and their scopes. *BMC bioinformatics*, 9(11), 1–9.

Wang, K. & Xia, R. 2016. A Survey on Automatical Construction Methods of Sentiment Lexicons. *ACTA AUTOMATICA SINICA*, 42(04):495–511.

Yang, J. & Zhang, Y. & Li, L. & Li, X. 2017. YEDDA: A lightweight collaborative text span annotation tool. *arXiv preprint arXiv*:1711.03759.

Zhang, X. & Feng, S. & Ding, E. 2020. Entity recognition and relation extraction model for coal mine. *Journal of Computer Applications*, 40(08):2182–2188.

Zheng. S. & Deng. Y. & Jiang. T. 2006. Kappa Analysis on Assessment of Urban Emergency Capability. *China Safety Science Journal*:69–72.

Zhou. H. & Yang. H. & Xu. J. & Zhang, J. & Kang, S. 2017. Construction of Chinese Hedge Scope Corpus. *Journal of chinese information processing*, 31(3):77–85.

Zou, B. & Zhu, Q. & Zhou, G. 2015. *Negation and speculation identification in Chinese language*. Proceedings of the 53rd Annual Meeting of the Association for Computational Linguistics and the 7th International Joint Conference on Natural Language Processing: 656–665.

*Smart cities and prospects for urban upgrading and transformation*

*Advances in Urban Construction and Management Engineering – Cha (Ed)*
 *ISBN 978-1-032-39018-5*

# Research on the mechanism of smart city's effect on urban innovation

Yunnong Jiang*
*Tianjin University of Technology, Tianjin, China*

ABSTRACT: Using smart cities to drive urban innovation and lead China to the forefront of innovative countries is an important issue in the high-quality development stage. Based on the panel data of 285 prefecture-level cities in China from 2006 to 2021, this paper uses the difference-in-difference method to test the net impact of smart cities on urban innovation. The research conclusions show that smart cities have a significant role in promoting urban innovation; smart cities mainly promote urban innovation by innovating information science and technology, gathering high-end human capital, and optimizing the institutional business environment. In the future, through the positive impact of smart cities on urban innovation, fully release the catalytic effect of information technology, human capital, and institutional environment on smart cities to promote urban innovation, and help China to become the forefront of innovative countries.

## 1 INTRODUCTION

Enhancing urban innovation capabilities and building an innovative city is an important starting point for building an innovative country, and it is also a major strategy for China to face the future. With the promotion and application of new-generation information technologies such as artificial intelligence, 5G, and big data, a new model for the integrated development of urbanization and information technology—the smart city has become an accelerator for the realization of a digital country and a smart country (Kar 2019). A smart city aims to analyze and integrate big data of urban operation by using a new generation of information technology, efficiently allocate urban resources, realize smart city management and operation, and promote the harmonious and sustainable development of the city (Wu 2016). Driven by the smart city development policy, China's urban informatization process has made great progress and the development of smart cities has achieved remarkable results.

The academic community has conducted useful explorations on the development effects of smart cities. Smart cities can drive urban innovation (Fu 2019; Liu 2018; Zhang 2020), but the selection of samples for the experimental group is one-sided and lacks in-depth discussions on internal mechanisms. It is necessary to conduct an in-depth discussion on the innovation effect of smart cities (Gu 2013). This will have important practical significance for the development of China's smart cities and the construction of an innovative country. In view of this, based on the panel data of 285 prefecture-level cities in China from 2006 to 2021, this paper uses the difference-in-difference method to test the net impact of smart cities on urban innovation. The objective of this paper is to examine the role of smart cities in promoting urban innovation and to answer theoretically and empirically how smart cities can promote urban innovation.

*Corresponding Author: jiangyunnong@163.com

DOI 10.1201/9781003348023-61

## 2 RESEARCH HYPOTHESIS

Technology, talents, and institutions are the three core factors in the development of smart cities (Nam 2011). On the one hand, the development and upgrading of smart cities will strengthen their positive impact on urban innovation through information technology effects, human capital effects, and institutional environmental effects; on the other hand, the inter-regional integration and network connection of smart cities will accelerate the spillover of urban innovation, thereby strengthening the positive impact of smart cities on urban innovation. From this, the following assumptions can be made:

H1: The establishment and development of smart cities play a significant role in promoting urban innovation.
H2a: Smart cities can drive urban innovation by revolutionizing information science and technology.
H2b: Smart cities can drive urban innovation by gathering high-end human capital.
H2c: Smart cities can promote urban innovation by optimizing the institutional business environment, but the effect may not be significant.

## 3 MODEL BUILDING

Based on the mechanism of smart cities promoting urban innovation, this paper will use the difference-in-difference model for empirical tests. The smart city pilots established in China correspond to 161 prefecture-level cities, which constitute the "experimental group" studied in this paper. Based on the research population of 285 prefecture-level cities in this paper, the remaining 124 prefecture-level cities constitute the "control group". If a prefecture-level city becomes a smart city pilot, it will be assigned a value of 1; otherwise, it will be assigned a value of 0; if a smart city pilot has been established in a certain year, it will be assigned a value of 1 for that year and its later, and a value of 0 before, which directly generates the smart city dummy variable is tested by setting the following two-way fixed effect model:

$$\text{cityinnovation}_{i,t} = \alpha_0 + \alpha_1 \text{smartcity}_{i,t} + \sum \delta x_{i,t} + \mu_i + \eta_t + \varepsilon_{i,t} \tag{1}$$

In formula (1), $\text{cityinnovation}_{i,t}$ is the explained variable, representing the urban innovation level of region i in period t. $\text{smartcity}_{i,t}$ is directly generated by the smartcity virtual variable directly generated. If an area is set as a smartcity pilot in a certain year, the value of the city in that year and later is 1, otherwise, it is 0. $\alpha_1$ is the estimation coefficient of policy effect and the core parameter of this paper. If $\alpha_1 > 0$, it indicates that a smart city can significantly enhance urban innovation ability; otherwise, a smart city has an adverse impact on urban innovation. $x_{i,t}$ is a series of control variables; $\mu_i$ is the individual fixed effect; $\eta_t$ is the time fixed effect, $\varepsilon_{i,t}$ is the error term.

The data used in the article are all from the "China Urban Statistical Yearbook" and "China Urban and Industrial Innovation Report" over the years. All variables measured in currency are based on 1990, excluding the impact of price factors.

## 4 BENCHMARK MODEL CHECKING

The benchmark model equation (1) is tested to get Table 1. In column (1) of Table 1, when no control variables are added, the estimated coefficient of the smart city on urban innovation is 11.6717, which is significant at the 1% level, indicating that the establishment of a smart city can significantly enhance the ability of urban innovation. Preliminary verification H1. After adding the control variables, it can be seen from column (2) that the net impact coefficient of smart cities on urban innovation is 10.2732, which is significant at the 1% level, which fully shows that smart cities can significantly promote urban innovation, thus effectively verifying H1.

Table 1. Smart city's benchmark return to urban innovation.

| Variable | cityinnovation (1) | cityinnovation (2) |
|---|---|---|
| Smart city | 11.6717*** | 10.2732*** |
| | (3.43) | (3.10) |
| Control variable | NO | YES |
| Time fixed | YES | YES |
| Individual fixed | YES | YES |
| N | 4544 | 4259 |
| $R^2$ | 0.1100 | 0.1821 |

In Table 1, t statistics are in parentheses; *, **, and *** represent being significant at 10%, 5%, and 1% levels respectively; all regressions use cluster-robust standard errors with the region as the clustering variable.

## 5 IDENTIFICATION OF THE MECHANISM OF ACTION

Whether the role of smart cities is transmitted through information technology effects, human capital effects and institutional environment effects need to be tested one by one using a mediation effect model. Here, we set the equation of the mediation effect model as:

$$\text{cityinnovation}_{i,t} = \alpha_0 + \alpha_1 \text{smartcity}_{i,t} + \mu_i + \eta_t + \varepsilon_{i,t} \tag{2}$$

$$\text{mediator}_{i,t} = \beta_0 + \beta_1 \text{smartcity}_{i,t} + \mu_i + \eta_t + \varepsilon_{i,t} \tag{3}$$

$$\text{cityinnovation}_{i,t} = \gamma_0 + \gamma_1 \text{smartcity}_{i,t} + \gamma_2 \text{mediator}_{i,t} + \mu_i + \eta_t + \varepsilon_{i,t} \tag{4}$$

In formula (4), $\text{mediator}_{i,t}$ represents mediating variables, including information technology effect, human capital effect, and institutional environment effect. If this three-way mediating transmission mechanism exists, $\alpha_1$, $\beta_1$, and $\gamma_2$ must satisfy statistical significance.

### 5.1 *Information technology effect test*

It can be seen from Table 2 that the influence coefficient of the smart city on urban innovation is 11.6717, which is significant at the 1% level. When smart city and Internet information service

Table 2. Mechanism test—information technology effect.

| Variable | urban innovation (1) | Internet information information service (2) | urban innovation (3) | software development (4) | urban innovation (5) |
|---|---|---|---|---|---|
| Smart city | 11.6717*** | 0.2112* | 6.3627*** | 0.2287* | 5.4746*** |
| | (3.43) | (1.92) | (3.47) | (1.79) | (4.17) |
| Internet information service | | | 37.8841*** | | |
| | | | (65.56) | | |
| Software development | | | | | 25.6459*** |
| | | | | | (83.84) |
| Time fixed | YES | YES | YES | YES | YES |
| Individual fixed | YES | YES | YES | YES | YES |
| N | 4544 | 999 | 999 | 1331 | 1331 |
| $R^2$ | 0.1100 | 0.0727 | 0.8871 | 0.0687 | 0.8964 |

innovation are added at the same time, this influence factor drops to 6.3627, which is significant at the 1% level, indicating that the promotion of smart cities to urban innovation is partly transmitted through the innovation of the Internet information services. The promotion effect of smart city on software development innovation is 0.2287, which is significant at the 10% level. When both smart city and software development innovation are included, the promotion effect of smart city on urban innovation drops to 5.4746, which is significant at the 1% level, indicating that the smart city promotes urban innovation partly through software development and innovation. The above two points confirm the view that smart cities promote urban innovation through the effects of information technology. H2a is validated.

## 5.2 *Human capital effect test*

As shown in Table 3, when the smart city and human capital quantity are added at the same time, the promotion effect of the smart city on urban innovation decreases from 11.6717 to 10.4236, indicating that this promotion is partly transmitted through the human capital quantity; when the smart city and human capital quality are added at the same time, the promotion effect of smart city on urban innovation drops from 11.6717 to 2.0873, and it is not statistically significant. It shows that the promotion effect of a smart city is completely realized through the transmission of human capital quality. Overall, smart cities can drive urban innovation through human capital effects.H2b is validated.

Table 3. Mechanism test—human capital effect.

| Variable | urban innovation (1) | human capital quantity (2) | urban innovation (3) | human capital quality (4) | urban innovation (5) |
|---|---|---|---|---|---|
| Smart city | 11.6717*** (3.43) | 35.6271*** (3.55) | 10.4236*** (2.90) | 1.7016*** (4.53) | 2.0873 (1.37) |
| Human capital quantity | | | 0. 0410*** (3.18) | | |
| Human capital quality | | | | | 5.6748*** (3.02) |
| Time fixed | YES | YES | YES | YES | YES |
| Individual fixed | YES | YES | YES | YES | YES |
| N | 4544 | 4409 | 4399 | 4504 | 4494 |
| $R^2$ | 0.1100 | 0.3310 | 0.1365 | 0.2716 | 0.5056 |

## 5.3 *Institutional environment effect test*

Smart cities can reduce institutional transaction costs, but they are not statistically significant. When smart cities and institutional transaction costs are added at the same time, institutional transaction costs significantly inhibit urban innovation, and smart cities can significantly promote urban innovation, and the promotion effect was significantly reduced from 11.6717 to 10.0898, indicating that reducing institutional transaction costs is an intermediary variable for smart cities to promote urban innovation, but this transmission mechanism is not statistically significant due to deep-seated reasons such as the immature development of smart cities. Smart cities promote marketization, but are not statistically significant, when both the smart city and Fan Gang marketization index are included, both marketization index and the smart city can significantly promote urban innovation, and the promotion effect of smart city on urban innovation decreased from 11.6717 to 9.4942. To sum up, optimizing the institutional business environment is an intermediary variable for smart cities to promote urban innovation, but this intermediary effect is not statistically significant, and H2c has been verified.

Table 4. Mechanism test—institutional environment effect.

| Variable | urban innovation (1) | institutional transaction costs (2) | urban innovation (3) | Fan Gang marketization index (4) | urban innovation (5) |
|---|---|---|---|---|---|
| Smart city | 11.6717*** | -0.0255 | 10.0898*** | 0.0146 | 9.4942*** |
| | (3.43) | (3.30) | (−0.87) | (0.70) | (3.39) |
| Institutional transaction costs | | | −5.7865** | | |
| | | | (−2.32) | | |
| Fan Gang marketization index | | | | | 4.7201*** |
| | | | | | (2.90) |
| Time fixed | YES | YES | YES | YES | YES |
| Individual fixed | YES | YES | YES | YES | YES |
| N | 4544 | 3990 | 3976 | 3420 | 3408 |
| $R^2$ | 0.1100 | 0.3373 | 0.1190 | 0.7083 | 0.1216 |

## 6 CONCLUSION

Based on 285 prefecture-level cities in China from 2006 to 2021, this paper uses the difference-in-difference model to systematically examine the net impact of smart cities on urban innovation. The study finds: (1) Smart cities can significantly promote urban innovation; (2) The mechanism of action shows: first, smart cities can promote urban innovation through information technology effects mainly based on Internet information service innovation and software development innovation; second, smart cities can enhance human capital effects from both the quantity and quality dimensions, thereby promoting urban innovation in China; third, smart cities have a clear tendency to optimize the institutional business environment by reducing institutional transaction costs and accelerating the process of marketization, thereby promoting urban innovation in China.

In the future, to further strengthen the positive impact of smart cities on urban innovation, we should start from the following aspects: First, speed up the development and construction of smart cities, and at the same time use intelligent service platforms to achieve resource sharing to enhance the role of smart cities in promoting urban innovation. Second, strengthen the level of new-generation information technology, improve the new-generation information technology service system, and enhance the positive role of information technology in the process of smart cities promoting urban innovation. Third, continuously improve the quality of China's human capital, continue to optimize the structure of human capital, especially to increase the stock of high-end human capital, and provide sufficient talent support for smart cities to drive urban innovation. Finally, accelerate the process of marketization, reduce institutional transaction costs in the process of economic development, continuously optimize the institutional business environment, provide a more open and inclusive environment, accelerate the role of smart cities in promoting urban innovation, and help build China into an innovative nation.


## ACKNOWLEDGMENTS

I am very grateful to my teachers for their guidance and my classmates for their help. In the following days, I will continue to study hard and strive for more academic achievements.

## REFERENCES


Fu Ping, Liu Dexue. Research on the effect of technological innovation in smart cities: An empirical analysis based on panel data of 282 prefecture-level cities in China [J]. *Economic Issues Exploration*, 2019(09): 72–81.

Gu Shengzu, Yang Jianwu, Liu Jiangri. Problems and countermeasures in the current construction of smart cities in my country [J]. *China Soft Science*, 2013(01):6–12.

Kar A K, Ilavarasan V, Gupta M P, et al. Moving beyond smart cities: Digital nations for social innovation & sustainability[J]. *Information Systems Frontiers*, 2019, 21(3):495–501.

Liu Qiao, Shi Daqian, Liu Jianjiang. The impact of smart city construction on urban technological innovation [J]. *Technical Economy*,2018, 37(05):81–85.

Nam T, Pardo T A. *Conceptualizing smart city with dimensions of technology, people, and institutions*[C]//Proceedings of the 12th annual international digital government research conference: digital government innovation in challenging times. 2011: 282–291.

Wu Biaobing, Lin Chengliang. The Open Governance Innovation Model of Smart City: The Practice and Enlightenment of the EU and South Korea [J]. *China Soft Science*, 2016(05):55–66.

Zhang Longpeng, Zhong Yilin, Tang Zhiwei. Research on the Impact of Smart City Construction on Urban Innovation Capability: A Quasi-Natural Experiment Based on China's Smart City Pilot [J]. *Soft Science*, 2020, 34(01):83–89.

*Advances in Urban Construction and Management Engineering – Cha (Ed)*


# Research progress and prospects of China's age-friendly city construction

Shi Wu, Jie Shen* & Feng Hu
*Wuhan University of Science and Technology, Wuhan, China*

ABSTRACT: This essay examines the development of Age-friendly cities in China today in light of the societal issue of population aging. CNKI is chosen as the database, Citespace-based bibliometric analysis is conducted, and the development of research on age-friendly city planning is researched from the perspectives of research hotspots, clustering maps, hotspot evolution, and major content. The findings indicate that research interest in the development of age-friendly cities is increasing, and the topics covered are becoming more extensive. However, there is little overlap between the various cluster themes, and the interdisciplinary scope needs to be expanded in both breadth and depth. Future urban research and practice are advised to concentrate on fundamental empirical studies in order to promote the development of urban settings better suited for the aging population.

## 1 INTRODUCTION

Building age-friendly communities is a crucial way to actively deal with the population aging as a result of the global aging crisis (Lu 2021). Using the Citespace visual analysis software, this paper explores the frontier hotspots from the perspectives of publishing institutions, author cooperation networks, keyword co-occurrence, and research evolution, and looks forward to the development trend to sort out the pertinent studies on the construction of age-friendly cities in my country over the past 30 years. References for urban construction research and practice should be provided.

## 2 BIBLIOMETRIC ANALYSIS

Retrieval time set from 1991 to 2021. The search keywords are "age-friendly city", "active aging" and "age-friendly city", and the search method is the subject. 4843 articles were retrieved and screened by CNKI. In the Citespace software, the time slice is set to 1 year. In order to obtain the 10% documents with the highest frequency each year, the threshold value is Top N%=10%.

### 2.1 *Analysis of authors and research institutions*

Based on the analysis of Citespace, it is concluded that there are 484 nodes in the author's cooperation network map, 324 times of connections, and the cooperation network density is 0.0028, indicating that the relevant scholars have not yet formed a close cooperation network. At present, scholars in various fields have carried out research on the construction of age-friendly cities,

*Corresponding Author: 675493824@qq.com

DOI 10.1201/9781003348023-62

including cooperative groups with one or more core figures. Scholars in such cooperative groups have a small number of participants but have stronger research capabilities, and the number of published papers is all 5 In addition, a multi-person cooperative network of cooperative research within the same institution has been formed; the rest is published by the author alone, which is the main force in the current domestic research on the construction of an elderly-friendly city.

The cooperative network of research institutions can reflect the strength of the connection between the authors' institutions and the degree of agreement between theory and practice at another level. From the perspective of institution type, the research institutions related to age-friendly cities are mainly colleges and universities. There are also a considerable number of urban planning and design institutes and research institutes. Judging from the number of published papers, there are 9 institutions that have issued documents more than 10 times, and all of them are universities except the China Research Center for Aging. From the perspective of the cooperation network structure, the participating institutions have the characteristics of "less aggregation and more dispersion". There is relatively little cooperation between universities, and there are certain connections within universities and with research institutions.

Figure 1. Research institutions.

## 3 ANALYSIS OF RESEARCH HOTSPOTS

The findings of the "Keyword" node analysis reveal 325 linked research hot words, 363 links between the nodes, dense connections between the keywords, and a significant correlation. A research trend centered on "aging", "appropriate aging" and "home care" is formed. Cluster analysis of the top 15 important keywords in terms of co-occurrence intensity, and three types of focus were obtained: First, the description of the problem solved by the construction of an elderly-friendly city: "Elderly Population", "Ageing Suitability", "Aging", is closely related to other keywords, with high betweenness centrality and frequency. The second is the practice direction represented by "home-based care", "community-based care" and "combination of medical care and elderly care". Elderly care services guarantee the quality of life of the elderly, and the entire process of serial research is the focus of research development, and the co-occurrence frequency of keywords is high. The third is the keywords of the research content of the construction of an elderly-friendly city, "elderly education", "social participation", etc., covering the development trend of different disciplines in this field (Table 1).

Table 1. Top 15 keywords in total occurrence frequency.

| order number | keyword | Co-occurrence frequency/times | Intermediary centrality |
|---|---|---|---|
| 1 | Elderly Population | 777 | 0.28 |
| 2 | Aging Suitability | 682 | 0.12 |
| 3 | aging | 398 | 0.15 |
| 4 | home-based care | 183 | 0.61 |
| 5 | combination of medical care and elderly care | 164 | 0.01 |
| 6 | community-based care | 141 | 0.05 |
| 7 | Pension services | 98 | 0.01 |
| 8 | social participation | 94 | 0.40 |
| 9 | elderly education | 93 | 0.05 |
| 10 | public space | 82 | 0.05 |
| 11 | influencing factor | 80 | 0.05 |
| 12 | Pension model | 66 | 0.03 |
| 13 | Pension real estate | 61 | 0.02 |
| 14 | transform | 58 | 0.07 |
| 15 | Strategy | 53 | 0.49 |

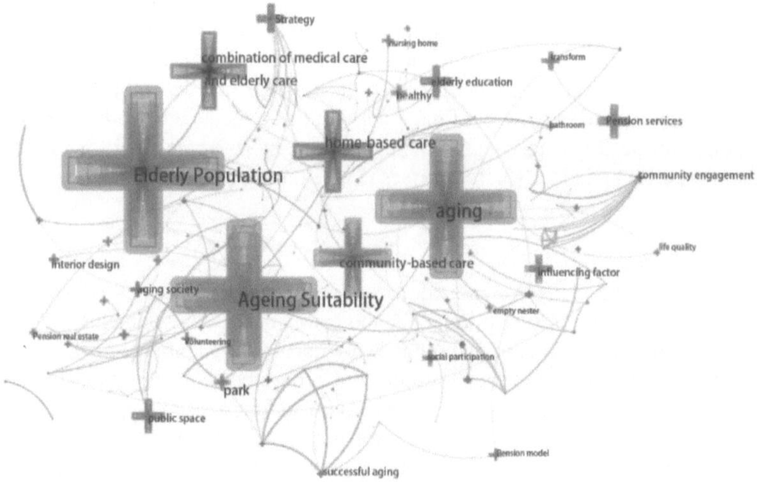


Figure 2. Co-occurrence of keywords.

## 4 ANALYSIS OF THE MAIN RESEARCH CONTENT

### 4.1 *Theoretical evolution*

From the state of "getting old before getting rich" in the early days to the current situation where the age structure of the population is inclined to decline, the exploration of the construction of an elderly-friendly city according to the actual situation in my country is also developing continuously (Huang 2022). From the perspective of the theoretical development on which the construction of an age-friendly city is based, it has experienced the evolution process of "productive aging", "successful aging", "healthy aging" and then "active aging".

"Productive aging" abandons the "negative" traditional concept of aging as a "burden" of social and economic development, and advocates for the elderly to be productive and actively participate in life, so as to achieve health, economic value, social and social benefits of ensuring the quality of life of the elderly for the purpose of the three levels of value. But letting the elderly re-engage in social work is contrary to the traditional Chinese concept of filial piety and cannot be tolerated by social morals (Shan 2008). Additionally, the involvement of seasoned seniors will surely create competition for young people's social resources. Therefore, the idea of "productive aging" has a limited impact in China.

On the basis of "productive aging", the concept of "successful aging" that pays more attention to the health of the elderly is derived. It is needed to pursue a lower disease risk, higher cognitive level, and healthy physiological state, emphasizing the positivity and plasticity of aging. Domestic-related research is mainly reflected in two aspects: one is to explore and analyze successful aging individuals from a medical point of view; the second is to explore and measure the influencing factors of Chinese successful aging from an empirical perspective. By comparing the influencing factors in the two models of Successful-aging, it is found that the use of different theoretical models has a greater impact on the measurement and analysis results of SA influencing factors.

From the perspective of function, "healthy aging" takes into account a wide range of social environments, and focuses on the development and maintenance of the functions required for the healthy life of the elderly, that is, the health influencing factors that the elderly can exercise according to their own concepts and preferences. The three dimensions covered by "active aging": participation, health, and security, are the inheritance and development of the first three concepts. The dimension of "participation" originates from "productive aging", which defines two types of "outputs" in productive aging: one is remunerative economic output, in which the elderly directly participate in market competition and create economic wealth; the two are unpaid non-economic outputs, such as participation in volunteer activities, family care, or independent study. Both types of "output" connotations lay a theoretical foundation for the social participation of the elderly. The dimension of "health" is expanded and deepened on the basis of "healthy aging". It not only focuses on the health, longevity, and happiness of the elderly but also meets the needs of the elderly care system and policy for healthy elderly and provides basic support for the elderly with diseases. The combination of medical and elderly care services is taken into consideration.

### 4.2 *Evaluation index system of age-friendly cities*

To explore and practice the evaluation indicators for the construction of an elderly-friendly city, its main functions are as follows: 1) to optimize the indicators of the core elements of the evaluation system to provide a reference for later planning; 2) to improve the accuracy of evaluating the city's friendliness to the elderly. Creating a social support and living environment suitable for the life of the elderly is conducive to maintaining the normal physiological functions of the elderly and improving their health (Tan 2010). Domestic research on the evaluation of elderly-friendly city construction mostly focuses on the two spatial scales of city and community. Most of the evaluations on the urban scale are based on the built environment, ecological environment, social environment, and social security. And medical coverage is a necessary feature for the elderly population. Research at the community scale is mostly carried out around the planning and design environment of the residential area and specific activity spaces and facilities. In the evaluation research based on the community scale, service and management operations are used by Jia Weiyang et al. (Jia 2016) as evaluation elements to explore the reasonable weight of each sub-item. Chen Tian et al. (Chen 2021) introduced factors that affect the number of visits to community elderly activity venues and constructed a clustering analytic hierarchy process evaluation model for community elderly activity venues to evaluate various types of community elderly activity venues.

At the same time, with the continuous improvement of material conditions and awareness levels, people's attention to the special group of the elderly is also rising, and related research has also completed a shift from focusing solely on the living environment to focusing on the coordination of material and spiritual needs. The elderly's education and social participation, which are included in the social environment, are all important factors in judging the friendliness of a city to the elderly. From the perspective of active aging, it is necessary to make changes in the understanding, concept, and evaluation of elderly education: it is regarded as a multi-evaluation method for potential development, promotion of health, participation, and security.

The construction and application of an age-friendly urban environmental assessment system involve many aspects, multiple disciplines, and multiple perspectives. Relevant research in my country has only started in recent years, and most of it focuses on built environment assessment.

### 4.3 *Empirical research on the construction of age-friendly cities*

Faced with the problem of aging, the discipline of urban planning mainly focuses on research on the impact of the built environment and social environment on the quality of life and health of the elderly. The built environment does not directly affect the health level of the elderly but affects the physical and mental health of the elderly by affecting the behavior and social activities of the population (Li 2021). With the continuous improvement of lifestyle and quality of life, the main threat to the health of the elderly has shifted from infectious diseases to chronic non-communicable diseases. The increase in the proportion of deaths and the economic burden caused by them have seriously threatened the lives, health, and property safety of elderly residents in my country. In the study of Mao Ying et al. (Mao 2015), it was confirmed that apart from individual factors, both the material environment and the complex social environment have different intensities on the prevalence and age of chronic diseases. Physical activity and social interaction are often used as mediating variables to explore the health effects of the built environment. Chen Chun et al. (Chen 2017) pointed out that the accessibility and availability of public leisure activities have a significant impact on the health quality of the elderly. For the special group of the elderly, considering the reduction of physical function, compared with other physical activities, walking activities are easier to carry out and have appropriate intensity, and are mostly used as research topics to explore the relationship between the built environment and the health status of the elderly. Scholars conduct empirical research on the impact of the built environment using statistical models or experimental analysis to identify significant environmental factors and investigate the impact mechanism. For example, Huang et al. (Huang 2020) constructed an "environment-health" mechanism with physical activity and community cohesion as mediating variables. The study found that greater facility accessibility has a significant positive impact on the physical and mental health of the elderly. From the perspective of research methods, some studies used the quantitative methods of MATLAB and GIS to explore the relationship between the elderly's going-out behavior and urban spatial structure and measure the spatiotemporal characteristics of the elderly's outdoor physical activity; Jiang Xiaoyan et al.(Jiang 2022) conducted the qualitative research method to analyze the behavioral characteristics of the old and young people on the main living streets of the community, and to plan a healthy community that is more suitable for the elderly to live in.

An elderly-friendly city should not only build a comfortable and livable environment but also meet the spiritual needs of the elderly. It is attempted to advocate for the elderly to have a healthy life and opportunities to participate and contribute to society. Incorporating new demand concepts at the level of the social environment, social participation, elderly education, and intergenerational integration are important influencing factors. Relevant scholars are mostly based on the characteristics of the elderly to explore the significance and practice of these concepts in planning and design. The social environment, institutional policies, and regional differences are the main factors affecting the social participation of the elderly. Studies have shown that good health status and active social participation have a cyclical effect of mutual influence. Good social participation can not only enrich their own lives but also meet the social needs of the elderly and enhance their social identity.

## 5 CONCLUSION AND PROSPECTS

Through the bibliometric analysis of the research field of age-friendly city construction in the past 30 years, it is found that: 1) At present, the research topic of this topic shows a state of continuous growth, and the degree of research attention is getting higher and higher, but the degree of cooperation among research subjects is low. The relationship between the author and the institution is not strong; 2) The research on the built environment, policy measures, and their possible health effects on the elderly group have a high degree of attention, reflecting the characteristics of "people-oriented", and the research topics involve a variety of disciplinary knowledge, showing a certain degree of an interdisciplinary phenomenon; 3) Research topics tend to be diversified, and there are more and more studies using big data analysis.

However, from the research content, we can see that 1) the existing research is mostly qualitative, and there is a lack of empirical research based on the current situation in my country. It is necessary to further expand and enrich relevant empirical research in the future. 2) The breadth of the research content needs to be widened, and the target population of most research is only based on the basic concept of "elderly people". There are few detailed studies on different regions, different socioeconomic attributes, and different cultures. 3) There is a scarcity of broad-based and diverse interdisciplinary collaboration. Although there is overlap between disciplines in existing research, it lacks a strong foundation for cooperation. Research on the construction of elderly-friendly cities is an interdisciplinary research, and the cooperation of related disciplines can form complementary advantages.

## ACKNOWLEDGEMENT

Fund Projects: 2021 Hubei Province College Philosophy and Social Science Research Project (No. 21Y031), 2021 Hubei Provincial Construction Science and Technology Plan Project (No. 202117).

## REFERENCES

Chen Chun, Chen Yong, Yu Li, et al. Planning for Healthy Cities: A Study on the Relationship between the Built Environment and the Body Mass Index of the Elderly [J]. *Urban Development Research*, 2017,24(04): 7–13.

Chen Tian, Wang Zheyu. Research on community-based elderly activity sites based on the analytic hierarchy process model: Taking the downtown area of Fuzhou as an example [J]. *Science and Technology Herald*, 2021, 39(24): 108–117.

Huang Fan, Duan Chengrong. From Demographic Dividend to Demographic Dividend—Based on the Analysis of the Seventh National Census Data [J]. *Population and Development*, 2022, 28(01): 117–126.

Huang Ling, Mu Yanchuan. The mechanism and empirical results of the impact of a community-built environment on the health of the elderly [J]. *Nanjing Social Sciences*, 2020(12): 51–58.

Jia Weiyang. A Preliminary Study on the Evaluation Index System of Community Aging Suitability [J]. *Urban Planning*, 2016,40(08): 65–70.

Jiang Xiaoyan, Zhu Xiaolei. A Study on the Interactive Behavior Scenarios of Old and Young in Community Commercial Streets—Three Cases of Guangzhou [J]. *Southern Architecture*, 2022(05), 91–99.

Li Kangkang, Yang Dongfeng. How the Urban Built Environment Affects the Physical Activity of the Elderly: Model Construction and Dalian Empirical Study [J]. *Human Geography*, 2021, 36(05): 111–120.

Lu Jiehua, Wu Xuqing. Changes in Population Age Structure: Main Features, Multiple Impacts and Coping Strategies [J]. *Youth Exploration*, 2021(04): 28–40.

Mao Ying, Zhu Bin, Liu Jinlin, et al. Empirical evidence on the influencing factors of chronic diseases in middle-aged and elderly people from the perspective of health ecology [J]. *Journal of Xi'an Jiaotong University* (Social Science Edition), 2015, 35(05): 15–24.

Shan Lingling, Chen Guopeng. A New Perspective of Analyzing Successful Aging—Lifetime Control Theory[J]. *Psychological Science*, 2008,31(04): 950–952.

Tan Shaohua, Guo Jianfeng, Jiang Yi. Active Intervention of Human Settlements on Health: A New Trend in Urban Planning Discipline[J]. *Urban Planning Journal*, 2010(04): 66–70.

*Advances in Urban Construction and Management Engineering – Cha (Ed)*


# Research and analysis on technological innovation of prefabricated building construction

Fei Du*
*Tianjin Vocational University, China*

ABSTRACT: With the continuous improvement and rapid development of China's economic strength, the construction industry has entered a new stage of development. This paper studies the application of precast concrete (PC) and aluminum formwork integrated construction in prefabricated buildings, PC connection joints, and a new system of ultra-high performance prefabricated structure (PCUS), the application of ultra-high performance concrete materials (UHPC), as well as the innovation and optimization of assembled monolithic composite concrete structure system and its industrial production technology. Compared with the traditional construction techniques, the new technology combined with the digital production mode can greatly improve the construction efficiency, shorten the construction period, save costs, and maximize the rationalization of resource allocation and utilization. At the same time, meeting the current social needs of low-carbon industrial production is conducive to promoting the transformation and upgrading of the construction industry, improving the level of construction quality and economic industrial efficiency. The above analysis can provide a reference for the research on the innovation of prefabricated building construction technology.

## 1 INTRODUCTION

With the rapid development of the modern industrial economy and the continuous improvement of people's living and economic standards, prefabricated buildings can better meet the development needs of society for building quality and concept. In order to promote the high-quality development of the construction industry, the state has continuously increased policy support, which marks that the development of prefabricated buildings in China has entered the stage of large-scale industrialization. Under the requirements of new science and technology development norms, we should change the traditional construction technology methods, expand the development path of prefabricated building innovative technology, reasonably allocate resources, improve the construction quality and improve the efficiency of construction management (Wang 2019). Through the analysis of a variety of new prefabricated building construction technologies that have been popularized and applied in the project practice, and through the project implementation cases, reflect their actual application effect in the project, so as to improve the utilization rate and application and promotion scope of prefabricated building construction technology. The application of new technology for prefabricated buildings lays a solid foundation for further improving the development quality and efficiency of the construction industry, accelerating the upgrading and transformation of building industrialization, digitization, and intelligence, and helps to continuously promote the high-quality development of the construction industry.

*Corresponding Author: susansm@qq.com

DOI 10.1201/9781003348023-63

## 2 INNOVATIVE CONSTRUCTION TECHNOLOGY OF PREFABRICATED BUILDINGS

### 2.1 *PC + aluminum formwork integrated construction technology*

The integrated process flow of PC + aluminum formwork is as follows: Measurement and setting out→shear wall reinforcement binding, installation and pipeline embeddin→concealed acceptance→wall formwork installation→beam formwork installation→aluminum formwork floor keel installation→PC laminated plate installation→formwork acceptance→beam slab reinforcement binding, installation and pipeline embedding→concealed acceptance→concrete pouring→rapid removal system formwork removal→transfer to the next floor. Although the process flow of the integrated construction of aluminum formwork and prefabricated buildings is relatively fixed, and the sequence is roughly the same, it should still be adjusted appropriately in combination with the actual situation in the actual construction environment (Ma, Wang, He, 2020).

Several key steps of this technology include deepening the design of the aluminum film, pasting foam double-sided adhesive tape on PC components, installing aluminum alloy beam formwork, installing aluminum alloy wall column formwork, processing the joints at the junction of PC components, strengthening PC components with aluminum film, and concrete pouring (Wang 2020). There are many quality problems between PC components and cast-in-place concrete during the combined construction of PC components and aluminum formwork processes, such as dislocation, slurry leakage, and cracks. In order to solve the quality defects at the joint between PC components and cast-in-place concrete, the construction technology of a local combination of aluminum formwork and PC exterior wall was developed. Technical measures such as optimizing the design of component size and filling with elastic sealing materials in the joint were taken to ensure the construction quality and improve the construction efficiency.

This technology has the following advantages: it meets the green development construction concept in the current low-carbon environment and is more in line with the development direction of modern industrialization. The aluminum mold has high strength and precision, and the concrete surface is flat and smooth after construction, which can achieve the effect of free plastering or thin plastering. The construction is efficient, which can reduce costs, reduce late maintenance, facilitate the management, and effectively improve the quality of concrete.

### 2.2 *Prefabricated connection joint + new system of ultra-high performance fabricated structure*

The new system of ultra-high performance prefabricated structure (PCUS) is also composed of precast beams, precast columns, and precast floors. Different from the traditional system, precast beams and columns are spliced into an integral frame through joints. The beam-column joints and column joints of the new system use post-cast joints based on ultra-high-performance concrete (UHPC).

The reinforcement of the beam-beam joint is lapped and poured with UHPC materials. When the critical anchorage length of the reinforcement at the connection point is 4 times the diameter of the reinforcement, and the lap length of the connection is 10 times the diameter of the reinforcement, the mechanical properties of each node connection meet the standard and have a considerable seismic capacity. The column node is supported on the lower column by 10 I-steel embedded in the upper column to complete the vertical positioning. At the same time, the column is supported by diagonal braces on both sides to complete the horizontal positioning. UHPC grouting is carried out through the shaped steel formwork with a dustpan opening to form a whole (Tang 2019).

This new set of matching structure systems has the characteristics of flexible form, efficient construction, and reliable structure, which solves the difficulty of uncontrollable strength and effect of prefabricated joints in the original prefabricated construction. It solves the problems of loose grouting and material shrinkage of fabricated joints, controls the pouring quality after joints, and forms a set of construction technology suitable for pouring after all kinds of joints. The joint connection quality is easy to detect, and the sleeve grouting is avoided. The bent-free construction can be realized, which reduces the construction difficulty, reduces the process, saves resources, and ensures the construction quality at the same time.

## 3 ASSEMBLED MONOLITHIC COMPOSITE CONCRETE STRUCTURE SYSTEM AND ITS INDUSTRIAL PRODUCTION TECHNOLOGY

Research on the system of "assembled monolithic composite concrete structure system and its industrial production technology", combined with the existing SPCS technology system and prestressed structure technology in the society, a new prefabricated building structure technology system is formed through research, development, and innovation (Wang, Liao, Liu,2019). This technology mainly includes two construction technologies: SPCS assembly integral composite structure process method and steel tube truss prestressed concrete composite beam (PKL composite beam for short); the cast-in-situ technology of prestressed concrete composite slab (PK3 composite slab for short).

The main innovations of the technical system are as follows:

Firstly, the combined application mode of "SPCS wall column and steel pipe truss prestressed composite beam slab" is proposed, the assembled integral composite concrete structure system is formed, and the corresponding design theory and calculation method are established.

Secondly, the efficient construction technology of assembled integral laminated concrete structure system has been developed, forming a less support and template-free construction process of structural engineering, and realizing efficient construction.

Thirdly, the SPCS wall column series complete equipment and the steel pipe truss prestressed composite beam and slab series complete equipment have been developed, realizing efficient and high-quality automatic production and transportation.

The assembled monolithic composite concrete structure system and its industrial production technology are applicable to the above-ground and underground buildings of the frame, frame shear wall, shear wall frame core tube, and other structural forms. It has the advantages of rapid construction, increasing the net height of buildings, improving the appearance quality of concrete structures, and has more obvious advantages in long-span and heavy-duty structures (Yu 2021).

## 4 APPLICATION CASE ANALYSIS OF PREFABRICATED BUILDING ENGINEERING

### 4.1 *Project overview*

A high-rise building project, with a total construction area of 107643 $m^2$, a basement of 2 floors, and the external wall and internal partition wall of the main structure are constructed with PC fabricated components. The total construction period of the civil works of the project is 30 months. The project passed the review of the "Luban Award" organized by the Ministry of housing and urban-rural development in a certain year.

### 4.2 *Deepen design and technological innovation measures*

The integrated design of civil engineering and decoration of the project reaches 80%, and a large number of new technologies and processes such as prefabricated component assembly installation and aluminum alloy formwork are applied. The production technology of assembled integral composite concrete structure system is applied, and the fabricated cavity composite concrete structure or formed reinforcement skeleton + shaped formwork is adopted, and the prestressed concrete composite members are adopted for horizontal members. In addition, the project also adopts the whole process of digital application of the platform, aiming at the design, production, transportation, construction, and data management of prefabricated components.

### 4.3 *Plasterless structure and high assembly rate*

The architectural structure design is optimized, and the external wall full concrete structure +pc wall + fabricated internal wall panel system is applied. The aluminum alloy formwork support system, the deepening of nodes, and the application of connectors and other technologies are adopted, and the structure meets the requirements of no plastering.

### 4.4 *Application of prefabricated building innovative structure technology*

The project adopts the assembled integral reinforced welded mesh laminated concrete structure technology (referred to as SPCS), which is more suitable for building industrialization than the traditional fabricated structure technology. Its innovation is conducive to production, transportation, and high production efficiency.

### 4.5 *Fine construction and general contracting management*

The total construction period is 5 months ahead of the same period of last year by adopting the technology of interleaved construction and municipal advance construction. Thanks to the high assembly rate, the prefabricated component installation and concrete pouring construction of the first-floor structure will be completed within 7 days of this project, and the later processes will be carried out alternately to meet the requirements of efficient flow operation (Sun, Fan, Qi, Ou, Wang 2018).

### 4.6 *Application effect analysis*

#### 4.6.1 *Industrialized prefabricated components*

The construction technology of PC component + aluminum mold cast-in-situ concrete connection is applied in the construction of standard floor structure, which has good quality and meets the requirements of no leakage and no plastering. PC components are connected by reliable technology, and the joints are glued fully and smoothly. The water tightness, air tightness, weather resistance, width depth ratio, and displacement adjustment ability meet the design requirements.

#### 4.6.2 *The technology of assembling integrally welded steel mesh to superimpose concrete structure*

This technology is more suitable for building industrialization. Its innovation is that components are conducive to production, transportation, and construction, and have good safety performance.

#### 4.6.3 *Digital production and construction*

The whole process digital application platform is adopted to complete the design of prefabricated components through intelligent deepening design software and generate drawings, lists, and processing data. The data is uploaded to the collaboration platform and associated with the project plan to carry out unified data management and transmission and complete the construction task of digital assembly.

### 4.7 *Cause result analysis of prefabricated building quality*

In the whole life cycle of prefabricated building design, production, transportation, and construction operations, in the process of quality management, the main causes of quality fluctuations include five factors: people, materials, machinery, construction methods, and environment. 24 influencing factors are selected, and the impact indicators of each influencing factor are calculated according to the DEMATEL method. According to the calculation results, the center degree is the horizontal axis and the cause degree is the vertical axis. The cause result diagram of the factors affecting the quality of prefabricated buildings is obtained, as shown in Figure 1.

According to the requirements of the DEMATEL method, the questionnaire is designed according to the influencing factors, and experts are invited to score to determine the relationship between the factors. The mic-mac method is used to determine the category of influencing factors, and calculate the driving force and dependence. With the help of the four quadrants of the dependency driving force classification diagram, 24 influencing factors are divided into four categories, as shown in Figure 2.

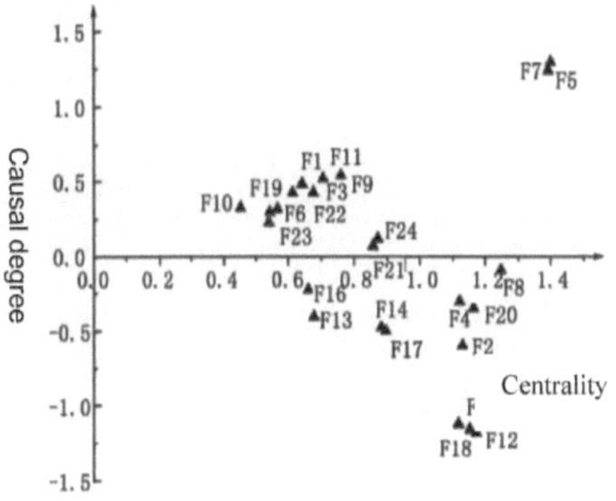


Figure 1. Cause result of quality factor.

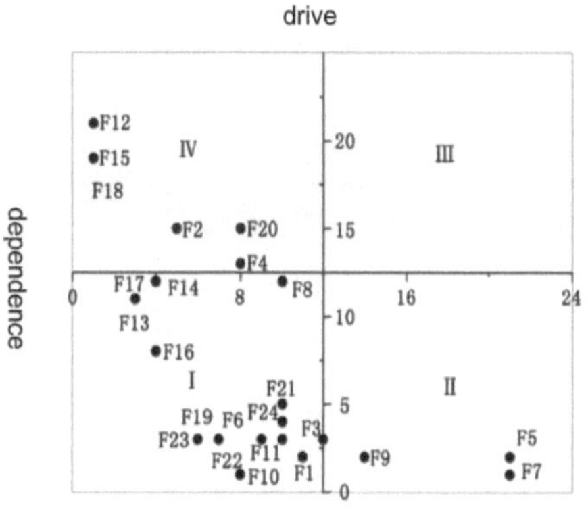


Figure 2. dependency driving force classification.

Through the cause result analysis of prefabricated building quality, this paper studies the structural hierarchical relationship between the influencing factors of prefabricated building quality, summarizes its dependence and driving force, and further improves the construction quality of prefabricated buildings through government support, formulating complete quality management specifications, strengthening communication and cooperation, and promoting technical improvement (Guo 2019).

## 5 CONCLUSION

To sum up, according to the development of China's prefabricated construction industry, new construction technology has been gradually promoted in the construction industry. Adopting innovative construction technology can optimize the on-site construction technology and structural construction technology, shorten the construction period, save costs and improve efficiency. Through the research and analysis of new technologies and processes related to prefabricated buildings, combined with a prefabricated building technology application case described in this paper, this paper makes an in-depth analysis of the project's technological innovation, structural innovation, application effect, and the cause result influencing factors of prefabricated building quality using the DEMATEL method, so as to further show the application of prefabricated technology in the quality control of construction projects and significant advantages in ensuring construction period and improving economic benefits. The promotion and application of prefabricated building innovative technology industry can achieve the goal of "double carbon" and optimize the environmental protection level of building construction, which is conducive to promoting the transformation and upgrading of the construction industry, improving the level of construction quality and improving the efficiency of the construction industry.

## REFERENCES


Guo Enji *Research on influencing factors and quality improvement schemes of prefabricated building construction quality* [D] Tianjin: Tianjin University, 2019.

Ma Xiao, Wang Chao, He Youzhen Research on construction technology and quality management of prefabricated buildings [J] *Famous city painting*, 2020 (3): 164.

Sun Hui, fan Zesen, Qi He, OuYaming, Wang long. Application Research on the combination of EPC mode and prefabricated building technology [J] *Construction Technology* 2018 (07).

Tang Xiaowei Research on construction technology of prefabricated buildings [J] *China's strategic emerging industries* (theoretical Edition), 2019 (14): 1.

Wang Qinglong. Key Points and Research on Construction Technology of Prefabricated Concrete Building Structure [J]. *Green Building Materials*. 2020(07):112–113+122.

Wang Ru, Liao Wentao, Liu Qingnan. Construction of Prefabricated Building Quality Information Model [J]. *Journal of civil engineering and management*, 2019, 36(6):8–16.

Wang Yingchun. Application status and Construction Quality control of prefabricated Buildings [J]. *Science and Technology Innovation*, 2019(33):110–111.

Yu Wubing *Research on cost control in the construction stage of construction projects* [d] Lanzhou: Lanzhou Jiaotong University, 2021.

*Advances in Urban Construction and Management Engineering – Cha (Ed)*


# Optimization of sponge city problems and measures based on water supply and drainage network in Beijing

Zikang Chu*
*Water supply and drainage science and Engineering, Heilongjiang University of China, China*

ABSTRACT: Sponge city construction involves implementing the concept of green development for a major change. The construction model consists of several modules and involves multiple fields and aspects. The paper takes the sponge city construction in Beijing as an example and discusses the effect assessment model for degree of fit, synthesis, and the bioretention construction problems. In addition, the paper also discusses the problems of visible pollution on the blockage of pipe networks and the infiltration of heavy metal pollutants, as well as the construction of relevant regulations and the establishment of advanced concepts in operation and maintenance management. The paper also proposes corresponding countermeasures to the above problems. Multi-model coupling for the construction of facilities in different areas, topsoil replacement of detention facilities, optimization of fillers and planting of enrichment plants, improvement of development concepts, and establishment of relevant regulations can effectively improve the current problems related to sponge cities.

## 1 INTRODUCTION

In recent years, China's cities have grown rapidly and the density of the urban population has increased. While high-intensity construction brings socioeconomic progress, the contradiction between it and resource and environmental bearing has also emerged. Problems such as urban black smelly water bodies and urban flooding are becoming increasingly severe. To solve this problem effectively, the country has issued a series of technical guidelines and guidance for sponge city construction. At present, there are more than 30 national pilot cities and about 100 provincial pilot cities for practice and exploration (Notice on the demonstration of systematic and healing promotion of sponge city construction [R]. Finance Office Construction [2021] No. 35.). Sponge cities have achieved remarkable results in "natural storage, natural infiltration and natural purification" of urban water bodies, and are about to enter a comprehensive promotion stage.

To effectively solve the traditional urban construction mode brought about by the "water problem," a new urban construction concept and urban development model—sponge city—is proposed. This model uses the concept and measures of Low-Impact Development to recycle and reuse rainwater. Using the city's underground pipe network, the characteristic way of rainfall surface runoff is changed to achieve the purpose of alleviating urban flooding, protecting surface water sources, and recharging underground water sources. In 2014, the Ministry of Housing and Construction prepared and released the "Technical Guide for Sponge City Construction" to clarify the content, requirements, and methods of urban planning, engineering design, construction, maintenance, and management processes affecting the construction of rainwater systems for development (Technical guide for construction of sponge cities.). The concept of "sponge city" was explicitly proposed in the official document, and since then the management and maintenance of urban ecological water resources and their reuse have moved from theory to practice.

*Corresponding Author: czk20000082020@163.com

DOI 10.1201/9781003348023-64

The paper focuses on "sponge city" and analyzes the problems faced by Beijing in promoting the normalization of sponge city construction based on the existing water supply and drainage network in Beijing. In the context of the reform of the ecological civilization system and sustainable ecological development, the paper compares different countries and cities with Beijing's sponge city to summarize the problems faced by Beijing's underground water pipeline network and proposes corresponding countermeasures and suggestions.

## 2 STUDY AREA OVERVIEW

Beijing is located in the northern part of China's North China Plain, adjacent to Tianjin in the east, about 150 km from the Bohai Sea in the southeast, and bordering Hebei Province in the north, east, and south. The city covers a land area of 16,410 sq. km. Among them, the plain area is 6,338 sq. km, accounting for 38.6% of the total area. The mountainous area is 10,072 sq. km, accounting for 61.4% of the total area. The urban area is 87.1 sq. km, accounting for 0.53% of the total area. The Beijing area includes large rivers such as the Yongding River, Chaobai River, and North Canal, and there are more than 80 large and small rivers. Among them, the Yongding River and Chaobai River are the two most important rivers in Beijing.

Beijing has a semi-temperate continental monsoon climate with high temperatures and rain in summer and cold and dry winters, with an uneven seasonal distribution of precipitation and frequent short periods of heavy rainfall. The spatial distribution of precipitation is localized, mainly concentrated in the southwestern part of Beijing as well as in the urban area of Beijing, which often faces short-duration heavy rainfall in summer. According to the rainfall observation data from 1950 to the present, 72.3% (39.5%–87.4%) of the annual precipitation is concentrated in summer (June–August), and 81.5% (60.1%–91.5%) of the annual rainfall in the June–September flood season. Beijing is prone to short-term, high-intensity rainstorm events. Urbanization increases the likelihood of storm centers in urban centers, such as the "7–21" rainfall event in 2012 and the "8–12" rainfall event in 2020.

## 3 THE MAIN PROBLEMS AND COUNTERMEASURES OF SPONGE CITY EFFECT EVALUATION

The effect assessment in the construction of a sponge city as a feedback mechanism can effectively respond to the achievements and shortcomings of the construction of the drainage system in the city, and different assessment methods should be adopted for different assessment objects. The effect evaluation of the sponge city has a vital role in promoting the improvement of the sponge city construction effect.

### 3.1 *Assessment points for the construction of sponge cities*

The construction of sponge cities is mainly assessed from multiple perspectives, such as water ecology, water environment, water resources, and water security. It is necessary to combine the natural background of the city, urban construction, implementation effectiveness, industrial development and other aspects of the construction of sponge cities to make an overall systematic assessment from multiple perspectives.

### 3.2 *Problems with the evaluation of the effectiveness of sponge cities*

For the effect and evaluation of sponge city construction, based on the collection and data monitoring, data reliability, and the subsequent data fitting are crucial and are also the basic work for the construction of smart cities and smart water.

Based on the current Beijing for sponge city construction effect, the current modeling software does not have a high degree of the fitting. The inability to source drainage from drainage pipes and drains, the proportion of black smelly water bodies, the proportion of waterlogging point elimination, and other aspects, such as multi-angle systemic comprehensive simulation are required to restore its true effect. That is, the utilization rate of a large number of pipeline data in the underground pipeline integrated management information system is not high. The main manifestation is that the mathematical model of rainfall runoff, drainage network, and inundation in the city needs to have more comprehensive real-time monitoring data and more accurate topographic data. The current single pipe network model and hydrodynamic model cannot simulate the urban flooding and drainage process comprehensively and accurately. The data fit is not high enough to comprehensively analyze the social, economic, and environmental benefits generated by sponge city construction and provide more comprehensive top-level design suggestions for the construction of sponge cities.

## 3.3 *Effectiveness evaluation method solution*

### 3.3.1 *SWMM model introduction*

SWMM, as one of the representative models for urban surface runoff studies at this stage, can effectively simulate urban stormwater runoff processes under short ephemeral rainfall or annual rainfall events (Ming & Xie 2021). WMM was developed by the United States Environmental Protection Agency (EPA) in the 1970s. It is mainly used for urban water and drainage network design, urban stormwater simulation, urban water pollution simulation, and urban low-impact development (LID). The model consists of four computational modules: runoff module, conveyance module, extended conveyance module, storage and treatment module, and a service component for statistical analysis and mapping. The software can simulate the complete urban rainfall-runoff process according to the precipitation of the city, mainly focusing on surface runoff, water flow in the drainage network, flood control and regulation of rainwater, as well as the impact evaluation of receiving water body patterns and water quality (Chen et al. 2015; Zhu et al. 2020).

The core simulation process of SWMM includes the surface flow production process, surface catchment process, pipe network catchment process, and water quality simulation process. Compared with other models this model possesses the ability to deal with the data generated by each sub-basin more comprehensively about precipitation, runoff, pollution load, and underground pipe network storage, and drainage and pumping sluice. SWMM can be simulated continuously at the time scale compared to other models. On the spatial scale, it is mainly applied to cities and is suitable for the simulation of pipe networks, rainfall, and surface runoff in large cities such as Beijing. The simulation of pollutant movement in SWMM is not only limited to the surface but also can simulate the pollution in the pipe network to achieve a comprehensive assessment. SWMM has a variety of function simulation methods, which can fit the data more closely to the actual.

### 3.3.2 *Coupling of SWMM models with other models for different modules*

Currently, SWMM has undergone several versions of optimization and upgrades, its operation pages are faster and more convenient, and the overall degree of fitting to the data is perfect. The modeling software can also be coupled with other modeling software to establish new coupled models, make up for the shortcomings of the modeling software, and expand the scope of evaluation of the sponge city system. It can be better applied to the fitting process of sponge city data monitoring in Beijing.

(1) SWMM-CCHE2D coupled model

The SWMM model and the CCHE2D model coupling combine the pipe network drainage system and the surface hydrodynamic model to simulate the urban drainage system comprehensively. The SWMM model simulates the surface runoff as a sink, and the surface-produced sink flows are fed into the pipe network through rainwater grates, inspection wells, and sponge facilities. The pipe network model is used to simulate the flow and water level of the pipe network, and the inspection wells and rainwater grates are used as overflow outlets to simulate the overflow points (Fang 2022). The model had been applied to Wuhan, combining high-precision topographic data and real-time monitoring data from multiple directions to simulate the corresponding mountainous sponge area with rapid confluence conditions, providing forecast and early warning and emergency management reference data support for urban flooding (Fang et al. 2021).

(2) SWMM-TELEMAC-2D coupled model
SWMM model has mature rainfall production, sink module, and pipe network hydrodynamic module, but it cannot provide important information about inundation such as inundation extent, inundation depth, and water velocity. The TELEMAC-2D model can simulate the two-dimensional hydrodynamics well and has a reserved water input interface. The combination of the SWMM model and TELEMAC-2D model can realize the whole process of simulation of rainfall production and confluence, pipe network flow, and 2D surface flooding. A new coupled model TSWM is constructed based on SWMM and TELEMAC-2D models in the Changjan area of Guangzhou. The numerical simulations are carried out for different periods of rainstorms through the validation of storm events. To improve the coupling computational efficiency and simplify the model operability, this model fitting can take one-way coupling without relying on the model source code for subsequent modification and development, and utilization (Wang et al. 2022).

(3) SWMM-MIKE11 coupling model
SWMM-MIKE11 model establishes the end management measures based on the sponge concept of source reduction and communication water system. The model entails three overall construction ideas: 1. The MIKE11 is used for regional generalization of the sub-water area and pipe network. 2. The MIKE11 model is used for the generalization of the river water system. 3. The pipe network discharge volume flow is added to flow boundary conditions to the MIKE11 model to obtain coupled simulation calculations, to realize the coupling of SWMM and MIKE11 model. The coupled model method has been applied to the Tonghui River water system and Tonglu County. For the Tonghui River system, after analyzing the current status of the water environment system, a hydrodynamic-water quality model was established and an optimal water scheduling scheme was proposed (Yang et al. 2021). For Tonglu County, urban flooding was predicted when encountering a 5-year, 10-year, and 20-year rainstorm, and flood risk maps were drawn in order (Luan et al. 2016). The model can be established to evaluate the degree of regional flooding risk mitigation and river water quality improvement effect under different rainfall recurrence periods respectively, and provide a reference for the decision of sponge city river construction work (Kong et al. 2021).

## 4 BIORETENTION FACILITY CONSTRUCTION PROBLEMS AND SOLUTIONS

### 4.1 *Introduction to bioretention*

Bioretention is a technology that fits into the construction of low-impact development of sponge cities through decentralization and ecology. Bioretention technology has become one of the most widely used low-impact development technologies due to its efficient natural purification and treatment characteristics of rainwater in the specific implementation process (Che et al. 2013; Du et al. 2017). Bioretention facilities are stormwater facilities that store and purify runoff through the plant, soil, and microbial systems in areas with low terrain. Bioretention facilities are divided into simple bioretention facilities and complex bioretention facilities, which are also known as rain gardens, bioretention strips, high-level flower beds, ecological tree ponds, etc. depending on the application location.

Bioretention mode of action: during the rainfall process, rainwater is retained and purified using the combined plant-microbial soil-filler processes of interception, infiltration, sedimentation, evaporation, transpiration, filtration, absorption, assimilation, adsorption, and microbial action (Li & Davis 2008).

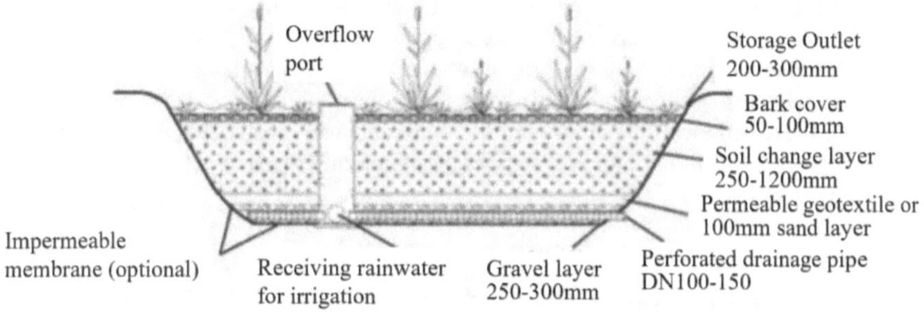


Figure 1. Schematic diagram of Bioretention facility.

### 4.2 *Bioretention facilities are problematic*

Long-term operation of bioretention subsequently brings the problems of internal clogging and accumulation of heavy metal pollutants such as gravel, leaves, dust, food residues, paper, plastic bags, and other pollutants, with rainwater runoff into the facility, resulting in bioretention facility clogging, increased treatment load, secondary pollution, and other problems. Soluble nitrogen, phosphorus, organic matter, heavy metals, and other pollutants continue to accumulate into the facility and gradually reach saturation and infiltration transferred to the fill inside. The accumulation of pollutants will not only affect the normal operation of bioretention facilities but also pose a potential threat to the surrounding soil and groundwater and affect human health (Chen et al. 2013).

Corresponding to different places sponge city bioretention facilities pollutant composition: buildings, residential areas, trade areas, retention facilities mainly set bioretention belt, high-level flower beds, ecological tree pools. Pollutants mainly from the surface of the building body, rainwater runoff. Their main pollutant composition: SS = 136 mg/L, chromium oxide = 123 to 328 mg/L, TN = 6.0 to 9.8 mg/L, TP = 0.2 to 0.94 mg/L, metal ions Hg = 0.000 05 to 0.067 mg/L, PAHs = 197 to 233 mg/L (Huang 2012; Ren 2019). Park green areas, squares, and playground detention facilities are mainly set up with high-level flower beds and rain gardens. The pollutants mainly come from atmospheric precipitation, and pesticides. The main components of its pollutants are polycyclic aromatic hydrocarbons. The content of this pollutant in different cities: Beijing = 0.72 to 40.5 mg/kg; Shanghai = 6.87 to 32.2 mg/kg; Guangzhou = 0.84 to 12.3 mg/kg (Li et al. 2020; Liu 2017). This shows that heavy metal contamination is an urgent problem for bioretention facilities at present.

### 4.3 *Solutions*

#### 4.3.1 *Filler optimization*

In bioretention facilities, the filler is the main factor that affects the adsorption effect and clogging problem of visible, i.e., large particle size pollutants. To improve the pollutant siltation and clogging problems in bioretention, the filler can be optimized according to its particle size, chemical composition, and porosity. For different types of fillers can be mixed and matched in a certain proportion, to ensure the effectiveness of the filler layer for the adsorption of pollutants to improve its infiltration capacity to alleviate the clogging problem.

#### 4.3.2 *Planting super-enriched plants*

Super-enriched plants can enrich heavy metal elements in the soil more than 100 times more than ordinary plants. Different plant species have different enrichment abilities for different heavy metals, and some plants have been found to enrich multiple metal pollutants at the same time. Some plants have also been found to enrich multiple metal pollutants at the same time. The planting of

Bioretention zone facilities should take into account the geographical location, precipitation, and temperature of the area. In the north, cold and drought-tolerant plants, such as horse rushes and reeds, are suitable; in the south, heat- and flood-loving plants, such as lamprey and celandine, are suitable. For the selection of super-enriched plants, the preferred growth cycle is short and the plant growth is large. Preferred local species are predominant and have good adaptability. The preferred ones are cheap-easy to maintain later and possess certaincertain ornamental value and are easy to match with the urban streetscape.

#### 4.3.3 *Replacement of topsoil*

In the process of bioretention of rainwater infiltration, heavy metal elements are also left in the soil, and with the increase of soil thickness, the content of heavy metal pollutants is decreasing with it. This means that heavy metal pollutants are mainly retained in the topsoil. Heavy metal contaminants in topsoil not only cause vegetation soil contamination but also have an impact on human health due to their exposure to humans. Regular replacement of topsoil can effectively reduce the pollution caused by heavy metal element retention in Bioretention facilities and reduce ecological risks and human health risks. The depth and frequency of soil replacement should be determined by taking into account the pollution level, precipitation density, and soil composition of different areas.

## 5 SPONGE CITY DRAINAGE NETWORK OPERATION AND MAINTENANCE MANAGEMENT ISSUES

Currently, the first and second batch of pilot cities of sponge cities in China have been completed and successfully put into use. Because of the current urban situation after the construction of sponge cities, it has a certain effect on urban flooding and pollution mitigation of runoff. Due to the high diversity of sponge city construction facilities, the base is large, and the division is too scattered. To make the sponge city in the later stage, give full play to its flood storage and drought resistance and regulate the role of urban runoff, the sponge city should strengthen its operation and maintenance management for pipe network facilities after completion. Therefore, the operation and maintenance management, and maintenance of drainage network facilities in sponge cities, so that they can play the function of "natural storage, natural infiltration, and natural purification" in a long-term and stable manner, has become a problem that needs urgent attention at present.

### 5.1 *Sponge city facilities operation and maintenance management status quo*

For the construction of sponge cities in different regions, there are differences in their construction subjects as well as socioeconomic attributes. Combined with the current situation of sponge city construction in Beijing, we find that there are some problems in the construction of its drainage network and the management mode of the later stage. First, there is a lack of clear operation and maintenance reference standards and legal provisions at the national policy level, and the lack of reference standards leads to less frequent maintenance or unreasonable maintenance methods, which reduces the life cycle of drainage network facilities in sponge cities (Shan et al. 2021). Second, the management system level lacks a clear responsibility for maintenance of the main body, maintenance management mode single (Chen et al. 2017). The completion of the operation of the sponge city pipe network facilities requires the collaboration of multiple departments and business units, and the unclear division of labor among city management departments will negatively affect the maintenance of the sponge city pipe network at a later stage. Third, the basic information on drainage networks lacks systematic management and dynamic update, and not enough attention is paid to the collection and management of the basic information of drainage networks (Lu et al. 2022). The aging and damage of the pipe network are likely to cause urban flooding, overflow of sewage, and ground subsidence. The current lack of attention to post-construction operation and maintenance management of pipe network facilities can cause serious problems for pipe networks.

Facing the problems of short life span and frequent repair brought by the damaged pipe network increases the cost.

### 5.2 *Solutions*

#### 5.2.1 *Adapt to local conditions and introduce advanced concepts and technologies from abroad*

Foreign developed countries in the process of building smart cities, smart water put forward many management concepts to solve the urban flooding and water ecological crisis, development of a more complete operational management mechanism, and achieved certain results (Zhang et al. 2017). For example, in the United States, Japan, Australia, and other countries in Europe, urban water use management practices have a relatively early start. The above-mentioned countries put forward the concepts of"low-impact development," "water-sensitive urban design," and "integrated water resources management." In some foreign developed countries sponge city design concept is shown in the following table, the design concept in urban planning can be well used, for China's sponge city construction to provide some reference. Some foreign developed countries' sponge city design concept is shown in the following table (Liao et al. 2008; Liu & Chang 2017; Wang 2018; National Research Council. Urban stormwater management in the United States [M]. New York: The National Academic Press 2009; Water Environment Federation. Design of urban stormwater controls [M] New York: Mc Graw Hill 2012).

Table 1. Foreign sponge city development concept in some developed countries.

| Country | Design concept | Application method |
|---|---|---|
| United States | Low impact development, LID | The concept developed based on BMP is mainly applied to the management and maintenance of various rainwater control technologies such as storage, infiltration, evaporation, filtration, and purification. |
| Japan | Japan Water Recycling Basic Law | Emphasizing the role of integrated watershed management, the development should adhere to a development approach that has minimal impact on the water cycle and maximizes the maintenance of a sound water cycle system. |
| Australia | Water sensitivity urban design, WSUD | From the perspective of urban planning and design, the urban water cycle is considered as a whole, and tap water, sewage, and rainwater discharge are managed comprehensively as constituent elements of the water cycle. |
| United Kingdom | The sustainable urban drainage system, SUDS | By considering water quality, water quantity, sewage recycling, economic development, and ecological values, we improve the overall water cycle of the city. |

#### 5.2.2 *Improve the rainwater pipe network management information construction*

The construction of sponge cities occupies an important part of the rainwater pipe network. After the completion of the construction of the network, the operation and maintenance can guarantee the sustainable development of the urban water cycle system. In order to improve the information construction of rainwater pipe network management, not only should we collect the information on the existing urban rainwater management measures and facilities, but also need to collect the information on facilities and observation information containing other industries. Comprehensive use of the currently existing planning is to built information system, check the gaps and fill the gaps, and improve the urban stormwater pipe network management information construction. Will all kinds of information storage, analysis, and maximum usage, over-planned water storage and emergency response plan is requiredto mitigate the damage caused by unexpected events.

### 5.2.3 *Improve related regulations*

Introduce specific regulations that are unified on a national and regional scale. Supporting policies and regulations are the basis to guarantee the effective use and scientific management of rainwater resources. China so far has no national rainwater management policy and regulations, although some water-scarce cities have successively introduced rainwater management incentives and guiding measures, most of them are not systematic enough, not to mention rising to the legal level (Li et al. 2017). In the context of Beijing, we can learn from the relevant policies and regulations in Germany. In 1989, the German government issued DIN 1989 "Standard for Rainwater Utilization Facilities," which is the first national standard for rainwater utilization facilities in Germany, symbolizing that the first generation of rainwater utilization technology in Germany has matured. In 1992, the second generation of rainwater utilization technology emerged. In 1995, the German government issued the first European standard EN 752-1 "Outdoor drains and drainage pipes." In 1996, the German government revised EN 752-1 and pointed out in the revised standard that rainwater harvesting systems should be set up to reduce the risk of flooding on the ground floor of buildings. At the beginning of this century, the third generation of rainwater utilization technology and related standards were formed (Schuetze 2013). Considering the regional characteristics of Beijing and the current situation of water resources and rainwater utilization, appropriate policies were introduced according to local conditions, such as the policy of rainwater reuse in water-scarce areas. In flood-prone areas, the policy of reducing flood peaks is introduced; the construction of sponge cities is strengthened to realize the recycling of water resources, etc. Make rainwater utilization play a bigger role in urban economic and social development.

## 6 CONCLUSION AND OUTLOOK

To sum up, the construction of a sponge city is related to the development of ecological civilization in China, but the problems exist. For example, in the assessment of the sponge city effect, the data fitting model currently used is too single, the fitting degree is not high, and the synthesis is not strong. In terms of Bioretention facility construction, visible pollutants enter the facility with rainwater runoff, causing Bioretention facility blockage. In contrast, heavy metal pollutants will seep down into the soil with surface runoff, causing heavy metal pollution. Policy on sponge city development concept and operation and maintenance management still needs to be improved, and relevant laws and regulations need to be improved.

The following solutions are proposed for the problems in sponge city construction: (1) The SWMM model coupled with CCHE2D, TELEMAC-2D, and MIKE11 models can be used to strengthen the fit and comprehensiveness of the model and realize the upgrade for the assessment system. (2) Solving the problem of the bioretention system for visible pollutant siltation and heavy metal infiltration can be achieved by combining methods such as optimizing filler, planting super-enriched plants, and replacing topsoil. (3) Drawing on advanced foreign concepts and technologies and relevant regulations, the current problems of operation and maintenance management of sponge city facilities can be improved.

## REFERENCES

Che W, Yan P, Li JQ, Zhao Y. Localized research and promotion of low-impact development[J]. *Construction Science and Technology*, 2013(23): 50–52. DOI:10.16116/j.cnki.jskj.2013.23.021. DOI:10.16116/j.cnki.jskj.2013.23.021. [14] Issued by the Ministry of Housing and Urban-Rural.

Chen H, Li JK, Li YJ, Xu Y, Shen B. Research and application progress of stormwater flood management model SWMM[J]. *Journal of Northwest Agriculture and Forestry University of Science and Technology* (Natural Science Edition), 2015, 43(12): 225–234. DOI:10.13207/j.cnki.jnwafu.2015.12.031.

Chen X L, Pelttier E, Sturm B S M, et al. Nitrogen removal and nitrifying and denitrifying bacteria quantification in a stormwater Bioretention system [J]. *Water Research*, 2013, 47(4);1691–1700.

Chen Yao, Duan Linghong, Xiong Yi, Qin Yu, Liu Mengyi, Wu Ming. Maintenance management of rain gardens in the construction of sponge cities [J]. *Water supply and drainage*, 2017, 53(S1):93–96. DOI:10.13789/j.cnki.wwe1964.2017.0441.

Development: Technical Guide for Sponge City Construction–Construction of Low-Impact Development Stormwater System [J]. *Construction Science and Technology*, 2015(01):10. DOI:10.16116/j.cnki.jskj.2015.01.002.

Du Xiaoli, Han Qiang, Yu Zhenya, Li Junqi. Analysis of some problems of Bioretention facilities application in the construction of sponge city[J]. *Water supply and drainage*, 2017, 53(01):54–58. DOI:10.13789/j.cnki.wwe1964.2017.0011.

Fang Yajun, Yu Chuanqi, Jin Xin, Li Jianhua, Xia Lijuan, Ren Mingxuan. Research on the effect of flood control in mountain sponge cities based on the SWMM-CCHE2D one-way coupling model [J]. *Journal of Wuhan University* (Engineering Edition), 2021, 54(10):898–906+941. DOI:10.14188/j.1671-8844.2021-10-003.

FANG Yajun, YU Chuanqi, JIN Xin. Evaluation of flood control effect in sponge cities based on SWMM-CCHE2D coupled model [J/OL]. *Journal of Jilin University* (Earth Science Edition):1–10 [2022-04-21]. DOI:10.13278/j.cnki.jjuese.20200136.

Huang Chenzhe. *Experimental and applied research on artificial soil infiltration technology for urban rainwater* [D]. Beijing University of Architecture, 2012.

Kong Y, Sun W, Li S, Xia Y, Xu H, Yang S. L. Construction and application ideas of SWMM-MIKE 11 coupling model in river sponge construction[J]. *Water Resources Conservation*, 2021, 37(06):74–79.

LI H, Davis A P. Capture and accumulation of heavy metals in Bioretention media [J] *Environmental Science and Technology, Environmental Science and Technology*, 2008, 42(14); 5247–5253.

Li JK, Zhang ZX, Jiang CHB, Wang SHM. Research progress of key technologies for Bioretention facilities in sponge cities[J]. *Water Resources Conservation*, 2020, 36(01):1–8+17.

Li, Chun, Xu, Yanling, Li, Mei. A preliminary study of foreign urban stormwater management policies and measures and their implications for the Beijing-Tianjin-Hebei region[J]. *Environmental Engineering*, 2017, 35(11):6–9. DOI:10.13205/j.hjgc.201711002.

Liao, Chao-Xuan, Zhan, Li-Mei, Chen, Jia-Liang, Qiu, Yi-Ru. Analysis of hydrological cycle mechanisms and improvement strategies in urban areas[J]. *Advances in Water Science*, 2008(01): 49–53.

Liu Liang. *Construction of genetically engineered bacteria to degrade polycyclic aromatic hydrocarbons in Bioretention ponds* [D]. Tsinghua University, 2017.

Liu, D.W., Chang, Y. Analysis of the Water Cycle Basic Act of Japan and its Implications for China [J]. *Water Resources Development Research*, 2017, 17(01):76–79. DOI:10.13928/j.cnki.wrdr.2017.01.018.

Lu Lu, Gao Feng, Guo Juan, Li Baotong, Li Guoqiang. Analysis of drainage network operation and maintenance management problems and countermeasures research [J]. *China Water Supply and Drainage*, 2022, 38(02): 8–13. DOI:10.19853/j.zgjsps.1000-4602.2022.02.002.

Luan Mu,Yuan Wenxiu,Liu Jun,Zhou Yantan,Kulejiang Dosjiang. Risk assessment of inundation in Tonglu County based on SWMM-MIKE11 coupled model[J]. Water Resources Conservation, 2016, 32(02): 57–61.

Ming, R. P., Xie, A. L. Research on the evaluation method of sponge city construction effect based on SWMM [J]. *Municipal Technology*, 2021, 39(06):110–112. DOI:10.19922/j.1009-7767.2021.06.110.

National Research Council. *Urban stormwater management in the United States* [M]. New York: The National Academic Press, 2009.

Notice on the demonstration of systematic and healing promotion of sponge city construction [R]. *Finance Office Construction* [2021] No. 35.

Ren Xiangyu. A multi-objective enhancement-based sponge city construction program for the northwest campus [J]. *Water purification technology*, 2019, 38(12):36–40+54. DOI:10.15890/j.cnki.jsjs.2019.12.008.

Schuetze T. *Rainwater harvesting and management-policy and regulations in Germany J l Water Science and Technology Water Supply*201313(2)376-385.

Shan Xihuan, Fang Zhida, Xie Wenxia, She Nian, Zhao Hongtao, Li Xuyong. Research on the technical system of operation and maintenance management of source engineering facilities in sponge cities [J]. *China Water Supply and Drainage*, 2021, 37(12):42–49. DOI:10.19853/j.zgjsps.1000-4602.2021.12.008.

Technical guide for construction of sponge cities.

Wang F. Australian water-sensitive city assessment practices and its inspiration[J]. *Ecological Economics*, 2018, 34(06):186–193.

WANG Zhaoli,CHEN Yuhong,LAI Chengguang. Numerical simulation of urban flooding based on TELEMAC-2D and SWMM models[J]. *Water Resources Conservation*, 2022, 38(01):117–124.

Water Environment Federation. *Design of urban stormwater controls* [M] New York: Mc Graw Hill, 2012.

Yang F, Wei Zunli, Liu Junmei, Wang Xiaoju. Research on the optimal scheduling scheme for improving water quality of urban rivers and lakes after rainfall based on SWMM and MIKE11 [J]. *Water Resources and Hydropower Technology* (Chinese and English), 2021, 52(07):73–84. DOI:10.13928/j.cnki.wrahe.2021.07.008.
Zhang B, Li JK, Li YJ. Advances in urban stormwater and pollution simulation under different development patterns[J]. *Environmental Science and Technology*, 2017, 40(08):87–95.
Zhu M, He Juntao, Li Jiang Shao, Feng Zhiting, Gao Fangjia, Fang Mian. Simulation analysis of the effect of different LID facilities on rainwater flood control based on SWMM model[J]. *Hydroelectricity*, 2020, 46(03):24–27+79.

*Advances in Urban Construction and Management Engineering – Cha (Ed)*


# Research on evaluation index of smart community construction level based on AHP and FCE

Ting Liu* & Yan Bao*
*Yangtze River Delta Urban Smart Management Institute, Shanghai Urban Construction Vocational College, Shanghai, China*

Liang Huang*
*School of Urban Operations and Management, Shanghai Urban Construction Vocational College, Shanghai, China*

ABSTRACT: Smart community is the "last mile" of urban fine governance and the basic unit of a smart city. This paper refers to the existing community evaluation index system structure in the smart community construction guide and constructs the evaluation index of the smart community construction level from the four dimensions of overall planning, infrastructure, smart application, and data platform. On this basis of the connotation of smart community, a quantitative analysis method is proposed by using the analytic hierarchy process (AHP) and fuzzy comprehensive evaluation (FCE) combining model, which provides a reference for community information construction.

## 1 INTRODUCTION

Since IBM put forward the concept of "smart city" in 2008, the wave of smart city construction has swept the world, and all parts of China are also constantly carrying out the pilot and promotion of smart cities. A community is a small cell unit in the urban system. Intelligence is the "urban brain" and the CPU of the community. With the development of 5G networks, IoT, big data, artificial intelligence, blockchain, and other new generation information technologies, an information-based community system will be built guided by the wisdom, greening, and human culture of the community. The system can maximize the sharing of resources in the community, effectively solve and improve people's livelihood, and create a happy and comfortable living space for residents.

At present, some research on the construction of smart communities has been carried out. Yang et al. (2009) constructed the evaluation index system of China's community information construction by using the analytic hierarchy process, providing some basis for the development of China's smart community construction. Zhang (2013) agreed that smart community evaluation is an important part of community management. Chen (2004) used fuzzy mathematics theory to build an evaluation model of a smart community. Wang (2018) built an index system for the smart community and evaluated the smart community. Majedi (2015) evaluated the effect of a smart community and believed that it was beneficial to the sustainable development of the city. Li (2015) has established an index system for evaluating the development of smart communities to evaluate the development of smart communities. Macke (2019) applied factor analysis and linear regression to evaluate

*Corresponding Authors: ltblueskyday@163.com, baoyan127@163.com and lianghuanguse@126.com

DOI 10.1201/9781003348023-65

the sustainable development of smart communities. Jarke (2017) indicated that the community performance indicator system should be practical. Ma (2018) used the index system to evaluate the construction of local smart communities. And Lou (2013) shows that the evaluation of community performance level is a tool for community development and progress. However, most of the existing literature has only established an index system for evaluating smart communities, without further quantitative analysis, and the conclusions obtained are difficult to have universal applicability.

Therefore, this paper refers to the existing community evaluation index system structure in the smart community construction guide and constructs the evaluation index of the smart community construction level from the four dimensions of overall planning, infrastructure, smart application, and data platform. On this basis of the connotation of smart community, a quantitative analysis method is proposed by using the analytic hierarchy process (AHP) and fuzzy comprehensive evaluation (FCE) combining model, which provides a reference for community information construction.

## 2 SMART COMMUNITY EVALUATION INDEX SYSTEM

### 2.1 *Selection of smart community evaluation indicators*

The purpose of formulating the smart community evaluation index system is to guide the planning, construction, and operation of smart communities, evaluate the effect of smart community construction and play a guiding role and quantitative evaluation role through a quantitative scientific evaluation system. The compilation of the indicator system follows three principles. First, it is forward-looking, and the indicators can represent the latest development level in all fields of the smart community. The second is operability. The selection of indicators should fully consider the scientificity and convenience of data collection. The third is expansibility. The index system can be supplemented, improved, and revised according to the actual situation. In 2014, the guidelines for smart community construction issued by the Ministry of housing and urban-rural development proposed a complete and detailed evaluation system. The system involves six areas, including a security system, infrastructure and building environment, community governance and public services, community management, convenience services, and theme communities. It includes six first-class indicators, 23 second-class indicators, and 87 third-class indicators. Due to the complicated evaluation structure in the guide, it is not easy to operate in the actual evaluation work. Herein, this paper integrates similar indicators to form a new smart community evaluation index system.

### 2.2 *Construction of smart community evaluation index system*

Based on the relevant research results and the principles of operability, objectivity, comprehensiveness, and systematization, a comprehensive evaluation index system for smart communities is constructed. The indicator system includes four primary indicators, including overall planning, infrastructure, smart application, and data platform, and 19 secondary indicators, as shown in Table 1.

Among them, the overall planning of the primary indicators is subdivided into four secondary indicators, including management mode, support system, operational development, and residents' satisfaction; the infrastructure is subdivided into 5 secondary indicators, including information infrastructure, intelligent ecological environment, integrated service facilities, IoT sensing device, and information data center; the smart application is subdivided into 6 secondary indicators, including community government affairs, property management, convenience services, smart healthcare, intelligent old-age care, and intelligent life; and the data platform is subdivided into four secondary indicators, including data security, standard interfaces, data empowerment, and share resources.

Table 1. Smart community evaluation index system.

| Target level | Criterion level | Index level |
|---|---|---|
| Smart community evaluation index system (A) | Overall planning (A1) | Management model (A11) |
| | | Support system (A12) |
| | | Operational development (A13) |
| | | Residents' satisfaction (A14) |
| | Infrastructure (A2) | Information infrastructure (A21) |
| | | Intelligent ecological environment (A22) |
| | | Integrated service facilities (A23) |
| | | IoT sensing device (A24) |
| | | Information data center (A25) |
| | Smart application (A3) | Community government affairs (A31) |
| | | Property management (A32) |
| | | Convenience services (A33) |
| | | Smart healthcare (A34) |
| | | Intelligent old-age care (A35) |
| | | Intelligent life (A36) |
| | Data platform (A4) | Data security (A41) |
| | | Standard interface (A42) |
| | | Data empowerment (A43) |
| | | Share resource (A44) |

## 3 SMART COMMUNITY EVALUATION MODEL

Fuzzy comprehensive evaluation (FCE) is to use the membership degree theory in fuzzy mathematics to quantitatively evaluate those complex multi-factor schemes that are difficult to describe with precise mathematical relationships. In the calculation process of the membership degree of each factor, whether the calculation of the index weight is objective or scientific determines the quality of the fuzzy comprehensive evaluation to a large extent. Compared with the method of subjective qualitative determination of weight, AHP, as a qualitative and quantitative multi-factor decision-making method, can calculate the index weight of fuzzy factors more objective and scientific. This paper adopts the combination of the AHP and FCE methods to evaluate the smart community. The specific calculation process is mainly divided into two parts: the first is to use the AHP method to determine the weight of the evaluation indicators of the smart community, and then based on the AHP method to determine the weight of the indicators, the multi-level FCE method is used to comprehensively evaluate the effect of the evaluation index system.

### 3.1 *Determine the index weight by the AHP method*

1. Constructing judgment matrix and scoring

   First, the smart community indicator evaluation system in Table 1 is constructed into a two-layer hierarchical structure. The first layer is the criterion layer, including four primary indicators of overall planning, infrastructure, smart applications, and data platforms; the second layer is the indicator layer, including 19 secondary indicators. Secondly, the 1-9 scale method proposed by Satty is used to compare and score the relative importance of the indicators. By judging the importance of each index, the judgment matrix of the corresponding index is constructed. According to the above method, the criterion layer index judgment matrix A is constructed, which is shown in Table 2.

Table 2. The criterion layer index judgment matrix A.

| | A1 | A2 | A3 | A4 |
|---|---|---|---|---|
| A1 | 1 | 1/3 | 1/3 | 1/2 |
| A2 | 3 | 1 | 1/3 | 2 |
| A3 | 3 | 3 | 1 | 2 |
| A4 | 2 | 1/2 | 1/2 | 1 |

2. Calculation of index weights and consistency check
   The indicator weight is the relative importance of the indicators at all levels relative to the upper-level factors. The specific calculation steps are as follows:
   (1) Calculate the geometric mean of the elements in each row of the judgment matrix to obtain the eigenvector $\omega$

$$\omega_i = \sqrt[n]{\prod_{j=1}^{n} a_{ij}} \quad i = 1, 2, \ldots, n \tag{1}$$

   (2) Normalize the feature vector

$$\omega_i = \frac{\omega_i}{\sum_{j=1}^{n} \omega_j} \quad i = 1, 2, \ldots, n \tag{2}$$

   (3) Calculate the maximum eigenvalue of the judgment matrix

$$\lambda_{\max} = \sum_{i=1}^{n} \frac{(A\omega)_i}{n\omega_i} \tag{3}$$

   where $(A\omega)_i$ is the element $i$ in the vector obtained by the product of the judgment matrix A and the eigenvector.
   (4) To ensure that experts have the same thinking in comparing the importance of indicators, it is also necessary to calculate the consistency ratio of the matrix. The calculation formula of the consistency ratio is:

$$CR = \frac{CI}{RI} \tag{4}$$

   where $CI$ is the consistency index and $RI$ is the average random consistency index.

$$CI = \frac{\lambda_{\max} - n}{n - 1} \tag{5}$$

   where $n$ is the order of the matrix.

$RI$ can be obtained by looking up the table according to the order of the matrix. The specific calculation results are shown in Table 3 and Figure 1, respectively.

We use the same calculation method to calculate the overall planning, infrastructure, smart applications, and data platforms in the criterion layer, and obtain the weight of each indicator in the indicator layer. Then, we multiply the weight of each indicator in the indicator layer by the weight of the indicator in the criterion layer to calculate the combined weight of the indicators in the indicator layer in the entire indicator system and then sort their importance accordingly. See Table 4 for details.

Table 3. Weight and consistency test of judgment matrix of primary indicators for smart community evaluation.

| Index | Eigenvector | Weight ($\omega$) | $\lambda_{max}$ | Consistency check |
|---|---|---|---|---|
| A1 | 0.425 | 0.1062 | | $CI = 0.055$ |
| A2 | 1.058 | 0.2644 | | $RI = 0.89$ |
| A3 | 1.779 | 0.4448 | 4.165 | $CR = 0.062 < 1$ |
| A4 | 0.738 | 0.1846 | | Pass |

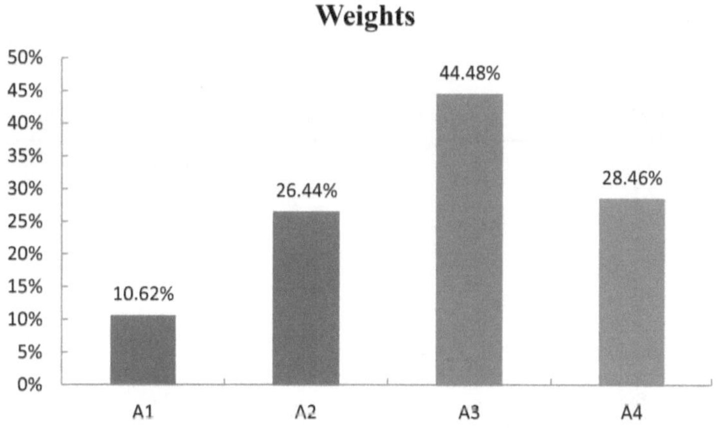


Figure 1. Weights of primary indicators for smart community evaluation.

## 3.2 *Quantitative analysis of FCE*

1. Constructing evaluation factor sets and comment sets
   The evaluation factor set is a set composed of various evaluation indicators that constitute the evaluation object, which is expressed as

$$U = \{u_1, u_2 ... u_n\}$$

The evaluation subsets of the comprehensive evaluation of the smart community are as follows.

$$A_1 = \{A_{11}, A_{12}, A_{13}, A_{14}\}$$

$$A_2 = \{A_{21}, A_{22}, A_{23}, A_{24}\}$$

$$A_3 = \{A_{31}, A_{32}, A_{33}, A_{34}\}$$

$$A_4 = \{A_{41}, A_{42}, A_{43}, A_{44}\}$$

The comment set is a set of grade standards for the change of the evaluation object when the evaluator designs the evaluation value questionnaire, describes the evaluation content in an appropriate language, and assigns points to different grade standards to form a corresponding numerical set. Generally, $V$ is used to represent the comment set. The comments and corresponding value sets of the smart community evaluation value questionnaire are:

$$V = \{V_1, V_2, V_3, V_4\} = \{\text{Excellent, Good, Moderate, Poor, Very Poor}\} = \{5, 4, 3, 2, 1\}$$

2. Build a weight set
   The weight coefficient vectors of the indicators at all levels are obtained from Table 4.

$$W = (0.1062, 0.2644, 0.4448, 0.1846)$$

$$W_1 = (0.4922, 0.2501, 0.1456, 0.1121)$$

$$W_2 = (0.3704, 0.2155, 0.2798, 0.0678, 0.0665)$$

$$W_3 = (0.0991, 0.3561, 0.1534, 0.1766, 0.1789, 0.0359)$$

$$W_4 = (0.3337, 0.2744, 0.2784, 0.1435)$$

Table 4. Weight and ranking of secondary indicators of a smart community evaluation system.

| Criterion layer | Indicator layer | Peer index weight | Overall indicator combination weight | Order of importance |
|---|---|---|---|---|
| Overall planning (A1) (0.1062) | Management model (A11) | 0.4922 | 0.0523 | 9 |
| | Support system (A12) | 0.2501 | 0.0266 | 13 |
| | Operational development (A13) | 0.1456 | 0.0155 | 18 |
| | Residents' satisfaction (A14) | 0.1121 | 0.0119 | 19 |
| Infrastructure (A2) (0.2644) | Information infrastructure (A21) | 0.3704 | 0.0979 | 2 |
| | Intelligent ecological environment (A22) | 0.2155 | 0.0570 | 8 |
| | Integrated service facilities (A23) | 0.2798 | 0.0740 | 5 |
| | IoT sensing device (A24) | 0.0678 | 0.0179 | 15 |
| | Information data center (A25) | 0.0665 | 0.0176 | 16 |
| Smart application (A3) (0.4448) | Community government affairs (A31) | 0.0991 | 0.0441 | 12 |
| | Property management (A32) | 0.3561 | 0.1584 | 1 |
| | Convenience services (A33) | 0.1534 | 0.0682 | 6 |
| | Smart healthcare (A34) | 0.1766 | 0.0786 | 4 |
| | Intelligent old-age care (A35) | 0.1789 | 0.0796 | 3 |
| | Intelligent life (A36) | 0.0359 | 0.0160 | 17 |
| Data platform (A4) (0.1846) | Data security (A41) | 0.3337 | 0.0616 | 7 |
| | Standard interface (A42) | 0.2744 | 0.0507 | 10 |
| | Data empowerment (A43) | 0.2484 | 0.0459 | 11 |
| | Share resource (A44) | 0.1435 | 0.0265 | 14 |

3. Establish a fuzzy relationship matrix and determine the fuzzy weight vector of the evaluation factors

   First, by choosing the appropriate fuzzy relationship matrix construction method, the membership degree subset of each factor of the evaluation object is calculated, and the fuzzy relationship matrix R is established. The second step is to select a suitable synthetic fuzzy operator and synthesize the fuzzy relation matrix R and the fuzzy weight vector $W$ into the fuzzy comprehensive evaluation result vector of each evaluation object. There are four types of common fuzzy operators. The research shows that the weighted average fuzzy operator makes the most use of the fuzzy matrix and can effectively reflect the characteristics of the evaluation object. The model of the weighted average fuzzy operator is M (•⊕). The calculation formula is:

$$b_j = \sum_{i=1}^{n} a_i r_{ij} \tag{6}$$

Therefore, this paper uses the weighted average fuzzy operator to obtain the membership of smart community evaluation indicators, as shown in Table 5.

Table 5. Membership in smart community evaluation indicators.

| Index | | Membership | | | | |
|---|---|---|---|---|---|---|
| | | Excellent | Good | Moderate | Poor | Very poor |
| Criterion layer | A1 | 0.3347 | 0.3795 | 0.2752 | 0.0106 | 0 |
| | A2 | 0.3473 | 0.3899 | 0.2506 | 0.0122 | 0 |
| | A3 | 0.3031 | 0.3349 | 0.3253 | 0.0367 | 0 |
| | A4 | 0.3565 | 0.2245 | 0.3792 | 0.0398 | 0 |
| Target layer | | 0.3280 | 0.3338 | 0.3102 | 0.0280 | 0 |

According to the principle of maximum membership degree of a fuzzy comprehensive evaluation, it can be seen from Table 5 that the membership degree of expert evaluation of indicators at all levels of the criterion layer except "A4" belongs to "moderate", and the other three indicators have the maximum weight (A1 is 0.3795, A2 is 0.3899, and A3 is 0.3349) are all "good". The maximum weight (0.3338) of the membership of the target layer smart community evaluation index belongs to "good".

Finally, multiply the result vector Z of the membership degree of the smart community evaluation index by the assigned comment set U to obtain a comprehensive evaluation score (3.96 points), and it can be judged that the overall evaluation of the evaluation object is "general and close to good" level.

## 4 CONCLUSIONS

In this paper, referring to the existing community evaluation index system structure in the smart community construction guide, the evaluation index of the smart community construction level is constructed from the four dimensions of overall planning, infrastructure, smart application, and data platform. On this basis of the connotation of smart community, a quantitative analysis method is proposed by using the analytic hierarchy process (AHP) and fuzzy comprehensive evaluation (FCE) combining model. The research results show that the overall evaluation of the evaluation index system is close to the "good" level, which can better reflect the characteristics of community intelligence.

Therefore, the evaluation index system for the smart community proposed in this paper can provide a reference for community information construction. In the next stage of smart community construction and development, we will pay more attention to people-oriented community construction, and improve the residents' participation. Then the construction of smart communities can better meet the needs of community residents.

## REFERENCES


Chen Z (2004). Research on evaluation method for the network systems performances in smart residential community. *Electrotechnical Journal*, 08, 52–56.

Li R., Kido A. et al (2015). Evaluation Index Development for intelligent transportation system in smart community based on big data. *Advances in Mechanical Engineering*, 7(2).

Lou M., Catherine B. et al (2013). The community development progress and evaluation tool: assessing community development fieldwork / Un outil pour evaluer les stages en developpement communautaire: Le community development progress and evaluation tool. *Canadian Journal of Occupational Therapy*, 80(4).

Jarke, J (2017). New findings on information technology described by investigators at University of Bremen. *Information Technology Newsweekly*.

Majedi H., Hasanvand S. et al (2015). The effectiveness of community engagement in public health interventions for disadvantaged groups: a meta-analysis. *BMC Public Health*, 15(1).

Macke J., Sarate J.R. et al (2019). Smart sustainable cities evaluation and sense of community. *Journal of Cleaner Production*, 239.

Ma L.H., Chiew K. et al (2018). Evaluation of local community metrics: from an experimental perspective. *Journal of Intelligent Information Systems*, 51(1).

Wang J., Ding S. et al (2018). Smart community evaluation for sustainable development using a combined analytical framework. *Journal of Cleaner Production*, 193.

Yang J. & Jiang H. (2009). Study on the evaluation index system of community informatization construction based on system analysis and AHP. *IEEE*, 364–368.

Zhang C.G. & Jiang C. (2013). Research on integrated management method for intelligent community. *Advanced Materials Research*, 614–615:1973-1978.

*Advances in Urban Construction and Management Engineering – Cha (Ed)*


# Asphalt pavement maintenance plans based on intelligent detection and maintenance decision technology

Zhi-hao Fu, Zhao-hui Ning & Xiao-xia Xu
*Beijing Highway Development Center, Beijing, China*

Cheng Liu*
*China-Road Transportation Verification & Inspection Hi-Tech Co., Ltd., Beijing, China*

ABSTRACT: The use of detection and maintenance decision technology has greatly developed in several engineering fields. This paper is aimed to provide a state-of-the-art survey of detection and maintenance decision technology applications in asphalt pavement. The reported studies are briefly divided into four categories: (1) detection technology of asphalt pavement, (2) intelligent identification of pavement diseases, (3) maintenance decision of asphalt pavement, and (4) conclusion. The conclusions drawn from the analysis show that: Asphalt pavement detection technologies such as fiber grating and acoustic vibration technology have improved the efficiency of pavement detection, but there are still some problems with sensors in terms of durability, systematic, and intelligence. Artificial intelligence and big data analysis have been widely used in intelligent maintenance decision-making of asphalt pavement. Algorithm-based intelligent recognition of pavement smoothness, anti-skid performance, and cracks has also made great progress. Multi-disease intelligent recognition technology can be used for the whole-process disease monitoring of the future road surface, and realize the automatic and efficient analysis of road service performance and road maintenance strategy optimization. However, further research is needed on intelligent decision-making training sets and deep learning algorithms.

## 1 INTRODUCTION

Asphalt pavement has the advantages of convenient maintenance and comfortable driving, accounting for more than 70% of the total road mileage in China. Asphalt pavement is affected by traffic loads and various external conditions during operation, which can easily lead to changes in pavement performance (Hou 2020; Ma 2017). The inspectors can evaluate the health status of the pavement by detecting and monitoring the performance and state of the pavement under the action of traffic load and climate, and by processing the collected data (Xu 2021). Based on the evaluation conclusions, pavement maintenance and maintenance strategies can be formulated scientifically. The intelligent detection technology of pavement technical conditions can timely detect early damage and disease of the structure, provide preventive maintenance, avoid large-scale renovation of the pavement, and effectively reduce the maintenance cost and the impact of traffic. This paper analyzes the research results and application status of intelligent detection and decision-making of asphalt pavement maintenance at home and abroad, analyzes different asphalt pavement detection and monitoring technologies, intelligent identification technology of pavement diseases, and asphalt pavement maintenance decision-making methods, and summarizes the research progress of asphalt pavement maintenance intelligent detection and decision-making, and pointed out the problems existing in the intelligent detection and decision-making of asphalt pavement maintenance,

*Corresponding Author: liucheng.bridge@foxmail.com

DOI 10.1201/9781003348023-66 

to provide guidance and reference for the intelligent detection and decision-making of asphalt pavement maintenance.

## 2 ASPHALT PAVEMENT DETECTION AND MONITORING TECHNOLOGY

### 2.1 *Visualization detection technology*

Pavement detection tools, techniques, and systems can convert large amounts of abstract data obtained experimentally or numerically into computer graphics images that can be directly perceived by human vision. Using the detection data to accurately calculate the pavement characteristics, make the evaluation results of the pavement technical condition more comprehensive and reliable. Asphalt pavement detection and monitoring visualization technologies mainly include cameras, line scan cameras, infrared cameras, photometric stereo technology, thermal infrared imaging, ground laser scanning, etc. The limitation of the technologies is that the amount of data processing and calculation is relatively large, which requires the support of specific image processing algorithms, and the detection accuracy is easily affected by the measurement environment, as shown in Table 1.

Table 1. Asphalt pavement visualization detection technology features.

| Visualization technologies | Features |
|---|---|
| Cameras | 3D raw information; road pavement geometric reconstruction |
| Line scan cameras | High data transmission efficiency and real-time recording; real-time road information |
| Infrared cameras | Automatic acquisition of image data; non-contact rapid non-destructive testing |
| Photometric stereo technology | The normal vector of the object surface; the reflectivity of object surface points; road pavement geometric reconstruction |
| Thermal infrared imaging | Infrared radiation intensity converted to visible light; non-contact rapid non-destructive testing |
| Ground laser scanning | 3D coordinates of road surface, surface reflectivity, and texture; road pavement geometric reconstruction |

### 2.2 *Fiber bragg grating monitoring technology*

The basic theory of the Fiber Bragg Grating (FBG) sensor is to measure strain and temperature changes by monitoring the offset of Bragg reflection center wavelength. When the FBG sensor has an external temperature change or is deformed by loading, the tension or compression of the grating may cause periodic variation in Bragg wavelength. FBG Sensing Technology can be used for quality control of the compaction of asphalt pavement, as well as monitoring the structural response of asphalt pavement and evaluating pavement performance.

Fiber Bragg grating sensor has a simple structure, small, and is easy to embed into composite components and large-scale buildings, providing real-time monitoring of the structure's integrity, safety, fatigue levels, damage degree, and other states. The measurement results have good repeatability, it is suitable for measuring static or quasi-static physical quantities based on strain and temperature changes, and is not suitable for measuring dynamic signals (such as vibration signals) and humidity, wind speed, and other signals.

## 3 INTELLIGENT RECOGNITION OF PAVEMENT DISEASES

Intelligent detection of road diseases is essential to realize two-dimensional or three-dimensional imaging and recognition of diseases. The imaging conditions are random, and image recognition

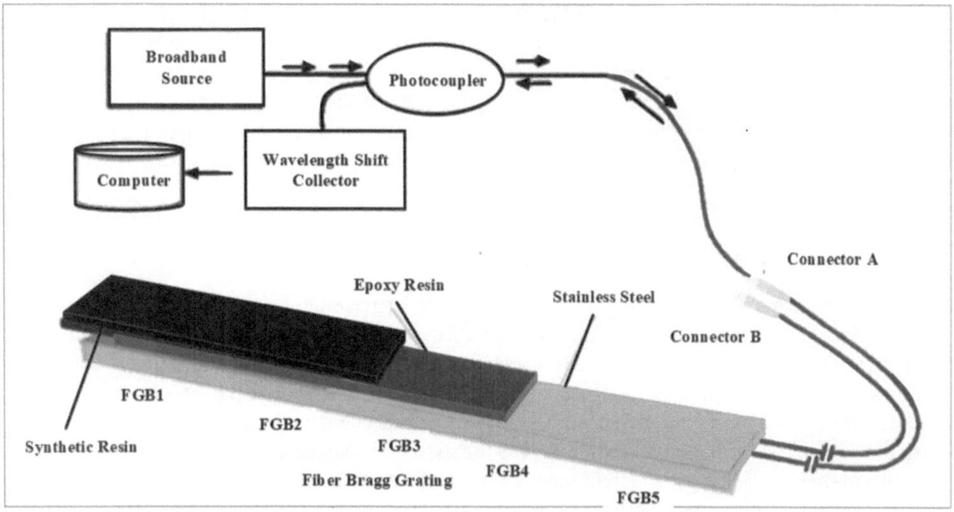


Figure 1. FBG sensing beam system.

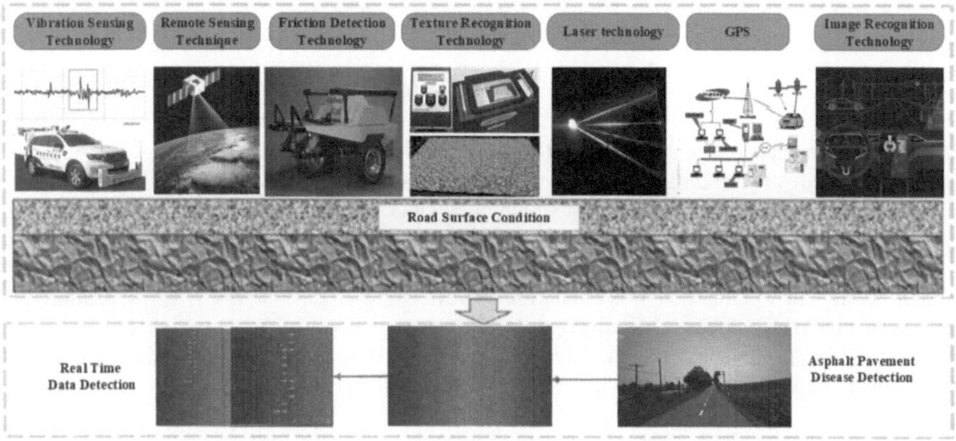


Figure 2. The intelligent detection process of pavement diseases.

depends on the algorithm. Measurements such as potholes require good imaging results in a reasonable space, while assessments such as cracks require very accurate imaging in a local area. These two extreme cases pose challenges for imaging techniques and recognition algorithms. Through the comprehensive application of relevant technologies, common diseases such as pavement flatness, skid resistance, rutting, and cracks are intelligently detected.

## 4 THE MAINTENANCE DECISION-MAKING ON ASPHALT PAVEMENT

The core of decision optimization of highway pavement maintenance management system lies in how to make the minimum budget fund scheme under the condition that the road performance and service level do not decline, and thus determine the specific maintenance decision so that the road pavement maintenance fund and maintenance quality reach the optimal balance. With the development of information technology, the intellectualization of asphalt pavement maintenance decisions has been improved. Based on the existing pavement maintenance management system, the introduction of preventive maintenance, artificial intelligence, big data analysis, and other advanced maintenance concepts and analysis methods, can effectively improve the routine maintenance management in the pavement disease identification, efficiency of road service performance evaluation and it plays a driving role in optimizing the multi-index decision-making system and carrying out systematic maintenance decisions.

### 4.1 *Decision tree method*

The decision tree method is based on the performance of pavement used to judge, through the introduction of a comprehensive evaluation index of the technical condition of asphalt pavement, combined with the special detection data of each road condition, traffic load, climatic environment, and construction quality and other factors grading weights, the use of the trigger value of the evaluation index to classify the disease, to achieve a comprehensive judgment and evaluation of the pavement test performance, and then divide the assessed road section into daily maintenance, preventive maintenance, repair and maintenance, special maintenance and emergency maintenance, etc. This decision-making method has the advantages of easy data collection, a simple evaluation model, and clear sub-indicators, but the method also has defects such as vague boundary values, large differences between different evaluation indicators, and difficulty in accurately defining the type of conservation.

### 4.2 *Preventive conservation decisions*

Preventive maintenance refers to the maintenance activities in that the original road is in good condition without structural damage and only strengthens its functionality or usability. Preventive maintenance measures for asphalt pavement mainly include the cost-benefit analysis method, life-cycle cost method, decision tree decision matrix method, comprehensive evaluation method, cluster analysis method, and analytic hierarchy process. Different methods have their characteristics and limitations, and their implementation effects are also affected by road conditions, traffic conditions, and natural climate. To effectively improve pavement performance, prolong pavement life, and save maintenance funds, preventive maintenance measures should be scientifically carried out according to pavement conditions, external environment, and financial status.

### 4.3 *Asphalt pavement maintenance based on big data statistical analysis*

The starting point of road maintenance based on big data analysis is to establish a management database based on large-scale road monitoring data, deeply excavate the internal relationship between various data, statistics, analysis and prediction of road service performance and service life, and provide better auxiliary services for road maintenance decision-making. The traditional project-level asphalt pavement maintenance decision-making method is usually made by maintenance engineers according to the road conditions on site, which is not only inefficient but also highly dependent on the experience of inspectors and subjective. With the rapid development and wide application of artificial intelligence, road maintenance and management have entered a new stage. Artificial intelligence belongs to a branch of computer science, its powerful feature learning and classification capabilities are suitable for image classification, object detection, image semantic segmentation, and other fields. The basic idea of road maintenance based on artificial intelligence is to apply neural networks, deep learning (convolutional neural network), expert system, and other artificial intelligence methods and means to fully analyze road diseases, road service performance, and road maintenance strategies to realize efficient road maintenance management and operation services.

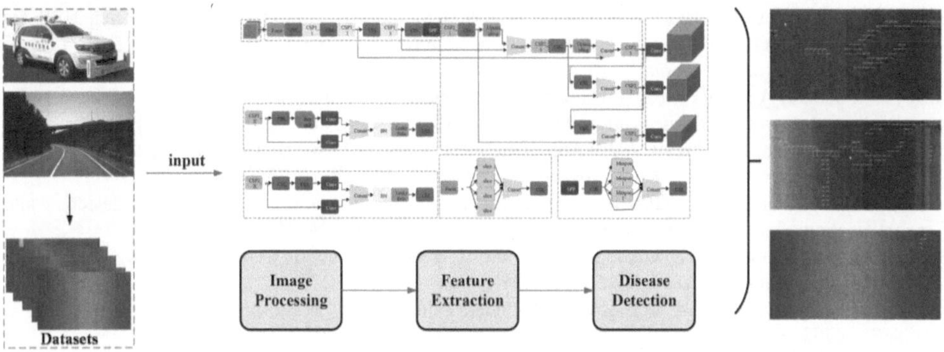


Figure 3. Applications of object detection in asphalt pavement disease detection.

## 5 CONCLUSIONS

The key to asphalt pavement maintenance is to measure and evaluate its road performance accurately. With the advancement of sensor technology, sensors with small sizes, high precision, low costs, and wide applications are increasingly used for road information perception. However, the sensor hardware technology is still insufficient in precision and sensitivity. Due to the insufficient durability of the road surface, the embedded sensor is prone to damage, which reduces the reliability of the use, the use efficiency of monitoring data is low, the information integration is not high, and the data mining is not deep enough. With the progress of science and technology, sensors will continue to develop in the direction of miniaturization, systematization, and intelligence, providing a strong technical guarantee for the detection and monitoring of asphalt pavement.

The highway is a trans-regional belt structure, and the road surface disease is a trans-temporal process from occurrence to development. At present, the intelligent road surface detection technology that realizes the whole process has rarely been reported. With the rapid development of micro-sensing technology, Beidou positioning system, and computer technology, it is expected to develop an integrated intelligent disease recognition machine to monitor the whole process of pavement disease and provide a scientific basis for the optimization design of pavement materials.

Method of artificial intelligence and big data analysis method has been widely used in the asphalt pavement maintenance decision, performance evaluation of road disease identification, road service, road maintenance strategy optimization in areas such as automatic efficient analysis, digging deeper into the intrinsic relation between various monitoring data, statistics, analysis, the road service performance, and service life prediction, to provide better auxiliary services for road maintenance decisions. At present, the common use of neural networks, machine learning, and other artificial intelligence technologies to establish pavement disease identification models requires many uniform formats, high-quality, open detection, and monitoring data databases to provide training samples. Moreover, it is important to develop a deep learning algorithm tailored to the characteristics of the industry for small data sets related to pavement maintenance, as well as a general-purpose rapid intelligent learning algorithm, and a platform for on-site maintenance. It is important to support and serve asphalt pavement maintenance intelligently in a country with a strong transportation infrastructure.

## REFERENCES


Alamdarlo M N, Hesami S. (2018). Optimization of the photometric stereo method for measuring pavement texture properties. *J. Measurement*, 127: 406–413.

Biscarini C, Catapano I, Cavalagli N.(2020). UAV photogrammetry, infrared thermography and GPR for enhancing structural and material degradation evaluation of the Roman masonry bridge of Ponte Lucano in Italy. *J. NDT & E International*, 115: 102287.

Gendy A E, Shalaby A.(2007). Mean profile depth of pavement surface macrotexture using photometric stereo techniques. *J. Journal of Transportation Engineering*, 133(7): 433–440.

Hou Yue, Chen Yihan, Gu Xingyu.(2020) Automatic identification of pavement objects and cracks using the convolutional auto-encoder. *J. China Journal of Highway and Transport*, 33(10): 288–303.

Ji AK, Xue XL, Wang YN. (2020). An integrated approach to automatic pixel-level crack detection and quantification of asphalt pavement. *J. Automation in Construction*, 114: 103176.

Ma Jian, Zhao Xiangmo, He Shuanhai.(2017) Review of pavement detection technology. *J. Journal of Traffic and Transportation Engineering*, 17(5): 121–137.

Ma WH, Dong T, Tian H. (2014) Line-scan CCD camera calibration in 2D coordinate measurement. *J. Optik*, 125(17): 4795–4798.

Xu Peng, Zhu Xuan, Yao Ding, Shi Changyun, Qian Guoping, Yu Huanan.(2021) Review on intelligent detection and decision-making of asphalt pavement maintenance. *J. Journal of Central South University* (Science and Technology), 52(7): 2100–2117.

*Advances in Urban Construction and Management Engineering – Cha (Ed)*


# Exploring the agile governance model of city clusters in the context of smart cities—Taking the Guangdong-Hong Kong-Macao Greater Bay Area as an example

Mengqi Du* & Yifen Yin*
*Macao Polytechnic University, Macao, China*

ABSTRACT: Digital transformation inspires the development of smart cities, as well as new requirements for the governance model of urban agglomerations. As an endogenous driving force for the development of urban agglomerations, smart cities are an important grip for the enhancement of regional collaborative governance capacity and governance modernization. Urban cluster governance is highly complex and uncertain, especially the cross-border and cross-system characteristics of urban cluster governance in the Guangdong-Hong Kong-Macao Greater Bay Area. It is necessary to promptly respond to the diverse demands of urban agglomeration governance, respond flexibly to the external governance environment full of uncertainties, and achieve integrated regional development. Agile governance of city clusters is an innovative idea of empowering agile thinking in the governance of city clusters. Thus, this paper proposes the "value-institution-synergy" system of agile governance in the Guangdong-Hong Kong-Macao Greater Bay Area city cluster. It also explores the path of agile thinking through inter-governmental coordination governance, common market governance, refined social governance, and cultural integration governance. to achieve agile governance of urban agglomerations and better promote the development of modernized urban agglomeration governance.

## 1 INTRODUCTION

"The 14th Five-Year Plan" proposes to strengthen the construction of digital society and digital government and to enhance the digital intelligence of urban services and social governance. The rapid economic growth of the Guangdong-Hong Kong-Macao Greater Bay Area has driven the extremely rapid expansion of total population and town size, bringing challenges and transformation burdens to the governance of urban agglomerations.

In the face of the challenges of mega-city cluster governance, such as shortage of governance resources, lagging support system, poor governance process, imbalance of governance structure, and decentralized governance (Yu & Dai 2018), we are facing the challenges of mega-city cluster governance. In the reform of urban cluster governance, intergovernmental cooperation is more helpful than market-driven and locational advantages to break the shackles of administrative boundaries and become the endogenous driving force of urban cluster development (Suo 2016). However, excessive transaction costs, risks of cooperation, and adverse effects from political tournaments have become the central dilemma facing the governance of urban agglomerations. There is an urgent need for city cluster governance systems to break out of their inherited paradigms and adapt to the challenges of the 21st century (Nuno et al. 2018).

*Corresponding Authors: P1909839@ipm.edu.mo and yfyin@ipm.edu.mo

DOI 10.1201/9781003348023-67

Digital technologies are profoundly influencing the innovation of governance systems and the transformation of governance in urban agglomerations. As a special global cross-border and cross-governance bay area, the Guangdong-Hong Kong-Macao Greater Bay Area, under the institutional environment of "one country, two systems", faces a series of urban cluster governance challenges such as emergency management, public service supply, and demand distribution, natural resource sustainability, and scientific urban spatial layout, etc. How to effectively improve the efficiency of cooperation through digital technology and establish a world-class urban cluster with institutional advantages. With the current rise of new-generation information technology such as big data, 5G, cloud computing, blockchain, and the change in the global governance system (Dai & Bao 2017), smart city development becomes the intersection of digital empowerment and governance-driven (Wen et al. 2022), in the context of smart cities, there is an urgent need for innovation in the governance model of urban agglomerations. Agile governance is an innovative theory that has emerged along with the reform of digital government and digital public services in various countries (Mergel et al. 2018, 2021), leading urban governance change with agile thinking, driving technological and institutional change, and taking advantage of synergies to promote the modernization of governance capabilities (Xue & Zhao 2019). The application of agile thinking in the digital development of the world-class city clusters in the Guangdong-Hong Kong-Macao Greater Bay Area is urgently needed to bring efficiency and convenience to the development of the city clusters and to unleash the innovative potential of city cluster governance.

## 2 THE REAL NEEDS OF GOVERNANCE IN THE GUANGDONG-HONG KONG-MACAO GREATER BAY AREA URBAN AGGLOMERATION

Smart city development is driving changes in the governance system of urban agglomerations, and digital technology provides technical support for innovation in the governance of urban agglomerations in the Guangdong-Hong Kong-Macao Greater Bay Area. At the same time, the diversity of governance needs, the uncertainty of the governance environment, and the integration of regional cooperation jointly drive agile thinking to empower the governance of urban agglomerations in the Greater Bay Area of Hong Kong and Macao and to realize the modernization of urban agglomerations.

First, it responds to the diversity of urban cluster governance needs promptly. The resident population of the Guangdong-Hong Kong-Macao Greater Bay Area has now basically reached over 70 million people. Smart cities bring more instant information and convenient life to residents, but the integration of large cities also brings new challenges to the governance of urban clusters. The complex governance system needs to address the individual needs of urban cluster residents for a wide range of local and cross-domain public services and infrastructure. In the face of the "big city disease", where the spatial distribution of population is not reasonable, and to make the spatial allocation of high-quality public resources more balanced, the government has been able to provide a more balanced distribution of public resources (Shi 2014), a faster and more convenient supply is an urgent issue for the current governance of urban agglomerations.

Second, it responds effectively to the complex and changing external environment of urban development. The unstable world development situation is prone to multiple crises such as social disorder, economic decline, and public sector failure in the country, affecting the sustainable development of the Guangdong-Hong Kong-Macao urban agglomeration. This includes the following two aspects. First, urban clusters are required to respond quickly to emergency events. Events such as COVID-19 and other public health events, as well as natural disasters such as typhoons, bring challenges to the emergency management linkage, economic cooperation, and social stability of the governments of Guangdong, Hong Kong, and Macao. Next, the counter-current of economic globalization has evolved the trend of unilateralism and local protectionism. Hong Kong and Macao create negative perceptions when cooperating with mainland cities to maximize their interests. Agile thinking and digital tools are needed to further stimulate market linkages, innovate city cluster governance, and overcome the changing external environment with a more agile city cluster governance model.

Finally, accelerate the iteration of the collaborative governance model for urban clusters. In the future, China's economy will move towards a more coordinated, higher quality, and more competitive regional integration, which needs to rely on the important vehicle of city clusters (Yun & Guo 2022). Guangdong, Hong Kong, Macao, and the Greater Bay Area to build a world-class city cluster goal to break through the traditional regional collaborative governance model. Under the "one country, two systems", the "9+2" cities interact with each other in a multidimensional and people-oriented manner. Break through the administrative boundaries, focus on grassroots interdepartmental cooperation, timely and effective response to the needs of the people, and feedback system effects, so that urban clusters have more intelligent governance. Cross-border government departments need to adapt to the requirements of integrated city cluster governance as soon as possible and apply agile thinking to accelerate the pace of transformation.

## 3 "VALUE-INSTITUTION-SYNERGY": AN INNER SYSTEM OF AGILE GOVERNANCE IN URBAN CLUSTERS

Since China's 14th Five-Year Plan, "Accelerating Digital, Development Building Digital China" has become the new goal and direction of the Guangdong-Hong Kong-Macao Greater Bay Area, the construction of digital urban agglomeration is also being stepped up. Agile governance is people-centered, with flexible iterations of governance driven by institutional innovation and smart cities responding collaboratively to diverse governance needs. The concept aims to "build a governance model that can respond quickly and responsively to public needs to improve the efficiency of organizational operations and the user experience (Fowler & Highsmith 2001; Mergel et al. 2018; Takeuchi & Nonaka 1986)".

### 3.1 *Putting people at the center is the core value of agile governance in urban clusters*

Putting people first and meeting the diverse needs of the people are the core values of agile governance in urban clusters. The characteristics of agile governance include participatory, resilient, inclusive, sustainable, responsive, and people-centered (Xie & Ren 2022). Agile governance can facilitate the ability of government departments to quickly identify social problems and respond to social needs promptly, scientifically predict their development trends and potential impacts, rapidly adjust organizational structures and workflows, and redesign public goods and services rapidly and iteratively (Yu 2022).

In an era of rapid technological progress, the public is receiving far more information than ever before and has a higher demand for public services and public product quality. Digitally driven city cluster governance must respond to the needs of the people in a timely and effective manner, transforming city governments into agile organizations with rapid responsiveness (Chen 2022), to alleviate the contradiction between the diversity of governance needs in the digital age and the lagging response of government departments, and to reshape the form and governance of government. Bringing agile thinking to public administration enables governments to respond to changing public needs in an effective manner (Ines et al. 2020). Smart cities are "user-centric" and focus more on the city's citizens and other stakeholders than just the concept of a "city" (Calderoni 2012). Then, in the context of smart cities, urban cluster governance should grasp the advantages of science and technology while also innovating and optimizing the governance system to move towards a more agile path of governance.

### 3.2 *Institutional innovation is an important driver of agile governance in urban agglomerations*

The continuous change of digital technology has led to the continuous optimization and change of the governance system of urban clusters to meet the requirements of the development of smart city clusters while satisfying the diversified needs of people for a better life. If institutional change cannot keep pace with technological innovation, then even the most advanced governance technology will not be as effective as it should be (Chen 2021). The governance system of urban agglomerations contains multi-disciplinary, multi-level, and cross-sectoral governance systems, especially in the Guangdong-Hong Kong-Macao Greater Bay Area, which also involves cross-border and cross-institutional mechanisms. Agile governance of urban agglomerations requires a high degree of cooperation among local governments to promote the development of diverse and integrated urban agglomerations with open-minded absorption and innovative reforms.

Inter-governmental cooperation is the cornerstone of urban cluster governance. However, the "top-down" hierarchy of government departments in urban agglomerations is not sufficient to meet the demand for efficient governance in urban agglomerations in terms of precise identification of people's needs or the use of technology in departmental programs. The innovation of the urban cluster governance system should grasp the opportunity of technological change, consider the multiple objectives of public interest, collaborative development, and improving governance effectiveness, and continuously promote urban cluster governance to agile governance driven by both technology empowerment and institutional empowerment. By using digital technology and creating an open governance system, we can build a mutually beneficial, policy-coordinated, and mechanism-coordinated urban cluster governance system through the extensive absorption and integration of regional development systems, and dovetail demand and services with a fit, fine, efficient, and convenient response mechanism.

### 3.3 *Smart city cooperation is the fundamental guarantee for agile governance of urban clusters*

Decentralized governance patterns are an obstacle to modernizing the governance of urban agglomerations. Smart cities advance the development process of smart city clusters. A smart city cluster is a high-level construction consortium of multiple cities in a specific regional context, which can be a "cluster" construction of a smart city or a "smart" upgrade of a city cluster (Hu et al. 2021).

On the one hand, enhance the synergistic governance effect of urban clusters. Unlike other governance models, the issues of inter-governmental coordination, social integration, and regional market activities are intertwined in the "synergy" of urban cluster governance. To realize the modernization of collaborative urban governance, we need to understand the basic rules and main characteristics of the intersection of regional government, society, market, and multiculturalism. With the help of modern digital technology tools and methods, playing a smart city in the development of ideas, management tools, efficiency transformation, and other aspects of innovation lead the role (Li & Liu 2020). On the other hand, promote the continuous evolution of smart city cluster governance to resilient governance and smart governance. The concept of toughness includes both the ability to withstand and adapt to pressure impacts and the ability to recover and regenerate after an impact (Shao & Xu 2015). Resilient governance of urban agglomerations emphasizes that when responding to uncertain events such as emergencies, urban governance can not only keep the original system continue to functioning but also restore the basic functions of the city as soon as possible under the impact, which is also the goal of agile governance of urban agglomerations. Smart governance places greater emphasis on the need for technology to be subordinated to the systemic governance of ideas, values, and other factors (Xia & Tan 2017), The finer satisfaction of the most fundamental needs of the people coincides with the concept of agile governance of urban clusters.

## 4 AGILE GOVERNANCE SYSTEM IN THE GUANGDONG-HONG KONG-MACAO BAY AREA CITY CLUSTER: THE WAY FORWARD

From a public management perspective, agile empowerment is a skillful tool that acts on the process of development to improve the efficiency of the subject. Thus, agile governance of urban agglomerations can be defined as the application of agile thinking to give urban agglomerations more creative space for development to stimulate a more refined governance model for urban subjects. Putting agile thinking throughout the inter-governmental coordination, common market, and social integration in the governance system of the Guangdong-Hong Kong-Macao Greater Bay Area city cluster, covering innovation in five areas: institutions, mechanisms, models, and policies and services (Figure 1).

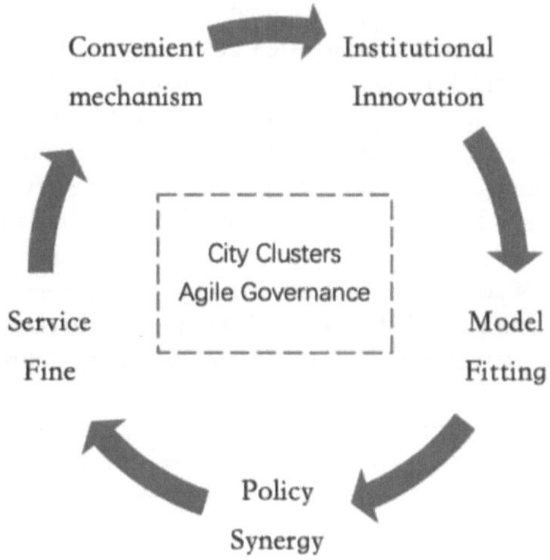


Figure 1. The Framework of the Agile Governance Model for the Guangdong-Hong Kong-Macao Greater Bay Area City Clusters.

### 4.1 *Inter-governmental coordination governance*

Building a common market is an important foundation for achieving market integration in the Guangdong-Hong Kong-Macao Greater Bay Area city clusters. City cluster governance depends on the level of economic development and the openness and vitality of the regional market. The effective cooperation between city governments is the key to the high-quality development of city clusters. The unified leadership of the central government, the autonomy of local governments, and the high degree of autonomy of the Hong Kong and Macao SAR governments in the governance process of the Guangdong-Hong Kong-Macao Greater Bay Area city cluster (Fang 2021), the multi-level subjects that make up the governments of Guangdong, Hong Kong, and Macau have some difficulty in interfacing and collaborating. In the development of smart cities, agile culture is embedded in the idea of institutional transformation, and a more modern urban agglomeration development strategy is developed based on a digital analysis of "cost-benefit-risk-assessment" by different urban agglomeration governance actors, such as local government departments, private sector, NGOs, and citizens, coordinated by the central government.

From the perspective of horizontal collaboration of city governments, information is effectively collected with the help of digital technology, and data is accurately analyzed through agile thinking. Translating complex urban agglomeration governance issues into targeted policy and administrative measures allows Guangdong, Hong Kong, and Macao to form a collaborative governance model under the premise of diversified development and improves the science of decision-making. From the perspective of vertical coordination of government departments, the difference between the administrative systems of Hong Kong, Macao, and the mainland makes the policy implementation effectiveness different and builds a "bottom-up" data platform to give policies according to the needs of classification. The application of digital technology can realize the implementation and feedback mechanism of "resident-organization-policy", and agile governance can effectively deal with the complex governance problems of the Guangdong-Hong Kong-Macao city cluster. However, local governments may conceal information for strategic purposes and derive personal benefit from it (Shresthaha 2008), a practical benefit distribution mechanism needs to be developed based on the data element system to ensure effective cooperation in inter-governmental governance.

### 4.2 *Common market governance*

Building a common market is an important foundation for achieving market integration in the Guangdong-Hong Kong-Macao Greater Bay Area city clusters. A city cluster's governance is affected by its level of economic development, and the openness and vitality of the regional market also have an impact. The Guangdong-Hong Kong-Macao Greater Bay Area market opening characteristics are not significant, in Hong Kong, Macao, and Guangdong under the influence and constraints of institutional mechanisms and economic systems, the development of regional market integration is still difficult. To promote the coordination of interests and win-win cooperation among cities by common market rules, and to form an economic community of city clusters with industrial clusters. The use of advanced mathematical technology, deep into the supply-side structure "inside" to reverse the imbalance between the financial and real economy (Chen 2022), drives the matching of supply and demand in the common market of city clusters, reduces transaction costs and transaction risks, and the free flow and rational allocation of regional resource factors.

In the current context of "reverse globalization" and the intention of Western countries to "decouple" from China's industries, large domestic city clusters should take on the responsibility of building a complete domestic industrial chain and creating a world-class regional industrial chain supply chain center (Wang 2020). As one of the three major city clusters in China, the Guangdong-Hong Kong-Macao Greater Bay Area has the development goal of building a world-class city cluster. The free and convenient flow of data elements provides data support for the regional business environment and realizes the common market governance of the Guangdong-Hong Kong-Macao Greater Bay Area city cluster.

Technology misuse, privacy protection, and failure risks associated with digital technologies increase the technical complexity of urban governance (Sancino & Hudson 2020), it also constrains the good economic development of urban clusters. Only by continuously promoting the integration of agile thinking in common market governance will we be able to anticipate potential risks and effectively address existing risks promptly. In addition, we should strengthen the regulation and protection of the common market between Guangdong, Hong Kong, and Macao, and develop an emergency linkage system between cities. We will work together to find the optimal solution and to build a community of destiny for the Guangdong-Hong Kong-Macao Greater Bay Area.

### 4.3 *Refined social governance*

The fundamental aim of social governance in urban agglomerations is to make life better for residents. At present, the mega-population size brings difficulties in the interface between social governance and inter-city government service systems in urban agglomerations. Infusing agile culture into urban grassroots governance docking, empowering city cluster government service system, making city clusters more efficient, convenient, and wise. Agile governance can respond to social needs more precisely, achieve refined social governance, and build a highly resilient governance structure of functional city clusters in the Guangdong-Hong Kong-Macao Greater Bay Area. The city should provide diverse, human-machine service channels and methods to multiple citizens, and then give the choice to the citizens to choose online or offline services according to their digital capabilities, specific service scenarios, and needs (Zheng 2021). Nurturing smart citizens, transforming them from consumers to participants (White et al. 2021), enhancing the ability and willingness of multiple subjects to participate in social governance, and promoting the formation of a social governance community (Yu & Fan 2022).

Through the integration of agile thinking into modern information technology, including artificial intelligence and the Internet of Things, a specific response mechanism for social governance is established, with residents' needs as the starting point and ultimate destination of social governance.

The purpose is to convert the "rough" governance thinking, to meet the expression of multiple interests, and enhance the level of personalized service (Xue & Wu 2022). Through data collection and analysis to refine community services, we can accurately sense and solve problems such as pre-school education, elderly services, community security, and community sanitation, making the city warmer and attracting more talents for the development of the Guangdong-Hong Kong-Macao Greater Bay Area city cluster. With big data platform and other technical means to promote the standardization of refined social governance in the Guangdong-Hong Kong-Macao Greater Bay Area city cluster, it can meet social needs, safeguard social rights and interests, coordinate social relations, guide social behavior, solve social problems, improve social security, promote social justice, prevent social risks, optimize the social environment, obtain the best social order, and promote social civilization and harmony and security and stability (Tian 2020).

## 5 DISCUSSION AND FURTHER RESEARCH

At this stage, the smart city construction in Guangdong, Hong Kong, and Macao are still under development, and there is a mismatch between urban cluster governance and digital governance structure. In the next phase, urban cluster governance should continue to play the role of big data, the Internet of Things, blockchain, and other technologies. At the same time, incorporate agile governance thinking as much as possible to promote regional social interaction in the Guangdong-Hong Kong-Macao Greater Bay Area, especially to strengthen the cooperation of institutions, mechanisms, policies, services, and models among the four central cities of Guangzhou, Shenzhen, Hong Kong, and Macao.

The agile governance model of the Guangdong-Hong Kong-Macao Greater Bay Area city cluster is the key to enhancing world competitiveness and influence. Digital technology, the most advanced governance tool available, can help urban agglomerations achieve deeper inter-governmental synergies, a regional common market, social refinement, and deep cultural integration. As the smart city is gradually developing, city cluster development is promoted with "agility" and "synergy" so that it meets the challenges of globalization and integration, while injecting new energy into the development of a world-class city cluster within the Guangdong-Hong Kong-Macao Greater Bay Area.

The following studies may be conducted in the future. First, the future innovation model of the world-class city cluster governance system for the Guangdong-Hong Kong-Macao Greater Bay Area needs to be explored in greater depth. Furthermore, this paper examines how agile thinking can be integrated into governance within the Guangdong-Hong Kong-Macao Greater Bay Area urban agglomeration considering smart cities, and further research is required on how to leverage intergovernmental synergies, common market development, and integration of social services through agile governance.

## REFERENCES

Calderoni L, Maio D and Palmieri P. (2012). Location-aware Mobile Services for a Smart City: Design, Implementation, and Deployment. *Journal of Theoretical and Applied Electronic Commerce Research* 7, 3, 74–87.

Chen Chang. (2022). Digital empowerment of financial supply-side structural reform in the new era: logic, characteristics, and path. *Journal of Xi'an University of Finance and Economics*, 35, 2, 50–61.

Chen Shuisheng. (2021). Technology, System, and Humanity: Orientation and Adaptation of Urban Refined Governance. *Journal of Shanxi University* (Philosophy and Social Science Edition), 3.

Chen Shuisheng. (2022). Digital transformation of urban governance: motivation, connotation, and path. *Theory and Reform*, 1, 33–46.

Dai Changzheng, Bao Jing. (2017). Digital government governance – an examination based on the evolution process of social forms. *China Public Administration*, 9, 21–27.

Fang Muhuan. (2021). Categorical docking and cross-level coordination: A new model of regional governance in the Guangdong-Hong Kong-Macao Greater Bay Area. *China Administration*, 3, 36–44.

Fowler, M., & Highsmith, J. (2001). *The Agile Manifesto*. Retrieved from HTTP:// agilemanifesto.org/.

Hu Guangwei, Zhao Siyu, Yao Min, Liu Jianxia. (2021). On the Construction of Smart City Cluster in China: Form, Architecture, and Path–Jiangsu Smart City Cluster as an Example. *E-GOVERNMENT*, 4, 2–15.

Ines Mergel, Sukumar Ganapati, Andrew B. (2020). Whitford. Agile: A New Way of Governing. *Public Administration Review*, 81, 1, 161–165.

Li Qing, Liu Haijun. (2020). Smart cities and modernization of urban governance: from conflict to empowerment. *Administrative Reform*, 4, 56–63.

Mergel I.,Ganapati S.,Whitford A. B. (2021). Agile: A New Way of Governing. *Public Administration Review*, 1.

Mergel I., Gong Y., Bertot J. (2018). Agile government: Systematic literature review and future research. *Government Information Quarterly*, 2.

Nuno F. da Cruz, Philipp Rode & Michael McQuarrie.(2018). New urban governance: A review of current themes and future priorities. *Journal of Urban Affairs*, 1–19.

Sancino A. and Hudson L. (2020). Leadership in, of, and for smart cities: case studies from Europe, America, and Australia. *Public Management Review*, 22, 5, 701–725.

Shao Yewen, Xu Jiang. (2015). Urban resilience: A conceptual analysis based on international literature review. *International Urban Planning*, 2.

Shi Yishai. (2014). The crux, root cause, and triggering force of "big city disease" and its obstacles. *Journal of Nantong University* (Social Science Edition, 3, 120–127.

Shrestha Has M K. (2008). *Decentralized Governments, Networks, and Interlocal Cooperation in Public Goods Supply*. Florida State University.

Suo Liming. (2016). Urban agglomeration governance for inter-governmental collaboration: trends, characteristics, and future directions. *Economic and social system comparison*, 6, 13–16.

Takeuchi, H., & Nonaka, I. (1986). The new. New product development game. *Harvard Business Review*, 1.

Tian Xiaoping. (2020). Opening up a new pattern of standardization of social governance in the new era. *China Quality and Standards Herald*, 1, 23–29.

Wang Bin. (2020). Building a governance mechanism compatible with the integrated development of urban agglomerations. *Open Herald*, 5, 34–39.

Wen Yating, Yu Jiang, Hong Zhisheng, Chen Feng. (2022). Research on the governance effect and governance process of smart cities in the context of digitalization. *Science and Technology Management*.

White G, Zink A, Codecà L, et al. (2021). A digital twin smart city for citizen feedback. *Cities*, 3, 1–16.

Xia Zhiqiang, Tan Yi. (2017). The basic logic of urban governance system and governance capacity building. *Journal of Shanghai Administrative College*, 5.

Xie Xiaoqin, Ren Shihui. (2022). Agile governance-driven mega-city governance in the digital economy: empirical evidence from the construction of a smart city in Chengdu. *Urban Issues*, 2, 86–95.

Xue Lan, Zhao Jing. (2019). Towards agile governance: An exploration of emerging industry development and regulatory model. *China Administration*, 8, 28–34.

Xue Zelin, Wu Chen. (2022). American practice and inspiration of digital transformation of urban governance. *E-Government*, 1, 251–260.

Yu Jianxing, Fan Liang. (2022). Digital Technology Empowered Social Governance and Its Limits: An Analysis of Hangzhou City Brain. *Comparative Economic and Social Systems*, 1, 117–126.

Yu Lu, Dai Xiangyu. (2018). Coordinated economic development, regional co-governance, and collaborative governance of local governments. *Hubei Social Science*, 7, 38–45.

Yu Wenxuan, (2022). Running elephants: Agile governance in mega cities. *Academia Bimestris*, 1, 139–149.

Yu Binbin, Guo Dong. (2022). Regional integration boosts the high-quality development of urban clusters. *China Social Science Net – China Social Science News*.

Zheng Lei. (2021). Content, path, and direction of digital transformation of cities. *Exploration and Controversy*, 4, 147–152.

*Advances in Urban Construction and Management Engineering – Cha (Ed)*
 *ISBN 978-1-032-39018-5*

# Research on countermeasures for the development of new intelligent cities in Jining City under the perspective of "double carbon" strategy

Luge Xing*
*Chengdu University of Technology, Chengdu, China*

Nailin Yang* & Wenhao Gai*
*Qufu Normal University, Qufu, China*

ABSTRACT: Throughout history, people have been searching for a more livable and better city. The "double carbon" strategy has put forward new requirements for the development of new smart cities, and the construction of new command cities is inevitable. The article firstly reviews the research on the development of smart cities at home and abroad and reviews the problems that exist in them. Then, it analyzes the current situation and problems in the development of new smart cities in Jining and the requirements for urban transformation and upgrading under the "double carbon" strategy, concerning the policy background of the development of new smart cities in Jining. Drawing on useful experiences from the construction of new smart cities at home and abroad, the article proposes countermeasures for the development of new smart cities in Jining in three dimensions: government, enterprises, and individuals, in the hope of promoting the development of new smart cities in Jining by taking active countermeasures. The article hopes to explore feasible countermeasures for the development of a new type of smart city at the prefectural level to provide a reference for the construction of smart cities in other small and medium-sized cities in China and to promote the urbanization and urban transformation process in China.

## 1 INTRODUCTION

In the 21st century, the problem of "urban disease" has become increasingly prominent, and there is an urgent need to find a new way out of urban construction. The concept of "Smart Earth" was proposed in the United States in 2008, and the "Smart Earth Winning in China" project was released in the same year, in which the concept of smart cities was proposed for the first time. As soon as the concept of "Smart Earth" was put forward, it led to a new wave of smart city construction worldwide (Cui 2021). In China, "smart city" was included in the national strategy in 2010, and the "14th Five-Year Plan" has put the "double carbon" target on the agenda, in this context, China's new smart city construction is imperative. Therefore, the most fundamental task in building the future society is to create a physical, natural, and institutional environment to realize the vision of a better future life for all people, which is also an important step toward the construction of smart cities in China (Wang 2022). However, small, and medium-sized cities should be the main force of urbanization construction and transformation, so this paper combines the experience of previous research and practice, reflects on the shortcomings in the research, and proposes relevant countermeasures for the construction of smart cities in Jining City to provide a reference for the exploration of the path of smart city construction in small and medium-sized cities in China.

*Corresponding Authors: ruger_xing@sina.com, 2390891730@qq.com and 2463819568@qq.com

DOI 10.1201/9781003348023-68

## 2 STATUS OF DOMESTIC AND INTERNATIONAL RESEARCH

### 2.1 *Status of foreign research*

Menniti (2014) studied the coordination of a virtual energy district mini-smart grid in a future smart city, arguing that in the future smart city, new information and communication technologies will allow for more efficient management of available resources; on the other hand, Schlingensiepen (2015) investigated the issue of framing automated transport systems in smart cities, arguing that a core function of smart cities is transport and proposing a design framework for an automated transport system that can provide personalized mobility services to users in a smart environment. Hashem (2015) discussed the feasibility of applying big data to smart city construction, arguing that big data can be used to derive potentially valuable clues from the vast amount of data from different data sources and use them to make decisions for the city. And Anthopoulos (2016) explored the interrelationship between smart cities and e-government. Sankowska (2018) showed the definition and ground rules of smart government from a European perspective, focusing on the key ICT components of smart government, and listing some possible challenges that may occur throughout the implementation process. challenges. Hana Kopackova (2020) suggested that citizen engagement is an important factor in smart city building. According to Cesar Ferro's research (2020), achieving the goal of sustainable urban development begins with improving smart city governance and quality of life.

### 2.2 *Current status of domestic research*

Liang (2012) explored the construction path of Ningbo's wisdom city, proposing that the path of Ningbo's wisdom city construction begins with promoting the construction of the core platform and strengthening the promotion of the application of Internet of Things technology; the construction mode of the wisdom city is government-led, with innovative cooperation and the implementation of the strategy of introducing talents, thereby promoting urban economic transformation. Han (2016) proposed that the construction of new smart cities vigorously develop the Internet + industry while invoking the PPP construction model to ease the pressure of government financial funds on construction. Xu (2016) studied the role of the government at three levels: macro, meso, and micro. Jia (2018) proposed that the difficulties are solved through refined governance means in the participation of multiple subjects in the common construction, using a digital platform grid-based governance model. Xu (2019) specified the importance of strengthening the innovation of government management mechanisms in the construction of smart cities. Zhang and Liu (2019) proposed that cities should be governed in a refined manner to break the fragmentation in government governance. Xie (2019) proposed that the construction of a smart government requires cooperation and cooperation among different subjects and levels. Dong (2019) stated in his study that this framework combined the *Technical Reference Model for Smart Cities* (Standard ID: GBT34678-2017) and the "*Six Ones*" architecture of new smart cities, and gave a schematic diagram of the top-level design framework. Yin (2019) analyzed the current problems of domestic smart city construction and proposes corresponding solutions, emphasizing the government's leadership, and strengthening top-level design awareness. Wu (2019) argued that strengthening the construction of smart cities will promote the upgrading of governance. With the development of information technology, smart cities will move from a new era of "looking for government work" to "government-initiated services". Smart cities are no longer a simple replacement for digital information cities, but a twist to make social governance more progressive.

### 2.3 *Review of the status of domestic and international research*

Overall, China's new smart city practice started a little later than abroad, but with the tilt in policy in recent years, there has been a marked increase in attention. In terms of research content, foreign countries attach great importance to the study of smart city-related application scenarios and technologies, while domestic research is gradually shifting from external technical-level research to focus on internal government governance levels. In terms of research regions, most studies have focused on first-tier cities such as North, Guangzhou, and Shenzhen or new first-tier cities like Ningbo, with a lack of attention to prefecture-level cities and a gap in research.

## 3 BACKGROUND OF THE CONSTRUCTION OF A NEW TYPE OF INTELLIGENT CITY IN JINING

### 3.1 *Policy support for the construction of new smart cities*

Since May 2012, China has issued many policy documents related to smart cities. In the past ten years, China has formed a good policy environment and a more favorable atmosphere for the construction of smart cities.

Table 1. China's smart city-related policies since 2012.

| Name of policy document | Issuing Unit | Release Time |
|---|---|---|
| Announced 18 pilot cities of intelligent tourism | China National Tourism Administration | May.2012 |
| Smart City Spatio-temporal Information Cloud Platform Construction Pilot Technical Guide | State Administration of Surveying, Mapping and Geographic Information of China | February.2013 |
| 15 cities selected for cooperation with the EU: Launch of the China-EU Green Smart Cities Cooperation Project | MIIT | November.2013 |
| National New Urbanisation Plan | State Council | March.2014 |
| Launched 80 national pilot cities of information for the people, Shenzhen, Foshan and Suzhou are among them | NDRC | June.2014 |
| The "Guidance on Promoting the Development of Intelligent Tourism" | China National Tourism Administration | January.2015 |
| Notice on promoting the transformation and upgrading of digital cities to smart cities | State Administration of Surveying, Mapping and Geographic Information of China | May.2015 |
| The "new smart city evaluation" notice, | NDRC | November.2016 |
| The Guide to the Construction of a Comprehensive Standardization System for the Smart Home | MIIT、Standardisation Management Committee | December.2016 |
| Notice on Matters Relating to the Use of 3300-3600MHz and 4800-5000MHz Frequency Bands for Fifth Generation Mobile Communication Systems" to determine the 5G IF spectrum. | MIIT | November.2017 |
| Notice on the in-depth development of "Internet + medical health" activities for the benefit of the people | National Health Planning and Development | April.2018 |
| Notice on continuing the evaluation of the construction of new intelligent cities to promote the healthy and rapid development of new intelligent cities | NDRC | December.2018 |
| Smart City Data Fusion Part 5: Municipal Infrastructure Data Elements | General Administration of Market Regulation | September.2019 |
| General technical requirements for the integrated service platform for smart city buildings and residential areas | General Administration of Market Regulation | October.2019 |
| Key Tasks of New Urbanization Construction and Urban-Rural Integration Development in 2020 | NDRC | October.2020 |
| The Outline of the Fourteenth Five-Year Plan for National Economic and Social Development of the People's Republic of China and its Vision 2035. | National People's Congress | April.2021 |

In 2016, the *Regulations on the Promotion of Smart Cities in Jining City* were issued, setting out requirements for the construction of smart cities in the municipality in several aspects, including the construction of information infrastructure, the pooling and sharing of information resources, information for the benefit of the people, smart government and social management, and industrial development.

In 2018, Jining launched the research work on the construction of a new type of smart city, and formed and issued the "Outline of Top-level Design Planning for a New Type of Smart City in Jining", which emphasizes the development concept of "green, efficient, low-carbon and smart" and provides a guide for the top-level design of smart cities by combining the characteristics of Jining.

At the end of 2020, Jining issued the Implementation Plan for the Construction of New Type of Intelligent City in Jining, proposing to closely follow the main line of "accelerating the construction of a new type of intelligent city and promoting the integrated development of digital economy and the real economy"; relying on the two-wheel drive of digital industrialization and industrial digitization; insisting on the combination of integrated promotion and key breakthrough, insisting on the combination of independent It is required to focus on breaking through key core technologies, strengthening infrastructure construction, upgrading industrial chains and building industrial ecology to eventually build intelligent Jining, fine Jining, happy Jining, vibrant Jining, and orderly Jining.

### 3.2 *New requirements for smart city construction under the "double carbon" strategy*

#### 3.2.1 *The traditional urban development model needs to be transformed urgently*

According to the Urban Air Quality Index and $PM_{2.5}$, $NO_2$, and CO indices compiled by the Ministry of Ecology and Environment, the months with good air quality in Jining are concentrated in the second half of 2021, and there are no months with excellent monthly air quality ratings (Table 2). Therefore, the air quality situation in Jining remains critical, and controlling carbon emissions will continue to have a significant impact on the future development of the city.

Table 2. 2021 Jining AQI monthly statistics historical data.

| Month | AQI | Scope | Quality grade | $PM_{2.5}$ | $PM_{10}$ | $NO_2$ | CO | $SO_2$ | $O_3$ |
|---|---|---|---|---|---|---|---|---|---|
| 2021-01 | 138 | 63~285 | Light pollution | 88 | 126 | 17 | 1.239 | 49 | 58 |
| 2021-02 | 78 | 46~129 | Good | 53 | 80 | 12 | 0.914 | 24 | 87 |
| 2021-03 | 128 | 35~500 | Light pollution | 51 | 77 | 11 | 0.942 | 31 | 97 |
| 2021-04 | 86 | 41~416 | Good | 39 | 57 | 10 | 0.827 | 23 | 113 |
| 2021-05 | 107 | 38~500 | Light pollution | 33 | 44 | 11 | 0.865 | 18 | 139 |
| 2021-06 | 119 | 28~202 | Light pollution | 30 | 60 | 10 | 0.777 | 15 | 178 |
| 2021-07 | 67 | 27~108 | Good | 18 | 36 | 7 | 0.642 | 11 | 117 |
| 2021-08 | 84 | 23~142 | Good | 26 | 48 | 6 | 0.726 | 19 | 136 |
| 2021-09 | 70 | 28~130 | Good | 22 | 44 | 7 | 0.667 | 19 | 117 |
| 2021-10 | 80 | 23~178 | Good | 51 | 77 | 9 | 0.748 | 37 | 96 |
| 2021-11 | 99 | 29~189 | Good | 68 | 110 | 12 | 0.763 | 46 | 74 |
| 2021-12 | 94 | 33~198 | Good | 65 | 104 | 14 | 0.894 | 48 | 54 |

The city of Jining has seven counties and two cities and two county-level cities, with a total of 11 counties and urban areas. These areas have markedly different levels of economic development, and public resources are often allocated according to the different administrative levels of the city, making more resources concentrated in the highest administrative level cities, thus inadvertently widening the imbalance in development between counties and urban areas. The problems of increasing contradictions between urban population growth and resource allocation, and the unscientific nature of the traditional economic growth model have prompted the need for cities to transform their development.

#### 3.2.2 *The development of a new type of smart city is necessary under the "double carbon" strategy*

Cities play an important role in the implementation of the "dual carbon" strategy, as they are currently home to more than half of the resident population and account for more than 70% of total carbon emissions. Cities have a range of problems that cannot be eliminated naturally and need to be transformed through urban transformation. The 'dual carbon' strategy is aimed at transforming cities, and there is a fit between the two objectives. The application of smart cities in multiple fields and scenarios also offers new ideas and possibilities for the low carbon development of cities.

### 3.3 *Status of the construction of a new type of wisdom city in Jining*

In terms of innovation leadership, Jining City grasps the position of enterprises as the mainstay of entrepreneurship and supports leading enterprises to build incubation bases with universities and research institutes; deepens the reform of the innovation system in Jining city, accelerates the construction of the Jining China Science and Technology Park, expands cooperation with research institutions and releases the effect of the city's industrial research institutes; optimizes the use of research funds, explores and optimizes new investment and financing models, and leverages more social capital to participate in science and technology innovation; optimizes the innovation ecological environment; strengthens the demonstration and leading role of Jining Innovation Valley, support the provincial laser research institute to accelerate its development, actively cultivates collaborative innovation platforms such as the Jining-Zhongguancun Innovation Center and the Big Data Industry Research Institute, and fully release the innovation platform function of Jining's renowned R&D center enterprises as the main engine to drive innovation in Jining.

In terms of industrial development, we are making every effort to build a strong manufacturing city. Vigorously develop the "four new" economy, accelerate the industrial city to the manufacturing city to move forward. Implementation of the cluster "three chain integration" plan. Promote enterprises to produce more high-end, greener, and smarter, guide enterprises to implement improved equipment, encourage them to increase investment in green and energy-saving transformation, and help them accelerate their transformation. A big data platform will be built for the city's industry to enable digital empowerment for industrial development; promote the deep integration of big data and other information technology into the manufacturing industry; achieve breakthroughs in the service industry.

### 3.4 *Deficiencies in the construction of a new type of wisdom city in Jining*

The analysis of the current situation of the new smart city construction in Jining reveals that there are still certain shortcomings in three aspects: top-level design, talent pool application, and infrastructure construction. The top-level design of the Jining government covers a wide range of areas and lacks a certain degree of relevance to the construction of smart cities. Most of the initiatives for the construction of smart cities are short-term, lacking a longer-term layout for new city construction planning, which is not conducive to sustainable development. In terms of talent, Jining's talent strategy is more focused on introducing talent, and there is still a certain lack of talent incubation and talent retention. Compared to first- and second-tier cities, Jining lacks the advantage of attracting talent and does not have enough talent incubation, making it difficult to win the initiative of applying talent reserves. In terms of infrastructure construction, Jining focuses on the optimal use and reuse of infrastructure, and there are still certain shortcomings in the field of new infrastructure based on high technology, which can slow down the pace of urban transformation.

## 4 PRACTICAL EXPERIENCE IN ADVANCED NEW SMART CITY CONSTRUCTION AT HOME AND ABROAD

### 4.1 *Practical experience of advanced new smart cities abroad*

The practice of smart cities in Europe and the US, as well as in developed countries such as South Korea, Japan, and Singapore in Asia, is generally earlier than in China. Among the representative, programs are the "Amsterdam City of the Future" in the Netherlands, the "Digital Birmingham" in the UK, the "Intelligent Transport System" in Stockholm, Sweden, the "Smart Island" in Singapore, the "I-JAPAN Strategy" in Japan and the "Green IT National Strategy" in Korea. The "Smart Island" in Singapore, the "I-Japan Strategy" in Japan, and the "Green IT National Strategy" in Korea (Table 3). As can be seen, many countries have adopted smart measures such as smart lighting systems, smart sewerage systems, smart transport, and smart grids to reduce energy consumption and pollution, which in effect reflects the beneficial value of smart city construction for low-carbon cities. Overall, these advanced foreign experiences of new smart city practices can provide some reference for China's practice.

Table 3. The practice of smart city building abroad.

| Country | Start time | Project Description |
|---|---|---|
| USA | 2008 | Boulder Smart Grid City Project: relies on advanced IT systems interconnected with the grid to enable remote monitoring and the collection and distribution of real-time information to reduce electricity consumption during transmission and distribution and to influence the way citizens use electricity(CARPENTER,2008) |
| | 2009 | San José Smart Street Lighting Project: aims to transform the city into a world centre for clean energy technology innovation, promoting environmental sustainability and reducing energy consumption. |
| | 2009 | he City of Dubuque's Smart Planet programme, which plans to digitise all of the city's resources, including water, electricity, oil, gas, transport and public services, to monitor, analyse and integrate data to intelligently respond to citizens' needs and reduce the city's energy consumption and costs"(Qin,2010) |
| | 2015 | The plan is to install mobile chips in existing streetlights and connect all streetlights in the city of Los Angeles to the cellular network, eventually allowing control of LED streetlights throughout the city with just a laptop(Sina,2016) |
| Germany | Early 21st century | The city of Hafen has launched more than forty smart city projects in the fields of health and education, such as telemedicine, oncology consultations, online education, online kindergartens and the "Independent Living" project to cope with an increasingly ageing society. |
| | 2012 | Milheim's energy monitoring system containing ICT systems in public buildings, such as the university community, intelligently counts and calculates data on all energy consumption in the building, such as air conditioning, lights and TVs, while the system uses a multifunctional energy-saving material in the building that can produce and store energy, as well as repair and clean itself, significantly reducing building costs and achieving zero greenhouse gas emissions (Chen,2014). In 2016, Mülheim received the ICF Top 7 Smart Cities of the World award |
| UK | 2006 | "Digital Birmingham: aims to promote the use of digital and information technology across all citizens and sectors to accelerate sustainable growth in the regional economy, while benefiting more citizens and improving quality of life(Wu,2014) |
| Sweden | 2006 | Stockholm Intelligent Transport System: automatic identification of all vehicles to reduce congestion and air pollution in the city(Peng,2012) In 2009, Stockholm was awarded the ICF Global Smart City of the Year for the theme "Smart Mobility". |
| Spain | 2009 | Barcelona values the role of the Internet of Things for smart cities and the city is covered with large areas of wireless sensors and routers. |
| Netherlands | 2009 | Two projects in Amsterdam, West Orange and Geuzenveld, have been launched to reduce CO2 emissions and energy consumption through energy-saving smart technologies to improve environmental issues |
| Vienna | 2011 | Vienna's efforts to build a smart city management system: "City heating and cooling schemes" to reduce energy consumption and CO2 emissions; "Smart sewage system" to monitor the indicators of sewage in the network in real time and dynamically(Chen,2014) |
| Finland | 2014 | The Helsinki Public Transport System Upgrade Project: aims to create an integrated, full-coverage, peer-to-peer, "on-demand mobility" public transport system built on the mobile internet, making public transport the number one travel option for citizens and thus making the local transport system smarter, safer and more sustainable. |
| Singapore | 1992 | "In 1999, Singapore was awarded the ICF Global Smart City Network of the Year for "High Speed Internet Connectivity for All". In 1999, Singapore was awarded the ICF Global Smart City Network of the Year for the theme "High Speed Internet Connectivity for All". |
| | 2006 | "Smart Nation 2015: a partnership between citizens, businesses and government to transform Singapore into an infocomm-driven metropolis using ubiquitous information and communications technology(Peng,2012) |
| Japan | 2001 | The "e-Japan" project: the creation of an IT strategy headquarters to focus on the national information strategy. |
| | 2004 | The "u-Japan" project: Promoting information and communication technology in Japan and the development of ubiquitous networks and related industries. |
| | 2009 | The T-Japan project focuses on the establishment of e-government and e-local self-government, as well as the electronic provision of healthcare, human resources and education. |
| Korea | 2004 | U-Korea Strategy: To digitize, network, visualize and smarten all of Korea's resources based on wireless sensor networks as a way to promote new economic development and social change in Korea, with the hope of bringing Korea into a smart society ahead of schedule(Jin,2009) |

### 4.2 *Practical experience of advanced new smart cities in China*

China's smart city practice began in large cities, thus leaving advanced exploration experience in first- and second-tier cities, with Shenzhen, Shanghai, and Ningbo having good exemplary exploration. Shenzhen is home to many domestic internet companies and has a unique technological development atmosphere and talent advantage, providing technical and talent support to promote the construction of smart cities. In addition, Shanghai's advanced smart city construction is also reflected in the application of multiple fields and scenarios, with smart communities, smart healthcare, smart education, and other fields having been widely applied in Shanghai. Ningbo, on the other hand, has developed the country's first city brain system, using the city brain as a general handle for smart city operations. The Ningbo government has worked hard to promote refined governance and collaborative multi-party governance to maximize the benefits to the people. Significantly, all three cities are among the first 5G application cities in China, reflecting the important role of innovative technology in smart city empowerment.

## 5 COUNTERMEASURES TO PROMOTE THE CONSTRUCTION OF A NEW TYPE OF WISDOM CITY IN JINING

### 5.1 *The government strengthens top-level design and improves the strategic layout of urban development*

The Jining government needs to strengthen its top-level design and lay out the construction of a smart city at a high level. It should also develop a refined standard system to break down the barriers to information and communication between different departments and authorities in terms of government services, and build up a database and platform for intelligent circulation to improve the efficiency and effectiveness of government services. The Jining municipal government should tap into the intrinsic needs of the people so that initiatives involving urban construction can truly address the needs of people's livelihoods and allow the dividends of future development to benefit the people in a greater and fairer way. The Jining government should respect the differences in the socio-spatial structure of each county and district, promote multi-party construction, respond to the demands of all parties, and explore the greatest possible convention for future smart city construction.

### 5.2 *Cultivate emerging leading enterprises to help build new smart cities*

Emerging high-tech enterprises can support the construction of smart cities in multiple scenarios and elements, adding to the architecture of high-end cities and stimulating more domestic demand potential. Emerging leading enterprises not only have a better lead role but also have unparalleled innovative vitality and vitality, providing an innovative backbone and forward momentum for sustainable urban competition. To promote the building of a smart enterprise within the enterprise, services at the application level require developers to take full account of the thinking habits and usage requirements of users. The smart enterprise, in turn, requires an innovative capability to truly access application technologies and business services based on the most granular information from the underlying infrastructure.

### 5.3 *The public responds positively to the call for integration into the urban development process*

Citizens are the beneficiaries of urban development, and the promotion of urban development cannot be achieved without their efforts. In the process of the future smart city enabling a "double carbon" strategy, citizens should also respond to the call to consciously practice low carbon living, and high-quality talents should actively integrate into the development of the city and contribute to the construction of the smart city. For the public, it is important to experience the benefits of a smart city and to provide feedback and collect information through the government's smart system promptly, thereby promoting the optimization of smart city services.

## 6 SUMMARY AND OUTLOOK

This paper finds that the construction of a new type of smart city in Jining involves multiple fields and levels, and requires multiple dimensions such as government, enterprises, and people to promote joint construction. There is still much room for improvement in the construction of Jining's new smart city, which needs to be continuously promoted in the future, without the government's top-level design and the use of advanced governance tools and innovative technology, and through the joint construction of multiple entities. We hope that in the future, Jining will be able to leverage the "double carbon" strategy to build a better new smart city.

## REFERENCES


Anthopoulos L G, Reddick C G. Understanding electronic government research and smart city: a framework and empirical evidence[J]. *Information Polity*, 2016, 21(1):99–117.

Carpenter S. It takes a city: boulder offers a proving ground for a smarter way to do a smart grid [J]. *Electric Light & Power*, 2008, 86(4):60–62.

Chen Weiqing, Qin Yun, Sun Luan. A review of domestic and international smart city research and practice[J]. *Guangxi Social Science*, 2014(11):141–145.

Cui Wei. Research on the construction path of new smart cities in the era of big data [J]. *Social Science Front*, 2021, (02):251–255.

Dong Hengchang, Zhang Peng, Wang Yuan. New smart city top-level design architecture [J]. *Intelligent Building and Smart City* 2019(09):21–24.

Han Zhaozhu, Ma Wenjuan. Exploration of the path of wisdom city construction in the context of "Internet+" [J]. *E-Government*, 2016(06): 89–96.

Hana Kopackova, Jitka Komarkova. Participatory technologies in smart cities: what citizens want and how to ask them [J]. *Telematics and Informatics, Telematics and Informatics*, 2020(47):17–20.

Hashemiat, Chang V, Anuar N B, et al. The role of big data in smart city[J]. *International Journal of Information Management*, 2016, 36(5):748–758.

Jia Xiaorong, Yang Ling. *Research on the construction mode and optimization path of smart city* [D]. Xi'an: Xi'an University of Architecture and Technology, 2018.

Julio CesarFerro De Guimaraes,Eliana Andréa Severo,Luiz Antonio FelixJúnior,et al.Govemance and quality of life in smart cities: Towards sustainable Govemance and quality of life in smart cities: Towards sustainable development goals[J].*Joumal of Cleaner Production*, 2020(253):86–89.

Kim Gang-jun. Korean cities enter the U-City era[J]. *Information Technology Construction*, 2009(10):9–10.

Komninos N, Bratsas C, Kakderi C, et al. Smart city ontologies: improving the effectiveness of smart city applications [J]. *Journal of Smart Cities*, 2015, 1(1):1–16.

Liang Jinlan. Exploring the path of building a "smart city" in Ningbo[J]. *Northern Economy*, 2012(07): 57–58.

Menniti D, Sorrention N, Pinnarellia, et al. *In the future smart cities: coordination of micro smart grids in a virtual energy district* [C] // Proceedings In the future smart cities: coordination of micro smart grids in a virtual energy district [C] // Proceedings of the International Symposium on Power Electronics, Electrical Drives, Automation & Motion. 2014:676–681.

Patrycja Sankowska. Smart Government: an European Approach toward Building Sustainable and Secure Cities of Tomorrow[J].*International Journal of Technology*, 2018, (9):7–9.

Peng Jidong. *Research on domestic and international smart city construction models* [D]. Changchun: Jilin University, 2012.

Qin Honghua, Li Hanqing, Zhao Xia. The development status of " smart city" at home and abroad[J]. *Information Technology Construction*, 2010(9):50–52.

Schlingensiepen J, Mehmood R, Nentanu F C. Framework for an autonomic transport system in smart cities [J]. *Cybernetics and Information Technologies, Cybernetics and Information Technologies*, 2015, 15(5):50–62.

Sina Technology. *New LA government plan: one laptop to control city's streetlights* [EB/OL]. [2016-08-06].http://tech.sina.com.cn/it/2015-04-13/doc-iawzuney3231062.shtml.

Wang Peng; Lai Meitong; Mao Zhehan. Patterns and ideas of smart city development under the perspective of "double carbon" strategy[J]. *Science and Technology*, 2022, (03):38–48.

Wu Xiaolin. Relying on smart cities to promote governance upgrading [N]. Taizhou Daily, 2018-06-06(005).

Wu, C. Yang Bo. The latest practice of smart cities in Europe[J]. *Journal of Urban Planning*, 2014(05):15–22.

Xie Xi. *Research on the construction of smart government from the perspective of collaborative governance - taking Z city as an example* [D]. Zhengzhou: Zhengzhou University, 2019.

Xu Aiping. Functional positioning and enhancement path of government governance in smart cities[J]. *E-Government*, 2016(04): 98–103.

Xu Zuozhong. *Research on the innovation of government management mechanism in the construction of smart city - taking Hangzhou Economic and Technological Development Zone as an example* [D]. Shanghai: East China University of Political Science and Law, 2019.

Yin Yi. Smart city construction: problems and cracking path[J]. *Journal of the Party School of the Guilin Municipal Committee of the Communist Party of China* 2019, 19(03):61–64.

Zhang Mingdou, Liu Yi. Research on the framework and path of urban refinement governance in the new era[J]. *E-Government*, 2019(09):76–84.

*Advances in Urban Construction and Management Engineering – Cha (Ed)*


# Digital application of traditional pattern elements in urban construction

Han Xiao*
*Changchun University of Architecture and Civil Engineering, China*

ABSTRACT: In recent years, the development of traditional culture has gradually entered the world vision, and the traditional pattern as an important part of traditional culture has gradually been widely used. In modern urban construction, traditional pattern elements account for most of them and are of great significance, while art and science collide and merge to produce digital industry. It can not only enrich urban design forms to a certain extent but also better inherit and carry forward traditional culture through the carrier of traditional patterns. Based on the integration of art and technology and the inheritance and innovation of traditional patterns, this paper expounds on the tradition on the basis of the characteristics of traditional Chinese patterns. The principles, rules, and expressions of the application of patterns in modern art design will be studied and discussed in terms of the digital application of traditional pattern elements in urban construction at home and abroad.

## 1 INTRODUCTION

Traditional pattern art has profound cultural base rhyme, rich connotation, and changeable form. Its development has a consistent vein, but also varied and consistent style, entrusted to the urban construction, can reflect far-reaching and extensive significance, so as to form a unique art system. Integrating it into the urban construction system can greatly enrich the ideological connotation of the shape in the landscape, and produce signs with aesthetic value, which is the precious wealth of people's art. Throughout urban development, digitalization is the future development trend, which also opens an important chapter of innovation and development of traditional patterns in China. Different nations and countries, regions, and histories will create different forms of culture. Inheriting and carrying forward traditional pattern elements and integrating their spirit into urban digital construction will make the design more unique in The Times.

## 2 THE TRACE OF TRADITIONAL PATTERN ELEMENTS

### 2.1 *Content*

Since ancient times, workers at home and abroad have created a magnificent and unique national culture with their hard work and wisdom, including a dazzling array of traditional handicrafts, large and obscure landscape gardens, unique ancient buildings, and so on (Huang 2015). Traditional pattern elements interspersed among them show diversity and variability, with unique national characteristics, and set up a distinct artistic banner. With the passage of time and the accumulation of culture, traditional pattern elements continue to increase with the accumulation of abundant, reflecting the unique profound culture at home and abroad in many ways.

*Corresponding Author: 359343335@qq.com

DOI 10.1201/9781003348023-69

### 2.2 *Cultural background*

Things have a certain historical background and reason. The emergence of traditional patterns is no exception, and it is also influenced by social and various realistic factors, spreading through the ages, and evolving. Since ancient times, all kinds of legends about myths have emerged in an endless stream. After artificial processing and high-level and in-depth cultural understanding in the later period, the types of traditional pattern elements evolved from them have been continuously enriched, which is the result of generations of diligent study. The world has a long and rich history, in the river of history, traditional pattern elements continue to develop, and people have endowed patterns with rich meanings, which are not only closely related to the traditional national culture of various places but also closely related to the development of various countries.

Table 1. Classification of traditional patterns.

| Name | Period | Content | | Demonstration pattern |
|---|---|---|---|---|
| Primitive social pattern | Primitive society | Painted pottery pattern | Figure |  |
| | | | Animal | 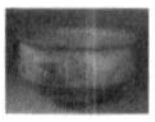 |
| | | | Plant |  |
| | | | The waves | 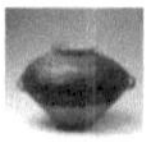 |
| | | | Geometric lines | 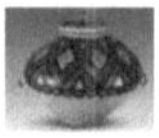 |
| | | | Primitive religious patterns |  |

### 2.3 *Meaning*

The ideological connotation of traditional pattern elements pursues a high degree of harmony between man and nature, emphasizes the pursuit of the spiritual realm and the performance of unique scenes, and emphasizes integrity and unity. Under the full use and understanding of it, later generations deeply study and extract the essence of traditional pattern elements. On the basis of the original research, it is needed to inherit and carry forward the traditional pattern elements completely, combine them with urban construction, and design the urban design highlighting the characteristics of the city from many aspects such as ideological connotation and humanistic consideration.

Traditional pattern elements are an epitome of the world's traditional culture, which not only can be simply displayed in graphics, but also contain rich emotional appeal and connotation of traditional culture, which is the rich carrier of ancient national culture.

## 3 THE PERFORMANCE OF TRADITIONAL PATTERN ELEMENTS IN DOMESTIC URBAN CONSTRUCTION

Traditional cultural elements in China emphasize the characteristics of endless growth and growth. Therefore, in urban design, different places will display patterns in different forms, which can not only achieve complementary effects but also do not affect individual characteristics.

In the development of modern urban design, traditional pattern elements are endowed with a positive attitude, which can get high attention from the public, so as to better balance and expand the ecological relationship between landscape and city, which plays a very significant positive role in the development of China's modern urban construction system. When it comes to concrete urban construction, we can refer to the lake district of Chengde Summer Resort, whose dike island is shaped in the shape of "Ruyi" and "Lucidum lucidum", the mascots of ancient China. Through the concept of "Zhijingyun Dike", the image of three islands in the lake forms a "Ruyi Lucidum tree", which is the icing on the cake of the traditional lake and island pattern of "one pool with three mountains" (Lu 2010). Among them, the traditional Chinese patterns "Ruyi" and "Lucidum lucidum" are endowed with the connotation of auspiciousness and auspiciousness, revealing the ancient emperor's vision for his great cause.

The application of traditional pattern elements in Chinese urban design is not only the understanding of traditional culture itself but also the attitude towards the historical connotation, artistic system, and development concept of different cities (Hao 2007). People have been pursuing the harmonious and common development of human and natural living environment, humanized design, "people-oriented principle" has become an important pursuit of the urban art design. "The Book of Rites · Zhongyong" said that "the joys and sorrows of not hair that, hair and are in the section that the sum of. In also, the world of this also; Harmony also, the world to reach the tao also." The beauty of harmony can show the best state of things. The style of urban architecture pays attention to coordination and unity, so as to achieve the same destination with the rapid development of urbanization and keep pace with The Times.

## 4 THE PERFORMANCE OF TRADITIONAL PATTERN ELEMENTS IN THE CONSTRUCTION OF FOREIGN CITIES

Foreign architecture pursues free and unrestrained style, lively lines, and dramatic presentation effect, and has gradually developed into a mature system containing rich design elements. This kind of architecture is not only more relaxed in space, but also increasingly rich in scenery.

In the Stowe Garden Valley in England, where the river Is called the Styx, man-made cave openings first appear. In ancient western mythology, the natural cave with spring water was regarded as the home of the noble goddess Muse, and this design is symbolic of the Muse. The Italian gardens of the 17th century took on a dramatic baroque style, characterized by a large number of sinuous curves, allegorical sculptures with strong impact, and various fountains and falls. It was built along the slopes of the mountain in layers of the platform, built on the stage of architecture and landscape axis-symmetric formation, layers level railing jade build by laying bricks or stones, trees have been cut to the rules of geometry, powerfully in the garden plant square altar, like the pool strict symmetry, a congested spring along the steep fall into which layer upon layer, called the "chain" waterfall, is a symbol of a raging river.

However, looking at the development of traditional patterns in landscape systems at home and abroad, according to the survey, the audience of this kind of landscape is more inclined to be middle-aged people. Due to the pursuit and understanding of modern art by modern and contemporary teenagers, traditional pattern landscape cannot cover their target leisure places. Therefore, a new landscape system is formed by combining traditional landscape elements with modern digitalization, showing the integration of tradition and modernity.

## 5 DIGITAL REPRESENTATION OF TRADITIONAL PATTERNS IN CONTEMPORARY URBAN CONSTRUCTION

In today's urban construction process, we record data through two-dimensional scanning, digital photography and other ways, and use graphics and image software to process relevant traditional patterns, creating many new patterns and design ideas, which provide a good digital reference for

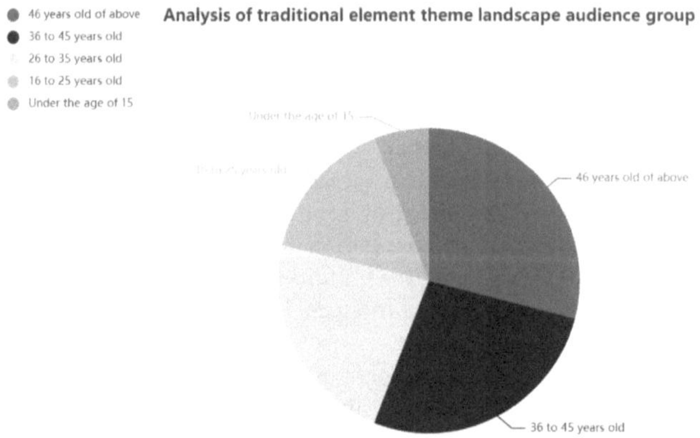


the study of the urbanization process. The application of digital technology in the inheritance and innovation of traditional patterns, on the one hand, will create a new style of artistic design; on the other hand, people's artistic design methods will also be greatly changed.

It is very convenient to use computer software to perform the digital performance of traditional patterns, and it is also easy to achieve the three-dimensional performance effect of patterns through digital technology. It is convenient to create patterns directly on the computer and show the works vividly and intuitively.

When the traditional pattern is digitalized, firstly, the traditional pattern is selected according to the standard, and then the software tools are used for digital modeling design, color design, structure and reconstruction, deformation and reorganization, aggregation, etc., and the extracted traditional pattern elements are processed, and then the new pattern is recombined. (Sun 2008). In addition, technical personnel can also use methods such as sorting, selection, segmentation, classification, and integration to perform digital representation of traditional patterns and establish a database of traditional pattern data, so as to better protect and inherit the artistic value of traditional patterns.

Among them, 3D printing is a new technology of contemporary popular, we can use 3 d printing to the traditional design and the physical performance of representational patterns, and we can innovate traditional patterns in the process of this part is very rich, such as model design aspects of early and late graphic materials performance and product exhibition, etc.

## 6 THE PERFORMANCE OF TRADITIONAL PATTERN ELEMENTS IN THE OPENING CEREMONY OF THE BEIJING OLYMPICS

Throughout the history of the modern Olympic Games, each opening ceremony is a cultural feast. The host country should not only fully show the characteristics of its own national culture and fully present its historical essence, but also be indispensable to the building of national image and the integration of world culture. In 2008, Beijing, as the host country of the Olympic Games, adopted new technologies to inherit and innovate the traditional pattern design from the perspective of two-dimensional and three-dimensional in the opening ceremony of the Olympic Games, combining the national classic symbols with the gorgeous high-tech stage, and jointly constructed a gorgeous modern art pageant.

In "Historical Footprints", 29 fireworks footprints walk along the central axis of Beijing towards the "Bird's Nest", which symbolizes the 29th Olympic Games step by step into China, the historical footprints of the 29th Olympic Games, China's hundred-year journey to pursue the Olympic dream. The footprints, made of flames, represent gunpowder, one of the four great inventions of ancient China. A huge scroll of 70 meters long, this is a Chinese painting scroll across time and space, the picture is showing the four treasures of the study that exudes the Chinese classical charm: ink, paper, and inkstone, which is a very representative cultural symbol of China. Movable type printing was on display in the text of The Analects of Confucius. Three thousand disciples of Confucius chanted the famous lines of the Analects of Confucius loudly and walked slowly into the venue. It is one of the greatest projects in the history of human civilization.

## 7 INHERITANCE AND DEVELOPMENT

Traditional pattern elements have become an indispensable part of modern urban construction. There are some differences between Chinese urban construction art and foreign urban construction art in their formation. The theme of Urban construction in China is the traditional philosophical thoughts of "harmony between man and nature" and "advocating nature". In the early period, foreign art was bound by European feudal thoughts and religious theocracy. In the later period, people began to oppose the feudal thoughts and system and re-examine the view system between nature and man. Only then did they get rid of the shackles of feudal thoughts and began the construction of natural urban architectural style art. However, both at home and abroad, traditional culture is diverse. In its long history, distinctive urban culture has been formed in each age.

In contemporary urban construction, digital representation of traditional patterns has completely changed the modeling thinking and corresponding methods of traditional pattern designers. Inheritance and innovation of traditional patterns are not to directly copy the traditional patterns but on the basis of preserving the soul of the traditional patterns, combining the elements of The Times and the trend of cultural development to innovate. The modern digital design shows the characteristics of The Times, while traditional pattern elements show the essence of history. As the youth of the new generation, we should pay attention to examining the most expressive sign language in local art. This unique cultural integration, is more in line with the law of modern design development. The history of traditional pattern elements connects the past and the future. We should work together to inherit and develop traditional pattern elements and urban culture.

## 8 CONCLUSION

Traditional pattern elements have a long history and gradually show distinct characteristics through generations of inheritance and accumulation. The traditional pattern culture is the art form that the masses grope out in life practice, and is the manifestation of the simplest artistic language in human labor life. As a result, magnificent traditional patterns appear in people's life, constantly improving people's aesthetics and improving people's pursuit of beauty. To some extent, this pattern culture spreads and carries forward the combination of the world's excellent traditional culture and modern design, which not only reflects the charm of art but also shows the extensive and profound world culture through excellent modern design and enriching people's vision.

## REFERENCES


Haoqin. (2007). Olympic dissemination: course, elements, characteristics-also discuss the inspiration of the Olympic Games to the Beijing Olympic Games. *J. Sports science*. 27, 2–8.

Huang Jingna. (2015). Discussion on the design of the new Chinese garden landscape. *Scientific cultivation*. 9, 199.

Lu Yunting. (2010). Pioneer art in the vision of netizens — the folk consideration of the art of the opening of the Beijing Olympic Games directed by Zhang Yimou. *J. Sports and science*. 31, 17.

Sun Haitao, Zhao Pengpeng. (2008). The Baroque symbol in the 17th-century Italian garden. *J. Shanxi architecture*. 3, 347–348.

Wang Shixiang. (2013). *Interpretation of the decoration book*. M. Beijing. Cultural Relics Publishing House. 82–85.

*Advances in Urban Construction and Management Engineering – Cha (Ed)*
*© 2023 Copyright the Author(s), ISBN 978-1-032-39018-5*

# Urban construction: Re-optimizing food systems

Zijiang Tan*
*Beijing Normal University, Beijing, China*

ABSTRACT: A proper food system is indispensable for the good development of cities. Prioritizing profitability and efficiency, our current global food system produces a large amount of food and brings high profits to producers every year. However, equity and sustainability are not given enough consideration. Thus, we develop a Multi-objective Programming Model, putting equity and sustainability in the first priority level as well as adding profitability and efficiency in the second priority level. In this way, we get a new food system named Ideal Food System. Then we discuss the benefits and cost of re-optimizing, and also make a comparison between developed countries and developing countries. In this new system, people in different areas will consume the same amount of food energy, the negative impact of the system on the environment will be reduced by 66.85%, but the efficiency of the food system will be reduced, and the total food production will be reduced.

## 1 INTRODUCTION

### 1.1 *Background*

It is necessary to change the current food system. Operating on economies of scale and utilizing economic models such as economic specialization and global trade, our global food system maximizes profitability and efficiency, but little consideration is taken to equity and environmental sustainability. While in some countries or regions, a large amount of food is wasted, being dumped into the garbage, there are some areas where people cannot get enough affordable and nutritious food, and even die due to hunger or food pollution. At the same time, our current food system is influencing the ecosystem in a negative way by emitting a lot of greenhouse gas, using much fresh water, causing biodiversity loss, etc.

This is not what we want to see in our cities. We should consider profitability and efficiency, but in order to meet everyone's basic right to exist and protect our environment, it is time to pay more attention to equity and sustainability. Therefore, it is very important to develop a new food system for urban construction around the world.

### 1.2 *Overview of our work*

By using the Multi-objective Programming Method, we build our food system model, giving different priority levels to equity, sustainability, profitability, and efficiency. In order to find the optimal solution, we use data and indicators such as coefficient of change and greenhouse gas emissions to quantify these four factors. Based on our model, the benefits and cost can be seen and they also differ between developed and developing countries. Finally, we discuss the scalability and adaptability of our model, which proves that our model is reliable.

*Corresponding Author: a1049534843@163.com

 DOI 10.1201/9781003348023-70

## 2 IDEAL FOOD SYSTEM

To re-optimize the food system, we should consider various objectives: equity, sustainability, profitability, and efficiency. Hence, we use the method of Multi-objective Programming, putting equality, sustainability, profitability, and efficiency in our objective function, and we give them different priority levels. In this way, we build a multi-objective programming model (MOP model) and develop a new food system called Ideal Food System. Unlike the current food systems used in urban development around the world, finding a new system is more focused on equity and sustainability.

### 2.1 *Equity*

The assessment of equity is very important for our model. Citizens' satisfaction with their city is the cornerstone of a city's steady development. The fair distribution of food greatly affects citizens' satisfaction with the city. In the current global food system, the amounts of food energy that citizens in different regions intake every day are very different, even extreme hunger and a large amount of food waste can be found. Hence, if we can balance Food Energy Consumed per Person in One Year (EC) in different regions, decreasing the large gap, we can solve the equity issue to some extent. Equity in the food system can be measured by EC.

To ensure that the areas of food scarcity in the current food system have sufficient affordable and nutritious food, we should restrict the purchasing behavior of people in regions with abundant food, prohibiting their over-consumption. Also, the study has shown that such over-consumption can lead to the problems of food waste and obesity. According to HHS and USDA, recommended daily intake is 2,600 kcal and 2,000 kcal for men and women (respectively) between 31 and 35. Assume that the ratio of the male population to the female population in the world is 1:1 (which is roughly equal to the real data). So, on average, a person needs 2300 kcal every day to live a healthy life. Thus, a regulation is made in our Ideal Food System: people should not buy and consume more than 2300 kcal per day (i.e. 839500 kcal per year).

In addition, considering that the unreasonableness of diet structure is an important factor contributing to food scarcity, our model will re-optimize the diet structure and give a recommendation concerning the proportion of each food type that people should consume, and we give such food combination a name: best food combination. Thus, in the Ideal Food System, everyone in the world buys and consumes food in accordance with the best food combination.

Table 1. Food type and their energy.

| | *Food Type* | *Energy (kcal/kg)* |
|---|---|---|
| 1 | Cereal | 3791 |
| 2 | Fruit & Vegetable | 563 |
| 3 | Milk | 423 |
| 4 | Beef | 2505 |
| 5 | Sheep/Goat | 2940 |
| 6 | Pork | 2421 |
| 7 | Chicken | 2390 |

### 2.2 *Sustainability*

Cities need to be sustainable, so our food system needs to be sustainable. The current food system is negatively influencing the environment, for example, by emitting a lot of greenhouse gas, leaving a water footprint (i.e., using fresh water), and using much energy in an inefficient way. Table 1 shows the impact of different types of food on the environment from the perspective of water footprint, greenhouse gas emissions, and energy efficiency (measured as Food Calories/Energy Used in Production).

Table 2. Impact of different food on the environment.

| | | Water Footprint (Liter per kilogram) | GHG Emissions (Kg equivalents per kg product) | Energy Efficiency |
|---|---|---|---|---|
| Cereal | | 1644 | 2.1 | 102% |
| Fruit & vegetable | | 642 | 0.7 | 20% |
| Milk | | 1020 | 3 | 24% |
| Meat | Beef | 15415 | 60 | 1.9% |
| | Sheep/goat | 8763 | 24 | 4.4% |
| | Pork | 5988 | 7 | 8.6% |
| | Chicken | 4325 | 6 | 13% |

According to Table 2, compared with other types of food, meat production uses more water, emits more greenhouse gas, and has smaller energy efficiency. To estimate the impact of food production on the environment, we introduce the Sustainability Indicator (SI) to our model. SI is a sustainability indicator for the food system.

### 2.3 *Profitability and efficiency*

We measure profitability by the gross income of the food industry. At the same time, a system is efficient when it maximizes desired outputs given available inputs. Thus, the efficiency of the food system is measured by the food energy produced. And the food energy produced must exceed the energy required to survive for the whole population in the world, otherwise, there are always some people who cannot consume enough food and will face hunger and even death.

### 2.4 *Result*

Based on the model described previously and the data we have searched, we get the following results with the help of MATLAB.

Figures 1 and 2 show the diet structure of the current food system and the Ideal Food System. We think everyone should buy and consume food in accordance with the best food combination shown in Figure 4. Compared with the current one, the new proportion of fruit and vegetable decreases to 18%, while the proportions of milk and meat increase to 20% and 12% respectively. We think such change is reasonable because fruit and vegetable contain relatively low energy. In order to reduce the

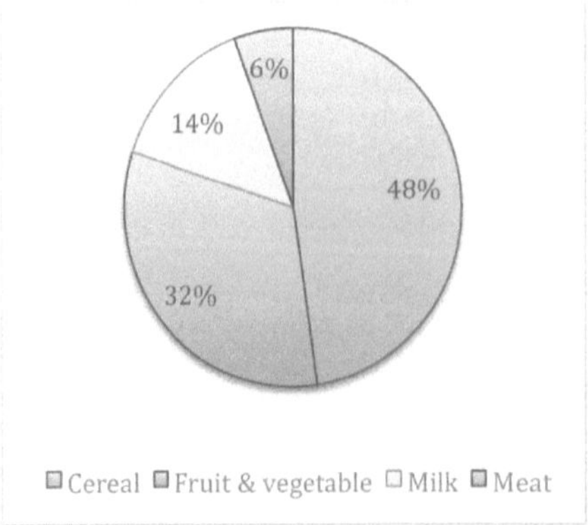


Figure 1. Diet structure of current food system.

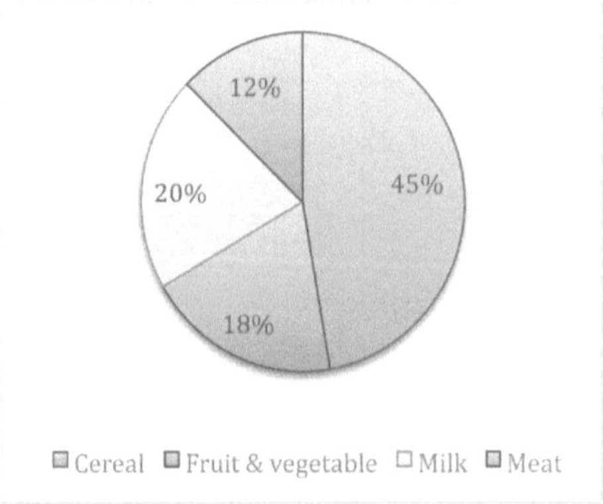


Figure 2. Diet structure of ideal food system.

environmental footprint, we should try to decrease food output (kg). However, we also need to ensure that people all over the world can survive in our food system, so the food energy (kcal) produced cannot be too low. In this situation, food with relatively high energy is a good choice to minimize the environmental footprint under the condition that food energy is enough for the whole population.

After re-optimization, the total grain output and the changes of the four indicators will change. Global food output decreases by 74.48% as a natural result of restraining food demand. And the reduction, along with the adjustment of the diet structure above, attributes to the change in equity, sustainability, revenue, and efficiency, which are −0.1237, −0.4295, 2.24 trillion dollars, and −1.6894 respectively. And one thing noticeable is that the new equity indicator is equal to 0, which means that people in different regions will consume the same amount of food energy. Thus, we achieve equity to some extent. Such changes will naturally have an impact on the development of the city. The impact of the new system will be described below.

## 3 BENEFITS AND COST

There are benefits and costs to moving from an existing food system to an ideal one. We start from a global perspective, and then consider the gains and losses from the perspective of developed and developing countries.

### 3.1 *World level*

Generally, Cities around the world as a whole have benefited in terms of equity, sustainability, and profitability. In the Ideal Food System, the equity indicator is equal to 0, which means that people in all regions intake the same amount of food energy every year. This corresponds with our goal of achieving equity in the food system. At the same time, according to our result on the sustainability indicator and the data of the current food system, we find that if we use the Ideal Food System, the negative impact on the environment will reduce by 66.85%, a very significant change. Surprisingly, the profitability will also improve although we do not put it in the first priority level when building our model. The total revenue of the food system will increase by 57.89%. This may be due to the increase in food prices.

However, there is also cost, which is mainly reflected by efficiency. Efficiency is a concept that pursues "the largest size of the cake", while equity is related to "how to divide the cake properly". The trade-off between equity and efficiency is quite common in many fields, and the food system is not an exception: the efficiency indicator will decrease by 73.49% if we move from the current food system to the Ideal Food System, namely, the annual global food energy produced will decrease by 73.49%.

We obtain the benefits and cost directly by reorganizing the proportion of each type of food and regulating the food prices. Hence the process of changing proportions and prices is always accompanied by the re-optimization of the food system, namely, a move toward a more equal, more sustainable, less productive, and less effective system. Therefore, the benefits and costs occur at the start of our re-optimization and gradually move to the maximum.

### 3.2 *Developed countries & developing countries*

The benefits and costs for developed countries and developing countries are different in many aspects.

The Ideal Food System will solve the problem of food insecurity and food scarcity to some extent. Since the phenomena of food insecurity and food scarcity are more frequently found in developing countries, they are benefited more with respect to equity. As for sustainability and profitability, the countries which produce large amounts of food will be benefited most, including developing countries such as China, Mexico, and developed countries like U.S. and Australia. Concerning efficiency, both developed countries and developing countries will bear some loss.

## 4 SCALABILITY AND ADAPTABILITY

In different sizes of food systems and in different regions, the food energy needed for a person may be different. Thus, we change the value of the Food Energy Needed to Live a Healthy Life Per Person Per Year (EN). Table 3 shows that after changing by 5%, total food output changes by less than 5%, and changes in other variables are all less than 0.3%, which indicates the good scalability and adaptability of our model.

Table 3. Maximum fluctuation after changing energy recommended by 5%.

| Variables | | Fluctuation |
|---|---|---|
| Total Food Output | | ±4.83% |
| Proportion | Cereal | ±0.23% |
| | Fruit & Vegetable | ±0.042% |
| | Milk | ±0.33% |
| | Meat | ±0.037% |
| Price | Cereal | ±0.00378% |
| | Fruit & Vegetable | ±0.00245% |
| | Milk | ±0.011% |
| | Meat | ±0.000026% |

## 5 CONCLUSION

By setting up the Multi-objective Programming Model, the current urban food system is eventually re-optimized for equity and sustainability. Specifically, the diet structure is modified in order to provide sufficient affordable and nutritious food to more people as well as reduce the negative impact of food production on the environment. Meanwhile, food prices are adjusted accordingly and total food output is decreased to some extent. As a result, the new system gains increased equity and sustainability but sacrifices some production and efficiency. Then this system is applied to the developed and developing countries, which can also get good enlightenment. Through the analysis of scalability and adaptability, it is proved that our food system has good scalability and can adapt to different areas. The new system we are building will help the city build better. At the same time, in the process of the rapid development of cities around the world, city builders should consider the difficulties and challenges they will encounter in urban construction in a more comprehensive and detailed way, and create the most suitable food system for urban development.

## REFERENCES


En.wikipedia.org. 2021. *Economic efficiency*. [online] Available at:<https://en.wikipedia.org/wiki/Economic_efficiency>

En.wikipedia.org. 2021. *Food system*. [online] Available at: <https://en.wikipedia.org/wiki/Food_system> [Accessed 6 February 2021].

Royalsocietypublishing.org. 2021. [online] Available at: <https://royalsocietypublishing.org/doi/pdf/10.1098/rstb.2010.0180>

U.S. Department of Health and Human Services and U.S. Department of Agriculture. *2015 – 2020 Dietary Guidelines for Americans*. 8th Edition. December 2015. Available at https://health.gov/our-work/food-nutrition/previous-dietary-guidelines/2015.

*Advances in Urban Construction and Management Engineering – Cha (Ed)*


# Analysis of real-time detection algorithm for dynamic performance of railway vehicles based on sensors

Zhidan Wu & Ying Huang*
*Liuzhou Railway Vocational Technical College, Liuzhou Guangxi, Guang Xi, China*

ABSTRACT: With the development of urbanization, inter-city and inner-city transportation are facing severe challenges. With the help of its intelligent system, railway transportation can solve the demand for personnel flow between regions and improve the level of economic development. Therefore, this paper studies the real-time detection algorithm of vehicle dynamic performance based on sensors, which provides theoretical support for the effective operation of railway vehicles.

## 1 INTRODUCTION

### 1.1 *Literature review*

Based on the sensor technology and the maintenance technology of railway vehicles, Okamura Jihuang and Jin Xianglin studied the detection technology of the bogie inside the vehicle to provide ideas for the detection of vehicle dynamic performance (Okamura et al. 2019). Huang Jingwu and others analyzed the current vehicle detection algorithm by virtue of the advantages of rapid sensor detection, proposed a vehicle detection algorithm based on the analysis of a track map, and verified the correctness of the algorithm by simulation (Huang et al. 2011). Based on the principle of virtual excitation, Zhou Jinsong and other researchers studied the stability algorithm of vehicle dynamic performance and found that the higher the running speed of the vehicle, the wider the wavelength generated by the stability and the greater the difficulty of railway maintenance (Zhou et al. 2008). Based on a wireless sensor tracking method, Xiao Shuo et al. used a GM model to predict the specific location of the vehicle and found that the model can effectively determine the direction of the vehicle (Xiao et al. 2008). Tan Wei and fan Keqing use the basic performance of the geomagnetic detector to extract the vehicle operation data, extract and process the relevant data features, and then get the vehicle-related data. And through the simulation test, it is further judged that the algorithm has high accuracy and good application effect (Tan et al. 2011). Based on the principle of laser scanning and ranging, Shi Hongmei proposed a subway gauge monitoring system. Through the experimental study, the specific data of the method can be judged, which can provide an effective reference for Metro dynamic performance monitoring (Shi et al. 2014).

### 1.2 *Research purpose*

At present, the intelligent transportation system adopts modern intelligent technology, digital technology, internet technology, control technology, and other advanced technologies to establish an accurate and efficient traffic management system, in order to fully solve various problems in the transportation industry. With the continuous development of global digital technology, the process of urbanization is accelerating, and the level of residents is significantly improved. For this reason, the traditional traffic management mode has been greatly challenged, which is mainly reflected in

*Corresponding Author: huangying800816@163.com

DOI 10.1201/9781003348023-71

railway vehicle management. In the process of railway vehicle driving, affected by the complex internal environment, it is difficult for the management department to accurately monitor the specific performance of the vehicle, greatly reducing the efficiency of vehicle operation and the level of department management. As a widely used vehicle detection technology, the sensor has become the main application technology of vehicle monitoring in the railway department because of its intelligent and convenient application. Therefore, it is of great significance to analyze the real-time detection algorithm of railway vehicle dynamic performance based on sensors.

## 2 THEORETICAL ANALYSIS

Vehicle dynamic detection is to correctly extract vehicle information, with the help of artificial intelligence, digital image processing, image recognition, communication and measurement, and control technology, identify specific vehicle information, and then detect specific vehicle speed, weight, driving track, vehicle spacing and road occupancy. In the railway intelligent traffic detection system, sensors are mainly used to collect specific vehicle information and convert it into electrical signals (Wang 2017). Then, according to the related vehicle dynamic detection principle, the output electric signal is processed, and then the specific signal is converted into traffic parameters.

The real-time detection algorithm of railway vehicle dynamic performance based on sensors mainly judges the specific traffic information according to the magnetic field disturbance of the site where the vehicle passes in the railway, so as to achieve the real-time monitoring and judgment of vehicle dynamic performance (Zhou 2018). In the specific work, relevant personnel needs to use sensor technology to extract more real and accurate vehicle signals and try to avoid interference from other factors inside the railway, as well as interference from relevant detection technology and personnel. In the process of processing the detection signal, it is necessary to fully consider the specific characteristics of the original driver without changing the vehicle, and simplify the detection algorithm as much as possible.

At present, most of the real-time vehicle dynamic performance detection algorithms mainly compare the vehicle signal with the signal passing by the vehicle, then standardize the comparison signal, and finally determine the specific vehicle dynamic performance. However, this detection method will change with the change in temperature and other vehicle signals, and the detection data is not accurate. The sensor detection method can not only avoid the influence of related factors inside the railway but also improve the accuracy of vehicle dynamic performance detection. The real-time detection algorithm of railway vehicle dynamic performance based on sensors mainly includes an adaptive threshold algorithm, fixed threshold algorithm, pattern recognition algorithm, and state machine detection algorithm. Among them, the most commonly used are fixed threshold algorithm and state locomotive and vehicle detection algorithm.

As far as the fixed threshold algorithm is concerned, it mainly deals with the detected vehicle signal by wavelet transform, and then compares it with the corresponding fixed threshold. Generally, in the interior of the railway, when the vehicle passes, its output signal will change significantly. In the actual monitoring process, due to the influence of ambient temperature and humidity, the background signal of the vehicle environment changes greatly. When using the fixed threshold algorithm, the correlation signal has been transformed into a stationary signal after wavelet processing. When there are no vehicles on the railway, the variance is basically constant. From this, we can judge the specific fluctuation variance when the vehicle passes by, and then get the vehicle dynamic-related data.

As far as the detection algorithm of state locomotive and the vehicle is concerned, it mainly deals with the specific output signal curve, and then judges whether there is a vehicle passing by according to the basic characteristics of the curve. The vehicle disturbance in the railway can be regarded as a bipolar magnet model, which can reflect the disturbance of the earth's magnetic field. When the sensor passes through the vehicle, the signal curve formed has the characteristics of two peaks, which can eliminate the interference of the track map and completely remove the dependence on the threshold.

## 3 PROCESS ANALYSIS OF REAL-TIME DETECTION ALGORITHM FOR DYNAMIC PERFORMANCE OF RAILWAY VEHICLES BASED ON SENSORS

In the existing intelligent transportation system, the monitor needs to be installed at the railway entrance. Generally, there are many vehicles passing through the railway intersection, the traffic situation is the most complex, and the detector needs to work continuously for a long time. Therefore, it is necessary for vehicle detectors to have good sensitivity and accuracy. Considering the data processing ability of the sensor node hardware system, it is necessary to simplify the design of the detection algorithm as much as possible for the application of relevant personnel.

The use of sensors to detect the dynamic performance of railway vehicles is the use of magnetic field characteristics. When the magnetic field material passes through the corresponding magnetic field, it will be affected by the distribution of the ground magnetic field, resulting in magnetic field disturbance. Generally speaking, the railway passing through vehicles can be regarded as a bipolar magnet model. When the two vehicles pass the detector, the detection signal inside the detector will fluctuate obviously, especially at the engine of the vehicle. Therefore, in the process of vehicle dynamic new energy monitoring, it is necessary to set a specific threshold value, when the detected signal mean value is greater than the threshold value, it can be judged that there is a vehicle passing. On the contrary, no vehicles pass by. After judging whether there is a vehicle passing by, according to the corresponding signal value, identify the specific signal value in the vehicle signal, and then accurately obtain the vehicle dynamic performance-related data. At the same time, according to the noise signal in the vehicle signal, the accuracy of the algorithm can be increased. The real-time detection algorithm flow of vehicle dynamic performance is shown in Figure 1.

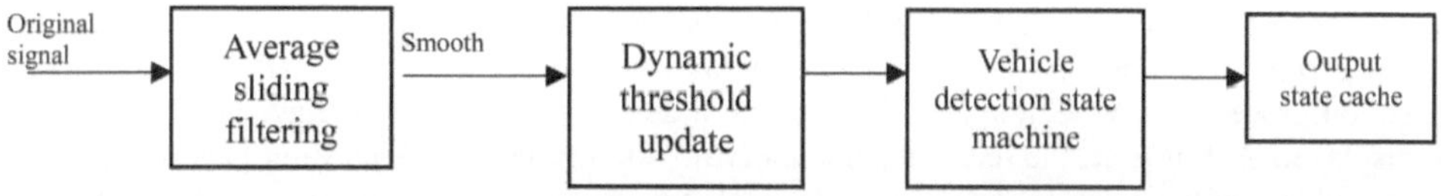


Figure 1. Real-time detection algorithm process of vehicle dynamic performance.

### 3.1 *Signal preprocessing*

After collecting the related data of vehicle dynamic performance, the collected signals are filtered. Considering the real-time requirement of the data, the average filtering method of the sliding window is adopted to remove the noise signal in the detection signal. In the actual calculation process, this method is relatively simple and can remove the noise signal in the vehicle's dynamic performance detection signal. The algorithm can also be used to distinguish vehicle models. The specific Equation 1 is as follows:

$$x(k) = \begin{cases} \dfrac{r(k) + r(k-1) + \ldots r(1)}{k} & fork < M \\ \dfrac{r(k) + r(k-1) + \ldots r(K-M+1)}{k} & fork \geq M \end{cases} \tag{1}$$

In the above formula, M is the specific length of the average sliding window, x (k) is the original vehicle data. In the application process of the algorithm, the noise signal is effectively removed, and the strength of the vehicle dynamic signal is weakened to some extent, but the performance is always good. Therefore, this algorithm can reduce the noise interference generated by adjacent vehicles to a maximum, and has high accuracy.

### 3.2 *Dynamic threshold acquisition*

Due to the influence of the temperature inside the railway, different shaped magnetoresistance sensors will produce a drift phenomenon in a period of time when detecting specific geomagnetic signals. Moreover, in the detection, this kind of drift phenomenon lasts for a long time and is unstable, so it can not be processed by digital filtering. Therefore, when using a dynamic threshold algorithm to detect vehicle dynamic performance, it is necessary to set a dynamic threshold according to drift law to maximize the practicability of the detection method. Under the effect of the sensor detection method, the collected drift signal has stability in a certain period of time, which is much longer than the vehicle passing time. Therefore, by setting the specific threshold value, the corresponding signal value is obtained, and then the vehicle's dynamic performance is judged. The dynamic threshold is calculated with Equation 2.

$$B_i(k)=\begin{cases}[x(k)+x(k-1)+\ldots x(x-I+1)]/t \\ B_{i-1}(k)\end{cases} \tag{2}$$

In the above formula, $x(k)$ means that the magnetic field signal stored in the buffer zone, without passing by the vehicle, $B_{i-1}(k)$ mainly represents the specific threshold judgment standard for the first time. When the dynamic threshold method is used to judge the vehicle dynamic performance, it mainly uses the dynamic threshold update standard to process the detected vehicle situation signal, and then obtains the corresponding parameters.

### 3.3 *State machine detection*

In general, the original vehicle signal will be converted to a certain discrete value after a series of filtering and threshold comparative analyses. According to the discrete value, the vehicle-specific information index can be determined. The state machine detection algorithm is to set the vehicle sensor state accurately, and then judge the specific signal visually according to the detection data. This algorithm is simple, but its accuracy needs to be considered. In the specific application and detection process, if the vehicle is parked next to a detector, the corresponding signal value will be displayed. At this time, the relevant personnel can initialize the threshold value according to the signal value, and the detector can detect the vehicle's dynamic performance and count the specific detection value. In addition, this algorithm can accurately judge the running time and speed of the vehicle and can reduce the impact of the railway environment.

### 3.4 *Dynamic performance test*

According to the parameters of lane occupancy, driving speed, and direction detected by the detector, the specific dynamic performance of the vehicle can be calculated. In the intelligent railway, two sensor nodes can be used to estimate the vehicle speed. In the specific calculation, first of all, we need to know the distance between the two sensor nodes s, and accurately calculate the time $t_1$ and $t_2$ when the vehicle passes between the two sensors. Then we can know the vehicle speed v, which is presented in Equation 3.

$$v=s/(t_2-t_1) \tag{3}$$

However, the distance between the two sensors is relatively close in the interior of the railway, and the vehicle runs faster. This will lead to the difference between the vehicle dynamic performance parameters and the actual data. Therefore, we need to take a series of methods to improve the accuracy of the algorithm and further analyze the vehicle signal.

## 4 APPLICATION OF REAL-TIME DETECTION ALGORITHM FOR DYNAMIC PERFORMANCE OF RAILWAY VEHICLES BASED ON SENSORS

### 4.1 *A general dynamic performance estimation algorithm*

The detection of vehicle dynamic performance is mainly based on the calculation of double coil, and the basic principle is roughly calculated by the ratio of distance and time. The estimation of the dynamic performance of double coil vehicles is mainly applicable to the ground induction coil speed measurement system. In general, in the railway interior, the speed measurement system of the ground induction coil mainly includes two or three coils, and the specific between each coil is constant. When the vehicle reaches the first coil, the speed detector will start to work, monitor the vehicle's pulse signal, and maintain the pulse signal in a stable state until the vehicle passes through the coil completely. When the vehicle passes through the second coil, the vehicle detector inside the coil starts to work and detects specific signals within the synchronization time. The second pulse signal will, to a certain extent, accurately calculate the time when the vehicle passes two coils. This dynamic performance estimation algorithm has better timeliness. However, when the internal environment of the railway changes greatly and the vehicle driving is changed by the complex environment, it may lead to signal overlapping and affect the accuracy of the short volume. The specific detection process is shown in Figure 2.

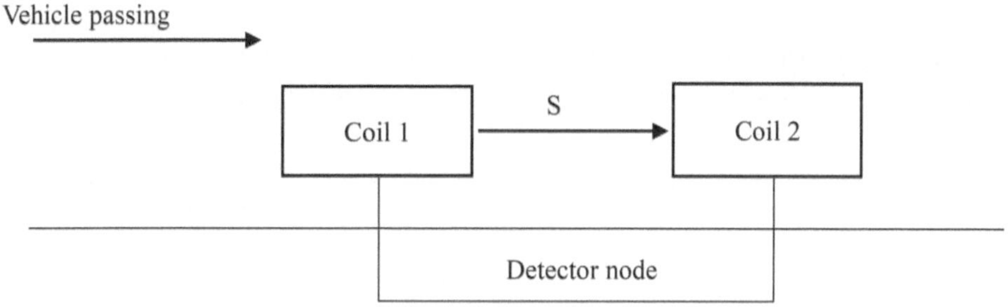


Figure 2. Detection process of ground induction coil.

### 4.2 *The system adopts a vehicle dynamic performance test and estimation method*

Generally, the estimation method of a double node vehicle detector is used to detect vehicle speed. When the vehicle is running on the railway, it is assumed that the relevant signal frequency is the same when the vehicle is always in the range of sensor detection. And it is assumed that the vehicle between the two sensors is traveling at a constant speed. With the help of the corresponding hardware equipment, the access point is synchronized between the two sensors every other period of time, which can eliminate the system detection error. Assuming that the specific length of the sensor node is $L_{AB}$, the actual representation of the vehicle arriving between sensor A and sensor B is $L_{AB}t_{B,up}$, and the departure time is $t_{A,down}$, $t_{B,down}$ respectively. Then the vehicle speed can be calculated with Equation 4 to Equation 8.

$$\Delta t_{up} = t_{B,up} - t_{A,up} \tag{4}$$

$$\Delta t_{down} = t_{B,down} - t_{A,down} \tag{5}$$

$$v_{up} = L_{AB}/\Delta t_{down} \tag{6}$$

$$V_{down} = L_{AB}/\Delta t_{down} \tag{7}$$

$$v = (v_{up} - v_{down})/2 \tag{8}$$

This method is simple, but its accuracy is low. The sensitivity is different between different sensors, simultaneous interpreting the performance and the environment. Therefore, there may be errors in determining the time of vehicle arrival and departure. However, digital filtering may cause detection delay and reduce detection sensitivity. Therefore, combined with the sensor performance, the sensor sensitivity is introduced, and the specific time of vehicle arrival is $\partial$ and departure is $\beta$. Equations 9 to Equation 11 are obtained.

$$\Delta t_{up} = \Delta t_{up} - \partial \tag{9}$$

$$\Delta t_{down} = \Delta t_{down} + \beta \tag{10}$$

$$v'_{up} + v'_{down} = v_{up} + v_{down} \tag{11}$$

In general, the correction function is different with different sensors. In order to ensure the accuracy of calculation, the same correction parameters can reduce the complexity of the algorithm.

### 4.3 *Time synchronization scheme of sensor nodes*

In the real-time detection of vehicle dynamic performance, most of the sensor technology used is ZigBee technology. This technology is mainly composed of sensor nodes and access points. All control commands can reach each sensor node from the access point to achieve time synchronization. Time synchronization means that the access point adjusts the sensor time to the standard time of detection within the set time. Only when the time between the two sensors is synchronized can the collected vehicle signal be analyzed systematically. When the access point sends specific messages to each sensor, although it may be affected by different sensor distances, is the network delay ignored, which can improve the accuracy of detection information (see Figure 3 for details).

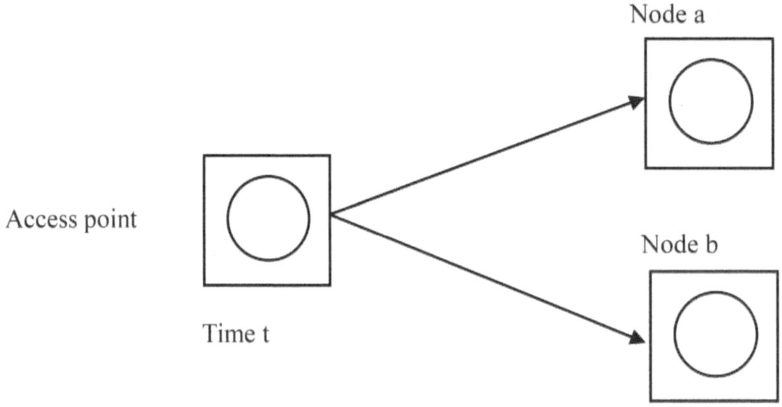


Figure 3. Time synchronization model of wireless sensor.

## 5 CONCLUSION

At present, railway transportation solves the problems of inconvenient traffic management, traffic jam, and frequent traffic accidents in China, which can accelerate the regional population flow, and is of great significance for the development of the transportation industry and the improvement of the economic level. Because of its intelligent and information-based characteristics, sensors provide the possibility for real-time detection of vehicle dynamic performance. This paper analyzes the real-time detection process of railway vehicle dynamic performance based on sensors and analyzes the real-time detection algorithm of vehicle dynamic performance, which has a good guiding role in improving vehicle operation efficiency and guiding the work of railway departments. In the future, the real-time detection algorithm of railway vehicle dynamic performance based on sensors will be widely used in related fields.


## ACKNOWLEDGMENTS

This research was supported by the Guangxi Department of Education Scientific Research Project (Research on Fault Information Extraction and Intelligent Diagnosis Algorithm of Gear Transmission System in Multi-source Strong Noise Environment of Metro) (No. 2022KY1401).

## REFERENCES

Huang J.W., Zou C.Y., Chen R.X. (2011). Vehicle Detection Algorithm Based on Track Map Magnetoresistance Sensor. *Communication Technology*, 44 (07), 89–91.

Jihuang Okamura., Xiang L.J. (2019). Bogie Maintenance Technology of Railway Vehicles Based on Sensors. *Foreign Rolling Stock Technology*, 55 (01), 19–23.

Shi H.M., Wu P., Yu Z.J. (2014). Research on Calibration Method of Non-contact Metro Gauge Detection System Based on Point Set Registration. *Journal of Railway*, 36 (11), 16–22

Tan W., Fan K.Q. (2011). Vehicle Detection Algorithm Based on Magnetoresistance Sensor. *Software*, 33 (11), 38–41.

Wang L.L. (2017). Application of Strain Type Force Sensor in Trackside Dynamic Monitoring System of Vehicle Operation Quality. *Harbin Railway Science and Technology*, 60 (03), 3–5.

Xiao S., Wei X.X., Wang Y. (2008). Design and Analysis of Vehicle Tracking Algorithm Based on Wireless Sensor Network. *Journal of Beijing Jiaotong University*, 7 (05), 64–67 + 76.

Zhou D.Y. (2018). Piezoelectric Rubber Sensors for Railway Vehicles. *Foreign Railway Vehicles*, 55 (002), 38–43.

Zhou J.S., Li D.G., Zhang X.W. (2008). Fast Algorithm of Stationarity and Its Application in Dynamic Analysis of High-speed Railway Vehicles. *Journal of Railways*, 30 (06), 38–41.

*Advances in Urban Construction and Management Engineering – Cha (Ed)*


# Study on the coordinated development of urbanization and ecological environment in Baoji City

Ke Ma*
*Institute of Human Geography, Xi'an International Studies University, China*

ABSTRACT: This paper takes Baoji City as an example through data collection and field research, based on the relevant data on urbanization development and ecological environment protection construction of Baoji City in 2009–2018, and analyzes the relevant indicators of urbanization and ecological environment in this city in recent years. By using the comprehensive evaluation model, coupling degree, and coupling coordination model, this paper discusses the coordinated development relationship and changes in urbanization and the ecological environment in Baoji City. The research results show that the urbanization comprehensive index of Baoji City has increased from 0.0941 in 2009 to 0.5871 in 2018, the urbanization level has gradually improved, the ecological environment quality has increased from 0.1855 in 2009 to 0.2525 in 2018, the ecological environment quality has steadily improved, the coupling coordination degree has increased from 0.36 in 2009 to 0.62 in 2019, the level of coupling coordination degree rises from medium-low level to medium-high level, and the overall coupling coordination degree is on the rise.

## 1 INTRODUCTION

At present, China's urbanization construction can be considered in a high-quality development stage, so various ecological and environmental problems are caused by social concerns. The level of urbanization in Baoji has been on the rise in recent years, and the urban ecological environment is relatively fragile. It is of great practical significance to study the coupling relationship between urbanization and the ecological environment in Baoji, so as to promote the coordinated and sustainable development of urbanization and the ecological environment.

## 2 INDEX SYSTEM CONSTRUCTION AND RESEARCH METHODS

### 2.1 *Establishment of index system*

The research fully absorbs the experience and achievements of domestic researchers in this field (Fang 2016). Overall consideration is given to the actual situation of the region. Among the four dimensions, social urbanization, population urbanization, spatial urbanization, and economic urbanization, 11 indicators are collected to reflect the urbanization level, including urbanization rate, urban population density, and per capita disposable income of urban residents. They also include per capita local fiscal revenue, per capita urban road area, per capita GDP, population density, the proportion of built-up area, number of doctors per 10,000 people, retail sales of consumer goods per 10,000 people, and Internet broadband access per 10,000 people; according to the "pressure-state-response" (PSR) model (Fang 2016), the overall quality of the ecological

*Corresponding Author: 1758348697@qq.com

 DOI 10.1201/9781003348023-72

environment is reflected based on the following three dimensions, namely ecological environmental governance, ecological environmental state, and ecological environmental pressure collection, namely built-up area green coverage rate, built-up area green area per capita, per capita industrial sulfur dioxide emissions, per capita smoke and dust emissions, sewage treatment rate, and harmless treatment rate of household garbage. The evaluation index system of coordinated development of urbanization and ecological environment in Baoji is established by integrating the above indexes.

### 2.2 *The research methods*

#### 2.2.1 *Standardization processing and determination of index weight*

The range method is adopted to implement unlimited tempering operations around the data. The coefficient of variation method is used to determine the weight of secondary indexes. Based on this, combined with the weighting and summary output matched with the weight of the first-level indicators (Grossman 1995). Its mathematical expression is as follows:

$$u_i = \frac{\sigma_i}{\overline{x_i}} \tag{1}$$

$$w_i = u_i / (\sum_{i=1}^{n} u_i) \tag{2}$$

Where, $u_i$ and $\sigma_i$ represent the coefficient of variation and standard deviation of index i, while x1 and $w_i$ represent the mean value and weight of index i.

#### 2.2.2 *Comprehensive evaluation model*

The linear weighting method is used in the synthetic evaluation model. Its mathematical expression is as follows:

$$U_i = \sum_{i=1}^{n} w_i v_i \tag{3}$$

$U_i$ reflects urbanization development index or ecological environment development index, $v_i$ reflects standardized data matching urbanization indicators or ecological environment indicators.

#### 2.2.3 *Coupling coordination degree model*

Coupling mainly refers to the influence of mutual influence and dependence of two or more systems simultaneously. The coupling degree reflects the measurement value to measure the actual effect of the mutual influence of systems (Adnan 2015), and the model is as follows:

$$C = 2\left(\frac{U_1 U_2}{(U_1 + U_2)^2}\right)^{\frac{1}{K}} \tag{4}$$

$C$ is the coupling, $U_1$ represents the urbanization development index, $U_2$ represents the ecological environment development index, and K is the adjustment coefficient. Since this study covers two system layers, the value of K is equal to 2.

Only depending on the coupling degree model can not completely show the degree of system correlation, so the coordination degree model needs to be perfected and improved (Dogan 2016). The mathematical expression is as follows:

$$H = \sqrt{C \times I} \tag{5}$$

$$I = aU_1 + bU_2 \tag{6}$$

$H$ is the degree of coordination; $C$ is the coupling; $I$ represents the composite harmonic index; $U_1$ represents the urbanization development index; $U_2$ represents the ecological environment development index; $a$ and $b$ are the coefficients. Based on the specific value of the coordination degree

$H$, the coordination development degree is subdivided into five different grades (Shi 2019) (See Table 1).

## 3 THE DATA SOURCE

Data from Baoji Statistical Yearbook and Baoji Statistical Bulletin of National Economy (2009-2018).

Table 1. Grade classification of coordinated development degree.

| Coordinate the range of development degrees | Coordinate the level of development | $u_1(x) > u_2(x)$ | $u_1(x) < u_2(x)$ |
|---|---|---|---|
| $0 \leq H < 0.2$ | Low-grade coordination | Ecological environment lag | Urbanization lag |
| $0.2 \leq H < 0.4$ | Medium and low degree of coordination | | |
| $0.4 \leq H < 0.6$ | Moderate coordination | | |
| $0.6 \leq H < 0.8$ | Medium-high coordination | | |
| $0.8 \leq H < 1$ | High coordination | | |

## 4 RESULTS AND ANALYSIS

### 4.1 *Evaluation of urbanization and ecological environment development*

The comprehensive evaluation index system of urbanization involves the following four first-level indicators, namely social urbanization, population urbanization, spatial urbanization, and economic urbanization, including 11 second-level indicators, such as urbanization rate, urban population density, and per capita GDP. The comprehensive evaluation index system of the eco-environmental system involves the following three first-level indicators, namely eco-environmental governance, eco-environmental state, and eco-environmental pressure, including seven second-level indicators, such as the green coverage rate of built-up urban areas and sewage treatment rate. According to the index weight calculation method. According to Formulae (1) and (2), the weights of urbanization and ecological environment indexes are calculated.

According to the comprehensive evaluation model (Formula 3), the comprehensive indexes of urbanization and ecological environment are calculated. As can be seen from Table 2 and Figure 1, the development speed of the comprehensive indexes of urbanization in Baoji slowed down in 2009–2010 and 2014–2015, and the overall fluctuation trend is obvious. The eco-environment index showed a gradual upward trend from 2009 to 2010, and a downward trend from 2010 to 2014. From 2014 to 2018, the trend was gradually increasing. In 2012, the eco-environment index began to be lower than the urbanization index. From 2013 to 2014, the urbanization level increased at a significantly faster pace, which contributed a lot to social urbanization. The eco-environment index increased significantly from 2014 to 2015, and the green coverage rate of the built-up area was 40.61% in 2015, witnessing an increase of 4.94 percentage points over the previous year.

### 4.2 *Analysis of the coupling and coordinated development of urbanization and ecological environment*

Based on the coupling coordination degree model (Formulae 4 and 5), this paper studies the coordination relationship between urbanization and the ecological environment in Baoji city in recent ten years. Based on the coupling degree and coupling coordination degree model and coordination degree development level standard (Table 1), studies are conducted on the coupling degree and coordinated development level of urbanization and ecological environment in this region (Table 3).

Table 2. Comprehensive index values of urbanization and ecological environment.

| Year | Comprehensive indicators for urbanization | Comprehensive index value of ecological environment |
|---|---|---|
| 2009 | 0.0941 | 0.1855 |
| 2010 | 0.0679 | 0.2223 |
| 2011 | 0.1490 | 0.2844 |
| 2012 | 0.2015 | 0.2227 |
| 2013 | 0.2262 | 0.1198 |
| 2014 | 0.3889 | 0.1115 |
| 2015 | 0.3633 | 0.1991 |
| 2016 | 0.4661 | 0.1856 |
| 2017 | 0.5691 | 0.2502 |
| 2018 | 0.5871 | 0.2525 |

The coupling degree of the system is higher than 0.9 in all years except 2010 and 2014, which is in the high coupling stage. In terms of the degree of coordinated development, the trend of continuous increase is obvious, from 0.36 in 2009 to 0.62 in 2018, indicating that the coordination relationship has been effectively optimized.

Table 3. Coupling coordination degree of urbanization and ecological environment in Baoji city.

| Year | The coupling | Degree of coordinated development | Coordinate the level of development |
|---|---|---|---|
| 2009 | 0.95 | 0.36 | Medium and low degree of coordination |
| 2010 | 0.85 | 0.35 | Medium and low degree of coordination |
| 2011 | 0.95 | 0.45 | Moderate coordination |
| 2012 | 1.00 | 0.46 | Moderate coordination |
| 2013 | 0.95 | 0.41 | Moderate coordination |
| 2014 | 0.83 | 0.46 | Moderate coordination |
| 2015 | 0.96 | 0.52 | Moderate coordination |
| 2016 | 0.90 | 0.54 | Moderate coordination |
| 2017 | 0.92 | 0.61 | Medium-high coordination |
| 2018 | 0.92 | 0.62 | Medium-high coordination |

The study shows that the coordinated development process of urbanization and the ecological environment in Baoji can be divided into three stages. In the first stage, the coordinated development degree of urbanization and ecological environment decreased from 0.36 in 2009 to 0.35 in 2010, which was within the middle and low level of coordination. In the second stage, the degree of coordinated development increased from 0.45 in 2011 to 0.54 in 2016, between the moderate level of coordination. In the third stage, the coordinated development degree of urbanization and ecological environment increased from 0.61 in 2017 to 0.62 in 2018, which is between the middle and high coordination levels.

## 5 CONCLUSION

After a comprehensive evaluation, it can be concluded that the ecological environment level is in a backward state, the coordinated development level of urbanization and the ecological environment is gradually improving, and the coupling coordination degree has changed to a medium-high coordination stage. The coupling degree and coordinated development degree of urbanization and ecological environment in this region have been improved.

## REFERENCES


Adnan K, Yavuz Sclman D. $CO_2$ emissions, economic growth, energy consumption, trade and urbanization in new EU member and candidate countries; A panel data analysis [J]. *Economic Modelling*, 2015, 44(96): 97–103.

Dogan E, Turkekul B. $CO_2$ emissions real output, energy consumption, trade, urbanization and financial development: testing the EKC hypothesis for the USA [J]. *Environmental Science and Pollution Research*, 2016, 23(2):1203–1213.

Fang Chuanglin, Zhou Chenghu, Gu Chaolin, et al. Theoretical framework and technical path for analysis of interactive coupling effect between urbanization and eco-environment in megalopolis [J]. *Acta Geographica Sinica*, 2016, 71(4):531–550.

Grossman G, Krucger A. Economic growth and the environment [J]. *Quarterly Journal of Economics*, 1995, 110(2):353–377.

Shi Yue, Qi Yijun. *Journal of Inner Mongolia normal university* (natural sciences in Chinese edition), 2019,48(4):331–341.

*Advances in Urban Construction and Management Engineering – Cha (Ed)*


# Evaluation of urban planning resilience from the perspective of urban epidemic prevention

Cong Peng* & Chunyang Guo*
*Faculty of Engineering, Technology & Built Environment, UCSI University, Kuala Lumpur, Malaysia*

ABSTRACT: The outbreak of the COVID-19 pandemic had a huge impact on the country and posed a challenge to the level of management of public health emergencies in the country. In the reflection on this epidemic situation, this paper analyzes the necessity of building resilient cities in urban epidemic prevention from the angle of urban planning, and puts forward the planning key points and measures pertinently, aiming to improve the professional scope of urban planning and similar events to predict and handle the ability to ensure urban safety.

## 1 INTRODUCTION

Since the beginning of the 2008 Chicago Climate Action Plan, the concept of "resilient cities" has been put forward globally, and the concept of "resilience planning" for cities has been put forward (Yan et al. 2020). By 2018, China was the first to propose "building a more sustainable and resilient ecological city" in Shanghai, which clearly defined the determination and direction of China's development of resilient cities. Urban resilience refers to the ability to maintain the original characteristic ability, adapt quickly, and restore balance in the face of external disturbance or disaster (Laxminarayan 2021). Due to the deepening and expansion of the research content, urban resilience research has become a hot topic in the prevention and resistance of various natural disasters, manufactured disturbances, and sudden diseases. In the concept of a resilient city, the vulnerability of the city comes from the level of economic growth, emergency management capability, and so on, and the destruction of the city is caused by various natural events (Domenico et al. 2014). Therefore, in the construction of a resilient city, the concept of overall planning should be emphasized, and the sustainable development of the city can be realized by combining the analysis of the resource carrying level within the city and the planning scheme according to local conditions (Ahern 2011).

Previous studies on urban resilience planning usually concentrated on the natural resource conditions of the city and the damage caused to the city by various unpredictable natural disasters (Li et al. 2020). However, during the pandemic, new problems have been raised regarding the spatial layout and resilience development of urban planning.

## 2 KEY POINTS OF RESILIENT CITY PLANNING FROM THE PERSPECTIVE OF URBAN EPIDEMIC PREVENTION

### 2.1 *Create a unitized pattern to form a fully functional community*

In ancient China, there were relevant urban planning schemes for the unitary pattern. For example, Chang'an City in Tang Dynasty, as the "first city in the world" with a population of more than 1 million at that time, its overall urban planning and layout comprehensively considered factors

*Corresponding Authors: 1057710640@qq.com and 444711360@qq.com

DOI 10.1201/9781003348023-73 

such as natural terrain and military defense. To facilitate the management, the city system was established to make the boundaries of each residential area clear, although, to some extent, its hindered population mobility has affected the development of social progress, it can be found that in the face of the outbreak of infectious diseases and other epidemics in modern cities, the urban pattern has its applicability.

When public health and safety incidents of infectious nature occur, unit management should be carried out quickly, which is similar to the community form in the current urban planning. However, the full functionality of the community should be paid attention to, and all the infrastructure necessary for people's lives should be built within the community to provide a basic living guarantee in case of an epidemic. To ensure various energy supplies in community units, such as electricity and gas, in addition to paying attention to the stable supply of water in the water supply and drainage facilities, we should also consider the diversion plan of sewage. In this transmission route of Wuhan COVID-19, fecal transmission is included. Therefore, in future urban pipe network design, we should also consider the diversion emergency plan of sewage pollutants to deal with the same type of public health safety emergencies. In addition to the above guarantees, at the beginning of the epidemic development, there was a phenomenon of patients seeking treatment across regions due to insufficient accommodation levels, which further expanded the scope of infection. Therefore, in the design of the unitized pattern of the community, the proportion of relevant functional facilities should be optimized to enable the community under the unitized pattern to meet the basic living guarantee within its district and reduce the expansion of adverse conditions (Ma et al. 2020).

### 2.2 *Form joint defense and joint control, and set up an emergency response system*

For the sake of improving management efficiency, besides unitized management, we should also pay attention to the linkage between units and form an effective joint defense and control mechanism. For urban planning, it is necessary to create an organic channel for the connection between units, which can become a public open place at ordinary times and a rigid node for the connection of unit patterns during the special period of urban epidemic prevention. Underground space can be fully utilized to make its functions more diversified. In addition to meeting the daily requirements, the underground space will also serve as an air defense and storage facility. Through this epidemic, new ideas can be developed in the underground space design in urban planning, an epidemic prevention space linking the ground and underground can be created, and a corresponding emergency linkage system can be established.

Just as Gundersen and Huolin put forward their adaptive ability for the change of resilient cities, they can keep the social ecosystem continuously adjusted through reasonable urban planning and have dynamic adaptability. In the joint defense and joint control of the unitized pattern, a dynamic connection can be maintained to realize the organic series connection of each unit, thus improving the city's emergency response capability, absorbing the impact of emergencies on the city, and realizing the sustainable development of the city. The characteristics of the urban epidemic are different from the urban damage caused by natural disasters. In this epidemic, there is a great contradiction between the ability of urban resources to respond to the epidemic. On the one hand, there is a lack of resource reserves, and on the other hand, there is a limited ability to allocate resources. The development of a unit pattern of urban planning allows all kinds of resource reserve standards to be established after it is established to balance the city's needs, meet regional usage and benign flow daily, to maintain dynamic equilibrium within each unit, as well as during epidemics. It can quickly reach the resource concentration to meet the emergency needs of a certain unit, realize the optimal management of resource reserves and deployment, and improve the city's ability to combat sudden epidemics. The behavioral governance model of resilient cities is shown in Figure 1.

### 2.3 *The Internet enables smart communities*

Since the disclosure of the "human-to-human" transmission of 2019-nCoV in January 2020, various industries have resumed work in March, and the closed management of provinces, cities, and

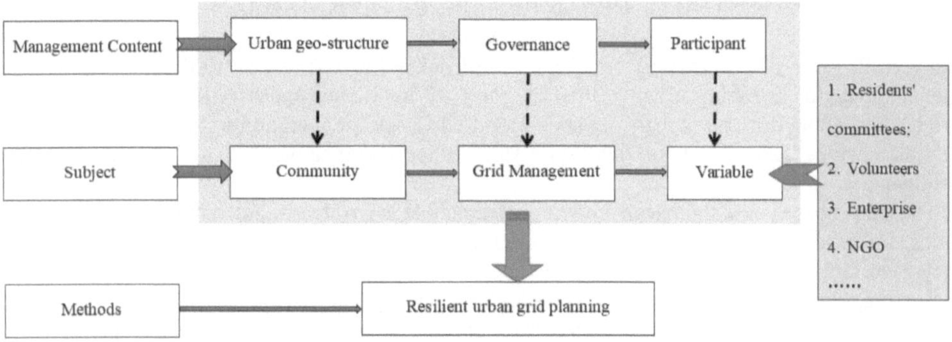


Figure 1. Behavioral governance model of resilient cities.

communities has undergone at least two 14-day closed cycles. In the event of an emergency, the essential lives of the people are guaranteed. On the one hand, it depends on the implementation of emergency plans and measures, and on the other hand, the management level is improved through the "Internet +" technology. The realization of smart communities within the city through Internet technology is the future trend. For example, in this epidemic, the closure of crucial infected areas, or the closure of buildings with confirmed patients, has affected the purchase of residents' daily necessities. The current response method is to utilize WeChat (a mobile application) groups to register and purchase, and property or community personnel to distribute, and requires each owner to submit procurement requirements within a certain time to facilitate collective procurement. After the realization of smart communities, 24 orders are placed every hour to improve convenience and meet diversified needs and minimize the impact of the closure on residents' lives.

Therefore, from the perspective of urban planning, it can be considered to set a center point in the community network to facilitate the full coverage of the smart community, provide storage, turnover, and allocation of various materials, and link property and information technology to improve community management and service level.

In addition to providing convenience in life, the smart community can also register and share information through the Internet. Through this epidemic situation, it can realize the collection of tracked information of people's moving about places, personally sensitive information, and so forth to effectively lock the whereabouts of relevant personnel when conducting close contact investigation and timely contact and isolation, effectively blocking further disease transmission. From the realization of the above smart community, we can also see the future urban layout of light volume, high turnover, and modular urban pattern.

### 2.4 *Risk prediction of resilient cities*

Resilient city management needs to conform to objective laws and realize dynamic and closed-loop management in the whole cycle. The idea of full-cycle management can be applied in different aspects of urban and rural governance. The risk management of megacities should carry out full-cycle closed-loop management including prevention in the early stage, containment in the early stage, control in the middle stage, and reflection in the late stage.

Especially in response to the urban epidemic situation, timely disposal is an essential content, but risk prediction is also important. Therefore, in the future of urban planning research, it is necessary to strengthen the safety risk assessment of the city, carry out risk prediction in advance in combination with different urban characteristics, make flexible management of the overall urban layout, and plan the places and land for emergency medical treatment, safety disposal, and personnel distribution in advance; for example, the site selection of Huoshenshan Hospital and Leishenshan Hospital in this epidemic is still imperfect. In future urban planning, especially in megacities, attention should be paid to the preservation of relevant infrastructure locations and land reserves.

## 3 EVALUATION SYSTEM OF URBAN RESILIENCE PLANNING

Between 2013 and 2018, Arup worked with the Rockefeller Foundation to develop a practical methodology for assessing urban resilience, the City Resilience Index (CRI). In addition, the project was applied to the global "Tenacity Hundred Cities" cooperation initiative. CRI consists of 4 dimensions, 12 resilient city goals, and 156 evaluation indicators that can be used to build a resilient city.

The CRI is based on a significant body of research that is firmly grounded in the experiences of cities, and evidence of what contributes to their ability to survive and thrive whatever disruption they have faced, or anticipated. It embraces the wide array of issues and complexity inherent in measuring resilience at a city scale, and considerable thought has been given to developing a basis of an assessment that is both comprehensive and manageable. The topics of indicators and 12 resilient city goals are shown in Table 1.

Table 1. Indicators' topics and 12 resilient city goals of CRI.

| Indicators | Topics |
|---|---|
| Health and wellbeing | Housing, utilities – energy, utilities – water supply, utilities – drainage & sanitation, food, employment & labor, education & training, business, finance & economy, employment & labor, health, disaster management |
| Economy and society | Support & welfare, citizen participation, and awareness, culture, citizen participation and awareness, crime, and policing, legal and justice, budget business, finance & economy |
| Infrastructure and ecosystems | Disaster management, urban planning, environment, protective infrastructure, environment, utilities, transport, ICT |
| Leadership and strategy | Governance, disaster management, education, citizen participation and awareness, city data, urban planning |
| | 12 Resilient city goals |
| I | Minimal human vulnerability |
| II | Diverse livelihood and employment |
| III | Adequate safeguards for human life and health |
| IV | Collective identity and mutual support |
| V | Social stability and security |
| VI | Economic security |
| VII | Reduced physical exposure and vulnerability |
| VIII | Continuity of critical services |
| IX | Reliable communications and mobility |
| X | Effective leadership and management |
| XI | Empowered stakeholders |
| XII | Integrated development planning |

Now 12 cities in the world are selected as samples (denoted as No. 1-12) and scored according to the goals of 12 resilient cities. The following original data can be obtained.

Correlation analysis was performed on the original data to find the degree of Correlation information between each indicator. The results of the data correlation analysis were shown in Figure 2.

From Figure 2, we can conclude that all indicators of the sample data in Table 2 have a good degree of correlation information, but the relationship is not simply linear but complex cross-coordination. In general, linear distributions have been assumed, but if the threshold midpoint is deemed inappropriate based on the median and mean of available data and professional judgment as to what global average performance is estimated to be, then an alternative distribution is assumed based on the best fit.

Table 2. Original data of representative samples of 12 resilient cities.

| | Cities | | | | | | | | | | | |
|---|---|---|---|---|---|---|---|---|---|---|---|---|
| Goals | 1 | 2 | 3 | 4 | 5 | 6 | 7 | 8 | 9 | 10 | 11 | 12 |
| 1 | 10 | 10 | 20 | 90 | 50 | 90 | 90 | 20 | 20 | 50 | 80 | 70 |
| 2 | 30 | 70 | 90 | 20 | 80 | 80 | 10 | 40 | 80 | 90 | 20 | 80 |
| 3 | 30 | 10 | 60 | 70 | 80 | 60 | 20 | 80 | 30 | 20 | 30 | 30 |
| 4 | 50 | 30 | 80 | 10 | 30 | 70 | 20 | 50 | 10 | 90 | 50 | 80 |
| 5 | 10 | 90 | 30 | 50 | 30 | 60 | 90 | 60 | 60 | 50 | 50 | 50 |
| 6 | 40 | 50 | 70 | 70 | 10 | 60 | 50 | 70 | 70 | 60 | 90 | 50 |
| 7 | 50 | 40 | 50 | 80 | 90 | 30 | 10 | 90 | 40 | 90 | 90 | 80 |
| 8 | 10 | 20 | 60 | 40 | 60 | 90 | 30 | 60 | 90 | 90 | 40 | 30 |
| 9 | 70 | 50 | 70 | 70 | 80 | 20 | 50 | 40 | 40 | 20 | 50 | 90 |
| 10 | 70 | 30 | 70 | 10 | 90 | 60 | 70 | 80 | 50 | 70 | 10 | 80 |
| 11 | 40 | 20 | 50 | 10 | 20 | 30 | 90 | 90 | 50 | 10 | 70 | 40 |
| 12 | 20 | 50 | 20 | 80 | 40 | 10 | 90 | 10 | 10 | 90 | 90 | 20 |

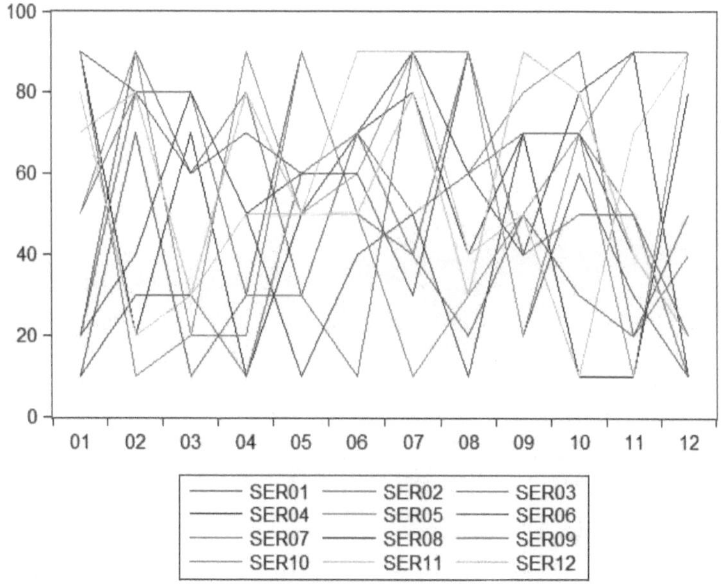


Figure 2. Analysis results of 12 goals and cities association information degree.

## 4 CONCLUSIONS

Through the reflection on the epidemic situation in COVID-19, there is still much research and analysis space for urban planning. The epidemic situation has tested China's national conditions. Wuhan has accepted various tests in this epidemic situation and has gradually entered the stage of repair and adaptation and grown into a more mature and resilient city. As a key link in the construction of a resilient city, urban planning is essential in maximizing the city's planning advantages in response to emergencies and disposal of public health and safety incidents, and in promoting the effective implementation of various emergency measures. Furthermore, to ensure urban safety and social stability, we should learn from the lessons of this epidemic and improve the handling capacity of our cities in the face of various public emergencies.

## REFERENCES


Ahern J (2011). From Fail-safe to Safe-to-Fail: Sustainability and Resilience in the New Urban World. *Landscape and Urban Planning*, 100(4):341–343.

Chao Ma, Yingxia Yun, Xiaosong Ma (2020). Study on the methods to promote community resilience in urban comprehensive disaster prevention and reduction planning. *City Planning Review*: 44(6). doi:10.11819/cpr20200610a.

Domenico Asprone, Gaetano Manfredi (2014).Linking Disaster Resilience and Urban Sustainability: A Global Approach for Future Cities. *Disasters*, 39(1):96–111.

Laxminarayan R, et al(2021). SARS-CoV-2 infection and mortality during the first epidemic wave in Madurai, South India: a prospective, active surveillance study[J]. *The Lancet Infectious Diseases*, 21(12): 1665–1676.

Shuiyu Yan, Jun Tang (2020). Progress on the Theory and Practice of Resilient City. *Journal of Human Settlements in West China*: 35(2). doi:10.13791/j.cnki.hsfwest.20200215.

Ye Li, Xinghua Liu, Qing He (2020). Urban Transportation System Resilience During the COVID-19 Pandemic. *Urban Transport of China*: 18 (3). doi:10.13813/j.cn11-5141/u.2020.0025.

*Advances in Urban Construction and Management Engineering – Cha (Ed)*


# Intelligent control of road lighting based on traffic flow

Feng Du*, Feiyang Zhang* & Zhiqiang Dai*
*Fujian University of Technology, China*

ABSTRACT: With the acceleration of urban modernization, the growth of urban road lighting scale, and the proposal of the national dual-carbon strategy, lighting energy saving has become a topic of increasing concern, and traditional nighttime dimming methods can no longer meet the needs of urban intelligent construction. This paper proposes an intelligent control method of road lighting combined with traffic flow. Firstly, the traffic flow and road surface illuminance are analyzed, and the DIALux software is used to simulate the road surface lighting conditions under different powers to verify the rationality of dimming. Secondly, microwave detection technology is used for the information collected by the traffic flow, and fuzzy control is used for the control method. The traffic flow and the rate of change of the traffic flow are used as the input of the controller to adjust the power of the LED street lights at night. Finally, using MATLAB software, the simulation model of the LED street light control system is built, and the simulation experiment of the dimming controller is carried out. The experimental results show the effectiveness of the traffic flow fuzzy control method, which saves 23.1% of electric energy compared with the traditional control method.

## 1 INTRODUCTION

In recent years, with the gradual improvement of urban road infrastructure, increased roads have been built and put into use. According to the statistical data in the "Statistical Bulletin on the Development of the Transportation Industry in 2021", by the end of 2021, the total mileage of roads nationwide will reach 5,280,700 kilometers. While road construction brings convenience to people's transportation, energy consumption cannot be ignored. At present, lighting electricity accounts for about 13% to 14% of China's total power generation. Among them, road lighting electricity accounts for about 20% to 30% of a country's total lighting power consumption, while the electricity utilization rate is less than 65%. The problem of power waste is serious (Qiao 2016). On urban roads, with the reduction of traffic flow at night, the traditional "all-night lighting" method causes a lot of waste of electric energy, while the "midnight lighting" method will bring safety hazards due to insufficient road illumination.

Therefore, this paper proposes an intelligent control method of road lighting combined with traffic flow, which adjusts the output power in real-time according to the current traffic flow value and the change rate of traffic flow and changes the brightness of LED street lights to achieve the purpose of energy saving. Its application in the urban road lighting system can achieve more safety and energy saving than traditional lighting regulation.

*Corresponding Authors: 59154586@qq.com, 1185796323@qq.com and 1768949893@qq.com

DOI 10.1201/9781003348023-74

## 2 OVERALL SYSTEM DESIGN

### 2.1 *Fuzzy control theory*

A fuzzy controller is the core part of the fuzzy system, which is composed of fuzzification, fuzzy reasoning, knowledge base, and clarity (Zhang 2020). Fuzzy control is to formalize and fuzzily human experience and turn it into a computer for effective control (Yang 2022).

In this study, in the process of lighting control, the collected traffic flow value and the change rate of traffic flow are transformed into fuzzy quantity as the input of the fuzzy controller. According to the fuzzy rules in the knowledge base, the reasoning is carried out. Finally, the output of the street lamp brightness grade adjustment is clarified, and the accuracy required for street lamp brightness is obtained. The block diagram of the fuzzy controller system for vehicle flow is shown in Figure 1.

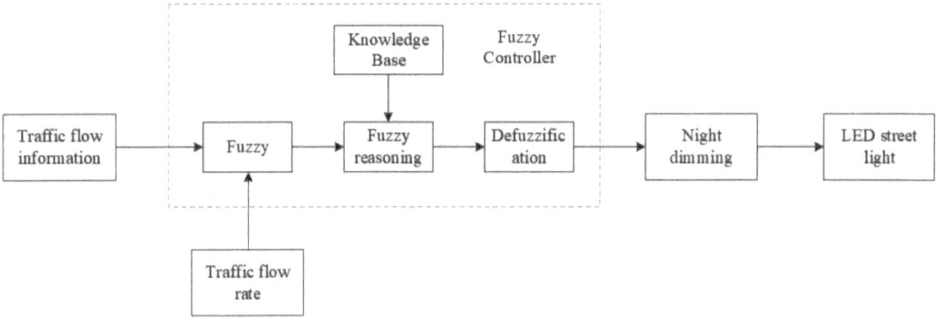


Figure 1. Block diagram of vehicle flow fuzzy controller system.

### 2.2 *Road lighting at different power levels*

According to the *Urban Road Lighting Design Standards* (Standards ID: CJJ 45-2015), the road illuminance standards for road lighting construction are shown in Table 1, but the average illuminance of the adjusted expressway, main road, and secondary road shall not be less than 10 lx, and the average illuminance of the branch shall not be less than 8 lx.

Table 1. Standard value of road illumination of motorway.

| Level | Road type | $E_{h,avmin}$/(lx) | $E_{h,avmax}$/(lx) |
|---|---|---|---|
| I | Expressway, main road | 20 | 30 |
| II | Secondary trunk road | 15 | 20 |
| III | Branch | 8 | 10 |

In this paper, the DIALux software is used for simulation, and the four-lane sub-main road is used as an example. The 3D simulation diagram is shown in Figure 2. The arrangement mode of bilateral arrangement is used. The street lamp spacing is 25 m, and the arrangement is symmetrical. The height of the luminescent point is 8 m, the inclination of the lamp arm is 10°, and the distance between the lamp pole and the lane is 0. 224 m. The design uses an LED street lamp, with a working voltage of 220/50 Hz and a rated power of 120 W.

The lighting brightness levels (i.e., lighting power) of the five LED street lamps are set as P1 (40%), P2 (60%), P3 (80%), P4 (90%), and P5 (100%) are simulated. The simulation results under each power are shown in Table 2, and the average illuminance of each power pavement meets the national standard.

Figure 2. DIALux road 3D simulation diagram.

Table 2. Simulation results of LED street lamp road illumination under different power.

| Lamp brightness grade | $E_{h,av}$ (lx) | $E_{h,min}$ (lx) | $E_{h,max}$ (lx) |
|---|---|---|---|
| P1 | 10.1 | 8.27 | 11.52 |
| P2 | 14.5 | 11.9 | 16.5 |
| P3 | 19.4 | 15.9 | 22.0 |
| P4 | 20.3 | 17.9 | 23.8 |
| P5 | 22.5 | 19.6 | 25.1 |

## 3 DETECTION OF TRAFFIC FLOW BY MICROWAVE TECHNOLOGY

Nowadays, there are many traffic flow detection technologies, such as induction coil detection, infrared detection, microwave detection, and video detection. Compared with other technologies, microwave detection technology is easy to install and maintain, and is not easy to affect by weather and other environmental factors. By adjusting the installation mode and installation angle, multi-lane traffic information can be detected at the same time.

Figure 3. Physical map of GH-719C.

Microwave sensor using GH-719C module as shown in Figure 3, the module pin 0-3 of these four connection points, according to different short-circuit mode, can adjust the sensitivity of 16, induction distance in 1-12 meters. Pins 4 and 5 are used to set the working mode. The specific working mode is as follows.

(1) 00 non-repeatable trigger mode;
(2) 01 repeatable trigger mode, delay time 2 seconds;
(3) 10 repeatable trigger mode, delay time 10 seconds;
(4) 11 Repeatable trigger mode, delay time 30 seconds.

The system adjusts the sensing distance to the maximum of 12 m, adopts the third repeatable trigger mode, and the delay time is 10 seconds.

## 4 DESIGN OF TRAFFIC FLOW FUZZY CONTROLLER

The traffic flow fuzzy controller uses the traffic flow information and the traffic flow rate of change information to control the brightness of the LED street lights.

(1) Fuzzy processing
Traffic flow F and traffic flow change rate FC. Traffic flow F is measured by the detector every 5 minutes, the traffic flow is divided into five fuzzy sets: traffic flow is small (NB), small (NS), general (O), large (PS), and large (PB). The range of vehicle flow F is (0, 80). The change rate of traffic flow is also divided into five fuzzy sets: rapid decrease (NB), slow decrease (NS), constant (O), slow increase (PS), and rapid increase (PB). The value range of the change rate of vehicle flow F is (−20, 20).

(2) Fuzzy rules
For the dual-input and single-output fuzzy controller, both input quantities have 5 language values, so the traffic flow fuzzy control rules have a total of 25 rules, as shown in Table 3 below,

Table 3. Traffic flow fuzzy control rule.

| | | FC | | | | |
|---|---|---|---|---|---|---|
| P | | NB | NS | O | PS | PB |
| F | NB | P1 | P1 | P1 | P2 | P3 |
| | NS | P1 | P1 | P2 | P2 | P3 |
| | O | P2 | P2 | P2 | P3 | P4 |
| | PS | P3 | P3 | P4 | P5 | P5 |
| | PB | P4 | P5 | P5 | P5 | P5 |

(3) Clarification
The output of the fuzzy control is the fuzzy amount of the brightness adjustment level of the LED street light, and the specific value is obtained by clearing it. The brightness level of LED street lighting is divided into 5 grades, namely P1 means 40%brightness, P2 means 60% brightness, P3 means 80% brightness, P4 means 90% brightness, and P5 means 100% brightness.

The above rules are based on the traffic flow to determine the output power of the street light. For example, when the traffic flow is small and the traffic flow rate decreases rapidly, it indicates that the traffic flow is further reduced, and the output power of the street light is relatively low at this time. P1; When the flow rate is small and the rate of change of traffic flow increases rapidly, it indicates that the number of vehicles on the road is increasing. At this time, the power of the street light will be adjusted from P1 to P2 and the street lamp brightness level is reduced from P4 to P3.

## 5 SIMULATION TEST AND ANALYSIS

To verify the energy-saving effect of the traffic flow dimming controller, MATLAB was used to test, record, and analyze the night lighting energy consumption. In this paper, combined with the traffic flow fuzzy controller designed above, a Simulink simulation model is built, as shown in Figure 4.

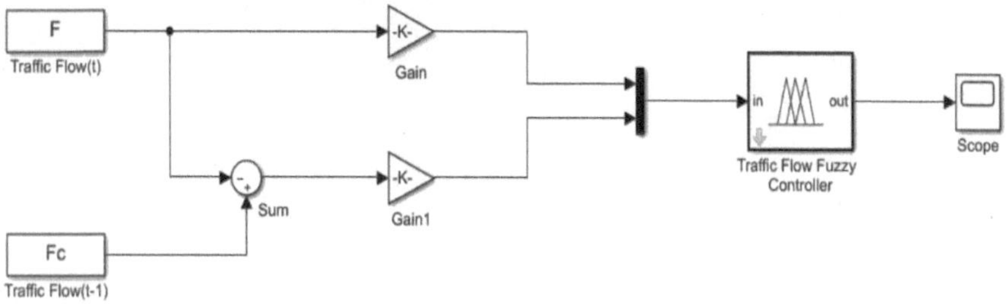


Figure 4. Diagram of the simulation model for Simulink.

This paper selects the variation trend of vehicles on a street in a typical city from 8:00 at night to 7:00 the next day, as shown in Figure 5. The horizontal axis is the time series number (increasing by one every five minutes), and the vertical axis is the traffic flow (Wang 2018).

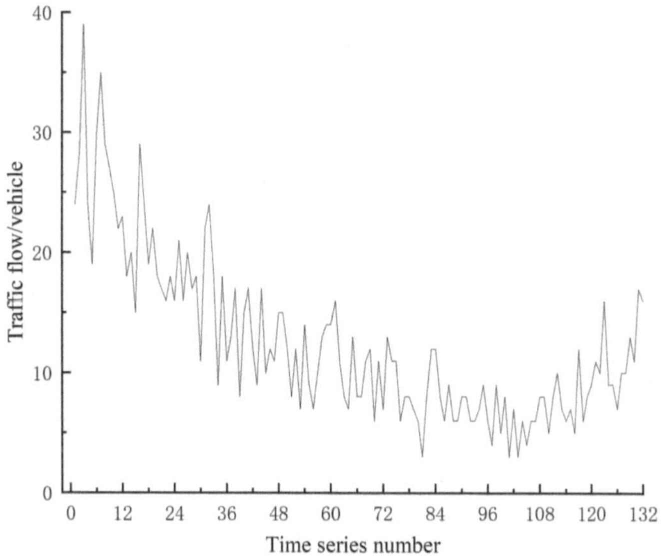


Figure 5. Trend chart of vehicle volume change.

The above-mentioned fuzzy control model of traffic flow is used for simulation, and the output power results are shown in Figure 6. The horizontal axis is the time series (increasing by one every five minutes), and the vertical axis is the dimming power multiplier of the LED street light.

The output power effect using the traditional control scheme is shown in Figure 7. The black curve is the full night light mode, and the red curve is the midnight light mode. The method is to reduce the output power of the street light by half after midnight. The horizontal axis is the time series (increasing by one every five minutes), and the vertical axis is the dimming power multiplier of the LED street light.

Since the abscissa of the LED street light dimming power diagram represents the time and the ordinate represents the dimming power, the total power consumption can be calculated from the area enclosed by the curve and the coordinate axis. The power consumption of a single lamp under the three control modes is calculated respectively. The full-night light mode: P1 = 13.1 kW; the midnight light mode: P2 = 9 kW; the traffic flow fuzzy control mode: P3 = 6.92 kW.

Therefore, choosing the fuzzy control mode of traffic flow from 8:00 p.m. to 11:00 p.m. the next day saves 47.2% of electricity compared to the full-night light mode and 23.1% energy compared

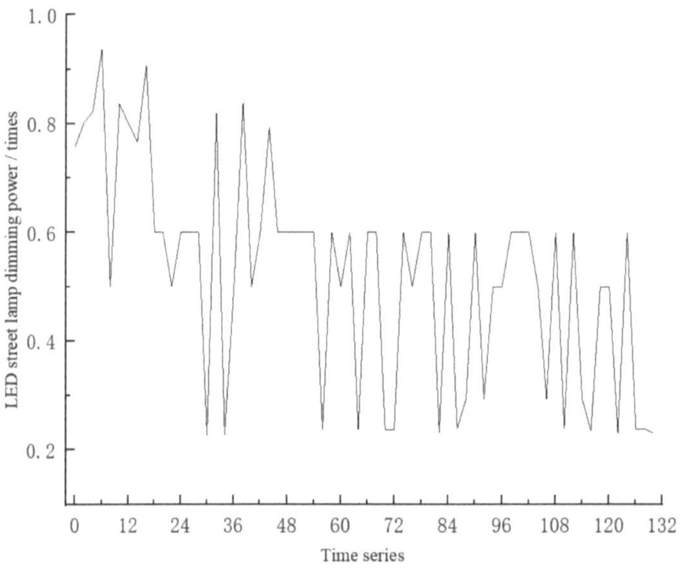


Figure 6. LED street lamp dimming power chart based on fuzzy control of vehicle flow.

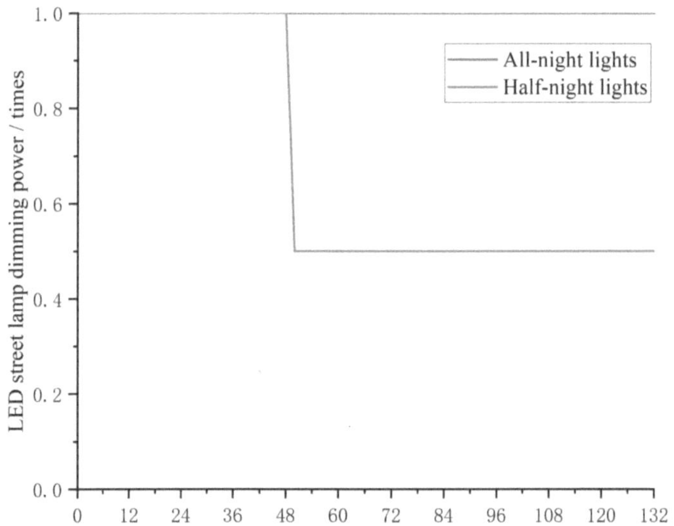


Figure 7. LED street light dimming power diagram under traditional control mode.

to the midnight light mode. From the above results, it can be concluded that the energy-saving efficiency of the fuzzy control method of traffic flow is significantly improved compared with the traditional method.

Urban street lighting accounts for more than 20% of the national lighting. In 2021, China's total power generation will be 8,112.18 billion kWh, of which 13%, or 1,054.58 billion kWh, will be used for lighting, of which 20% will be used for road lighting, about 210.9 billion kWh when the intelligent control method is used to save electricity up to 48.51 billion kWh.

## 6 CONCLUSIONS

Through the analysis of traffic flow changes and road brightness, this paper proposes a road lighting control method based on traffic flow changes. Using DIALux software to simulate the road illuminance to verify the safety of the road illuminance of LED street lamps under different powers, and use MATLAB software to simulate the rationality of the control method, and finally draw the following conclusions:

(1) We propose a dynamic dimming control method based on the change of road traffic flow, which provides a theoretical basis for energy saving of road lighting;
(2) In this method, the brightness of the street lights is high when the traffic flow is heavy, and the brightness of the street lights is low when the traffic flow is small. On the premise of ensuring the safety of road traffic, the energy consumption is saved by 23.1% compared with the traditional lighting method;
(3) The use of fuzzy control of traffic flow can save electricity up to 48.51 billion kWh;
(4) With the further expansion of the city scale, the scale of urban lighting equipment will also increase. It is believed that this system can play a certain practical value in this field; this study did not consider the impact of weather conditions on road lighting. The weather variables will be added to the research, which will further improve the degree of intelligence.

## ACKNOWLEDGEMENTS

This paper was supported by the Natural Science Foundation of Fujian Science and Technology Department (Grant No. 2019J01787) and Fujian Tonglian intelligent illumination intellectual property research and development (Grant No. GY-H-21203).

## REFERENCES

Jian Wang. *Research on Fuzzy Control of Urban Intelligent Lighting System Based on Illuminance and Traffic Flow* [D] Chengdu: University of Electronic Science and Technology of China, 2018.

Ministry of Transport of the People's Republic of China. *Statistical Bulletin on the Development of the Transportation Industry in* 2021 [R]. 2021.

Qiao Zhu, Yisheng Liu, Yashuang Liao, Xiaowen Wang. Research on energy saving potential of urban road lighting [J]. *Power Electronics Technology*, 2016, 50(12):38–41.

Qihang Yang, Jiachun Li, Yuewei Chen, Biaobiao Hu. Application of Fuzzy Control Technology in Phosphate Mine Wastewater Treatment System[J]. *Water Purification Technology*, 2022, 41(02):81–86.

Suwen Zhang. *Research and realization of urban smart street lamp construction* [D]. East China Jiaotong University, 2020.

Urban Road Lighting Design Standard: CJJ 45-2015 [S]. Beijing: China Construction Industry Press, 2015: 8–9.

*Advances in Urban Construction and Management Engineering – Cha (Ed)*


# Subway track bed foreign body detection based on deep learning in subway inspection

Jie He*, Yuanhui Ning*, Ming Geng* & Xiaohua Luo*
*China Railway Siyuan Survey and Design Group Co., Ltd., China*

Li Yan*
*Ningbo Rail Transit Group Co., Ltd., Smart Operation Branch, China*

Qin Yin*
*China Railway Siyuan Survey and Design Group Co., Ltd., China*

ABSTRACT: In view of the current subway patrol inspection, the patrol inspection manual has not only a heavy workload but also is a low efficient way to finish subway patrol inspection. In this paper, a method for subway track bed foreign body detecting is proposed and carried by the subway inspection electric trolly. This method only uses normal image data without foreign bodies to make up the training dataset, which overcomes the difficulty of collecting different foreign body samples. At the same time, this paper uses the depth information obtained by the 3D camera module to reprocess the output results of the foreign body detection algorithm and eliminate false alarms caused by illumination and water stains, and improve the positive rate of the system. Experiments have proved that this method can effectively detect the specific position of abnormal objects in images.

## 1 INTRODUCTION

Foreign body detection is an important task in subway inspection, which is related to the safety of subway trains. At the present, foreign body detection is mainly carried out by manual inspection. Because the position of foreign bodies is random, workers need to check foreign bodies on the whole subway track bed. This method is inefficient, time-consuming, and labor-intensive (Chen 2019; Lv 2014; Liao 2020; Tian 2019; Wen 2019).

In recent years, with the development of artificial intelligence, deep learning technology leads the rapid development of the machine vision field and computer vision. Especially the ability to label a large amount of data and feature learning based on a convolution neural network. Target detection based on supervised deep learning has been greatly improved, such as the RCNN series, YOLO series, and SSD series (Bochkovskiy 2004; Girshick 2015; He 2019; Liu 2016; Ren 2016; Redmon 2016). However, when supervised target detection is used for subway track bed foreign body detection, there are the following disadvantages: in actual scenes, it is very difficult to collect foreign bodies, and it is impossible to provide enough samples for network model training. There are many kinds of foreign bodies, so it is difficult to define each kind of foreign body exhaustively. Foreign body samples are unbalanced, and the number of samples without foreign bodies is far larger than samples with foreign bodies, which is an important challenge in foreign body detection.

Therefore, this paper proposes an unsupervised deep learning-based subway track bed foreign body detection algorithm, which uses the GANomaly network to train the normal data of the track

*Corresponding Authors: 627787180@qq.com, 289738668@qq.com, mynameisdc@yeah.net, sclead315@yeah.netlongbo2019@qq.com and 1149092572@qq.com

 DOI 10.1201/9781003348023-75

bed (Akcay 2019). After training, the subway track bed pictures with foreign bodies, and the trained network model will reconstruct the subway track bed data without foreign bodies. After training, the network model will still reconstruct the normal track bed picture without foreign bodies. Therefore, by comparing the difference between the original image and the image reconstructed by the network, we can judge whether there is a foreign body.

At the same time, to overcome it, 2D images are easily affected by illumination and water stains. The detection results of the network and the 3D information of the corresponding position are jointly judged to avoid false positives caused by illumination and water stains.

## 2 TRACK BED ACQUISITION MODULE

The track bed acquisition module is mainly composed of three 3D laser line scanning cameras, which are distributed at the left, middle and right positions of the track (Figure 1). The images captured by the three cameras are shown in Figure 2.

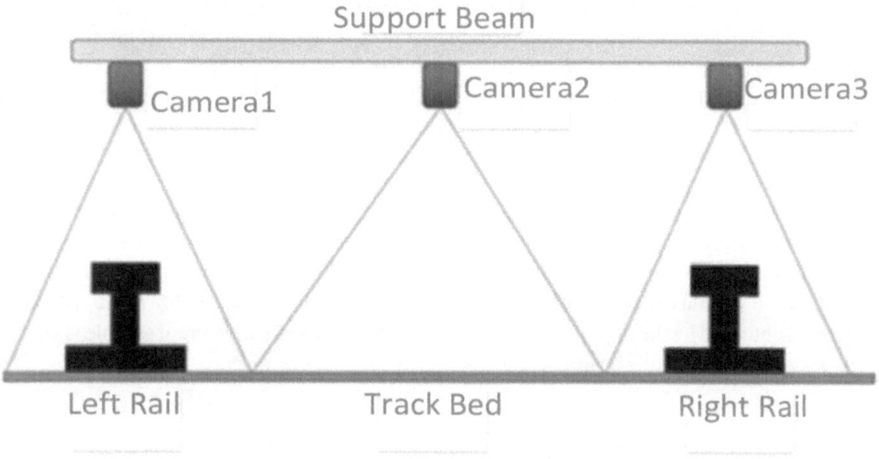


Figure 1. Track bed acquisition module.

It can be seen from Figure 2 that the images collected by the three cameras cover the left and right rails and the middle track bed area completely.

## 3 FOREIGN BODY DETECTION IN TRACK BED BASED ON DEEP LEARNING

### 3.1 *GANomaly network architecture and principles*

The basic idea of the GANomaly model is that the self-encoder is trained by using normal class images in the training stage, and the main function of this self-encoder is to reconstruct normal class images. In the test stage, if the image belongs to the type of no foreign matter, the reconstruction error between the reconstructed image and the original image is smaller. If the test image belongs to the type of foreign matter, then the reconstructed image has a larger reconstruction error than the original image. According to the difference of reconstruction error between a normal image and an abnormal image, it can be judged whether the image contains foreign matter or not by setting an appropriate threshold value.

The breakdown diagram of the GANomaly model framework is shown in Figure 3. The GANomaly model framework consists of a generator, reconstruction encoder, and discriminator. In Figure 3, the generator is composed of an encoder $G_E(x)$ and a decoder $G_D(\mathbf{Z})$. The input

(a) Images of Camera 1 (b) Images of Camera 2 (c) Images of Camera 3

Figure 2. Image captured by track bed module.

data X passes through the encoder to obtain its potential space vector **Z**, and **Z** passes through the decoder $G_D(\mathbf{Z})$ to obtain the reconstructed data $\hat{x}$ of X; The reconstruction encoder $E(\hat{x})$ is used for re-encoding the reconstruction data $\hat{x}$ to obtain the potential space vector $\hat{\mathbf{Z}}$ after the reconstruction data is encoded; The discriminator $D(x, \hat{x})$ introduces the idea of antagonistic learning to distinguish the difference between the original input data $x$ and the reconstructed data $\hat{x}$. When the original input data $x$ is judged to be true and the reconstructed data is judged to be false, its structure is designed as the same network as the decoder structure in the generator to reduce the gap between the reconstructed data and the original input data, and then judge the abnormal state or fault state of the system.

In the training stage, the model is learned by using positive samples, that is, the encoder $G_E(x)$, decoder $G_D(\mathbf{Z})$, and reconstruction encoder $E(\hat{x})$ are all suitable for positive samples. After training, when an abnormal sample is an input in the test stage because the decoder is trained by positive samples, there is a certain difference between the encoded potential space vector **Z** and the potential space vector $\hat{\mathbf{Z}}$ obtained by the reconstructed encoder, and then whether the input sample is abnormal is judged according to this difference.

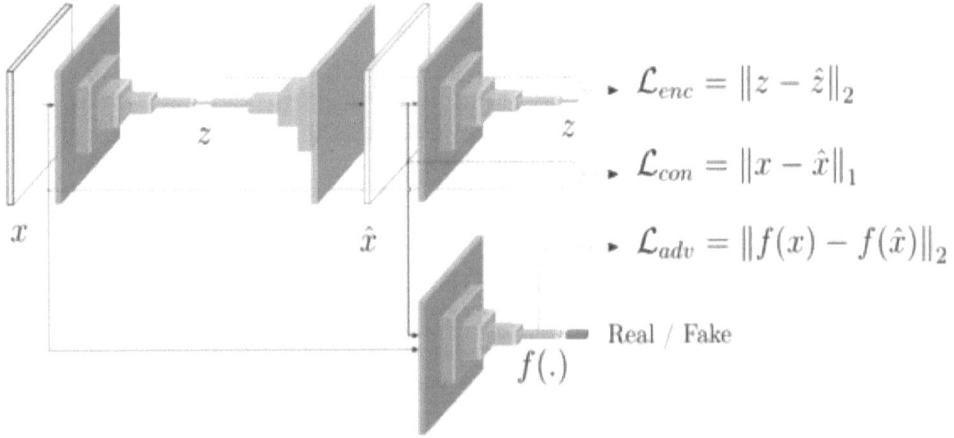


Figure 3. GANomaly.

The net loss of this model is divided into three parts. In the generation network, the reconstruction error loss is defined to reduce the gap between the original input data and the reconstructed data.

$$L_{con} = |x - \hat{x}| \tag{1}$$

where $L_{con}$ is the reconstruction error loss in the generated network. A feature matching error is set in the discriminator to optimize the data feature layer.

$$L_{adv} = |f(x) - f(\hat{x})| \tag{2}$$

where $L_{adv}$ is the loss in the discriminant network, specifically refers to the loss of confrontation between the generator and the discriminator; $f(\cdot)$ is the transfer function of the model.

Ideally, for normal data, the difference between the latent space vector encoded by the reconstructed data $\hat{\mathbf{Z}}$ and the latent space vector $\mathbf{Z}$ encoded by the original input data is very small. To quantify and optimize this difference in the training stage, the error between potential vectors is proposed.

$$L_{enc} = ||\mathbf{Z} - \hat{\mathbf{Z}}||_2 \tag{3}$$

where $L_{enc}$ is the reconstruction potential space vector error loss in GANomaly.

Because only positive samples participate in the training process, the model can achieve better coding and decode only for positive samples. If negative samples are input, the potential space vectors obtained by coding will be greatly different under coding and decoding, so it is judged as abnormal.

For the model, the entire loss function $L_{total}$ can be expressed as:

$$L_{total} = \omega_{adv} L_{adv} + \omega_{con} L_{con} + \omega_{enc} L_{enc} \tag{4}$$

where $\omega_{adv}, \omega_{con}, \omega_{enc}$ are the parameters for judging network loss, reconstruction error loss in generating network, and reconstruction potential space vector error loss in GANomaly.

The whole GANomaly foreign body detection algorithm flow is as follows:

1. Segmenting the images in the sample set to form a training set;
2. GANomaly network is trained on the training set;
3. Segmenting the images in the sample set to form a test set;
4. Using the trained generator network to test the test set;
5. Splicing and restoring the measured images to obtain the generated images;
6. Difference between the restored image and the test original image to obtain L1 distance;
7. Segmenting the obtained L1 distance according to the threshold to obtain the final foreign body detection result;

### 3.2 *Results filtering based on 3D information*

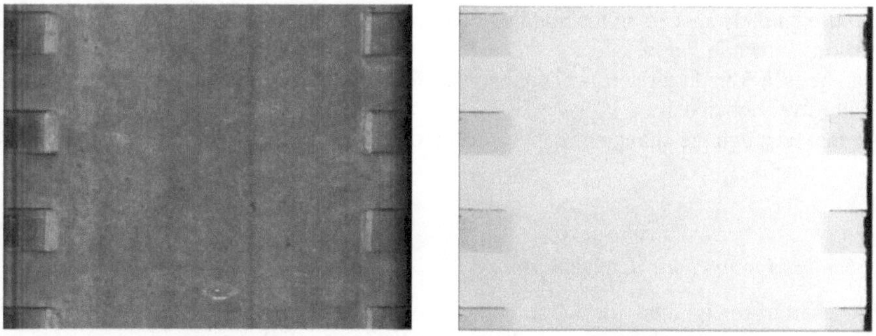

(a) 2D image (b) 3D depth map

Figure 4. 3D images captured by the camera.

When stains appear on the surface of the track bed, the 3D information of the track bed surface will not change, as shown in Figure 5.

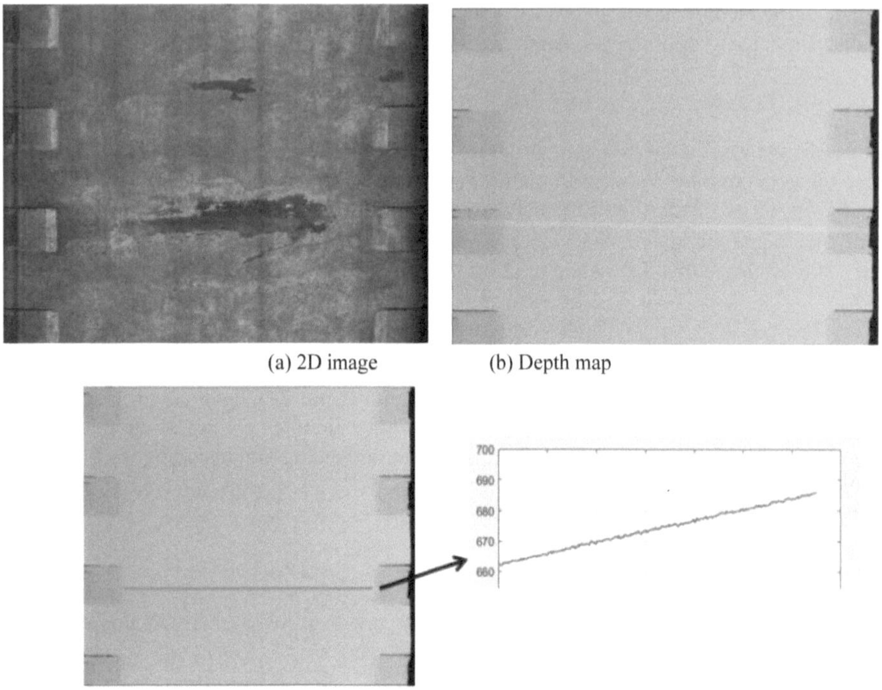


(a) 2D image (b) Depth map

(c) Height change of stain area

Figure 5. Stain area.

When there is a large area of stain on the track bed, it is easy to judge it as a foreign body only by 2D images, because the large area of stain makes the texture difference of 2D images obvious. Therefore, this paper introduces 3D information of the image to screen the results to avoid false positives caused by stains. There are two kinds of images obtained by the 3D camera: one is the gray image and the other is a depth image. As shown in Figure 4, there are two kinds of images acquired by the 3D camera.

As can be seen from Figure 5(c), a section of 3D information of the stained area is selected for analysis, and the stain does not cause the change of 3D information of the track bed surface. When a foreign body appears in the middle of the track bed, the 3D depth information will change obviously, as shown in Figure 6.

As can be seen from Figure 6(c), when there is a foreign body, the depth information of the track bed changes very obviously.

Therefore, through the change of depth information, the stain and other areas can be judged to avoid false positives.

### 3.3 *Track bed foreign body detection process*

The flow chart of foreign body detection in the whole track bed is shown in Figure 7.

The process steps of track bed foreign body detection are as follows:

Step 1: Input a 2D image;

Step 2: The 2D image is reconstructed by the GANomaly network model, and whether there are foreign bodies is judged according to the reconstruction error threshold.

Step 3: If there is no foreign body, output the result. Otherwise, go to Step 4.

Step 4: Enter the 3D depth image and the location of the foreign body.

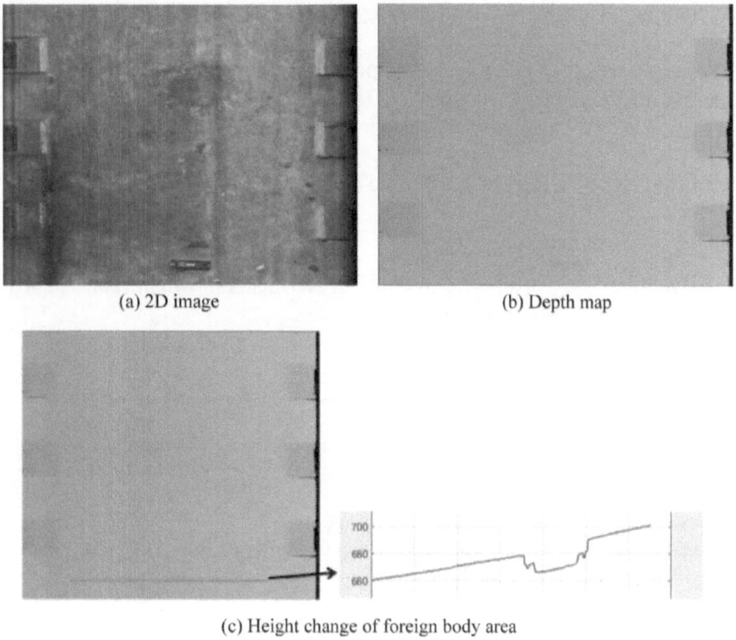


Figure 6. Foreign body area.

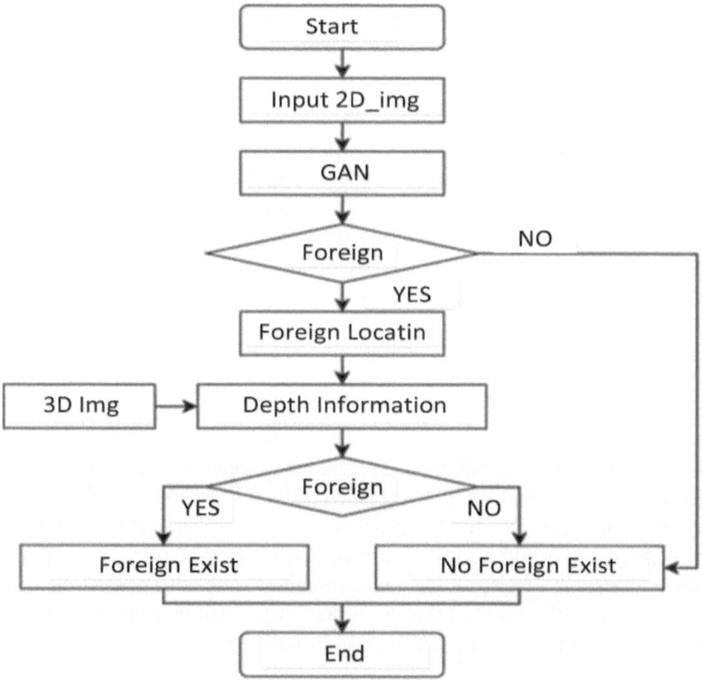


Figure 7. Flow chart of foreign body detection in the track bed.

Step 5: Combining with the depth image, judging whether the corresponding position is a foreign body.

Step 6: If it is judged as a foreign body, the output detection result is that there is a foreign body; if it is judged as not a foreign body, the output detection result is that there is no foreign body.

## 4 EXPERIMENTAL RESULTS AND ANALYSIS

### 4.1 *Experimental environment*

The hardware environment used in all experiments in this paper is shown in Table 1, and the software environment used in experiments is shown in Table 2.

Table 1. Hardware environment.

| Hardware name | Model | Parameter |
|---|---|---|
| CPU | Intel core i7 | 2.4 GHz |
| GPU | RTX3090 | 16 GB |
| Memory | Double data rate v4 | 16 GB |
| Hard disk | SAMSUNG 850 EVO SATA3 | 512 GB |

Table 2. Software environment.

| Software name | Version |
|---|---|
| Operating system | Windows 10 |
| Python | 3.5. 2 |
| PyTorch | 1.3. 1 |
| CUDA | Release 10.0 |

### 4.2 *Experimental data set*

The data set used in this study is 2,000 images of track with no beds collected by subway inspection vehicles on a subway line. Among them, there are 100 foreign bodies, and the image size is 2,000 × 2,560. There are 1,600 training sets and 400 test sets. Because the size of the original image is relatively large, it cannot be directly sent to the network for training. The image is segmented by blocks and then sent to the network for training. After trying, set the segmentation block size of the image to 200 × 256.

### 4.3 *Experimental results*

The change curve of training loss value in the training process is shown in Figure 9. There is a gradual decrease in loss after about 500 iterations, and the rapid decline in loss value indicates that the network has converged evidently and has learned a significant amount of existing information.

In the experiment, the recall rate and false alarm rate are used as evaluation indexes.

The recall rate is for the original sample, indicating how many positive classes in the sample have been predicted correctly. There are also two situations, one is to predict the original positive class as a positive class (TP), and the other is to predict the original positive class as a negative class (FN). The definition of recall rate is shown in Equation (5).

$$Recall = \frac{TP}{TP + FN} \tag{5}$$

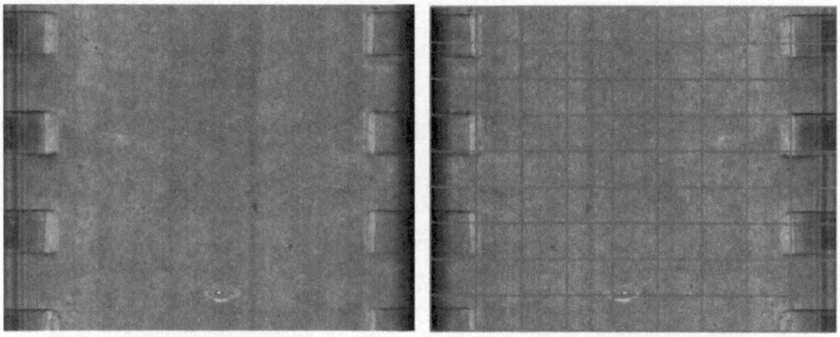

(a) Original image (b) Segmentation result

Figure 8. Image segmentation.

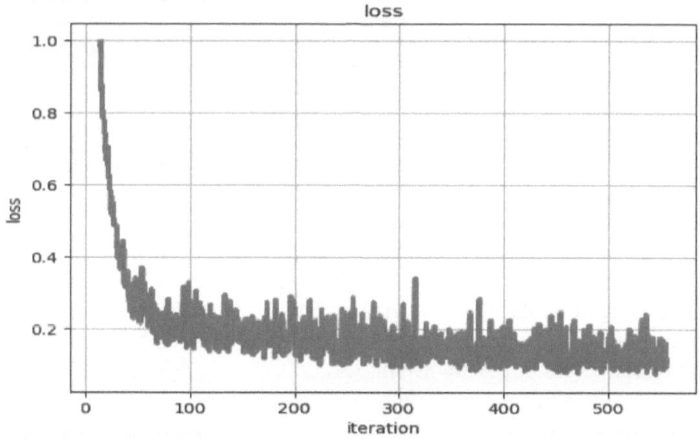


Figure 9. Variation curve of training loss.

The false alarm rate represents the proportion of negative samples divided into positive samples, which is defined as Equation (6).

$$FPR = \frac{FP}{TN + FP} \tag{6}$$

where $FP$ represents the number of samples that predict negative classes as positive classes, and TN represents the number of samples that predict negative classes as negative classes.

The results of basic GANomaly were compared with those of GANomaly with 3D information processing. Comparative information is shown in Table 3.

Table 3. Comparison table of result information.

| Method | Recall rate (%) | False alarm rate (%) |
|---|---|---|
| GANomaly | 96.5 | 10 |
| GANomaly plus 3D | 96.5 | 1 |

From the data in the table, GANomaly with the 3D method can reduce the false alarm rate to a very low level while ensuring a high recall rate, and effectively reducing the impact of false positives on the system.

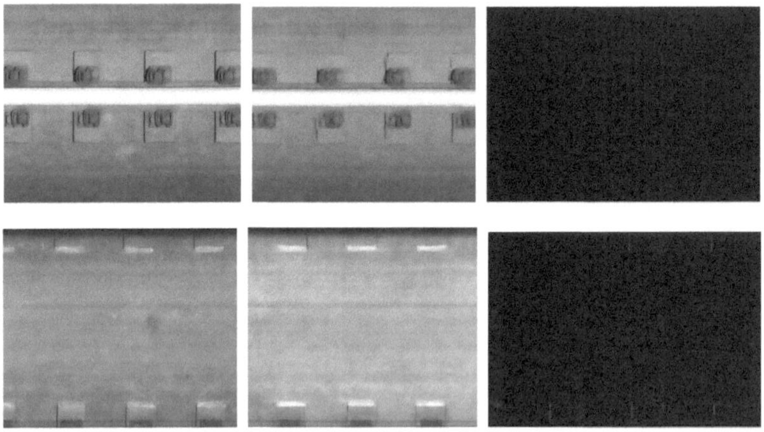

(a) Original map (b) Reconstructed map (c) Difference map

Figure 10. Results without foreign matter.

The reconstruction diagram and difference diagram of foreign body-free samples in the data set are shown in Figure 10.

From Figure 10, the difference between the reconstructed foreign body-free image and the original image is very small, and the foreign body-free data can be judged in this way. In Figure 11, the reconstruction diagram and difference diagram of some samples with foreign bodies are shown.

It can be seen from Figure 11 that there is a great difference between the reconstructed map and the original map in the position of foreign bodies, and foreign bodies can be effectively distinguished, which shows the effectiveness of the algorithm.

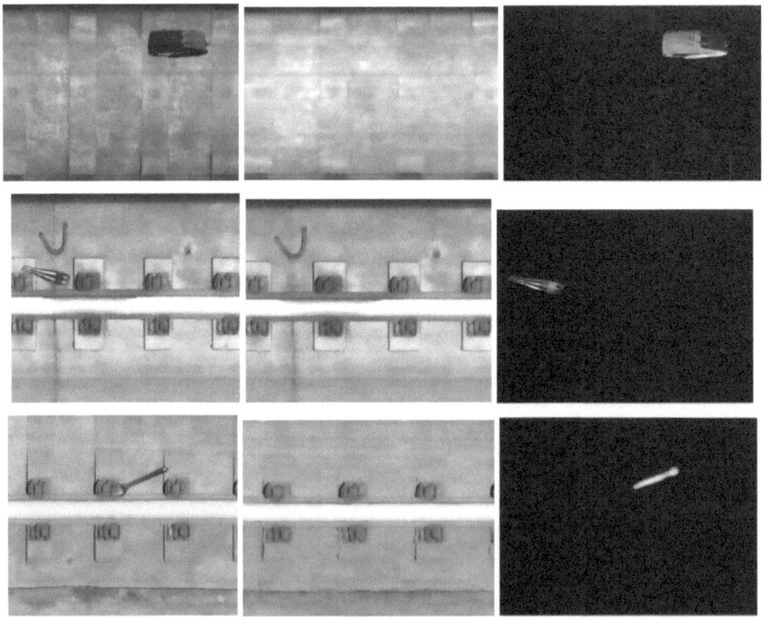

(a) Original map (b) Reconstructed map (c) Difference map

Figure 11. Foreign body results.

## 5 CONCLUSIONS

In this paper, a track bed foreign body detection method based on deep learning is proposed, which uses normal image data without foreign bodies to participate in training, and overcomes the difficulty in collecting different foreign body samples. At the same time, we use the depth information obtained by the 3D camera module to reprocess the output results of foreign body detection, eliminating false alarms caused by illumination and water stains, and improving the positive report rate of the system. Experimental results show that this method can reduce the false alarm rate while satisfying a high recall rate, and has high practical value.


## ACKNOWLEDGEMENTS

This research was supported by the National Key R&D Program of China (Grant No. 2018YFB2100903) and the Service Capacity Building of Hubei Innovative Method Promotion and Application Base (Grant No. 2020IM020800).


## REFERENCES


Akcay S, Atapour-Abarghouei A, Breckon T P. *GANomaly: Semi-supervised Anomaly Detection via Adversarial Training* [J]. 2019.

Anguelov D, Erhan D, Liu W, et al. *Ssd: Single shot multibox detector[C]//European conference on computer vision*. Springer, Cham, 2016: 21–37.

Bochkovskiy A, Liao H Y M, Wang C Y. Yolov4: Optimal speed and accuracy of object detection[J]. *arXiv preprint arXiv*:2004.10934, 2020.

Chen Lin, Lv Yifeng,. Metro Operation Tunnel Safety Inspection System and Application[J]. *Urban Rail Transit Research*, 2014, 17(10): 126–128. DOI:10.16037/j.1007-869x.2014.10.071.

Chen Zhuoxiong, Yu Xiaoli. Application Research of Intelligent Inspection Robot in Subway Communication and Signal Room[J]. *China New Communication*, 2019, 21(11):7–8.

Divvala S, Girshick R, Redmon J, et al. *You only look once: Unified, real-time object detection*[C]//Proceedings of the IEEE conference on computer vision and pattern recognition. 2016: 779–788.

Girshick R, He K, Ren S, et al. *Faster R-CNN: Towards Real-Time Object Detection with Region Proposal Networks*[C]// Advances in Neural Information Processing Systems. 2015: 91–99.

Girshick R. Fast r-cnn [C] *//Proceedings of the IEEE international conference on computer vision*. 2015: 1440–1448.

He Z, Zhang L. *Multi-adversarial faster-rcnn for unrestricted object detection*[C]//Proceedings of the IEEE/CVF International Conference on Computer Vision. 2019: 6668–6677.

Liao Jin, Liu Dong, Wu Songrong, Yang Ping. Design of Vehicle-mounted Subway Track Defect Inspection System[J].*Modern Urban Rail Transit*, 2020(04):11–16.

Liu Liang, Sun Kun, Wen Xiaokai. Conception of Subway Inspection Information System Based on the Internet of technology[J].*Modern Tunnel Technology*, 2019, 56(03):59–64. DOI:10.13807/j.cnki.mtt.2019.03.009.

Tian Jun. Exploration of automatic inspection of Shencheng subway lines [J]. *Urban express rail transit*, 2019, 32(05):19–24+32.

*Advances in Urban Construction and Management Engineering – Cha (Ed)*
 *ISBN 978-1-032-39018-5*

# Research on the development of new R&D institutions in Guangxi autonomous region based on PSR framework model

Xing Jin* & Hong Cui*
*Guilin University of Electronic Technology, Guilin, China*

Yi Guo
*Guilin University of Aerospace Technology, Guilin, China*

ABSTRACT: The report of the 19th National Congress points out that China has entered a new era. This historical positioning also puts forward new requirements and gives new missions to new R&D institutions. In the process of comprehensively promoting innovation development, the linear research model of traditional R&D institutions can no longer meet the demand for innovation, and new R&D institutions are the unique products of China-born under this demand. Compared with the construction of Guangdong and Jiangsu provinces, the development of new R&D institutions in the Guangxi Autonomous Region started late, and there are huge gaps in policy support, construction and operation, talent gathering, and transformation of research results. To analyze the impact of various pressure factors on new R&D organizations into "responsive" behavior, this paper constructs a framework model based on the concept of "pressure-state-response" for the development of new R&D institutions in the Guangxi Autonomous Region. Moreover, it offers constructive suggestions for the future development of new R&D institutions in the Guangxi Autonomous Region.

## 1 INTRODUCTION

For a long time, under the linear research model of "basic research-applied research-product development", universities and traditional research institutes have long performed research functions and produced results in the form of papers and experimental products, while product development and industrialization belonged to enterprises. This research model has undoubtedly deepened the gap between research results and product industrialization. In the process of promoting innovation development, the biggest dilemma facing China is that technology and economy are "two skins", and the traditional research institutions can no longer meet the requirements of innovation. To overcome this problem, Guangdong, Jiangsu, and other provinces have taken the lead in building many new R&D institutions such as Shenzhen Tsinghua University Research Institute and Nano Industry Technology Research Institute. The special feature of these new R&D institutions is that they give full play to the advantages of market-oriented operation and have strong power to transform scientific and technological achievements.

Guangxi Autonomous Region is located on the southwest border of China, and besides the local economic development level is at a lower level in the country, its scientific and technological innovation vitality, and power are also inferior to those in the east and southeast regions of China. And as one of the innovation subjects, the record construction of new R&D institutions in Guangxi started late (the first batch was recognized in 2019), and their development is still in the exploration stage, which is important to research to promote the development of new R&D institutions in Guangxi and accelerate the pace of science and technology system reform in Guangxi. Accordingly,

*Corresponding Authors: Shiny521@naver.com, 1784127878@qq.com and ddgybb@163.com

 DOI 10.1201/9781003348023-76

this paper will take new R&D institutions in the Guangxi region as the main body, explore their development and problems based on the PSR framework model, and then give corresponding suggestions to better promote the construction and development of new R&D institutions in the Guangxi region.

## 2 REVIEW OF THE STUDY

### 2.1 *Concept of new R&D institutions*

As innovation and practice of science and technology system reform, there is no unified and recognized concept of new R&D institutions, except that their functional positioning and institutional model are recognized to be different from traditional research institutes. Xia et al. (2014) argue that new R&D institutions have the following characteristics compared with traditional institutions: functional orientation reflecting government orientation, de-administrative governance model, flexible and innovative institutional mechanism, and high collaboration between government, industry, academia, and research. According to (Wu & Chen 2018), new R & D institutions aim at the forefront of Science and Technology, involve multiple investors, are market-oriented, and are driven by innovation, it is a new type of R&D mode that technology invention and industry development coordinate and promote each other. In short, new R&D institutions should be independent legal entities with diversified investment subjects, internationalized construction mode, market-oriented operation mechanism, enterprise management system, combined incubation and innovation and entrepreneurship, sustainable development capability, and collaborative innovation between industry, and university and research.

### 2.2 *Research on the motivation for the establishment of new R&D institutions*

The development of new R&D institutions cannot be separated from the change in the background situation at home and abroad. With the nourishment of a series of background factors, the pressure of building new R&D institutions - an innovation system - has arisen in China. Based on the research on the motivation of new R&D institutions, this paper sums up these pressure factors as follows: from the macro-environment point of view, China has low innovation vitality (Wu & Chen 2018) and market demand, in particular, the urgent need to solve the "two skins" problem of technology and economy (Gao & Zhang 2015), top-level design of our country, such as innovation-driven development strategy, the call of supply-side structural reform policy (Liu et al. 2019), the rising status of enterprise innovation as the main body (Liao 2016), and industrial transformation and upgrading (Liao 2016; Ou et al. 2019), are the macro-pressure to build new R&D institutions in our country. In terms of micro-environment, the pressure factors that exist among the participating subjects of new R&D institutions and produce positive effects include resources, learning, and collaboration (Gao & Zhang 2015; Wu & Chen 2018; Xia et al. 2014). As mentioned earlier, the birth of new R&D institutions in China is contemporary and urgent, which can be used to some extent to solve the part of market failure to promote the integration of science and technology and industry and promote the transfer and transformation of industrial application technology.

### 2.3 *Research on the establishment and development status of new R&D institutions*

The establishment and emergence of new R&D institutions in China were mainly concentrated after 2005 (Ou et al. 2019), and in terms of spatial distribution, the newly established new R&D institutions show an uneven trend. At the regional level, they are mostly concentrated in the southeast coastal areas, such as Guangdong, Jiangsu, and Zhejiang provinces; at the provincial level, the new R&D institutions in each province are mainly concentrated in the regions or core cities with a more developed industrial economy. At present, the world is still during a new round of scientific and technological revolution and industrial change, with disruptive technologies emerging and industrialization accelerating, and the development of new R&D institutions in the ascendant

(Wang & Wang 2020). In recent years, the development of new R&D institutions has received national attention, and the state and relevant provinces have introduced a series of policies to encourage and support their construction, and the number has shown rapid growth (Zhang et al. 2019), aiming at major international frontier issues, and the fields involved include new energy, new materials, new generation information technology, biomedicine, energy conservation, and environmental protection, and other strategic emerging industries that the country needs to develop urgently (Wang & Wang 2020). However, from the study (Mao 2020) and other scholars, it is easy to produce a direct connection between institutions in the same region, and there are few connections between institutions in different regions, which confirms from the side that there is territoriality in the development of new R&D institutions, suggesting that new R&D institutions should pay more attention to regional industrial development characteristics and promote local industrial technology progress. In addition, in terms of development mode, new R&D institutions in different regions also have their characteristics. At present, there are widely recognized classification views in China: according to the different construction subjects, they can be divided into government-led, university-led, research institution-led, enterprise-led, social organization-led, or individual-led types (Li 2014; Tan & Chen 2015; Zhu et al. 2013); according to the nature of the institution, they can be divided into private non-enterprise, institution, share enterprise way, non-profit research institutions and other types (Chen & Wu 2018; Zhang et al. 2018). Based on the above studies, this paper obtains data information from local science and technology departments and related literature and reports (data as of early June 2022), screens out the top 5 regions with better development of new R&D institutions, and comes up with the following Table 1 as an overview of the development status of new R&D institutions nationwide.

Table 1. List of the development status of new R&D institutions nationwide.

| Province | Registration time | Better developed cities | Industry sectors involved |
|---|---|---|---|
| Jiangsu | 2017 | Nanjing, Suzhou, Xuzhou, and Wuxi | Biomedicine, intelligent manufacturing, energy conservation, environmental protection, new electronic information, new materials, artificial intelligence, etc. |
| Hubei | 2019 | Wuhan and Shiyan | Integrated circuits, a new generation of information technology, intelligent manufacturing, automotive, biological, and new energy, and new materials |
| Shandong | 2019 | Qingdao, Weifang, Jinan, and Yantai | Intelligent equipment and manufacturing, information technology, biomedicine, new materials, agriculture, etc. |
| Guangdong | 2015 | Guangzhou, Foshan, and Shenzhen | Artificial intelligence, integrated circuits, biomedicine, marine engineering, communications, networks, etc. |
| Fujian | 2017 | Xiamen, Quanzhou, and Fuzhou | New materials, intelligent manufacturing, biomedicine, a new generation of information technology, measurement, testing, etc. |

## 3 CONSTRUCTION OF PSR MODEL

Drawing on the “Pressure-State-Response” (PSR) model, which is a causality model widely used to evaluate the impact of human activities on the ecological environment, this paper proposes a model for the construction and development of new R&D institutions applicable to the Guangxi

Autonomous Region (Figure 1). The PSR model was first proposed by two Canadian statisticians, Tony Friend, and David Rapport, in 1979 as a conceptual model of causality, and later revised as a framework system for studying environmental problems, and is widely used in environmental, land, and agricultural research (Hughey et al. 2004; Wolfslehner & Vacik 2008). The difference between the PSR model and other models is that it embodies a logical process of change and development, such as "why does change occur-what kind of change occurs-what should be done". The model proposed in this paper is combined with the current situation of the development of new R&D institutions in Guangxi, and is built according to the idea of "why new R&D institutions are built - the current situation of the construction of new R&D institutions - the choice of measures to promote the development of new R&D institutions in the future".

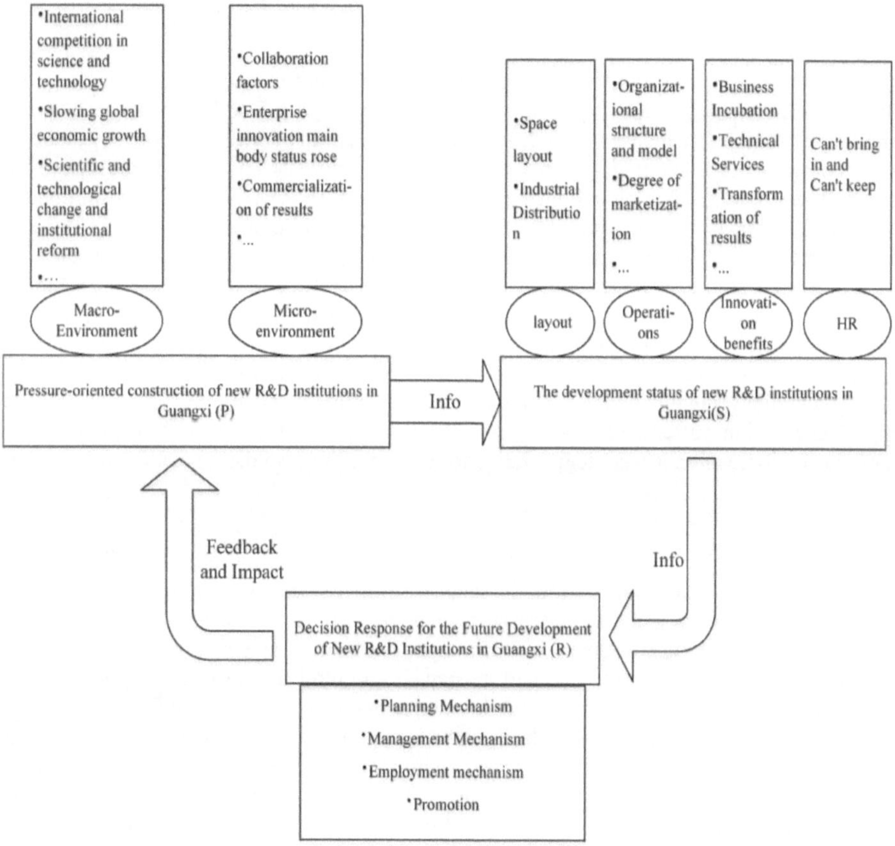


Figure 1. Model diagram of the development of the construction of new R&D institutions in Guangxi.

## 4 PRESSURE - REASON ORIENTATION OF BUILDING NEW R&D INSTITUTIONS IN GUANGXI

### 4.1 *Respond to policy calls*

Innovation is a system project that requires not only the intertwining and mutual support of innovation, industry, and capital chains, but also the policy chain to play a pushing effect. To open the channel between technological innovation and economic and social development, China issued the Implementation Plan for Deepening the Reform of Science and Technology System in 2015 and put forward the plan to promote the development of new R&D institutions and form cross-regional and cross-industry R&D and service networks. So far, new R&D institutions have gradually received attention at the national level. Under the guidance of national policies, in May 2019, to accelerate the construction of new R&D institutions in Guangxi and promote the high-quality development of innovation and entrepreneurship, the Department of Science and Technology and the Department of Finance of Guangxi Autonomous Region issued the Management Measures for New Industrial Technology R&D Institutions (Trial)", which for the first time openly selected new R&D institutions for the whole country and clarified the declaration and identification, policy support, management and assessment of new R&D institutions.

### 4.2 *The vision of improving the innovation and creation vitality of Guangxi*

From the basic indicators of the 2020 China Regional Innovation Capacity Evaluation Report, Beijing, Guangdong, Shanghai, and Jiangsu account for more than 50% of the national basic research investment, while the Guangxi region shows a weak foundation for science and technology innovation. In 2019, Guangxi's R&D expenditure investment is only 16.71 billion yuan, and the R&D expenditure investment intensity is 0.79%, ranking 27th place. In addition, enterprises as the main body of innovation have not given full play to their leading role, nearly 90% of enterprises above the size of the project are completed independently and have no R&D activities. In summary, the Guangxi region shows the limitation of innovation thinking and innovation ecology, whether it is R&D investment or technology accumulation. According to Browne, the choice of appropriate capital structure can promote the improvement of corporate governance structure and enhance the efficiency of corporate governance structure. As a new type of innovation consortium integrating R&D investment, innovation platform, and open cooperation between industry, academia, and research, new R&D institutions can further optimize the allocation of innovation resources, strengthen industrial technology supply and lead the development of strategic emerging industries, thus stimulating Guangxi's innovation and creativity vitality and improving the innovation capacity of the autonomous region, which is a strategic choice for Guangxi to grasp the innovation opportunities.

### 4.3 *The need to break the "two skins" of scientific research and industrial development*

During the "13th Five-Year Plan" period, the inactivity of enterprise innovation and creation, and the lack of enthusiasm of scientific researchers in transforming their achievements are still the major problems that Guangxi's scientific research and industrial development are in the state of "two skins", mainly reflected in the lack of awareness and motivation of the deep integration of industry, university, and research. From the horizontal point of view, the number of enterprises completing projects independently is as high as 90%; from the vertical point of view, universities get less than 30% of R&D funds from enterprises, while scientific research institutions get no more than 10% of R&D funds from enterprises. Such "skin-pulling" development eventually led to an extremely serious problem - the front-end innovation research results of universities and research institutions were put on the high shelf, and the enterprises were unable to industrialize the needed back-end technology. From the indexes of scientific and technological activities and marketization of technical achievements in Guangxi and the whole country, few high-quality scientific and technological achievements meet the market demand in Guangxi, as shown in Table 2.

Table 2. A comparison of technological achievements between Guangxi and the national level of scientific and technological activities*.

| Category | Guangxi | Nationwide | National ranking |
|---|---|---|---|
| Science and technology activity output level index | 36.04% | 75.42% | 22 |
| Marketization index of technological achievements | 21.07% | 77.37% | 27 |

*Source: Guangxi Intelligence Bureau.

## 5 STATUS - THE DEVELOPMENT STATUS OF NEW R&D INSTITUTIONS IN GUANGXI

In 2019, following the "Management Measures of New Industrial Technology R&D Institutions in Guangxi (Trial)", Guangxi Autonomous Region started the declaration of new R&D institutions. As determined by the meeting evaluation of relevant experts, 23 institutions eventually passed the accreditation in 2019; in 2020, 11 institutions eventually passed the accreditation. According to the statistics of Guangxi Information Bureau, during the "13th Five-Year Plan" period, the total number of R&D personnel of 34 new R&D institutions in the region reached 3,453, accounting for 60% of the total number of employees, with 655 valid invention patents, 1,928 transformations, and 1.632 billion yuan of transformation income, becoming an important regional innovation system in the autonomous region. It has become an important new force in the regional innovation system. At present, in the new round of the "14th Five-Year Plan", 12 new R&D institutions will be identified in Guangxi in 2021, and they are mainly concentrated in the Nanning area.

Table 3. Overview of the construction of new R&D institutions in Guangxi Autonomous Region*.

| No. | Location | Number | Main industry fields |
|---|---|---|---|
| 1 | Nanning | 29 | Intelligent transportation, new materials, biotechnology, testing technology and services, energy conservation and environmental protection, information technology, construction engineering, remote sensing technology |
| 2 | Liuzhou | 8 | Artificial intelligence, big data, the Internet of Things, new materials, systems, and software development |
| 3 | Guilin | 4 | New materials, geological minerals, artificial intelligence, special tires, and rubber products |
| 4 | Beihai | 3 | Biotechnology, new materials, system integration |
| 5 | Qinzhou | 1 | Biomedicine |
| 6 | Yulin | 1 | New materials, material recycling, and environmental protection |

*Source: Guangxi Science and Technology Department.

### 5.1 *Development layout*

From the statistics of the "Thirteenth Five-Year Plan" period, there are 34 new R&D institutions in Guangxi identified by the record. From the viewpoint of spatial distribution, six of the 14 cities in Guangxi have established new R&D institutions, of which 18 are in Nanning, the provincial capital alone, accounting for 53% of the number of new R&D institutions in the autonomous region, while the remaining eight cities are in blank construction, which shows that the construction of new R&D institutions in Guangxi is extremely unbalanced in terms of geographical distribution. From the perspective of construction form and operation, new R&D institutions in Guangxi are mainly of enterprise type, and many of them are established only in recent years, while the number of private non-enterprise types and career types is very small, and only Guilin Huigu Artificial Intelligence Industry Technology Research Institute and Beibu Gulf Institute of New Marine Materials are involved in the construction of two institutions. In addition, there are only five high-quality new R&D institutions that can establish continuously updated portals, have relatively independent operating conditions, and offer outstanding industrial services. This indicates that the construction quality of new R&D institutions in Guangxi is not high at present and lacks enthusiasm in the institutional filing. From the viewpoint of industrial distribution, the fields involved in new R&D institutions in Guangxi include information technology, advanced manufacturing, and new materials, which better comply with the strategic industrial layout actively developed by China.

### 5.2 *Operational construction*

Most of the traditional R&D institutions adopt the administrative management mode, follow the fixed "iron rice bowl" salary system, and implement the input mode with the government as the main input body in the organizational input, which is more homogeneous in operation (Browne 1999). A new generation of research and development institutions has changed their traditions, established market-oriented strategies, built a diversified and modernized management system, standardized the management of institutions, improved their scientific and technological innovation capacity and market competitiveness, and effectively enhanced the power and vitality of new research and development institutions through scientific and technological innovation. In addition, in terms of input subjects, new R&D institutions are also actively incorporating multiple resources to produce synergistic operational effects. For example, the Beibu Gulf Institute of New Marine Materials is a new R&D institution jointly built by Beihai Marine Industry Science and Technology Park, Beihai Science and Technology Bureau, and Zhejiang California International Institute of Nanotechnology of Zhejiang University, which is a self-supporting institution approved by Beihai Municipal People's Government. The Institute has become one of the new R&D institutions with a diversified construction mode, market-oriented operation mechanism, and modernized management system in Guangxi, with the development mode of inline and outline cooperation.

### 5.3 *Scientific research benefits*

Under the support of new policies and new models, the new R&D institutions have played a powerful role in science and technology innovation: in carrying out incubation and technical services, as of the first quarter of 2020, the first batch of 23 new R&D institutions have incubated and cultivated 66 science and technology-based enterprises and served 8938 enterprises, which greatly promoted the adjustment and transformation and upgrading of Guangxi's regional industrial structure and promoted the innovative development of Guangxi region. In the cooperation between government, industry, academia, and research, according to incomplete statistics, the first batch of 23 new R&D institutions have cooperated with 52 colleges and universities and 11 research institutes, carrying out 20 national science and technology projects, 96 provincial science and technology projects and 87 municipal science and technology projects. In the transformation of scientific and technological achievements, the new R & D institutions not only focus on scientific research results but also pay more attention to the use and transformation of scientific and technological achievements in the form of enterprise incubation and school-enterprise cooperation as mentioned above. Based on the combination of existing technologies and achievements, new R&D institutions take the initiative to investigate the market demand, accept the commission of enterprise projects and carry out joint basic, applied, and frontier research in this way, and this R&D process is the supply-side reform in the field of science and technology R&D (He 2019).

### 5.4 *Human resources*

According to the annual report on the employment quality of the 2021 ordinary college graduates in Guangxi, as of September 1, 2021, the number of 2021 ordinary college graduates in the region reached 30.09 million, including 10.5 thousand postgraduates and 127.7 thousand undergraduates. And from the perspective of the employment area chosen by the employed graduates, nearly half of the postgraduates and undergraduates chose to be employed outside the region, and the loss of higher-level talents is serious, which is undoubtedly a heavy blow to the cultivation and development of high-level talents in Guangxi region. Innovative talents are the most asset of new R&D institutions and the core power source for the development of new R&D institutions. At present, the region's science and technology talent, especially high-level talent and team scarcity, this talent cannot attract the embarrassing dilemma of retention also led to new R&D institutions as a carrier to build a high-level science and technology innovation platform is still scarce.

## 6 RESPONSE – THE FUTURE DEVELOPMENT OF NEW R&D INSTITUTIONS IN GUANGXI

### 6.1 *Optimize the development environment and plan the overall layout*

At present, the most obvious problem with the construction of new R&D institutions in Guangxi is the uneven quantity and quality of the development of institutions in each region horizontally, which is not only in Guangxi but also in most of the construction of new R&D institutions in the country. Therefore, as a government department, firstly, it should formulate more localized encouragement policies to clarify the development positioning and identity certification of new R&D institutions; secondly, as a highland for the collection of technology, knowledge, talents, capital and other resources, new R&D institutions need a variety of factors to work together, and at present, most of the new R&D institutions built in Guangxi are of enterprise type, while there are few of other types. Relevant departments need to optimize their allocation of resources following the collection of resources in different cities and districts, particularly by combining the development of local universities and leading enterprises, and by mixing and reforming the existing career-based new research to continue enriching the construction of new research and development institutions in Guangxi.

### 6.2 *Operation management*

#### 6.2.1 *Increase government support*

At present, Guangxi Autonomous Region provides financial support of no more than five million to the newly identified new R&D institutions, and after their construction and operation, they are assessed and evaluated with the help of third-party evaluation agencies to give subsequent financial subsidies of no more than two million. Compared with the government financial support in Guangdong and Jiangsu, Guangxi has more in-depth and stronger measures to support the new R&D institutions.

#### 6.2.2 *Building universities- and institutes-based platforms*

From the results of the author's current statistics, most of the new R&D institutions built in Guangxi are of the enterprise type, while the new R&D institutions built by universities, research institutes, and enterprises are not. The ultimate purpose of enterprise type is to make a profit, which may lead to weak R&D capability and ignore the long-term R&D benefits due to the lack of strong technology or talented team. For a viable new R&D institution, what it pursues is the continuous precipitation and development of technology and R&D capability. For example, the most representative German Franhof Association, which is a tripartite management body composed of enterprises, academia, and the government, is responsible for the overall research direction of the association and has been conducting market-oriented applied research since its establishment in 1949.

#### 6.2.3 *Increase information disclosure and broaden innovation channels*

At present, the construction of new R&D institutions around the world presents a thriving scene. However, as a new thing that has developed rapidly in China only in recent years, new R&D institutions have been lacking in terms of information integration and disclosure. Provinces such as Guangdong and Jiangsu, as the leaders in the development of new R&D institutions in China, have also failed to establish information databases about new R&D institutions (Zhao & Dai, 2017), not to mention those new R&D institutions that have been built only in recent years. Therefore, at present, in the construction of new R&D institutions, in addition to grasping the current offline physical development in various places, it is also necessary for each institution to do a good job of publishing the relevant online information and timely registering the development of institutions for the record, to form the effect of synergistic operation between online and offline.

### 6.3 *Improve the employment mechanism*

New R&D institutions as the assembly of various elements of the highland, especially the establishment of the early urgent demand for talent elements, while most of the high-end talents are a flexible introduction, mid-range talents are more inclined to go to well-known enterprises, research institutes or institutions, resulting in the talent gathering effect is not high. For this reason, the new R & D institutions in the employment mechanism must break the traditional concept of "only education, only thesis", with a more inclusive, open-minded to accept different levels of talent, emphasizing that "only talent is used".

### 6.4 *Promotion and publicity, committed to high-quality development*

As the leader of industrial R&D and application, the new R&D institutions should not only have the consciousness of taking other high-quality new R&D institutions as the development "benchmark" but also have the goal of developing themselves into the professional field "benchmark". To this end, Guangxi's new R&D institutions are in the process of continuous improvement and development to find their foothold, grasp the main direction, while learning from the development and operation mode of other institutions to form their advantageous features, and then take advantage of the launch to create their brand.

## 7 CONCLUSION

Based on the PSR framework structure, this paper constructs a model diagram of the construction and development of new R&D institutions in Guangxi, which provides new ideas for systematically sorting out the development of new R&D institutions. From the perspective of development status, the construction of new R&D institutions has undoubtedly brought positive results to the science and technology and economic development of the Guangxi region, but it still faces some difficulties and problems, such as the small number of institutions and uneven development level. At present, the research on new R&D institutions is mostly based on the internal development of institutions, but as a multi-subject institution integrating production, academia, and research, the development of new R&D institutions is bound to involve more stakeholders. Therefore, in the future, scholars can also explore how new R&D institutions can use their characteristics to achieve value co-creation based on the perspective of the innovation ecosystem to solve the current development confusion.

## REFERENCES

Browne, F.X. 1999. Corporate Financial Intermediaries And Firm Valuation. *The Journal of Finance* (11): 24–25.

Chen, S.Y. & Wu, H.B. 2018. Problems and Countermeasures for the Development of New R&D Institutions under Innovation-driven Strategy. Macroeconomic Management (06): 43–49.

Gao, R.H. & Zhang, W.W. 2015. Research on the development status of new R&D institutions in China. *Value Engineering* 34(33): 45–47.

He, X. 2019. Research on the development dilemma and policy support path of new R&D institutions. *Scientific Management Research* 37(06): 41–47.

Hughey, K. F., Cullen, R., Kerr, G. N. & Cook, A. J. 2004. Application of the pressure–state–response framework to perceptions reporting of the state of the New Zealand environment. *Journal of Environmental Management* 70(1): 85–93.

Li, D.L. 2014. Exploring the development mode and characteristics of new R&D institutions in Guangdong. *Guangdong Science & Technology* 23(23): 77–80.

Liao, Y.N. 2016. Exploration of new R&D institutions in China: Guangdong as an example. *Science & Technology Industry of China* (08): 70–76.

Liu, Y.X., Ou, C.R., Zhang, G.Y. & Yang, S.W. 2019. The growth path of new R&D institutions and their characteristics: A science and technology innovation perspective. *Journal of Guangdong University of Technology* 36(05): 94–101+110.

Mao, Y.H., Shui, Y.Y., Fang, Y.L. & Li, S.M. 2020. Research on the development of new R&D institutions and their development trend: a visual analysis based on Citespace. *Management and Administration* (05): 10–17.

Ou, C.R., Liu, Y.X., Zhang, G.Y. & Yang, S.W. 2019. Research on the Background and Realistic Opportunity of New R&D Institutions for Science and Technology Innovation. *Journal of Guangdong University of Technology* 36(05): 102–110.

Tan, L. & Chen, Y.S. 2015. Research and Suggestions on the Construction Model of New R&D Institutions in Guangdong. *Science and Technology Management Research* 35(20): 45–49.

Wang, S.G. & Wang, B.B. 2020. Reflections and Suggestions on the Development of China's New R&D Institutions in the Background of the New Era. *Science and Technology &Innovation* (01): 9–13.

Wolfslehner, B. & Vacik, H. 2008. Evaluating sustainable forest management strategies with the Analytic Network Process in a Pressure-State-Response framework. *Journal of environmental management* 88(1): 1–10.

Wu, W. & Chen, L.T. 2018. Research on the mechanism of formation and role of new R&D institutions under different information conditions: based on the perspective of innovation-driven development. *Science and Technology Management Research* 38(11): 87–94.

Xia, Y.T., Zhang, Y.F., Gao, R.H., Zhou, W.K. & Wang, C.L. 2014. Research on Collaborative Innovation Model and Mechanism of New R&D Institutions in China: An Example of 6 New R&D Institutions in Suzhou, Guangdong and Shaanxi. *Science & Technology Progress and Policy* 31(14): 13–18.

Zhang, Y.L., Li, R.Y., Liu, Y.X., Xu, Z.H. & Zhang, G.Y. 2018. Analysis and Research on the Current Situation of New R&D Institutions in Guangdong Province. *Science and Technology Management Research* 38(13): 124–132.

Zhang, Y.L., Zhen, Y., Luo, J.W. & Zhang, G.Y. 2019. Research and Prospects of New R&D Institutions Based on Systematic Literature Review Method. *Science and Technology Management Research* 39(15): 83–91.

Zhao, J.D. & Dai, Q.Y. 2017. Data Analysis of New R&D Institutions in Guangdong Province and its System Construction. *Science and Technology Management Research* 37(20): 82–87.

Zhu, J.J., Cai, J.W., Liu, S.F., Fang, Z.G. & Guan, Y.Q. 2013. Research on the Operation Mechanism and Construction Strategy of New R&D Institutions in Jiangsu. *Science & Technology Progress and Policy* 30(14): 36–39.

*Advances in Urban Construction and Management Engineering – Cha (Ed)*


# Research progress, hotspots and prospects of low-carbon city in China: Visual analysis based on CiteSpace

Xing Jin* & Tingting Qiao*
*School of Business, Guilin University of Electronic Technology, China*

Xuebin Huang*
*Department of Evaluation, Guilin Institute of Aerospace Industry, China*

ABSTRACT: The construction of a low-carbon city is one of the important measures to face the increasingly serious carbon emission problem. Visual analysis of relevant literature in the research field of urban low-carbon development can provide a basis and reference for further in-depth research on low-carbon city construction and development. Based on 1,360 pieces of literature collected from the CNKI database from 2008 to 2021, this paper analyzes the research progress, hot topics, and the evolution of hot topics in the field of urban low-carbon development in China by using CiteSpace visualization software and then puts forward research prospects.

## 1 INTRODUCTION

Although the development in recent years has created a good external environment for the development of low-carbon construction work, it is still facing a huge contradiction between economic and social development and carbon emissions, and the development of the low-carbon city is imminent. In general, research related to low-carbon city development focus on concept and connotation (Wang 2014), measurement (Li 2012), and evaluation (Chen 2016; Li 2011), mainly analyzing temporal and spatial differences (Zhu 2015), regional disparities (Zhong 2013) and endogenous dynamics (Zhou 2016). A few scholars discussed the planning and realization path of low-carbon city construction (Wang 2016), and put forward policy suggestions on improving the system design of urban green development, encouraging, and supporting green technology innovation (Lu 2013; Meng 2015). However, traditional literature review mainly relies on reading literature and making systematic comments on it, which is limited and inevitably subjective. This research tried to use CiteSpace software, the extracted from CNKI in 2008-2021 low carbon city development research literature related to the study of visual analysis, clear, clear to show the research progress of this field, heat evolution trend and future development direction, strive for our country provides the reference for the further research of low carbon city development and reference.

## 2 RESEARCH DESIGN

### 2.1 *Research methods*

The analysis tool used in this study is CiteSpace, a visual analysis software designed and developed by Chen Chaomei, version 6.1.R2. The focus of an area of research changes over time, and trends in scientific knowledge can be tracked over time by analyzing the academic literature (Saheb T 2019). Therefore, the scientific knowledge graph analysis method was adopted. In CiteSpace, the time interval was selected from 2008 to 2021, and the time slice was one year. Node Types were set as

*Corresponding Authors: Shiny521@naver.com, 1349837048@qq.com and bin2022028@163.com


DOI 10.1201/9781003348023-77

Author, Institution, and Keyword, and Selection Criteria are set as Top 20, 20, and 50, respectively, and author collaboration atlas, institutional collaboration atlas, keyword co-occurrence and cluster atlas can be drawn respectively. Based on keyword co-occurrence and cluster atlas, burstiness can further be used to generate keywords to explore low-carbon city research progress, hotspots, and evolution.

### 2.2 *Data sources and research methods*

The data of this study came from CNKI with the function of advanced retrieval, and the title of the article was set as "low-carbon city". The publication time of the retrieved literature was from 2008 to 2021, with a period of 14 years. Other retrieval conditions were all default values. After retrieval, to ensure the rationality of the selected papers, the data were checked item by item, and invalid records such as journal conference solicitation, communication, no author, and irrelevant to the subject were canceled. Finally, a total of 1,360 valid papers were obtained.

## 3 RESEARCH PROGRESS OF LOW-CARBON CITY

### 3.1 *Number of publications*

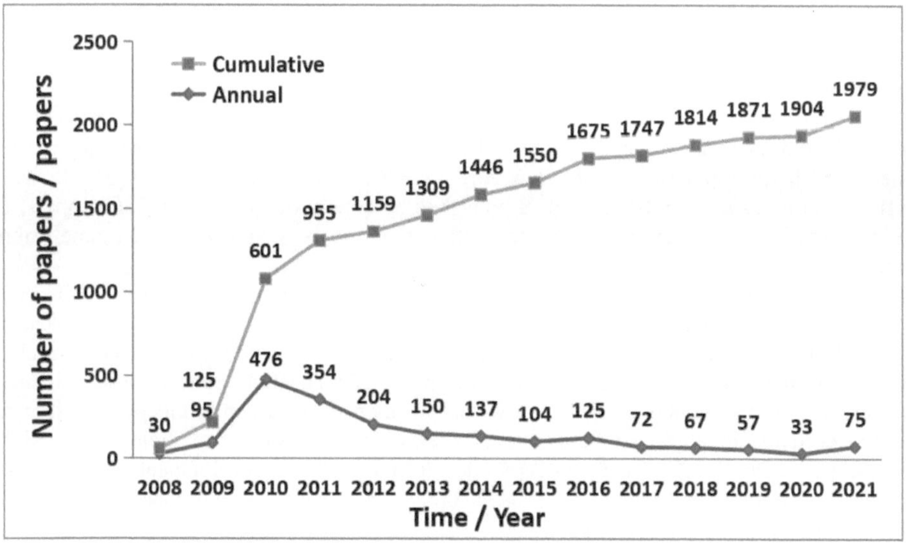


Figure 1. Annual and cumulative publications of low-carbon city research.

Figure 1 shows the number of "low-carbon cities" retrieved on an annual and cumulative basis. The cumulative number of documents has shown a continuous increase from 2008 to 2021. From 2008 to 2010, the number of annual documents has shown a rapid growth trend, and the annual number of documents in 2010 reached the maximum (476); from 2010 to 2015, the annual number of publications decreased, but remained at a high level of more than 100; in 2016, the annual number of documents issued increased slightly, and after 2017, the annual number of documents issued showed a steady downward trend, that is, taking 2010 as the critical point, the annual number of documents issued before 2010 increased steadily. After 2010, the research intensity weakened, and the annual number of documents issued continued to decline, but after 2020, the annual number of documents issued has an increasing trend, which may be related to the national "double carbon" goals and policies.

### 3.2 *Analysis of core authors and cooperative relationships*

Figure 2. Author collaboration map of low-carbon city research.

Core authors reflect the author's influence in a certain field, while author collaboration networks can reveal the research interests and links of different authors in that field (Ma 2018). In the author collaboration atlas (Figure 2), there are 159 lines among 421 authors, representing a certain cooperative relationship between authors. According to Price's law in bibliometrics, M = 0.749 (N max)1/2 determines the number of papers published by core authors. Among 1,360 papers, Du Dong published the most papers with 8 papers, that is, N max = 8, and M = 2.12 (namely, three papers).

It can be seen from Figure 2 that the size of nodes reflects the number of papers published by authors. Based on the number of papers published by authors, scholars such as Du Dong, Zhuang Guiyang, Li Kexin, and Li Ruiqing are the core authors in the field of low-carbon city research. In addition, from the perspective of connections, the number of connections between authors is small, indicating that most authors in the field of low-carbon city research are in an independent research state. It is currently apparent that there are only a few obvious author cooperation teams, and this decentralized research state among scholars does not facilitate the development of a consensus on theoretical research and practical results. Therefore, the exchange and cooperation between scholars still need to be strengthened.

### 3.3 *Distribution of institutions and cooperation*

From the perspective of the distribution of the core research forces in the field of low-carbon city research in China, statistics on the institutions where the authors of the main literature work can reveal the main research institutions in this field. Based on the literature data, the cooperation map of institutions was drawn (Figure 3). The analysis shows that there are 130 links between 367 research institutions, which reflects that there is a certain research cooperation relationship between research institutions, but the degree of cooperation needs to be further improved at present.

Analysis of Figure 3 further indicates that research on low-carbon cities is concentrated in scientific research institutes or comprehensive universities located in the major cities of Beijing, Shanghai, and Guangzhou. From the size of nodes and the connection between nodes, Tongji University, Tsinghua University, Huazhong University of Science and Technology, Peking University, and other institutions have formed a large cooperation group, but have not formed a close cooperation network. Other agencies are sub-departmental or relatively independent of the same agency. Therefore, in the follow-up research, the cooperation between various institutions, especially trans-institutional and trans-regional cooperation, needs to be strengthened.

Figure 3. Collaboration map of institutions in low-carbon city research.

## 4 ANALYSIS OF RESEARCH HOTSPOTS OF LOW-CARBON CITY

### 4.1 *Overview of research hotspots based on keyword co-occurrence*

Keywords co-occurrence and cluster visualization atlas (Figure 4). The analysis shows that 446 keywords have 669 links with each other, and the module value is 0.6885, greater than 0.30, indicating that there are significant differences among research topics of urban low-carbon development, and the average contour value is high, reaching 0.9607, indicating that the clustering results are convincing. According to the analysis, keywords such as low-carbon economy, low-carbon city pilot policy, low-carbon city construction, and sustainable development are closely related, and key clusters such as climate change, carbon sink, carbon emission, and evaluation index system overlap each other, which indicates that these themes blend and influence each other logically, and also reflects that low-carbon city research is inseparable from the low-carbon economy, carbon sink, carbon emissions, and other topics.

Further analysis found that the three keywords with the highest frequency in the research related to urban low-carbon development were "low carbon city", "low carbon economy" and "carbon emission" (Table 1), reaching 416, 73, and 39 times, respectively. On this basis, further cluster analysis was conducted on keywords, and a total of 23 clusters were generated. Among them, low-carbon economy, evaluation index system, sustainable development, climate change, low-carbon city pilot policy, carbon sink, and low-carbon city construction have formed 9 relatively large clusters. And the planning and construction composed of low-carbon buildings, urban planning, and public policies are also hot topics. According to the results of keyword frequency and cluster analysis, there are three major research hotspots in low-carbon city research since 2008.

### 4.2 *Research hotspots and development dynamic analysis based on keyword co-occurrence*

#### 4.2.1 *Connotation and development concept of low-carbon city*

The UK is the first country to attach importance to the low-carbon city. In 2003, the UK government's energy white paper our Energy Future: Building a Low-carbon Economy mentioned

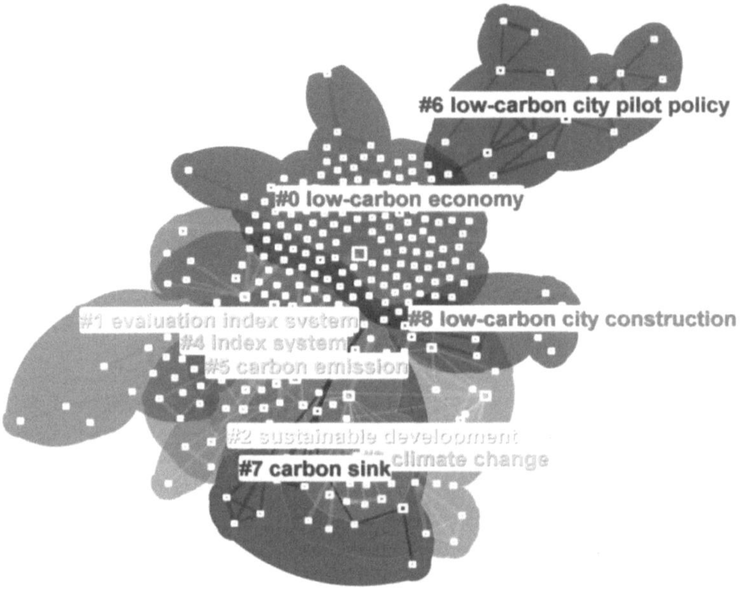


Figure 4. Co-occurrence and clustering of keywords in China's low-carbon city research field.

Table 1. The top 10 frequency keywords of the urban low-carbon development field.

| ID | Count | Keywords | ID | Count | Keywords |
|---|---|---|---|---|---|
| 1 | 416 | Low-carbon city | 6 | 22 | Sustainable development |
| 2 | 73 | Low-carbon economy | 7 | 18 | Index system |
| 3 | 39 | Carbon emission | 8 | 14 | Evaluation index system |
| 4 | 33 | Climate change | 9 | 10 | Low-carbon city construction |
| 5 | 27 | Urban planning | 10 | 8 | Analytical hierarchy process |

"low-carbon economy" for the first time, which means to create higher production value with less consumption of natural resources and pollution emissions. The low-carbon city comes from a low-carbon economy. It is a city with a low-carbon economy as its development mode and direction, citizens with low-carbon life as their concept and behavior characteristics, and government officials with a low-carbon society as its construction sample and blueprint. Subsequently, Japan put forward the idea of building a "low-carbon society", and the concept of "low-carbon" has gradually attracted wide attention around the world. In general, a low-carbon city is to achieve the all-around low-carbon transformation to minimize greenhouse gas emissions by developing a low-carbon economy and innovating low-carbon technologies, and forming a sustainable development path of harmonious coexistence between man and nature. Although China's low-carbon city construction started relatively late, with the continuous strengthening of low-carbon awareness, different regions of China have carried out low-carbon city development exploration, and actively promote low-carbon city construction. In early 2008, WWF selected Shanghai and Baoding as the first batch of the low-carbon pilot city to launch the Low-carbon City Development Project in China, which helps to promote building energy efficiency, renewable energy, and the manufacturing and application of energy-saving products (Jin 2008).

#### 4.2.2 *Low-carbon city development planning and construction*

Urban planning, as a means of national macro-control, determines the future development direction of cities. Integrating low-carbon concepts into urban planning and realizing low-carbon construction through urban planning means is particularly important. First, the government plays a leading role in the process of low-carbon city construction, guiding and promoting the transformation of the traditional city into low-carbon cities through policy guidance and system design (Jiang 2013). With the release of the national low-carbon pilot provinces and city notice, pilot regions began to prepare low-carbon city development plans according to local conditions (National Development and Reform Commission 2012), the academia has carried out extensive research, mainly including the form of urban spatial structure, land use type, low carbon community planning, urban traffic system, and urban industry. Based on this, we should advance the compact urban layout, multi-center group, land function mix, green building design, construction of the comprehensive traffic patterns, and urban green space system low carbon development strategy. Secondly, in the context of addressing climate change and realizing energy conservation and emission reduction, the implementation of the carbon tax and carbon emission right trading is an effective policy means for low-carbon city construction (Zhang 2010). At present domestic scholars in the connotation of the carbon tax and carbon tax division, the carbon tax system construction has made the corresponding research, and most scholars think city industrial, transportation, construction is the main source of carbon emissions, reducing carbon emissions as the core of low carbon city construction goal and key measures of response to climate change (Qin 2011), domestic scholars proposed to implement building energy efficiency standards, establish urban public transport and rapid rail transit systems, improve energy efficiency and other measures to reduce CO from the scope 2 emissions.

#### 4.2.3 *Low-carbon city development evaluation*

At present, the evaluation of urban low-carbon development in China mainly involves two aspects: a comprehensive evaluation of a low-carbon economy and an evaluation index system of low-carbon cities. The comprehensive evaluation of a low-carbon economy includes economy, society, environment, resources, technology, policy, and other aspects. Domestic research on the evaluation index system mainly follows the ideas of many researchers including Edward L. Glaeser, Matthew E. Kahn, and Chris Goodall (Wang 2012). The weight and standardized treatment are determined according to the importance of indicators, and the core evaluation indicators are consistent with those of foreign countries. However, the research on the indicator system of Domestic scholars is relatively complex, with specific indicators ranging from more than 10 to more than 40, and quantitative indicators are the majority, covering different fields and evaluation standards (Liu 2013). In recent years, new measurement methods and models have also entered the quantitative research of low-carbon city construction. Yuan et al. (2017) introduced basic data and advanced technical means in urban meteorology, GIS, and other related fields into the natural ventilation evaluation method, and established a relatively complete evaluation and research framework through interdisciplinary research with urban microclimate; Xu et al. (2020) through the integration of ecological methods, put forward countermeasures of building low-carbon cities and ecological corridors according to the current situation of urban development, which is an important attempt to study the interdisciplinary research of low-carbon city construction evaluation system. In general, the study of the low-carbon city evaluation system is still in the exploratory stage in China, and most of the corresponding studies are mainly low-carbon pilot cities, lacking a unified measurement standard, and its applicability needs to be proved by several practical cases.

## 5 EVOLUTION OF LOW-CARBON CITY RESEARCH TOPICS

Emergent keywords refer to the high-frequency keywords in a certain period, which constitute the main keyword emergent map and reflect the research hot spots and development trends in a specific period to some extent (Sun 2018). The top 15 emerging keywords in the field of low-carbon city research are sorted out (Figure 5). First, from the perspective of emergent intensity, the keyword with the highest emergent intensity is "low-carbon economy", whose emergent value is four and whose first occurrence time is 2009. It shows that the theme of the urban low-carbon economy has been rising since 2009, and it still has high research heat and development potential,

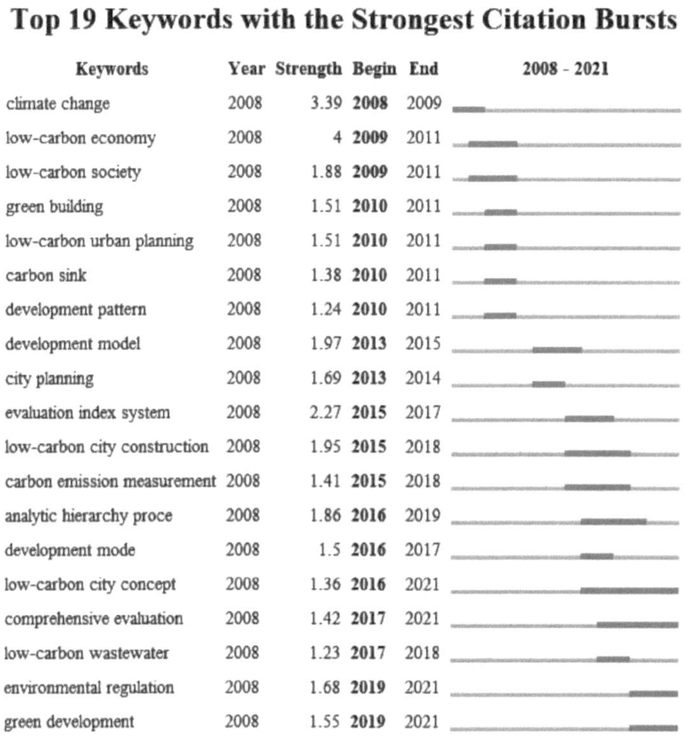


Figure 5. The map of keywords emerging in the research field of low-carbon city.

followed by "climate change", with a sudden value of 3.39, indicating high academic attention to it. Secondly, in terms of the duration of emergence, the "low carbon city concept", "low carbon city construction", "carbon emission measurement" and "comprehensive evaluation" have the longest influence cycle (4 to 6 years), and all of them have appeared since 2015, indicating that these themes have been the hot spots of scholars' continuous attention for many years, and are also significant turning points in the early stage of urban low-carbon development. Finally, from the perspective of time series changes, the emerging keywords have changed from "green building", "low carbon urban planning" to "development model", "low carbon city construction", and finally to "green development", "comprehensive evaluation", which shows the evolution of the connotation and form of urban low-carbon development research. Urban low-carbon development is not only reflected in the physical structure but also endowed with green, ecological, sustainable, measurable, and other multidimensional and deep-seated connotations.

## 6 CONCLUSIONS AND PROSPECTS

### 6.1 *Main conclusions*

Using CiteSpace to draw a knowledge map, this paper makes a visual analysis of 1,360 documents on the study of low-carbon cities from 2008 to 2021 from the aspects of the distribution of documents, research hotspots and research path evolution, major institutions, and key authors, and the following conclusions are drawn:

(1) From the time of publication, the number of annual publications of low-carbon city research in China has experienced a rapid growth stage, gradual decline stage, and fluctuation increase stage; The cumulative number of documents issued shows a steady increasing trend.

(2) From the perspective of authors, Du Dong, Zhuang Guiyang, and Li Ruiqing constitute the core authors of low-carbon city research in China. The contributions of the core authors have laid the foundation of low-carbon city research and are the backbone of research in the field. Non-core author groups publish fewer articles and lack cooperation with each other.
(3) From the perspective of published institutions, Tongji University, Tsinghua University, Peking University, and other universities are the main research institutions, and the internal cooperation between the teams is relatively close; In addition, the cooperation among all research institutions is mainly regional, and the inter-institutional cooperation across regions is weak.
(4) From the results of keyword co-occurrence, keywords such as low carbon economy, urban planning, carbon emission, and index system have the highest frequency and become a hot research direction; at the same time, the keyword emergence results show that the research frontier points to the development mode, comprehensive evaluation, environmental regulation, green development, and other fields, which are also the focus of our attention in the future.

### 6.2 *Prospects*

Based on the above analysis, the research field of low-carbon cities in China is constantly expanding, from basic theoretical research to technological innovation, which helps promote the faster and better development of low-carbon cities in China. However, there is still a large gap between China's low-carbon city research and foreign countries in terms of basic theory, research depth, and combination with national policies. In the future, the following aspects should be paid more attention to.

(1) We should strengthen the cooperation among researchers and promote the regional cooperation of research units. Domestic researchers in low-carbon cities and related fields should strengthen cooperation and realize cross-regional cooperation through policy encouragement and overall funding support and other ways to promote the high-quality development of domestic low-carbon city research field.
(2) The development and improvement of a carbon accounting system can provide more scientific and comprehensive greenhouse gas emission data inventory support for the formulation of urban low-carbon policy and the development of a low-carbon society, which is an important content that cannot be ignored in the expansion of urban low-carbon development research field. At present, the SPD method has been used as a cutting-edge analytical tool to analyze carbon emission pathways in foreign academic circles. However, the research started late, and there is no systematic study on the carbon emission pathway in China with the SPD method, so the application of the SPD method in the study of low-carbon urban development in China needs to be further expanded.
(3) Although all cities in China are deepening the development of a low-carbon economy, and three batches of the low-carbon pilot city have been set up at the national level, there are only a few studies involving the basic work of performance, monitoring, and evaluation of China's low-carbon city construction. Most academic circles attribute the difference in the data analysis results of low-carbon cities before and after the implementation of the policy to the effect of low-carbon policies or measure the implementation effect of low-carbon policy through econometric modeling. However, there is uncertainty in the causal relationship between policy and effect, further research needs to adopt scientific empirical methods to separate the influencing factors other than low-carbon policy actions to scientifically measure the effect of low-carbon policy implementation.

## FUNDING

This research was supported by the Improvement of Basic Scientific Research Ability of Young and Middle-Aged Teachers in Guangxi Universities in 2022 (Project title: A Study on The

Impact Mechanism of Green Innovation Performance of Guangxi Enterprises Based on Knowledge Behavior).

## REFERENCES


Chen F. & Zhu D.J. (2009). Theoretical methods of low-carbon city research and empirical analysis of Shanghai. *J. Urban development research*. 16(10): 71–79.

Chen J.G. (2016). Research on evaluation index system and method of low-carbon city based on sustainable development. *J. Engineering Construction and Design*. 8, 163.

Fu Y. & Wang Y.L. & Li D. (2008). Research on the development path of low-carbon city. *J. The Impact of Science on Society*. 2, 5–10.

Jiang R.L. (2013). Logic program, function and policy orientation of low-carbon city construction. *J. Business Times*. 21, 13–15.

Jin S. WWF launches Low-carbon Urban Development project in China. *J. Environmental Protection*. 3, 22.

Li G.D. & Zhao Z.S. & Zhang J.H. (2012). Measurement of urban low-carbon development level under the background of climate change. *J. Meteorology and disaster reduction research*. 35(04):1–8.

Li K.X. (2011). Low carbon strategy of urbanization in China. *J. Journal of the Chinese Academy of Sciences*. 26 (01): 49–55.

Liu Q.P. (2013). Thoughts on the construction of low carbon city evaluation index system in China. *J. China population, resources and environment*. 23(11): 280–283.

Lu J. (2013). *Economic exploration of Low-carbon city construction in China*. D. Jilin University.

Ma T. & Jia R.Y. & Liu Q.L. et al. (2018). The evolution of China's innovation network research and analysis of the hotspots. *J. Science and technology progress and countermeasures*. 35 (3):22–28.

Meng X.M. (2015). *Research on low-carbon City development in China*. D. Shandong University.

Qin B. & Shao R. (2011). Low-carbon city and spatial structure optimization: Concept, demonstration and practice. *J. Urban planning international*. 26(3): 72–77.

Saheb T. & Saheb M. (2019). Analyzing and visualizing knowledge structures of health informatics from 1974 to 2018:A bibliometric and Social network analysis. *J. Health care Informatics Research*. 25(2): 61–7.

Sun W. & Mao L.X. (2018). Evolution of Beijing-Tianjin-Hebei Coordinated Development based on Citespace method. *J. Acta Geographica Sinica*. 73(12): 2378–2391.

Wang B.B. & Du S.H. (2012). ECO2: A third model for low-carbon city? *J. Journal of Sichuan Normal University* (Social Science Edition). 39(3): 45–50.

Wang B.J. & Wan Y.H. (2014). *Concept connotation of low-carbon city and development enlightenment of Low-carbon city in China*. C. (9th) Urban Development and Planning Conference proceedings – S11 Standard Discussion and Future prospect of Low-carbon Ecological City with Chinese characteristics. 37–41.

Wang D.D. (2016). Research on the construction mode and implementation path of low-carbon city.*J. Ecological economy*. 32(09): 47–51.

Yuan L. & Zhang Y.X. & Guo Y.Y. et al. (2017). Research on urban planning and design strategy for improving urban microclimate: A case study of Natural ventilation assessment in Shenzhen. *J. City planning review*. 41(09): 87–91.

Zhang Z.T. (2010). Some thoughts on China's carbon tax legislation. *J. Chinese Journal of Law*. 2, 98–100.

Zhong X. (2013). *Regional differences of urban low-carbon level in Guangdong Province*. D. Guangzhou University.

Zhou Z.N. (2016). Empirical research on driving factors of urban residents' low-carbon consumption behavior. *J. Science and Technology Square*. 1, 173–178.

Zhu Y. (2015). *Spatial-temporal coupling coordination between urban land intensive use and low-carbon city*. D. Ocean University of China.

*Advances in Urban Construction and Management Engineering – Cha (Ed)*
*© 2023 Copyright the Author(s), ISBN 978-1-032-39018-5*

# Development of a high-quality strategy based on SWOT analysis—Taking the *Populus euphratica* area as an example

Zhongyan Zhu[1,*]
*Research Institute of Human Geography, Xi'an International Studies University, Xi'an, China*

ABSTRACT: This paper investigates a *Populus euphratica* scenic area using field research, ArcGIS10.7, SWOT analysis, and various scenic tourism resources from four aspects to make a high-quality development strategy. There are many advantages (or strengths) to developing the poplar forest scenic area, including its location, resources, and policy, but there are also many disadvantages (or weaknesses), including strong tourism seasonality, single products, and the construction of infrastructure. A development strategy is proposed for the scenic area in terms of products, prices, channels, and publicity to promote high-quality development.

## 1 INTRODUCTION

On both sides of the Ejina River, a large population of euphratica grows. The Ejina Banner Populus Forest is one of the only three original poplar forests in the world, and the Populus Euphratica Scenic Area is one of its components. The scenic spot integrates Populus euphratica, Tamarix Chinensis Lour, desert, and Mongolian folklore, and is called the "pearl of the desert".

SWOT stands for strengths (S), weaknesses (W), opportunities (O), and threats (T). SWOT analysis is based on internal and external competitive environments and competitive conditions. In the 1980s, American scholars proposed SWOT analysis, which was first practiced in the field of management and then widely used in various industries. China began to apply it to tourism planning in the 1990s (Demir et al. 2016; Lin 2004; Li 2021; Ye 2005).

In the new era, high-quality development has become a theme in China's economic and social development. Since the 19th National Congress of the Communist Party of China (CPC), China's economy has shifted from high-speed growth to a stage of high-quality development (He 2018). Promoting high-quality development of tourism is an important driving force for the high-quality development of China's economy. Tourism is the core element of development, and high-quality development of scenic spots is a top priority (Li & Wang 2021).

This study considers the Populus euphratica scenic spot as the research object, introduces the SWOT analysis model, analyzes the strengths, weaknesses, opportunities, and threats of the development of scenic spots, and combines it with field research to judge the development direction of scenic spots and provide development strategies for the high-quality development of scenic spots.

## 2 OVERVIEW OF THE RESEARCH AREA AND RESEARCH METHODS

Ejina Banner is located in the westernmost part of the Inner Mongolia Autonomous Region. It is connected to the Alxa Right Banner in the southeast, Gansu Province in the west and southwest, and Mongolia in the north, with a border of 507 km. The Populus Euphratica Scenic Spot (national 4A level scenic spot) is located 2 km south of Dalaihubu Town in Ejina Banner (Figure 1), covering an area of 1,500 mu. It is a complex tourism project that integrates sightseeing, cultural teaching, scientific practice, and commerce.

*Corresponding Author: zhuzhongyan0225@163.com

DOI 10.1201/9781003348023-78

Through data collection, field investigation, and SWOT analysis (Markovska et al. 2009), a high-quality development strategy for scenic spots is proposed.

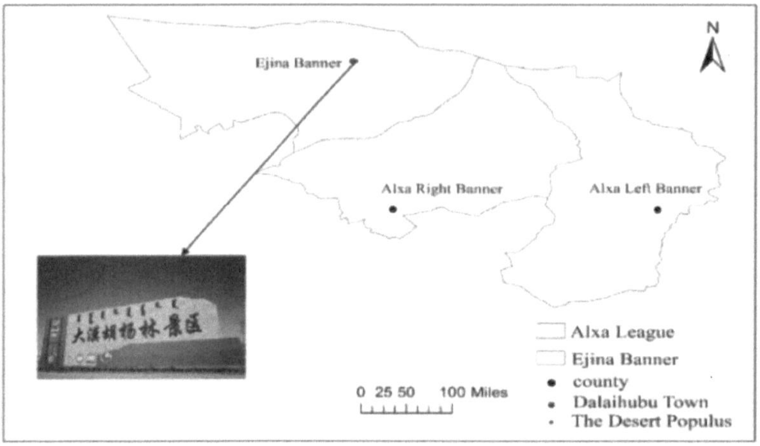


Figure 1. Location diagram of the study area.

## 3 SWOT ANALYSIS OF THE DESERT POPULUS EUPHRATICA SCENIC SPOT

To understand the opinions of different groups of people on scenic spots, the author conducted a field survey of tourists at scenic spots. Online questionnaires were distributed through a questionnaire star. A total of 215 valid questionnaires were collected. On this basis, the SWOT analysis method was adopted to analyze and determine the advantages, disadvantages, opportunities, and threats of the development of Desert Populus scenic spots.

### 3.1 *Analysis of strengths*

#### 3.1.1 *Advantages*

Desert Populus euphratica scenic spots have rich natural and cultural resources. In terms of natural resources, the natural landscape of the scenic spot is unique, among which the most characteristic scenic spots water mirror populous euphratica, here is also the best scenery of the reflection of this scenic spot; for human resources, there is a unique ethnic tribe in the scenic spot – Mongolian Tuerhute. The glorious history of Tuerhute's return as the pioneer of eastward return and the traditional culture of their tribe attracts many tourists. As shown in Figure 2, tourists are very fond of water mirror populous euphratica, Tuerhute family, desert countryside, Ruoshui River riverside, and other scenic spots. They also have a strong interest in camel culture, Tuerhute culture, folk performances, and other cultural displays (Figure 3).

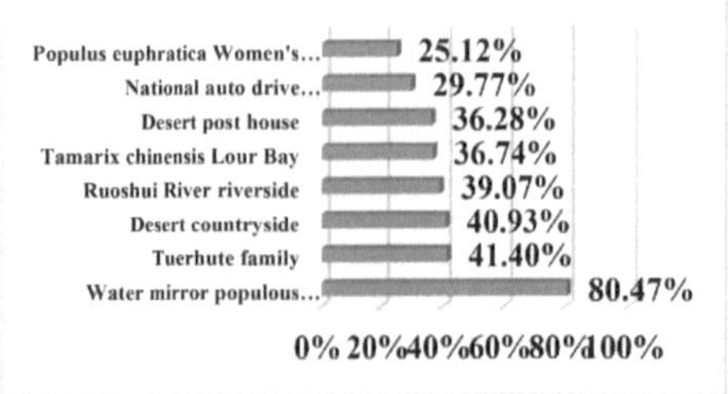


Figure 2. Attractions of interest to tourists.

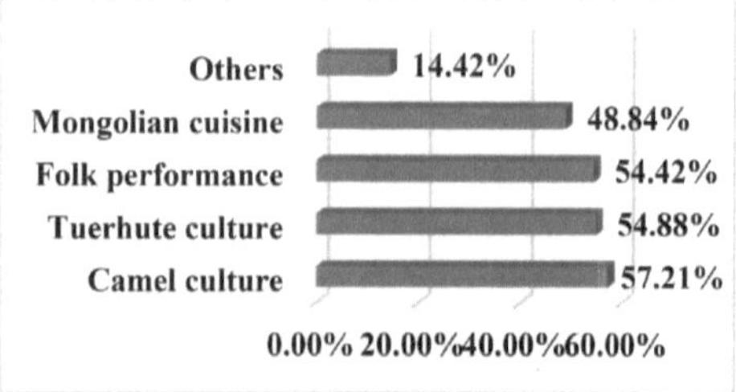


Figure 3. The favorite pastime for tourists.

### 3.1.2 *Location*

Desert Populus euphratica scenic spot is located in Dalaihubu Town, Ejina Banner. The town's railways are Jiuquan-Ejina railway, Linhe-Hami railway and Jiayuguan-Ceke railway, and Xi'an direct flights to Ejina Banner. In recent years, the Ejina Banner has been continuously strengthening the construction of transportation infrastructure, road traffic conditions have significantly improved, and people's travel has become more convenient.

## 3.2 *Analysis of weaknesses*

### 3.2.1 *Tourism products are single and homogenized seriously*

Although scenic spots are rich in natural resources and profound in cultural heritage, they lack tourism products that can highlight their characteristics, and existing tourism products lack diversity. The survey found that 30.23% of tourists thought that the features of scenic spots were not outstanding. Tourists for this purpose are mainly sightseeing tourists; most of the tourists visiting Populus euphratica, and Populus euphratica in Ejina Banner have several similar scenic spots, even Ejina Populus euphratica scenic spot.

### 3.2.2 *Network marketing means lagging behind*

With the rapid development of modern network technology, information transmission has increasingly relied on it. However, as shown in Figure 4, the marketing mode of the scenic spot lags, mainly relying on the introduction of family members or friends, while TV, the Internet, WeChat public numbers, and we-media promotion account for only 5.58%, 11.16%, 7.91%, and 5.58%, respectively. This shows that there is a serious lag in network marketing, the development time of the scenic spot is not long, and scenic spot operations have not been formed.

Figure 2. Channels for tourists to obtain information on the Populus Euphratica Scenic Spot.

### 3.2.3 *Infrastructure construction is inadequate*

To reduce costs, the scenic spot adopts an outsourcing shuttle bus to reduce capital investment in scenic spot traffic. However, the outsourced shuttle bus lacks coordination with the scenic spot, causing inconvenience to tourists. Regarding accommodation, the seasonal limitation of viewing Populus euphratica causes tourists to concentrate in autumn. The rooms in Ejina are in short supply, resulting in a sharp increase in housing prices. As can be seen from Figure 5, 42.79% of tourists are not satisfied with the traffic conditions of the scenic spot, 32.09% are dissatisfied with the catering and accommodation of the scenic spot, and 22.79% think the supporting facilities of the scenic spot are incomplete.

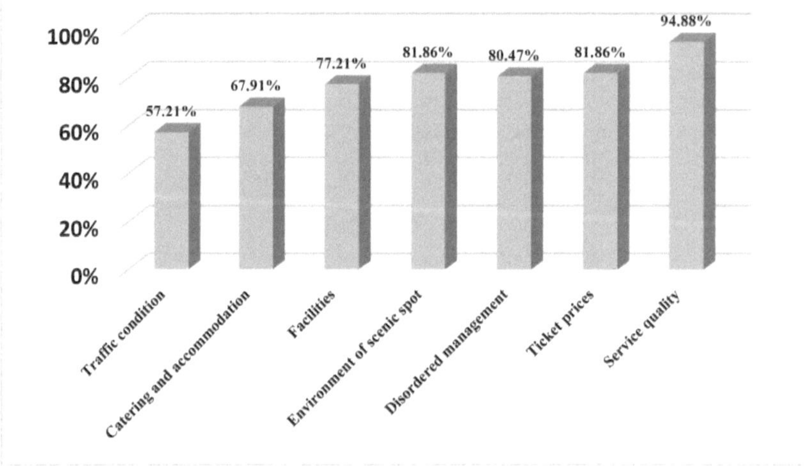


Figure 3. The satisfaction of tourists.

### 3.3 *Analysis of opportunities*

In 2020, the COVID-19 epidemic spread across the world, plunging the tourism industry into a trough. With the normal development of the pandemic, new modes of tourism have emerged (Lin 2021). The rise of domestic and peripheral tourism has also provided new development opportunities for regional tourism, which has brought opportunities for the development of the Desert Populus euphratica scenic spot.

#### 3.3.1 *National policies*

The state has implemented coordinated regional development and rural revitalization. The state stressed the need to vigorously support the accelerated development of ethnic minority areas, poor areas, and border areas, and strengthen measures to promote the development of the western region to form a new pattern, which provides opportunities for the development of scenic spots.

#### 3.3.2 *Local policies*

Ejina Banner tourism has been blowing from the new economic growth point towards a new pillar industry, and actively developing all-for-one tourism, which has an important role in promoting the economy and culture of Ejina Banner. In the face of the scenic spot "off-season crisis" at the same time, the Ejina Banner set issued the "Ejina Banner off-season tourism development incentive subsidy scheme", to further optimize the environment of Ejina Banner cultural tourism development.

### 3.4 *Analysis of threats*

#### 3.4.1 *Domestic competition intensifies*

The enthusiasm for tourism development is increasing worldwide, leading to intensified competition in the tourism industry and rising development costs. Simultaneously, competition for the domestic tourist source market is becoming fiercer.

#### 3.4.2 *Regional competitive development*

The Populus Euphratica Scenic Spot faces the challenge of many scenic spots, such as the Ejina Populus euphratica scenic spot and the Black City Ruins site. The landscape and tourism market are similar; therefore, the Desert Populus euphratica scenic spot faces huge challenges.

## 4 HIGH-QUALITY DEVELOPMENT STRATEGY OF THE POPULUS EUPHRATICA SCENIC SPOT

Based on the SWOT analysis of the enhanced Populous Scenic Spot, it can be found that the development of the scenic spot has a certain advantage, but there is still a backward way of publicity. Populus euphratica scenic area lacks infrastructure construction to better grasp development opportunities and meet the challenges of the future. In this regard, we propose the following constructive suggestions and development strategies to achieve high-quality development of the scenic area.

### 4.1 *Strengthen development strengths*

#### 4.1.1 *Vigorously develop cultural tourism products*

Based on the existing, it can develop the story of the legend of the Tuerhute tribe, Ruoshui River three thousand live action deduction. Populus euphratica theme cultural festival, music festival, food street, and other activities can be held in the case of more tourists. We can design and produce a series of exclusive souvenirs for sale at scenic spots, such as imitation poplar leaf ornaments (earrings, necklaces, rings), poplar-style clothing, poplar-related stationery, and supplies, which can increase the income of the scenic spot while allowing tourists to have a deep understanding of its culture.

#### 4.1.2 *Develop leisure and vacation tourism products*

Populus euphratica is the scenic spot of the main body, but cannot completely populate euphratica repeated development; it should be to join other element extension tourism products, such as the rich cultural connotation of poplars as the woman and the family of Tuerhot, giving them new meaning, which will stimulate the tourist's imagination and memory to the greatest extent. In addition, traditional routes should be constantly updated to develop new routes, attract, new activities, and give new meaning to tourism products.

### 4.2 *Make up the weak points and promote high-quality development*

#### 4.2.1 *Strengthen infrastructure construction*

In terms of transportation, first, scenic spots outside the traffic problem should contact the local transportation department, and the railway station and urban area should be opened to tourist buses for scenic spots; second, the internal traffic problems of scenic spots should be taken into consideration to withdraw the outsourcing system, and staff of scenic spots should manage and coordinate scenic traffic vehicles. In terms of accommodation, firstly, innovative accommodation should be provided based on the specific situation of the scenic spot to increase the number of rooms. Secondly, the relevant local departments should adopt some preferential conditions that will divert tourists from Ejina Banner, resulting in a shortage of rooms there.

#### 4.2.2 *Develop various marketing methods*

First, scenic spots can cooperate with WeChat, Meituan, Ctrip, Qunar, and other travel network companies to promote smart ticketing systems that provide more convenience for tourists. Second, scenic spots can sign cooperative agreements with other local scenic spots, hotels, and travel agencies in various provinces and regions to achieve resource sharing and mutual benefit. Additionally, official social media platforms such as WeChat, Weibo, TikTok, and others should also be utilized to provide timely information on tourism resources and activities. In addition, advertising, and marketing directly within tourists' moments in WeChat can help attract tourists to other tourists by increasing investment in network information.

### 4.3 *Seize policy opportunities and speed up development*

All departments of the scenic spot should study carefully and deeply understand the spirit of the superior documents, to accurately understand the requirements, tasks, policy support, and other content, especially for the specific content related to the development of the scenic spot and tourism, we should carefully study and grasp, find the right direction, earnestly seize, and use the policy opportunities, and make breakthroughs. It is also important to speed up the development of scenic spots to create growth points, poles, and belts, as well as to promote the high-quality development of scenic spots.

### 4.4 *Strengthen regional cooperation and jointly meet challenges*

To strengthen regional cooperation, we should strengthen cooperation with other tourist attractions, such as "Black City Ruins" and "Strange Woods", to enrich tourism content, extend the stay time of tourists and meet their various tourism needs. The integration of local tourism resources promotes the formation of an Ejina banner tourism system, making Ejina Banner an important travel destination.

## 5 CONCLUSIONS AND DISCUSSIONS

During the 14th Five-Year Plan period, the high quality remained the main theme in the development of tourist attractions. Through a combination of questionnaires and field research, the development status of the Populus Euphratica Scenic Spot has been analyzed the SWOT method.

(1) It was found that the scenic spot has rich tourism resources and obvious location strength, but there are still weaknesses such as a single product, serious homogenization, and insufficient publicity.
(2) Facing opportunities and challenges, combined with the unique Ruoshui River area, the ethnic culture of the Tuerhute tribe vigorously develops, highlights the tourism products with the characteristics of the scenic spot, and strengthens the infrastructure construction.
(3) Relevant departments and governments should combine related industries and pay attention to marketing across all channels. They should also make full use of modern promotional means for publicity and promotion and provides research support for the development of scenic spots to achieve high-quality development of Populus Euphratica Scenic Spots.

With the continuous development of the economy, the tourism industry is also facing challenges, but also opportunities. First, the premise of all development is the sustainable development of scenic spots. It is very important to ensure the stable development of the ecological environment in scenic spots. Additionally, the scenic spot should follow the development trend, make good use of the new media platform, and introduce changes and innovations to attract tourists and maintain a sustainable income.

## REFERENCES

Demir S Esbah H Akgün A A (2016) Quantitative SWOT analysis for prioritizing ecotourism-planning decisions in protected areas: Igneada case Int. *J. of Sustainable Development & World Ecology*. 23: 456–468.
He J M (2018) *Research on the System and Strategy of China's tourism High quality Development in the New era Tourism Tribune*. 33: 9–11.
Li T Wang X T (2021) *Research on high-quality Development of Yanmenguan Scenic Spot based on Network comment 2021 Proceedings of China Tourism Science Annual Conf.* 593–603.
Li X X (2021) Study on Rural Tourism Development Strategy Based on SWOT Quantitative*Analysis Urbanism and Architecture*. 18: 51–53.
Lin Y (2021) Reflections on regional tourism development under the new mode of tourism development in the post-epidemic era. *J. of Anshan Normal University*. 23: 47–49.
Lin Z H (2004) *SWOT Analysis of Tourism Planning Forest Inventory and Planning*. 3 L 64–69.
Markovska N Taseska V Pop-Jordanov J (2009) SWOT analyses of the national energy sector for sustainable energy development. *Energy*. 34: 752–756.
Ye Y H (2005) SWOT strategy analysis of tourism planning *J. of Hunan College of Finance and Economics*. 53–55.

*Advances in Urban Construction and Management Engineering – Cha (Ed)*
*© 2023 Copyright the Author(s), ISBN 978-1-032-39018-5*

# A phase sequence optimization method oriented by ideal bidirectional green wave

Liang Zhu*, Jin-xin Wang* & Sheng Dai*
*Lianyungang JARI Electronics Co., Ltd., Lianyungang Jiangsu China*

Ju-yuan Wu*
*Shanghai JARI Zhaoxin Information Technology Co., Ltd., Shanghai, China*

ABSTRACT: The conventional green wave design is to unify the signal control cycle of all the intersections on the arterial street and set a reasonable phase offset for each intersection to ensure the traffic flow moves on the street, and the signal light is always green when they arrive at each intersection. It is proved that only adjusting the phase offset cannot solve the problem of bidirectional green wave compatibility in many cases. Given this, a phase sequence optimization method oriented by the ideal bidirectional green wave is proposed. Firstly, the signal engineer designs a unidirectional green wave scheme for the arterial street based on the background phase sequence, which is a time-distance graph. Secondly, the engineer finds the intersection in the figure with the minimum green time of another coordinated phase and draws the ideal green wave for another direction according to the designed green wave speed. Thirdly, the engineer draws the adjustable green time of another coordinated phase at each intersection in the figure. Then, the engineer moves up and down the ideal green wave to make it as far as possible, located in another adjustable green time with a coordinated phase at each intersection to maximize the actual green wave bandwidth. Finally, the engineer adjusts the phase sequence of each intersection according to the final ideal green wave position. The example shows that the proposed method is simple and effective, and a reasonable bidirectional green wave can be designed quickly.

## 1 INTRODUCTION

Coordinated control can effectively reduce the number of stops and delay time of traffic flow by giving a unified signal control cycle to each intersection of the arterial street, setting a reasonable phase offset for each intersection, and relying on clock calibration to make traffic flow run on the arterial street and arrive at each intersection in a green traffic state (Wang et al. 2006). To ensure the maximum overall traffic capacity of the arterial street, we usually select the cycle of the key intersection with the largest load on the arterial street as the public cycle time (Lu et al. 2009).

There is much research on how to design green wave bands. Chen Yao et al. studied the phase offset optimization of green wave band based on Synchro (Chen et al. 2016). Pei Yulong et al. established a delay equation based on parallelogram type to calculate the phase offset that minimizes the delay of bidirectional straight vehicles (Pei et al. 2007). In addition, Jing Binbin et al. proposed a numerical solution to design bidirectional maximum green wave band bandwidth (Jing et al. 2017). The result of designing bidirectional green waves merely by adjusting the common period time and phase offset is not satisfactory in many cases. In addition, the numerical method often simplifies and solidifies the model of intersection and arterial streets in the calculation. It sets some boundary conditions and excludes some seemingly non-major, which actually may be important influencing factors. This may lead to the miss of the expected effect. This paper proposes a phase sequence

*Corresponding Authors: zhuliang200894@163.com, 462378754@qq.com, daisheng716@163.com and celine_jari@yeah.net

DOI 10.1201/9781003348023-79

optimization method that takes the bidirectional green wave design as the goal orientation. This method does not need a complex theoretical model or complicated calculation, but can design a reasonable bidirectional green wave signal scheme only according to the basic planes and traffic flows of the intersections on the arterial street.

## 2 BIDIRECTIONAL GREEN WAVE DESIGN

The arterial street green wave with maximum bandwidth gain is called the ideal green wave. Furthermore, for the upstream and downstream directions of the arterial street, the concept of the ideal green wave is further divided into the upstream ideal green wave and downstream ideal green wave. The upstream and ideal downstream bandwidth is

$$\begin{cases} Gb_{up}^{\max} = \min\{g_{up}^{1}, g_{up}^{2}, \ldots, g_{up}^{n}\} \\ Gb_{down}^{\max} = \min\{g_{down}^{1}, g_{down}^{2}, \ldots, g_{down}^{n}\} \end{cases} \tag{1}$$

Where, $n$ is the number of intersections on the arterial street, $g_{up}^{i}$ is the upstream coordinated phase green time of the $i$ intersection on the arterial street, $g_{up}^{i}$ is the downstream coordinated phase green time of the $i$ intersection on the arterial street.

In Figure 1 the green time of upstream and downstream coordinated phases of intersection $I_i$ and intersection $I_j$ are equal, so the ideal upstream and downstream bandwidths are equal. By adjusting the phase offset, it is easy to get the unidirectional ideal green wave shown in Figure 1. However, the downstream green wave bandwidth is far from reaching the ideal green wave. The reason is that the upstream and downstream coordinated phases are started and ended simultaneously, which limits the formation of green waves in the downstream direction. At this point, the start time or end time of the upstream or downstream coordinated phase needs to be adjusted to design an ideal bidirectional green wave.

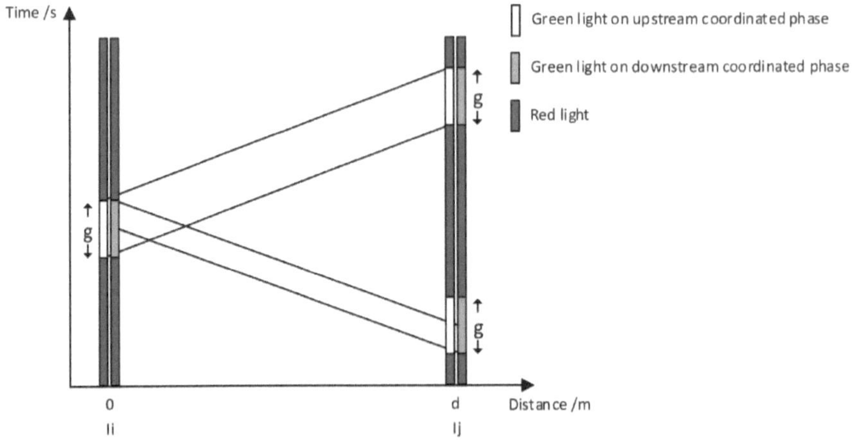


Figure 1. Schematic diagram of the ideal upstream green wave.

It can be intuitively concluded from the figure that if the green time of the downstream coordinated phase of $I_j$ at the intersection in the figure will be moved downstream and forward for the period, and the downstream green wave bandwidth of the arterial street can be effectively increased. Therefore, the phase sequence of the intersection $I_j$ will be adjusted by the engineer from the phase alignment mode in Figure.1 to the phase overlap mode in Figure 2. With the adjustments described above, the ideal green wave will be designed for upstream and downstream coordinated phases.

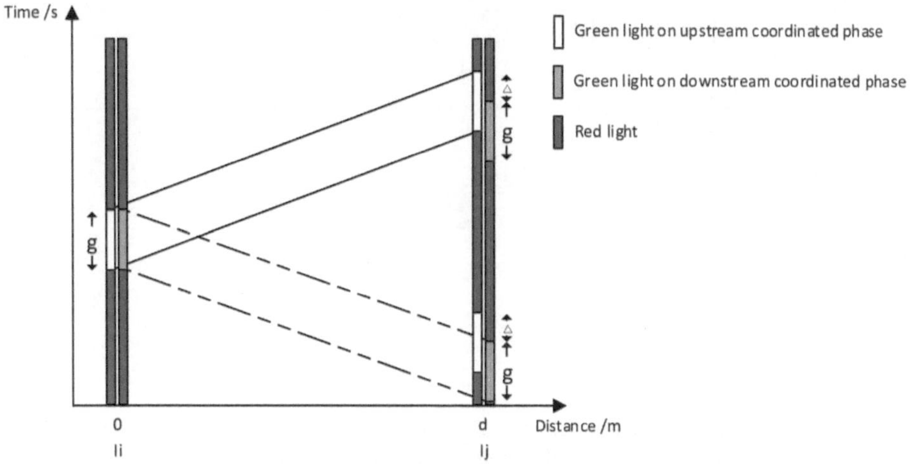


Figure 2. Schematic diagram of the bidirectional ideal green wave.

Whether the phase sequence of intersection $I_j$ can be adjusted to the phase overlap shown in Figure 2 is a real problem. Taking Figure 3 as an example, when the green time for going straight is longer than that for the left turn, the phase sequence can be adjusted to overlap mode 1 and overlap mode 2. When the green time for going straight is equal to that for the left turn, the phase sequence can be adjusted to single side mode 1 and single side mode 2. The phase sequence can be adjusted to overlap mode 3 and overlap mode 4 when the green time for the straight is shorter than that for the left turn. It can be concluded that the phase overlap mode in Figure 2 can be adjusted when and only when the green time for the left turn phase that following the coordination phase of the intersection $I_j$ is Δ.

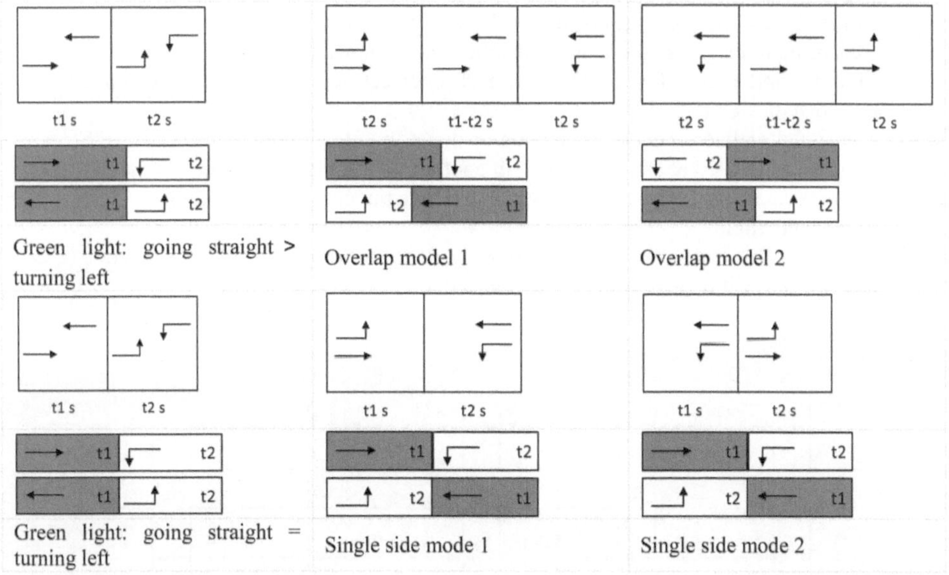


Figure 3. Phase sequence adjustment.

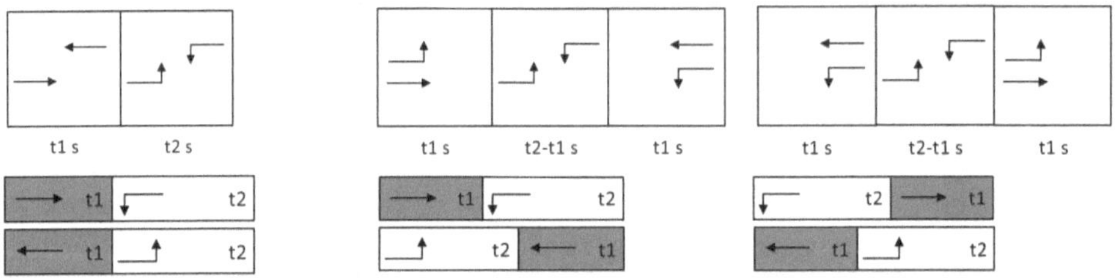


Green light: going straight < turning left

Overlap model 3

Overlap model 4

Figure 3. Continued.

Therefore, in many cases, the effect of the bidirectional ideal green wave cannot be achieved by adjusting the phase sequence. At this point, maximizing the bandwidth of the green wave in the other direction becomes the goal of signal engineers on the premise that the main coordination direction is the ideal green wave. According to Formula (1), the maximum value of the ideal green wave bandwidth of the arterial street is the minimum green time of the coordinated phase at each intersection. Therefore, if the main coordination direction is the ideal green wave, the closer the green wave bandwidth in another direction is to the minimum green time of another coordinated phase at each intersection of the arterial street, the better.

Given this, a phase sequence optimization method oriented by an ideal bidirectional green wave was proposed. The specific steps are described as follows:

Step 1: The signal engineer needs to draw the arterial street time-distance graph based on the background phase sequence scheme and adjust each intersection's phase offset to design a unidirectional ideal green wave in the main coordination direction for the arterial street.

Step 2: The signal engineer needs to find the intersection with the minimum green time from another direction in the figure. At the beginning and end of the green light, the engineer needs to draw two oblique lines according to the designed green wave speed. The area formed by two oblique lines is the ideal green wave in another direction.

Step 3: According to the phase sequence adjustment rules shown in Figure.3, the adjustable green time of another coordinated phase at each intersection was analyzed and plotted in the time-distance graph.

Step 4: The signal engineer needs to move up or down the ideal green wave drawn in Step 2 to ensure it is located as far as possible in another adjustable green time with a coordinated phase at each intersection, so that we can achieve the goal of maximizing the actual green wave bandwidth.

Step 5: Based on the ideal green wave position obtained in Step 4, the phase sequence of each intersection will be determined and adjusted.

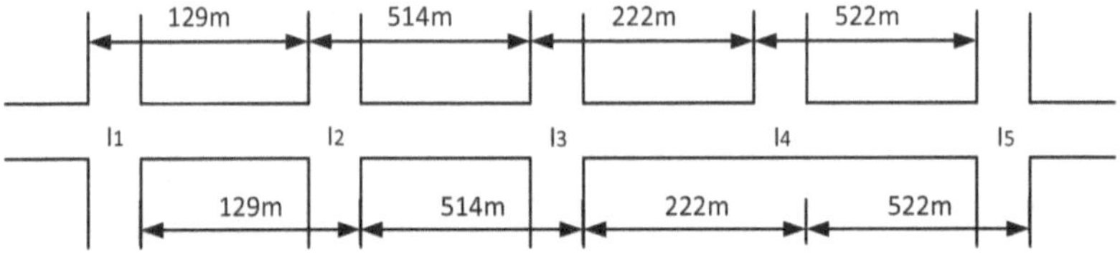


Figure 4. Layout of intersections on the arterial street.

## 3 INSTANCE ANALYSIS

Take an arterial street with five intersections in Figure 5 as an example. It is defined as the upstream direction of the arterial street from west to east and the downstream direction of the arterial street from east to west. Among them, the upstream direction is the main coordinating direction. The bidirectional design green wave velocity is 11m/s. The basic phase sequences for each intersection are shown in Figure 6.

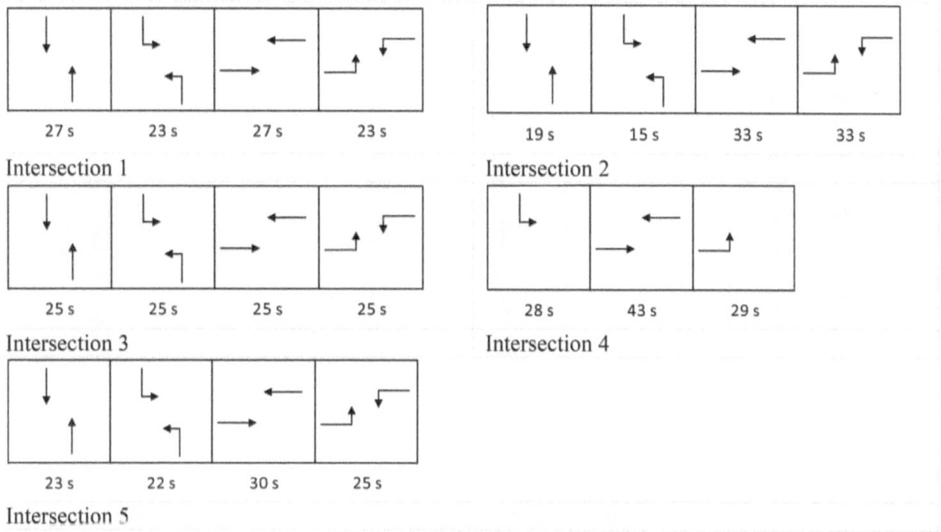


Figure 5. Timing scheme of the basic signal at the intersection.

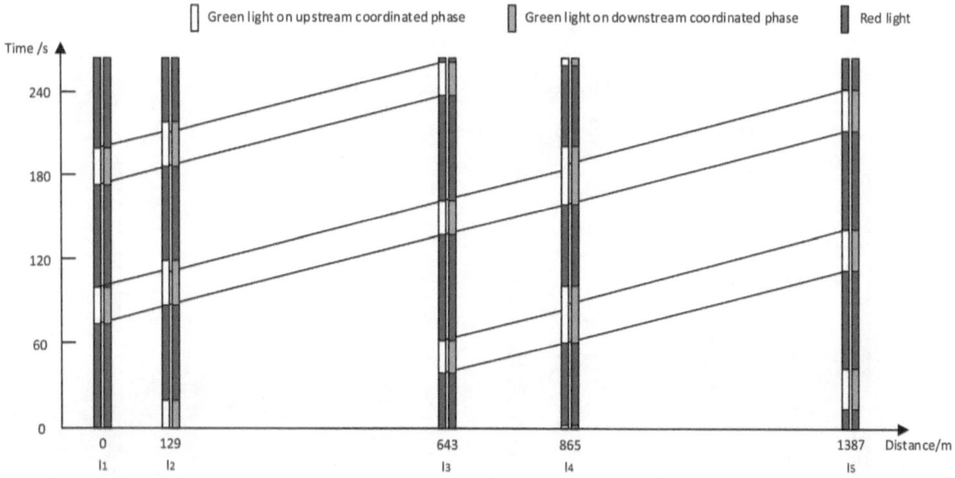


Figure 6. The ideal green wave of arterial street upstream.

Details of how the proposed method was applied to the bi-directional green wave design in this example are as follows:

Step 1: By adjusting the phase offset, the bidirectional ideal green wave from west to east in the main coordination direction of the arterial street was designed, and the result is shown in Figure 6;

Step 2: Centered on $I_3$, the minimum green time of the downstream coordinated phase in the arterial street, two oblique lines are drawn at the beginning and end of the green light according to the designed green wave speed to form the downstream ideal green wave region, as shown in Figure 7

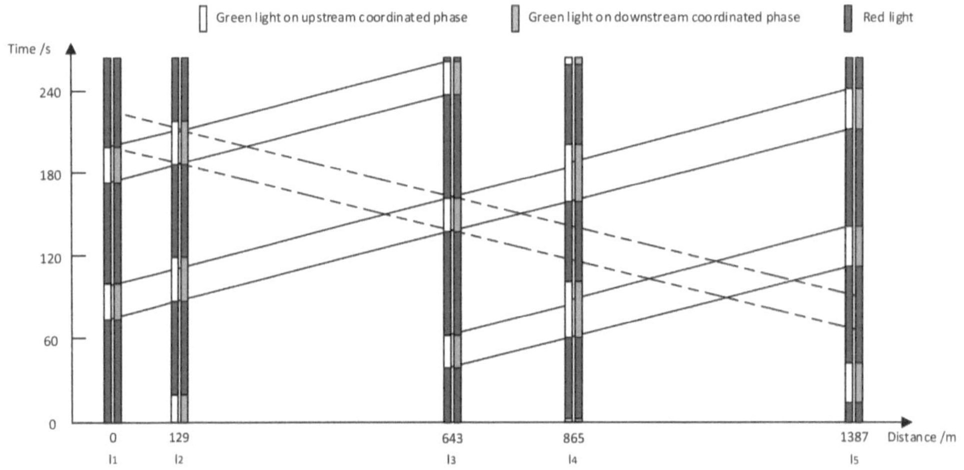


Figure 7. The ideal green wave area of arterial street downstream.

Step 3: According to the phase sequence adjustment rules shown in Figure 3, the adjustable green time of the downstream coordinated phase at each intersection was analyzed successively and plotted in the time-distance graph, as shown in Figure 8.

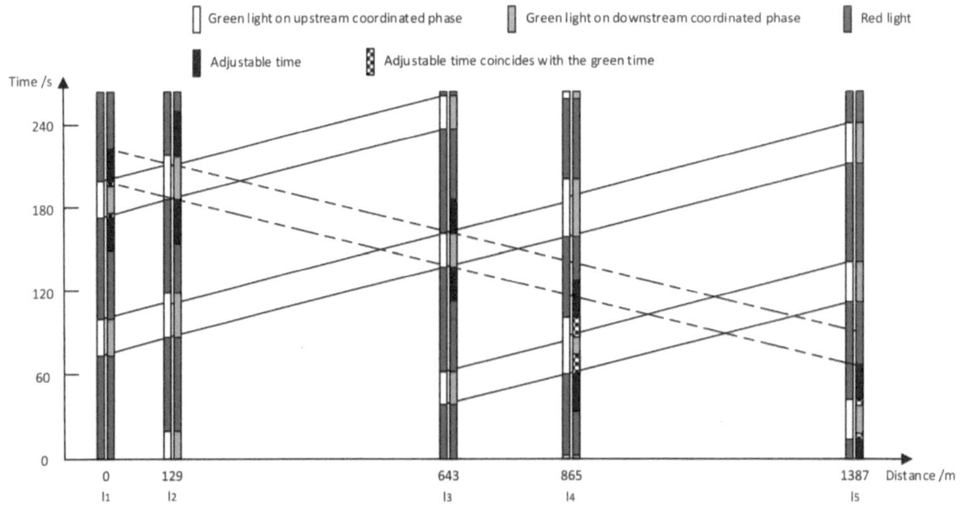


Figure 8. Downstream coordinated phase green light adjustment area.

Step 4: The downstream ideal green wave was moved up or down to make it be located as far as possible in the adjustable green time of the downstream coordinated phase at each intersection to search the position to maximize the actual downstream green wave bandwidth. The result is shown in Figure 9. In this case, the downstream ideal green wave was founded.

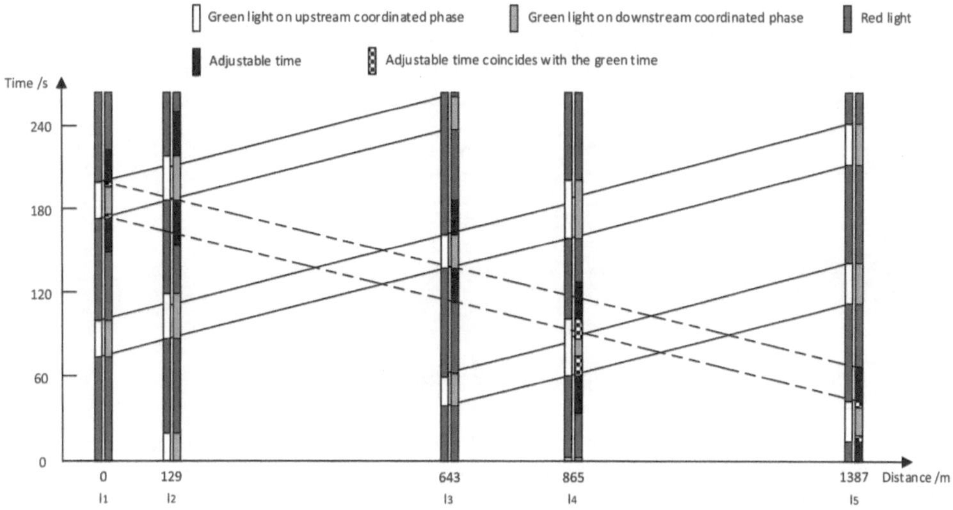


Figure 9. Maximum location of downstream actual green wave bandwidth.

Step 5: The phase sequence of each intersection was adjusted based on the downstream ideal green wave position obtained in the previous steps, as shown in Figure 10. Signal timing scheme after phase sequence adjustment at each intersection is shown in Figure 11. At this step, the arterial street bidirectional green wave was designed ended.

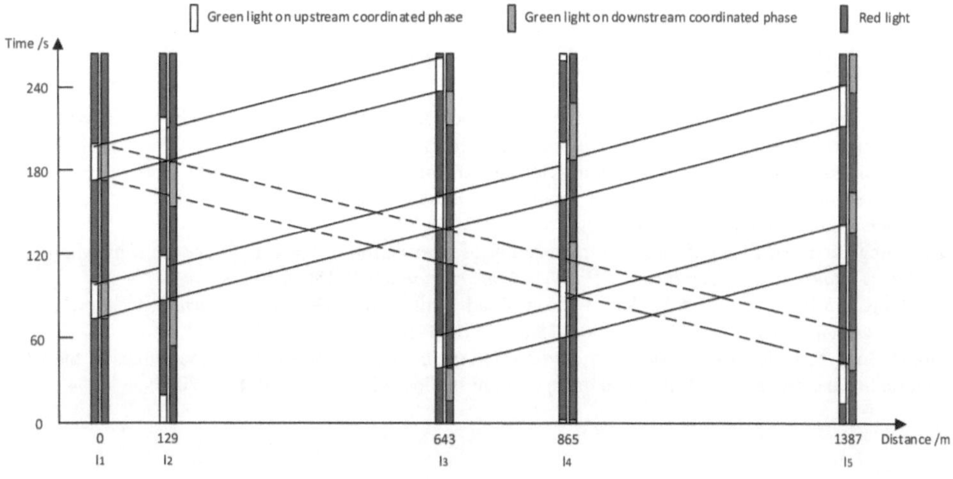


Figure 10. The final bidirectional green wave design.

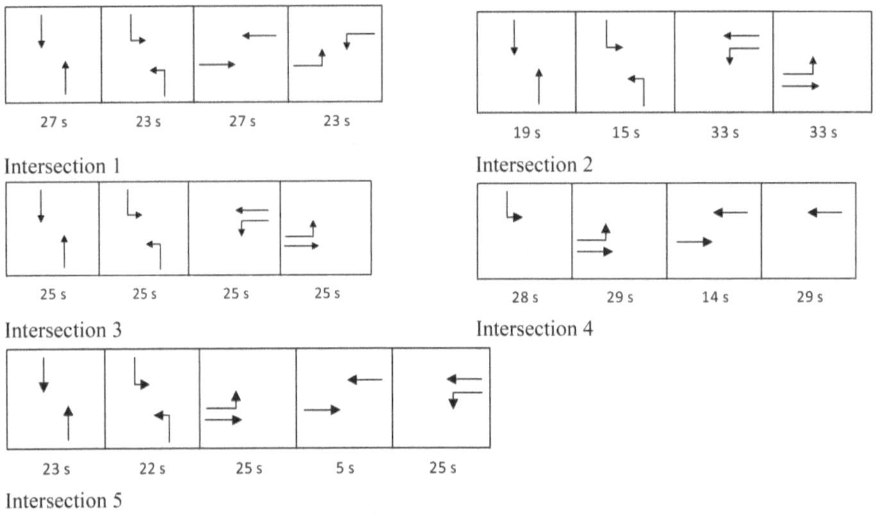


Figure 11. Timing scheme of the basic signal at the intersection.

In the above green wave application, the signal engineer can make fine-tune according to the actual situation, such as selecting the beginning of green, the ending of green, or the ending of yellow to coordinate the position of the green wave.

## 4 CONCLUSION

The proposed method takes the bidirectional ideal green wave as the target and optimizes the phase sequence of each intersection under the premise that the main coordination direction of the arterial street is the ideal green wave, so that another coordination direction will converge to the ideal green wave as much as possible. This method does not need a complex theoretical model and tedious calculation. If there is auxiliary software that can draw the time-distance graph, it can make the design work more efficient. This method can also design reasonable and suitable bidirectional green waves for the arterial street using a simple paper ruler and pen.

## REFERENCES

Chen Yao, Liu Shasha, Li Lingli, et al. Synchro-based Offset Optimization Method: A Case Study of Intersections in Chang'an Avenue [J]. *Journal of Transport Information and Safety*, 2012, 30(6): 115–117.

Jing Binbin, Yan Xiaowen, Wu Huan, et al. General Algebraic Algorithm for Arterial Coordination Control Based on Maximum Bidirectional Progression Bandwidth [J]. *Journal of Transportation Systems Engineering and Information Technology*, 2017, 17(2): 76–82.

Lu Lanping, Li Yijie, Zhang Zhongda. Optimization of Intersection Cycle Length and Split by Minimum Delay. [J]. *Journal of Tianjin Institute of Urban Construction*, 2009, 15(1): 16–18.

Pei Yulong, Liu Bohang. Offset Model for Urban Road Optimization Method[J]. *Journal of Central South University: Natural Science Edition*, 2007, 38(add1):1023–1027.

Wang D, Song X, Li F. *Research on offset optimization of signalized arterials* [C]//Proceedings of the fifth international conference on traffic and transportation studies. Xi an, China: ICTTS. 2006: 584–594.

*Advances in Urban Construction and Management Engineering – Cha (Ed)*


# Some thoughts on urban planning under the low-carbon concept caused by the COVID-19 pandemic

Haiying Yang*
*Changchun University of Architecture and Civil Engineering, Changchun, China*

ABSTRACT: In the 21st century, with extreme weather conditions and pandemic, human survival faces a great threat. The reason is inseparable from the destruction of nature by human activities. The concept of low carbon has long been put forward by scholars, more and more recognized by people, and gradually put into action. The Chinese government has done a lot of work in low-carbon development, which can be seen in the 5-year plan for national economic and social development. "Climate change is not only an environmental problem, but also a development problem. In the final analysis, it is a development problem." Urban planning is serving for urban development. It is imperative to achieve the low-carbon development goals. The author summarizes the existing low-carbon planning documents and cases and the planning methods suitable for low-carbon development and puts forward some thoughts on the content of future low-carbon planning.

## 1 INTRODUCTION

In the 21st century, as we are marching towards the Chinese dream of national rejuvenation, the COVID-19 pandemic has slowed down our pace. Heavy rains, droughts, locust plagues, extreme cold, various virus epidemics, and more are recorded. "Brazil is frequent in heavy rains, and climate change is the main culprit." On July 20, 2021, a heavy rainstorm was recorded in Zhengzhou, and the rainfall within an hour was equal to 106 West Lake water storage capacity. The roads of Zhengzhou became rivers, cars became boats, subways became vast oceans, and people's lives were threatened by floods. According to China Climate Communique, in 2021, extreme weather events frequently occurred in China, and the national annual average temperature hit a record high. Cold increased beyond the normal range. Now, in addition to COVID-19, unexplained childhood hepatitis and the Monkeypox virus are spreading rapidly.

When we lament the nonsensical infinite self-replication of the COVID-19 virus, leaving hosts dead, they themselves will not exist. Can we think of the relationship between man and the earth?

Driven by the "self-interest gene," human beings plunder resources indefinitely, leading to the irreversible destruction of the living environment and finally dying together. Is it really the final fate of the human beings known for having wisdom?

As early as 1896, Svante·Arrhenius (Nobel Prize winner in Chemistry) predicted an increased $CO_2$ due to fossil fuel combustion, which will lead to global warming, but humans ignored.

According to the IPCC Fourth Scientific Assessment report, the last 50 years (more than 90%) of global warming is likely due to human activities (Gu &Tan 2010).

James Hansen, director of the Goddard Institute for Space Studies (NASA), called for $CO_2$ concentrations in the atmosphere to reach dangerous levels (385 ppm is the "tipping point," 383.1 ppm in 2007).

Global climate change and continuous warming will lead to disruption in the earth's ecosystem, and cause a great disaster to human society. Controlling the $CO_2$ concentration in the atmosphere has become an urgent need.

*Corresponding Author: 13944119585@139.com

DOI 10.1201/9781003348023-80

## 2 RESEARCH BACKGROUND

Systematic talks of low-carbon economies go back to the United Nations Framework Convention on Climate Change in 1992 and the Kyoto Protocol in 1997.

The 2003 UK Energy White Paper, "The Future of Our Energy: Creating a Low-carbon Economy," wrote the concept of a low-carbon economy in a government document. The book says that a "low-carbon economy" means getting more economic output through less natural resource consumption and less environmental pollution, and is the way to create higher living standards and a better quality of life. The low-carbon economy creates opportunities for the development, application, and export of advanced technologies, as well as creating new business opportunities and more jobs. Its overall goal is to cut $CO_2$ emissions by 60% from 1990 levels by 2050, fundamentally turning the UK into a low-carbon economy (Chen & CAI, 2015).

At the 2005 ("8 + 5") summit, Chinese President Jintao Hu made it clear that "climate change is both an environmental issue and a development issue, and in the final analysis, it is a development issue." (Chen & CAI 2015).

At the Copenhagen conference in 2009, the "convention" of the department director was broken. Not only did the ministers participate in the COCA negotiations, but also the leaders came to the front line for negotiations. In addition to the UN platform, many major international bilateral and multilateral platforms have also themed climate change (Chen & CAI 2015).

In the face of increasingly severe environmental problems, various countries have also changed from the initial attention to the present action. The Chinese government has always attached great importance to the sustainable development of mankind. And the 5-year plan for social and economic development has always focused on controlling greenhouse gas emissions from the 11th Five-Year Plan period to the 14th Five-Year Plan period. During the 10th Five-Year Plan period in 2006, China reversed the previous trend of the rapid rise in energy intensity. In 2011, during the 12th Five-Year Plan, basic policies for low-carbon communities were laid out. The "Five-Year Plan for the 13th Greenhouse Gas Emission Control" in 2016 set specific requirements for low-carbon communities and became the general outline of the country's response to climate change. In 2014, the National Development and Reform Commission issued the Notice on Carrying out the Pilot Work of Low-carbon Communities, proposing the goal of 1,000 pilot construction. In 2015, the National Development and Reform Commission proposed some specific quantitative indicators for constructing low-carbon communities. During the 14th Five-Year Plan of 2020, a broad green way of production and life was proposed. Carbon emissions fell steadily, the ecological environment was fundamentally improved, and the goal of building a beautiful China was achieved (Zhang 2016). At the 75th session of the United Nations General Assembly in September of the same year, China officially proposed to achieve a carbon peak by 2030 and achieve carbon neutrality by 2060. Although the goal is clear, the road is still difficult.

## 3 SUMMARY OF THE CONTENT OF URBAN PLANNING UNDER THE CURRENT LOW-CARBON CONCEPT

Although all sectors of society have studied how to achieve energy conservation, emission reduction, and low-carbon development, due to the limitations of the field, it seems that blind people cannot cover the whole picture. Due to its comprehensive, dynamic, practical, and democratic characteristics, the urban and rural planning field is duty-bound in the practice of low-carbon city goals. The reading literature is summarized as follows:

### 3.1 *"Low-carbon" planning at the regional level*

At the regional level, the loose layout of urban agglomeration (model 1) leads to the increased energy consumption of transportation and infrastructure, which needs to be adjusted. The goal of the adjustment is to make the urban agglomeration more "compact" and conducive to low-carbon development (Pan 2008). Urban agglomerations or a single city under the new form needs a new balance.

Currently, the motor vehicle regional traffic mode with expressway as the carrier may be replaced by high-speed rail transit. The balanced urban relationship within urban agglomerations will be classified in different grades, such as satellite cities and central cities, thus reducing the carbon emission brought about by networked traffic (Model 2).

Model 1. Motor vehicle transit.

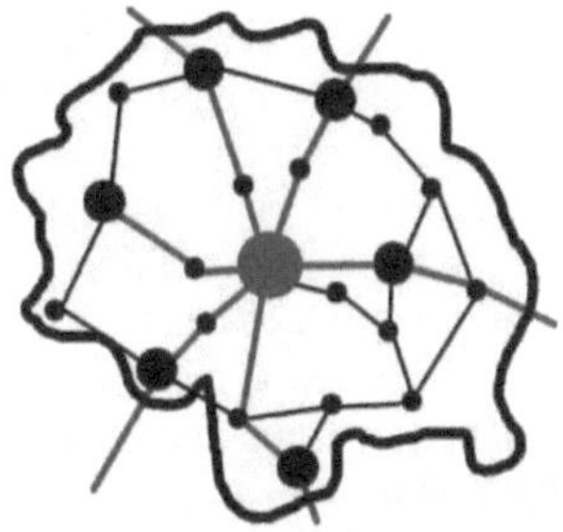

Model 2. High-speed rail transit.

### 3.2 *"Low-carbon" planning at the city level*

First, the compact urban spatial structure improves the efficiency of public transport and makes the space use more efficient. At the same time, the "Infinite extension" type of land layout and the "pendulum" traffic functional zoning should be readjusted. With low-carbon development and rail public transportation, compact and unbalanced intensity is more suitable for future development (Pan 2008).

Second, what has become the protagonist of urban transportation is that public transportation is efficient, convenient, safe, and comfortable. The urban transportation system also provides a variety of travel options, builds low-carbon travel systems for cycling and walking, and explores the application of personal rapid transit systems (PRT).

Third, in terms of land use, moderate and reasonable mixing of functions in planning does not mean a complete denial of the principle of "functional zoning" in modern urban planning, but replacing the rigid understanding and implementation of "functional zoning." Since the 1950s, China's various "courtyards" formed due to historical conditions have gained the opportunity to be reviewed and evaluated again.

Fourth, the application of various "low-carbon" technologies is mainly focused on urban infrastructures, such as the new (non-fossil-dependent) energy supply system, the regional energy supply system, the garbage recycling and incineration system, the water treatment, and reuse system.

Finally, the optimized urban greening system can absorb some $CO_2$.

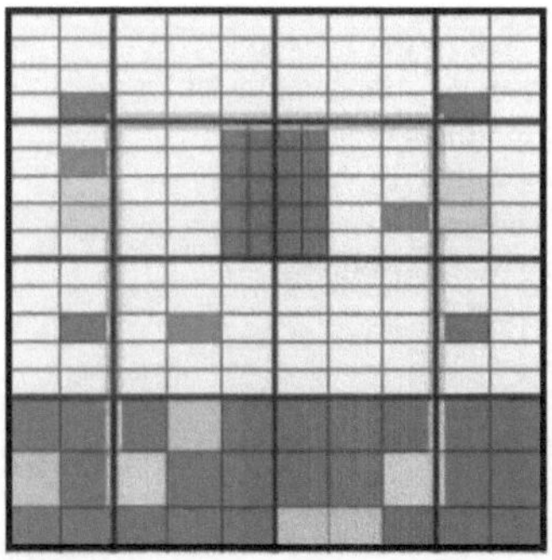

Model 3. General city space.

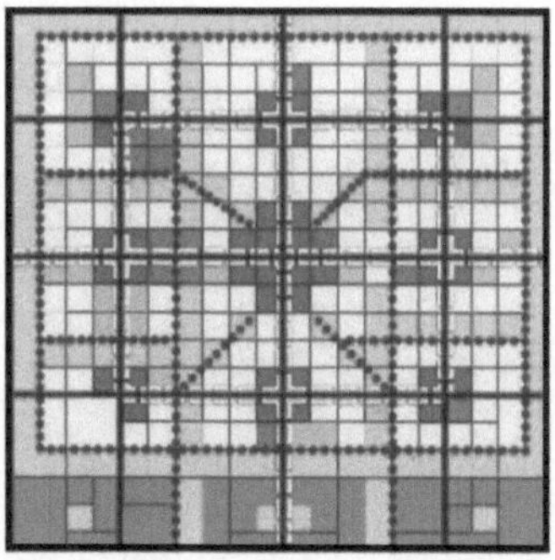

Model 4. Low carbon city space.

### 3.3 *Discussion on "low-carbon" planning at the community level*

Building communities with moderately mixed functions and moderate density is the key to achieving the "low-carbon" goal. High community density is not conducive to using renewable energy sources such as solar energy, and too low community density can cause problems for the efficiency of transportation and urban infrastructure. We must plan low-carbon target priorities according to different types of communities. For example, the urban center (CBD) should avoid excessive development intensity and make efforts in indoor and outdoor environment adjustment, vertical transportation, and other aspects (Pan 2008).

The residential area should have appropriate population density; green energy-saving buildings; a compact, convenient, and dynamic community center; and comfortable and safe outdoor venues. Cycling and walking can be alternative traveling modes, and rainwater collection and utilization, water utilization, and garbage classification recycling system need to be considered.

In the industrial parks dominated by industrial production, in addition to achieving the purpose of "energy conservation and emission reduction" through the reform of the production process itself, the unified energy supply system, the upstream and downstream connection of the production process, the treatment and recycling of wastewater and waste system, and the carbon capture system are the key pointers (Pan & Zhuang 2018).

## 4 FOREIGN LOW-CARBON CITY PLANNING CASES

### 4.1 *Field City Railway Station Area Project Development Planning, California, USA*

This project was completed under the framework of the California Environmental Quality Act. The planning team proposed it according to different design ideas and several schemes, as shown in Figure 5 (Li 2011), with different characteristics. And the deepening plan is selected through the comparison of key performance indicators, as shown in Tables 1 and 2 (Li 2011), while showing the determination of the project to pursue low carbon.

Table 1. Key performance indicators of the low-carbon city form.

| Evaluation | Metric (KPI) |
|---|---|
| Parks and open spaces | Park/open space area per person<br>Coverage (percentage of population reaching the park within 5 minute walk)<br>Comprehensive accessibility<br>Possible carbon summary area<br>Community service facilities |
| Community service supporting facilities | Coverage (proportion of the population reaching at least one service point within 5 minute walk)<br>Comprehensive accessibility |
| Transportation | Comprehensive accessibility of bus station coverage (proportion of the population who can reach bus stops within 5 minutes of walk)<br>VKT: Vehicle Kilometer Travelled (preliminary estimate) |
| Building energy | Basic per person total building energy demand<br>Energy demand per square meter of the basic building |
| Water resource | Basic per person demand for drinking water resources<br>Basic per person landscape water demand |
| Garbage generation | Basic per person garbage generation |

The design combines the railway station with a community bus connection line to achieve the purpose of bus-oriented development. At the same time, the development density of the core area of the railway station should be improved to strengthen its central position in the whole community service system. The service population density and the level of bus stations are superimposed to improve the comprehensive accessibility of public transportation. It is hoped that a comprehensive bus system will encourage people to use buses more, thus reducing travel in cars.

The plan reduces the types of homes, mainly independent villas, and turns them into more compact structures and high and low density, to reduce energy consumption in heating, gas supply, lighting, and water demand.

Community service facilities (business, retail, services, schools, etc.) also try to shift people from car travel to walking or cycling travel within walking distance, thus reducing carbon emissions from car travel.

Design schemes to improve the coverage of parks and open spaces to build a park system within a few minutes' walk.

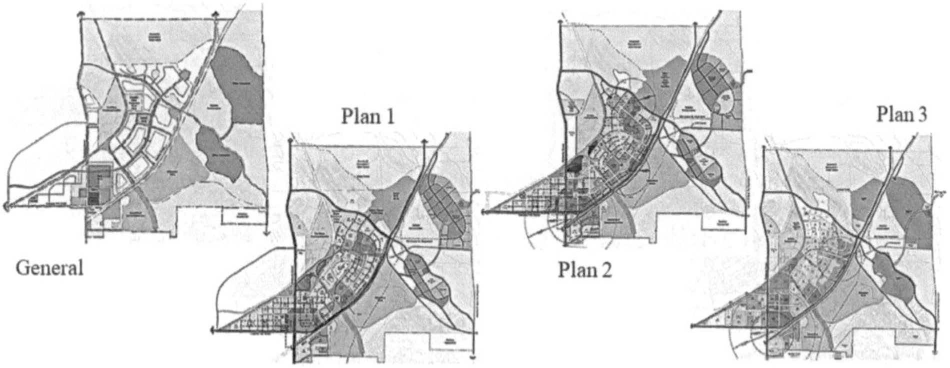


Figure 5. Fairford railway station area concept scheme.

Table 2. Draft basic carbon emission results.

| Scheme | Annual per person (ton) |
|---|---|
| General development model | 6.2 |
| Plan one | 5.6 |
| Plan 2 | 5.3 |
| Plan 3 | 5.5 |

The designer compared four schemes and six main indicators (park and open space, community service supporting facilities, transportation, construction energy, water resources, and garbage generation) to determine the plan with the lowest carbon emissions.

## 4.2 *Hammarby, Sjostad, Stockholm, Sweden*

The design is to achieve the low carbon goal of the project from public transportation, block scale, natural elements, and energy resource utilization.

First, the design increases the supply of public transport to reduce the dependence on cars.

Second, the urban spatial form emphasizes small blocks: the block scale is about 60~200 m, the building height is controlled within 2~8 floors, the main street has service functions, and the commercial interface reaches 1800 m, so as to reduce carbon emission in terms of traffic and form design.

Third, the design focuses on the natural elements to create a high-quality public open space. There is more than one green space above 30 $m^2$ within 300 m of any apartment, with ventilation requirements considered, and this is designed to increase the carbon sink and reduce energy consumption.

Finally, the community garbage and solid waste energy conversion are designed to meet the community's energy needs. Residents will classify the garbage first and then collect it. Convert combustible waste into electricity and heat energy through the cogeneration energy center. That energy accounts for about 20% of the community's normal energy supply. The designers also tried turning the biogas from the silt into natural gas to supply city buses and improve their carbon reduction capacity.

### 4.3 *Vauban, Freiburg, Germany*

The designer achieved the low carbon reduction goals through the five public transport systems, renewable energy, water utilization, low carbon community, and a green landscape.

Public transportation and travel. The community controls the number of private cars through street control measures. If there is no street parking space on the road, only temporary parking is allowed, and public garages charge high parking fees. Currently, 70% of the residents in the community no longer own private cars, and 80% rely on "T-shaped" public transportation.

Renewable energy sources. Eighty-one percent of the community's energy supply comes mainly from solid waste conversion, which is achieved by sending wood waste from the wood industry around Vauban to the community's cogeneration energy center. Wind power and 1,200 m two solar panels also contribute 4% of the region's energy. The current average energy consumption is about 80 kwh/year/$m^2$, about 24% below the regional average of 105 kwh.

Water system. All the green landscape in the community has high water permeability, including road separation greening has a water storage function. After simple treatment, the rainwater can be used back for toilets in residents or public buildings. The data show that Vauban has left 90% of its rainwater in the region and achieved self-digestion.

## 5 CONCLUSION

Cities are only one part of climate change, but they are the key to solving the issues. Conscious urban planning and management can help reduce the risk of adverse effects of climate change and promote the transformation of cities to a low-carbon economy.

At present, the field of urban planning under the low-carbon target mainly covers urban form, block scale, functional layout, construction density, construction, transportation, energy utilization, and others. Under the new trends such as 5G, big data, and smart city, there is room for further exploration in urban planning targeting low-carbon development.

In addition, the impact of the gap between rich and poor on people's values and low-carbon development is also noteworthy. Whether we can look back at the people's values and life attitudes in the new China period and absorb the nutrition suitable for contemporary society is what we can discuss in future studies.

In addition to urban planning, various countries are exploring and practicing in different fields, such as legislation, evaluation standards, and the carbon trading market, which play a significant role in realizing the low-carbon target. We hope humanity can achieve sustainable development goals early and coexist in harmony with the earth.

## REFERENCES

Chaolin, Gu. & Zong bo, Tan. (2010). *Urban and regional planning research: Low-carbon city Research on Urban and Regional Planning: Low Carbon Cities* (Volume 3). M. Commercial Press Published Press, 101–195.

Haixiao, Pan. (2008). Spatial planning strategy of "Low-carbon city" in China. *J. Journal of Urban Planning*. 178, 57–64.
Jiahua, Pan. & Guiyang, Zhuang. (2018). *Research on Low-carbon of Xiamen City*. M. CitySocial Sciences Academic Press, 94–109
Meigu, Chen. & Hai, CAI. (2015). *Low carbon Economics*. M. Tsinghua University Press, 10–23
Xiaoduan, Zhang. (2016). *Strong and sustainable urban development*. M. Southeast University Press, 23–53
Xin, Li. (2011). Low Carbon Planning Strategy Under AB32: the Case of Fairfield Train Station Development Plan. *J. Urban Planning International*. 28–35

*Advances in Urban Construction and Management Engineering – Cha (Ed)*


# Study on the renewal of the ancient city of Suocheng, Zhifu District, Yantai City based on big data

Hongmei Li & Xiangtian Ma
*The School of Fine Arts, University of Jinan, Jinan, China*

Aiping Jia*
*Yishui County Administrative Approval Service Bureau, Linyi, China*

ABSTRACT: The ancient city carries the memory of a city's historical development and is the carrier of the city's historical storage. The renewal of ancient urban areas is a living display of urban history, which is important to the development of contemporary cities. With the development of information technology in the new era, big data brings real-time data support for a comprehensive grasp of the state of conservation and renewal of the ancient urban area. With human activities and needs as the core, the data makes the study more practical. Taking the ancient city of Suocheng, Zhifu District, Yantai City as an example, we use preliminary research data and open data (Baidu POI data collection, etc.) to establish different groups' demand indicators, and then we may combine numerical analysis method, preliminary selection of gold data for analysis and selection of SPSS software for data statistics, so as to quantify the description and explore the study of the renewal of Suocheng ancient city supported by big data. The data eventually boosted the local economy.

## 1 INTRODUCTION

### 1.1 *The status of the city in Yantai's urban development*

Qishan Suocheng is located in the middle of Zhifu District, YantaiCity. It was built in 1398 as a defensive city to prevent Japanese invaders (Li 2016). It has traditional folk characteristics. It is a relatively well-preserved defensive castle of the Ming and Qing dynasties and important evidence for studying the history of Yantai's urban development (Zhang 2013). It can be said that the history of Yantai's founding originated from Qishan Suocheng; thus, it can be seen that Suocheng plays an important role in Yantai's urban development (Li 2015).

### 1.2 *The situation facing the development of the city in the new period*

Despite the complete layout of the city's streets and alleys and the richness of its historical architectural heritage, the city has inevitably encountered the challenges of sustainable development in the new era. Overtourism, commercial overdevelopment, lack of popularity, one-size-fits-all, and cold "showrooms" of ancient buildings are also plaguing this ancient city, thus making it develop healthily and sustainable development has become a key issue of research, as shown in Figure 1.

### 1.3 *Big data-driven development transformation in the city*

With the advent of the cloud era, big data is widely valued. The sustainable development of Suocheng needs more accurate data support in the management and design to make precise policies. With

*Corresponding Author: 1024407356@qq.com


DOI 10.1201/9781003348023-81

Figure 1. The old town of Suocheng after renovation.

human activities as the core, big data can quantitatively describe human needs and provide dynamic, accurate, and scientific descriptions of the development of the ancient city of Suocheng. Therefore, big data helps transform the development of Suocheng (Zhang 2019).

## 2 STUDY ON THE UPDATE OF THE CITY SUPPORTED BY BIG DATA

### 2.1 *Data idea architecture*

Preliminary research data and open data (Baidu POI data collection, etc.) are used to initially establish the demand indicators of the consumer groups in the ancient city of Suocheng (Zhang 2021). In combination with numerical analysis, we used software such as Golden Data to analyze the data and SPSS software to conduct data statistics, so as to quantitatively describe the needs of the consumer groups in the ancient city of Suocheng and provide a reference basis for the construction of evaluation indicators. Thus big data can drive the development of the local economy (Xue 2016).

### 2.2 *Analysis of indicators*

#### 2.2.1 *Data analysis for the respondent group*

From the data results, it can be seen that there are more women than men among the respondents, as women tend to engage in commercial consumption behaviors such as shopping, strolling, and gourmet food, while men tend to appear as their escorts. Similarly, women tend to have a higher aesthetic threshold than men and are more discerning about the form, function, and quality of consumer goods, as shown in Figure 2.

Respondents in the 18-25 age group tend to pursue the new trend of fashion and have a love for spending. The emergence of a new tourist hotspot often attracts mainly young people. The emergence of the renovated Suocheng Old Town attracts many young people from Yantai City and the surrounding area. They constitute the main force of tourism and consumption in the ancient city, as shown in Figure 3.

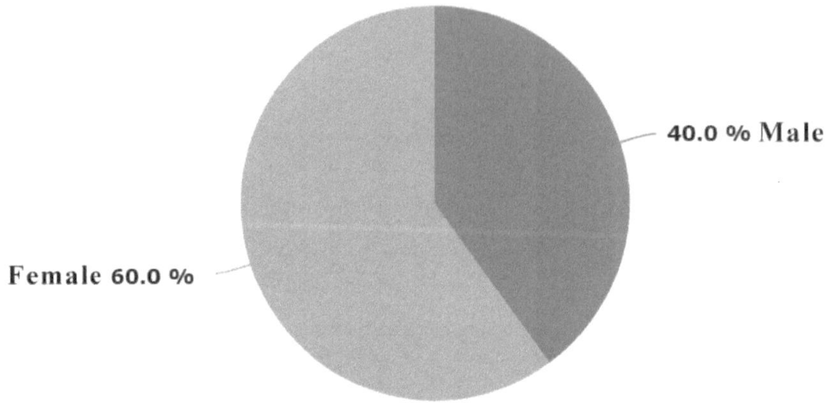


Figure 2. Gender distribution of respondents.

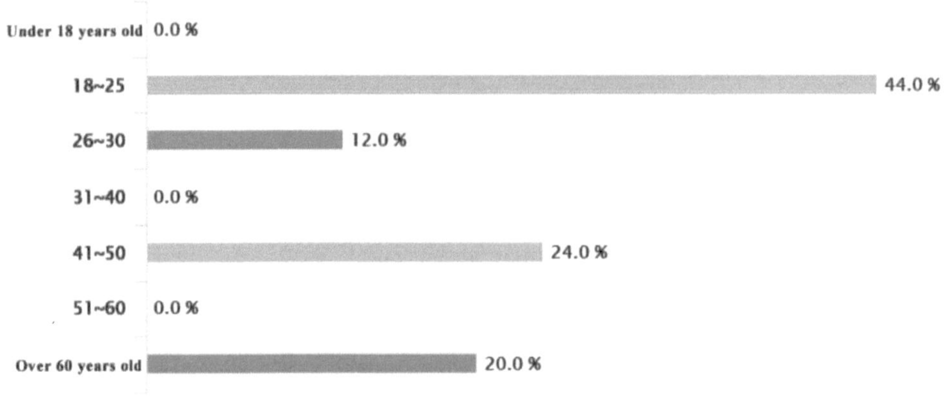


Figure 3. Respondents' age distribution.

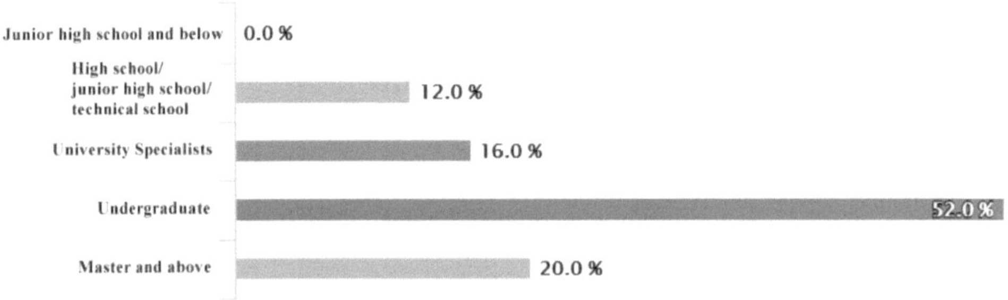


Figure 4. Distribution of respondents' education.

Among all the respondents' education segments, the largest group of people has a bachelor's degree, which focuses on the group of 18 to 25 years old. This indicates that there are more people with higher education among the visitors in Suocheng. A higher education level of people will lead to a higher level of aesthetics. This also puts a better demand on the overall level of renovation design in Suocheng, as shown in Figure 4.

### 2.2.2 *Satisfaction survey of the renovation projects in the city*

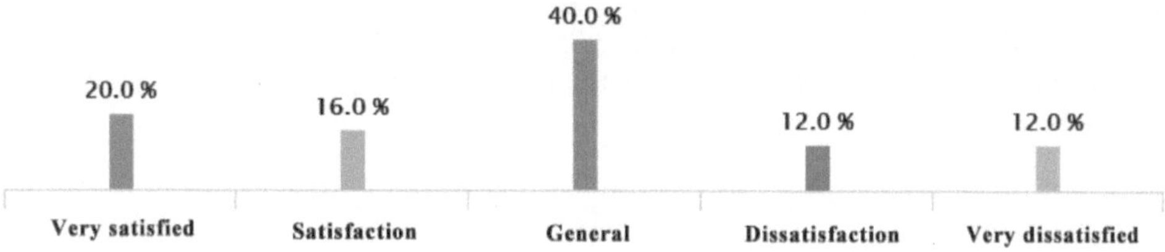


Figure 5. Respondents' satisfaction with the design of the renovation of the ancient city of Suocheng.

The survey on the satisfaction with the renovation design of the old town of Suocheng shows that, compared with the dissatisfiers, satisfiers are larger; however, the number of people with a neutral attitude is the highest. To a certain extent, it shows that the renovated Suocheng can not fully satisfy the needs of all customer groups in some aspects. The design and renovation party should widely solicit and investigate the renovation needs of different customer groups before the design and renovation and reasonably integrate opinions and suggestions based on the local historical background, superior and inferior policies, history and humanities, folk customs, natural resources, architectural styles, and business models of Zhifu District, Yantai City, as shown in Figure 5.

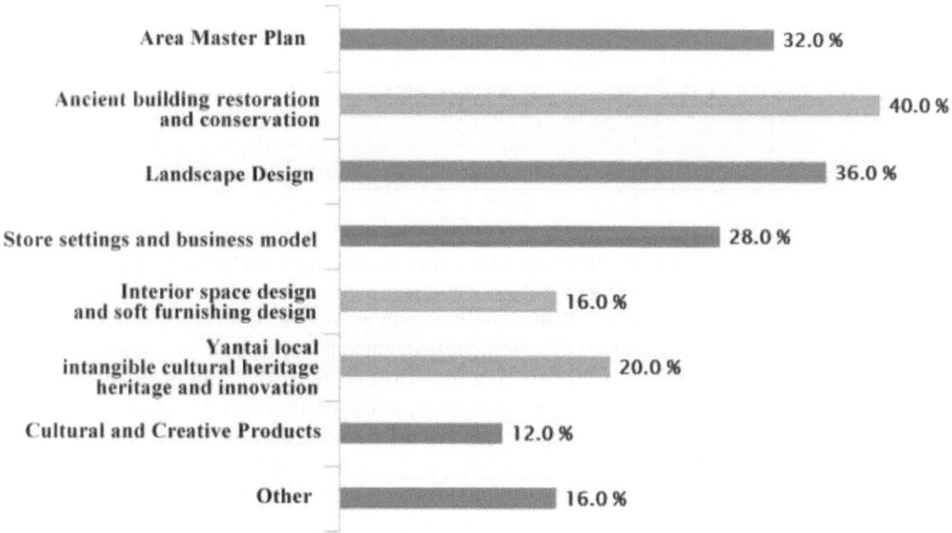


Figure 6. Which aspect of the renovation project in the city was more successful for the respondent?

The survey covers many aspects, such as the ancient building restoration and preservation, landscape design, store settings, and business model, interior space design and soft furnishing design, local intangible cultural heritage heritage heritage and innovation, IP cultural and creative products belonging to the city, and other eight aspects of satisfaction. Forty percent of the respondents thought that the ancient building restoration and protection in Suocheng was the most successful. The renovation project in Suocheng renovated and repaired the facades of the main buildings on both sides of the central axis, basically maintaining the original architectural appearance and preserving the style features of the local ancient buildings in the details of the ancient buildings, as shown in Figure 6.

The satisfaction with the overall planning of the area in the city ranks second. The overall layout of the city is symmetrical in the central axis, the city set cross street divided into four equal areas (Wang 2018), and the rearrangement of the city's commercial street is based on the original cross street in the city is set, so as not to destroy the original layout of the city, but also to separate the noisy commercial area and quiet residential areas. It is reasonable planning of the dynamic and quiet areas.

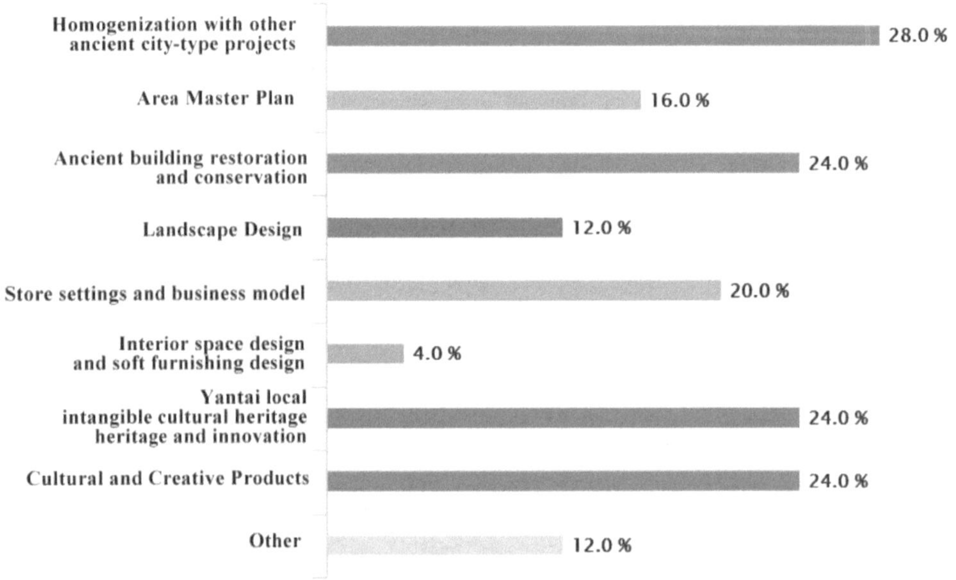


Figure 7. What do respondents think needs to be improved in the renovation project in their city?

The highest percentage of the survey data is that the renovation of Suocheng is homogeneous with other ancient cities, indicating that the renovation of Suocheng is not fundamentally different from other projects in China that focus on ancient city tourism and does not highlight the maritime cultural attributes and characteristics of Yantai on the Jiaodong Peninsula. It also does not show the cultural characteristics of the city as a military fortress in ancient China to defend against foreign invasion (Wang 2018), which leads to the tourists may have some aesthetic fatigue. Secondly, in terms of the repair and protection of ancient buildings, although it is mentioned above that the protection of ancient buildings in Suocheng is reasonable and tourists are satisfied, in the process of field research, it was found that the gate building at the entrance of the city was too new and too simple, which led to the incompatibility with the surrounding ancient architectural style, so the transformation of the entrance of the city gate here is a failure, as shown in Figure 7.

In terms of intangible culture inheritance and innovation, Yantai's rich and diversified local marine fishery culture and special recreational projects such as Lan Guan opera, oceanic rice-song, Tong Bei Quan, Dajiao, Bagua drumming, Zhifu whip, and paper-cutting are not well utilized (Li 2010). Through the field research, we found that only a few stalls in the city demonstrated the process of making and selling Chinese sugar paintings and paper-cutting, and the selling price was too high, which affected visitors' enthusiasm to experience Yantai's intangible cultural heritage.

Similarly, the level of dissatisfaction of visitors is also high in terms of IP cultural and creative products in Suocheng. There are no creative products with historical and cultural characteristics of Suocheng that can be derived from the unique marine and ancient military culture IPs in Suocheng.

It can be seen from the data that the degree of dissatisfaction of tourists is not low in terms of the setting and business model of the stores in the city. The business model is too single, most of which only contains two business forms: ordinary commodity sales and the catering industry. Of course, the catering stores are also some ordinary popular cuisine restaurants, which do not reflect the rich local food culture of Yantai. There are many special snacks, such as Shandong oil whirlpool, Jiaodong buns, Qing blanching oysters, mixed grain pancakes, etc., which can set up special stores for sale and add Yantai specialties. So we can experience links, integrating the concept of user experience design, in order to improve the experience of local tourists (Wang 2021).

## 3 CONCLUSION

The study has shown that data collection and analysis of the consumer population can provide early warning of management policies, service facilities, and environmental design of Suocheng Old Town, and timely adjustment of related businesses to avoid over-tourism and over-development, which can contribute to the sustainable development of Suocheng Old Town and the preservation of traditional culture and big data also brings economic benefits to the city.

## ACKNOWLEDGMENTS

The thesis is the stage result of the Shandong Social Science Planning Research Project (No. 17CWYJ18).

## REFERENCES

Huan, L. (2016) Basic research on the preservation planning of Suochengli, a historical area in Yantai city. *Architectural Journal*, S1: 71–76.

Jiaqi, Z., Yannan, L. & Binbin, S. (2021) *Spatial distribution of living service industry in the main city of Zhengzhou based on POI data Spatial Distribution Characteristics of Life Service Industry in Main City of Zhengzhou*. World Regional Studies.

Jun, W., Ying, Q. & Gang, W., (2018) Talking about the Continuation and Vitality Renaissance of the Historic District of Qishansuo City in Yantai. *Architecture & Culture*, 11: 218–219.

Ling, L. (2010) *A Preliminary Study of Yantai Folk Culture Resources and Their Development and Utilization*, Ludong University Journal (Philosophy and Social Sciences Edition). 27: 39–42.

Runxia, W. (2018) Survey and Thinking on Historic and Cultural Streets Tourism Development, *Shanghai Urban Management*. 27: 77–81.

Shasha, L. (2015) *Analysis on the Historic District's Protection and Renewal of Yantai Suo Cheng Li Based on the Urban Catalytic Theory*. Shenyang Jianzhu University.

Wei, Z. (2013) Objectives of tourism development in historic and cultural districts. *Modern Business*, 31: 179–180.

Weiliang, Z. & Hui, C. (2019) Analysis and Research of Suzhou Open Block Based on Big Data. Huazhong Architecture, *Huazhong Architecture*. 02: 63–67.

Yujie, W. & Yan, Xian. (2021) Research on user needs of red culture digital tourism from the perspective of experience design. *Design*, 34: 132–135.

Zhenlin, X. & Minmin, H. (2016) Study of Outdoor Advertising Setting in Ancient City. *Journal of Hefei University of Technology* (Social Sciences), 30: 83–88.

*Advances in Urban Construction and Management Engineering – Cha (Ed)*
 *ISBN 978-1-032-39018-5*

# A study on the siting of a comprehensive domestic waste treatment center based on an immune algorithm

Zhenhua Yan*
*Public Utility Service Center of Liaobu Town, Dongguan City, Dongguan, Guangdong, China*

Yawei Hu
*School of Electronic Engineering and Intelligence, Dongguan University of Technology, Dongguan, Guangdong, China*

Baihui Luo
*Public Utility Service Center of Liaobu Town, Dongguan, Guangdong, China*

ABSTRACT: The problem of waste disposal is gradually becoming the focus of attention in several fields, and the optimal siting plan is derived by taking the siting of a comprehensive domestic waste treatment center as the research object. The characteristics of domestic waste are obtained, the daily per capita domestic waste production in the region is calculated, a production prediction model is constructed using an immune algorithm, and the regional space is regarded as a limited resource to improve the siting model of the integrated treatment center. The analysis results: Under the same daily domestic waste production conditions, the average values of waste collection distance of the four candidate addresses are 12.448 km, 13.112 km, 9.600 km, and 13.131 km, respectively, and candidate address C has the shortest collection distance, so candidate address C is taken as the best site selection option.

## 1 INTRODUCTION

The standard of living in today's world is increasing, and the total amount of household waste generated is also increasing. According to research, all current waste disposal methods are unable to dispose of all domestic waste, and some waste that cannot be recycled requires additional treatment. As existing landfill sites become saturated, larger-scale integrated waste treatment centers are created (Alcaraz-Londoo et al. 2022; Pu & Yu 2021). In practice, in addition to traditional means such as sanitary landfills and composting, the planning and construction of a corresponding number of integrated domestic waste treatment centers have become an emerging waste treatment method. In order to improve the pollution problems caused by traditional waste disposal methods to the environment, the application of immune algorithms to the study of the siting of comprehensive domestic waste disposal centers can improve the efficiency of waste disposal to a certain extent. Foreign research on landfills started relatively early, dating back to around 3000 BC, with the emergence of landfills as evidence. In the United States, the initial form of waste disposal was basically a simple pile. At that time, the impact of rubbish piling on vehicle traffic was considered more than the environmental pollution.

The turning point from a focus on traffic only to a focus on the environment came in the late 1970s, prompting the development of landfills to new heights. Although China's waste disposal research is somewhat distant from some developed countries, detailed measures have been developed concerning our specific national conditions, and fruitful research results have been achieved. In the course

*Corresponding Author: yanzhenhua@126.com

 DOI 10.1201/9781003348023-82

of subsequent developments, countries worldwide have enacted laws and regulations to ensure the regularisation of waste disposal. One of the landmark initiatives was the separation of waste. Firstly, construction waste was separated from domestic waste, then the waste generated was divided in detail, and finally, a resource recovery plan was developed that corresponded to the separation policy (Abbasi et al. 2020; Ren & Yang 2020; Wang et al. 2020). Along with the development of science and technology, the facilities for waste treatment have been upgraded, from simple mechanized facilities at the beginning to intelligent equipment today, all relying on the results of advanced technological developments. The focus of the research selected for the comprehensive domestic waste treatment center has shifted from a single waste treatment method to a study of the balance between multiple resources and environmental management. From the current research results available in academia, the combination of immunization algorithms with the study of the siting of integrated domestic waste treatment centers is not yet complete, and there is still room for progress.

## 2 OBTAINING HOUSEHOLD WASTE CHARACTERISTICS

Domestic waste is mainly solid waste generated in production and life, including ordinary residents and enterprises, institutions, and schools. The final production of domestic waste is mainly governed by both economic development and the living environment, but the disposal of domestic waste can have an incalculable impact on many aspects. In a period of relative material backwardness, the volume of all kinds of waste was relatively small, and at the same time, the disposal methods adopted by social agents were relatively simple. This proves that there is a close relationship between the volume of domestic waste and the material standard of living, and that the various characteristics presented are not obvious (Blaise, Issock, Roberts-Lombard, et al. 2020). The characteristics of domestic waste are latent, time-sensitive, spatial, and complex. The latent nature is mainly reflected in the long self-degradation cycle of domestic waste, while the disadvantages of the simple landfill method are not very obvious in the short term. In particular, in some food waste, in the absence of dry and wet waste separation, the accumulation of large amounts of domestic waste can produce a build-up of liquid, accompanied by an unpleasant stench, which poses a serious threat to the atmosphere and water sources. All these aspects directly or indirectly reflect the spatial characteristics of domestic waste. It is well known that the classification of domestic waste is the most complex of all waste types in terms of components. Faced with the classification of dry and wet waste, recyclables and non-recyclables, the subsequent treatment is more difficult, and the processes are more complex. In particular, the complexity of domestic waste is becoming more prominent in the face of increasingly refined methods of domestic waste disposal. Based on the above description, the steps to obtain the characteristics of domestic waste are completed.

## 3 IMMUNOLOGICAL ALGORITHMS TO CONSTRUCT YIELD PREDICTION MODELS

The immune algorithm originates from the immune system of living organisms, and the main principle of action is the process of seeking the optimal solution according to the corresponding problem to be optimized. In selecting the location of a comprehensive domestic waste treatment center, the surrounding area that can be covered by the waste removal link is a very important influencing factor. On a societal level, better-educated residents are more receptive to the policy prescriptions of waste separation and have higher implementation power. Combining the above variables affecting domestic waste production, the coverage radius in the immunization algorithm is used as the characteristic value to derive the formula for calculating the daily per capita domestic waste production within a region. This is specified as follows.

$$\bar{D} = \frac{k_m}{L_m} \times \beta \tag{1}$$

In equation (1), $k$ represents the total amount of domestic waste, $L$ represents the total population of the area, $\beta$ represents the coverage radius of the domestic waste removal service, and $m$ represents the domestic waste removal cycle. Based on equation (1), the expression for the production of food waste in the production of domestic waste in the region is calculated as:

$$Q = \frac{|\beta - 1|}{m} \times \frac{k_n}{\sum_{n=1}^{N} g_n} \tag{2}$$

In equation (2), $g$ represents the estimated population of the region, $n$ represents the percentage of domestic wastewater content, and the rest of the variables have the same meaning as in equation (1). With the known capacity of the domestic waste transfer station, the expression of the yield prediction model is derived as follows.

$$R = \sum_{p \in M} \sum_{j \in U} q_j z_{jp} \cdot q_i z_{ip} \tag{3}$$

In equation (3), $p$ represents the garbage compression station, $M$ represents the collection of garbage compression stations. $j$ represents a candidate transit center, $q_j$ represents the garbage compression station at the traffic center $j$, $U$ denotes the set of candidate traffic centers, and R represents a set of candidate transit centers, $q_i$ represents the garbage capacity of the $i$ transfer station, and $z_{ip}$ represents the change coefficient of the transfer volume. In addition, in the immune algorithm, the coverage of the service point needs to meet the corresponding waste removal demand for the service area where waste removal is required. Based on the above calculations, the steps to construct a yield prediction model using the immune algorithm are completed.

## 4 IMPROVED SITING MODEL FOR INTEGRATED TREATMENT CENTERS

Considering the operational model of the integrated domestic waste treatment center and the related influencing factors, the siting model needs to be improved in conjunction with the immunization algorithm. In essence, the problem of siting an integrated domestic waste treatment center is also a dynamic resource allocation problem. In addition to the common explicit elements, some intrinsic implicit elements are also of great value in the site selection process, as shown in Figure 1.

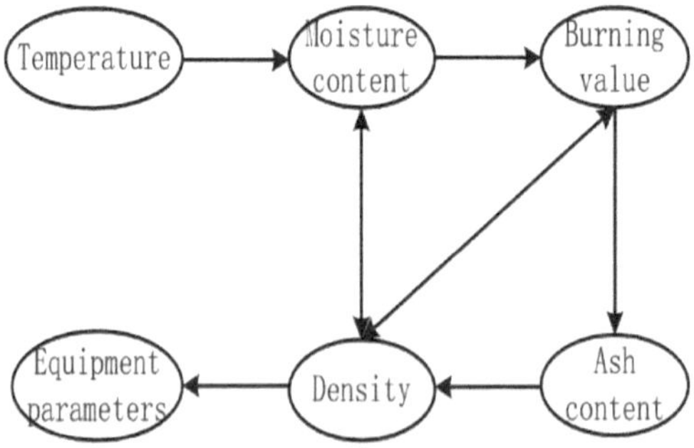


Figure 1. Implicit elements of site selection.

As can be seen from Figure 1, the implicit elements to be considered in an integrated domestic waste treatment center include temperature, moisture content, combustion value, ash content, equipment parameters, and density. In addition to the temperature of the domestic waste itself, the indoor temperature of the integrated domestic waste treatment center must also meet the temperature needs of waste incineration. According to research data, if the smoke produced by burning domestic waste cannot be discharged in time within an environment of 200–400°C, the higher the concentration of harmful gases produced. At the same time, the temperature range for high-temperature composting should be set between 45 and 55°C. Domestic waste is usually un-compacted waste with high water content and can affect the combustion value of domestic waste under certain circumstances. For domestic waste, the higher the combustion value, the higher the recycling value, generally judged at 5000 KJ/kg, while waste below 3500 KJ/kg is more suitable for landfill. Ash is the solid inorganic material in domestic waste. The higher the proportion of ash, the slower the burning rate and the lower the corresponding combustion value. Once the density of domestic waste exceeds 800 kg/m$^3$, the best disposal method is a sanitary landfill, and when the density is between 350 and 650 kg/m$^3$, composting is an option. Based on the above description, the steps to improve the siting pattern of the integrated treatment center are completed.

## 5 CASE STUDY

### 5.1 *Regional overview*

A township in Dongguan City was selected as the research object to study the location of its domestic waste treatment center. The villages and towns have a favorable geographical location and are surrounded by many high-tech industrial parks, so it is representative to study the site selection of domestic waste in this area. At present, most of the garbage collection equipment in this area is in the form of semi-open trucks and agricultural tricycles, which not only have poor sealing performance, but also easily cause secondary pollution to the ground. In addition, the frequency of use of domestic waste collection equipment in villages and towns is low, and the garbage bins are densely distributed, making them difficult to handle. It is understood that by the beginning of 2022, there will be 39 local domestic waste collection stations, but the construction of existing stations is gradually lagging behind, resulting in an increase in the amount of domestic waste transshipment, and a high breeding tendency. When the temperature reaches 25°C and above, the waste will produce an odor. Therefore, the site selection of the village's comprehensive garbage disposal center is imperative.

### 5.2 *Analysis of site selection options*

The total amount of domestic waste generated locally from 2020-2022 has shown a year-on-year increase, with a growth rate of 3% from 600 (t/d), 618 (t/d) to 637 (t/d). The main types of domestic waste include food waste, paper, textiles, glass, ash and rubber, and plastic. Food and plastic waste account for a relatively large proportion of the waste, at 41% and 29%, respectively. In addition to the 39 landfill sites mentioned above, four additional sites have been selected as candidates for the comprehensive domestic waste treatment center, considering the water distribution in the area. The proposed candidate address A is located in the southeast of the village, candidate address B is located in the south of the village, candidate address C is located in the north of the village, and candidate address D is located in the southwest of the village. At the same time, the distance between the four candidate addresses for domestic waste collection and transportation under different daily domestic waste production conditions in the area is simulated, and the address with the smallest collection and transportation distance between the four candidate addresses is set as the best site selection option. The specific results are as follows.

The results in Table 1 show that when the daily production of domestic waste in the town is less than or equal to 20 (t/d), the average values of the collection distances between the candidate addresses of the four comprehensive domestic waste treatment centers are not very different: 4.211 km, 4.771 km, 3.743 km, and 4.287 km respectively.

The results in Table 2 show that when the daily production of domestic waste in the town is 40 (t/d), the average values of the collection distances between the candidate addresses of the four integrated domestic waste treatment centers are 7.403 km, 8.423 km, 6.217 km, and 7.156 km respectively.

As can be seen from the results in Table 3, when the daily production of domestic waste in the town is 80 (t/d), the average difference in the collection and transportation distances between the

Table 1. Waste collection distances for a daily domestic waste output of 20t (km).

| Candidate addresses/ daily waste production (t) | Candidate Address A | Candidate Address B | Candidate Address C | Candidate Address D |
|---|---|---|---|---|
| 2 | 3.615 | 3.557 | 3.004 | 3.115 |
| 4 | 3.748 | 3.698 | 3.378 | 3.203 |
| 6 | 3.825 | 4.263 | 3.499 | 3.226 |
| 8 | 3.997 | 4.521 | 3.514 | 3.334 |
| 10 | 4.006 | 4.696 | 3.678 | 4.498 |
| 12 | 4.145 | 4.835 | 3.814 | 4.516 |
| 14 | 4.337 | 5.248 | 3.906 | 4.955 |
| 16 | 4.668 | 5.362 | 4.112 | 5.226 |
| 18 | 4.744 | 5.553 | 4.223 | 5.307 |
| 20 | 5.022 | 5.978 | 4.301 | 5.492 |

Table 2. Waste collection distances for a daily domestic waste output of 40t (km).

| Candidate addresses/ daily waste production (t) | Candidate Address A | Candidate Address B | Candidate Address C | Candidate Address D |
|---|---|---|---|---|
| 22 | 6.006 | 7.115 | 5.436 | 6.331 |
| 24 | 6.347 | 7.233 | 5.697 | 6.425 |
| 26 | 6.645 | 7.249 | 5.702 | 6.466 |
| 28 | 7.103 | 8.366 | 5.855 | 6.519 |
| 30 | 7.209 | 8.554 | 5.971 | 6.523 |
| 32 | 7.337 | 8.667 | 6.410 | 7.227 |
| 34 | 7.416 | 8.802 | 6.526 | 7.403 |
| 36 | 8.537 | 9.331 | 6.733 | 7.619 |
| 38 | 8.649 | 9.419 | 6.829 | 8.228 |
| 40 | 8.776 | 9.522 | 7.015 | 8.817 |

Table 3. Waste collection distances for a daily domestic waste output of 80t (km).

| Candidate addresses/ daily waste production (t) | Candidate Address A | Candidate Address B | Candidate Address C | Candidate Address D |
|---|---|---|---|---|
| 62 | 12.121 | 11.366 | 8.221 | 13.265 |
| 64 | 12.348 | 11.518 | 8.317 | 13.457 |
| 66 | 12.506 | 11.645 | 8.456 | 15.633 |
| 68 | 13.296 | 13.713 | 9.812 | 15.849 |
| 70 | 14.374 | 13.825 | 9.944 | 15.901 |
| 72 | 15.445 | 13.916 | 9.986 | 15.933 |
| 74 | 15.636 | 14.326 | 10.012 | 15.978 |
| 76 | 16.224 | 15.877 | 11.314 | 16.305 |
| 78 | 18.599 | 17.224 | 11.401 | 16.479 |
| 80 | 18.710 | 17.561 | 11.530 | 19.511 |

candidate addresses of the four comprehensive domestic waste treatment centers is progressively more pronounced: 14.9226km, 14.097km, 9.899km, and 15.831km respectively.

From the results in Table 4, it can be seen that when the daily production of domestic waste in the town is 120 (t/d), the average values of the collection and transportation distances between the four candidate addresses of the comprehensive domestic waste treatment center are: 23.251 km, 25.153 km, 18.539 km, and 25.249 km respectively. From the results in Table 1–Table 4, it can be seen that the average values of the waste collection and transportation distances of the four candidate addresses. Therefore, among the four options for the location of the comprehensive waste treatment

Table 4. Waste collection distances for a daily domestic waste output of 120t (km).

| Candidate addresses/ daily waste production (t) | Candidate Address A | Candidate Address B | Candidate Address C | Candidate Address D |
|---|---|---|---|---|
| 102 | 20.144 | 23.207 | 16.122 | 22.271 |
| 104 | 21.206 | 23.316 | 16.256 | 22.332 |
| 106 | 21.337 | 23.425 | 16.337 | 23.458 |
| 108 | 22.448 | 23.528 | 18.215 | 24.513 |
| 110 | 22.469 | 25.310 | 18.336 | 24.554 |
| 112 | 22.523 | 25.455 | 18.348 | 25.622 |
| 114 | 24.589 | 25.609 | 18.442 | 25.719 |
| 116 | 25.644 | 25.883 | 20.311 | 26.734 |
| 118 | 25.780 | 27.274 | 20.405 | 28.518 |
| 120 | 26.372 | 28.521 | 22.618 | 28.773 |

center in the area, candidate address C, which is located directly north of the town, is the most suitable for the location requirements and can always maintain the minimum waste collection and transportation distance under different waste production conditions, which can effectively improve the efficiency of waste collection and transportation in the area. This can effectively improve the efficiency of waste collection and transportation in the area. Therefore, candidate address C is selected as the comprehensive waste treatment center for the area.

## 6 CONCLUSION

The article integrates various factors of domestic waste disposal methods and uses immune algorithms to study the site selection scheme that best meets the actual needs. To a certain extent, it alleviates the contradiction between land resources and waste disposal and provides an effective reference model for subsequent waste disposal studies. Through the case study, a more objective implementation approach is selected, overcoming the subjective tendencies of decision-makers. At the same time, the application scenarios of the immunization algorithm are enriched. More fuzzy variables can be introduced into future studies of the siting problem to produce more accurate results.

## REFERENCES


Abbasi A, Araban M, Heidari Z, et al. Comparing the impact of educational messages based on an extended parallel process model on solid waste separation behaviors in female students: A four-group randomized trial. *Waste Management*, 2020, 117(1):1–8.

Alcaraz-Londoo L M, Ortiz-Clavijo L F, Duque C, et al. Review on techniques of automatic solid waste separation in domestic applications. *Bulletin of Electrical Engineering and Informatics*, 2022, 11(1):128–133.

Blaise P, Issock I, Roberts-Lombard M, et al. Normative Influence on Household Waste Separation: The Moderating Effect of Policy Implementation and Sociodemographic Variables. *Social Marketing Quarterly*, 2020, 26(2):93–110.

Pu M, Yu X. The Research on the Regulation Mechanism of Garbage Classification and Recycling Based on the Analysis of Subject Game. *Open Access Library Journal*, 2021, 08(6):1–20.

Ren Yue, Yang Junjie. Eco-efficiency assessment of municipal solid waste sorting treatment mode. *China Environmental Science*, 2020, 40(3): 1166–1175.

Wang Dandan, Jian lirong, Fu Shuaishuai. Incentive and supervision mechanism of municipal solid waste separation and recycling. *China Environmental Science*, 2020, 40(7): 3188–3195.

*Advances in Urban Construction and Management Engineering – Cha (Ed)*


# Prospects for the application of UAVs in the field of port security

Weiwei Sun*
*China Waterborne Transport Research Institute, Beijing, China*

Zhiyan Zhao
*Shuikeyuanda Traffic Design Institute (Beijing) Co., Ltd., Beijing, China*

Xiaoshu Shao
*Zhoushan Port and Shipping Development Center, Zhejiang, China*

ABSTRACT: Boasting the advantages of low cost, high accuracy, mobility and flexibility, and the ability to carry equipment, unmanned aerial vehicles (UAVs) have been widely used in various industries. But due to natural conditions, salt spray corrosion, signal interference, and other reasons, the application of UAVs in ports has been restricted. This paper studies the functions of UAVs and relevant policies and regulations and puts forward application prospects from the aspects of security inspections in the port, inspection of large equipment in ports, intra-port traffic management, personnel tracking, emergency rescue, and so on. A development model of "UAV + port security" is constructed from the aspects of simplified operation by personnel, the durability of battery life, standardization of safety performance, and integration with high-tech, which promotes the application of UAV technology in the field of port security and facilitates the safe and stable development of ports.

## 1 INTRODUCTION

UAVs, which were only used in the military in the past, are being widely used in different industries, such as transportation, public security, logistics, agriculture, etc., by combining the latest technology with the improvement of modern science and technology and economic development. Through the research on the existing technology of UAVs, the application functions suitable for the field of port security are explored, and the UAVs are transformed from "aerial photography artifacts" to "aerial police," which further improves the overall security level of the port.

## 2 ANALYSIS OF THE STATUS QUO OF UAV DEVELOPMENT

### 2.1 *UAV system composition and function*

A UAV is a powered, controllable, unmanned aerial vehicle that can carry multiple mission equipment, perform multiple tasks, and can be reused (Wu 2020). It integrates navigation, communication, power, chip, and other technologies, with many advantages such as small size, low cost, convenient use, and high flexibility. UAV comprises an aircraft body, control system, navigation system, camera system, etc. (Li 2014), with the composition shown in Figure 1.

UAV originated in the military field and is gradually applied to civilian applications and equipped with functions such as image acquisition, sensor detection, communication relay, and material

*Corresponding Author: sww@wti.ac.cn

 DOI 10.1201/9781003348023-83

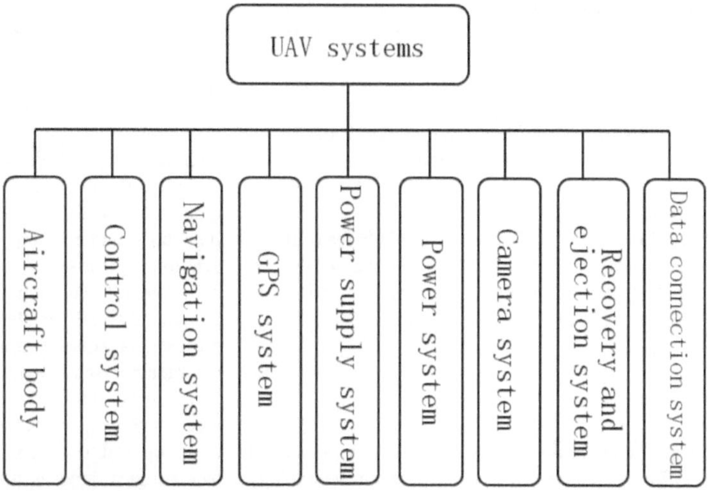

Figure 1. Composition of UAV system.

transportation, with the continuous improvement and optimization of UAV hardware and software technology. The flying altitude of UAVs is generally not more than 1,000 m, the flight speed is generally less than 200 km/h, and radar reflection is not more than 2 $m^2$ (Jiang et al. 2022). UAVs have been widely used in aerial photography, plant protection, map surveying and mapping, power inspection, fire rescue, express transportation, traffic supervision, wildlife observation, news reporting, film and television shooting, and other fields.

### 2.2 *UAV-related policies and regulations*

Currently, the State Council and various ministries and commissions have issued departmental rules and regulations, policies, and measures related to the development and safety supervision of the UAV industry (Jin & Shang 2019), which mainly include two aspects. One is incentive policies to promote the development of the emerging UAV industry, improve the industry-standard system, and enhance the application level of the UAV industry as well as the overall competitiveness of the industry (Table 1). The other is incentive policies to strengthen the supervision of UAVs, standardize the development of the industry, and put forward relevant management requirements for UAVs from the perspective of the management responsibilities of various departments.

## 3 APPLICATION ANALYSIS OF UAVS IN THE FIELD OF PORT SECURITY

Currently, China's port security management is still in the mode of "cameras + security guards," which is mainly based on civil defense and supplemented by the technical defense. Except for face recognition and vehicle recognition, other security equipment or functions are rarely used in ports. UAVs are characterized by high efficiency, low cost, and wide adaptability. They are not constrained by the terrain environment, which will guide the transformation of port security supervision to a more efficient and convenient management method. The main application directions of UAVs in port security include security inspection, equipment inspection, port traffic management, personnel tracking, emergency rescue, and so on.

Table 1. Incentive policies for UAVs.

| Promulgation Time | Policy | Promulgating Body | Gist |
|---|---|---|---|
| November 2016 | *The 13th Five-Year Plan for the Development of National Strategic Emerging Industries* | State Council of the People's Republic of China | It propounds to vigorously develop civil helicopters, multi-purpose aircraft, special aircraft, and industrial-grade UAVs with high market demand |
| August 2017 | *The 13th Five-Year Plan for the Construction of the National Emergency Response System* | General Office of the State Council of the People's Republic of China | It supports and encourages general aviation enterprises to increase helicopters, fixed-wing aircraft, unmanned aerial vehicles, and related professional equipment with emergency rescue capabilities to play their roles in rescue and disaster relief, medical rescue, and other fields |
| December 2017 | *Guiding Opinions on Promoting and Regulating the Development of Civil UAV Manufacturing* | Ministry of Industry and Information Technology of the People's Republic of China | It is proposed to vigorously carry out technological innovation, expand service application fields, establish and improve the standard system, and promote the development of China's civil UAV manufacturing industry |
| September 2019 | *Outline for Building a Strong Transportation Country* | State Council of the Central Committee of the Communist Party of China | It advocates actively developing UAV logistics and delivery, and the development of the UAV industry has become an important part of the construction of a strong transportation country |

### 3.1 *Security inspections in the port*

From a commanding perspective, UAVs can monitor the situation in the port area in a wide range, quickly reach places that are difficult for people to reach, expand the patrol range, and build a three-dimensional prevention and control network in the air and on the ground. In addition, limited by factors such as the range of personnel activities, the arrangement of monitoring probes at the front of the wharf, and the occlusion of berthing ships, the patrolling and surveillance at the frontier waters of the wharf have always been relatively weak link in the security field. After UAVs are equipped with cameras, the patrolling at the frontier waters of the wharf can be realized. As cameras' night vision, fog penetration, and other functions continue to be powerful, real-time monitoring and video return can also be performed in bad weather such as night or heavy fog.

Dalian Customs uses UAVs for supervision and security inspections in special areas. Using advanced technologies such as wireless remote control, high-altitude hovering, digital image return, navigation and positioning, and video tracking, the processing efficiency of video monitoring centers in special areas is effectively improved. Relying on modern scientific and technological means, the traditional manual "one inspection a day" is upgraded to "multiple inspections a day."

### 3.2 *Inspection of large equipment in ports*

The key point of port security protection is the safety of personnel, equipment, and goods in the port. Therefore, preventing the equipment from being damaged in the port is also one of the contents of port security. Due to the large size and high height of mechanical equipment in the port, such as gantry cranes and quay cranes, the traditional manual inspection method has risks and disadvantages such as high-altitude operation, blind vision, and long machine downtime. UAVs can use GIS + Beidou satellite navigation technology to independently plan flight routes, freely set inspection tasks, and meet the needs of efficient patrols, which can greatly shorten equipment inspection time, reduce inspection costs, and decrease personnel operation risks.

Zhenhua Heavy Industry took the lead in applying UAV technology in port machinery maintenance projects (Sohu 2019), which can collect health data of all large machinery and equipment in the port and use professional image detection technology to automatically form health evaluation reports after uploading, so as to cover blind spots of manual detection fully. The system can automatically inspect forty detection areas of a quay crane within about two hours, which is more than eight times more efficient than manual detection.

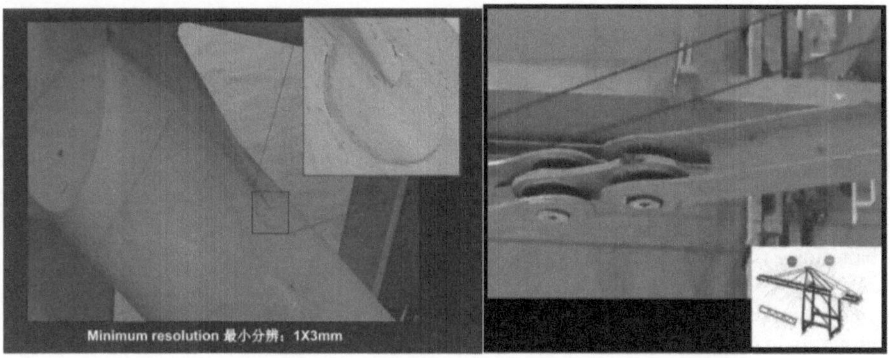


Figure 2. Utilizing UAVs for inspection and data collection of large port equipment.

### 3.3 *Intra-port traffic management*

For ports with a large number of vehicles for the collection and distribution of containers, bulk cargo, etc., the traffic management in the port can be carried out by UAVs, and the real-time positioning and tracking of vehicles entering and leaving the port can be carried out by using image recognition and other technologies to prevent vehicles from running out of the prescribed route or entering restricted areas.

In the meantime, traffic management can identify and capture traffic violations such as vehicle speeding and use the shouting system to warn people and vehicles when they approach a dangerous area to discourage and correct them in time to eliminate traffic hazards.

### 3.4 *Personnel tracking*

Personnel is the key target of prevention and control in port security. UAVs have the function of tracking target personnel; that is, for the personnel who carry out sabotage activities in the port, after a UAV locks the target, the personnel can be automatically tracked and followed through technologies such as face recognition, so as to deter and track saboteurs and assist capture. In addition, UAVs can match the detected images of suspicious persons with the background image library to obtain more detailed profiles and information about the persons.

### 3.5 *Emergency rescue*

UAVs can lock and reach the emergency rescue area in a short time and provide multi-angle and large-view monitoring images, the flight trajectory of which is generally insusceptible to terrain constraints. Therefore, UAVs can be used in emergency rescue, display the accident scene comprehensively and macroscopically, and provide decision-making for decision-makers to deploy personnel and materials. Through the cooperation of the video conference system and UAVs, the local operator can share the observation scene with multiple participants in real-time, and emergency experts worldwide can quickly participate in the rescue command online.

In addition, UAVs can also carry simple emergency supplies and drop food, small life-saving buoys, GPS buoys, etc., for initial rescue. At present, in the field of emergency rescue, UAVs have the functions of search and positioning, video transmission, material delivery, firefighting, etc.

## 4 CONSTRAINTS AND IMPROVEMENT DIRECTIONS FOR THE DEVELOPMENT OF UAVS IN THE FIELD OF PORT SECURITY

### 4.1 *Constraints*

#### 4.1.1 *Natural conditions*

In severe weather such as rainy days, windy days, and thunderstorms, the stability and safety of UAVs will be restricted to a certain extent. In windy weather, to maintain a balanced attitude and fly normally, UAVs will consume more power, which greatly shortens the battery life and reduces flight stability.

Furthermore, the high-temperature environment will affect the thermal conductivity of UAVs and accelerate the aging of UAV materials. Severe cold environments can reduce the efficiency of UAV batteries, resulting in accidents such as motor stalls.

#### 4.1.2 *Salt spray corrosion*

Port terminals in coastal areas are subjected to some form of salt spray environment throughout their life cycle due to high humidity and salt spray deposition. The deposition of salt spray will form a liquid film on the surface of the UAV, which constitutes an environmental medium that corrodes the components of the UAV and affects its life cycle. Metal materials in UAVs are more sensitive to salt spray, which easily damages electronic equipment and corrode conductive layers and metal materials.

#### 4.1.3 *Signal interference*

There are many ships at the port terminal, and there are radars and high-frequency signal interference on the ship, as well as the relationship between the metal material of the ship itself, which will affect or even block the signal controlled by UAVs. If terrorists use radar, radio, and other methods to track and lock, interfere with the data link and positioning systems of UAVs, or intercept the communication and navigation between UAVs and their remote control, they can prompt the UAV to descend on its own and control them to descend and return.

### 4.2 *Lifting direction*

#### 4.2.1 *Simplified operation by personnel*

As of the end of 2019, the number of Chinese civil UAV licenses was 67,218. The employment distribution of those holding UAV licenses is mainly in the production and research and development enterprises, related application units, and universities of various types of civil UAVs. The security of port facilities is mainly in the charge of the internal staff of the company. The UAV operation interface should be convenient and simple to facilitate the operation and use of a wider range of staff.

#### 4.2.2 *The durability of battery life*

Due to the large area of the port area, the endurance of UAVs should be improved, the flight time of UAVs should be extended, and their application functions should be maximized. The automatic take-off and landing power exchange function is applied, and the battery is automatically replaced to support high-frequency operations in the UAV port. With the development of UAV battery performance and the research of wireless charging technology, the endurance of UAVs is constantly being improved.

#### 4.2.3 *Standardization of safety performance*

Due to a large number of people and vehicles in the port and the relatively high equipment in the port, during the flight of the UAV, the environment perception and obstacle avoidance technology should be used to achieve effective observation, assessment, and threat judgment of the air traffic environment, strengthen avoidance, and prevent collisions, so as not to affect normal production operations. In addition, UAVs use high-definition camera aerial photography of port data; thus, it is necessary to improve the information security protection capabilities of UAV systems to ensure the production, storage, and transmission of data security in the process of interacting with sensors.

#### 4.2.4 *Integration with high-tech*

With the rapid development of high and new technologies such as intelligent monitoring, 5G, and Beidou satellite navigation systems, UAVs will be further improved in terms of data transmission, shooting pixels, flight distance, control range, and information processing. Besides, the development of new equipment, technologies, and energy will also promote UAVs into a more intelligent, lightweight, and professional era.

## 5 CONCLUSION

UAV technology has rapidly developed due to the advancement of artificial intelligence; however, its application in port security is still in its infancy. "5G+UAV" can realize a wide range of face recognition, vehicle recognition, and intelligent tracking, which meets the structured application of video (Che 2022). Equipped with cameras and emergency relief supplies, UAVs have the advantages of on-site monitoring and rescue, playing an irreplaceable role. With the development of the UAV inertial navigation system, waterproof pressure relief film, and other technologies, UAVs will be further developed in the port field, The application of UAVs in port security is a major transformation in the port security management model, which can promote the development of port security systems from "civil defense" to "technical defense," enhance the intelligence, information technology, and modernization level of the port security system, and meet the development needs of "smart port" and "safe port."

## REFERENCES


Che, H.2022 Design of three-dimensional intelligent security construction scheme based on 5G. *Radio & Television Network*, Issue 1: 52–55.

Jiang, D.T., Fan, C.J., Yong, Q.R., Qu, C.X., Liu, S., Zhang, Y.J.2022 Research on UVA detection and counter technologies for security in key areas. *Journal of applied sciences*, Issue 1: 167–178.

Jin, W., Shang Y. 2019 Safety supervision of UAVs in China. *Science and Technology Herald*, 37(14):12.

Li, C.2014 Application prospect and implementation of security surveillance system based on unmanned aerial vehicle in smart city. Science Mosaic, Issue 6: 72–75.

Sohu. 2019. Open a new era of intelligence! The world's first intelligent UAV inspection system in the dock has been successfully applied. https://www.sohu.com/a/341637100_237762.

Wu, J.2020 Application prospect and realization of UAV-based security monitoring system in smart cities. *Wireless Internet Technology*, Issue 24: 108–109.

*Advances in Urban Construction and Management Engineering – Cha (Ed)*


# A review of the cost efficiency and environmental sustainability of cross-laminated timber compared with concrete for multi-storey buildings in Australia

Yun Sun*
*The University of New South Wales, Australia*

ABSTRACT: The construction of buildings has always been closely involved in our modern life, and a large amount of energy consumption and waste generated by it has led to environmental degradation. In terms of reducing the energy consumption and construction waste caused by the common building structure of reinforced concrete, the use of Cross Laminated Timber (CLT) new materials can potentially alleviate such problems.

The prior literature review has shown that although domestic and foreign researchers have conducted much research on CLT, the application of CLT materials in construction is still very few. In addition, the specific energy saved and carbon emissions reduced by using CLT to compare with reinforced concrete structures, the expected findings will be of great significance to the environment. In the Australian residential construction market, CLT buildings are conceptually novel and worthy of promotion.

The project will use Revit to model a 6-story residential building with a reinforced concrete structure and CLT structure to analyze using the Life Cycle Costing (LCC) method and Life Cycle Impact Assessment (LCIA) method. The results of this research will demonstrate whether CLT is more economically and environmentally sustainable than reinforced concrete structures. The conclusion drawn can be considered to alleviate the negative impact of construction activities on the environment. Therefore, the costs and environmental impact assessment throughout the whole life cycle of both materials will be analyzed and calculated with the specific formulas during different stages. The outcome of this research will indicate which construction material performs better in environmental sustainability and cost efficiency. In addition, the final accurate comparison between these two materials can positively contribute to the current construction market, and the Australian government may encourage CLT applications on multi-level buildings.

## 1 INTRODUCTION

The current construction industry shows significant energy consumption and waste production, which leads to environmental deterioration. In the current market, the concrete structure is a common type of construction product that not only represents enormous costs but also leads to huge carbon dioxide ($CO_2$) emissions, as the consumption of concrete is approximately 1 ton of $CO_2$ per person per year in the atmosphere (Şanal 2018). Compared to the large consumption of concrete, Cross Laminated Timber (CLT) is a sustainable building material that can be made of renewable wood and isolates carbon, and be produced without fossil fuels, which results in the reduction of 26.5% of global warming in terms of CLT buildings rather than reinforced concrete constructions (Lewis et al. 2021). In addition, the production of CLT does not aggravate the deterioration of environmental harm, like global warming. On the other hand, CLT has an improvement in strength-to-weight ratio, durability, and dimensional stability compared to reinforced concrete.

*Corresponding Author: z5243598@zamil.unsw.edu.au

 DOI 10.1201/9781003348023-84

This project will examine whether CLT is more sustainable than steel reinforced concrete structures or not, specifically the cost efficiency in the life cycle stages of buildings, its recyclability, and the effect on carbon emission. By identifying the beneficial outcomes, the more affordable construction material, CLT, can be used profitably in sustainable building, minimize waste production and reduce carbon emissions compared to the usage of reinforced concrete.

## 2 LITERATURE REVIEW

This literature review aims to provide a background of the problems and challenges of CLT material. Although a lot of research on CLT and comparison between CLT and concrete material has been done by researchers, there have still been few applications of CLT material in construction buildings. Therefore, based on the prior work and research, analyzing whether the CLT can be an alternative material compared with reinforced concrete, specifically in terms of cost efficiency and environmental impacts, is covered in this review.

### 2.1 *CLT materials introduction*

CLT is a type of composite timber used for structural load bearing. CLT is available as a two-way element to form shear walls and is used in high-rise pure timber structures and mixed-frame buildings (Table 1) (Li 2019).

Table 1. CLT characteristics and scope of application.

| Material | Component Type | Mechanical properties | Appearance | Size | Applications |
|---|---|---|---|---|---|
| Cross Laminated Timber (CLT) | Tow-way element: wall, floor, roof | Load bearing | Natural wood texture | Length: Max. 20m<br>Width: 50~250mm<br>Thickness: Max. 508mm<br>timber construction | High-rise residential and office buildings; Large span |

CLT plates are made of solid sawn timber or structural composite in odd numbers (at least three layers), which are pressed through longitudinal and horizontal stacks and then pressed by adhesive.

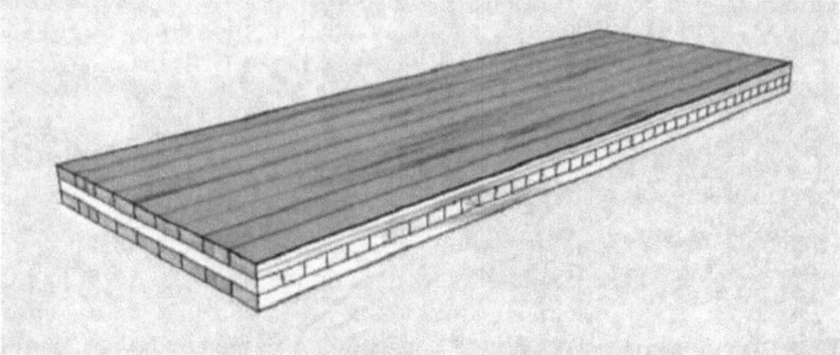

Figure 1. Cross laminated timber.

### 2.2 *Cost efficiency*

#### 2.2.1 *Sourcing raw material and manufactured products*

At present, the possibilities of CLT are embodied in the multi-storey timber structures used in offices and homes, thus driving a renaissance in urban timber engineering. Local forestry and other components of the construction supply chain have great opportunities to develop products for Mass timber construction (MTC) in the construction industry. The recovery and removal of poor-quality materials and the opportunity to recover CLT production wood from other waste sources offer financial and political advantages in an increasingly carbon-constrained world (Evison et al. 2018). In addition, compared with reinforced concrete materials, CLT is highly suitable for reconstructing and upgrading existing buildings and is resistant to special loads, such as earthquakes (Brandner 2013).

In a survey conducted in 2019, over 75% of CLT manufacturers agreed that high raw material cost is a barrier to further growth of the CLT industry (Larasatie et al. 2021). However, two-thirds of the respondents were neutral or against the opinion that high production costs would be a barrier to industry growth. The survey also showed prevailing optimism for the CLT market.

No matter whether the cost of excavating raw materials is high, the usage of CLT should be compliant with the Building Codes. As indicated in the Global CLT Industry survey (Muszynski et al. 2020), 90-100% of the CLT fabrication were custom order/project machined panels, so there is room to decline in the cost of sourcing CLT when products are standardized. By contrast, the uncommonly sized products result in higher transportation costs, especially if imported (Forest and Wood Products Australia, Massive timber construction systems: cross-laminated timber (CLT). 2019.). There are several CLT importers in Australia but only one locally established plant (Evison et al. 2018). Considering this situation, to reduce the cost, CLT manufacturers can update their supply chain from importing to integrating elements, including material sourcing, engineering, and even building construction.

#### 2.2.2 *Construction and maintenance*

One of the obstacles in CLT construction is the higher material cost, on average 2% to 6%, than concrete construction. A study concluded that a three-story commercial building in New Zealand could become more expensive with a 4% higher cost using CLT material instead of using concrete (Smith et al. 2009). A study shows that the cost of using CLT for a single-family residential building in the United States can be 23% higher than that of traditional light-framed wood options (Burback & Pei 2021). In addition, Ahmed and Arocho analyzed the construction cost of wood and concrete based on a residential building project in Canada, and they concluded that mass timber building has a higher construction cost of 6.43% than concrete building (Ahmed & Arocho 2021).

Although the capital cost of CLT (£240/m$^2$) is higher than steel and concrete (£190/m$^2$), when the project is built up to 8 storeys, the reduced project time can offset the material cost of CLT (Ahmed & Arocho 2021). The speed of timber construction is much quicker at the construction site than in conventional buildings, with almost no waiting time. In addition, Forest and Wood Products Australia published a report which indicates that compared to traditional structure buildings (office, residential, healthcare, and industrial), the overall cost saving of CLT construction can be 2.2 to 13.9% cheaper.

### 2.3 *Environmental sustainability*

Research has found that timber structure buildings have better environmental performance at all stages of their life cycle than other common types of building materials. According to the results of energy assessments in North America and Australia, wood construction has advantages over concrete and steel structures in terms of energy consumption, carbon emissions, and water and air pollution during the full life cycle of the materials (Table 2).

#### 2.3.1 *Waste production and recyclability of reinforced concrete and CLT*

Reinforced concrete structures are the preferred choice when designing high risers. However, the industry is facing waste production and environmental sustainability challenges. U.S. Environmental Protection Agency conducted that in America, concrete is the largest contributor by weight to construction and demolition waste, with a figure of 220 tones generated in 2014Tm. In addition, Tam indicated that only 54% of concrete waste was recycled into a new construction project in Australia each year, and the rest of the waste was sent to the landfill (Tam 2009).

Table 2. Comparison of energy consumption between the timber structure and common structure in the life cycle.

| | Timber frame | Concrete structure | Steel structure |
|---|---|---|---|
| Energy consumption | 100% | 120% | 112% |
| Carbon emissions | 100% | 129% | 115% |
| Water pollution | 100% | 225% | 300% |
| Air pollution | 100% | 112% | 110% |

Note: The life assessment cycle is 20 years.
(Source: *CLT Handbook* (Karacabeyli & Douglas 2013))

Compared with concrete, CLT has a huge advantage in environmental sustainability due to its strong energy storage and easy installation and removal. At the end of the life cycle, CLT can be recycled easily because of its high performance in the strength-to-weight ratio of the material. Another research shows that with the incineration of CLT components in the end-of-life stage, the recovery of energy can be saved significantly in greenhouse gas (GHG) emissions (Jayalath et al. 2020), which means that CLT requires the least amount of energy to be recovered at its end-of-life stage compared to concrete and steel.

### 2.3.2 *Carbon emission*

According to the United Nations Environment program (2020), 38% of global carbon emissions in 2019 came from the building industry, and the amount of carbon emissions is growing yearly. And due to the carbon emission level, the environment might be affected.

The definition of the life cycle to define the scope of building materials' service life from the start to the end includes five stages: transportation, manufacturing, construction, operation, and demolition (Zhang & Wang 2015). A case study in China found that each square meter of reinforced concrete structure causes at least 3 thousand kilograms of CO2 in its life cycle. In addition, the amount of carbon emission for reinforced concrete structures was calculated by an equation simplified as E=Q*EF, which means the emission is equal to the quantities of the material multiplied by the emission factors (Zhang & Wang 2015). Therefore, by knowing the quantities and emission factors, the carbon emission of each stage of the life cycle can be calculated.

According to research about Australian softwood and hardwood forests, each cubic meter of hardwood causes an amount of 982 kg of CO2, which is the total average of its life cycle, including deforesting and transportation, etc. The emission for hardwood is estimated by collecting previous data from each case study region and scaled up based on assumptions (England et al. 2013). This means the carbon emission for reinforced concrete of each cubic meter is higher than CLT by comparing the amount of carbon emission.

However, the calculation of carbon emission between CLT and reinforced concrete is different as well; therefore, using the same method of estimating the emission is required for further accurate comparison.

## 2.4 *Conclusion*

Prefabrication of CLT structure can significantly reduce construction time, and the offsite factory manufacturer can maintain high-quality control. Before processing and integrating with each penetration, connection, and interface, a comprehensive 3D digital documentation process can ensure the success of the timber prefabrication.

Compared to the concrete and steel manufacturing process, the production of CLT has a high-efficient performance in terms of energy consumption and carbon emission. Research about life cycle analysis (LCA) reveals that CLT may emit less greenhouse gas (GHG), store more carbon, and consume less fossil energy during raw material extraction, transportation, and manufacture phases when compared with concrete, steel, and brick (Guo et al. 2017).

However, some areas also require further studies and research on CLT. The degradation of this material is one of the areas which requires additional investigation and experiments. There is no publicly available evidence regarding the end-of-life fate of CLT, including potential degradation in landfill (Bart et al. 2013). Because the research on CLT degradation requires a certain amount of time, there is no research currently available in this area.

## 3 RECOMMENDED METHODOLOGIES

The scope of this research is to identify and compare the reinforced concrete and CLT building material in terms of cost efficiency and sustainability. To address the problems, this research conducted a comparative life cycle cost analysis (LCC) for the cost efficiency of CLT and reinforced concrete and conducted a comparative life cycle impact analysis (LCIA) for the sustainability of these two materials. This project assumes that the analysis object is a 6-storey residential building with a duration of 50 years in Australia, and a monetary discount rate is applied to the LCC assessment. Meanwhile, the costs of land acquisition are not included in this study.

### 3.1 *Life cycle cost analysis*

LCC is a commonly used assessment for analyzing the cost performance of a product or a system. This research proposes to use comparative LCC for the two building materials identified. The life cost arising from the development, construction, operation, maintenance, decommissioning, and disposal/recycling of the constructed asset will be studied in the analysis. The capital cost of land for development will not be considered in this research.

The LCC analysis will follow the ISO 15686-5:2017 standards and refer to the LCC Estimation Model: A Construction Material Perspective by Biolek et al. 2019. Cost efficiency will be compared using present value analysis. All costs incurring at different times of the studied life will be discounted correspondingly to present values.

LCC indicator will be calculated according to ISO 15686-5:2017 standards:

$$LCC = \sum_{t=0}^{T} \frac{C_t}{(1+r)^t}$$

Where:
Ct – all costs as equivalent cash flows in year t
r – discount rate (to be selected for economic analysis for the private sector in Australia)
t – analyzed year (t = 0, 1, 2..., T)
T – studied life cycle in years (T=50 for this research)

Collection of local data is preferred as the study looks at buildings in Australia. Data collection will mainly focus on current market prices in both the private and public sectors to enhance the application of results in the current economy. Local cost figures for reinforced concrete are expected to be readily available. For the CLT alternative, local cost figures can be limited. The research aims to collect data from local suppliers, designers, and developers. For the missing data entries, figures from similar projects overseas will be collected.

### 3.2 *Life cycle impact assessment*

LCIA is one of the most useful assessments to analyze the environmental impacts of production activities. The LCIA will quantify the required construction materials and assess the volume of global warming impact by calculations. Therefore, the environmental impacts of CLT and reinforced concrete as multi-story building construction materials can be analyzed and compared.

The first step of this assessment will be quantifying how much CLT or reinforced concrete is required for a multi-level building. The global building material statistics can be accessed through the Statista website under its building construction section. Then, a Life Cycle Inventory table can be concluded, and the quantities of the construction materials will be summarized under the table.

Then, the second step is to identify which substances involved in negative environmental impacts will be produced during the life cycle. The life cycle begins with the raw materials (timber, aggregates, ore) to the end of the building life. Therefore, all the toxic and global warming impacts substances are accounted for in this part.

The third step will be calculating the greenhouse gas (GHG) emission for the construction and operation stages of the building. Jayalath introduced calculation methods to estimate the GHG emission volume by different construction materials throughout the life cycle (Jayalath, Navaratnam, Ngo, Mendis, Hewson, Aye, 2020). The following series of formulas can estimate the GHG emission for different materials during the construction stage:

$$GHG_C = GHG_M + GHG_T + GHG_S$$

$$GHG_M = \sum_{m=1}^{M} Q_m GHG_S$$

$$GHG_T = \sum Q_m D_i f_E f_{GHG-Diesel}$$

Where $GHG_C$ (kg $CO_2$-e) is the GHG emission during the construction, $GHG_M$ (kg $CO_2$-e) is the GHG by the extraction and manufacturing of the building materials, $GHG_T$ (kg $CO_2$-e) is the GHG emission during the transportation of materials to the site, $GHG_S$ (kg $CO_2$-e) is the energy consumption of equipment, $Q_m$ (kg, m3, m2) is the quantity of material m, $D_i$ (km) is the distance traveled, $f_E$ (MJ/kg.km) is the fuel energy factor for trucks, $f_{GHG-Diesel}$ is the GHG emission factor for diesel.

After estimating the GHG emission for the material manufacturing and construction stage, the energy consumption and GHG emission can be calculated using the following formula:

$$GHG_{OE} = OE * f_{GHG-electricity}$$

Where $GHG_{OE}$ (kg $CO_2$-e) is the GHG emission and energy consumption during the operational stage, OE (kWh) is the operation energy requirements of the building, $f_{GHG-electricity}$ (kg $CO_2$-e/kWh) is the state-specific GHG emission factor.

Finally, in the demolition phase of the building, the recycling rate of waste materials for both reinforced concrete structures and CLT structures buildings can be estimated based on the open data from the Australian Bureau of Statistics under the construction section.

## 4 CONCLUSION

This review consists of calculation processes throughout the LCC and LCIA analysis. The LCC indicator shows the expected monetary value of the object, and all the GHG indicators ($GHG_C GHG_M$, $GHG_T$, etc.) shows the expected Greenhouse Gas emission in different phases of the building. Therefore, a building material with a smaller value of LCC and GHG indicators possesses better cost efficiency and sustainability performance. However, the application of CLT in multi-storey buildings is still at the early stage, and the detailed life cycle cost and environmental impact assessments were not analyzed by other academic peers. Although CLT brings benefits to the current environmental problems, the expense of the material is still high due to the lack of manufacturers in Australia and the expense of raw materials. In the next step, using BIM technologies to carry out experimental modeling is essential. The data from Australian government agencies, which include statistical data, the factors of data diversity, and variability, must be considered in this experiment to reduce the error of the experimental results of the CLT model and the reinforced concrete model. The outcome of this experiment will accurately reflect the performance of environmental sustainability and cost efficiency of CLT and reinforced concrete structures. Therefore, CLT boards mixed with recycled timber could potentially reduce the cost, and this CLT can be a huge innovation, so it will potentially encourage local governments to promote CLT application on high rises.

## REFERENCES


A. Jayalath, S. Navaratnam, T. Ngo, P. Mendis, N. Hewson, and L. Aye, 'Life cycle performance of Cross Laminated Timber mid-rise residential buildings in Australia,' *Energy Build.*, vol. 223, p. 110091, Sep. 2020, doi: 10.1016/j.enbuild.2020.110091.

B. Burback and S. Pei, 'Cross-Laminated Timber for Single-Family Residential Construction: Comparative Cost Study,' *Archit. Eng.*, vol. 23, no. 3, Apr. 2021, [Online]. Available: https://doi.org/10.1061/(ASCE)AE.1943-5568.0000267

D. Bart, E. Crossin, and J. Wong, '*Life cycle assessment of a cross-laminated timber building,*' For. Wood Prod. Aust. Ltd., Jun. 2013, [Online]. Available: https://www.researchgate.net/publication/274964059_Life_Cycle_Assessment_of_a_cross_laminated_timber_building

D. C. Evison, P. D. Kremer, and J. Guiver, 'Mass timber construction in Australia and New Zealand—status, and economic and environmental influences on adoption,' *Wood Fiber Sci.*, vol. 50, no. Special, pp. 128–138, Aug. 2018, doi: 10.22382/wfs-2018-046.

E. Karacabeyli, B. Douglas, FPInnovations (Institute), and Binational Softwood Lumber Council, CLT handbook: cross-laminated timber. 2013.

Forest and Wood Products Australia, *Massive timber construction systems: cross-laminated timber* (CLT). 2019.

H. Guo, Y. Liu, W.-S. Chang, Y. Shao, and C. Sun, 'Energy Saving and Carbon Reduction in the Operation Stage of Cross Laminated Timber Residential Buildings in China,' *Sustainability*, vol. 9, no. 2, p. 292, Feb. 2017, doi: 10.3390/su9020292.

İ. Şanal, 'Discussion on the effectiveness of cement replacement for carbon dioxide (CO2) emission reduction in concrete: Original Research Articles: Discussion on the effectiveness of cement replacement for carbon dioxide (CO2) emission reduction', *Green. Gases Sci. Technol.*, vol. 8, no. 2, pp. 366–378, Apr. 2018, doi: 10.1002/ghg.1748.

J. R. England, B. May, R. J. Raison, and K. I. Paul, 'Cradle-to-gate inventory of wood production from Australian softwood plantations and native hardwood forests: Carbon sequestration and greenhouse gas emissions,' *For. Ecol. Manag.*, vol. 302, pp. 295–307, Aug. 2013, doi: 10.1016/j.foreco.2013.03.010.

K. Lewis, R. Shrestha, and K. I. Crews, 'Introduction to Cross Laminated Timber and Development of Design Procedures for Australia and New Zealand', presented at the 23rd Australasian Conference on the Mechanics of Structures and Materials (ACMSM23), *Byron Bay*, Australia, Apr. 2021.

L. Muszynski, P. Larasatie, J. E. Guerrero, and R. Albee, 'Global CLT Industry in 2020: Growth beyond the Alpine Region', presented at the 63rd International Convention of Society of Wood Science and Technology, *virtual conference*, Jul. 2020.

P. Larasatie, R. Albee, L. Muszyński, and J. E. M. Guerrero, '*Global CLT industry survey: the 2020 updates*', p. 9, Apr. 2021.

R. Brandner, '*Production and Technology of Cross Laminated Timber (CLT): A state-of-the-art Report,*' vol. 1, May 2013.

S. Ahmed and I. Arocho, 'Analysis of cost comparison and effects of change orders during construction: Study of a mass timber and a concrete building project,' *J. Build. Eng.*, vol. 33, p. 101856, Jan. 2021, doi: 10.1016/j.jobe.2020.101856.

S. Li, '*Research on Tectonics of Cross-Laminated Timber Walls,*' Jun. 2019.

T. Smith, M. Fragiacomo, S. Pampanin, and A. H. Buchanan, 'Construction time and cost for post-tensioned timber buildings,' *Proc. Inst. Civ. Eng. – Constr. Mater.*, vol. 162, no. 4, pp. 141–149, Nov. 2009, doi: 10.1680/coma.2009.162.4.141.

V. W. Y. Tam, 'Comparing the implementation of concrete recycling in the Australian and Japanese construction industries,' *J. Clean. Prod.*, vol. 17, no. 7, pp. 688–702, May 2009, doi: 10.1016/j.jclepro.2008.11.015.

X. Zhang and F. Wang, 'Life-cycle assessment and control measures for carbon emissions of typical buildings in China,' *Build. Environ.*, vol. 86, pp. 89–97, Apr. 2015, doi: 10.1016/j.buildenv.2015.01.003.

*Advances in Urban Construction and Management Engineering – Cha (Ed)*


# The design of an intelligent utility tunnel based on BIM

Linlin Ma, Pooya Saffari* & Boxiao Li
*Qingdao City College, Qingdao, Shandong, China*

Qiuying Sun
*Qingdao Institute of Technology, Qingdao, Shandong, China*

ABSTRACT: In view of the increasing number of underground utility tunnel corridors and the subsequent problems such as equipment status monitoring, fault prevention, and treatment, combined with the operational requirements of smart cities, this paper introduces the realization method of intelligent pipe corridors system based on BIM and optical fiber sensing technology. In this study, the establishment of a 3D model of pipeline based on BIM, the construction of an optical fiber sensing system, system networking, and data uploading are introduced.

## 1 INSTRUCTION

Due to the expansion of the urban area in China, the number of urban underground pipelines is also gradually increasing. The underground comprehensive pipe networks are centralized with important livelihood pipe networks such as heating power, water supply, electricity, and communication. However, problems such as the aging of underground pipeline networks and difficulties in maintenance are becoming increasingly prominent. Therefore, one of the important factors in the construction of a smart city is to establish a standardized, networked and intelligent underground pipeline project, so as to carry out real-time and effective management of the comprehensive underground pipeline and improve the comprehensive operational capacity of the city. The construction of a smart pipeline needs to have relevant general functions such as environmental detection, equipment detection, data analysis, safety detection, etc., among which the difficulty lies in environmental and equipment detection. The traditional pipeline systems are equipped with temperature, humidity, smoke, and other detection equipment. These detectors are large in number, modeling them is messy, and uploading data is difficult. Therefore, the traditional pipeline systems do not have the conditions to integrate into the unified intelligent pipeline system (Dahai 2020).

In order to overcome the above problems, a smart pipeline system with high intelligence, high standardization, and simple and reliable monitoring measures is needed to improve the safety and efficiency of the urban, comprehensive pipeline.

## 2 ESTABLISHMENT OF THE BIM MODEL

Based on Revit software, a model for the construction and operation of the pipeline was established to manage and control the operation process of the pipeline. Revit is one of the most widely used software in the BIM system in China, and its main feature is parametric components. Revit provides an open schematic system, which can create a pipeline model according to the requirements of the pipeline, simulates the local conditions, section, and three-dimensional information of

*Corresponding Author: pooya_saffari@qdc.edu.cn

DOI 10.1201/9781003348023-85

the pipeline according to the actual needs, and facilitates the communication and understanding between departments.

(1) Pipeline structure modeling
Combined with the CAD drawing software, the model was established in the component set mode, as shown in Figure 1. The materials were defined, and attention was paid to the chamfering treatment in the structure and the setting method of pipe support.
(2) Pipe and cable modeling
The component set mode was carried out to arrange the heating pipes, cables, and cable supports. Special attention was paid to the selection of materials. The monitoring instruments were placed at the predetermined position.

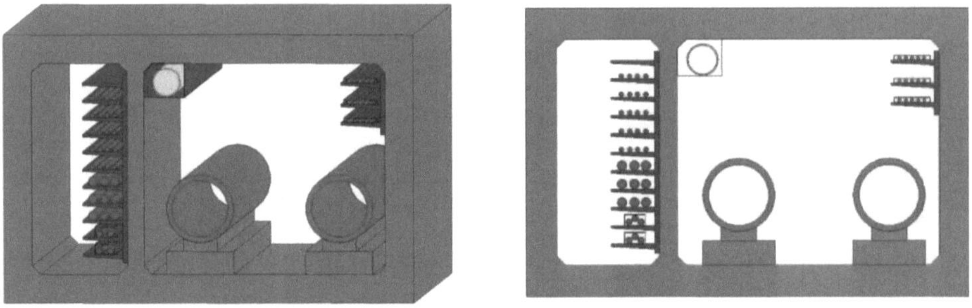

Figure 1. Pipeline model.

## 3 OPTICAL FIBER SENSING SYSTEM

If the fiber Bragg grating sensor is affected by external factors (such as temperature, vibration, and displacement), some characteristics of the transmitted light in the fiber will change. By configuring a special demodulation device, through signal acquisition and analysis, the characteristics of the light can be detected, including changes in attenuation, phase, wavelength, polarization, mode field distribution, and propagation time. These characteristic changes are analyzed and processed by the optical fiber alarm host and combined with special algorithms, so that the type of external interference can be distinguished and the abnormal position can be located to realize the accurate alarm function.

Taking the temperature grating as an example, the optical fiber is irradiated with a light pulse of a specific frequency. When the light pulse is shifted downward along the optical fiber, various scattering phenomena such as Rayleigh scattering, Brillouin scattering, and Raman scattering will occur. Among them, Raman scattering is the most sensitive to temperature interference, and Raman scattering will exist in the whole optical path (Yang 2019).

Raman scattering is caused by the energy exchange between the thermal vibration of optical fiber molecules and the interaction of photons. When a part of light energy is converted into thermal vibration kinetic energy, it will produce a light with a longer wavelength than the light source, called Stokes light. On the contrary, it will produce a light with a shorter wavelength than the light source, i.e., anti-Stokes light.

The ratio of the intensity of anti-Stokes light to Stokes light is a function related to temperature. During the transmission process, a part of Raman scattering light will return along the optical fiber, and the demodulator will receive and demodulate this part of the light and calculate the intensity ratio of anti-Stokes light and Stokes light, so as to calculate the temperature change at the abnormal interference point. By using OTDR technology, the transmission rate of light in the optical fiber is known. (Nuochen 2020) According to the time difference between incident light and

Raman scattering light, different temperature points can be determined. As a result, the abnormal temperature points along the whole optical fiber can be distinguished.

The overall structure scheme of the optical fiber detection system is shown in Figure 2.

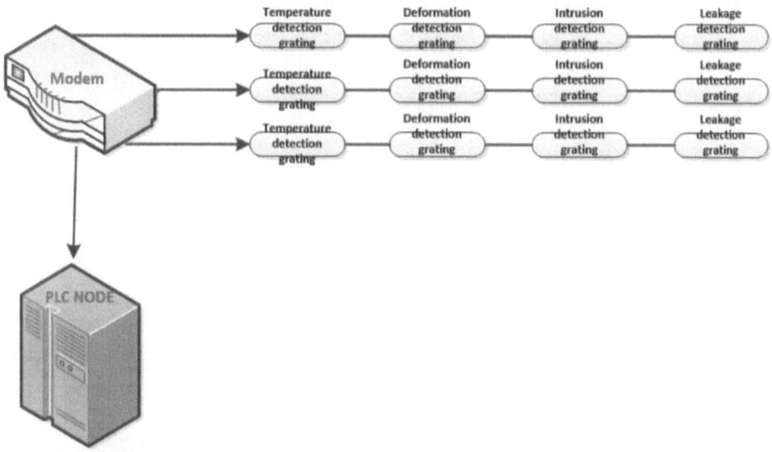


Figure 2. Overall structure scheme of the optical fiber detection system.

The whole system is composed of fiber demodulator, transmission fiber, and various detection gratings with different functions. The demodulator also supports Modbus Communication Protocol and can be used as a MODBUS slave of the PLC node. In addition, the PLC node also needs to be connected to other equipment controlled by this node, such as ventilation fans, sewage pumps, etc.

## 4 PLC DATA ACQUISITION AND MONITORING SYSTEM

Number In this study, the CPU of the PLC node is 1511-1pn of Siemens 1500 series. The main functional modules include distributed I/O node IM155 module, 16-point DI module *2, 8-point DO module *2, 8-point AI module *1, 4-Point AO module *1, MODBUS communication module *2. There are abundant standardized function blocks in Siemens series PLC, among which MODBUS function block is widely used. (Hao 2019)

The 14 PLC nodes planned for this project, as shown in Figure 3, are divided into two branches: A and B, with 7 PLC nodes per branch. Each PLC is set with an IP address in the same network segment, and the IP address of each PLC is not repeated, as shown in Figure 4.

### 4.1 *Modbus communication between PLC node and demodulator*

The Modbus_Comm_Load instruction in the Siemens PLC system configures the communication module for communication through the Modbus RTU protocol. When the Modbus_Comm_Load instruction is added to the program, the system automatically assigns an instance background data block to it. When configuring Modbus communication between PLC node and demodulator, MODBUS _ Comm_ Load instruction needs to be called first and assigned relevant parameters, such as module address, transmission baud rate, parity bit, and MODB address bits, as shown in Figure 5. The PLC node in this project is the Modbus master station, so the relevant parameters of the Modbus master block should be configured accordingly, as shown in Figure 6.

### 4.2 *System networking*

Due to the long distance between the nodes of the comprehensive underground pipeline, it is difficult to use optical fiber or other physical means for networking. This study adopts a public

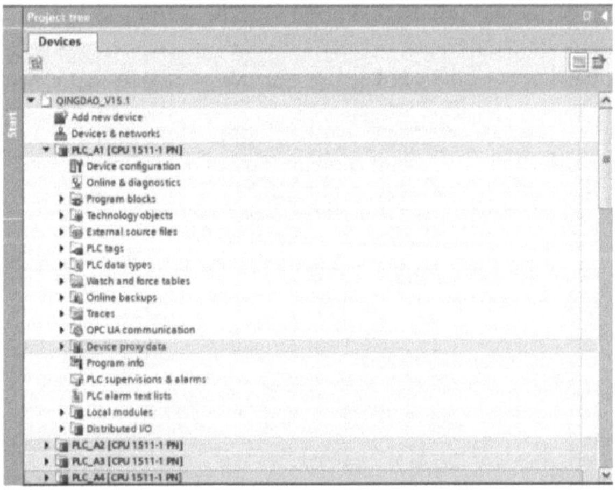


Figure 3. Device configuration.

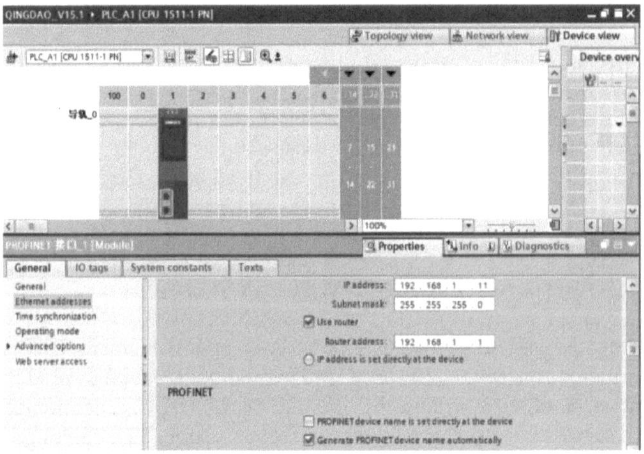


Figure 4. Device configuration continued.

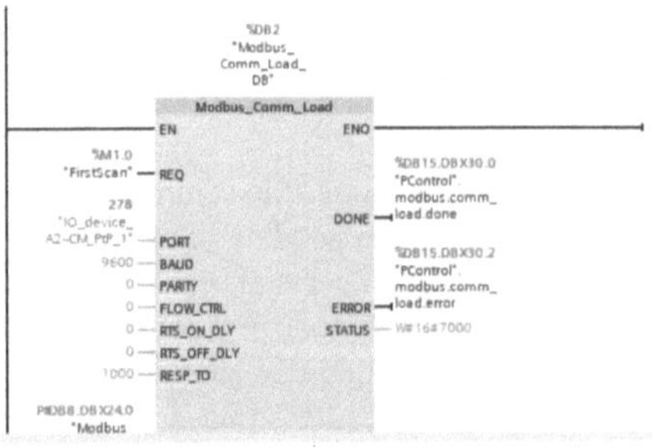


Figure 5. MODBUS program initialization.

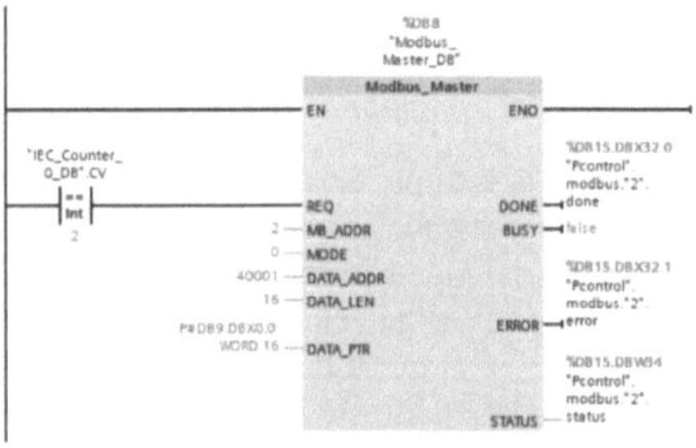


Figure 6. MODBUS master block.

network networking method provided by the mobile network operator to introduce mobile network signals (broadband network or 4G network card) at each node unit. The gateway communication equipment adopts Huawei AR101GW-Lc-S. The gateway module can support broadband network and 4G network card simultaneously and support various SSL VPN, L2TP, and other networking methods. In this study, the L2TP method is used for networking all distributed nodes. The system scheme is shown in Figure 7.

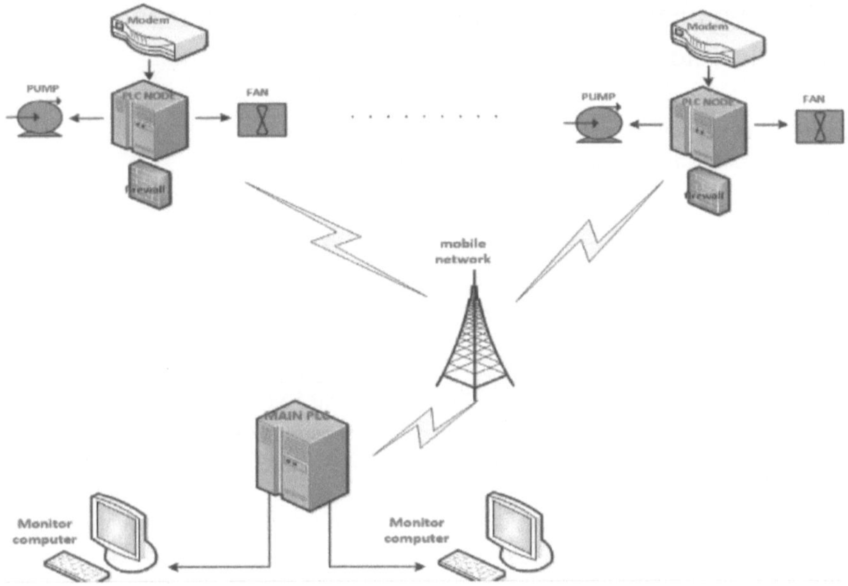


Figure 7. System structure diagram.

## 5 CONCLUSION

This research is an intelligent pipeline system based on BIM technology, which has the characteristics of equipment data exchange interface standard, a stable detection unit, a high degree of automation, and a high degree of integration. It is conducive to the intelligent operation of the underground integrated pipeline system, which can greatly reduce the labor intensity. Additionally, using a standardized interface protocol will help reduce the difficulty of connecting each unit in the smart city (Vodák 2021).

## REFERENCES


Dahai, B. (2020). Research on information construction of integrated pipe gallery based on BIM (in Chinese). *Municipal Technology*, 38 (02): 215–220.

Hao, L. (2019). *Research on the construction of intelligent management and control system of urban integrated pipe gallery based on BIM+GIS* (in Chinese) [D] Shenyang Architecture University.

Nuochen, L. (2020). *Research on operation and maintenance management platform of smart pipe gallery based on BIM+GIS* (in Chinese) [D] Hebei Institute of Architecture and Engineering.

Vodák, J., Šulyová, D. & Kubina. M. (2021). Advanced technologies and their use in smart city management[J] *Sustainability*, 13(10).

Yang, B. (2019). *Development and application of optical fiber security monitoring information management system for pipe gallery* (in Chinese) [D] Jilin University.

*Advances in Urban Construction and Management Engineering – Cha (Ed)*
*© 2023 Copyright the Author(s), ISBN 978-1-032-39018-5*

# Study on the outpatient layout optimization based on medical process simulation technology

Yu Wang & Guoqing Peng*
*Hualan Design & Consulting Group, Nanning, Guangxi, China*

Yanya Huang
*Guilin University of Electronic Technology, Guilin, Guangxi, China*

Rongjing Li & Jian Cao
*Hualan Design & Consulting Group, Nanning, Guangxi, China*

ABSTRACT: Medical process simulation method can provide the evaluation of the rationality of the medical process, visible, dynamic, and quantitative in the design stage, supporting the normal and epidemic situation. However, this technology is still seldom used in the hospital streamline design. So the simulation method applied in this paper used a complex mathematical model and heterogeneous peer groups theory, to simulate the pedestrian flow. In this paper, taking the outpatient department of Jingxi Public Health Emergency Center as a case, we simulated the operation of the medical building during the epidemic time. Based on the medical process simulation method, we establish an outpatient simulation model to visualize, quantitatively analyze, and evaluate the outpatient clinic. After the optimizing pre-examination and triage, modular general design of outpatient service the density and maximum queuing number of each department decreased, extraordinary for Respiratory Dept.2. The optimization strategy proposed in this paper has reference significance for the hospital design of normalized epidemic prevention and control.

## 1 INTRODUCTION

Outpatient is the first station of receiving patients, the spatial layout and streamlined design rationality are the main factors affecting patients' medical efficiency. At the beginning of the epidemic, there were a large number of patients with internal respiratory diseases, and the outpatient service was saturated (Petrone 2021).

Riquelme (2021) et al. Simulated COVID-19 infection process in the air transportation network. Pan (2021) et al. simulated deployment arrangements within hospitals during the COVID-19 pandemic in New York City, promoting the timely deployment of hospitals in public health emergencies by proposing solutions to identified problems. In terms of pedestrian microsimulation, Moussaïd (2009) and Kuang (2011) et al. have carried out the traditional pedestrian flow simulation, but they lack the description of heterogeneous individuals in the peer group. Patients and their families are the objects of pedestrian simulation in this paper, they are typical heterogeneous peer groups. So the traditional pedestrian simulation is not applicable.

To sum up, research on medical process simulation during epidemic time is endless, but there is still a lack of research on the streamlined layout of the medical process in peacetime and epidemic time. Therefore, this paper takes the outpatient department of Jingxi Public Health Emergency Treatment Center as a case, based on the heterogeneous peer group theory, and uses the medical

*Corresponding Author: 243564846@qq.com


DOI 10.1201/9781003348023-86

process simulation technology to simulate, analyze, and evaluate the outpatient pedestrian flow. Then it puts forward the optimization strategy and epidemic prevention plan for the outpatient bottleneck unit, which provides a reference for the outpatient medical technology design combined peacetime and epidemic.

## 2 MEDICAL PROCESS SIMULATION

### 2.1 *Medical process simulation method introduction*

The medical process simulation method based on an improved social force model considers the influence of pedestrians walking together and the differences between heterogeneous individuals so that the pedestrian simulation is closer to reality. This method uses the complex mathematical model as the logical basis and has the characteristics of multi-dimensional parameterization, visualization, and digitization. It aims to solve the problems that cannot be evaluated quantitatively, visually, and digitally in the design process of medical building streamline and service process, and to solve the problem that the general simulation software has only demonstration effect but can not generate data.

### 2.2 *The process of medical process simulation*

The research steps of this paper are as follows:

a) Gather basic requirements. Including the number of patients, the process of diseases, etc.
b) Outpatient simulation model construction.
c) Simulation in epidemic time.
d) Simulation analysis. Analyze indicators such as visit density, waiting number, etc.
e) Put forward the design optimization strategy and spatial layout optimization scheme.
f) Simulation again and verify the effectiveness of the optimization strategy.

## 3 OUTPATIENT DEPARTMENT USING STATE SIMULATION

### 3.1 *Project overview*

This paper takes the outpatient department of Jingxi Public Health Emergency Treatment Center as a case. It is located in Jingxi City, Guangxi Zhuang Autonomous Region, and on the border between China and Vietnam. The outpatient department is used as the infectious disease area of Jingxi people's Hospital in peacetime, and as a professional infectious disease hospital during the epidemic.

### 3.2 *Medical process simulation model construction*

The simulation model is divided into the space environment model and the medical behavior model, which is based on the improved social force model. In the process of building the simulation model, the function module needs to be dragged intothe designated area for parameter calibration, and input the parameters to the logic module at the same time.Second, it is required to connect the logic modules according to the medical treatment process; finally, the data set is created and the chart is generated. To focus on the implementation process of the medical process simulation method, this paper mainly focuses on the patient treatment process, and temporarily ignores the research on medical staff and other medical processes.

#### 3.2.1 *Space environment model construction*

Environment modeling includes the hospital building's internal and external space environment. In this paper, the length, width, height, and dimension of the space are calibrated based on the design parameters. The internal space environment includes public service facilities and medical units. The external space environment includes roadways, squares, parking lots, buildings, obstacles on pedestrian passages, and other traffic facilities.

### 3.2.2 *Visit behavior model construction and parameter calibration in epidemic time*

In this paper, the respiratory infectious disease is the simulated case of epidemic time. According to the project design concept, other departments are used for emergency response after the outbreak, and the "pre-hospital screening and triage" process is added to screen fever patients. The main links are as follows:

- Routine respiratory patient visits.
- Patient arrival → screening (no fever) → registration → waiting → diagnose → payment → treatment/blood drawing/medical examination/taking medicine → leaving (end);
- Nucleic acid testing personnel visits.
  Patient arrival → screening (no fever) → registration → nucleic acid sampling → leaving (end);
- Fever patient visits.
  Patient arrival → screening (fever) → diagnosis and treatment examination → medical technology examination → observation (End the simulation process in this step).

Figure 1 and Table 1 show the logic diagram and logic module parameter settings in the epidemic.

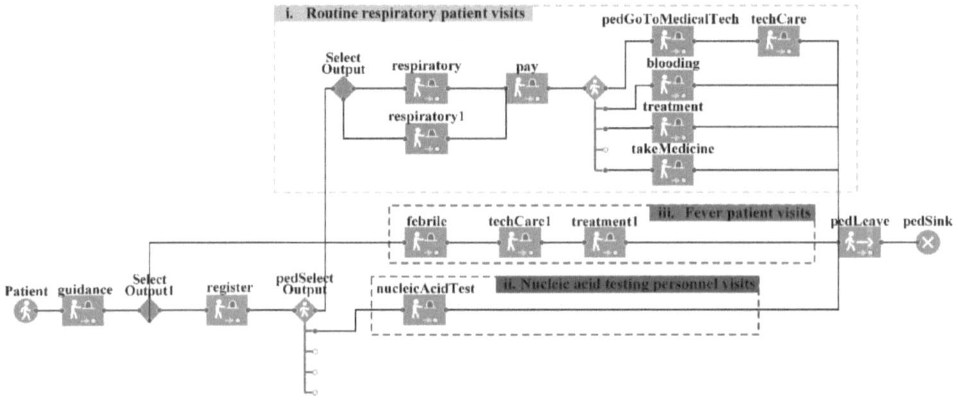


Figure 1. Patient visit process logic in the epidemic.

## 4 SPACE LAYOUT OPTIMIZATION STRATEGY BASED ON SIMULATION

### 4.1 *Epidemic time simulation data analysis*

The simulation period in this paper is 8:00–12:00 on working days. Figure 2 shows the simulation operation interface, which generates data every minute. The spatial density map reflects the density in each region. Red indicates high density, and blue indicates low density. During the epidemic time, only two clinics were open, including the Respiratory Dept. (Record as respiratory room 1, red circle in Figure 2) and Intestinal Dept. (Transformed into Respiratory room 2, yellow circle in Figure 2). AIDS Dept. and Hepatitis Dept. stopped taking care of patients (Transformed into Standby Dept.). Meanwhile, a pre-hospital triage channel and nucleic acid detection area were built.

During the simulation period, the total amount of patients entering is 503, and the number of patients leaving is 292. Among them, the time of discharged patients is mainly distributed in 8 to 39 minutes, accounting for about 0.78%, which is due to the fact that 236 discharged patients that underwent nucleic acid testing accounted for 80.8% of the discharged patients.

Table 1. Logic module and parameter calibration table.

| Module name | Parameter meaning | Calibration value | Measurement |
|---|---|---|---|
| Patient | Average patient arrival rate | Poisson (70, 150, 120, 0, 1) | Person/hour |
| | Patient walking speed | Uniform (0.6, 1.2) | M / S |
| Guidance | Triage time | Uniform (5, 15) | Second |
| Register | Registration Time | Uniform (1, 3.0) | Minute |
| pedSelectOutput | Patients proportion in various departments | Respiratory department patients proportion—0.4 | — |
| | | Nucleic acid testing personnel proportion—0.6 | — |
| Respiratory | Respiratory Department waiting time | Uniform (5, 12) | Minute |
| Febrile | Fever patients' diagnosis and treatment time | Uniform (8, 20) | Minute |
| Selectoutput | Routine patients (respiratory patients, nucleic acid testing personnel): Fever patients | 8:2 | — |
| NucleicAcidTest | Nucleic acid sampling time | Uniform (10, 20) | Second |
| techCare1 | Medical technical examination time (fever patients) | Uniform (7, 18) | Minute |
| Pay | Payment time | Uniform (1.0, 3.0) | Minute |
| techCare | Medical technology diagnosis and treatment time | uniform(7, 18) | Minute |
| Blooding | Blood drawing time | Uniform (2.0, 3.0) | Minute |
| Treatment | Treatment time | Uniform (11, 30) | Minute |
| hospitalized | Inpatient registration (fever patients) | Uniform (3, 10) | Minute |

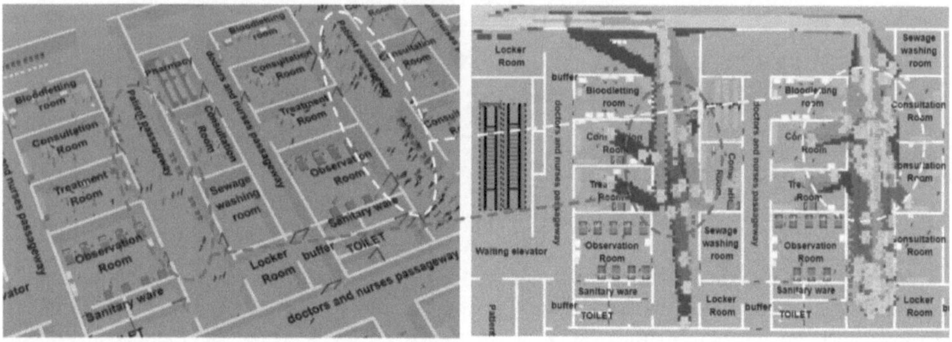


Figure 2. Epidemic time simulation operation interface.

Figure 3 shows that the regional density of Respiratory room 2 continues to rise. There are two reasons: on the one hand, the overall patients' number increases; on the other hand, during the simulation, it also undertakes the diagnosis of fever patients and suspected confirmed cases, which makes its diagnosis and treatment services overloaded, and waiting room seriously crowded (as shown in the yellow circle in Figure 2).

Figure 4 shows presence of a large number of waiting queues in the Respiratory room1 and Respiratory room 2, which is consistent with the regional density trend in Figure 3. After the

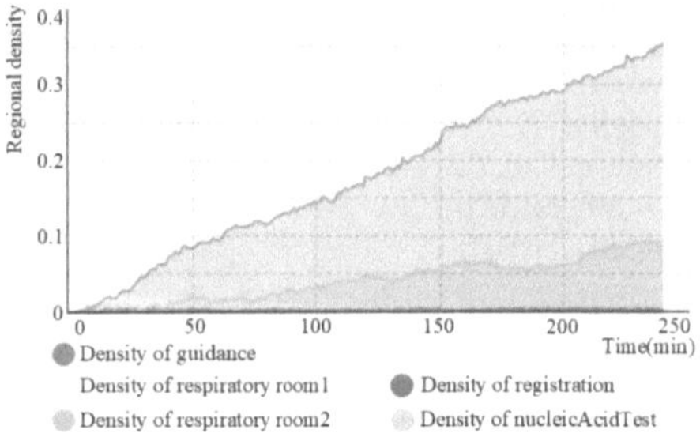


Figure 3. Department density-time variation curve.

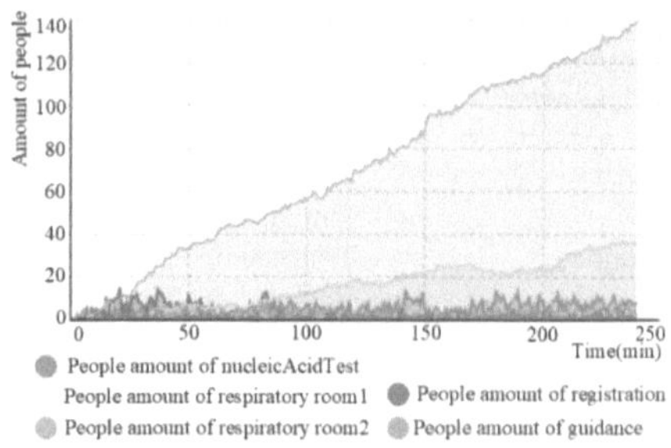


Figure 4. Graph of the number of patients waiting to be seen in the department during the epidemic.

simulation, the queue included 140 people; There is less queuing between registration and other departments.

Figure 5 shows the patient passageway of Respiratory Dept. room 1(red circle) and room 2(yellow circle) in the simulation operation. It can be seen that the number of patients and the per capita density of patients in Respiratory Dept. room 2 are high, which is the same as the analysis results in Figures 3 and 4.

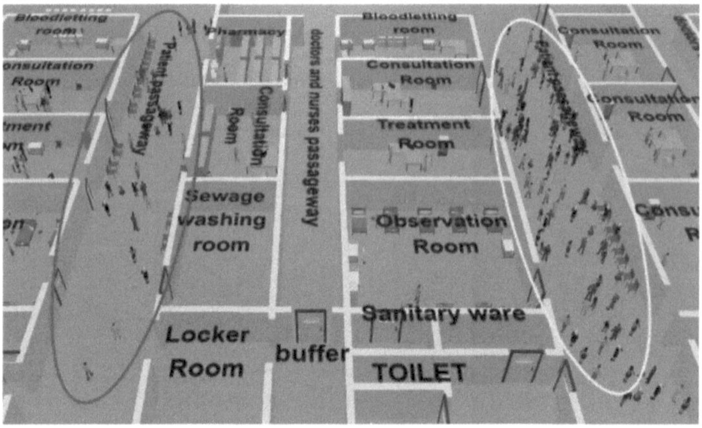


Figure 5. Patient passageway in Respiratory Dept. room 1 and 2.

Generally speaking, the outpatient treatment process and plane layout have not been reasonably and carefully adjusted in the epidemic time. Patients gather in some areas of the hospital, which not only affects the medical treatment service experience but also has potential medical and health safety hazards such as cross infection and increasing disease transmission. Therefore, to meet the diagnosis and treatment needs, hospitals need to have corresponding optimization strategies, to make full preparations to deal with the epidemic and improve the level of medical services.

### 4.2 *Optimization strategy*

#### 4.2.1 *Front triage table, separate diagnosis, treatment, and collection*

Changing the function of the triage table is one of the optimization methods. It is a triage of disease causes in peacetime, and triage of single nucleic acid sampling and the diagnosis of fever patients in the epidemic, so the triage table should be in front. At the same time, increase the number of

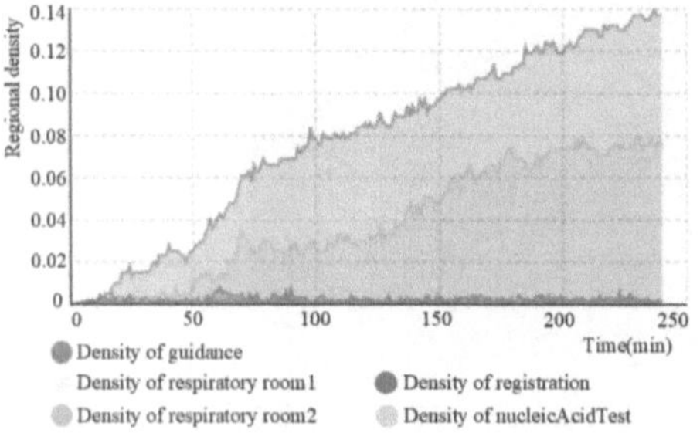


Figure 6. Department density-time variation curve after optimization.

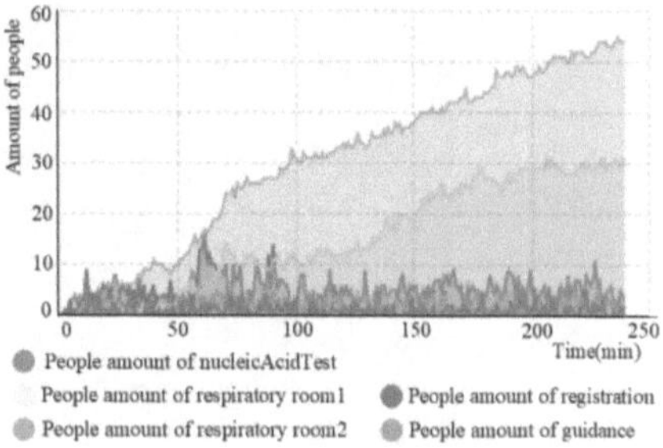


Figure 7. Diagram of the time variation of the number of patients waiting in the department after optimization.

registration machines in the treatment unit, to complete registration, charging, and treatment in a small area.

#### 4.2.2 *The dimensions of the outpatient consultation room and observation room of the infectious diseases department were designed in general*

After the epidemic outbreak, the number of patients visiting hospital outpatient clinics declined, and the infectious disease department increased a large number of suspected patients for treatment. Given this phenomenon, the observation room is set up for suspected patients. It needs to be noted that patients cannot leave the hospital or enter other areas without authorization to avoid increasing the risk of infection. During the epidemic time, the consulting room close to the Infectious Diseases Dept. will be converted into a temporary observation room, to alleviate the space pressure of the Infectious Diseases Dept. At the same time, it is considered to set an isolation roller shutter to isolate the observation room from other parts of the hospital.

#### 4.2.3 *Improve the guiding effect of traffic space on patients*

Unify the style and color of the traffic space in the same area to form spatial integration. The identification system is unified with the spatial structure, spatial interface division, and decorative materials in the traffic space.

### 4.3 *Optimization results and simulation*

Based on the optimization strategy, the adjusted hospital layout is verified by simulation. Using the control variable method, the input parameters in the simulation remain unchanged. Figure 6 and Figure 7 show that the regional density of departments decreased, especially Respiratory room 2, which decreased from the highest value of 0.35 to 0.14, indicating that the space congestion has been alleviated. Meanwhile, the maximum queuing number in room 2 decreased from about 140 to 54, indicating that the service level has been improved. Comprehensive simulation results show that the optimization strategy proposed in this paper is effective.

As shown in Figure 8, after optimization, the number and density of patients in Respiratory Dept. rooms 1 and 2 decreased, which is consistent with the above analysis. It shows that the optimization strategy proposed in this paper for the simulation case is effective, reduces the phenomenon of personnel aggregation to a certain extent, and reduces cross-infection at the same time.

## 5 CONCLUSION

The research shows that the medical process simulation method makes the hospital susceptible to unit identification and the medical process visible evaluation, dynamic, and quantitative. At the

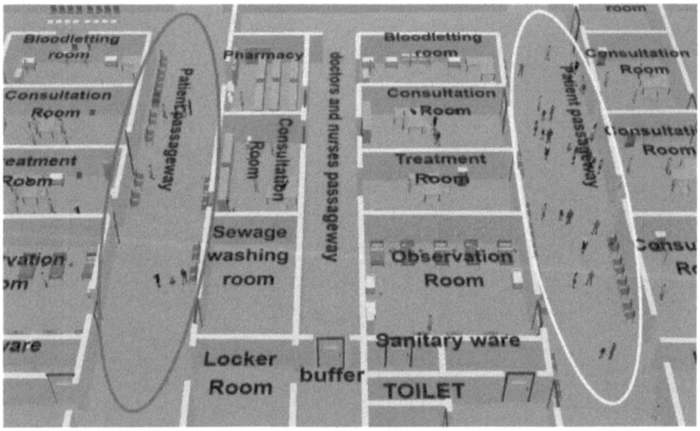


Figure 8. Patient passageway in Respiratory Dept. rooms 1 and 2 after optimization.

same time, it provides data support for hospital design and decision-making, which is in the hospital design stage combination of peacetime and epidemic period.

The results show that the medical process simulation method has great potential and practicality, and can be used to guide hospital design. It can assist designers to judge in advance where the node is prone to problems in the "combination of peacetime and epidemic period" hospital. It can improve the medical process, improve the efficiency of diagnosis/ treatment services and reduce the risk of infection by formulating corresponding plans and measures in advance in epidemic time. Meanwhile, this paper also verifies the effectiveness of optimization strategies such as pre-diagnosis of epidemic disease, separation of diagnosis and treatment and collection, the addition of self-service equipment, and modular general design of outpatient service.

## ACKNOWLEDGMENT

The authors appreciate the funding support from the scientific research project of the National Ministry of Housing and Urban-Rural Construction (Entry name: Research on medical process simulation and spatial layout optimization method for infection control in hospital ICU, 2019-K-151), and the science and technology plan project of Guangxi Housing and Urban-Rural Construction (Entry name: Research on Key Technologies of airport terminal space layout optimization and management and control evaluation under normalized epidemic prevention and control).

## REFERENCES

Kuang H, Fan Y H, Li X L, et al. (2011). Heterogeneous Effect and Pattern Formation on Pedestrian Queue Dynamics. *J. Applied Mechanics & Materials*, 94–96:2043-2046.

Moussaïd M, Helbing D, Garnier S, et al. (2009). Experimental study of the behavioral mechanisms underlying self-organization in human crowds. *J. Proceedings of the Royal Society Biological Sciences*, 276:2755–2762.

Pan D, Rajwani K. (2021). Implementation of simulation training during the COVID-19 pandemic: a New York hospital experience. *J. Simulation in Healthcare*, 16(1): 46.

Petrone G, Brown L, Binder W, et al. (2021) The Calm Before the Storm: Utilizing In Situ Simulation to Evaluate for Preparedness of an Alternative Care Hospital During COVID-19 Pandemic. *J. Disaster Medicine and Public Health Preparedness*, 1–17.

Riquelme F, Aguilera A, Inostrosa-Psijas A. (2021) Contagion modeling and simulation in transport and air travel networks during the COVID-19 pandemic: a survey. *J. IEEE Access*, 9: 149529–149541.

*Advances in Urban Construction and Management Engineering – Cha (Ed)*


# Talking about the application and development of laser ranging telescope in overhead line survey

Jingliang Zhang*
*South China University of Technology, China*

ABSTRACT: A laser-ranging telescope is one of the main tools for line measurement. With the continuous development and promotion of electronic technology and computer technology, the telescope cannot be limited to the measurement between the measurement point and the target point, but can be extended to the target point (Feng 2008). The measurement between space then extends to the measurement between any point in the space of the measurement range, and finally, a three-dimensional topographic map can be formed, which provides data support for engineering measurement, unmanned driving, and bionic robots (Guo 2012).

## 1 INTRODUCTION

Overhead line measurement is mainly used in the fields of power grid engineering and telecommunication and communication engineering (He 2022). Telecommunication communication engineering is gradually moving towards the form of wireless transmission and underground laying of optical fiber cables, and overhead line engineering is still one of the main ways of power transmission (Li 2016). The measurement of overhead lines has also become one of the main links in the power overhead line project. The measurement of overhead lines in power engineering mainly focuses on the distance and angle between overhead lines, electrical equipment and devices, obstacles, and the ground, providing necessary data materials for power line design, implementation, and operation and maintenance (Wang 2020). The characteristics of the measurement of overhead lines in power engineering are complex outdoor terrain, large line spans, and lines often crossing rivers, highways, buildings, etc. These are the difficulties that survey engineers need to face in their work (Wang 2009).

## 2 COMPARISON OF DISTANCE MEASURING TOOLS

The main purpose of engineering measurement is to obtain data on the distance and angle between the measurement objects and provide the basis for the project implementation (Wu 2015). The commonly used distance measuring tools are tape measures, measuring wheelbarrows, infrared rangefinders, laser rangefinder telescopes, total stations, electrographic software, etc. (Yang 2010). The selection of distance measurement tools should be based on the working environment, engineering stage, and measurement accuracy requirements (Zuo 2016).

The measuring method of the tape measure is the simplest and most traditional, but there are drawbacks such as flat terrain and low work efficiency.

The engineering surveying trolley is pushed on the ground by the small wheel, the path length of the ground is transmitted to the recording module of the surveying trolley, and the reading is

*Corresponding Author: 1175102843@qq.com

DOI 10.1201/9781003348023-87 

displayed through the diode. The measuring wheelbarrow is not only simple in structure and low in cost, but also easy to operate, and its data accuracy can basically meet the requirements of power line measurement. It is one of the most commonly used tools for measuring the length of outdoor power underground pipelines. However, since the towers of the power overhead line and the ground projection line of the power line usually do not fall on flat ground, the engineer and the surveying BMX cannot walk normally, so the instrument has considerable limitations.

The handheld infrared rangefinder emits infrared light through the instrument, generates light reflection by touching the target object with light, and then receives the reflected light through the sensor in the instrument, and transmits the signal to the calculator module for calculation, so as to obtain the measuring instrument to the target and the straight-line distance to the target. The instrument is small in size, light in weight, accurate in measurement data, realizes non-contact measurement, and has been widely used and promoted. However, due to the serious attenuation of infrared rays under the refraction of strong natural light outdoors and the limited human vision, the difficulty of aiming the target with the handheld infrared rangefinder is generally only about 40-50 meters. Therefore, it is mainly used in places where the working environment is dark, the measurement distance is short, and the target is clear and easy to see, such as electric rooms, cable trenches, indoor equipment, and other places. Its scope of application generally does not include measuring outdoor power overhead lines.

## 3 THE APPLICATION OF LASER RANGING TELESCOPE

A laser ranging telescope is a combination of laser range finder and telescopic system that integrates observation and measurement. The module consists of several parts. The laser ranging telescope generally uses the laser pulse method to measure the distance. The laser emitted by the rangefinder is reflected by the object to be measured and then received by the rangefinder. The rangefinder simultaneously records the time of the laser round trip. By calculating half the product of the speed of light and the round-trip time, the distance between the rangefinder and the measured object can be found. In order to take advantage of the relatively strong laser penetration, a telephoto system is set up inside. Through the refraction principle of the lens, the measuring distance of the laser rangefinder can generally reach 400-3000 meters, and the accuracy can reach $\pm 1/1000$. The demand for survey and measurement of overhead lines has improved greatly compared to the previous measuring tools such as tape measures, measuring wheelbarrows, and infrared rangefinders.

The laser emission module of the laser ranging telescope is usually a one-way laser emission, and can only detect a single measured target; so, only the data of a single measured target can be obtained, that is, the straight-line distance between the telescope and the measured target. The elevation angle between the measured target and the horizontal line cannot be directly read from the straight-line distance between two or more measurement targets. In the measurement project of electric overhead lines, when the engineer needs to measure the span of the tower, the engineer must stand on the side of one of the towers in person and then use the telescope to aim at the target tower to trigger the laser, so as to measure the straight-line distance between the two towers. When measuring the vertical height of the overhead wire to the ground, the engineer needs to stand on the ground projection point of the overhead line before using the telescope. The positioning of the projection foot will directly affect the accuracy of the data.

However, due to the unsatisfactory geographical conditions such as the location of the poles and towers of the outdoor power overhead lines and the direction of the overhead lines, they often fall in steep mountains, rivers, ponds, highways, schools, hospitals, government agencies, restricted military areas, etc. The surveying engineer cannot visit the position of the tower in person, cannot determine the ground projection point of the overhead line, and cannot enter the outer edge of buildings and obstacles for measurement. To obtain the span between two towers or the safe net distance of an overhead line, the engineer only needs to measure the measurement target at two parallel positions, obtain the straight-line distance between the two positions and the target, and then go through the trigonometric function cosine theorem. Using the equation below, we obtained the straight-line distance between two measurement targets:

$L^2 = a^2 + b^2 - 2ab \bullet \cos(\arccos \frac{a^2+c^2-x^2}{2ac} - \arccos \frac{a^2+c^2-y^2}{2bc})$, and measure the vertical height of the target to the ground: $H = x \bullet \sin \arccos \frac{x^2+c^2-a^2}{2xc}$. Although the whole conversion process is complicated, the operation process is cumbersome, the positioning of the measurement points is inaccurate, and deviations are likely to occur, the engineer does not need to visit the position of the pole and tower in person, nor does it need to find the projection point of the overhead wire and cannot measure it, and it can also be avoided. If the data is inaccurate due to the angle measurement error, the straight-line distance between any two points in space can be obtained. To improve the efficiency of measurement work, engineers will sometimes import the calculation equations into a spreadsheet and store them on a mobile phone or tablet computer. As long as the measurement data are recorded on the spot (a,b,c,x,y), the calculation results of L and H can be obtained, which provides great help for the measurement work. It is widely used in the measurement field of power overhead lines.

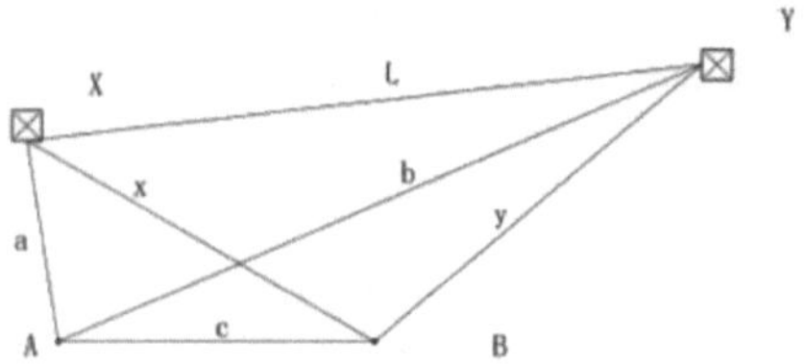

(一)

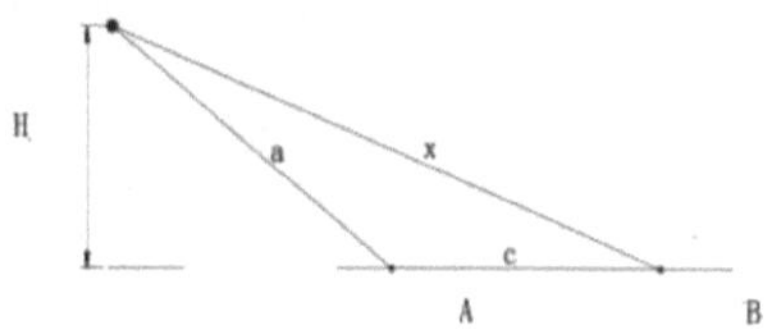

(二)

## 4 EXPANSION OF LASER RANGING TELESCOPES

The integrated circuits industry is developing rapidly with the rapid development of science and technology. Users only need to adjust the interior of the telescope and integrate the micro-calculator to help solve the complex calculation process of engineers and even get a higher level of improvement.

According to the above calculation formula, as long as the distance between the two laser emission points and the straight-line distance between the laser point and the two measured targets are known, the straight-line distance between the two measurement targets can be calculated. Therefore, as long as a set of laser emission modules and laser detection modules are added to the telescope, two sets of data (a, b, x, y) can be obtained. Combined with the distance (c) between the two laser emission points, the above-mentioned data can be used. The calculation formula obtains the distance between the two measurement targets and the height above the ground (L, H). In order to improve the operation speed of the measuring instrument, the data operation processing module can be replaced with a micro calculator, and the measurement software can be developed on the Visual Studio platform using the computer C# language, and parameters such as atmospheric refraction coefficient can be taken into account to improve the measurement accuracy. The operation formula is imported into the program software, and the single-chip microcomputer system is integrated through digital circuit integration technology. After the single-chip computer system collects two sets of data, and then through the operation of the internal program of the microcomputer, the calculation results can be obtained, and the measurement readings can be displayed. Such a rangefinder can complete the measurement more quickly without the need for other equipment or software, is convenient to carry, has higher work efficiency, and has more complete functions.

Suppose this measuring instrument is further improved and expanded, and the straight-line distance between any two points in space and the height of any point to the ground is extended. In that case, a three-dimensional topographic map of the measuring area can be constructed, including the relationship between objects on the terrain. To achieve this goal, it is necessary to expand the measuring instrument further. First, an optical shooting system is introduced into the instrument, and the terrain in the measurement range is optically scanned through the shooting system to obtain a terrain data model and establish an image platform. Then, the laser of the measuring instrument is used. The emission point and detection module are expanded into a laser emission surface scanning and detection data processing system, and the basic data between the instrument and the measurement target within the measurement range is obtained through laser measurement. Then, a large amount of basic data is transmitted to the computer processing system through the communication system. The system classifies, calculates, and integrates a large amount of data and assigns values ??to all the measurement targets on the topographic map, thus forming a three-dimensional spatial topographic map with data relationships, and can perform data query, display, and analysis on any target operations such as printing.

With the development of the computer age, engineering survey software has also emerged. Some outdoor topographic surveys are realized by some computer application software developed on the platform of satellite maps. Then latitude and longitude coordinates and elevation are attached through computer programming, and some simple tools such as measurement and line drawing are added. However, the measurement data provided by the map software is inaccurate due to the interval of satellite scanning, the long lag of terrain data update, and the limited resolution of satellite photos. In addition, because the overhead line measurement is usually performed outdoors in remote mountains or under the jungle, the network signal is often interrupted, making the application software inoperable and limited to a certain extent.

By comparing the above-mentioned various measurement methods, the measurement methods of various instruments have their advantages and disadvantages. However, scientific development always expands past, existing, and envisaged products to form new products to meet the needs of greater demand. We also improve, extend and integrate various measuring instruments to make up for their respective shortages to obtain the greatest improvement and make more suitable products to meet the greater needs of engineers.

## 5 CONCLUDING REMARKS

With the continuous development of science and technology and the economy, the future will be the era of artificial intelligence technology, and laser measurement will continue to be developed into intelligent measurement and intelligent data acquisition. Intelligent measurement and data acquisition are indispensable in intelligent products, such as computer graphics, unmanned aerial vehicles, unmanned vehicles, and bionic robots. The development of science is derived from countless tiny innovative ideas. We expect this idea will also promote scientific and technological progress and look forward to the birth of new products and engineering applications.

## REFERENCES


Feng Qibo et al. edited "*Optical Measurement Technology and Application*" ISBN9787302171362, Tsinghua University Press, 2008.05.

Guo Yuzhen et al. edited "*Measurement Data Processing and Software Application*" ISBN9787550902954, Yellow River Water Conservancy Press, 2012.08.

He Lenian edited "*Analog Integrated Circuit Design and Simulation*" ISBN9787030214270, Science Press, 2022.1.

Li Zhenghang and Huang Jinsong edited *GPS Measurement and Data Processing* (Third Edition) ISBN9787307176805, Wuhan University Press, 2016.05.

Wang Peng, Fu Luhua, Sun Changku edited "*Laser Measurement Technology*" ISBN 978711168932, Machinery Industry Press, 2020.07.01.

Wang Tong edited "*The Application of C Language in Measurement and Control*" ISBN9787111261902, Machinery Industry Press, 2009.04.

Wu Anzhuang edited "*Development Technology of C Language Measurement Software Based on VS2012 Platform*" ISBN9787550911949, Yellow River Water Conservancy Press, 2015.08.
Yang Daihua et al. edited "*The Principle and Application of Single Chip Microcomputer and Embedded System*" ISBN9787562524397, China University of Geosciences Press, 2010.2.
Zuo Xiangang, Liu Yanchang, Jia Meng edited "*The Principle of Single-chip Microcomputer and Its C Language Programming*" ISBN9787560633411, Xidian University Press, 2016.9.

*Advances in Urban Construction and Management Engineering – Cha (Ed)*
*© 2023 the Editor(s), ISBN 978-1-032-39018-5*

# Author index